1000 Solved Problems in Modern Physics

Ahmad A. Kamal

1000 Solved Problems in Modern Physics

Springer

Dr. Ahmad A. Kamal
425 Silversprings Lane
Murphy, TX 75094, USA
anwarakamal@yahoo.com

ISBN 978-3-642-43390-0 SBN 978-3-642-04333-8 (eBook)
DOI 10.1007/978-3-642-04333-8
Springer Heidelberg Dordrecht London New York

Cover design: eStudio Calamar Steinen

Springer is part of Springer Science+Business Media (www.springer.com)

Dedicated to my parents

Preface

This book is targeted mainly to the undergraduate students of USA, UK and other European countries, and the M.Sc of Asian countries, but will be found useful for the graduate students, Graduate Record Examination (GRE), Teachers and Tutors. This is a by-product of lectures given at the Osmania University, University of Ottawa and University of Tebrez over several years, and is intended to assist the students in their assignments and examinations. The book covers a wide spectrum of disciplines in Modern Physics, and is mainly based on the actual examination papers of UK and the Indian Universities. The selected problems display a large variety and conform to syllabi which are currently being used in various countries. The book is divided into ten chapters. Each chapter begins with basic concepts containing a set of formulae and explanatory notes for quick reference, followed by a number of problems and their detailed solutions.

The problems are judiciously selected and are arranged section-wise. The solutions are neither pedantic nor terse. The approach is straight forward and step-by-step solutions are elaborately provided. More importantly the relevant formulas used for solving the problems can be located in the beginning of each chapter. There are approximately 150 line diagrams for illustration.

Basic quantum mechanics, elementary calculus, vector calculus and Algebra are the pre-requisites. The areas of Nuclear and Particle physics are emphasized as revolutionary developments have taken place both on the experimental and theoretical fronts in recent years. No book on problems can claim to exhaust the variety in the limited space. An attempt is made to include the important types of problems at the undergraduate level.

Chapter 1 is devoted to the methods of Mathematical physics and covers such topics which are relevant to subsequent chapters. Detailed solutions are given to problems under Vector Calculus, Fourier series and Fourier transforms, Gamma and Beta functions, Matrix Algebra, Taylor and Maclaurean series, Integration, Ordinary differential equations, Calculus of variation Laplace transforms, Special functions such as Hermite, Legendre, Bessel and Laguerre functions, complex variables, statistical distributions such as Binomial, Poisson, Normal and interval distributions and numerical integration.

Chapters 2 and 3 focus on quantum physics. Chapter 2 is basically concerned with the old quantum theory. Problems are solved under the topics of deBroglie waves, Bohr's theory of hydrogen atom and hydrogen-like atoms, positronium and mesic atoms, X-rays production and spectra, Moseley's law and Duan–Hunt law, spectroscopy of atoms and molecules, which include various quantum numbers and selection rules, and optical Doppler effect.

Chapter 3 is concerned with the quantum mechanics of Schrodinger and Hesenberg. Problems are solved on the topics of normalization and orthogonality of wave functions, the separation of Schrodinger's equation into radial and angular parts, 1-D potential wells and barriers, 3-D potential wells, Simple harmonic oscillator, Hydrogen-atom, spatial and momentum distribution of electron, Angular momentum, Clebsch–Gordon coefficients ladder operators, approximate methods, scattering theory-phase-shift analysis and Ramsuer effect, the Born approximation.

Chapter 4 deals with problems on Thermo–dynamic relations and their applications such a specific heats of gases, Joule–Thompson effect, Clausius–Clapeyron equation and Vander waal's equation, the statistical distributions of Boltzmann and Fermi distributions, the distribution of rotational and vibrational states of gas molecules, the Black body radiation, the solar constant, the Planck's law and Wein's law.

Chapter 5 is basically related to Solid State physics and material science. Problems are covered under the headings, crystal structure, Lattice constant, Electrical properties of crystals, Madelung constant, Fermi energy in metals, drift velocity, the Hall effect, the Debye temperature, the intrinsic and extrinsic semiconductors, the junction diode, the superconductor and the BCS theory, and the Josephson effect.

Chapter 6 deals with the special theory of Relativity. Problems are solved under Lorentz transformations of length, time, velocity, momentum and energy, the invariance of four-momentum vector, transformation of angles and Doppler effect and threshold of particle production.

Chapters 7 and 8 are concerned with problems in low energy Nuclear physics. Chapter 7 covers the interactions of charged particles with matter which include kinematics of collisions, Rutherford Scattering, Ionization, Range and Straggiling, Interactions of radiation with matter which include Compton scattering, photoelectric effect, pair production and nuclear resonance fluorescence, general radioactivity which includes problems on chain decays, age of earth, Carbon dating, alpha decay, Beta decay and gamma decay.

Chapter 8 is devoted to the static properties of nuclei such as nuclear masses, nuclear spin and parity, magnetic moments and quadrupole moments, the Nuclear models, the Fermi gas model, the shell model, the liquid drop model and the optical model, problems on fission and fusion and Nuclear Reactors.

Chapters 9 and 10 are concerned with high energy physics. Chapter 9 covers the problems on natural units, production, interactions and decays of high energy unstable particles, various types of detectors such as ionization chambers, proprortional and G.M. counters, Accelerators which include Betatron, Cyclotron, Synchro-Cyclotron, proton and electron Synchrotron, Linear accelerator and Colliders.

Chapter 10 deals with the static and dynamic properties of elementary particles and resonances, their classification from the point of view of the Fermi–Dirac and Bose–Einstein statistics as well as the three types of interactions, strong, Electromagnetic and weak, the conservation laws applicable to the three types of interactions, Gell-mann's formula, the properties of quarks and classification into supermultiplets, the types of weak decays and Cabibbo's theory, the neutrino oscillations, Electro–Weak interaction, the heavy bosons and the Standard model.

Acknowledgements

It is a pleasure to thank Javid for the bulk of typing and suggestions and Maryam for proof reading. I am indebted to Muniba for the line drawings, to Suraiya, Maqsood and Zehra for typing and editing. I am grateful to the Universities of UK and India for permitting me to use their question papers cited in the text to CERN photo service for the cover page to McGraw-Hill and Co: for a couple of diagrams from Quantum Mechanics, L.I. Schiff, 1955, to Cambridge University Press for using some valuable information from Introduction to High Energy Physics, D.H. Perkins and to Ginn and Co: and Pearson and Co: for access to Differential and Integral Calculus, William A. Granville, 1911. My thanks are due to Springer-Verlag, in particular Claus Ascheron, Adelheid Duhm and Elke Sauer for constant encouragement.

Murphy, Texas
February 2010

Ahmad A. Kamal

Contents

Chapter 1
Mathematical Physics

1.1 Basic Concepts and Formulae

Vector calculus

Angle between two vectors, $\cos\theta = \frac{\boldsymbol{A}.\boldsymbol{B}}{|\boldsymbol{A}||\boldsymbol{B}|}$
Condition for coplanarity of vectors, $\boldsymbol{A}.\boldsymbol{B} \times \boldsymbol{C} = 0$

Del

$$\nabla = \frac{\partial}{\partial x}\hat{i} + \frac{\partial}{\partial y}\hat{j} + \frac{\partial}{\partial z}\hat{k}$$

Gradient

$$\nabla\phi = \left(\frac{\partial\phi}{\partial x}\hat{i} + \frac{\partial\phi}{\partial y}\hat{j} + \frac{\partial\phi}{\partial z}\hat{k}\right)$$

Divergence

If $V(x, y, z) = V_1\hat{i} + V_2\hat{j} + V_3\hat{k}$, be a differentiable vector field, then $\nabla.\boldsymbol{V} = \frac{\partial}{\partial x}V_1 + \frac{\partial}{\partial y}V_2 + \frac{\partial}{\partial z}V_3$

Laplacian

$$\nabla^2 = \frac{\partial^2}{\partial x^2} + \frac{\partial^2}{\partial y^2} + \frac{\partial^2}{\partial z^2} \quad \text{(Cartesian coordinates } x, y, z)$$

$$\nabla^2 = \frac{1}{r^2}\frac{\partial}{\partial r}\left(r^2\frac{\partial}{\partial r}\right) + \frac{1}{r^2\sin\theta}\frac{\partial}{\partial\theta}\left(\sin\theta\frac{\partial}{\partial\theta}\right) + \frac{1}{r^2\sin^2\theta}\frac{\partial^2}{\partial\Phi^2}$$

(Spherical coordinates r, θ, Φ)

$$\nabla^2 = \frac{\partial^2}{\partial r^2} + \frac{1}{r}\frac{\partial}{\partial r} + \frac{1}{r^2}\frac{\partial^2}{\partial\theta^2} + \frac{\partial^2}{\partial z^2} \quad \text{(Cylindrical coordinates } r, \theta, z)$$

Line integrals

(a) $\int_C \phi \, \mathrm{d}\boldsymbol{r}$

(b) $\int_C \boldsymbol{A} \,.\, \mathrm{d}\boldsymbol{r}$
(c) $\int_C \boldsymbol{A} \times \mathrm{d}\boldsymbol{r}$

where ϕ is a scalar, $\boldsymbol{A}$ is a vector and $\boldsymbol{r} = x\hat{i} + y\hat{j} + z\hat{k}$, is the positive vector.

Stoke's theorem

$$\oint_C \boldsymbol{A} \,.\, d\boldsymbol{r} = \iint_S (\nabla \times \boldsymbol{A}) \,.\, n \,\mathrm{d}s = \iint_S (\nabla \times \boldsymbol{A}) \,.\, \mathrm{d}\boldsymbol{s}$$

The line integral of the tangential component of a vector $\boldsymbol{A}$ taken around a simple closed curve C is equal to the surface integral of the normal component of the curl of $\boldsymbol{A}$ taken over any surface S having C as its boundary.

Divergence theorem (Gauss theorem)

$$\iiint_V \nabla \,.\, \boldsymbol{A} \,\mathrm{d}v = \iint_S \boldsymbol{A}.\hat{n} \,\mathrm{d}s$$

The volume integral is reduced to the surface integral.

Fourier series

Any single-valued periodic function whatever can be expressed as a summation of simple harmonic terms having frequencies which are multiples of that of the given function. Let $f(x)$ be defined in the interval $(-\pi, \pi)$ and assume that $f(x)$ has the period 2π, i.e. $f(x + 2\pi) = f(x)$. The Fourier series or Fourier expansion corresponding to $f(x)$ is defined as

$$f(x) = \frac{1}{2}a_0 + \sum_{n=1}^{\infty} (a_0 \cos nx + b_n \sin nx) \tag{1.1}$$

where the Fourier coefficient a_n and b_n are

$$a_n = \frac{1}{\pi}\int_{-\pi}^{\pi} f(x) \cos nx \,\mathrm{d}x \tag{1.2}$$

$$b_n = \frac{1}{\pi}\int_{-\pi}^{\pi} f(x) \sin nx \,\mathrm{d}x \tag{1.3}$$

where $n = 0, 1, 2, \ldots$

If $f(x)$ is defined in the interval $(-L, L)$, with the period $2L$, the Fourier series is defined as

$$f(x) = \frac{1}{2}a_0 + \sum_{n=1}^{\infty} (a_n \cos(n\pi x/L) + b_n \sin(n\pi x/L)) \tag{1.4}$$

where the Fourier coefficients a_n and b_n are

$$a_n = \frac{1}{L}\int_{-L}^{L} f(x)\cos(n\pi x/L)\,\mathrm{d}x \tag{1.5}$$

$$b_n = \frac{1}{L}\int_{-L}^{L} f(x)\sin(n\pi x/L)\,\mathrm{d}x \tag{1.6}$$

Complex form of Fourier series

Assuming that the Series (1.1) converges at $f(x)$,

$$f(x) = \sum_{n=-\infty}^{\infty} C_n e^{in\pi x/L} \tag{1.7}$$

with

$$C_n = \frac{1}{L}\int_{C}^{C+2L} f(x)e^{-i\pi nx/L}\,\mathrm{d}x = \begin{cases} \frac{1}{2}(a_n - ib_n) & n > 0 \\ \frac{1}{2}(a_{-n} + ib_{-n}) & n < 0 \\ \frac{1}{2}a_o & n = 0 \end{cases} \tag{1.8}$$

Fourier transforms

The Fourier transform of $f(x)$ is defined as

$$\Im(f(x)) = F(\alpha) = \int_{-\infty}^{\infty} f(x)e^{i\alpha x}\,\mathrm{d}x \tag{1.9}$$

and the inverse Fourier transform of $F(\alpha)$ is

$$\Im^{-1}(f(\alpha)) = F(x) = \frac{1}{2\pi}\int_{-\infty}^{\infty} F(\alpha)e^{i\alpha x}\mathrm{d}\alpha \tag{1.10}$$

$f(x)$ and $F(\alpha)$ are known as Fourier Transform pairs. Some selected pairs are given in Table 1.1.

Table 1.1

$f(x)$	$F(\alpha)$	$f(x)$	$F(\alpha)$
$\frac{1}{x^2+a^2}$	$\frac{\pi e^{-a\alpha}}{a}$	e^{-ax}	$\frac{a}{\alpha^2+a^2}$
$\frac{x}{x^2+a^2}$	$-\frac{\pi i\alpha}{a}e^{-a\alpha}$	e^{-ax^2}	$\frac{1}{2}\sqrt{\frac{\pi}{a}}e^{-\alpha^2/4a}$
$\frac{1}{x}$	$\frac{\pi}{2}$	xe^{-ax^2}	$\frac{\sqrt{\pi}}{4a^{3/2}}\alpha e^{-\alpha^2/4a}$

Gamma and beta functions

The gamma function $\Gamma(n)$ is defined by

$$\Gamma(n) = \int_0^\infty e^{-x} x^{n-1}\,\mathrm{d}x \quad (Re\, n > 0) \tag{1.11}$$

$$\Gamma(n+1) = n\Gamma(n) \tag{1.12}$$

If n is a positive integer

$$\Gamma(n+1) = n! \tag{1.13}$$

$$\Gamma\left(\frac{1}{2}\right) = \sqrt{\pi}; \Gamma\left(\frac{3}{2}\right) = \frac{\sqrt{\pi}}{2}; \Gamma\left(\frac{5}{2}\right) = \frac{3}{4}\sqrt{\pi} \tag{1.14}$$

$$\Gamma\left(n + \frac{1}{2}\right) = \frac{1.3.5\ldots(2n-1)\sqrt{\pi}}{2^n} \quad (n = 1, 2, 3, \ldots) \tag{1.15}$$

$$\Gamma\left(-n + \frac{1}{2}\right) = \frac{(-1)^n 2^n \sqrt{\pi}}{1.3.5\ldots(2n-1)} \quad (n = 1, 2, 3, \ldots) \tag{1.16}$$

$$\Gamma(n+1) = n! \cong \sqrt{2\pi n}\, n^n e^{-n} \quad \text{(Stirling's formula)} \tag{1.17}$$

$$n \to \infty$$

Beta function $B(m, n)$ is defined as

$$B(m, n) = \frac{\Gamma(m)\Gamma(n)}{\Gamma(m+n)} \tag{1.18}$$

$$B(m, n) = B(n, m) \tag{1.19}$$

$$B(m, n) = 2\int_0^{\pi/2} \sin^{2m-1}\theta \cos^{2n-1}\theta\,\mathrm{d}\theta \tag{1.20}$$

$$B(m, n) = \int_0^\infty \frac{t^{m-1}}{(1+t)^{m+n}}\,\mathrm{d}t \tag{1.21}$$

Special funtions, properties and differential equations

Hermite functions:

Differential equation:

$$y'' - 2xy' + 2ny = 0 \tag{1.22}$$

when $n = 0, 1, 2, \ldots$ then we get Hermite's polynomials $H_n(x)$ of degree n, given by

$$H_n(x) = (-1)^n e^{x^2} \frac{\mathrm{d}^n}{\mathrm{d}x^n}\left(e^{-x^2}\right) \quad \text{(Rodrigue's formula)}$$

First few Hermite's polynomials are:

$$H_o(x) = 1,\ H_1(x) = 2x,\ H_2(x) = 4x^2 - 2$$
$$H_3(x) = 8x^3 - 12x,\ H_4(x) = 16x^4 - 48x^2 + 12 \tag{1.23}$$

Generating function:

$$e^{2tx-t^2} = \sum_{n=0}^{\infty} \frac{H_n(x)t^n}{n!} \tag{1.24}$$

Recurrence formulas:

$$H_n'(x) = 2nH_{n-1}(x)$$
$$H_{n+1}(x) = 2xH_n(x) - 2nH_{n-1}(x) \tag{1.25}$$

Orthonormal properties:

$$\int_{-\infty}^{\infty} e^{-x^2} H_m(x)H_n(x)\,\mathrm{d}x = 0 \quad m \neq n \tag{1.26}$$
$$\int_{-\infty}^{\infty} e^{-x^2} \{H_n(x)\}^2\,\mathrm{d}x = 2^n\, n!\sqrt{\pi} \tag{1.27}$$

Legendre functions:

Differential equation of order n:

$$(1 - x^2)y'' - 2xy' + n(n+1)y = 0 \tag{1.28}$$

when $n = 0, 1, 2, \ldots$ we get Legendre polynomials $P_n(x)$.

$$P_n(x) = \frac{1}{2^n n!}\frac{\mathrm{d}^n}{\mathrm{d}x^n}(x^2 - 1)^n \tag{1.29}$$

First few polynomials are:

$$P_o(x) = 1,\ P_1(x) = x,\ P_2(x) = \frac{1}{2}(3x^2 - 1)$$
$$P_3(x) = \frac{1}{2}(5x^3 - 3x),\ P_4(x) = \frac{1}{8}(35x^4 - 30x^2 + 3) \tag{1.30}$$

Generating function:

$$\frac{1}{\sqrt{1 - 2tx + t^2}} = \sum_{n=0}^{\infty} P_n(x)t^n \tag{1.31}$$

Recurrence formulas:

$$x P'_n(x) - P'_{n-1}(x) = n P_n(x)$$
$$P'_{n+1}(x) - P'_{n-1}(x) = (2n+1) P_n(x) \tag{1.32}$$

Orthonormal properties:

$$\int_{-1}^{1} P_m(x) P_n(x)\, \mathrm{d}x = 0 \quad m \neq n \tag{1.33}$$
$$\int_{-1}^{1} \{P_n(x)\}^2\, \mathrm{d}x = \frac{2}{2n+1} \tag{1.34}$$

Other properties:

$$P_n(1) = 1,\ P_n(-1) = (-1)^n,\ P_n(-x) = (-1)^n P_n(x) \tag{1.35}$$

Associated Legendre functions:

Differential equation:

$$(1-x^2)y'' - 2xy' + \left\{ l(l+1) - \frac{m^2}{1-x^2} \right\} y = 0 \tag{1.36}$$
$$P_l^m(x) = (1-x^2)^{m/2} \frac{\mathrm{d}^m}{\mathrm{d}x^m} P_l(x) \tag{1.37}$$

where $P_l(x)$ are the Legendre polynomials stated previously, l being the positive integer.

$$P_l^o(x) = P_l(x) \tag{1.38}$$
$$\text{and } P_l^m(x) = 0 \quad \text{if } m > n \tag{1.39}$$

Orthonormal properties:

$$\int_{-1}^{1} P_n^m(x) P_l^m(x)\, \mathrm{d}x = 0 \quad n \neq l \tag{1.40}$$
$$\int_{-1}^{1} \{P_l^m(x)\}^2\, \mathrm{d}x = \frac{2}{2l+1} \frac{(l+m)!}{(l-m)!} \tag{1.41}$$

Laguerre polynomials:

Differential equation:

$$xy'' + (1-x)y' + ny = 0 \tag{1.42}$$

if $n = 0, 1, 2, \ldots$ we get Laguerre polynomials given by

$$L_n(x) = e^x \frac{d^n}{dx^n}(x^n e^{-x}) \quad \text{(Rodrigue's formula)} \tag{1.43}$$

The first few polynomials are:

$$\begin{aligned} &L_o(x) = 1,\, L_1(x) = -x + 1,\, L_2(x) = x^2 - 4x + 2 \\ &L_3(x) = -x^3 + 9x^2 - 18x + 6,\, L_4(x) = x^4 - 16x^3 + 72x^2 - 96x + 24 \end{aligned} \tag{1.44}$$

Generating function:

$$\frac{e^{-xs/(1-s)}}{1-s} = \sum_{n=0}^{\infty} \frac{L_n(x)s^n}{n!} \tag{1.45}$$

Recurrence formulas:

$$\begin{aligned} &L_{n+1}(x) - (2n + 1 - x)L_n(x) + n^2 L_{n-1}(x) = 0 \\ &xL_n'(x) = nL_n(x) - n^2 L_{n-1}(x) \end{aligned} \tag{1.46}$$

Orthonormal properties:

$$\int_0^{\infty} e^{-x} L_m(x) L_n(x)\, dx = 0 \quad m \neq n \tag{1.47}$$

$$\int_0^{\infty} e^{-x} \{L_n(x)\}^2\, dx = (n!)^2 \tag{1.48}$$

Bessel functions: $(J_n(x))$

Differential equation of order n

$$x^2 y'' + xy' + (x^2 - n^2)y = 0 \quad n \geq 0 \tag{1.49}$$

Expansion formula:

$$J_n(x) = \sum_{k=0}^{\infty} \frac{(-1)^k (x/2)^{2k-n}}{k!\Gamma(k + 1 - n)} \tag{1.50}$$

Properties:

$$J_{-n}(x) = (-1)^n J_n(x) \quad n = 0, 1, 2, \ldots \tag{1.51}$$

$$J_o'(x) = -J_1(x) \tag{1.52}$$

$$J_{n+1}(x) = \frac{2n}{x} J_n(x) - J_{n-1}(x) \tag{1.53}$$

Generating function:

$$e^{x(s-1/s)/2} = \sum_{n=-\infty}^{\infty} J_n(x)t^n \tag{1.54}$$

Laplace transforms:

Definition:
A Laplace transform of the function $F(t)$ is

$$\int_0^\infty F(t)e^{-st}\,\mathrm{d}t = f(s) \tag{1.55}$$

The function $f(s)$ is the Laplace transform of $F(t)$. Symbolically, $\mathcal{L}\{F(t)\} = f(s)$ and $F(t) = \mathcal{L}^{-1}\{f(s)\}$ is the inverse Laplace transform of $f(s)$. $\mathcal{L}^{-1}$ is called the inverse Laplace operator.

Table of Laplace transforms:

$F(t)$	$f(s)$
$aF_1(t) + bF_2(t)$	$af_1(s) + bf_2(s)$
$aF(at)$	$f(s/a)$
$e^{at}F(t)$	$f(s-a)$
$F(t-a)\ t > a$ $0\quad t < a$	$e^{-as}f(s)$
1	$\dfrac{1}{s}$
t	$\dfrac{1}{s^2}$
$\dfrac{t^{n-1}}{(n-1)!}$	$\dfrac{1}{s^n} n = 1, 2, 3, \ldots$
e^{at}	$\dfrac{1}{s-a}$
$\dfrac{\sin at}{a}$	$\dfrac{1}{s^2+a^2}$
$\cos at$	$\dfrac{s}{s^2+a^2}$
$\dfrac{\sinh at}{a}$	$\dfrac{1}{s^2-a^2}$
$\cosh at$	$\dfrac{s}{s^2-a^2}$

Calculus of variation

The calculus of variation is concerned with the problem of finding a function $y(x)$ such that a definite integral, taken over a function shall be a maximum or minimum. Let it be desired to find that function $y(x)$ which will cause the integral

$$I = \int_{x_1}^{x_2} F(x, y, y')\,\mathrm{d}x \tag{1.56}$$

to have a stationary value (maximum or minimum). The integrand is taken to be a function of the dependent variable y as well as the independent variable x and $y' = \mathrm{d}y/\mathrm{d}x$. The limits x_1 and x_2 are fixed and at each of the limits y has definite value. The condition that I shall be stationary is given by Euler's equation

$$\frac{\partial F}{\partial y} - \frac{\mathrm{d}}{\mathrm{d}x}\frac{\partial F}{\partial y'} = 0 \tag{1.57}$$

When F does not depend explicitly on x, then a different form of the above equation is more useful

$$\frac{\partial F}{\partial x} - \frac{\mathrm{d}}{\mathrm{d}x}\left(F - y'\frac{\partial F}{\partial y'}\right) = 0 \tag{1.58}$$

which gives the result

$$F - y'\frac{\partial F}{\partial y'} = \text{Constant} \tag{1.59}$$

Statistical distribution

Binomial distribution

The probability of obtaining x successes in N-independent trials of an event for which p is the probability of success and q the probability of failure in a single trial is given by the binomial distribution $B(x)$.

$$B(x) = \frac{N!}{x!(N-x)!}p^x q^{N-x} = C_x^N p^x q^{N-x} \tag{1.60}$$

$B(x)$ is normalized, i.e.

$$\sum_{x=0}^{N} B(x) = 1 \tag{1.61}$$

It is a discrete distribution.
The mean value,

$$\langle x \rangle = Np \tag{1.62}$$

The S.D.,

$$\sigma = \sqrt{Npq} \tag{1.63}$$

Poisson distribution

The probability that x events occur in unit time when the mean rate of occurrence is m, is given by the Poisson distribution $P(x)$.

$$P(x) = \frac{e^{-m} m^x}{x!} \quad (x = 0, 1, 2, \ldots) \tag{1.64}$$

The distribution $P(x)$ is normalized, that is

$$\sum_{x=0}^{\infty} p(x) = 1 \tag{1.65}$$

This is also a discrete distribution.

When NP is held fixed, the binomial distribution tends to Poisson distribution as N is increased to infinity.

The expectation value, i.e.

$$\langle x \rangle = m \tag{1.66}$$

The S.D.,

$$\sigma = \sqrt{m} \tag{1.67}$$

Properties:

$$p_{m-1} = p_m \tag{1.68}$$

$$p_{x-1} = \frac{x}{m} p_m \; and \; p_{x+1} = \frac{m}{m+1} p_x \tag{1.69}$$

Normal (Gaussian distribution)

When p is held fixed, the binomial distribution tends to a Normal distribution as N is increased to infinity. It is a continuous distribution and has the form

$$f(x)\,\mathrm{d}x = \frac{1}{\sqrt{2\pi}\sigma} e^{-(x-m)^2/2\sigma^2} \mathrm{d}x \tag{1.70}$$

where m is the mean and σ is the S.D.

The probability of the occurrence of a single random event in the interval $m - \sigma$ and $m + \sigma$ is 0.6826 and that between $m - 2\sigma$ and $m + 2\sigma$ is 0.973.

Interval distribution

If the data contains N time intervals then the number of time intervals n between t_1 and t_2 is

$$n = N(e^{-at_1} - e^{-at_2}) \tag{1.71}$$

where a is the average number of intervals per unit time. Short intervals are more favored than long intervals.

Two limiting cases:

$$\text{(a) } t_2 = \infty; N = N_o e^{-\lambda t} \quad \text{(Law of radioactivity)} \tag{1.72}$$

This gives the number of surviving atoms at time t.

$$\text{(b) } t_1 = 0; N = N_o(1 - e^{-\lambda t}) \tag{1.73}$$

For radioactive decays this gives the number of decays in time interval 0 and t.

Above formulas are equally valid for length intervals such as interaction lengths.

Moment generating function (MGF)

$$\begin{aligned} \text{MGF} &= Ee^{(x-\mu)t} \\ &= E\left[1 + (x-\mu)t + (x-\mu)^2\frac{t^2}{2!} + \ldots\right] \\ &= 1 + 0 + \mu_2\frac{t^2}{2!} + \mu_3\frac{t^3}{3!} + \ldots \end{aligned} \tag{1.74}$$

so that μ_n, the nth moment about the mean is the coefficient of $t^n/n!$.

Propagation of errors

If the error on the measurement of $f(x, y, \ldots)$ is σ_f and that on x and y, σ_x and σ_y, respectively, and σ_x and σ_y are uncorrelated then

$$\sigma_f^2 = \left(\frac{\partial f}{\partial x}\right)^2 \sigma_x^2 + \left(\frac{\partial f}{\partial y}\right)^2 \sigma_y^2 + \cdots \tag{1.75}$$

Thus, if $f = x \pm y$, then $\sigma_f = \left(\sigma_x^2 + \sigma_y^2\right)^{1/2}$

And if $f = \frac{x}{y}$ then $\frac{\sigma f}{f} = \left(\frac{\sigma_x^2}{x^2} + \frac{\sigma_y^2}{y^2}\right)^{1/2}$

Least square fit

(a) Straight line: $y = mx + c$

It is desired to fit pairs of points $(x_1, y_1), (x_2, y_2), \ldots, (x_n, y_n)$ by a straight line

Residue: $S = \sum_{i=1}^{n}(y_i - mx_i - C)^2$

Minimize the residue: $\frac{\partial s}{\partial m} = 0; \frac{\partial s}{\partial c} = 0$

The normal equations are:

$$m\sum_{i=1}^{n} x_i^2 + C\sum_{i=1}^{n} x_i - \sum_{i=1}^{n} x_i y_i = 0$$

$$m\sum_{i=1}^{n} x_i + nC - \sum_{i=1}^{n} y_i = 0$$

which are to be solved as ordinary algebraic equations to determine the best values of m and C.

(b) Parabola: $y = a + bx + cx^2$

Residue: $S = \sum_{i=1}^{n}(y_i - a - bx_i - cx_i^2)^2$
Minimize the residue: $\frac{\partial s}{\partial a} = 0; \frac{\partial s}{\partial b} = 0; \frac{\partial s}{\partial c} = 0$

The normal equations are:

$$\sum y_i - na - b\sum x_i - c\sum x_i^2 = 0$$
$$\sum x_i y_i - a\sum x_i - b\sum x_i^2 - c\sum x_i^3 = 0$$
$$\sum x_i^2 y_i - a\sum x_i^2 - b\sum x_i^3 - c\sum x_i^4 = 0$$

which are to be solved as ordinary algebraic equations to determine the best value of a, b and c.

Numerical integration

Since the value of a definite integral is a measure of the area under a curve, it follows that the accurate measurement of such an area will give the exact value of a definite integral; $I = \int_{x_1}^{x_2} y(x)\mathrm{d}x$. The greater the number of intervals (i.e. the smaller Δx is), the closer will be the sum of the areas under consideration.

Trapezoidal rule

$$\text{area} = \left(\frac{1}{2}y_0 + y_1 + y_2 + \cdots y_{n-1} + \frac{1}{2}y_n\right)\Delta x \tag{1.76}$$

Simpson's rule

$$\text{area} = \frac{\Delta x}{3}(y_0 + 4y_1 + 2y_2 + 4y_3 + 2y_4 + \cdots y_n),\ n\ being\ even. \tag{1.77}$$

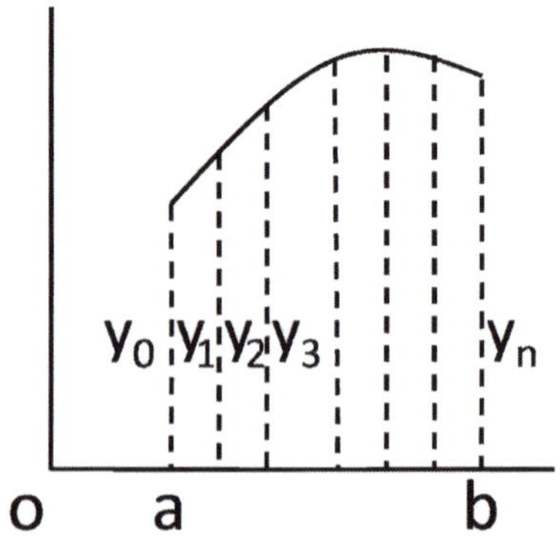

Fig. 1.1 Integration by Simpson's rule and Trapezoidal rule

Matrices

Types of matrices and definitions

Identity matrix:

$$I_2 = \begin{pmatrix} 1 & 0 \\ 0 & 1 \end{pmatrix}; I_3 = \begin{pmatrix} 1 & 0 & 0 \\ 0 & 1 & 0 \\ 0 & 0 & 1 \end{pmatrix} \tag{1.78}$$

Scalar matrix:

$$\begin{pmatrix} a_{11} & 0 \\ 0 & a_{22} \end{pmatrix}; \begin{pmatrix} a_{11} & 0 & 0 \\ 0 & a_{22} & 0 \\ 0 & 0 & a_{33} \end{pmatrix} \tag{1.79}$$

Symmetric matrix:

$$\left(a_{ji} = a_{ij}\right); \begin{pmatrix} a_{11} & a_{12} & a_{13} \\ a_{12} & a_{22} & a_{23} \\ a_{13} & a_{23} & a_{33} \end{pmatrix} \tag{1.80}$$

Skew symmetric:

$$\left(a_{ji} = -a_{ij}\right); \begin{pmatrix} a_{11} & a_{12} & a_{13} \\ -a_{12} & a_{22} & a_{23} \\ -a_{13} & -a_{23} & a_{33} \end{pmatrix} \tag{1.81}$$

The *Inverse of a matrix* $B = A^{-1}$ (B equals A inverse):
if $AB = BA = I$ and further, $(AB)^{-1} = B^{-1}A^{-1}$
A commutes with B if $AB = BA$
A anti-commutes with B if $AB = -BA$

The *Transpose* (A') *of a matrix* A means interchanging rows and columns.

$$\begin{aligned} Further, (A+B)' &= A' + B' \\ (A')' &= A, (kA)' = kA' \end{aligned} \tag{1.82}$$

The *Conjugate of a matrix*. If a matrix has complex numbers as elements, and if each number is replaced by its conjugate, then the new matrix is called the conjugate and denoted by A^* or $\overline{A}$ (A conjugate)

The *Trace (Tr) or Spur* of a matix is the num of the diagonal elements.

$$Tr = \sum a_{ii} \tag{1.83}$$

Hermetian matrix

If $\overline{A'} = A$, so that $a_{ij} = \bar{a}_{ji}$ for all values of i and j. Diagonal elements of an Hermitian matrix are real numbers.

Orthogonal matrix

A square matrix is said to be orthogonal if $AA' = A'A = I$, i.e. $A' = A^{-1}$

The column vector (row vectors) of an orthogonal matrix A are mutually orthogonal unit vectors.

The inverse and the transpose of an orthogonal matrix are orthogonal.
The product of any two or more orthogonal matrices is orthogonal.
The determinant of an Orthogonal matrix is ± 1.

Unitary matrix

A square matrix is called a unitary matrix if $(\overline{A})'A = A(\overline{A})' = I$, i.e. if $(\overline{A})' = A^{-1}$.

The column vectors (row vectors) of an n-square unitary matrix are an orthonormal set.

The inverse and the transpose of a unitary matrix are unitary.
The product of two or more unitary matrices is unitary.
The determinant of a unitary matrix has absolute value 1.

Unitary transformations

The linear transformation $Y = AX$ (where A is unitary and X is a vector), is called a unitary transformation.

If the matrix is unitary, the linear transformation preserves length.

Rank of a matrix

If $|A| \neq 0$, it is called non-singular; if $|A| = 0$, it is called singular.

A non-singular matrix is said to have rank r if at least one of its r-square minors is non-zero while if every $(r + 1)$ minor, if it exists, is zero.

Elementary transformations

(i) The interchange of the ith rows and jth rows or ith column or jth column.
(ii) The multiplication of every element of the ith row or ith column by a non-zero scalar.
(iii) The addition to the elements of the ith row (column) by k (a scalar) times the corresponding elements of the jth row (column). These elementary transformations known as row elementary or row transformations do not change the order of the matrix.

Equivalence

A and B are said to be equivalent ($A \sim B$) if one can be obtained from the other by a sequence of elementary transformations.

The adjoint of a square matrix

If $A = [a_{ij}]$ is a square matrix and α_{ij} the cofactor of a_{ij} then

$$\text{adj } A = \begin{bmatrix} \alpha_{11} & \alpha_{21} & \cdots & \alpha_{n1} \\ \alpha_{12} & \alpha_{22} & \cdots & \cdots \\ \cdots & \cdots & \cdots & \cdots \\ \cdots & \cdots & \cdots & \alpha_{nn} \end{bmatrix}$$

The cofactor $\alpha_{ij} = (-1)^{i+j} M_{ij}$
where M_{ij} is the minor obtained by striking off the ith row and jth column and computing the determinant from the remaining elements.

Inverse from the adjoint

$$A^{-1} = \frac{adj\ A}{|A|}$$

Inverse for orthogonal matrices

$$A^{-1} = A'$$

Inverse of unitary matrices

$$A^{-1} = (\overline{A})'$$

Characteristic equation

$$\text{Let } AX = \lambda X \tag{1.84}$$

be the transformation of the vector X into λX, where λ is a number, then λ is called the eigen or characteristic value.
From (1.84):

$$(A - \lambda I)X = \begin{bmatrix} a_{11} - \lambda & a_{12} & \cdots & a_{1n} \\ a_{21} & a_{22} - \lambda & \cdots & a_{2n} \\ \vdots & \cdots & \cdots & \cdots \\ a_{n1} & \cdots & \cdots & a_{nn} - \lambda \end{bmatrix} \begin{bmatrix} x_1 \\ x_2 \\ \vdots \\ x_n \end{bmatrix} = 0 \tag{1.85}$$

The system of homogenous equations has non-trivial solutions if

$$|A - \lambda I| = \begin{bmatrix} a_{11} - \lambda & a_{12} & \cdots & a_{1n} \\ a_{21} & a_{22} - \lambda & \cdots & a_{2n} \\ \vdots & \cdots & \cdots & \cdots \\ a_{n1} & \cdots & \cdots & a_{nn} - \lambda \end{bmatrix} = 0 \tag{1.86}$$

The expansion of this determinant yields a polynomial $\phi(\lambda) = 0$ is called the characteristic equation of A and its roots $\lambda_1, \lambda_1, \ldots, \lambda_n$ are known as the characteristic roots of A. The vectors associated with the characteristic roots are called invariant or characteristic vectors.

Diagonalization of a square matrix

If a matrix C is found such that the matrix A is diagonalized to S by the transformation

$$S = C^{-1}AC \tag{1.87}$$

then S will have the characteristic roots as the diagonal elements.

Ordinary differential equations

The methods of solving typical ordinary differential equations are from the book "Differential and Integral Calculus" by William A. Granville published by Ginn & Co., 1911.

An ordinary differential equation involves only one independent variable, while a partial differential equation involves more than one independent variable.

The order of a differential equation is that of the highest derivative in it.

The degree of a differential equation which is algebraic in the derivatives is the power of the highest derivative in it when the equation is free from radicals and fractions.

Differential equations of the first order and of the first degree

Such an equation must be brought into the form $M\mathrm{d}x + N\mathrm{d}y = 0$, in which M and N are functions of x and y.

Type I variables separable

When the terms of a differential equation can be so arranged that it takes on the form

$$(A) \quad f(x)\,\mathrm{d}x + F(y)\,\mathrm{d}y = 0$$

where $f(x)$ is a function of x alone and $F(y)$ is a function of y alone, the process is called separation of variables and the solution is obtained by direct integration.

$$(B) \quad \int f(x)\,\mathrm{d}x + \int F(y)\,\mathrm{d}y = C$$

where C is an arbitrary constant.

Equations which are not in the simple form (A) can be brought into that form by the following rule for separating the variables.

First step: Clear off fractions, and if the equation involves derivatives, multiply through by the differential of the independent variable.

Second step: Collect all the terms containing the same differential into a single term. If then the equation takes on the form

$$XY\,\mathrm{d}x + X'Y'\,\mathrm{d}y = 0$$

where X, X' are functions of x alone, and Y, Y' are functions of y alone, it may be brought to the form (A) by dividing through by $X'Y'$.

Third step: Integrate each part separately as in (B).

Type II homogeneous equations

The differential equation

$$M\mathrm{d}x + N\mathrm{d}y = 0$$

is said to be homogeneous when M and N are homogeneous function of x and y of the same degree. In effect a function of x and y is said to be homogenous in the variable if the result of replacing x and y by λx and λy (λ being arbitrary) reduces to the original function multiplied by some power of λ. This power of λ is called the degree of the original function. Such differential equations may be solved by making the substitution

$$y = vx$$

This will give a differential equation in v and x in which the variables are separable, and hence we may follow the rule (A) of type I.

Type III linear equations

A differential equation is said to be linear if the equation is of the first degree in the dependent variables (usually y) and its derivatives. The linear differential equation of the first order is of the form

$$\frac{\mathrm{d}y}{\mathrm{d}x} + Py = Q$$

where P, Q are functions of x alone, or constants, the solution is given by

$$ye^{\int P\mathrm{d}x} = \int Qe^{\int P\mathrm{d}x}\,\mathrm{d}x + C$$

Type IV equations reducible to linear form

Some equations that are not linear can be reduced to the linear form by a suitable substitution, for example

$$(A)\quad \frac{dy}{dx} + Py = Qy^n$$

where P, Q are functions of x alone, or constants. Equation (A) may be reduced to the linear form (A), Type III by means of the substitution $x = y^{-n+1}$.

Differential equations of the nth Order and of the nth degree

Consider special cases of linear differential equations.

Type I – The linear differential equation

$$(A)\quad \frac{d^n y}{dx^n} + P_1\frac{d^{n-1}y}{dx^{n-1}} + P_2\frac{d^{n-2}y}{dx^{n-2}} + \cdots + P_n y = 0$$

in which coefficients $P_1, P_2, \ldots P_n$ are constants.

Consider the differential equation of third order

$$(B)\quad \frac{d^3 y}{dx^3} + P_1\frac{d^2 y}{dx^2} + P_2\frac{dy}{dx} + P_3 y = 0$$

where P_1, P_2 and P_3 are constants. The corresponding auxiliary equation is

$$r^3 + P_1 r^2 + P_2 r + P_3 = 0$$

Let the roots be r_1, r_2, r_3.

If r_1, r_2, r_3 are real and distinct,

$$y = C_1 e^{r_1 x} + C_2 e^{r_2 x} + C_3 e^{r_3 x}$$

If r_1, r_2, r_3 are real and equal

$$y = C_1 e^{-r_1 x} + C_2 x e^{-r_2 x} + C_3 x^2 e^{-r_3 x}$$

In case $a + bi$ and $a - bi$ are each multiple roots of the auxiliary equation occurring s times, the solutions would be

$$C_1 e^{ax}\cos bx,\ C_2 x e^{ax}\cos bx,\ C_3 x^2 e^{ax}\cos bx, \ldots C_s x^{s-1} e^{ax}\cos bx$$
$$C_1' e^{ax}\sin bx,\ C_2' x e^{ax}\sin bx,\ C_3' x^2 e^{ax}\sin bx, \ldots C_s' x^{s-1} e^{ax}\sin bx$$

Summary for the rule for solving differential equations of the type

$$\frac{d^n y}{dx^n} + P_1\frac{d^{n-1}y}{dx^{n-1}} + P_2\frac{d^{n-2}y}{dx^{n-2}} + \cdots + P_n y = 0$$

where $P_1, P_2, \ldots P_n$ are constants.

First step: Write down the corresponding auxiliary equation

$$D^n + p_1 D^{n-1} + p_2 D^{n-2} + \cdots + p_n = 0$$

Second step: Solve completely the auxiliary equation.

Third step: From the roots of the auxiliary equation, write down the corresponding particular solutions of the differential equation as follows

Auxiliary equation	Differential equation
(a) Each distinct real root r_1	Gives a particular solution $e^{r_1 x}$
(b) Each distinct pair of imaginary roots $a \pm bi$	Gives two particular solutions $e^{ax}\cos bx$, $e^{ax}\sin bx$
(c) A multiple root occurring s times	Gives s particular solutions obtained by multiplying the particular solutions (a) or (b) by $1, x, x^2, \ldots, x^{n-1}$

Fourth step: Multiple each of the n independent solutions by an arbitrary constant and add the results. This gives the complete solution.

Type II

$$(I) \quad \frac{d^n y}{dx^n} + P_1 \frac{d^{n-1} y}{dx^{n-1}} + P_2 \frac{d^{n-2} y}{dx^{n-2}} + \cdots + P_n y = X$$

where X is a function of x alone, or constant, and $P_1, P_2, \ldots P_n$ are constants.

When $X = 0$, (I) reduces to (A) Type I.

$$(J) \quad \frac{d^n y}{dx^n} + P_1 \frac{d^{n-1} y}{dx^{n-1}} + P_2 \frac{d^{n-2} y}{dx^{n-2}} + \cdots + P_n y = 0$$

The complete solution of (J) is called the complementary function of (I).

Let u be the complete solution of (J), i.e. the complementary function of (I), and v any particular solution of (I). Then

$$\frac{d^n v}{dx^n} + P_1 \frac{d^{n-1} v}{dx^{n-1}} + P_2 \frac{d^{n-2} v}{dx^{n-2}} + \cdots + P_n v = X$$

$$\text{and} \quad \frac{d^n u}{dx^n} + P_1 \frac{d^{n-1} u}{dx^{n-1}} + P_2 \frac{d^{n-2} u}{dx^{n-2}} + \cdots + P_n u = 0$$

Adding we get

$$\frac{d^n (u+v)}{dx^n} + P_1 \frac{d^{n-1} (u+v)}{dx^{n-1}} + P_2 \frac{d^{n-2} (u+v)}{dx^{n-2}} + \cdots + P_n (u+v) = X$$

showing that $u + v$ is a solution of I.

Procedure:

First step: Replace the RHS member of the given equation (I) by zero and solve the complimentary function of I to get $y = u$.

Second step: Differentiate successively the given equation (I) and obtain, either directly or by elimination, a differential equation of a higher order of type I.

Third step: Solving this new equation by the previous rule we get its complete solution

$$y = u + v$$

where the part u is the complimentary function of (I) already found in the first step, and v is the sum of additional terms found

Fourth step: To find the values of the constants of integration in the particular solution v, substitute

$$y = u + v$$

and its derivatives in the equation (I). In the resulting identity equation equate the coefficients of like terms, solve for constants of integration, substitute their values back in

$$y = u + v$$

giving the complete solution of (I).

Type III

$$\frac{\mathrm{d}^n y}{\mathrm{d}x^n} = X$$

where X is a function of x alone, or constant

Integrate n times successively. Each integration will introduce one arbitrary constant.

Type IV

$$\frac{\mathrm{d}^2 y}{\mathrm{d}x^2} = Y$$

where Y is a function of y alone

Multiply the LHS member by the factor $2\frac{\mathrm{d}y}{\mathrm{d}x}\,\mathrm{d}x$ and the RHS member by k equivalent factor $2\mathrm{d}y$

$$2\frac{\mathrm{d}y}{\mathrm{d}x}\frac{\mathrm{d}^2 y}{\mathrm{d}x^2}\,\mathrm{d}x = d\left(\frac{\mathrm{d}y}{\mathrm{d}x}\right)^2 = 2Y\,\mathrm{d}y$$

$$\int d\left(\frac{\mathrm{d}y}{\mathrm{d}x}\right)^2 = \left(\frac{\mathrm{d}y}{\mathrm{d}x}\right)^2 = \int 2Y\,\mathrm{d}y + C_1$$

Extract the square root of both members, separate the variables, and integrate again, introducing the second arbitrary constant C_2.

Complex variables

Complex number $z = r(\cos\theta + i\sin\theta)$, where $i = \sqrt{-1}$
$z^n = \cos n\theta + i\sin n\theta$

Analytic functions

A function f of the complex variable z is analytic at a point z_o if its derivative $f'(z)$ exists not only at z_o but at every point z in some neighborhood of z_o. As an example if $f(z) = \frac{1}{z}$ then $f'(z) = -\frac{1}{z^2}(z \neq 0)$. Thus f is analytic at every point except the point $z = 0$, where it is not continuous, so that $f'(0)$ cannot exist. The point $z = 0$ is called a singular point.

Contour

A contour is a continuous chain of finite number of smooth arcs. If the contour is closed and does not intersect itself, it is called a closed contour. Boundaries of triangles and rectangles are examples. Any closed contour separates the plane into two domains each of which have the points of C as their only boundary points. One of these domains is called the interior of C, is bounded; the other, the exterior of C, is unbounded.

Contour integral is similar to the line integral except that here one deals with the complex plane.

The Cauchy integral formula

Let f be analytic everywhere within and on a closed contour C. If z_o is any point interior to C, then

$$f(z_o) = \frac{1}{2\pi i}\int_C \frac{f(z)\,dz}{z - z_o}$$

where the integral is taken in the positive sense around C.

1.2 Problems

1.2.1 Vector Calculus

1.1 If $\phi = \frac{1}{r}$, where $r = (x^2 + y^2 + z^2)^{1/2}$, show that $\nabla\phi = \frac{r}{r^3}$.

1.2 Find a unit vector normal to the surface $xy^2 + xz = 1$ at the point $(-1, 1, 1)$.

1.3 Show that the divergence of the Coulomb or gravitational force is zero.

1.4 If $\mathbf{A}$ and $\mathbf{B}$ are irrotational, prove that $\mathbf{A}\times\mathbf{B}$ is Solenoidal that is div $(\mathbf{A}\times\mathbf{B}) = 0$

1.5 (a) If the field is centrally represented by $\boldsymbol{F} = f(x, y, z), \boldsymbol{r} = f(r)\boldsymbol{r}$, then it is conservative conditioned by curl $\boldsymbol{F} = 0$, that is the field is irrotational.
(b) What should be the function $F(r)$ so that the field is solenoidal?

1.6 Evaluate $\int_c \boldsymbol{A}\,.\,\mathrm{d}\boldsymbol{r}$ from the point $P(0, 0, 0)$ to $Q(1, 1, 1)$ along the curve $\boldsymbol{r} = \hat{i}t + \hat{j}t^2 + \hat{k}r^3$ with $x = t,\ y = t^2,\ z = t^3$, where $\boldsymbol{A} = y\hat{i} + xz\hat{j} + xyz\hat{k}$

1.7 Evaluate $\oint_c \boldsymbol{A}\,.\,\mathrm{d}\boldsymbol{r}$ around the closed curve C defined by $y = x^2$ and $y^2 = 8x$, with $\boldsymbol{A} = (x + y)\hat{i} + (x - y)\hat{j}$

1.8 (a) Show that $\boldsymbol{F} = (2xy + z^2)\hat{i} + x^2\hat{j} + xyz\hat{k}$, is a conservative force field.
(b) Find the scalar potential.
(c) Find the work done in moving a unit mass in this field from the point $(1, 0, 1)$ to $(2, 1, -1)$.

1.9 Verify Green's theorem in the plane for $\int_c (x + y)\mathrm{d}x + (x - y)\,\mathrm{d}y$, where C is the closed curve of the region bonded by $y = x^2$ and $y^2 = 8x$.

1.10 Show that $\int_s \boldsymbol{A}\,.\,\mathrm{d}s = \frac{12}{5}\pi R^2$, where S is the sphere of radius R and $\boldsymbol{A} = \hat{i}x^3 + \hat{j}y^3 + \hat{k}z^3$

1.11 Evaluate $\int_r \boldsymbol{A}\,.\,\mathrm{d}\boldsymbol{r}$ around the circle $x^2 + y^2 = R^2$ in the xy-plane, where $\boldsymbol{A} = 2y\hat{i} - 3x\hat{j} + z\hat{k}$

1.12 (a) Prove that the curl of gradient is zero.
(b) Prove that the divergence of a curl is zero.

1.13 If $\phi = x^2y - 2xz^3$, then:
(a) Find the Gradient.
(b) Find the Laplacian.

1.14 (a) Find a unit vector normal to the surface $x^2y + xz = 3$ at the point $(1, -1, 1)$.
(b) Find the directional derivative of $\phi = x^2yz + 2xz^3$ at $(1, 1, -1)$ in the direction $2\hat{i} - 2\hat{j} + \hat{k}$.

1.15 Show that the divergence of an inverse square force is zero.

1.16 Find the angle between the surfaces $x^2 + y^2 + z^2 = 1$ and $z = x^2 + y^2 - 1$ at the point $(1, +1, -1)$.

1.2.2 Fourier Series and Fourier Transforms

1.17 Develop the Fourier series expansion for the saw-tooth (Ramp) wave $f(x) = x/L,\ -L < x < L$, as in Fig. 1.2.

1.18 Find the Fourier series of the periodic function defined by:
$f(x) = 0,\ \text{if } -\pi \le x \le 0$
$f(x) = \pi,\ \text{if } 0 \le x \le \pi$

Fig. 1.2 Saw-tooth wave

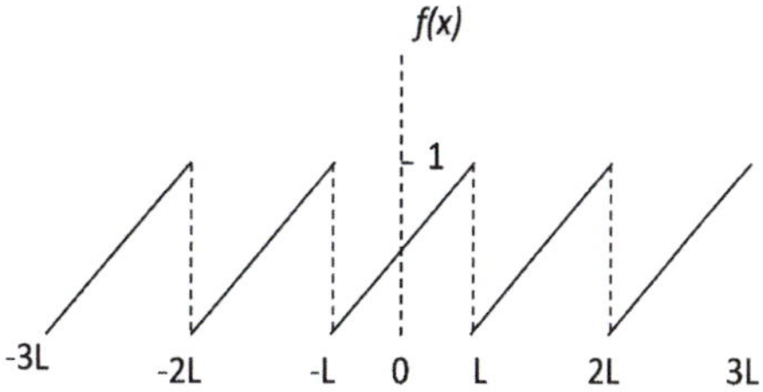

1.19 Use the result of Problem 1.18 for the Fourier series for the square wave to prove that:

$$1 - \frac{1}{3} + \frac{1}{5} - \frac{1}{7} + \cdots = \frac{\pi}{4}$$

1.20 Find the Fourier transform of $f(x) = \begin{cases} 1, & |x| < a \\ 0, & |x| > a \end{cases}$

1.21 Use the Fourier integral to prove that:

$$\int_0^\infty \frac{\cos ax \mathrm{d}x}{1 + a^2} = \frac{\pi}{2} e^{-x}$$

1.22 Show that the Fourier transform of the normalized Gaussian distribution

$$f(t) = \frac{1}{\tau\sqrt{2\pi}} e^{\frac{-t^2}{2\tau^2}}, \quad -\infty < t < \infty$$

is another Gaussian distribution.

1.2.3 Gamma and Beta Functions

1.23 The gamma function is defined by:

$$\Gamma(z) = \int_0^\infty e^{-x} x^{z-1} \,\mathrm{d}x, (Re\, z > 0)$$

(a) Show that $\Gamma(z+1) = z\Gamma(z)$
(b) And if z is a positive integer n, then $\Gamma(n+1) = n!$

1.24 The Beta function $B(m, n)$ is defined by the definite integral:

$$B(m, n) = \int_0^1 x^{m-1}(1-x)^{n-1} \,\mathrm{d}x$$

and this defines a function of m and n provided m and n are positive. Show that:

$$B(m, n) = \frac{T(m)T(n)}{T(m+n)}$$

1.25 Use the Beta functions to evaluate the definite integral $\int_0^{\pi/2} (\cos\theta)^r \, d\theta$

1.26 Show that:

(a) $\Gamma(n)\Gamma(1-n) = \frac{\pi}{\sin(n\pi)}; 0 < n < 1$
(b) $|\Gamma(in)|^2 = \frac{\pi}{n \sin h(n\pi)}$

1.2.4 Matrix Algebra

1.27 Prove that the characteristic roots of a Hermitian matrix are real.

1.28 Find the characteristic equation and the Eigen values of the matrix:

$$\begin{pmatrix} 1 & -1 & 1 \\ 0 & 3 & -1 \\ 0 & 0 & 2 \end{pmatrix}$$

1.29 Given below the set of matrices:

$$A = \begin{pmatrix} -1 & 0 \\ 0 & -1 \end{pmatrix}, B = \begin{pmatrix} 0 & 1 \\ 1 & 0 \end{pmatrix}, C = \begin{pmatrix} 2 & 0 \\ 0 & 2 \end{pmatrix}, D = \begin{pmatrix} \frac{\sqrt{3}}{2} & \frac{1}{2} \\ -\frac{1}{2} & \frac{\sqrt{3}}{2} \end{pmatrix}$$

what is the effect when A, B, C and D act separately on the position vector $\begin{pmatrix} x \\ y \end{pmatrix}$?

1.30 Find the eigen values of the matrix:

$$\begin{pmatrix} 6 & -2 & 2 \\ -2 & 3 & -1 \\ 2 & -1 & 3 \end{pmatrix}$$

1.31 Diagonalize the matrix given in Problem 1.30 and find the trace ($T_r = \lambda_1 + \lambda_2 + \lambda_3$)

1.32 In the Eigen vector equation $A\mathbf{X} = \lambda\mathbf{X}$, the operator A is given by $A = \begin{bmatrix} 3 & 2 \\ 4 & 1 \end{bmatrix}$.

Find:

(a) The Eigen values λ
(b) The Eigen vector $\mathbf{X}$
(c) The modal matrix C and it's inverse C^{-1}
(d) The product $C^{-1}AC$

1.2.5 Maxima and Minima

1.33 Solve the equation $x^3 - 3x + 3 = 0$, by Newton's method.

1.34 (a) Find the turning points of the function $f(x) = x^2 e^{-x^2}$.
(b) Is the above function odd or even or neither?

1.2.6 Series

1.35 Find the interval of convergence for the series:

$$x - \frac{x^2}{2^2} + \frac{x^3}{3^2} - \frac{x^4}{4^2} + \cdots$$

1.36 Expand $\log x$ in powers of $(x - 1)$ by Taylor's series.

1.37 Expand $\cos x$ into an infinite power series and determine for what values of x it converges.

1.38 Expand $\sin(a + x)$ in powers of x by Taylor's series.

1.39 Sum the series $s = 1 + 2x + 3x^2 + 4x^3 + \cdots$, $|x| < 1$

1.2.7 Integration

1.40 (a) Evaluate the integral:

$$\int \sin^3 x \cos^6 x \, \mathrm{d}x$$

(b) Evaluate the integral:

$$\int \sin^4 x \cos^2 x \, \mathrm{d}x$$

1.41 Evaluate the integral:

$$\int \frac{1}{2x^2 - 3x - 2} \, \mathrm{d}x$$

1.42 (a) Sketch the curve in polar coordinates $r^2 = a^2 \sin 2\theta$
(b) Find the area within the curve between $\theta = 0$ and $\theta = \pi/2$.

1.43 Evaluate:

$$\int \frac{(x^3 + x^2 + 2)}{(x^2 + 2)^2} \, \mathrm{d}x$$

1.44 Evaluate the definite integral:

$$\int_0^{+\infty} \frac{4a^3}{x^2 + 4a^2} \, \mathrm{d}x$$

1.45 (a) Evaluate:

$$\int \tan^6 x \sec^4 x \, \mathrm{d}x$$

(b) Evaluate:

$$\int \tan^5 x \sec^3 x \, \mathrm{d}x$$

1.46 Show that:

$$\int_2^4 \frac{2x+4}{x^2-4x+8}\,dx = \ln 2 + \pi$$

1.47 Find the area included between the semi-cubical parabola $y^2 = x^3$ and the line $x = 4$

1.48 Find the area of the surface of revolution generated by revolving the hypocycloid $x^{2/3} + y^{2/3} = a^{2/3}$ about the x-axis.

1.49 Find the value of the definite double integral:

$$\int_0^a \int_0^{\sqrt{a^2-x^2}} (x+y)\,dy\,dx$$

1.50 Calculate the area of the region enclosed between the curve $y = 1/x$, the curve $y = -1/x$, and the lines $x = 1$ and $x = 2$.

1.51 Evaluate the integral:

$$\int \frac{dx}{x^2 - 18x + 34}$$

1.52 Use integration by parts to evaluate:

$$\int_0^1 x^2 \tan^{-1} x\,dx$$

[University of Wales, Aberystwyth 2006]

1.53 (a) Calculate the area bounded by the curves $y = x^2 + 2$ and $y = x - 1$ and the lines $x = -1$ to the left and $x = 2$ to the right.
(b) Find the volume of the solid of revolution obtained by rotating the area enclosed by the lines $x = 0$, $y = 0$, $x = 2$ and $2x + y = 5$ through 2π radians about the y-axis.

[University of Wales, Aberystwyth 2006]

1.54 Consider the curve $y = x \sin x$ on the interval $0 \le x \le 2\pi$.
(a) Find the area enclosed by the curve and the x-axis.
(b) Find the volume generated when the curve rotates completely about the x-axis.

1.2.8 Ordinary Differential Equations

1.55 Solve the differential equation:

$$\frac{dy}{dx} = \frac{x^3 + y^3}{3xy^2}$$

1.56 Solve:

$$\frac{d^3y}{dx^3} - 3\frac{d^2y}{dx^2} + 4y = 0 \text{ (Osmania University)}$$

1.57 Solve:

$$\frac{d^4y}{dx^4} - 4\frac{d^3y}{dx^3} + 10\frac{d^2y}{dx^2} - 12\frac{dy}{dx} + 5y = 0 \text{ (Osmania University)}$$

1.58 Solve:

$$\frac{d^2y}{dx^2} + m^2y = \cos ax$$

1.59 Solve:

$$\frac{d^2y}{dx^2} - 5\frac{dy}{dx} + 6y = x$$

1.60 Solve the equation of motion for the damped oscillator:

$$\frac{d^2x}{dt^2} + 2\frac{dx}{dt} + 5x = 0$$

subject to the condition, $x = 5$, $dx/dt = -3$ at $t = 0$.

1.61 Two equal masses are connected by means of a spring and two other identical springs fixed to rigid supports on either side (Fig. 1.3), permit the masses to jointly undergo simple harmonic motion along a straight line, so that the system corresponds to two coupled oscillations.

Assume that $m_1 = m_2 = m$ and the stiffness constant is k for both the oscillators.

(a) Form the differential equations for both the oscillators and solve the coupled equations and obtain the frequencies of oscillations.
(b) Discuss the modes of oscillation and sketch the modes.

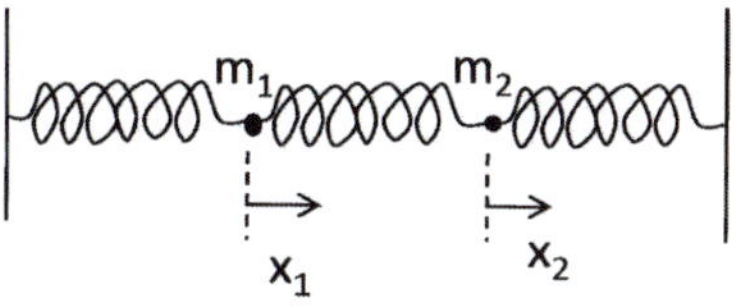

Fig. 1.3 Coupled oscillator

1.62 A cylinder of mass m is allowed to roll on a spring attached to it so that it encounters simple harmonic motion about the equilibrium position. Use the energy conservation to form the differential equation. Solve the equation and find the time period of oscillation. Assume k to be the spring constant.

1.63 Solve:

$$\frac{d^2y}{dx^2} - 8\frac{dy}{dx} = -16y$$

1.64 Solve:

$$x^2\frac{dy}{dx} + y(x+1)x = 9x^2$$

1.65 Find the general solution of the differential equation:

$$\frac{d^2y}{dx^2} + \frac{dy}{dx} - 2y = 2\cosh(2x)$$

[University of Wales, Aberystwyth 2004]

1.66 Solve:

$$x\frac{dy}{dx} - y = x^2$$

1.67 Find the general solution of the following differential equations and write down the degree and order of the equation and whether it is homogenous or in-homogenous.

(a) $y' - \frac{2}{x}y = \frac{1}{x^3}$
(b) $y'' + 5y' + 4y = 0$

[University of Wales, Aberystwyth 2006]

1.68 Find the general solution of the following differential equations:

(a) $\frac{dy}{dx} + y = e^{-x}$
(b) $\frac{d^2y}{dx^2} + 4y = 2\cos(2x)$

[University of Wales, Aberystwyth 2006]

1.69 Find the solution to the differential equation:

$$\frac{dy}{dx} + \frac{3}{x+2}y = x + 2$$

which satisfies $y = 2$ when $x = -1$, express your answer in the form $y = f(x)$.

1.70 (a) Find the solution to the differential equation:

$$\frac{d^2y}{dx^2} - 4\frac{dy}{dx} + 4y = 8x^2 - 4x - 4$$

which satisfies the conditions $y = -2$ and $\frac{dy}{dx} = 0$ when $x = 0$.

(b) Find the general solution to the differential equation:

$$\frac{d^2y}{dx^2} + 4y = \sin x$$

1.71 Find a fundamental set of solutions to the third-order equation:

$$\frac{d^3y}{dx^3} - \frac{d^2y}{dx^2} + \frac{dy}{dx} - y = 0$$

1.2.9 Laplace Transforms

1.72 Consider the chain decay in radioactivity $A \overset{\lambda_A}{\rightarrow} B \overset{\lambda_B}{\rightarrow} C$, where λ_A and λ_B are the disintegration constants. The equations for the radioactive decays are:

$$\frac{dN_A(t)}{dt} = -\lambda_A N_A(t), \text{ and } \frac{dN_B(t)}{dt} = -\lambda_2 N_B(t) + \lambda_A N_A(t)$$

where $N_A(t)$ and $N_B(t)$ are the number of atoms of A and B at time t, with the initial conditions $N_A(0) = N_A^0$; $N_B(0) = 0$. Apply Laplace transform to obtain $N_A(t)$ and $N_B(t)$, the number of atoms of A and B as a function of time t, in terms of N_A^0, λ_A and λ_B.

1.73 Consider the radioactive decay:

$A \overset{\lambda_A}{\rightarrow} B \overset{\lambda_B}{\rightarrow} C$ (Stable)

The equations for the chain decay are:

$$\frac{dN_A}{dt} = -\lambda_A N_A \quad (1)$$

$$\frac{dN_B}{dt} = -\lambda_B N_B + \lambda_A N_A \quad (2)$$

$$\frac{dN_C}{dt} = +\lambda_B N_B \quad (3)$$

with the initial conditions $N_A(0) = N_A^0$; $N_B(0) = 0$; $N_C(0) = 0$, where various symbols have the usual meaning. Apply Laplace transforms to find the growth of C.

1.74 Show that:

(a) $\pounds(e^{ax}) = \frac{1}{s-a},\ if\ s > a$
(b) $\pounds(\cos ax) = \frac{s}{s^2+a^2},\ s > 0$
(c) $\pounds(\sin ax) = \frac{a}{s^2+a^2}$

where $\pounds$ means Laplacian transform.

1.2.10 Special Functions

1.75 The following polynomial of order n is called Hermite polynomial:

$H_n'' - 2\xi H_n' + 2n H_n = 0$

Show that:

(a) $H_n' = 2n H_{n-1}$
(b) $H_{n+1} = 2\xi H_n - 2n H_{n-1}$

1.76 The Bessel function $J_n(x)$ is given by the series expansion

$$J_n(x) = \frac{\sum(-1)^k (x/2)^{n+2k}}{k!\Gamma(n+k+1)}$$

Show that:

(a) $\frac{d}{dx}[x^n J_n(x)] = x^n J_{n-1}(x)$
(b) $\frac{d}{dx}[x^{-n} J_n(x)] = -x^{-n} J_{n+1}(x)$

1.77 Prove the following relations for the Bessel functions:

(a) $J_{n-1}(x) - J_{n+1}(x) = 2\frac{d}{dx} J_n(x)$
(b) $J_{n-1}(x) + J_{n+1}(x) = 2\frac{n}{x} J_n(x)$

1.78 Given that $\Gamma\left(\frac{1}{2}\right) = \sqrt{\pi}$, obtain the formulae:

(a) $J_{1/2}(x) = \sqrt{\frac{2}{\pi x}} \sin x$

(b) $J_{-1/2}(x) = \sqrt{\frac{2}{\pi x}} \cos x$

1.79 Show that the Legendre polynomials have the property:

$$\int_{-l}^{l} P_n(x) P_m(x)\,dx = \frac{2}{2n+1}, \text{ if } m = n$$
$$= 0, \text{ if } m \neq n$$

1.80 Show that for large n and small θ, $P_n(\cos\theta) \approx J_0(n\theta)$

1.81 For Legendre polynomials $P_l(x)$ the generating function is given by:

$$T(x,s) = (1 - 2sx + s^2)^{-1/2}$$
$$= \sum_{l=0}^{\infty} P_l(x) s^l, \; s < 1$$

Use the generating function to show:

(a) $(l+1)P_{l+1} = (2l+1)x P_l - l\, P_{l-1}$
(b) $P_l(x) + 2x P_l'(x) = P_{l+1}'(x) + P_{l-1}'(x)$, Where prime means differentiation with respect to x.

1.82 For Laguerre's polynomials, show that $L_n(0) = n!$. Assume the generating function:

$$\frac{e^{-xs/(1-s)}}{1-s} = \sum_{n=0}^{\infty} \frac{L_n(x) s^n}{n!}$$

1.2.11 Complex Variables

1.83 Evaluate $\oint_c \frac{dz}{z-2}$ where C is:

(a) The circle $|z| = 1$
(b) The circle $|z + i| = 3$

1.84 Evaluate $\oint_c \frac{4z^2-3z+1}{(z-1)^3}\,dz$ when C is any simple closed curve enclosing $z = 1$.

1.85 Locate in the finite z-plane all the singularities of the following function and name them:

$$\frac{4z^3 - 2z + 1}{(z-3)^2(z-i)(z+1-2i)}$$

1.86 Determine the residues of the following function at the poles $z = 1$ and $z = -2$:

$$\frac{1}{(z-1)(z+2)^2}$$

1.87 Find the Laurent series about the singularity for the function:

$$\frac{e^x}{(z-2)^2}$$

1.88 Evaluate $I = \int_0^\infty \frac{dx}{x^4+1}$

1.2.12 Calculus of Variation

1.89 What is the curve which has shortest length between two points?

1.90 A bead slides down a frictionless wire connecting two points A and B as in the Fig. 1.4. Find the curve of quickest descent. This is known as the Brachistochrome, discovered by John Bernoulli (1696).

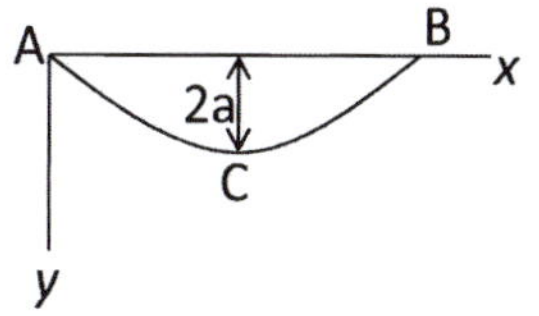

Fig. 1.4 Brachistochrome

1.91 If a soap film is stretched between two circular wires, both having their planes perpendicular to the line joining their centers, it will form a figure of revolution about that line. At every point such as P (Fig. 1.5), the horizontal component of the surface of revolution acting around a vertical section of the film will be constant. Find the equation to the figure of revolution.

1.92 Prove that the sphere is the solid figure of revolution which for a given surface area has maximum volume.

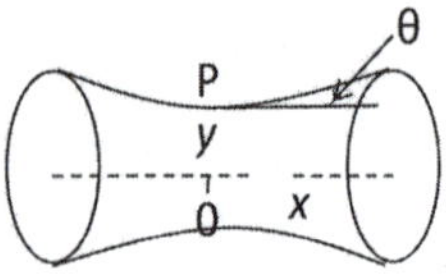

Fig. 1.5 Soap film stretched between two parallel circular wires

1.2.13 Statistical Distributions

1.93 Poisson distribution gives the probability that x events occur in unit time when the mean rate of occurrence is m.

$$P_x = \frac{e^{-m}m^x}{x!}$$

(a) Show that P_x is normalized.
(b) Show that the mean rate of occurrence or the expectation value $< x >$, is equal to m.
(c) Show that the S.D., $\sigma = \sqrt{m}$
(d) Show that $P_{m-1} = P_m$
(e) Show that $P_{x-1} = \frac{x}{m} P_m$ and $P_{x+1} = \frac{m}{x+1} P_x$

1.94 The probability of obtaining x successes in N-independent trials of an event for which p is the probability of success and q the probability of failure in a single trial is given by the Binomial distribution:

$$B(x) = \frac{N!}{x!(N-x)!} p^x q^{N-x} = C_x^N p^x q^{N-x}$$

(a) Show that $B(x)$ is normalized.
(b) Show that the mean value is Np
(c) Show that the S.D. is $\sqrt{Npq}$

1.95 A G.M. counter records 4,900 background counts in 100 min. With a radioactive source in position, the same total number of counts are recorded in 20 min. Calculate the percentage of S.D. with net counts due to the source.

[Osmania University 1964]

1.96 (a) Show that when p is held fixed, the Binomial distribution tends to a normal distribution as N is increased to infinity.
(b) If Np is held fixed, then binomial distribution tends to Poisson distribution as N is increased to infinity.

1.97 The background counting rate is b and background plus source is g. If the background is counted for the time t_b and the background plus source for a time t_g, show that if the total counting time is fixed, then for minimum statistical error in the calculated counting rate of the source(s), t_b and t_g should be chosen so that $t_b/t_g = \sqrt{b/g}$

1.98 The alpha ray activity of a material is measured after equal successive intervals (hours), in terms of its initial activity as unity to the 0.835; 0.695; 0.580; 0.485; 0.405 and 0.335. Assuming that the activity obeys an exponential decay law, find the equation that best represents the activity and calculate the decay constant and the half-life.

[Osmania University 1967]

1.99 Obtain the interval distribution and hence deduce the exponential law of radioactivity.

1.100 In a Carbon-dating experiment background counting rate = 10 C/M. How long should the observations be made in order to have an accuracy of 5%? Assume that both the counting rates are measured for the same time.

^{14}C + background rate = 14.5 C/M

1.101 Make the best fit for the parabola $y = a_0 + a_1 x + a_2 x^2$, with the given pairs of values for x and y.

x	−2	−1	0	+1	+2	+3
y	9.1	3.5	0.5	0.8	4.6	11.0

1.102 The capacitance per unit length of a capacitor consisting of two long concentric cylinder with radii a and b, $(b > a)$ is $C = \frac{2\pi\varepsilon_0}{\ln(b/a)}$. If $a = 10 \pm 1$ mm and $b = 20 \pm 1$ mm, with what relative precision is C measured?

[University of Cambrdige Tripos 2004]

1.103 If $f(x)$ is the probability density of x given by $f(x) = xe^{-x/\lambda}$ over the interval $0 < x < \infty$, find the mean and the most probable values of x.

1.2.14 Numerical Integration

1.104 Calculate $\int_1^{10} x^2\,\mathrm{d}x$, by the trapezoidal rule.

1.105 Calculate $\int_1^{10} x^2\,\mathrm{d}x$, by the Simpson's rule.

1.3 Solutions

1.3.1 Vector Calculus

1.1 $$\nabla\Phi = \left(\hat{i}\frac{\partial}{\partial x} + \hat{j}\frac{\partial}{\partial y} + \hat{k}\frac{\partial}{\partial z}\right)(x^2+y^2+z^2)^{-1/2}$$

$$= \left(-\frac{1}{2}.2x\hat{i} - \frac{1}{2}.2y\hat{j} - \frac{1}{2}.2z\hat{k}\right)(x^2+y^2+z^2)^{-3/2}$$

$$= -(x\hat{i} + y\hat{j} + z\hat{k})(x^2+y^2+z^2)^{-\frac{3}{2}} = -\frac{r}{r^3}$$

1.2 $\nabla(xy^2 + xz) = \left(\hat{i}\frac{\partial}{\partial x} + \hat{j}\frac{\partial}{\partial y} + \hat{k}\frac{\partial}{\partial z}\right)(xy^2 + xz)$

$= (y^2 + z)\hat{i} + (2xy)\hat{j} + x\hat{k}$

$= 2\hat{i} - 2\hat{j} - \hat{k}, at(-1, 1, 1)$

A unit vector normal to the surface is obtained by dividing the above vector by its magnitude. Hence the unit vector is

$(2\hat{i} - 2\hat{j} - \hat{k})[(2)^2 + (-2)^2 + (-1)^2]^{-1/2} = \frac{2}{3}\hat{i} - \frac{2}{3}\hat{j} - \frac{1}{3}\hat{k}$

1.3 $F \propto 1/r^2$

$\nabla \,.\, (r^{-3}\mathbf{r}) = r^{-3}\nabla \,.\, \boldsymbol{r} + \boldsymbol{r} \,.\, \nabla r^{-3}$

But $\nabla \,.\, \mathbf{r} = \left(\hat{i}\frac{\partial}{\partial x} + \hat{j}\frac{\partial}{\partial y} + \hat{k}\frac{\partial}{\partial z}\right) . \left(\hat{i}x + \hat{j}y + \hat{k}z\right)$

$= \frac{\partial x}{\partial x} + \frac{\partial y}{\partial y} + \frac{\partial z}{\partial z} = 3$

$\boldsymbol{r} \,.\, \nabla r^{-3} = (x\hat{i} + y\hat{j} + z\hat{k}) \,.\, \left(\hat{i}\frac{\partial}{\partial x} + \hat{j}\frac{\partial}{\partial y} + \hat{k}\frac{\partial}{\partial z}\right)(x^2 + y^2 + z^2)^{-3/2}$

$= (x\hat{i} + y\hat{j} + z\hat{k}) \,.\, \left(-\frac{3}{2}\right) . (2x\hat{i} + 2y\hat{j} + 2z\hat{k})(x^2 + y^2 + z^2)^{-5/2}$

$= -3(x^2 + y^2 + z^2)(x^2 + y^2 + z^2)^{-\frac{5}{2}} = -3r^{-3}$

Thus $\nabla \,.\, (r^{-3}\boldsymbol{r}) = 3r^{-3} - 3r^{-3} = 0$

1.4 By problem $\nabla \times \boldsymbol{A} = 0$ *and* $\nabla \times \boldsymbol{B} = 0$, it follows that

$\boldsymbol{B} \,.\, (\nabla \times \boldsymbol{A}) = 0$

$\boldsymbol{A}. \,(\nabla \times \boldsymbol{B}) = 0$

Subtracting, $\boldsymbol{B} \,.\, (\nabla \times \boldsymbol{A}) - \boldsymbol{A} \,.\, (\nabla \times \boldsymbol{B}) = 0$

Now $\nabla \,.\, (\boldsymbol{A} \times \boldsymbol{B}) = \boldsymbol{B} \,.\, (\nabla \times \boldsymbol{A}) - \boldsymbol{A} \,.\, (\nabla \times \boldsymbol{B})$

Therefore $\nabla \,.\, (\boldsymbol{A} \times \boldsymbol{B}) = 0$, so that $(\boldsymbol{A} \times \boldsymbol{B})$ is solenoidal.

1.5 (a) Curl $\{\boldsymbol{r} f(r)\} = \nabla \times \{\boldsymbol{r} f(r)\} = \nabla \times \{xf(r)\hat{i} + yf(r)\hat{j} + zf(r)\hat{k}\}$

$$= \begin{vmatrix} \hat{i} & \hat{j} & \hat{k} \\ \frac{\partial}{\partial x} & \frac{\partial}{\partial y} & \frac{\partial}{\partial z} \\ xf(r) & yf(r) & zf(r) \end{vmatrix} = \left(z\frac{\partial f}{\partial y} - y\frac{\partial f}{\partial z}\right)\hat{i} + \left(x\frac{\partial f}{\partial z} - z\frac{\partial f}{\partial x}\right)\hat{j} + \left(y\frac{\partial f}{\partial x} - x\frac{\partial f}{\partial y}\right)\hat{k}$$

But $\frac{\partial f}{\partial x} = \left(\frac{\partial f}{\partial r}\right)\left(\frac{\partial r}{\partial x}\right) = \frac{\partial f}{\partial r}\frac{\partial (x^2+y^2+z^2)^{1/2}}{\partial x} = \frac{xf'}{r}$

Similarly $\frac{\partial f}{\partial y} = \frac{yf'}{r}$ and $\frac{\partial f}{\partial z} = \frac{zf'}{r}$, where prime means differentiation with respect to r.

Thus,

$$\text{curl}\{\boldsymbol{r} f(r)\} = \left(\frac{zyf'}{r} - \frac{yzf'}{r}\right)\hat{i} + \left(\frac{xzf'}{r} - \frac{zxf'}{r}\right)\hat{j} + \left(\frac{yxf'}{r} - \frac{xyf'}{r}\right)\hat{k} = 0$$

(b) If the field is solenoidal, then, $\nabla.\boldsymbol{r}F(r) = 0$

$$\frac{\partial(xF(r))}{\partial x} + \frac{\partial(yF(r))}{\partial y} + \frac{\partial(zF(r))}{\partial z} = 0$$

$$F + x\frac{\partial F}{\partial x} + F + y\frac{\partial F}{\partial y} + F + z\frac{\partial F}{\partial z} = 0$$

$$3F(r) + x\frac{\partial F}{\partial r}\frac{x}{r} + y\frac{\partial F}{\partial r}\frac{y}{r} + z\frac{\partial F}{\partial r}\frac{z}{r} = 0$$

$$3F(r) + \left(\frac{\partial F}{\partial r}\right)\left(\frac{x^2 + y^2 + z^2}{r}\right) = 0$$

But $(x^2 + y^2 + z^2) = r^2$, therefore, $\frac{\partial F}{\partial r} = -\frac{3F(r)}{r}$

Integrating, $\ln F = -3\ln r + \ln C$ where $C =$ constant

$$\ln F = -\ln r^3 + \ln C = \ln \frac{C}{r^3}$$

Therefore $F = C/r^3$. Thus, the field is $\boldsymbol{A} = \frac{r}{r^3}$ (inverse square law)

1.6 $x = t,\ y = t^2,\ z = t^3$

Therefore, $y = x^2, z = x^3,\ \mathrm{d}y = 2x\mathrm{d}x,\ \mathrm{d}z = 3x^2\,\mathrm{d}x$

$$\int_c A.dr = \int (y\hat{i} + xz\hat{j} + xyz\hat{k}).(\hat{i}\mathrm{d}x + \hat{j}\mathrm{d}y + \hat{k}\mathrm{d}z)$$

$$= \int_0^1 x^2\,\mathrm{d}x + 2\int_0^1 x^5\mathrm{d}x + 3\int_0^1 x^8\,\mathrm{d}x$$

$$= \frac{1}{3} + \frac{1}{3} + \frac{1}{3} = 1$$

1.7 The two curves $y = x^2$ and $y^2 = 8x$ intersect at (0, 0) and (2, 4). Let us traverse the closed curve in the clockwise direction, Fig. 1.6.

$$\int_c \boldsymbol{A}.\mathrm{d}r = \int_c [(x + y)\hat{i} + (x - y)\hat{j}].(\hat{i}\ \mathrm{d}x + \hat{j}\ \mathrm{d}y)$$

$$= \int_c [(x + y)\mathrm{d}x + (x - y)\mathrm{d}y]$$

$$= \int_2^0 [(x + x^2)\mathrm{d}x + (x - x^2)2x\mathrm{d}x] \qquad \text{(along } y = x^2\text{)}$$

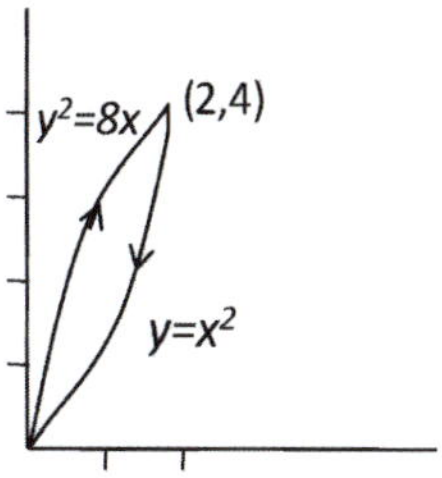

Fig. 1.6 Line integral for a closed curve

$$+\int_0^4 \left(\frac{y^2}{8}+y\right)\frac{y\mathrm{d}y}{4}+\left(\frac{y^2}{8}-y\right)\mathrm{d}y \qquad (\text{along } y^2=8x)$$
$$=+\frac{16}{3}$$

1.8 (a) It is sufficient to show that Curl $\boldsymbol{F}=0$

$$\nabla\times\boldsymbol{F}=\begin{vmatrix} i & j & k \\ \dfrac{\partial}{\partial x} & \dfrac{\partial}{\partial y} & \dfrac{\partial}{\partial z} \\ 2xy+z^2 & x^2 & 2xz \end{vmatrix}=\hat{\imath}.0-\hat{\jmath}(2z-2z)+\hat{k}(2x-2x)=0$$

(b) $\mathrm{d}\Phi=F.\,\mathrm{d}\boldsymbol{r}=((2xy+z^2)\hat{\imath}+x^2\hat{\jmath}+2xz\hat{k}).(\hat{\imath}\mathrm{d}x+\hat{\jmath}\mathrm{d}y+\hat{k}\mathrm{d}z)$

$=(2xy+z^2)\,\mathrm{d}x+x^2\,\mathrm{d}y+2xz\mathrm{d}z$

$=(2xy\mathrm{d}x+x^2\,\mathrm{d}y)+(z^2\,\mathrm{d}x+2xz\mathrm{d}z)$

$=\mathrm{d}(x^2y)+\mathrm{d}(z^2x)=\mathrm{d}(x^2y+xz^2)$

Therefore $\Phi=x^2y+xz^2+\text{constant}$

(c) Work done $=\Phi_2-\Phi_1=5.0$

1.9 Let $U=x+y; V=x-y$

$$\frac{\partial U}{\partial x}=1;\ \frac{\partial V}{\partial y}=-1$$

The curves $y=x^2$ and $y^2=8x$ intersect at (0, 0) and (2, 4).

$$\iint\left(\frac{\partial U}{\partial x}-\frac{\partial V}{\partial x}\right)\mathrm{d}x\,\mathrm{d}y=\iint_S(1-(-1))\mathrm{d}x\,\mathrm{d}y=2\int_{x=0}^{2}\int_{y=x^2}^{2\sqrt{2x}}\mathrm{d}x\mathrm{d}y$$

$$=2\int_0^2\left[\int_{x^2}^{2\sqrt{2x}}\mathrm{d}y\right]\mathrm{d}x=2\int_0^2(2\sqrt{2}\sqrt{x}-x^2)\,\mathrm{d}x=2\left[\frac{4\sqrt{2}}{3}x^{3/2}-\frac{x^3}{3}\right]_0^2=\frac{16}{3}$$

This is in agreement with the value obtained in Problem 1.7 for the line integral.

1.10 Use the divergence theorem

$$\iint \boldsymbol{A}\,.\,\mathrm{ds}=\iiint\nabla.\,\boldsymbol{A}\,\mathrm{d}v$$

But $\nabla.\,A=\dfrac{\partial}{\partial x}x^3+\dfrac{\partial}{\partial y}y^3+\dfrac{\partial}{\partial z}z^3$

$=3x^2+3y^2+3z^2=3(x^2+y^2+z^2)=3R^2$

$$\iint \boldsymbol{A}\,.\,\mathrm{ds}=\iiint 3R^2\mathrm{d}v=\int(3R^2)(4\pi R^2\,\mathrm{d}R)$$

$$=12\pi\int R^4\,\mathrm{d}R=\frac{12}{5}\pi R^5$$

1.11 $\displaystyle\int_c \boldsymbol{A}\,.\,\mathrm{d}\boldsymbol{r}=\int(2y\hat{\imath}-3x\hat{\jmath}+z\hat{k}).(\mathrm{d}x\hat{\imath}+\mathrm{d}y\hat{\jmath}+\mathrm{d}z\hat{k})$

$$=\int(2y\mathrm{d}x-3x\mathrm{d}y+z\mathrm{d}z)$$

Put $x = R\cos\theta,\ \mathrm{d}x = -R\sin\theta\ \mathrm{d}\theta,\ y = R\sin\theta,\ \mathrm{d}y = R\cos\theta, z = 0, 0 < \theta < 2\pi$

$$\int \mathbf{A}\,.\,\mathrm{d}\mathbf{r} = -2R^2\int \sin^2\theta\ \mathrm{d}\theta - R^2\int \cos^2\theta\ \mathrm{d}\theta$$
$$= -2\pi R^2 - \pi R^2 = -3\pi R^2$$

1.12 (a) $\nabla \times (\nabla\Phi) = \begin{vmatrix} i & j & k \\ \frac{\partial}{\partial x} & \frac{\partial}{\partial y} & \frac{\partial}{\partial z} \\ \frac{\partial\Phi}{\partial x} & \frac{\partial\Phi}{\partial y} & \frac{\partial\Phi}{\partial z} \end{vmatrix}$

$$= i\left(\frac{\partial^2\Phi}{\partial y\partial z} - \frac{\partial^2\Phi}{\partial z\partial y}\right) - j\left(\frac{\partial^2\Phi}{\partial x\partial z} - \frac{\partial^2\Phi}{\partial z\partial x}\right) + k\left(\frac{\partial^2\Phi}{\partial x\partial y} - \frac{\partial^2\Phi}{\partial y\partial x}\right) = 0$$

because the order of differentiation is immaterial and terms in brackets cancel in pairs.

(b) To show $\nabla.(\nabla \times \boldsymbol{V}) = 0$

$$\left(\hat{i}\frac{\partial}{\partial x} + \hat{j}\frac{\partial}{\partial y} + \hat{k}\frac{\partial}{\partial z}\right) \cdot \begin{vmatrix} i & j & k \\ \frac{\partial}{\partial x} & \frac{\partial}{\partial y} & \frac{\partial}{\partial z} \\ V_x & V_y & V_z \end{vmatrix}$$

$$= \left(\hat{i}\frac{\partial}{\partial x} + \hat{j}\frac{\partial}{\partial y} + \hat{k}\frac{\partial}{\partial z}\right) \cdot \left[\hat{i}\begin{vmatrix} \frac{\partial}{\partial y} & \frac{\partial}{\partial z} \\ V_y & V_z \end{vmatrix} - \hat{j}\begin{vmatrix} \frac{\partial}{\partial x} & \frac{\partial}{\partial z} \\ V_x & V_z \end{vmatrix} + \hat{k}\begin{vmatrix} \frac{\partial}{\partial x} & \frac{\partial}{\partial y} \\ V_x & V_y \end{vmatrix}\right]$$

$$= \frac{\partial}{\partial x}\begin{vmatrix} \frac{\partial}{\partial y} & \frac{\partial}{\partial z} \\ V_y & V_z \end{vmatrix} - \frac{\partial}{\partial y}\begin{vmatrix} \frac{\partial}{\partial x} & \frac{\partial}{\partial z} \\ V_x & V_z \end{vmatrix} + \frac{\partial}{\partial z}\begin{vmatrix} \frac{\partial}{\partial x} & \frac{\partial}{\partial y} \\ V_x & V_y \end{vmatrix}$$

$$= \begin{vmatrix} \frac{\partial}{\partial x} & \frac{\partial}{\partial y} & \frac{\partial}{\partial z} \\ \frac{\partial}{\partial x} & \frac{\partial}{\partial y} & \frac{\partial}{\partial z} \\ V_x & V_y & V_z \end{vmatrix} = 0$$

The value of the determinant is zero because two rows are identical.

1.13 $\Phi = x^2y - 2xz^3$

(a) $\nabla\Phi = \left(\hat{i}\frac{\partial}{\partial x} + \hat{j}\frac{\partial}{\partial y} + \hat{k}\frac{\partial}{\partial z}\right)(x^2y - 2xz^3)$

$$= 2(xy - z^3)\hat{i} + x^2\hat{j} + 6xz^2\hat{k}$$

(b) $\nabla^2\Phi = \left(\frac{\partial^2}{\partial x^2} + \frac{\partial^2}{\partial y^2} + \frac{\partial^2}{\partial z^2}\right)(x^2y - 2xz^3)$

$$= 2y - 12xz$$

1.14 (a) $\nabla(x^2y + xz) = \left(\hat{i}\frac{\partial}{\partial x} + \hat{j}\frac{\partial}{\partial y} + \hat{k}\frac{\partial}{\partial z}\right)(x^2y + xz)$

$$= (2xy + z)\hat{i} + x^2\hat{j} + x\hat{k}$$
$$= -\hat{i} + \hat{j} + \hat{k}$$

A unit vector normal to the surface is obtained by dividing the above vector by its magnitude. Hence the unit vector is

$$\frac{-\hat{i}+\hat{j}+\hat{k}}{[(-1)^2+1^2+1^2]^{1/2}} = -\frac{\hat{i}}{\sqrt{3}}+\frac{\hat{j}}{\sqrt{3}}+\frac{\hat{k}}{\sqrt{3}}$$

(b) $\nabla\Phi = \left(\hat{i}\frac{\partial}{\partial x}+\hat{j}\frac{\partial}{\partial y}+\hat{k}\frac{\partial}{\partial z}\right)(x^2yz+2xz^2)$

$= (2xyz+2z^2)\hat{i}+x^2z\hat{j}+4xz\hat{k}$

$= -\hat{j}-4\hat{k}$ at $(1, 1, -1)$

The unit vector in the direction of $2\hat{i}-2\hat{j}+\hat{k}$, is

$$\hat{n} = \frac{2\hat{i}-2\hat{j}+\hat{k}}{[2^2+(-2)^2+1^2]^{1/2}} = 2\hat{i}/3-2\hat{j}/3+\hat{k}/3$$

The required directional derivative is

$$\nabla\Phi.n = (-\hat{j}-4\hat{k}).\left(\frac{2\hat{i}}{3}-\frac{2\hat{j}}{3}+\frac{2\hat{k}}{3}\right) = \frac{2}{3}-\frac{4}{3} = -\frac{2}{3}$$

Since this is negative, it decreases in this direction.

1.15 The inverse square force can be written as

$f\alpha\frac{\boldsymbol{r}}{r^3}$

$\nabla.\boldsymbol{f} = \nabla.r^{-3}\boldsymbol{r} = r^{-3}\nabla.\boldsymbol{r}+\boldsymbol{r}.\nabla r^{-3}$

But $\nabla.\boldsymbol{r} = \left(\hat{i}\frac{\partial}{\partial x}+\hat{j}\frac{\partial}{\partial y}+\hat{k}\frac{\partial}{\partial z}\right)\cdot(\hat{i}x+\hat{j}y+\hat{k}z)$

$= \frac{\partial x}{\partial x}+\frac{\partial y}{\partial y}+\frac{\partial z}{\partial z} = 3$

Now $\nabla r^n = nr^{n-2}\boldsymbol{r}$

so that $\nabla r^{-3} = -3r^{-5}\boldsymbol{r}$

$\therefore \nabla.(r^{-3}\boldsymbol{r}) = 3r^{-3}-3r^{-5}\boldsymbol{r}.\boldsymbol{r} = 3r^{-3}-r^{-3} = 0$

Thus, the divergence of an inverse square force is zero.

1.16 The angle between the surfaces at the point is the angle between the normal to the surfaces at the point.

The normal to $x^2+y^2+z^2 = 1$ at $(1, +1, -1)$ is

$\nabla\Phi_1 = \nabla(x^2+y^2+z^2) = 2x\hat{i}+2y\hat{j}+2z\hat{k} = 2\hat{i}+2\hat{j}-2\hat{k}$

The normal to $z = x^2+y^2-1$ or $x^2+y^2-z = 1$ at $(1, 1, -1)$ is

$\nabla\Phi_2 = \nabla(x^2+y^2-z) = 2x\hat{i}+2y\hat{j}-\hat{k} = 2\hat{i}+2\hat{j}-\hat{k}$

$(\nabla\Phi_1).(\nabla\Phi_2) = |\nabla\Phi_1||\nabla\Phi_2|\cos\theta$

where θ is the required angle.

$(2\hat{i}+2\hat{j}-2\hat{k}).(2\hat{i}+2\hat{j}-\hat{k}) = (12)^{1/2}(9)^{1/2}\cos\theta$

$\therefore \cos\theta = \frac{10}{6\sqrt{3}} = 0.9623$

$\theta = 15.780$

1.3.2 Fourier Series and Fourier Transforms

1.17 $$f(x) = \frac{1}{2}a_0 + \sum_{n=1}^{\infty}\left(a_n \cos\left(\frac{n\pi x}{L}\right) + b_n \sin\left(\frac{n\pi x}{L}\right)\right) \quad (1)$$

$$a_n = (1/L)\int_{-L}^{L} f(x)\cos\left(\frac{n\pi x}{L}\right) \mathrm{d}x \quad (2)$$

$$b_n = (1/L)\int_{-L}^{L} f(x)\sin\left(\frac{n\pi x}{L}\right) \mathrm{d}x \quad (3)$$

As $f(x)$ is an odd function, $a_n = 0$ for all n.

$$b_n = (1/L)\int_{-L}^{L} f(x)\sin\left(\frac{n\pi x}{L}\right) \mathrm{d}x$$

$$= (2/L)\int_{0}^{L} x\sin\left(\frac{n\pi x}{L}\right) \mathrm{d}x$$

$$= -\left(\frac{2}{n\pi}\right)\cos n\pi = -\left(\frac{2}{n\pi}\right)(-1)^n = \left(\frac{2}{n\pi}\right)(-1)^{n+1}$$

Therefore,

$$f(x) = (2/\pi)\sum_{1}^{\infty}\frac{(-1)^{n+1}}{n}\sin\left(\frac{n\pi x}{L}\right)$$

$$= (2/\pi)[\sin\left(\frac{\pi x}{L}\right) - \frac{1}{2}\sin\left(\frac{2\pi x}{L}\right) + \left(\frac{1}{3}\right)\sin\left(\frac{3\pi x}{L}\right) - \cdots]$$

Figure 1.7 shows the result for first 3 terms, 6 terms and 9 terms of the Fourier expansion. As the number of terms increases, a better agreement with the function is reached. As a general rule if the original function is smoother compared to, say the saw-tooth function the convergence of the Fourier series is much rapid and only a few terms are required. On the other hand, a highly discontinuous function can be approximated with reasonable accuracy only with large number of terms.

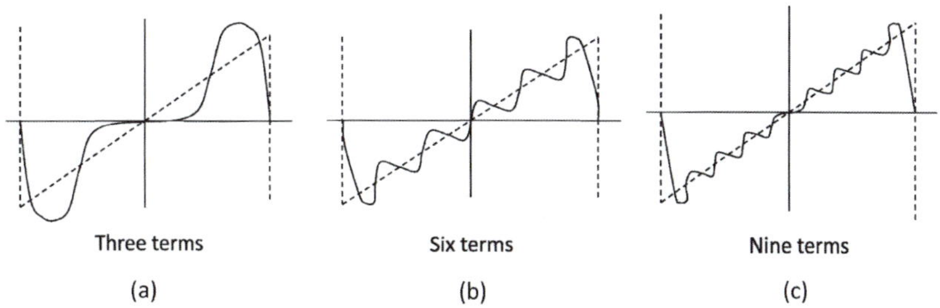

Fig. 1.7 Fourier expansion of the saw-tooth wave

1.18 The given function is of the square form. As $f(x)$ is defined in the interval $(-\pi, \pi)$, the Fourier expansion is given by

$$f(x) = \frac{1}{2}a_0 + \sum_{n=1}^{\infty} (a_n \cos nx + b_n \sin nx) \quad (1)$$

$$\text{where } a_n = (1/\pi) \int_{-\pi}^{\pi} f(x) \cos nx \, \mathrm{d}x \quad (2)$$

$$a_0 = (1/\pi) \int_{-\pi}^{\pi} f(x) \, \mathrm{d}x \quad (3)$$

$$b_n = \left(\frac{1}{\pi}\right) \int_{-\pi}^{\pi} f(x) \sin nx \, \mathrm{d}x \quad (4)$$

By (3)

$$a_0 = (1/\pi) \left(\int_{-\pi}^{0} 0 \mathrm{d}x + \int_{0}^{\pi} \pi \mathrm{d}x \right) = \pi \quad (5)$$

By (2)

$$a_n = (1/\pi) \int_{0}^{\pi} \cos nx \, \mathrm{d}x = 0, n \geq 1 \quad (6)$$

By (4)

$$b_n = (1/\pi) \int_{0}^{\pi} \pi \sin nx \, dx = \left(\frac{1}{n}\right) (1 - \cos n\pi) \quad (7)$$

Using (5), (6) and (7) in (1)

$$f(x) = \frac{\pi}{2} + 2 \left(\sin(x) + \left(\frac{1}{3}\right) \sin 3x + \left(\frac{1}{5}\right) \sin 5x + \cdots \right)$$

The graph of $f(x)$ is shown in Fig. 1.8. It consists of the x-axis from $-\pi$ *to* 0 and of the line AB from 0 *to* π. A simple discontinuity occurs at $x = 0$ at which point the series reduces to $\pi/2$.

Now,

$\pi/2 = 1/2[f(0-) + f(0+)]$

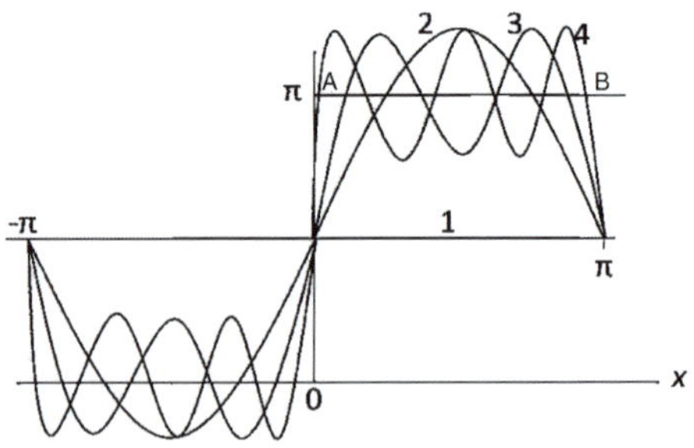

Fig. 1.8 Fourier expansion of a square wave

which is consistent with Dirichlet's theorem. Similar behavior is exhibited at $x = \pi, \pm 2\pi \ldots$ Figure 1.8 shows first four partial sums with equations

$y = \pi/2$
$y = \pi/2 + 2\sin x$
$y = \pi/2 + 2(\sin x + (1/3)\sin 3x)$
$y = \pi/2 + 2(\sin x + (1/3)\sin 3x + (1/5)\sin 5x)$

1.19 By Problem 1.18,

$$y = \frac{\pi}{2} + 2\left(\sin x + \left(\frac{1}{3}\right)\sin 3x + \left(\frac{1}{5}\right)\sin 5x + \left(\frac{1}{7}\right)\sin 7x + \cdots\right)$$

Put $x = \pi/2$ in the above series

$$y = \pi = \frac{\pi}{2} + 2\left(1 - \frac{1}{3} + \frac{1}{5} - \frac{1}{7} + \cdots\right)$$

Hence $\frac{\pi}{4} = 1 - \frac{1}{3} + \frac{1}{5} - \frac{1}{7} + \cdots$

1.20 The Fourier transform of $f(x)$ is

$$T(u) = \frac{1}{\sqrt{2\pi}} \int_{-a}^{a} e^{iux} f(x)\mathrm{d}x$$

$$= \frac{1}{\sqrt{2\pi}} \int_{-a}^{a} 1.e^{iux}\,\mathrm{d}x = \frac{1}{\sqrt{2\pi}} \frac{e^{iux}}{iu}\bigg|_{-a}^{a} = \frac{1}{\sqrt{2\pi}}\left(\frac{e^{iua} - e^{-iua}}{iu}\right)$$

$$= \sqrt{\frac{2}{\pi}}\frac{\sin ua}{u}, \quad u \neq 0$$

For $u = 0$, $T(u) = \sqrt{\frac{2}{\pi}}u$.

The graphs of $f(x)$ and $T(u)$ for $u = 3$ are shown in Fig. 1.9a, b, respectively

Note that the above transform finds an application in the FraunHofer diffraction.

$\tilde{f}(\omega) = A\sin\alpha/\alpha$

This is the basic equation which describes the Fraunhofer's diffraction pattern due to a single slit.

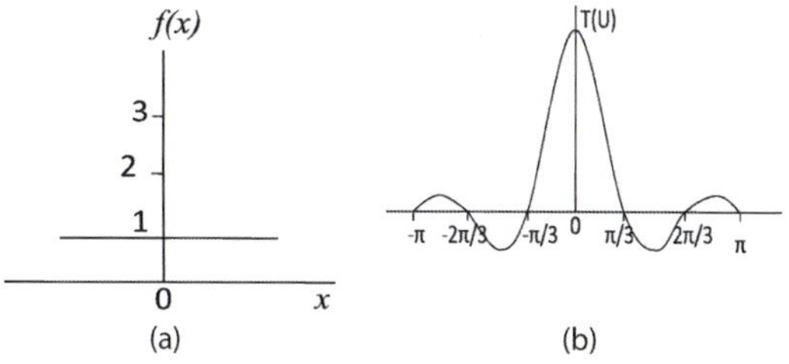

Fig. 1.9 Slit function and its Fourier transform

1.21 Consider the Fourier integral theorem

$$f(x) = \frac{2}{\pi}\int_0^\infty \cos ax \, da \int_0^\infty e^{-u} \cos au \, du$$

Put $f(x) = e^{-x}$. Now the definite integral

$$\int_0^\infty e^{-bu} \cos(au) \, du = \frac{b}{b^2 + a^2}$$

Here $\int_0^\infty e^{-u} \cos au \, du = \frac{1}{1+a^2}$

$$\therefore \frac{2}{\pi}\int_0^\infty \frac{\cos ax}{1+a^2} dx = f(x) \text{ or } \int_0^\infty \frac{\cos ax}{1+a^2} = \frac{\pi}{2} e^{-x}$$

1.22 The Gaussian distribution is centered on $t = 0$ and has root mean square deviation τ.

$$\tilde{f}(\omega) = \frac{1}{\sqrt{2\pi}}\int_{-\infty}^\infty f(t) e^{-i\omega t} dt$$

$$= \frac{1}{\sqrt{2\pi}}\int_{-\infty}^\infty \frac{1}{\tau\sqrt{2\pi}} e^{-t^2/2\tau^2} e^{-i\omega t} dt$$

$$= \frac{1}{\sqrt{2\pi}}\int_{-\infty}^\infty \frac{1}{\tau\sqrt{2\pi}} e^{-[t^2+2\tau^2 i\omega t + (\tau^2 i\omega)^2 - (\tau^2 i\omega)^2]/2\tau^2} dt$$

$$= \frac{1}{\sqrt{2\pi}} e^{-\frac{\tau^2\omega^2}{2}} \left\{ \frac{1}{\tau\sqrt{2\pi}} \int_{-\infty}^\infty e^{\frac{-(t+i\tau^2\omega^2)^2}{2\tau^2}} dt \right\}$$

The expression in the Curl bracket is equal to 1 as it is the integral for a normalized Gaussian distribution.

$$\therefore \tilde{f}(\omega) = \frac{1}{\sqrt{2\pi}} e^{-\frac{\tau^2\omega^2}{2}}$$

which is another Gaussian distribution centered on zero and with a root mean square deviation $1/\tau$.

1.3.3 Gamma and Beta Functions

1.23 $\Gamma(z+1) = \lim_{T\to\infty} \int_0^T e^{-x} x^z dx$

Integrating by parts

$$\Gamma(z+1) = \lim_{T\to\infty} [-x^z e^{-x}|_0^T + z\int_0^T e^{-x} x^{z-1} dx]$$

$$= z \lim_{T\to\infty} \int_0^T e^{-x} x^{z-1} dx = z\Gamma(z)$$

because $T^z e^{-T} \to 0$ as $T \to \infty$

Also, since $\Gamma(1) = \int_0^\infty e^{-x} dx = 1$

If z is a positive integer n,

$\Gamma(n+1) = n!$

Thus the gamma function is an extension of the factorial function to numbers which are not integers.

1.24 $$B(m,n) = \int_0^1 x^{m-1}(1-x)^{n-1}\mathrm{d}x \tag{1}$$

With the substitution $x = \sin 2\Phi$ (1) becomes

$$B(m,n) = 2\int_0^{\pi/2} (\sin \Phi)^{2m-1}(\cos \Phi)^{2n-1}\mathrm{d}\Phi \tag{2}$$

Now $\Gamma(n) = 2\int_0^\infty y^{2n-1}e^{-y^2}\mathrm{d}y$

$$\Gamma(m) = 2\int_0^\infty y^{2m-1}e^{-x^2}\mathrm{d}x$$

$$\therefore \Gamma(m)\Gamma(n) = 4\int_0^\infty \int_0^\infty x^{2m-1}y^{2n-1} \exp -(x^2+y^2)\mathrm{d}x\mathrm{d}y \tag{3}$$

The double integral may be evaluated as a surface integral in the first quadrant of the *xy-plane*. Introducing the polar coordinates $x = r\cos\theta$ and $y = r\sin\theta$, the surface element $\mathrm{d}s = r\mathrm{d}r\mathrm{d}\theta$, (3) becomes

$$\Gamma(m)\Gamma(n) = 4\int_0^{\pi/2}\int_0^\infty r^{2m-1}(\cos\theta)^{2m-1}(\sin\theta)^{2n-1}e^{-r^2}r\mathrm{d}r\mathrm{d}\theta$$

$$\Gamma(m)\Gamma(n) = 2\int_0^{\pi/2}(\cos\theta)^{2m-1}(\sin\theta)^{2n-1}\mathrm{d}\theta . 2\int_0^\infty r^{2(m+n)-1}e^{-r^2}\mathrm{d}r \tag{4}$$

In (4), the first integral is identified as $B(m,n)$ and the second one as $\Gamma(m+n)$. It follows that

$$B(m,n) = \frac{\Gamma(m)\Gamma(n)}{\Gamma(m+n)}$$

1.25 One form of Beta function is

$$2\int_0^{\pi/2}(\cos\theta)^{2m-1}(\sin\theta)^{2n-1}\mathrm{d}\theta = B(m,n) = \frac{\Gamma(m)\Gamma(n)}{\Gamma(m+n)} \quad (m>0, n>0) \tag{1}$$

Letting $2m-1 = r$, that is $m = \frac{r+1}{2}$ and $2n-1 = 0$, that is $n = 1/2$, (1) becomes

$$\int_0^{\pi/2}(\cos\theta)^r\,\mathrm{d}\theta = \frac{1}{2}\frac{\Gamma\left(\frac{r+1}{2}\right)\Gamma\left(\frac{1}{2}\right)}{\Gamma\left(\frac{r}{2}+1\right)} \tag{2}$$

Now $\Gamma(n) = \int_0^\infty x^{n-1}e^{-x}\mathrm{d}x$, put $x = y^2$, $\mathrm{d}x = 2y\mathrm{d}y$, so that

$$\Gamma(n) = 2\int_0^\infty y^{2n-1}e^{-y^2}\mathrm{d}y$$

$$\Gamma(1/2) = 2\int_0^\infty e^{-y^2}\mathrm{d}y = \frac{2\sqrt{\pi}}{2} = \sqrt{\pi}$$

So that

$$\int_0^{\frac{\pi}{2}} (\cos\theta)^r \mathrm{d}\theta = \frac{\sqrt{\pi}}{2}\frac{\Gamma\left(\frac{r+1}{2}\right)}{\Gamma\left(\frac{r}{2}+1\right)}$$

1.26 (a) $$B(m,n) = \int_0^1 x^{m-1}(1-x)^{n-1}\mathrm{d}x \quad (1)$$

Put $$x = \frac{y}{1+y} \quad (2)$$

$$B(m,n) = \int_0^\infty \frac{y^{n-1}\mathrm{d}y}{(1+y)^{m+n}} = \frac{\Gamma(m)\Gamma(n)}{\Gamma(m+n)}$$

Letting $m = 1-n$; $0 < n < 1$

$$\int_0^\infty \frac{y^{n-1}}{(1+y)}\mathrm{d}y = \frac{\Gamma(1-n)\Gamma(n)}{\Gamma(1)}$$

But $\Gamma(1) = 1$ and

$$\int_0^\infty \frac{y^{n-1}}{(1+y)}\mathrm{d}y = \frac{\pi}{\sin(n\pi)}; 0 < n < 1$$

$$\Gamma(n)\Gamma(1-n) = \frac{\pi}{\sin(n\pi)} \quad (3)$$

(b) $|\Gamma(in)|^2 = \Gamma(in)\Gamma(-in)$

Now $\Gamma(n) = \frac{\Gamma(n+1)}{n}$

$$\Gamma(-in) = \frac{\Gamma(1-in)}{-in}$$

$$\therefore |\Gamma(in)|^2 = \frac{\Gamma(in)\Gamma(1-in)}{-in(\sin\, i\pi n)}$$

by (3)

Further $\sinh(\pi n) = i \sin i\pi n$

$$\therefore |\Gamma(in)|^2 = \frac{\pi}{n\sinh(\pi n)}$$

1.3.4 Matrix Algebra

1.27 Let H be the hermitian matrix with characteristic roots λ_i. Then there exists a non-zero vector X_i such that

$$HX_i = \lambda_i X_i \quad (1)$$

$$\text{Now } \bar{X}_i' H X_i = \bar{X}_i' \lambda_i X_i = \lambda_i \bar{X}_i' X_i \quad (2)$$

is real and non-zero. Similarly the conjugate transpose

$$\bar{X}_i' H X_i = \bar{\lambda}_i \bar{X}_i' X_i \quad (3)$$

Comparing (2) and (3),

$\bar{\lambda}_i = \lambda_i$

Thus λ_i *is real*

1.28 The characteristic equation is given by

$$\begin{vmatrix} 1-\lambda & -1 & 1 \\ 0 & 3-\lambda & -1 \\ 0 & 0 & 2-\lambda \end{vmatrix} = 0$$

$$(1-\lambda)(3-\lambda)(2-\lambda) + 0 + 0 = 0 \quad (1)$$

$$\text{or } \lambda^3 - 6\lambda^2 + 11\lambda - 6 = 0 \text{ (characteristic equation)} \quad (2)$$

The eigen values are $\lambda_1 = 1$, $\lambda_2 = 3$, and $\lambda_3 = 2$.

1.29 Let $X = \begin{pmatrix} x \\ y \end{pmatrix}$

$$AX = \begin{pmatrix} -1 & 0 \\ 0 & 1 \end{pmatrix} \begin{pmatrix} x \\ y \end{pmatrix} = \begin{pmatrix} -x \\ -y \end{pmatrix}$$

It produces reflection through the origin, that is inversion. A performs the parity operation, Fig. 1.10a.

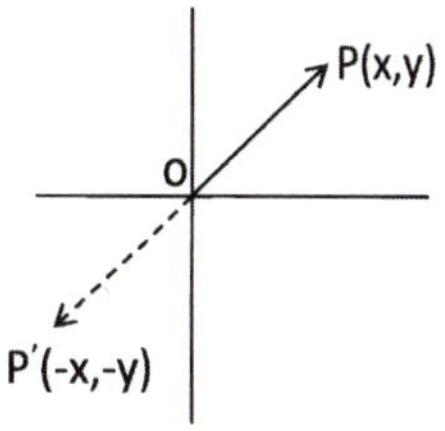

Fig. 1.10a Parity operation (inversion through origin)

$$BX = \begin{pmatrix} 0 & 1 \\ 1 & 0 \end{pmatrix} \begin{pmatrix} x \\ y \end{pmatrix} = \begin{pmatrix} y \\ x \end{pmatrix}$$

Here the x and y coordinates are interchanged. This is equivalent to a reflection about a line passing through origin at $\theta = 45°$, Fig. 1.10b

$$CX = \begin{pmatrix} 2 & 0 \\ 0 & 2 \end{pmatrix} \begin{pmatrix} x \\ y \end{pmatrix} = \begin{pmatrix} 2x \\ 2y \end{pmatrix}$$

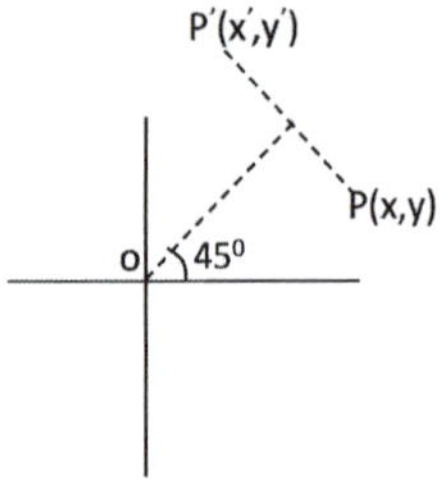

Fig. 1.10b Reflection about a line passing through origin at 45°

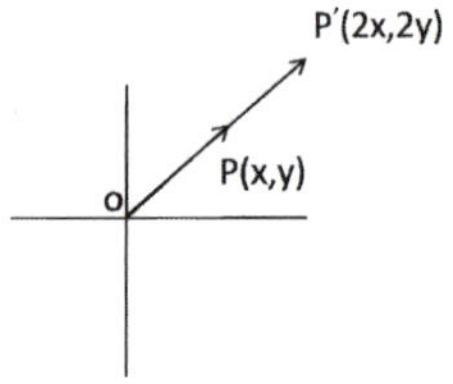

Fig. 1.10c Elongating a vector in the same direction

Here the magnitude becomes double without changing its orientation.

$$DX = \begin{pmatrix} \frac{\sqrt{3}}{2} & \frac{1}{2} \\ -\frac{1}{2} & \frac{\sqrt{3}}{2} \end{pmatrix} \begin{pmatrix} x \\ y \end{pmatrix} = \begin{pmatrix} \cos 30^\circ & \sin 30^\circ \\ -\sin 30^\circ & \cos 30^\circ \end{pmatrix} \begin{pmatrix} x \\ y \end{pmatrix} = \begin{pmatrix} \frac{\sqrt{3}}{2}x + \frac{y}{2} \\ -\frac{x}{2} + \frac{\sqrt{3}}{2}y \end{pmatrix}$$

The matrix D is a rotation matrix which rotates the vector through 30° about the z-axis,. Fig.1.10d.

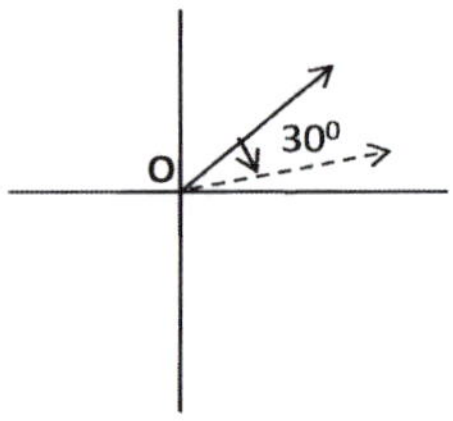

Fig. 1.10d Rotation of a vector through 30°

1.30 The matrix $A = \begin{pmatrix} 6 & -2 & 2 \\ -2 & 3 & -1 \\ 2 & -1 & 3 \end{pmatrix}$

The characteristic equation is

$$|A - \lambda I| = \begin{vmatrix} 6-\lambda & -2 & 2 \\ -2 & 3-\lambda & -1 \\ 2 & -1 & 3-\lambda \end{vmatrix} = 0$$

This gives $-\lambda^3 + 12\lambda^2 - 36\lambda + 32 = 0$
or $(\lambda - 2)(\lambda - 2)(\lambda - 8) = 0$
The characteristic roots (eigen values) are
$\lambda_1 = 2, \lambda_2 = 2$ and $\lambda_3 = 8$

1.31 $A = \begin{pmatrix} 6 & -2 & 2 \\ -2 & 3 & -1 \\ 2 & -1 & 3 \end{pmatrix}$

In Problem 1.30, the characteristic roots are found to be $\lambda = 2, 2, 8$. With $\lambda = 2$, we find the invariant vectors.

$$\begin{pmatrix} 6-2 & -2 & 2 \\ -2 & 3-2 & -1 \\ 2 & -1 & 3-2 \end{pmatrix} \begin{pmatrix} x_1 \\ x_2 \\ x_3 \end{pmatrix} = 0$$

The two vectors are $X_1 = (1, 1, -1)'$ and $X_2 = (0, 1, 1)'$. The third vector can be obtained in a similar fashion. It can be chosen as $X_3 = (2, -1, 1)'$. The three column vectors can be normalized and arranged in the form of a matrix. The matrix A is diagnalized by the similarity transformation.

$$S^{-1} AS = \text{diag } A$$

$$S = \begin{pmatrix} \frac{1}{\sqrt{3}} & 0 & \frac{2}{\sqrt{6}} \\ \frac{1}{\sqrt{3}} & \frac{1}{\sqrt{2}} & -\frac{1}{\sqrt{6}} \\ -\frac{1}{\sqrt{3}} & \frac{1}{\sqrt{2}} & \frac{1}{\sqrt{6}} \end{pmatrix}$$

As the matrix S is orthogonal, $S^{-1} = S'$. Thus

$$\begin{pmatrix} \frac{1}{\sqrt{3}} & \frac{1}{\sqrt{3}} & -\frac{1}{\sqrt{3}} \\ 0 & \frac{1}{\sqrt{2}} & \frac{1}{\sqrt{2}} \\ \frac{2}{\sqrt{6}} & -\frac{1}{\sqrt{6}} & \frac{1}{\sqrt{6}} \end{pmatrix} \begin{pmatrix} 6 & -2 & 2 \\ -2 & 3 & -1 \\ 2 & -1 & 3 \end{pmatrix} \begin{pmatrix} \frac{1}{\sqrt{3}} & 0 & \frac{2}{\sqrt{6}} \\ \frac{1}{\sqrt{3}} & \frac{1}{\sqrt{2}} & -\frac{1}{\sqrt{6}} \\ -\frac{1}{\sqrt{3}} & \frac{1}{\sqrt{2}} & \frac{1}{\sqrt{6}} \end{pmatrix} = \begin{pmatrix} 2 & 0 & 0 \\ 0 & 2 & 0 \\ 0 & 0 & 8 \end{pmatrix}$$

1.32 $H = \begin{pmatrix} a_{11} & a_{12} \\ a_{21} & a_{22} \end{pmatrix}$

$$A = \begin{bmatrix} 3 & 2 \\ 4 & 1 \end{bmatrix}$$

(a) $\begin{vmatrix} 3-\lambda & 2 \\ 4 & 1-\lambda \end{vmatrix} = 0$, characteristic equation is

$$(3-\lambda)(1-\lambda) - 8 = 0$$

$$\lambda^2 - 4\lambda - 5 = 0, (\lambda - 5)(\lambda + 1) = 0$$

The eigen values are $\lambda_1 = 5$ and $\lambda_2 = -1$

(b) and (c) The desired matrix has the form

$$C = \begin{pmatrix} C_{11} & C_{12} \\ C_{21} & C_{22} \end{pmatrix}$$

The columns which satisfy the system of equations

$$(a_{ij} - \delta_{ij}\lambda_k)C_{jk} = 0, \text{ no sum on } k \tag{1}$$

yielding

$$(a_{11} - \lambda_k)C_{1k} + a_{12}C_{2k} = 0, \text{ no sum on } k$$
$$a_{21}C_{1k} + (a_{22} - \lambda_k)C_{2k} = 0, \ k = 1, 2$$

Since $a_{11} = 3$, $a_{21} = 4$, $a_{12} = 2$, $a_{22} = 1$, we get on setting $k = 1$ and $\lambda_1 = 5$

$$-2C_{11} + 2C_{21} = 0 \quad (2)$$
$$4C_{11} - 4C_{21} = 0 \quad (3)$$

Thus $C_{21} = C_{11} = \text{a} = 1$
The substitution of $k = 2$ and $\lambda_2 = -1$ yields

$$(3 + 1)C_{12} + 2C_{22} = 0$$
$$4C_{12} + 2C_{22} = 0$$

or $C_{22} = -2C_{12}$
We may set $C_{12} = 1$ so that $C_{22} = -2$
Thus $C = \begin{pmatrix} 1 & 1 \\ 1 & -2 \end{pmatrix}$ (modal matrix)
The inverse of C is easily found to be

$$C^{-1} = \begin{pmatrix} \frac{2}{3} & \frac{1}{3} \\ \frac{1}{3} & -\frac{1}{3} \end{pmatrix}$$

Eigen vectors: $\begin{pmatrix} a_{11} - \lambda & a_{12} \\ a_{21} & a_{22} - \lambda \end{pmatrix} \begin{pmatrix} x_1 \\ x_2 \end{pmatrix} = 0$

Put $\lambda = \lambda_1 = 5$; $\begin{pmatrix} 3 - 5 & 2 \\ 4 & 1 - 5 \end{pmatrix} \begin{pmatrix} x_1 \\ x_2 \end{pmatrix} = 0 \rightarrow -2x_1 + 2x_2 = 0 \rightarrow x_1 = x_2$

The normalized invariant vector is $\frac{1}{\sqrt{2}} \begin{pmatrix} 1 \\ 1 \end{pmatrix}$

Put $\lambda = \lambda_2 = -1$; $\begin{pmatrix} 3 - (-1) & 2 \\ 4 & 1 - (-1) \end{pmatrix} \begin{pmatrix} x_1 \\ x_2 \end{pmatrix} = 0 \rightarrow 4x_1 + 2x_2 = 0 \rightarrow$
$x_2 = -2x_1$

The second invariant eigen normalized eigen vector is $\frac{1}{\sqrt{5}} \begin{pmatrix} 1 \\ -2 \end{pmatrix}$

(d) $C^{-1}AC = \begin{pmatrix} 2/3 & 1/3 \\ 1/3 & -1/3 \end{pmatrix} \begin{pmatrix} 3 & 2 \\ 4 & 1 \end{pmatrix} \begin{pmatrix} 1 & 1 \\ 1 & -2 \end{pmatrix} = \begin{pmatrix} 5 & 0 \\ 0 & -1 \end{pmatrix}$

1.3.5 Maxima and Minima

1.33 Let $f = y = x^3 - 3x + 3 = 0$
Let the root be a

If $x = a = -2$, $y = +1$
If $x = a = -3$, $y = -15$

Thus $x = a$ lies somewhere between -2 and -3.
For $x = a - 2.1$, $y = 0.039$, which is close to zero.
Assume as a first approximation, the root to be $a = v + h$
Put $v = -2.1$

$$h = -\frac{f(v)}{f'(v)};\ f(v) = f(-2.1)$$

$$f'(v) = \frac{dy}{dx}|_v;\ f'(v) = 10.23$$

$$h = -\frac{0.039}{10.23} = -0.0038$$

To a first approximation the root is $-2.1 - 0.0038123$ or -2.1038123. As a second approximation, assume the root to be

$a = -2.1038123 + h,$

Put $v_1 = -2.1038123$

$h_1 = -f(v_1)/f'(v_1) = -0.000814/6.967 = -0.0001168$

The second approximation, therefore, gives $a = -2.1039291$.

The third and higher approximations can be made in this fashion. The first approximation will be usually good enough in practice.

1.34 $y(x) = x^2 \exp(-x^2)$ (1)

Turning points are determined from the location of maxima and minima. Differentiating (1) and setting $\mathrm{d}y/\mathrm{d}x = 0$

$\mathrm{d}y/\mathrm{d}x = 2x(1 - x^2)\exp(-x^2) = 0$

$x = 0, +1, -1$. These are the turning points.

We can now find whether the turning points are maxima or minima.

$$\frac{\mathrm{d}y}{\mathrm{d}x} = 2(x - x^3)e^{-x^2}$$

$$y'' = 2(2x^4 - 5x^2 + 1)e^{-x^2}$$

For $x = 0$, $\frac{\mathrm{d}^2y}{\mathrm{d}x^2} = +2 \rightarrow$ minimum

For $x = +1$, , $\frac{\mathrm{d}^2y}{\mathrm{d}x^2} = -4e^{-1} \rightarrow$ maximum

For $x = -1$, , $\frac{\mathrm{d}^2y}{\mathrm{d}x^2} = -4e^{-1} \rightarrow$ maximum

$y(x) = x^2e^{-x^2}$ is an even function because $y(-x) = +y(x)$

1.3.6 Series

1.35 The given series is $x - \frac{x^2}{2^2} + \frac{x^3}{3^2} - \frac{x^4}{4^2} + \cdots$ (A)

The series formed by the coefficients is

$$1 - \frac{1}{2^2} + \frac{1}{3^2} - \frac{1}{4^2} + \cdots \quad \text{(B)}$$

$$\lim_{n=\infty}\left(\frac{a_{n+1}}{a_n}\right) = \lim_{n=\infty}\left[-\frac{n^2}{(n+1)^2}\right] = \frac{\infty}{\infty}$$

Apply L'Hospital rule.

Differentiating, $\lim_{n=\infty}\left(-\frac{2n}{2(n+1)}\right)=\frac{\infty}{\infty}$

Differentiating again, $\lim_{n=\infty}\left(-\frac{2}{2}\right)=-1(=L)$

$$\left|\frac{1}{L}\right|=\left|\frac{1}{-1}\right|=1$$

The series (A) is

I. Absolutely convergent when $|Lx|<1$ or $|x|>\left|\frac{1}{L}\right|$ i.e. $-\left|\frac{1}{L}\right|<x<+\left|\frac{1}{L}\right|$

II. Divergent when $|Lx|>1$, or $|x|>\left|\frac{1}{L}\right|$

III. No test when $|Lx|=1$, or $|x|=\left|\frac{1}{L}\right|$.

By I the series is absolutely convergent when x lies between -1 and $+1$
By II the series is divergent when x is less than -1 or greater than $+1$
By III there is no test when $x=\pm 1$.

Thus the given series is said to have $[-1, 1]$ as the interval of convergence.

1.36 $f(x)=\log x;\ f(1)=0$

$f'(x)=\frac{1}{x};\ f'(1)=1$

$f''(x)=-\frac{1}{x^2};\ f''(1)=-1$

$f'''(x)=\frac{2}{x^3};\ f'''(1)=2$

Substitute in the Taylor series

$$f(x)=f(a)+\frac{(x-a)}{1!}f'(a)+\frac{(x-a)^2}{2!}f''(a)+\frac{(x-a)^3}{3!}f'''(a)+\cdots$$

$$\log x=0+(x-1)-\frac{1}{2}(x-1)^2+\frac{1}{3}(x-1)^2-\cdots$$

1.37 Use the Maclaurin's series

$$f(x)=f(0)+\frac{x}{1!}f'(0)+\frac{x^2}{2!}f''(0)+\frac{x^3}{3!}f'''(0)+\cdots \tag{1}$$

Differentiating first and then placing $x=0$, we get

$f(x)=\cos\ x,\ f(0)=1$
$f'(x)=-\sin\ xf'(0)=0$
$f''(x)=-\cos\ x,\ f''(0)=-1$
$f'''(x)=\sin\ x,\ f'''(0)=0$
$f^{iv}(x)=\cos\ x,\ f^{iv}(0)=1$

etc.

Substituting in (1)

$$\cos\ x=1-\frac{x^2}{2!}+\frac{x^4}{4!}-\frac{x^6}{6!}+\cdots$$

The series is convergent with all the values of x.

1.38 $f(a+x) = \sin(a+x)$
Put $x = 0$

$f(a) = \sin\ a$
$f'(a) = \cos\ a$
$f''(a) = -\sin\ a$
$f'''(a) = -\cos\ a$

Substitute in

$$f(x) = f(a) + \frac{(x-a)}{1!} f'(a) + \frac{(x-a)^2}{2!} f''(a) + \frac{(x-a)^3}{3!} f'''(a) + \cdots$$

$$\sin(a+x) = \sin a + \frac{x}{1} \cos a - \frac{x^2}{2!} \sin a - \frac{x^3}{3!} \cos a + \cdots$$

1.39 We know that

$y = 1 + x + x^2 + x^3 + x^4 + \cdots = 1/(1-x)$

Differentiating with respect to x,

$dy/dx = 1 + 2x + 3x^2 + 4x^3 + \cdots = 1/(1-x)^2 = S$

1.3.7 Integration

1.40 (a) $\int \sin^3 x \cos^6 x dx = \int \sin^2 x \cos^6 x \sin x dx$

$= -\int (1-\cos^2 x) \cos^6 x\ d(\cos x) = \int \cos^8 x\ d(\cos x) - \int \cos^6 x\ d(\cos x)$

$$= \frac{\cos^9 x}{9} - \frac{\cos^7 x}{7} + C$$

(b) $$\int \sin^4 x \cos^2 x dx = \int (\sin x \cos x)^2 \sin^2 x dx$$

$$= \int \frac{1}{4} \sin^2 2x (\frac{1}{2} - \frac{1}{2} \cos 2x) dx$$

$$= \frac{1}{8} \int \sin^2 2x dx - \frac{1}{8} \int \sin^2 2x \cos 2x\ dx$$

$$= \frac{1}{8} \int (\frac{1}{2} - \frac{1}{2} \cos 4x) dx - \frac{1}{8} \int \sin^2 2x \cos 2x dx$$

$$= \frac{x}{16} - \frac{\sin 4x}{64} - \frac{\sin^3 2x}{48} + C$$

1.41 Express the integrand as sum of functions.

Let $$\frac{1}{2x^2 - 3x - 2} = \frac{1}{(2x+1)(x-2)} = \frac{A}{2x+1} + \frac{B}{x-2} = \frac{A(x-2) + B(2x+1)}{(2x+1)(x-2)}$$

$B - 2A = 1$
$A + 2B = 0$

Solving, $A = -\frac{2}{5}$ and $B = \frac{1}{5}$

$$I = -\frac{2}{5} \int \frac{dx}{2x+1} + \frac{1}{5} \int \frac{dx}{x-2} + C$$

$$= -\frac{1}{5}\ln(2x+1) + \frac{1}{5}\ln(x-2) + C$$

$$= \frac{1}{5}\ln\left(\frac{x-2}{2x+1}\right) + C$$

1.42 $r^2 = a^2 \sin 2\theta$

Elementary area

$$\mathrm{d}A = \frac{1}{2}r^2\mathrm{d}\theta$$

$$A = \frac{1}{2}\int_0^{\pi/2} r^2\mathrm{d}\theta = \frac{a^2}{2}\int_0^{\pi/2} \sin 2\theta\mathrm{d}\theta =$$

$$a^2\int_0^{\pi/2} \sin\theta\mathrm{d}(\sin\theta) = \frac{a^2}{2}\sin^2\theta|_0^1 = \frac{a^2}{2}$$

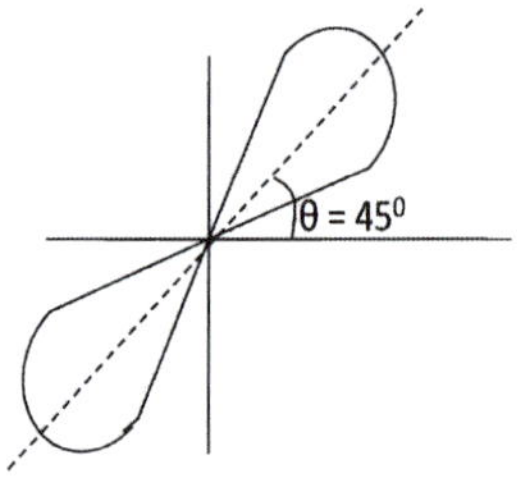

Fig. 1.11 Polar diagram of the curve $r^2 = a^2 \sin 2\theta$

1.43 Since $x^2 + 2$ occurs twice as a factor, assume

$$\frac{x^3+x^2+2}{(x^2+2)^2} = \frac{Ax+B}{(x^2+2)^2} + \frac{Cx+D}{x^2+2}$$

On clearing off the fractions, we get

$$x^3 + x^2 + 2 = Ax + B + (Cx + D)(x^2 + 2)$$

or $x^3 + x^2 + 2 = Cx^3 + Dx^2 + (A + 2C)x + B + 2D$

Equating the coefficients of like powers of x

$C = 1, D = 1, A + 2C = 0, B + 2D = 2$

This gives $A = -2, B = 0, C = 1, D = 1$

Hence,

$$\frac{x^3+x^2+2}{(x^2+2)^2} = -\frac{2x}{(x^2+2)^2} + \frac{x}{x^2+2} + \frac{1}{x^2+2}$$

$$\int\frac{(x^3+x^2+2)\mathrm{d}x}{(x^2+2)^2} = -\int\frac{2x\mathrm{d}x}{(x^2+2)^2} + \int\frac{x\mathrm{d}x}{x^2+2} + \int\frac{\mathrm{d}x}{x^2+2}$$

$$= \frac{1}{x^2+2} + \frac{1}{2}\ln(x^2+2) + \frac{1}{\sqrt{2}}\tan^{-1}\left(\frac{x}{\sqrt{2}}\right) + C$$

1.44 $$\int_0^\infty \frac{4a^3\mathrm{d}x}{x^2+4a^2} = \lim_{b=\infty}\int_0^b \frac{4a^3\mathrm{d}x}{x^2+4a^2} = \lim_{b=\infty}\left[2a^2\tan^{-1}\left(\frac{x}{2a}\right)\right]_0^b$$

$$= \lim_{b=\infty}\left[2a^2\tan^{-1}\left(\frac{b}{2a}\right)\right] = 2a^2.\frac{\pi}{2} = \pi a^2$$

1.45 (a) $\int \tan^6 x \sec^4 x dx = \int \tan^6 x(\tan^2 x + 1) \sec^2 x dx$

$$= \int (\tan x)^8 \sec^2 x dx + \int \tan^6 x \sec^2 x dx$$

$$= \int (\tan x)^8 d(\tan x) + \int (\tan x)^6 d(\tan x)$$

$$= \frac{\tan^9 x}{9} + \frac{\tan^7 x}{7} + C$$

(b) $\int \tan^5 x \sec^3 x dx = \int \tan^4 x \sec^2 x \sec x \tan x \, dx$

$$= \int (\sec^2 x - 1)^2 \sec^2 x \sec x \tan x \, dx$$

$$= \int (\sec^6 x - 2 \sec^4 x + \sec^2 x) d(\sec x)$$

$$= \frac{\sec^7 x}{7} - 2\frac{\sec^5 x}{5} + \frac{\sec^3 x}{3} + C$$

1.46 $\int_2^4 \frac{2x + 4}{x^2 - 4x + 8} dx = \int_2^4 \frac{2x - 4 + 8}{(x - 2)^2 + 4} dx$

$$= \int_2^4 \frac{2x - 4}{(x - 2)^2 + 4} dx + 8 \int_2^4 \frac{dx}{(x - 2)^2 + 4}$$

$$= \ln [(x - 2)^2 + 4]_2^4 + (8/2) \tan^{-1} 1$$

$$= \ln 2 + \pi$$

1.47 Let us first find the area OMP which is half of the required area OPP′. For the upper branch of the curve, $y = x^{3/2}$, and summing up all the strips between the limits $x = 0$ and $x = 4$, we get
Area OMP $= \int_0^4 y dx = \int_0^4 x^{3/2} dx = \frac{64}{5}$.
Hence area OPP′ $= 2x \frac{64}{5} = 25.6$ units.
Note: for the lower branch $y = x^{3/2}$ and the area will be $-64/5$. The area will be negative simply because for the lower branch the y-coordinates are negative. The result for the area OPP′ pertains to total area regardless of sign.

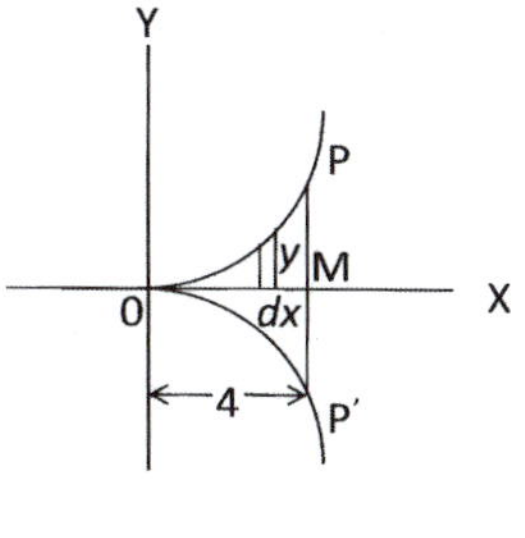

Fig. 1.12 Semi-cubical parabola

1.48 $x^{2/3} + y^{2/3} = a^{2/3}$ (1)

The arc AB generates only one half of the surface.

$$\frac{S_x}{2} = 2\pi \int_a^b y \left[1 + \left(\frac{\mathrm{d}y}{\mathrm{d}x}\right)^2\right]^{1/2} \mathrm{d}x \quad (2)$$

From (1) we find

$$\frac{\mathrm{d}y}{\mathrm{d}x} = -\frac{y^{1/3}}{x^{1/3}};\ y = \left(a^{\frac{2}{3}} - x^{\frac{2}{3}}\right)^{3/2} \quad (3)$$

Substituting (3) in (2)

$$\frac{S_x}{2} = 2\pi \int_0^a (a^{2/3} - x^{2/3}) \left[1 + \frac{y^{2/3}}{x^{2/3}}\right]^{1/2} \mathrm{d}x$$

$$= 2\pi \int_0^a (a^{2/3} - x^{2/3})^{3/2} \left(\frac{a^{2/3}}{x^{2/3}}\right)^{1/2} \mathrm{d}x$$

$$= 2\pi a^{1/3} \int_0^a (a^{2/3} - x^{2/3})^{3/2} x^{-1/3}\, \mathrm{d}x$$

$$= \frac{6\pi a^2}{5}$$

$$\therefore S_x = \frac{12\pi a^2}{5}$$

Fig. 1.13 Curve of hypocycloid $x^{2/3} + y^{2/3} = a^{2/3}$

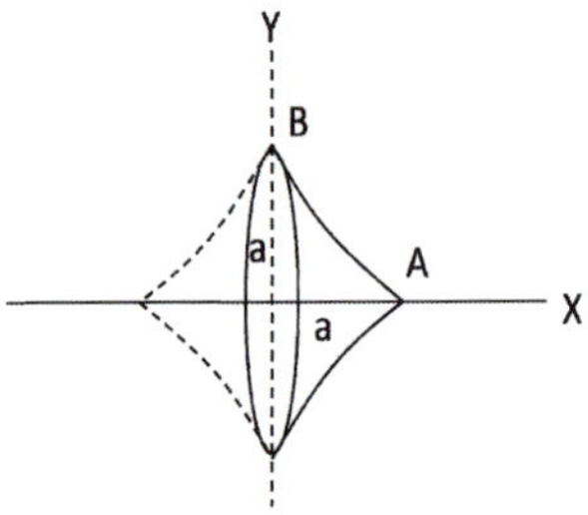

1.49 $$\int_0^a \int_0^{\sqrt{a^2-x^2}} (x+y)\mathrm{d}y\, \mathrm{d}x = \int_0^a \left[\int_0^{\sqrt{a^2-x^2}} (x+y)\mathrm{d}y\right] \mathrm{d}x$$

$$= \int_0^a \left[\left(xy + \frac{y^2}{2}\right) \mathrm{d}x\right]_0^{\sqrt{a^2-x^2}}$$

$$= \int_0^a \left(x\sqrt{a^2-x^2} + \frac{a^2-x^2}{2}\right) \mathrm{d}x$$

$$= \frac{2a^3}{3}$$

1.50 Area to be calculated is

$A = \mathrm{ACFD} = 2 \times \mathrm{ABED}$

$$= 2\int y\mathrm{d}x$$

$$= 2\int_1^2 \frac{1}{x}\,\mathrm{d}x = 2\ln 2$$

$$= 1.386 \text{ units}$$

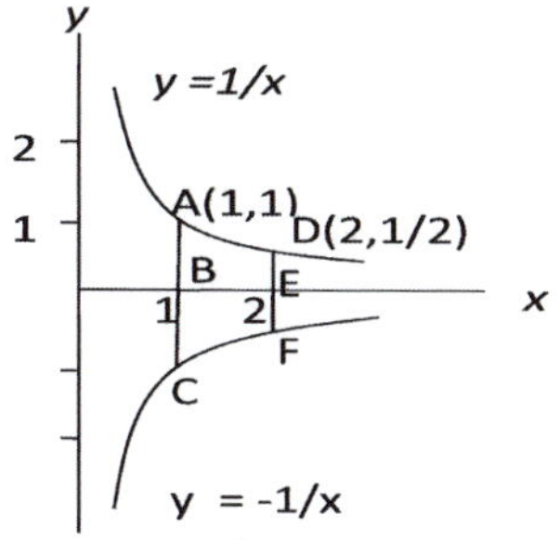

Fig. 1.14 Area enclosed between the curves $y = 1/x$ and $y = -1/x$ and the lines $x = 1$ and $x = 2$

1.51 $\int \frac{1}{x^2 - 18x + 34}\mathrm{d}x = \int \frac{1}{(x-3)^2 + 25}\mathrm{d}x$

$$= (1/5)\tan^{-1}\left(\frac{x-3}{5}\right)$$

1.52 $\int_0^1 x^2 \tan^{-1} x\, \mathrm{d}x = (x^3/3)\tan^{-1} x|_0^1 - 1/3\int_0^1 \frac{x^3}{(x^2+1^2)}\mathrm{d}x$

$$= \frac{\pi}{12} - \frac{1}{3}\int_0^1 \left[x - \frac{x}{(x^2+1)}\right]\mathrm{d}x$$

$$= \frac{\pi}{12} - \frac{x^2}{6}\bigg|_0^1 + \left(\frac{1}{6}\right)\ln(x^2+1)\bigg|_0^1$$

$$= \frac{\pi}{12} - \frac{1}{6} + \frac{1}{6}\ln 2$$

1.53 (a) The required area is for the figure formed by ABDGEFA. This area is equal to the area under the curve $y = x^2 + 2$, that is ACEFA, minus ΔBCD, plus ΔDGE (Fig 1.15a)

$$= \int_{-1}^{2} y\mathrm{d}x - \frac{1}{2}\mathrm{BC}\,.\,\mathrm{CD} + \frac{1}{2}\mathrm{DE}\,.\,\mathrm{EG}$$

$$= \int_{-1}^{2} (x^2+2)\mathrm{d}x - \frac{1}{2}\,.\,2.2 + \frac{1}{2}\,.\,1.1$$

$$= 7.5 \text{ units}$$

(b) The required volume V = Volume of cylinder BDEC of height H and radius r and the cone ABC. (Fig 1.15b)

$$V = \pi r^2 H + \frac{1}{3}\pi r^2 h = \pi r^2\left(H + \frac{h}{3}\right)$$

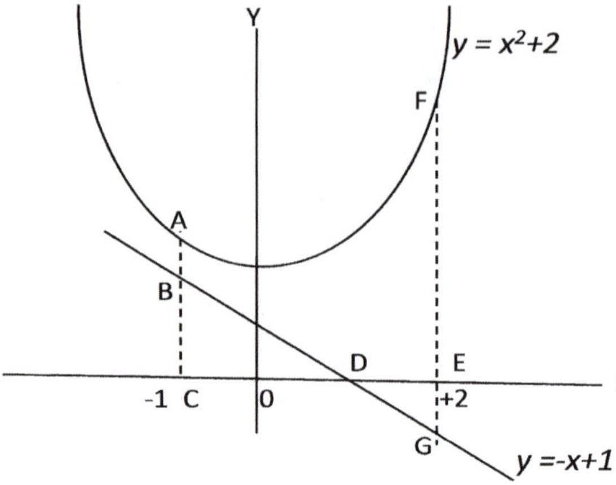

Fig. 1.15a Area bounded by ABGFA (see the text, Prob 1.53a)

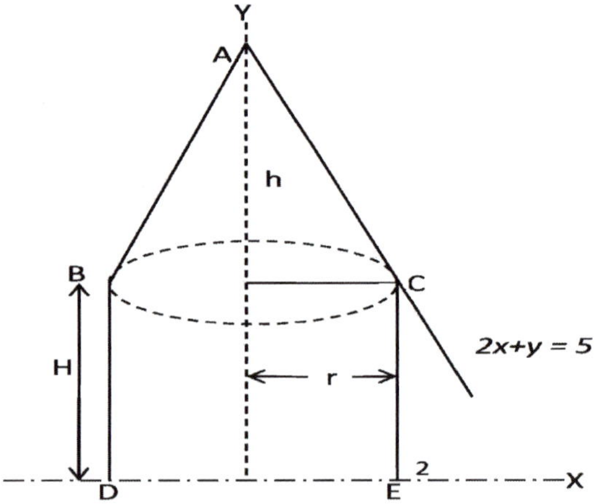

Fig. 1.15b Volume of the cylinder plus the cone (See Prob 1.15b)

$$= \pi . 2^2 \left(5 + \frac{4}{3}\right) = 25\frac{1}{3}\pi \text{ units}$$

1.54 (a) Area $= \int_0^{2\pi} y\mathrm{d}x = \int_0^{\pi} x \sin x \,\mathrm{d}x + \int_{\pi}^{2\pi} x \sin x \,\mathrm{d}x$

$= -x\cos x + \sin x|_0^{\pi} - x\cos x + \sin x|_{\pi}^{2\pi} = \pi + 3\pi = 4\pi$

The area refers to the magnitude

(b) Volume, $V = 2\pi \int y^2 \,\mathrm{d}x = 2\pi \int x^2 \sin^2 x \,\mathrm{d}x$

$$= \frac{4\pi^4}{3} - \frac{\pi^2}{2}$$

1.3.8 Ordinary Differential Equations

1.55 $\dfrac{dy}{dx} = \dfrac{x^3 + y^3}{3xy^2}$

The equation is homogenous because $f(\lambda x, \lambda y) = f(x, y)$. Use the transformation

$y = Ux, dy = U\ dx + x\ dU$

$\dfrac{U dx + x du}{dx} = \dfrac{x^3 + U^3 x^3}{3x.U^2 x^2} = \dfrac{1 + U^3}{3U^2}$

$3U^3\ dx + 3xU^2\ du = (1 + U^3) dx$

or $(2U^3 - 1) dx + 3xU^2\ du = 0$

Dividing by $x(2U^2 - 1)$,

$\dfrac{dx}{x} + \dfrac{3U^2 du}{2U^3 - 1} = 0$

Integrating, $\ln x + \frac{1}{2} \ln(2U^3 - 1) = C$

$2 \ln x + \ln \left(\dfrac{2y^3}{x^3} - 1 \right) = C$

or $2y^3 - x^3 = Cx$

1.56 $\dfrac{d^3 y}{dx^3} - 3\dfrac{d^2 y}{dx^2} + 4y = 0$

The auxiliary equation is

$D^3 - 3D^2 + 4 = 0$

Solving, the roots are $-1, 2, 2$.

The root -1 gives the solution e^{-x}.

The double root 2 gives two solutions e^{2x}, $x\ e^{2x}$.

The general solution is

$y = C_1 e^{-x} + C_2 e^{2x} + C_3 x e^{2x}$

1.57 $\dfrac{d^4 y}{dx^4} - 4\dfrac{d^3 y}{dx^3} + 10\dfrac{d^2 y}{dx^2} - 12\dfrac{dy}{dx} + 5y = 0$

The auxiliary equation is

$D^4 - 4D^3 + 10D^2 - 12D + 5 = 0$

Solving, the roots are $1, 1, 1 \pm 2i$

The pair of imaginary roots $1 \pm 2i$ gives the two solutions $e^x \cos 2x$ and $e^x \sin 2x$.

The double root gives the two solutions e^x, xe^x.

The general solution is

$Y = C_1 e^x + C_2 x e^x + C_3 e^x \cos 2x + C_4 e^x \sin 2x$

or, $y = (C_1 + C_2 x + C_3 \cos 2x + C_4 \sin 2x) e^x$.

1.58 $$\frac{d^2y}{dx^2} + m^2y = \cos bx \tag{1}$$

Replacing the right-hand member by zero,

$$\frac{d^2y}{dx^2} + m^2y = 0. \tag{2}$$

Solving, we get the complimentary function

$$y = C_1 \sin mx + C_2 \cos mx = U. \tag{3}$$

Differentiating (1) twice, we get

$$\frac{d^4y}{dx^4} + m^2\frac{d^2y}{dx^2} = -b^2 \cos bx. \tag{4}$$

Multiply (1) by b^2 and adding the result to (4) gives

$$\frac{d^4y}{dx^4} + (m^2 + b^2)\frac{d^2y}{dx^2} + b^2m^2y = 0. \tag{5}$$

The complete solution of (5) is

$$y = C_1 \sin mx + C_2 \cos mx + C_3 \sin bx + C_4 \cos\ bx$$

or $y = U + C_3 \sin bx + C_4 \cos bx = U + V$

We shall now determine C_3 and C_4 so that $C_3 \sin bx + C_4 \cos bx$ shall be a particular solution V of (1)

Substituting

$$y = C_3 \sin bx + C_4 \cos bx, \frac{dy}{dx} = C_3\, b \cos bx - C_4 b \sin bx,$$

$\frac{d^2y}{dx^2} = -C_3b^2 \sin \mathrm{bx} - C_4b^2 \cos \mathrm{bx}$ in (1), we get

$$C_4(m^2 - b^2) \cos bx + C_3(m^2 - b^2) \sin bx = \cos bx$$

Equating the coefficients of like terms in this identity we get

$$C_4(m^2 - b^2) = 1 \to C_4 = \frac{1}{m^2 - b^2}$$

$$C_3(m^2 - b^2) = 0 \to C_3 = 0$$

Hence a particular solution of (1) is

$$V = \frac{\cos bx}{m^2 - b^2}$$

and the complete solution is

$$y = 0 + V = C_1 \sin mx + C_2 \cos mx + \frac{\cos bx}{m^2 - b^2}$$

1.59 $$\frac{d^2y}{dx^2} - 5\frac{dy}{dx} + 6y = x \tag{1}$$

Replace the right-hand member by zero to form the auxiliary equation

$$D^2 - 5D + 6 = 0 \tag{2}$$

The roots are $D = 2$ and 3. The solution is

$$y = C_1e^{2x} + C_2e^{3x} = 0 \tag{3}$$

The complete solution is

$$y = U + V \tag{4}$$

where $V = C_3x + C_4$ (5)

In order that V be a particular solution of (1), substitute $y = C_3x + C_4$, in (1) in order to determine C_3 and C_4.

$$-5C_3 + 6(C_3x + C_4) = x$$

Equating the like coefficients

$6C_3 = 1 \rightarrow C_3 = 1/6$
$6C_4 - 5C_3 = 0 \rightarrow C_4 = 5/36$

Hence the complete solution is

$$y = C_1e^{2x} + C_2e^{3x} + \frac{x}{6} + \frac{5}{36}$$

1.60 $$\frac{d^2x}{dt^2} + 2\frac{dx}{dt} + 5x = 0 \tag{1}$$

Put $x = e^{\lambda t}$, $\frac{dx}{dt} = \lambda e^{\lambda t}$, $\frac{d^2x}{dt^2} = \lambda^2 e^{\lambda t}$ *in* (1)

$$\lambda^2 + 2\lambda + 5 = 0 \tag{2}$$

its roots being, $\lambda = -1 \pm 2i$

$x = Ae^{-t(1-2i)} + Be^{-t(1+2i)}$
$x = e^{-t}[C\cos 2t + D\sin 2t]$

where C and D are constants to be determined from the initial conditions.
At $t = 0$, $x = 5$. Hence $C = 5$. Further

$$\frac{dx}{dt} = -e^{-t}[C\cos 2t + D\sin 2t + 2C\sin 2t - 2D\cos 2t]$$

At $t = 0$, $dx/dt = -3$

$$-3 = -C + 2D = -5 + 2D$$

whence $D = 1$. Therefore the complete solution is

$$x = e^{-t}(5\cos 2t + \sin 2t)$$

1.61 (a) Let the mass 1 be displaced by x_1 and mass m_2 by x_2. The force due to the spring on the left acting on mass 1 is $-kx_1$ and that due to the coupling is $-k(x_1 - x_2)$.

The net force

$$F_1 = -kx_1 - k(x_1 - x_2) = -k(2x_1 - x_2)$$

The equation of motion for mass 1 is

$$m\ddot{x}_1 + k(2x_1 - x_2) = 0 \tag{1}$$

Similarly, for mass 2, the spring on the right exerts a force $-kx_2$, and the coupling spring exerts a force $-k(x_2 - x_1)$. The net force

$$F_2 = -kx_2 - k(x_2 - x_1) = -k(2x_2 - x_1)$$

The equation of motion for mass 2 is

$$m\ddot{x}_2 + k(2x_2 - x_1) = 0 \tag{2}$$

The two Eqs. (1) and (2) are coupled equations.

$$\text{Let } x_1 = A_1 \sin \omega t \tag{3}$$

$$x_2 = A_2 \sin \omega t \tag{4}$$

$$\ddot{x}_1 = -\omega^2 A_1 \sin \omega t = -\omega^2 x_1 \tag{5}$$

$$\ddot{x}_2 = -\omega^2 A_2 \sin \omega t = -\omega^2 x_2 \tag{6}$$

Inserting (5) and (6) in (1) and (2)

$$-m\omega^2 x_1 + k(2x_1 - x_2) = 0$$

$$-\, m\omega^2 x_2 + k(2x_2 - x_1) = 0$$

Rearranging

$$(2k - m\omega^2)x_1 - kx_2 = 0 \tag{7}$$

$$-kx_1 + (2k - m\omega^2)x_2 = 0 \tag{8}$$

In order that the above equations may have a non-trivial solution, the determinant formed from the coefficients of x_1 and x_2 must vanish.

$$\begin{vmatrix} 2k - m\omega^2 & -k \\ -k & 2k - m\omega^2 \end{vmatrix} = 0 \tag{9}$$

$$(2k - m\omega^2)^2 - k^2 = 0$$

$$\text{or } (m\omega^2 - k)(m\omega^2 - 3k) = 0$$

The solutions are

$$\omega_1 = \sqrt{\frac{k}{m}} \tag{10}$$

$$\omega_2 = \sqrt{\frac{3k}{m}} \tag{11}$$

(b) The two solutions to the problem are

$$x_1 = A_1 \sin \omega_1 t;\ x_2 = A_2 \sin \omega_1 t \tag{12}$$

$$x_1 = B_1 \sin \omega_2 t;\ x_2 = B_2 \sin \omega_2 t \tag{13}$$

In (12) and (13) the amplitudes are not all independent as we can verify with the use of (7) and (8). Substituting (10) and (12) in (7), yields $A_2 = A_1$. Substitution of (11) and (13) in (7), gives $B_2 = -B_1$.

Dropping off the subscripts on $A's$ and $B's$ the solutions can be written as

$$x_1 = A \sin \omega_1 t = x_2 \tag{14}$$

$$x_1 = B \sin \omega_2 t = -x_2 \tag{15}$$

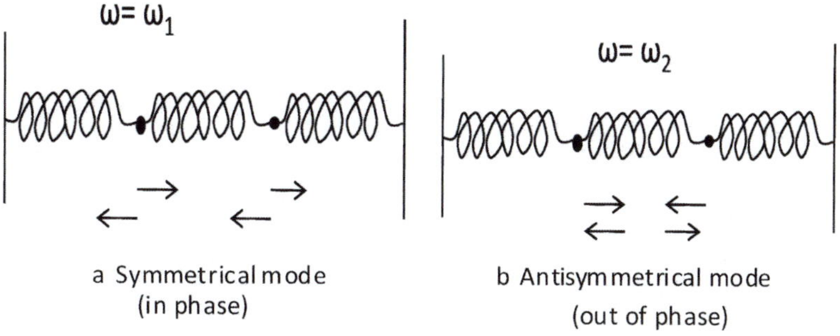

Fig. 1.16 Two modes of Oscillation

If initially $x_1 = x_2$, the masses oscillate in phase with frequency ω_1 (symmetrical mode) as in Fig. 1.16(a). If initially $x_2 = -x_2$ then the masses oscillate out of phase (asymmetrical) as in Fig. 1.16(b)

1.62 Sum of translational + rotational + potential energy = constant

$$\frac{1}{2}mv^2 + \frac{1}{2}I\omega^2 + \frac{1}{2}kx^2 = \text{const.}$$

But $I = \frac{1}{2}mR^2$ and $\omega = v/R$

Therefore $\frac{3}{4}mv^2 + \frac{1}{2}kx^2 = \text{const.}$

$$\frac{3}{4}m(\mathrm{d}x/\mathrm{d}t)^2 + \frac{1}{2}kx^2 = \text{const.}$$

Differentiating with respect to time,

$$\left(\frac{3}{2}\right)\left(\frac{m\mathrm{d}^2x}{\mathrm{d}t^2}\right)\cdot\frac{\mathrm{d}x}{\mathrm{d}t} + kx.\frac{\mathrm{d}x}{\mathrm{d}t} = 0$$

Cancelling $\mathrm{d}x/\mathrm{d}t$ through and simplifying $\mathrm{d}^2x/\mathrm{d}t^2 + (2k/3m)x = 0$. This is an equation to SHM.

Writing $\omega^2 = \frac{2k}{3m}$, time period $T = \frac{2\pi}{\omega} = 2\pi\sqrt{\frac{3m}{2k}}$

1.63 $\frac{\mathrm{d}^2y}{\mathrm{d}x^2} - 8\frac{\mathrm{d}y}{\mathrm{d}x} = -16y$

$$\frac{\mathrm{d}^2y}{\mathrm{d}x^2} - 8\frac{\mathrm{d}y}{\mathrm{d}x} + 16y = 0$$

Auxiliary equation:

$$D^2 - 8D + 16 = 0$$
$$(D - 4)(D - 4) = 0$$

The roots are 4 and 4.

Therefore $y = C_1e^{4x} + C_2xe^{4x}$

1.64 $x^2\frac{\mathrm{d}y}{\mathrm{d}x} + y(x + 1)x = 9x^2$ (1)

Put the above equation in the form

$$\frac{dy}{dx} + Py = Q \tag{2}$$

$$\frac{dy}{dx} + \frac{y(x+1)}{x} = 9 \tag{3}$$

$$\text{Let } y = Uz \tag{4}$$

$$\frac{dy}{dx} = \frac{U dz}{dx} + \frac{z dU}{dx} \tag{5}$$

Substituting (4) and (5) in (3)

$$\frac{U dz}{dx} + \left(\frac{dU}{dx} + \frac{U(x+1)}{x}\right) z = 9 \tag{6}$$

Now to determine U, we place the coefficients of z equal to zero. This gives

$$\frac{dU}{dx} + \frac{U(x+1)}{x} = 0$$

$$\frac{dU}{U} = -\left(1 + \frac{1}{x}\right) dx$$

Integrating, $\ln U = -x - \ln x$ or

$$U = e^{-x}/x \tag{7}$$

As the term in z drops off, Eq. (6) becomes

$$U \frac{dz}{dx} = 9 \tag{8}$$

Eliminating U between (7) and (8)

$$dz = 9x\, e^x\, dx$$

$$\text{Integrating } z = 9 \int x e^x\, dx = 9e^x(x-1) \tag{9}$$

Substituting U and z in $y = Uz$,

$$y = \frac{9(x-1)}{x}$$

1.65 $$\frac{d^2y}{dx^2} + \frac{dy}{dx} - 2y = 2\cosh 2x \tag{1}$$

The complimentary solution is found from

$$\frac{d^2y}{dx^2} + \frac{dy}{dx} - 2y = 0$$

$$D^2 + D - 2 = 0$$

$$(D-1)(D+2) = 0$$

$$D = 1, -2$$

$$Y = U = C_1 e^x + C_2 e^{-2x} \tag{2}$$

Differentiating (1) twice

$$\frac{d^4y}{dx^4} + \frac{d^3y}{dx^3} - 2\frac{d^2y}{dx^2} = 8\cosh 2x \tag{3}$$

Multiply (1) by (4) and subtract the resulting equation from (3)

$$\frac{d^4y}{dx^4} + \frac{d^3y}{dx^3} - \frac{6d^2y}{dx^2} - \frac{4dy}{dx} + 8y = 0 \tag{4}$$

$$D^4 + D^3 - 6D^2 - 4D + 8 = 0$$

$$(D-1)(D-2)(D+2)^2 = 0$$

$$D = 1, 2, -2, -2$$

The complete solution of (4) is

$$y = C_1e^x + C_3e^{2x} + C_2e^{-2x} + C_4xe^{-2x}$$
$$= U + C_3e^{2x} + C_4xe^{-2x}$$
$$= U + V$$
$$V = C_3e^{2x} + C_4xe^{-2x} \tag{5}$$

Inserting (5) in (1), writing $2\cosh 2x = e^{2x} + e^{-2x}$ and comparing the coefficients of e^{2x} and e^{-2x}, we find $C_3 = \frac{1}{4}$ and $C_4 = -\frac{1}{3}$. Thus the complete solution of (1) is

$$y = C_1e^x + C_2e^{-2x} + \frac{1}{4}e^{2x} - \frac{1}{3}x\,e^{-2x}$$

1.66 $\frac{xdy}{dx} - y = x^2$

$$\frac{dy}{dx} - \frac{y}{x} = x$$

The standard equation is

$$\frac{dy}{dx} + Py = Q$$

$$\therefore P = -\frac{1}{x}; Q = x$$

$$y\exp\left(\int p\mathrm{d}x\right) = \left[\int Q\exp\left(\int p\mathrm{d}x\right)\right]\mathrm{d}x + C$$

$$y\exp\left(-\int \frac{1}{x}\mathrm{d}x\right) = \left[\int x\exp\left(-\int \frac{1}{x}\mathrm{d}x\right)\right] + C$$

$$y\exp(-\ln x) = \left[\int x\exp(-\ln x)\right] + C$$

$$yx^{-1} = \int x\,x^{-1}\mathrm{d}x + C$$

$$y = x^2 + Cx$$

1.67 (a) $y' - \dfrac{2y}{x} = \dfrac{1}{x^3}$ (1)

Let $y = px$, $y' = p + xp'$

Then (1) becomes

$xp' - p = 1/x^3$

Now $\frac{d}{dx}\left(\frac{p}{x}\right) = \frac{xp-p}{x^2}$

$\therefore xp' - p = x^2 \dfrac{d}{dx}\left(\dfrac{p}{x}\right) = \dfrac{1}{x^3}$

$\dfrac{d}{dx}\left(\dfrac{p}{x}\right) = \dfrac{1}{x^5}$ or $d\left(\dfrac{p}{x}\right) = \dfrac{dx}{x^5}$

Integrating

$\dfrac{p}{x} = -\dfrac{1}{4x^4} + C$

or $\dfrac{y}{x^2} = -\dfrac{1}{4x^4} + C$

$y = -\dfrac{1}{4x^2} + Cx^2$

It is inhomogeneous, first order.

(b) $y'' + 5y' + 4y = 0$

$D^2 + 5D + 4 = 0$

$(D + 4)(D + 1) = 0$

$D = -4, -1$

$y = A\,e^{-4x} + B\,e^{-x}$

It is inhomogeneous, second order.

1.68 (a) $\dfrac{dy}{dx} + y = e^{-x}$

Compare with the standard equation

$\dfrac{dy}{dx} + py = Q$

$P = 1; Q = e^{-x}$

$y \exp\left(\int p dx\right) = \left[\int Q \exp\left(\int p\, dx\right)\right] dx + C$

$y \exp\left(\int 1\, dx\right) = \left[\int e^{-x} \exp\left(\int 1\, dx\right)\right] dx + C$

$ye^x = x + C$

$y = xe^{-x} + Ce^{-x}$

(b) $\dfrac{d^2y}{dx^2} + 4y = 2\cos(2x)$ (1)

The complimentary function is obtained from $y'' + 4y = 0$

$y = U = C_1 \sin 2x + C_2 \cos 2x$

Differentiate (1) twice

$$\frac{d^4y}{dx^4} + \frac{4d^2y}{dx^2} = -8\cos(2x) \quad (2)$$

Multiply (1) by 4 and add to (2),

$$\frac{d^4y}{dx^4} + \frac{8d^2y}{dx^2} + 16y = 0$$

$$D^4 + 8D^2 + 16 = 0$$

$$(D^2 + 4)^2 = 0$$

$$D = \pm 2i$$

$$y = C_1 \sin 2x + C_3\, x \sin 2x + C_2 \cos 2x + C_4 x \cos 2x$$

$$= U + C_3 x \sin 2x + C_4 x \cos 2x$$

$$= U + V$$

$$Y = V = C_3 x \sin 2x + C_4 x \cos 2x \quad (3)$$

Use (3) in (1) and compare the coefficients of $\sin 2x$ and $\cos 2x$ to find $C_3 = 1/2$ and $C_4 = 0$. Thus the complete solution is

$$y = C_1 \sin 2x + C_2 \cos 2x + \frac{1}{2} x \sin 2x$$

1.69 $\frac{dy}{dx} + \frac{3y}{x+2} = x + 2$

This equation is of the form

$$\frac{dy}{dx} + yp(x) = Q(x)$$

with $P = \frac{3}{x+2}$ and $Q = x + 2$

The solution is obtained from

$$y \exp\left(\int p(x)dx = \left[\int Q(x) \exp\left(\int p(x)dx\right)\right] dx + C\right.$$

$$\text{Now} \int p(x)dx = 3\int \frac{dx}{x+2} = 3\ln(x+2) = \ln(x+2)^3$$

$$\therefore y \exp(\ln(x+2)^3) = \left[\int (x+2) \exp(\ln(x+2)^3)\right] dx + C$$

$$y(x+2)^3 = \int (x+2)^4 dx + C$$

$$= \frac{(x+2)^5}{5} + C$$

$$\text{or } y = \frac{(x+2)^2}{5} + C$$

$y = 2$ when $x = -1$

Therefore $C = \frac{9}{5}$

The complete solution is

$$y = \frac{(x+2)^2}{5} + \frac{9}{5} = \frac{x^2 + 4x + 13}{5}$$

1.70 (i) $$\frac{d^2y}{dx^2} - \frac{4dy}{dx} + 4y = 8x^2 - 4x - 4 \quad (1)$$

Replace the RHS member by zero to get the auxiliary solution.

$D^2 - 4D + 4 = 0$

The roots are $D = 2$ and 2. Therefore the auxiliary solution is

$$y = Ae^{2x} + Bxe^{2x} \quad (2)$$

Complete solution is

$$y = (A + Bx)e^{2x} + Cx^2 + Dx + E \quad (3)$$

The derivatives are

$$\frac{dy}{dx} = (2A + 2Bx + B)e^{2x} + 2Cx + D \quad (4)$$

$$d^2y/dx^2 = 4(A + B + Bx)e^{2x} + 2C \quad (5)$$

Use (3), (4) and (5) in (1) and compare the coefficients of like terms. We get three equations. Two more equations are obtained from the conditions $y = -2$ and $\frac{dy}{dx} = 0$ when $x = 0$.

Solving the five equations we get, $A = -3$, $B = 3$, $C = 2$, $D = 3$ and $E = 1$. Hence the complete solution is $y = 3(x - 1)e^{2x} + 2x^2 + 3x + 1$

(ii) $$\frac{d^2y}{dx^2} + 4y = \sin x \quad (1)$$

Replace the RHS member by zero and write down the auxiliary equation $D^2 + 4 = 0$

The roots are $\pm 2i$. The auxiliary solution is

$Y = A\cos 2x + B\sin 2x$

The complete solution is

$$Y = A\cos 2x + B\sin 2x + C\sin x \quad (2)$$

$$\frac{d^2y}{dx^2} = -4(A\cos 2x + B\sin 2x) - C\sin x \quad (3)$$

Substitute (2) and (3) in (1) to find $C = 1/3$. Thus

$$y = A\cos 2x + b\sin 2x + \frac{1}{3}\sin x$$

1.71 $y''' - y'' + y' - y = 0$

Auxiliary equation is

$D^3 - D^2 + D - 1 = 0$

$(D - 1)(D^2 + 1) = 0$ The roots are $D = 1, \pm i$

The solution is

$y = A\sin x + B\cos x + ce^x$

1.3.9 Laplace Transforms

1.72 $$\frac{\mathrm{d}N_A(t)}{\mathrm{d}t} = -\lambda_A N_A(t) \tag{1}$$

$$\frac{\mathrm{d}N_B(t)}{\mathrm{d}t} = -\lambda_B N_B(t) + \lambda_A N_A(t) \tag{2}$$

Applying Laplace transform to (1)

$$s\mathcal{L}(N_A) - N_A(0) = -\lambda_A \mathcal{L}(N_A)$$

$$\text{or } \mathcal{L}(N_A) = \frac{N_A^0}{s + \lambda_A} = \frac{N_A^0}{s - (-\lambda_A)} \tag{3}$$

$$\therefore N_A = N_A^0 \exp(-\lambda_A t) \tag{4}$$

Applying the Laplace transform to (2)

$$s\mathcal{L}(N_B) - N_B(0) = -\lambda_B \mathcal{L}(N_B) + \lambda_A \mathcal{L}(N_A) \tag{5}$$

Using (3) in (4) and putting $N_2(0) = 0$

$$\mathcal{L}(N_B)(s + \lambda_B) = \frac{\lambda_A N_A^0}{s + \lambda_A}$$

$$\text{or } \mathcal{L}(N_B) = \frac{\lambda_A N_A^0}{(s + \lambda_A)(s + \lambda_B)} = \frac{\lambda_A N_A^0}{\lambda_B - \lambda_A}\left[\frac{1}{s + \lambda_A} - \frac{1}{s + \lambda_B}\right]$$

$$= \frac{\lambda_A N_A^0}{\lambda_B - \lambda_A}\left[\frac{1}{s - (-\lambda_A)} - \frac{1}{s - (-\lambda_B)}\right]$$

$$\therefore N_B = \frac{\lambda_A N_A^0}{\lambda_B - \lambda_A}\left[e^{-\lambda_A t} - e^{-\lambda_B t}\right]$$

1.73 $$\frac{\mathrm{d}N_A}{\mathrm{d}t} = -\lambda_A N_A \tag{1}$$

$$\frac{\mathrm{d}N_B}{\mathrm{d}t} = -\lambda_B N_B + \lambda_A N_A \tag{2}$$

$$\frac{\mathrm{d}N_C}{\mathrm{d}t} = +\lambda_B N_B \tag{3}$$

Applying the Laplace transform to (3)

$$s\mathcal{L}\{N_C\} - N_C(0) = \lambda_B \mathcal{L}\{N_B\} = \frac{\lambda_B \lambda_A N_A^0}{(s + \lambda_A)(s + \lambda_B)}$$

Given $N_c(0) = 0$

$$\mathcal{L}\{N_c\} = \frac{\lambda_A \lambda_B N_A^0}{s(s + \lambda_A)(s + \lambda_B)}$$

$$= \frac{\lambda_A \lambda_B N_A^0}{(\lambda_B - \lambda_A)s}\left[\frac{1}{s + \lambda_A} - \frac{1}{s + \lambda_B}\right]$$

$$= \frac{\lambda_A \lambda_B N_A^0}{(\lambda_B - \lambda_A)} \left[\frac{1}{\lambda_A} \left\{ \frac{1}{s} - \frac{1}{s + \lambda_A} \right\} - \frac{1}{\lambda_B} \left\{ \frac{1}{s} - \frac{1}{s + \lambda_B} \right\} \right]$$

$$= N_1^0 \left[\frac{1}{s} - \frac{\lambda_B}{(\lambda_B - \lambda_A)} \frac{1}{(s + \lambda_A)} + \frac{\lambda_A}{(\lambda_B - \lambda_A)} \frac{1}{(s + \lambda_B)} \right]$$

$$\therefore N_c = N_A^0 \left[1 + \frac{1}{\lambda_B - \lambda_A} (\lambda_A \exp(-\lambda_B t) - \lambda_B \exp(-\lambda_A t)) \right]$$

1.74 (a) $\mathcal{L}\{e^{ax}\} = \int_0^\infty e^{-sx} e^{ax}\, dx = \int_0^\infty e^{-(s-a)x}\, dx$

$$= \frac{1}{s - a}, \text{ if } s > a$$

(b) and (c). From part (a), $\mathcal{L}(e^{ax}) = \frac{1}{s-a}$ Replace a by ai

$$\mathcal{L}(e^{iax}) = \mathcal{L}\{\cos ax + i \sin ax\}$$
$$= \mathcal{L}\{\cos ax\} + i\mathcal{L}\{\sin ax\}$$
$$= \frac{1}{s - ai} = \frac{s + ai}{s^2 + a^2} = \frac{s}{s^2 + a^2} + \frac{ia}{s^2 + a^2}$$

Equating real and imaginary parts:

$$\mathcal{L}\{\cos ax\} = \frac{s}{s^2 + a^2}; \mathcal{L}\{\sin ax\} = \frac{a}{x^2 + a^2}$$

1.3.10 Special Functions

1.75 Express H_n in terms of a generating function $T(\xi, s)$.

$$T(\xi, s) = \exp[\xi^2 - (s - \xi)^2] = \exp[-s^2 + 2s\xi]$$

$$= \sum_{n=0}^{\infty} \frac{H_n(\xi) s^n}{n!} \tag{1}$$

Differentiate (1) first with respect to ξ and then with respect to s.

$$\frac{\partial T}{\partial \xi} = 2s \exp(-s^2 + 2s\xi) =$$

$$\sum_n \frac{2s^{n+1} H_n(\xi)}{n!} = \sum_n \frac{s^n H_n'(\xi)}{n!} \tag{2}$$

Equating equal powers of s

$$H_n' = 2n H_{n-1} \tag{3}$$

$$\frac{\partial T}{\partial s} = \xi(-2s + 2\xi) \exp(-s^2 + 2s\xi) = \sum_n (-2s + 2\xi) s^n H_n(\xi) = \sum_n \frac{s^{n-1} H_n(\xi)}{(n-1)!} \tag{4}$$

Equating equal powers of s in the sums of equations

$$H_{n+1} = 2\xi H_n - 2n H_{n-1} \tag{5}$$

It is seen that (5) satisfies the Hermite's equation

$$H_n'' - 2\xi H_n' + 2nH_n = 0$$

1.76 $J_n(x) = \sum_k \frac{(-1)^k \left(\frac{x}{2}\right)^{n+2k}}{k!\Gamma(n+k+1)}$

(a) Differentiate

$$\frac{d}{dx}[x^n J_n(x)] = J_n(x) n x^{n-1} + x^n \frac{dJ_n(x)}{dx}$$

$$= \sum_k \frac{(-1)^k \left(\frac{x}{2}\right)^{n+2k} n x^{n-1}}{k!\Gamma(n+k+1)} + \sum \frac{x^n (n+2k)(-1)^k x^{n+2k-1}}{k!\Gamma(n+k+1).2^{n+2k}}$$

$$= \sum_k \frac{(-1)^k \left(\frac{x}{2}\right)^{n+2k-1} (n+k) x^n}{k!\Gamma(n+k+1)} = \sum \frac{(-1)^k \left(\frac{x}{2}\right)^{n+2k-1} x^n}{\Gamma(n+k)} = x^n J_{n-1}(x)$$

(b) A similar procedure yields

$$\frac{d}{dx}[x^{-n} J_n(x)] = -x^{-n} J_{n+1}(x)$$

1.77 From the result of 1.76(a)

$$\frac{d}{dx}[x^n J_n(x)] = J_n(x) n x^{n-1} + x^n \frac{d\,J_n(x)}{dx} = x^n J_{n-1}(x)$$

Divide through out by x^n

$$\frac{n}{x} J_n(x) + \frac{dJ_n(x)}{dx} = J_{n-1}(x)$$

Similarly from (b)

$$-\frac{n}{x} J_n(x) + \frac{dJ_n(x)}{dx} = -J_{n+1}(x)$$

Add and subtract to get the desired result.

1.78 $J_n(x) = \sum_{k=0}^{\infty} \frac{(-1)^k \left(\frac{x}{2}\right)^{n+2k}}{k!\Gamma(n+k+1)}$

(a) Therefore, with $n = 1/2$

$$J_{\frac{1}{2}}(x) = \frac{\left(\frac{x}{2}\right)^{1/2}}{\Gamma\left(\frac{3}{2}\right)} - \frac{\left(\frac{x}{2}\right)^{5/2}}{1.\Gamma\left(\frac{5}{2}\right)} + \frac{\left(\frac{x}{2}\right)^{9/2}}{2!\Gamma\left(\frac{7}{2}\right)} - \cdots$$

Writing $\Gamma\left(\frac{3}{2}\right) = \frac{\sqrt{\pi}}{2}$, $\Gamma\left(\frac{5}{2}\right) = \frac{3\sqrt{\pi}}{4}$, $\Gamma\left(\frac{7}{2}\right) = \frac{15}{8}\sqrt{\pi}$

$$J_{\frac{1}{2}}(x) = \sqrt{\frac{2}{\pi x}}\left[x - \frac{x^3}{3!} + \frac{x^5}{5!} + \cdots\right]$$

$$= \sqrt{\frac{2}{\pi x}} \sin x$$

(b) With $n = -1/2$

$$J_{-\frac{1}{2}}(x) = \frac{\left(\frac{x}{2}\right)^{-1/2}}{\Gamma\left(\frac{1}{2}\right)} - \frac{\left(\frac{x}{2}\right)^{3/2}}{1.\Gamma\left(\frac{3}{2}\right)} + \frac{\left(\frac{x}{2}\right)^{7/2}}{2!\Gamma\left(\frac{5}{2}\right)} - \cdots$$

$$= \sqrt{\frac{2}{\pi x}}\left[1 - \frac{x^2}{2!} + \frac{x^4}{4!} - \cdots\right]$$
$$= \sqrt{\frac{2}{\pi x}} \cos x$$

1.79 The normalization of Legendre polynomials can be obtained by l – *fold* integration by parts for the conventional form

$$P_l(x) = \frac{1}{2^l l!}\frac{\mathrm{d}^l}{\mathrm{d}x^l}(x^2 - 1)^l \qquad \text{(Rodrigues's formula)}$$
$$\int_{-1}^{+1} [P_l(x)]^2 \, \mathrm{d}x = \left(\frac{1}{2^l l!}\right)^2 \int_{-1}^{+1} \left[\frac{\mathrm{d}^l (x^2 - 1)^l}{\mathrm{d}x^l}\right]\left[\frac{\mathrm{d}^l (x^2 - 1)^l}{\mathrm{d}x^l}\right] \mathrm{d}x$$
$$= (-1)^l (\frac{1}{2^l l!})^2 \int_{-1}^{+1} \left[\frac{\mathrm{d}^{2l} (x^2 - 1)}{\mathrm{d}x^{2l}}\right] (x^2 - 1)^l \, \mathrm{d}x$$
$$= (-1)^l \left(\frac{(2l)!}{2^l l!}\right)^2 \int_{-1}^{+1} (x^2 - 1)^l \, \mathrm{d}x = \frac{2}{2l + 1}$$

Put $l = n$ to get the desired result.

The orthogonality can be proved as follows. Legendre's differential equation

$$\frac{\mathrm{d}}{\mathrm{d}x}\left[(1 - x^2)\frac{\mathrm{d}P_n(x)}{\mathrm{d}x}\right] + n(n + 1)P_n(x) = 0 \qquad (1)$$

can be recast as

$$[(1 - x^2)P_n']' = -n(n + 1)P_n(x) \qquad (2)$$
$$[(1 - x^2)P_m']' = -m(m + 1)P_m(x) \qquad (3)$$

Multiply (2) by P_m and (3) by P_n and subtract the resulting expressions.

$$P_m[(1 - x^2)P_n']' - P_n[(1 - x^2)P_m']' = [m(m + 1) - n(n + 1)]P_m P_n \qquad (4)$$

Now, LHS of (4) can be written as

$$P_m[(1 - x^2)P_n']' - P_n[(1 - x^2)P_m']'$$
$$= P_m[(1 - x^2)P_n']' + P_m'[(1 - x^2)P_n'] - P_n[(1 - x^2)P_m'] - P_n[(1 - x^2)P_m']'$$

(4) can be integrated

$$\frac{\mathrm{d}}{\mathrm{d}x}[(1 - x^2)(P_m P_n' - P_n P_m') = [m(m + 1) - n(n + 1)]P_m P_n$$
$$(1 - x^2)\left(P_m P_n' - P_n P_m'\right)|_{-1}^{1} = [m(m + 1) - n(n + 1)] \int_{-1}^{1} P_m P_n \mathrm{d}x$$

Since $(1 - x^2)$ vanishes at $x = \pm 1$, the LHS is zero and the orthogonality follows.

$$\int_{-1}^{1} P_m(x)P_n(x)\mathrm{d}x = 0; m \neq n$$

1.80 Legendre's equation is

$$(1-x^2)\frac{\partial^2 P_n(x)}{\partial x^2} - 2x\frac{\partial P_n(x)}{\partial x} + n(n+1)P_n(x) = 0 \quad (1)$$

Put $x = \cos\theta$, Eq. (1) then becomes

$$\sin^2\theta\frac{\partial^2 P_n}{\partial\cos^2\theta} - 2\cos\theta\frac{\partial P_n}{\partial\cos\theta} + n(n+1)P_n = 0 \quad (2)$$

For large n, $n(n+1) \to n^2$, and $\cos\ \theta \to 1$ for small θ,

$$\sin^2\theta\frac{\partial^2 P_n}{\partial\cos^2\theta} - 2\frac{\partial P_n}{\partial\cos\theta} + n^2 P_n = 0 \quad (3)$$

Now, Bessel's equation of zero order is

$$x^2\frac{\mathrm{d}^2\ J_0(x)}{\mathrm{d}x^2} + x\frac{\mathrm{d}}{\mathrm{d}x}J_0(x) + x^2 J_0(x) = 0 \quad (4)$$

Letting $x = 2n\sin\theta/2 = n\sin\theta$, in (4) for small θ, and noting that $\cos\ \theta \to 1$, after simple manipulation we get

$$\sin^2\theta\frac{d^2 J_0(n\sin\theta)}{d\cos^2\theta} - 2d\frac{dJ_0(n\sin\theta)}{d\cos\theta} + n^2 J_0(n\sin\theta) = 0 \quad (5)$$

Comparing (5) with (3), we conclude that

$$P_n(\cos\ \theta) \to J_0(n\sin\theta) \quad (1)$$

1.81 $T(x,s) = (1-2sx+s^2)^{-1/2} = \sum p_l(x)s^l$

(a) Differentiate (1) with respect to s.

$$\frac{\partial T}{\partial s} = (x-s)(1-2sx+s^2)^{-\frac{3}{2}}$$

$$= \sum(x-s)(1-2sx+s^2)^{-1}p_l(x)s^l$$

$$= \sum l p_l(x)s^{l-1}$$

Multiply by $(1-2sx+s^2)$

$$\sum(x-s)p_l s^l = \sum l p_l s^{l-1}(1-2sx+s^2)$$

Equate the coefficients of s^l

$$xp_l - p_{l-1} = (l+1)p_{l+1} - 2xlp_l + (l-1)p_{l-1}$$

or $(l+1)p_{l+1} = (2l+1)xp_l - lp_{l-1}$

(b) Differentiate with respect to x

$$\frac{\partial T}{\partial x} = s(1 - 2sx + s^2)^{-\frac{3}{2}}$$

$$= \sum (1 - 2sx + s^2)^{-1} p_l s^{l+1}$$

$$= \sum p_l' s^l$$

Multiply by $(1 - 2sx + s^2)$

$$\sum s^{l+1} p_l = \sum (s^l - 2xs^{l+1} + s^{l+2}) p_l'$$

Equate coefficients of s^{l+1}

$$p_l = p_{l+1}' - 2xp_l' + p_{l-1}'$$

or $p_l(x) + 2xp_l'(x) = p_{l+1}' + p_{l-1}'$

1.82 $\dfrac{e^{-\frac{xs}{1-s}}}{1-s} = \sum_{n=0}^{\infty} \dfrac{L_n(x)s^n}{n!}$

Put $x = 0$

$$\sum_{n=0}^{\infty} L_n(0)\frac{s^n}{n!} = \frac{1}{1-s}$$

$$= 1 + s + s^2 + \cdots s^n + \cdots$$

$$= \sum_{n=0}^{\infty} s^n$$

Therefore $L_n(0) = n!$

1.3.11 Complex Variables

1.83 (a) Since the pole at $z = 2$ is not interior to $|z| = 1$, the integral equals zero
(b) Since the pole at $z = 2$ is interior to $|z + i| = 3$, the integral equals $2\pi i$.

1.84 Method 1

$$\oint_c \frac{4z^2 - 3z + 1}{(z-1)^3} \mathrm{d}z = \oint_c \frac{4(z-1)^2 + 5(z-1) + 2}{(z-1)^3} \mathrm{d}z$$

$$= 4\oint_c \frac{\mathrm{d}z}{z-1} + 5\oint_c \frac{\mathrm{d}z}{(z-1)^2} + 2\oint_c \frac{\mathrm{d}z}{(z-1)^3}$$

$$= 4(2\pi i) + 5(0) + 6(0) = 8\pi i$$

where we have used the result

$$\oint_c \frac{dz}{(z-a)^n} = 2\pi i \text{ if } n = 1$$
$$= 0 \text{ if } n > 1$$

Method 2
By Cauchy's integral formula

$$f(n)(a) = \frac{n!}{2\pi i}\oint_c \frac{f(z)}{(z-a)^{n+1}}dz$$

If $n = 2$ and $f(z) = 4z^2 - 3z + 1$, then $f''(1) = 8$. Hence

$$8 = \frac{2!}{2\pi i}\oint_c \frac{4z^2-3z+1}{(z-1)^3} \text{or} \int \frac{4z^2-3z+1}{(z-1)^3} = 8\pi i$$

1.85 $z = 3$ is a pole of order 2 (double pole);
$z = i$ and $z = -1 + 2i$ are poles of order 1 (simple poles).

1.86 $z = 1$ is a simple pole, $z = -2$ is a pole of order 2 or double pole.
Residue at $z = 1$ is $\lim_{z\to 1}(z-1)\left\{\frac{1}{(z-1)(z+2)^2}\right\} = \frac{1}{9}$
Residue at $z = -2$ is $\lim_{z\to -2}\frac{d}{dz}\left\{\frac{(z+2)^2}{(z-1)(z+2)^2}\right\}$

$$= \frac{d}{dz^2}\left\{\frac{1}{z-1}\right\} = \frac{2}{(z-1)^2} = \frac{2}{9}$$

1.87 The singularity is at $z = 2$
Let $z - 2 = U$. Then $z = 2 + U$.

$$\frac{e^z}{(z-1)^2} = \frac{e^{2+U}}{U^2} = e^2.\frac{e^U}{U^2}$$
$$= \frac{e^2}{U^2}\left[1 + U + \frac{U^2}{2!} + \frac{U^3}{3!} + \cdots\right]$$
$$= \frac{e^2}{(z-2)^2} + \frac{e^2}{z-2} + \frac{e^2}{2!} + \frac{e^2(z-2)}{3!} + \frac{e^2(z-2)^2}{4!} + \cdots$$

The series converges for all values of $z \neq 2$

1.88 Consider $\oint_C \frac{dz}{z^4+1}$, where C is the closed contour consisting of line from $-R$ to R and the semi-circle Γ, traversed in the counter clock-wise direction. The poles for $Z^4 + 1 = 0$, are $z = \exp(\pi i/4), \exp(3\pi i/4), \exp(5\pi i/4), \exp(7\pi i/4)$. Only the poles $\exp(\pi i/4)$ and $\exp(3\pi i/4)$ lie within C. Using L'Hospital's rule

$$\text{Residue at } \exp(\pi i/4) = \lim_{z\to\exp(\frac{\pi i}{4})}\left\{z - \exp\left(\frac{\pi i}{4}\right)\frac{1}{z^4+1}\right\}$$

$$= \frac{1}{4z^3} = \frac{1}{4}\exp\left(-\frac{3\pi i}{4}\right)$$

$$\text{Residue at } \exp(3\pi i/4) = \lim_{z\to\exp(\frac{3\pi i}{4})}\left\{z - \exp\left(\frac{3\pi i}{4}\right)\frac{1}{z^4+1}\right\}$$

$$= \frac{1}{4z^3} = \frac{1}{4}\exp\left(-\frac{3\pi i}{4}\right)$$

Thus

$$\oint_c \frac{\mathrm{d}z}{z^4+1} = 2\pi i\left\{\frac{1}{4}\exp\left(-\frac{3\pi i}{4}\right) + \frac{1}{4}\exp\left(-\frac{3\pi i}{4}\right)\right\} = \frac{\pi}{\sqrt{2}}$$

Thus

$$\int_{-R}^{R}\frac{\mathrm{d}x}{x^4+1} + \int\frac{\mathrm{d}z}{z^4+1} = \frac{\pi}{\sqrt{2}}$$

Taking the limit of both sides as $R\to\infty$

$$\lim_{R\to\infty}\int_{-R}^{+R}\frac{\mathrm{d}x}{x^4+1} = \int_{-\infty}^{\infty}\frac{\mathrm{d}x}{x^4+1} = \frac{\pi}{\sqrt{2}}$$

It follows that

$$\int_0^{\infty}\frac{\mathrm{d}x}{x^4+1} = \frac{\pi}{2\sqrt{2}}$$

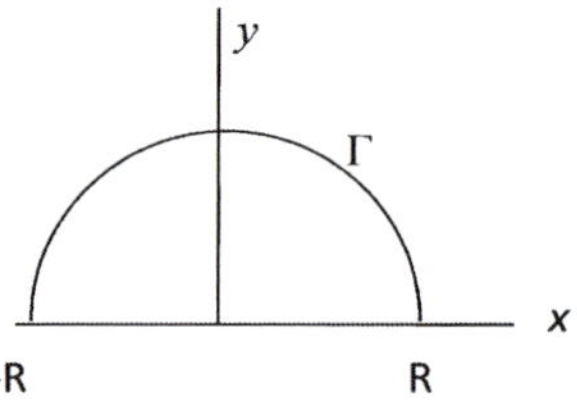

Fig. 1.17 Closed contour consisting of line from −R to R and the semi-circle Γ

1.3.12 Calculus of Variation

1.89 Let $I = \int_{x_0}^{x_1} F(x, y, y')\mathrm{d}x$ (1)

Here $I = \int_{x_0}^{x_1}\sqrt{1+\left(\frac{\mathrm{d}y}{\mathrm{d}x}\right)^2}\,\mathrm{d}x$ (2)

Now the Euler equation is

$$\frac{\partial F}{\partial y} - \frac{\mathrm{d}}{\mathrm{d}x}\left(\frac{\partial F}{\partial y'}\right) = 0 \quad (3)$$

But in (2), $F = F(y')$.
Hence

$$\frac{\partial F}{\partial y} = 0$$

$$\frac{\partial F}{\partial y'} = \frac{\partial}{\partial y'}(1 + y'^2)^{\frac{1}{2}} = y'(1 + y'^2)^{-1/2}$$

$$\frac{\mathrm{d}}{\mathrm{d}x}\left[y'(1 + y'^2)^{-1/2}\right] = 0$$

or $y'(1 + y'^2)^{-1/2} = C = \text{constant}$

or $y'^2(1 - C^2) = C^2$

or $y' = \frac{\mathrm{d}y}{\mathrm{d}x} = a = \text{constant}$

Integrating $y = ax + b$ which is the equation to a straight line. The constants a and b can be found from the coordinates $P_0(x_0, y_0)$ and $P_1(x_1, y_1)$

1.90 The velocity of the bead which starts from rest is

$$\frac{\mathrm{d}s}{\mathrm{d}t} = \sqrt{2gy} \tag{1}$$

The time of descent is therefore

$$I = t = \int \frac{\mathrm{d}s}{\sqrt{2gy}} = \frac{1}{\sqrt{2g}}\int\sqrt{\frac{\mathrm{d}x^2 + \mathrm{d}y^2}{y}} = \frac{1}{\sqrt{2g}}\int\sqrt{\frac{1 + y'^2}{y}}\mathrm{d}x \tag{2}$$

$$F = \sqrt{\frac{(1 + y'^2)}{y}} \tag{3}$$

Here F involves only y and y'. The Euler equation is

$$\frac{\mathrm{d}F}{\mathrm{d}x} - \frac{\mathrm{d}}{\mathrm{d}x}(\frac{\partial F}{\partial y'}) = 0 \tag{4}$$

which does not contain x explicitly. In that case $F(y, y')$ is given by

$$\frac{\mathrm{d}F}{\mathrm{d}x} = \frac{\partial F}{\partial y}\frac{\mathrm{d}y}{\mathrm{d}x} + \frac{\partial F}{\partial y'}\frac{\mathrm{d}y'}{\mathrm{d}x} \tag{5}$$

Multiply (4) by $\frac{\mathrm{d}y}{\mathrm{d}x}$

$$\frac{\mathrm{d}y}{\mathrm{d}x}.\frac{\mathrm{d}F}{\mathrm{d}y} - \frac{\mathrm{d}y}{\mathrm{d}x}\frac{\mathrm{d}}{\mathrm{d}x}\left(\frac{\mathrm{d}F}{\mathrm{d}y'}\right) = 0 \tag{6}$$

Combining (5) and (6)

$$\frac{\mathrm{d}F}{\mathrm{d}x} = \frac{\mathrm{d}}{\mathrm{d}x}\left(\frac{\mathrm{d}F}{\mathrm{d}y'}\frac{\mathrm{d}y}{\mathrm{d}x}\right) \tag{7}$$

Integrating $F = \frac{\mathrm{d}F}{\mathrm{d}y'}\frac{\mathrm{d}y}{\mathrm{d}x} + C$

$$F = \sqrt{\frac{1+y'^2}{y}} = \frac{y'^2}{\sqrt{1+y'^2}} + C$$

Simplifying $\frac{1}{\sqrt{y(1+y'^2)}} = C$ where we have used (3)

Or $y(1+y'^2) = \text{constant, say } 2a$

$$\therefore \left(\frac{dy}{dx}\right)^2 = \frac{2a-y}{y}$$

$$\therefore \frac{dx}{dy} = \left(\frac{y}{2a-y}\right)^{1/2} = \frac{y}{(2ay-y^2)^{1/2}}$$

This equation can be easily solved by a change of variable $y = 2a\sin^2\theta$ and direct integration. The result is

$$x = 2a\sin^{-1}\left(\frac{y}{2a}\right) - \sqrt{2ay-y^2} + b$$

which is the equation to a cycloid.

1.91 Irrespective of the function y, the surface generated by revolving y about the x-axis has an area

$$2\pi\int_{x_1}^{x_2} y\,ds = 2\pi\int y(1+y'^2)^{1/2}\,dx \tag{1}$$

If this is to be minimum then Euler's equation must be satisfied.

$$\frac{\partial F}{\partial x} - \frac{d}{dx}\left(F - y'\frac{\partial F}{\partial y'}\right) = 0 \tag{2}$$

Here

$$F = y(1+y'^2)^{1/2} \tag{3}$$

Since F does not contain x explicitly, $\frac{\partial F}{\partial x} = 0$. So

$$F - y'\frac{\partial F}{\partial y'} = a = \text{constant} \tag{4}$$

Use (3) in (4)

$$y(1+y'^2)^{\frac{1}{2}} - yy'^2(1+y'^2)^{-\frac{1}{2}} = a$$

Simplifying

$$\frac{y}{(1+y'^2)^{1/2}} = a$$

$$\text{or } \frac{dy}{dx} = \sqrt{\frac{y^2}{a^2} - 1},\ y = a\cosh\left(\frac{x}{a} + b\right)$$

This is an equation to a Catenary.

1.92 The area is

$$A = 2\pi\int y\,ds = 2\pi\int_0^a y(1+y'^2)^{1/2}dx$$

The volume:

$$V = \pi \int_0^a y^2 \mathrm{d}x$$

Therefore, dropping off the constant factors

$$K = y^2 + \lambda y(1 + y'^2)^{1/2}$$

which must satisfy the Euler's equation

$$\frac{\partial K}{\partial x} - \frac{\mathrm{d}}{\mathrm{d}x}(K - y'\frac{\partial K}{\partial y'}) = 0$$

It is convenient to use the above form as K does not explicitly contain x, and $\frac{\partial K}{\partial x} = 0$. Therefore,

$$K - y'\frac{\partial K}{\partial y'} = y^2 + \lambda y(1 + \lambda y'^2)^{\frac{1}{2}} - \lambda y y'^2 (1 + y'^2)^{-\frac{1}{2}} = 0$$

Now $y = 0$ at $x = 0$ and at $x = a$ which can be true if $C = 0$. Hence

$$y^2 + \lambda y(1 + y'^2)^{-1/2} = 0$$

Or $y = -\lambda(1 + y'^2)^{-1/2}$

Solving for y',

$$\frac{\mathrm{d}y}{\mathrm{d}x} = \frac{1}{y}(\lambda^2 - y^2)^{1/2}$$

Integrating,

$$-(\lambda^2 - y^2)^{\frac{1}{2}} = x - x_0$$

Or $(x - x_0)^2 + y^2 = \lambda^2$

This is the equation to a sphere with the centre on the x-axis at x_0, and of radius λ.

1.3.13 Statistical Distribution

1.93 (a) $\sum_{x=0}^{\infty} P_x = \sum_{x=0}^{\infty} \frac{e^{-m} m^x}{x!}$

$$= e^{-m}\left(1 + \frac{m}{1!} + \frac{m^2}{2!} + \cdots\right)$$

$$= e^{-m} \times e^{+m} = 1$$

Thus the distribution is normalized.

(b) $< x >= \sum_{x=0}^{\infty} x P_x = \sum_{x=0}^{\infty} \frac{x e^{-m} m^x}{x!} = \sum_{x=0}^{\infty} \frac{e^{-m} m^x}{(x-1)!}$

$$= e^{-m}\left(m + \frac{m^2}{1!} + \frac{m^3}{2!} + \cdots\right) (\because (-1)! = \infty)$$

$$= m\, e^{-m}\left(1 + \frac{m}{1!} + \frac{m^2}{2!} + \cdots\right)$$

$$= m\, e^{-m} \times e^m = m$$

(c) $< x^2 >= \sum x^2 \frac{e^{-m} m^x}{x!} = \sum [x(x-1)+x] \frac{e^{-m} m^x}{x!}$

$$= \sum_{x=0}^{\infty} \frac{e^{-m} m^x}{(x-2)!} + \sum_{x=0}^{\infty} x\, e^{-m} \frac{m^x}{x!}$$

$$= e^{-m} \left(m^2 + \frac{m^3}{1!} + \frac{m^4}{2!} + \cdots \right) + m$$

$$= m^2 e^{-m} e^m + m = m^2 + m$$

$$\sigma^2 =< (x - \bar{x})^2 >=< x^2 > -2 < x > \bar{x} + < \bar{x} >^2 =< x^2 > -m^2$$

$$\sigma^2 = m \text{ or } \sigma = \sqrt{m}$$

(d) $P_{m-1} = \frac{e^{-m} m^{m-1}}{(m-1)!} = \frac{e^{-m} m^m}{(m-1)!m} = \frac{e^{-m} m^m}{m!} = P_m$

That is the probability for the occurrence of the event at $x = m - 1$ is equal to that at $x = m$

(e) $P_{x-1} = \frac{e^{-m} m^{x-1}}{(x-1)!} = \frac{e^{-m} m^x}{x!} \frac{x}{m} = \frac{x}{m} P_x$

$$P_{x+1} = \frac{e^{-m} m^{x+1}}{(x+1)!} = \frac{m\, e^{-m} m^x}{x!(x+1)} = \frac{m}{x+1} P_x$$

1.94 (a) $(q + p)^N = q^N + N q^{N-1} P + \frac{N(N-1) q^{N-2}}{2!} P^2$

$$+ \cdots \frac{N!}{x!(N-x)!} P^x q^{N-x} + \cdots P^N$$

$$= \sum_{x=0}^{N} \frac{N!}{x!(N-x)!} P^x q^{N-x} = 1 (\because q + p = 1)$$

(b) We can use the moment generating function $M_x(t)$ about the mean μ which is given as

$$M_x(t) = E e^{(x-\mu)t}$$

$$= E \left[1 + (x - \mu)t + (x - \mu)^2 \frac{t^2}{2!} + \cdots \right]$$

$$= 1 + 0 + \mu_2 \frac{t^2}{2!} + \mu_3 \frac{t^3}{3!} + \cdots$$

So that μ_n is the coefficient of $\frac{t^n}{n!}$

$$M_x(t) = Ee^{xt} = \sum_{x=0}^{\infty} e^{xt} B(x)$$

$$= \sum_{x=0}^{\infty} \binom{N}{r} (pe^t)^r (1-p)^{N-r}$$

$$= (pe^t + 1 - p)^N \sum \binom{N}{r} p^r q^{N-r}$$

$$= (pe^t + 1 - p)^N$$

$$\mu_n^0 = \frac{\partial^n M_x(t)}{\partial t^n}|_{t=0}$$

Therefore $\mu_1^0 = \frac{\partial M}{\partial t}|_{t=0} = Npe^t(q + pe^t)^{N-1}|_{t=0} = Np$
Thus the mean $= Np$

(c) $$\mu_2^0 = \frac{\partial^2 M}{\partial t^2} = [N(N-1)p^2 e^t (q + pe^t)^{N-2} + Npe^t(q + pe^t)^{N-1}]_{t=0}$$

$$= N(N-1)p^2 + Np$$

But $\mu_2 = \mu_2^0 - (\mu_1^0)^2 = N(N-1)p^2 + Np - N^2 p^2$

$$= Np - Np^2 = Np(1-p) = Npq$$

or $\sigma = \sqrt{Npq}$

1.95 Total counting rate/minute, $m_1 = 245$
Background rate/minute, $m_2 = 49$
Counting rate of source, $m = m_1 - m_2 = 196$

$$m_1 = \frac{n_1}{t_1} \pm \frac{\sqrt{n_1}}{t_1}; m_2 = \frac{n_2}{t_2} \pm \frac{\sqrt{n_2}}{t_2}; \text{ Net count } m = m_1 - m_2$$

$$\sigma = (\sigma_1^2 + \sigma_2^2)^{1/2} = \left(\frac{n_1}{t_1^2} + \frac{n_2}{t_2^2}\right)^{1/2} = \left(\frac{m_1}{t_1} + \frac{m_2}{t_2}\right)^{1/2}$$

$$\sigma = \left(\frac{m_1}{t_1} + \frac{m_2}{t_2}\right)^{1/2}$$

$$= \left(\frac{49}{100} + \frac{245}{20}\right)^{1/2} = 3.57$$

Percentage S.D. $= \frac{\sigma}{m} \times 100 = \frac{3.57}{196} \times 100 = 1.8\%$

1.96 (a) $$B(x) = \frac{N!}{x!(N-x)!} p^x q^{N-x}$$

Using Sterling's theorem

$$N! \to \sqrt{2\pi N} N^N e^{-N}$$

$$x! \to \sqrt{2\pi x} x^x e^{-x}$$

$$(N-x)! \to \sqrt{2\pi(N-x)}(N-x)^{N-x} e^{-N+x}$$

$$B(x) \to f(x) = \sqrt{\frac{N}{2\pi x(N-x)}} \frac{p^x q^{N-x} N^N}{x^x (N-x)^{N-x}}$$

$$= \sqrt{\frac{N}{2\pi x(N-x)}} \left(\frac{Np}{x}\right)^x \left(\frac{Nq}{(N-x)}\right)^{N-x}$$

Let δ denote the deviation of x from the expected value Np, that is

$$\delta = x - Np$$

then $N - x = N - Np - \delta = Nq - \delta$

$$f(x) = \left[2\pi Npq \left(1 + \frac{\delta}{Np}\right)\left(1 - \frac{\delta}{Nq}\right)\right]^{-1/2}$$
$$\left(1 + \frac{\delta}{Np}\right)^{-(Np+\delta)} \cdot \left(1 - \frac{\delta}{Nq}\right)^{-(Nq-\delta)}$$

Assume that $|\delta| \ll Npq$ so that

$$\left|\frac{\delta}{Np}\right| \ll 1 \text{ and } \left|\frac{\delta}{Nq}\right| \ll 1$$

The first bracket reduces to $(2\pi Npq)^{-\frac{1}{2}}$. Take $\log_e$ on both sides.

$$\ln\left[f(x)(2\pi Npq)^{\frac{1}{2}}\right] = -(Np+\delta)\ln\left(1+\frac{\delta}{Np}\right) - (Nq-\delta)\ln\left(1-\frac{\delta}{Nq}\right)$$

$$= -(Np+\delta)\left(\frac{\delta}{Np} - \frac{\delta^2}{2N^2p^2} + \frac{\delta^3}{3N^3p^3} - \cdots\right)$$
$$-(Nq-\delta)\left(-\frac{\delta}{Nq} - \frac{\delta^2}{2N^2q^2} - \frac{\delta^3}{3N^3q^3} - \cdots\right)$$
$$= -\frac{\delta^2}{2Npq} - \frac{\delta^3(p^2-q^2)}{6N^2p^2q^2}$$

Neglect the δ^3 term and putting $\sigma = \sqrt{Npq}$ and $\sigma = x - Np = \bar{x}$.

$$f(x) = \frac{1}{\sigma\sqrt{2\pi}} \exp\left[\frac{-(x-\bar{x})^2}{2\sigma^2}\right]$$

(Normal or Gaussian distribution)

$$\text{(b) } B(x) = \frac{N!p^x q^{N-x}}{x!(N-x)!}$$

$$= \frac{N(N-1)\ldots(N-x-1)p^x(1-p)^{N-x}}{x!}$$

$$= \frac{N(N-1)\ldots(N-x+1)(Np)^x(1-p)^{N-x}}{N^x x!}$$

$$= \frac{m^x}{x!}\left(1-\frac{1}{N}\right)\left(1-\frac{2}{N}\right)\cdots\left(1-\frac{x-1}{N}\right)(1-p)^{N-x}$$

$$= \frac{m^x}{x!}\frac{\left(1-\frac{1}{N}\right)\left(1-\frac{2}{N}\right)\cdots\left(1-\frac{x-1}{N}\right)(1-p)^N}{(1-p)^x}$$

The poisson distribution can be deduced as a limiting case of the binomial distribution, for those random processes in which the probability of occurrence is very small, $p \ll 1$, while the number of trials N becomes very large and the mean value $m = pn$ remains fixed. Then $m \ll N$ and $x \ll N$, so that approximately

$$(1-p)^{N-x} \approx e^{-p(N-x)} \approx e^{-pN} = e^{-m}$$

Thus $B(x) \to P(x) = \frac{e^{-m}m^x}{x!}$

1.97 $S = (g-b) \pm \sqrt{\dfrac{g}{t_g} + \dfrac{b}{t_b}}$

$$t = t_b + t_g = \text{constant}$$

$$t_g = t - t_b$$

$$\sigma^2 = \sigma_g^2 + \sigma_b^2 = \frac{g}{t-t_b} + \frac{b}{t_b}$$

must be minimum. Therefore, $\frac{\partial(\sigma^2)}{\partial t_b} = 0$

$$\frac{g}{(t-t_b)^2} - \frac{b}{t_b^2} = 0$$

$$\frac{t_b^2}{(t-t_b)^2} = \frac{b}{g} \to \frac{t_b}{t_g} = \sqrt{\frac{b}{g}}$$

1.98 $A = A_0 e^{-\lambda t}$

$$\ln\left(\frac{A_0}{A}\right) = \lambda t$$

$$y = \lambda t$$

$$S = y - \lambda t$$

Normal equation gives

$$\lambda = \frac{\sum t_n y_n}{\sum t_n^2}$$

$$\sum_{n=1}^{6} t_n y_n = 1 \times \ln(1/0.835) + 2 \times \ln(1/0.695)$$

$$+ 3 \times \ln(1/0.58) + 4 \times \ln(1/0.485)$$
$$+ 5 \times \ln(1/0.405) + 6 \times \ln(1/0.335)$$
$$= 16.5175$$

$$\sum_{n=1}^{6} t_n^2 = 1^2 + 2^2 + 3^2 + 4^2 + 5^2 + 6^2 = 91$$

$$\therefore \lambda = \frac{16.5175}{91} = 0.1815\,\mathrm{h}^{-1}$$

$$T_{1/2} = \frac{0.693}{\lambda} = \frac{0.693}{0.1815} = 3.82\,\mathrm{h}$$

1.99 We determine the probability $P(t)$ that a given counter records no pulse during a period t. We divide the interval t into two parts $t = t_1 + t_2$. The probabilities that no pulses are recorded in either the first or the second period are given by $P(t_1)$ and $P(t_2)$, respectively, while the probability that no pulse is recorded in the whole interval is $P(t) = P_1(t_1 + t_2)$.
Since the two events are independent,

$$P(t_1 + t_2) = P(t_1)p(t_2)$$

The above equation has the solution

$$P(t) = e^{-at}$$

where a is a positive constant. The reason for using the minus sign for a is that $P(t)$ is expected to decrease with increasing t.
The probability that there will be an event in the time interval dt is c dt. The combined probability that there will be no events during time interval t, but one event between time t and $t + \mathrm{d}t$ is $c\, e^{-at}$ dt where $c =$ constant. It is readily shown that $c = a$. This follows from the normalization condition

$$\int_0^\infty P(t)dt = c\int_0^\infty e^{-at}\mathrm{d}t = 1$$

Thus $\mathrm{d}p(t) = ae^{-at}\,\mathrm{d}t$
Clearly, small time intervals are more favoured than large time intervals amongst randomly distributed events. If the data have large number N of intervals then the number of intervals greater than t_1 but less than t_2 is

$$n = N\int_{t_1}^{t_2} ae^{-at}dt = N(e^{-at_1} - e^{-at_2}) \text{ (Interval distribution)}$$

a represents the average number of events per unit time.

Two limiting cases

(a) $t_2 = \infty$. We find that the number of intervals greater than any duration is Ne^{-at} in which $at = average\ number\ of\ events\ in\ time\ t$. In the case of radioactivity at time $t = 0$, let $N = N_0$.
Then the radioactive decay law becomes

$$N = N_0 e^{-\lambda t}$$

where N is the number of surviving atoms at time t, and $a = \lambda$ is the decay constant, that is the number of decays per unit time.

(b) $t_1 = 0$, implies that the number of events shorter than any duration t is $N_0(1 - e^{-at})$
For radioactive decay the above equation would read for the number of decays in time interval 0 to t.

$$N = N_0(1 - e^{-\lambda t})$$

1.100 $N_s = N_0 - N_B = 14.5 - 10 = 4.5$

$$\sigma_s = \sqrt{\frac{10}{t} + \frac{14.5}{t}} = \sqrt{\frac{24.5}{t}}$$

$$\frac{\sigma_s}{N_s} = \frac{5}{100} = \frac{1}{4.5}\sqrt{\frac{24.5}{t}}$$

$$t = 484\,\text{min}$$

1.101 The best values of a_0, a_1 and a_2 are found by the Least square fit. The residue S is given by

$$S = \sum_{n=1}^{6} (y_n - a_0 - a_1 x_n - a_2 x_n^2)^2$$

Minimize the residue.

$$\frac{\partial S}{\partial a_0} = 0, \text{ gives}$$

$$\sum_{n=1}^{6} y_n = na_0 + a_1 \sum x_n + a_2 \sum x_n^2 \quad (1)$$

$$\frac{\partial S}{\partial a_1} = 0 \text{ gives}$$

$$\sum x_n y_n = a_0 \sum x_n + a_1 \sum x_n^2 + a_2 \sum x_n^3 \quad (2)$$

$$\frac{\partial S}{\partial a_2} = 0 \text{ gives}$$

$$\sum x_n^2 y_n = a_0 \sum x_n^2 + a_1 \sum x_n^3 + a_2 \sum x_n^4 \quad (3)$$

Equations (1), (2) and (3) are the so-called normal equations which are to be solved as ordinary simultaneous equations.

$$N = 6; \sum y_n = 29.5; \sum x_n = 3; \sum x_n^2 = 19$$

$$\sum x_n^3 = 27; \sum x_n^4 = 115; \sum x_n y_n = 21.3; \sum x_n^2 y_n = 158.1$$

Solving (1), (2) and (3) we find $a_0 = 0.582; a_1 = -1.182; a_2 = 1.556$

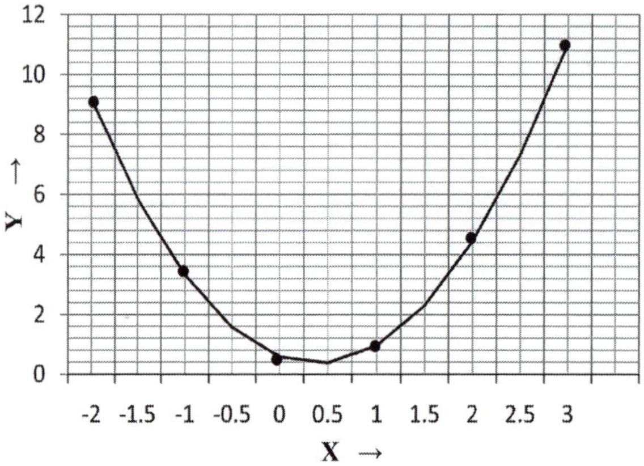

Fig. 1.18 Least square fit of the parabola

1.102 $$C = \frac{2\pi\varepsilon_0}{\ln\left(\frac{b}{a}\right)} \tag{1}$$

From propagation of errors

$$\sigma_c = \left[\left(\frac{\partial c}{\partial b}\right)^2 \sigma_b^2 + \left(\frac{\partial c}{\partial a}\right)^2 \sigma_a^2\right]^{1/2} \tag{2}$$

$$\frac{\partial c}{\partial b} = -\frac{c}{b \ln \frac{b}{a}}; \frac{\partial c}{\partial a} = \frac{c}{a \ \ln \frac{b}{a}} \tag{3}$$

Using (1), (2) and (3) and simplifying

$$\frac{\sigma_c}{c} = \left[\left(\ln\frac{b}{a}\right)\right]^{-1/2} \left[\frac{\sigma_a^2}{a^2} + \frac{\sigma_a^2}{b^2}\right]^{1/2}$$

Substituting $a = 10\,\text{mm}$, $b = 20\,\text{mm}$, $\sigma_a = 1\,\text{mm}$ and $\sigma_b = 1\,\text{mm}$, $\sigma_c/c = 0.16$

1.103

$$
\begin{aligned}
< x > &= \int_0^\infty x\, f(x)\mathrm{d}x \Big/ \int_0^\infty f(x)\mathrm{d}x \\
&= \int_0^\infty x^2 e^{-\frac{x}{\lambda}}\mathrm{d}x \Big/ \int_0^\infty x e^{-\frac{x}{\lambda}}\mathrm{d}x \\
&= \frac{2\lambda^3}{\lambda^2} = 2\lambda
\end{aligned}
$$

Most probable value of x is obtained by maximizing the function $xe^{-x/\lambda}$

$$
\begin{aligned}
&\frac{\mathrm{d}}{\mathrm{d}x}(xe^{-x/\lambda}) = 0 \\
&e^{-\frac{x}{\lambda}}\left(1 - \frac{x}{\lambda}\right) = 0 \\
&\therefore x = \lambda \\
&x(\text{most probable}) = \lambda
\end{aligned}
$$

1.3.14 Numerical Integration

1.104 The trapezoidal rule is

$$
\text{Area} = \left(\frac{1}{2}y_0 + y_1 + y_2 + \cdots + y_{n-1} + \frac{1}{2}y_n\right)\Delta x
$$

Given integral is $\int_1^{10} x^2\mathrm{d}x$. Divide $x = 1$ to $x = 10$ into 9 intervals.
Thus $\frac{b-a}{n} = \frac{10-1}{9} = 1 = \Delta x$
Substituting the abscissas in the equation $y = x^2$, we get the ordinates $y = 1, 4, 9, 16, \cdots 100$.

$$
\text{area} = \left(\frac{1}{2} + 4 + 9 + 25 + 36 + 49 + 64 + 81 + \frac{1}{2} \times 100\right) = 334.5
$$

This may be compared with the value obtained from direct integration, $\left[\frac{x^3}{3}\right]_1^{10} = 333$.
The error is 0.45%.

1.105 For Simpson's rule
take 10 intervals
Here $\frac{b-a}{n} = \frac{10-0}{10} = 1 = \Delta x$
The area under the curve $y = x^2$ is given by

$$
\frac{\Delta x}{3}(y_0 + 4y_1 + 2y_2 + 4y_3 + 2y_4 + \cdots + 4y_{n-1} + y_n)
$$

The ordinates are found by substituting $x = 0, 1, 2 \cdots 10$ in the equation $y = x^2$. Thus

$$\text{area} = \frac{1}{3}(0+4+8+36+32+100+72+196+128+324+100) = 333.3$$

In this case Simpson's rule happens to give exactly the same result as that from direct integration.

Chapter 2
Quantum Mechanics – I

2.1 Basic Concepts and Formulae

Wave number

$$\tilde{\nu} = \frac{1}{\lambda} \tag{2.1}$$

1 fm = 10^{-15} m; 1 Å = 10^{-10} m; 1 nm = 10^{-9} m; 1 μm = 10^{-6} m; $\hbar c$ = 197.3 Mev – fm

Photon energy

$$E = h\nu \tag{2.2}$$

Photon momentum

$$p = h\nu/c$$

Photon energy – wavelength conversion

$$\lambda(\text{nm}) = \frac{1241}{E(ev)} \tag{2.3}$$

de Broglie wavelength

$$\lambda = h/p \tag{2.4}$$

$$\lambda(\text{electron}) : \lambda(\text{Å}) = (150/V)^{1/2} \tag{2.5}$$

$$\lambda(\text{neutron}) : \lambda(\text{Å}) = 0.286\ E^{-1/2} \quad (E \text{ in ev}) \tag{2.6}$$

Atomic units

The Bohr radius $\hbar^2/m_e e^2$ is frequently used as the unit of length in atomic physics. In atomic units the energy is measured in multiples of the ionization energy of hydrogen atom that is $m_e e^4/2\hbar^2$. In these units $\hbar^2 = 1$, $e^2 = 2$ and $m_e = \frac{1}{2}$ in all equations.

Natural units

$$\hbar = c = 1 \tag{2.7}$$

Mosley's law (for characteristic X-rays)

$$\sqrt{\nu} = A(Z - b) \tag{2.8}$$

where Z is the atomic number, A and b are constants.
For K_α line

$$\lambda = \frac{1{,}200}{(Z-1)^2} \text{Å} \tag{2.9}$$

X-rays absorption

$$I = I_0 e^{-\mu x} \tag{2.10}$$

Duane–Hunt law (for continuous X-rays)

$$\lambda_c = \frac{c}{\nu_{\max}} = \frac{hc}{eV} = \frac{1{,}240}{V} \text{ pm} \tag{2.11}$$

where e is the electron charge and V is the P.D through which the electrons have been accelerated in the X-ray tube.

Doppler effect (*Non-relativistic*)

$$\nu = \nu_0(1 + \beta c \cos\theta^*) \tag{2.12}$$

where ν is the observed frequency, ν_0 the frequency of light in the rest frame of source emitted at angle θ^*, $v = \beta c$ is the source velocity. The inverse transformation is

$$\nu_0 = \nu(1 - \beta c \cos\theta) \tag{2.13}$$

Hydrogen atom (Bohr's model)

Angular momentum

$$L = n\hbar, \quad n = 1, 2, 3 \ldots\ldots \tag{2.14}$$

Energy of photon emitted from energy level E_i to final level E_f.

$$hv = E_\text{i} - E_\text{f} \tag{2.15}$$

Radius of the nth orbit

$$r_n = \frac{\varepsilon_0 h^2 n^2}{\pi \mu e^2 z} \tag{2.16}$$

where μ is the reduced mass given by

$$\mu = \frac{m_e m_p}{m_e + m_p} \tag{2.17}$$

and $Z = 1$ for H-atom.
The radius of the smallest orbit, called the Bohr radius,

$$a_0 = r_1 = \frac{\varepsilon_0 h^2}{\pi m e^2} = 0.529\,\text{Å} \tag{2.18}$$

Orbital velocity in the nth orbit

$$v_n = \frac{ze^2}{2\varepsilon_o nh} \tag{2.19}$$

Fine structure constant

$$\alpha = \frac{v}{c} = \frac{e^2}{2\varepsilon_{ohc}} = \frac{1}{137} \tag{2.20}$$

α is a measure of the electromagnetic interaction
Kinetic Energy of electron

$$K_n = \frac{1}{2}\, m\, v_n^2 = \frac{ze^2}{8\pi\,\varepsilon_o r_n} \tag{2.21}$$

Potential energy of electron

$$U_n = -\frac{ze^2}{4\pi\,\varepsilon_o r_n} \tag{2.22}$$

Total Energy of electron

$$E_n = K_n + U_n = -\frac{ze^2}{8\pi\,\varepsilon_o r_n} = -\frac{mz^2e^4}{8\varepsilon_o^2 h^2 n^2} = \frac{-13.6}{n^2}(\text{ev}) \tag{2.23}$$

$$\tilde{\nu}_{\text{if}} = \frac{1}{\lambda_{\text{if}}} = \text{R}\left(\frac{1}{n_{\text{f}}^2} - \frac{1}{n_{\text{i}}^2}\right) \tag{2.24}$$

Mesic atom

A negatively charged muon or pion when captured by the nucleus forms a bound system called mesic atom before absorption or decay system.

i. $r_n \propto 1/\mu$; therefore radii are shrunk by a factor of ~200 for muonic atom compared to H-atom.
ii. $E_n \propto \mu$; therefore energy levels are spaced 200 times greater than those of H-atom. X-rays are emitted instead of visible light, when the muon cascades down.

Table 2.1 Hydrogen spectrum

	Series	Region	First line (Å)	Series limit (Å)
1.	*Lyman*	Ultraviolet	1,215	911
	$\tilde{\nu} = \frac{1}{\lambda} = \mathrm{R}\left(\frac{1}{1^2} - \frac{1}{n_i^2}\right);\ n_{i=2,3,4\ldots}$			
2.	*Balmer*	Visible	6,561	3,645
	$\tilde{\nu} = \mathrm{R}\left(\frac{1}{2^2} - \frac{1}{n_i^2}\right);\ n_i = 3, 4, 5\ldots$			
3.	*Pashen*	Infrared	18,746	8,201
	$\tilde{\nu} = \mathrm{R}\left(\frac{1}{3^2} - \frac{1}{n_i^2}\right);\ n_i = 4, 5, 6\ldots$			
4.	*Brackett*	Infrared	40,501	14,580
	$\tilde{\nu} = \mathrm{R}\left(\frac{1}{4^2} - \frac{1}{n_i^2}\right);\ n_i = 5, 6, 7\ldots$			
5.	*Pfund*	Far infra-red	74,558	22,782
	$\tilde{\nu} = \mathrm{R}\left(\frac{1}{5^2} - \frac{1}{n_i^2}\right);\ n_i = 6, 7, 8\ldots$			

Positronium

A system of $e^+ - e^-$ is called positronium, the reduced mass, $\mu = 0.5m_e$. Therefore, the radii are expanded but energy levels are reduced by a factor of 2, compared to the H-atom, and the entire spectrum is shifted toward the longer wavelength.

Uncertainty principle (Heisenberg)

$$\Delta x \Delta p_x \gtrsim \hbar \tag{2.25a}$$

$$\Delta E \Delta t \gtrsim \hbar \tag{2.25b}$$

$$L_z \Delta \Phi \gtrsim \hbar \tag{2.25c}$$

Restricted uncertainty principle

$$\Delta x \Delta p_x \gtrsim \hbar/2 \tag{2.26}$$

Bohr magneton

$$\mu_B = \frac{e\hbar}{2m} \tag{2.27}$$

Zeeman effect

The splitting of spectrum lines in a magnetic field is known as Zeeman effect.

Normal Zeeman effect (Strong magnetic field)

Each term is split up into $2J + 1$ terms by the magnetic field. When observed transversely (magnetic field at right angle to the light path), the lines are observed to be split up into three, the middle line linearly polarized parallel to the field, and the outside lines at right angles to the field; but when observed longitudinally (field parallel to the light path), they are split into two, which are circularly polarized in the opposite directions.

The selection rule is $\Delta m = 0, \pm 1$, where m is the magnetic quantum number. We thus get a simple triplet or doublet. In the former they are equally spaced.

Anamolous Zeeman effect

For not too strong field, one observes splitting into more than three components, unequally spaced. The additional magnetic energy is given by

$$E_{\text{mag}} = -\frac{e\hbar Bmg}{2\mu c} \tag{2.28}$$

where g is Lande's g-factor. The undisturbed term again splits up into $2J+1$ equidistant terms but the lines will not be equidistant because the g-factor for the upper and lower levels would be different.

$$g = 1 + \frac{j(j+1) + s(s+1) - l(l+1)}{2j(j+1)} \tag{2.29}$$

Degeneracy of H-atom energy levels

$$\text{The degeneracy} = 2\sum_{l=0}^{n-1}(2l+1) = 2n^2 \tag{2.30}$$

where n is the principal quantum number.

Broadening of spectral lines

The observed spectral lines are not perfectly sharp. The broadening is due to

(i) Natural width explained by the uncertainty principle for time and energy.
(ii) Thermal motion of atoms.
(iii) Molecular collisions.

Spectroscopic notation

${}^{2S+1}L_J$,where S is the total electron spin, L is the orbital angular momentum, and J the total angular momentum.

Stern–Gerlah experiment

In this experiment a collimated beam of neutral atoms emerging from a hot oven is sent through an inhomogeneous magnetic field. The beam is split up into $2J + 1$ components. The experiment affords the determination of the spin of the atoms.

2.2 Problems

2.2.1 de Broglie Waves

2.1 (a) Write down the equation relating the energy E of a photon to its frequency f. Hence determine the equation relating the energy E of a photon to its wavelength.

(b) A π^0 meson at rest decays into two photons of equal energy. What is the wavelength (in m) of the photons? (The mass of the π^0 is 135 MeV/c)

[University of London 2006]

2.2 Calculate the wavelength in nm of electrons which have been accelerated from rest through a potential difference of 54 V.

[University of London 2006]

2.3 Show that the deBroglie wavelength for neutrons is given by $\lambda = 0.286\,\text{Å}/\sqrt{E}$, where E is in electron-volts.

[Adapted from the University of New Castle upon Tyne 1966]

2.4 Show that if an electron is accelerated through V volts then the deBroglie wavelength in angstroms is given by $\lambda = \left(\frac{150}{V}\right)^{1/2}$

2.5 A thermal neutron has a speed v at temperature $T = 300\,\text{K}$ and kinetic energy $\frac{m_n v^2}{2} = \frac{3kT}{2}$. Calculate its deBroglie wavelength. State whether a beam of these neutrons could be diffracted by a crystal, and why?
(b) Use Heisenberg's Uncertainty principle to estimate the kinetic energy (in MeV) of a nucleon bound within a nucleus of radius 10^{-15} m.

2.6 The relation for total energy (E) and momentum (p) for a relativistic particle is $E^2 = c^2p^2 + m^2c^4$, where m is the rest mass and c is the velocity of light. Using the relativistic relations $E = \hbar\omega$ and $p = \hbar k$, where ω is the angular frequency and k is the wave number, show that the product of group velocity (v_g) and the phase velocity (v_p) is equal to c^2, that is $v_\text{p}v_\text{g} = c^2$

2.2.2 Hydrogen Atom

2.7 In the Bohr model of the hydrogen-like atom of atomic number Z the atomic energy levels of a single-electron are quantized with values given by

$$E_n = \frac{Z^2 m_\text{e} e^4}{8\varepsilon_0^2 h^2 n^2}$$

where m is the mass of the electron, e is the electronic charge and n is an integer greater than zero (principal quantum number)

What additional quantum numbers are needed to specify fully an atomic quantum state and what physical quantities do they quantify? List the allowed quantum numbers for $n = 1$ and $n = 2$ and specify fully the electronic quantum numbers for the ground state of the Carbon atom (atomic number $Z = 6$)

[Adapted from University of London 2002]

2.8 Estimate the total ground state energy in eV of the system obtained if all the electrons in the Carbon atom were replaced by π^- particles. (You are given that the ground state energy of the hydrogen atom is $-13.6\,\text{eV}$ and that the π^- is a particle with charge -1, spin 0 and mass $270\,m_e$

[University of London]

2.9 What are atomic units? In this system what are the units of (a) length (b) energy (c) $\hbar^2$ (d) e^2 (e) m_e ? (f) Write down Schrodinger's equation for H-atom in atomic units

2.10 (a) Two positive nuclei each having a charge q approach each other and electrons concentrate between the nuclei to create a bond. Assume that the electrons can be represented by a single point charge at the mid-point between the nuclei. Calculate the magnitude this charge must have to ensure that the potential energy is negative.

(b) A positive ion of kinetic energy 1×10^{-19} J collides with a stationary molecule of the same mass and forms a single excited composite molecule. Assuming the initial internal energies of the ion and neutral molecule were zero, calculate the internal energy of the molecule.

[Adapted from University of Wales, Aberystwyth 2008]

2.11 (a) By using the deBroglie relation, derive the Bohr condition $mvr = n\hbar$ for the angular momentum of an electron in a hydrogen atom.

(b) Use this expression to show that the allowed electron energy states in hydrogen atom can be written

$$E_n = -\frac{me^4}{8\varepsilon_0^2 h^2 n^2}$$

(c) How would this expression be modified for the case of a triply ionized beryllium atom $Be(Z = 4)$?

(d) Calculate the ionization energy in eV of Be^{+3} (ionization energy of hydrogen $= 13.6\,\text{eV}$)

[Adapted from the University of Wales, Aberystwyth 2007]

2.12 When a negatively charged muon (mass $207\,m_e$ is captured in a Bohr's orbit of high principal quantum number (n) to form a mesic atom, it cascades down to lower orbits emitting X-rays and the radii of the mesic atom are shrunk by a factor of about 200 compared with the corresponding Bohr's atom. Explain.

2.13 In which mu-mesic atom would the orbit with $n = 1$ just touch the nuclear surface. Take $Z = A/2$ and $R = 1.3\,A^{1/3}$ fm.

2.14 Calculate the wavelengths of the first four lines of the Lyman series of the positronium on the basis of the simple Bohr's theory

[Saha Institute of Nuclear physics 1964]

2.15 (a) Show that the energy E_n of positronium is given by $E_n = -\alpha^2 m_e c^2/4n^2$ where m_e is the electron mass, n the principal quantum number and α the fine structure constant

(b) the radii are expanded to double the corresponding radii of hydrogen atom

(c) the transition energies are halved compared to that of hydrogen atom.

2.16 A non-relativistic particle of mass m is held in a circular orbit around the origin by an attractive force $f(r) = -kr$ where k is a positive constant

(a) Show that the potential energy can be written

$U(r) = kr^2/2$

Assuming $U(r) = 0$ when $r = 0$

(b) Assuming the Bohr quantization of the angular momentum of the particle, show that the radius r of the orbit of the particle and speed v of the particle can be written

$$v^2 = \left(\frac{n\hbar}{m}\right)\left(\frac{k}{m}\right)^{1/2}$$

$$r^2 = \left(\frac{n\hbar}{k}\right)\left(\frac{k}{m}\right)^{1/2}$$

where n is an integer

(c) Hence, show that the total energy of the particle is

$$E_n = n\hbar\left(\frac{k}{m}\right)^{1/2}$$

(d) If $m = 3 \times 10^{-26}$ kg and $k = 1180\,\mathrm{N\,m^{-1}}$, determine the wavelength of the photon in nm which will cause a transition between successive energy levels.

2.17 For high principle quantum number (n) for hydrogen atom show that the spacing between the neighboring energy levels is proportional to $1/n^3$.

2.18 In which transition of hydrogen atom is the wavelength of 486.1 nm produced? To which series does it belong?

2.19 Show that for large quantum number n, the mechanical orbital frequency is equal to the frequency of the photon which is emitted between adjacent levels.

2.20 A hydrogen-like ion has the wavelength difference between the first lines of the Balmer and lyman series equal to 16.58 nm. What ion is it?

2.21 A spectral line of atomic hydrogen has its wave number equal to the difference between the two lines of Balmer series, 486.1 nm and 410.2 nm. To which series does the spectral line belong?

2.2.3 X-rays

2.22 (a) The $L \rightarrow K$ transition of an X-ray tube containing a molybdenum ($Z = 42$) target occurs at a wavelength of 0.0724 nm. Use this information to estimate the screening parameter of the K-shell electrons in molybdenum.

[Osmania University]

2.23 Calculate the wavelength of the $Mo(Z = 42)K_\alpha$ X-ray line given that the ionization energy of hydrogen is 13.6 eV

[Adapted from the University of London, Royal Holloway 2002]

2.24 In a block of Cobalt/iron alloy, it is suspected that the Cobalt ($Z = 27$) is very poorly mixed with the iron ($Z = 26$). Given that the ionization energy of hydrogen is 13.6 eV predict the energies of the K absorption edges of the constituents of the alloy.

[University of London, Royal Holloway 2002]

2.25 Calculate the minimum wavelength of the radiation emitted by an X-ray tube operated at 30 kV.

[Adapted from the University of London, Royal Holloway 2005]

2.26 If the minimum wavelength from an 80 kV X-ray tube is 0.15×10^{-10} m, deduce a value for Planck's constant.

[Adapted from the University of New Castle upon Tyne 1964]

2.27 If the minimum wavelength recorded in the continuous X-ray spectrum from a 50 kV tube is 0.247 Å, calculate the value of Plank's constant.

[Adapted from the University of Durham 1963]

2.28 The wavelength of the K_α line in iron ($Z = 26$) is known to be 193 pm. Then what would be the wavelength of the K_α line in copper ($Z = 29$)?

2.29 An X-ray tube has nickel as target. If the wavelength difference between the K_α line and the short wave cut-off wavelength of the continuous X-ray spectrum is equal to 84 pm, what is the voltage applied to the tube?

2.30 Consider the transitions in heavy atoms which give rise to L_α line in X-ray spectra. How many allowed transitions are possible under the selection rule $\Delta l = \pm 1, \Delta j = 0, \pm 1$.

2.31 When the voltage applied to an X-ray tube increases from 10 to 20 kV the wavelength difference between the K_α line and the short wave cut-off of the continous X-ray spectrum increases by a factor of 3.0. Identify the target material.

2.32 How many elements have the K_α lines between 241 and 180 pm?

2.33 Moseley's law for characteristic x-rays is of the form $\sqrt{v} = a(z-b)$. Calculate the value of a for K_α

2.34 The K_α line has a wavelength λ for an element with atomic number $Z = 19$. What is the atomic number of an element which has a wavelength $\lambda/4$ for the K_α line?

2.2.4 Spin and μ and Quantum Numbers – Stern–Gerlah's Experiment

2.35 Evidence for the electron spin was provided by the Sterrn–Gerlah experiment. Sketch and briefly describe the key features of the experiment. Explain what was observed and how this observation may be interpreted in terms of electron spin.

[Adapted from University of London 2006]

2.36 (i) Write down the allowed values of the total angular momentum quantum number j, for an atom with spin s and l, respectively (ii) Write down the quantum numbers for the states described as $^2S_{1/2}$, $^3D_2$ and $^5P_3$ (iii) Determine if any of these states are impossible, and if so explain why.

[Adapted from the University of London, Royal Holloway 2003]

2.37 (a) show that an electron in a classical circular orbit of angular momentum $\boldsymbol{L}$ around a nucleus has magnetic dipole moment given by $\boldsymbol{\mu} = -e\,\boldsymbol{L}/2m_e$ (b) State the quantum mechanical values for the magnitude and the z-component of the magnetic moment of the hydrogen atom associated with (i) electron orbital angular momentum (ii) electron spin

[Adapted from the University of London, Royal Holloway 2004]

2.38 In a Stern-Gerlach experiment a collimated beam of hydrogen atoms emitted from an oven at a temperature of 600 K, passes between the poles of a magnet for a distance of 0.6 m before being detected at a photographic plate a further 1.0 m away. Derive the expression for the observed mean beam separation, and determine its value given that the magnetic field gradient is 20 $\mathrm{Tm^{-1}}$ (Assume the atoms to be in the ground state and their mean kinetic energy to be 2 kT; Bohr magneton $\mu_B = 9.27 \times 10^{-24}\,\mathrm{J\,T^{-1}}$

[Adapted from the University of London, Royal Holloway 2004]

2.39 State the ground state electron configuration and magnetic dipole moment of hydrogen ($Z = 1$) and sodium ($Z = 11$)

2.40 In a Stern–Gerlah experiment a collimated beam of sodium atoms, emitted from an oven at a temperature of 400 K, passes between the poles of a magnet for a distance of 1.00 m before being detected on a screen a further 0.5 m away. The mean deflection detected was 0.14°. Assuming that the magnetic field gradient was 6.0 $\mathrm{T\,m^{-1}}$ and that the atoms were in the ground state and their mean kinetic energy was $2kT$, estimate the magnetic moment.

[Adapted from the University of London, Royal Holloway and Bedford New College, 2005]

2.41 If the electronic structure of an element is $1s^2\,2s^2\,2p^6\,3s^2\,3p^6\,3d^{10}\,4s^2\,4p^5$, why can it not be (a) a transition element (b) a rare-earth element?

[Adopted from the University of Manchester 1958]

2.42 In a Stern–Gerlach experiment the magnetic field gradient is 5.0 V s m^{-2} mm^{-1}, with pole pieces 7 cm long. A narrow beam of silver atoms from an oven at 1, 250 K passes through the magnetic field. Calculate the separation of the beams as they emerge from the magnetic field, pointing out the assumptions you have made (Take $\mu = 9.27 \times 10^{-24}\,JT^{-1}$]

[Adapted from the University of Durham 1962]

2.43 (a) The magnetic moment of silver atom is only 1 Bohr magneton although it has 47 electrons? Explain.

(b) Ignoring the nuclear effects, what is the magnetic moment of an atom in the $3p_0$ state?

(c) In a Stern–Gerlah experiment, a collimated beam of neutral atoms is split up into 7 equally spaced lines. What is the total angular momentum of the atom?

(d) what is the ratio of intensities of spectral lines in hydrogen spectrum for the transitions

$2^2p_{1/2} \to 1^2s_{1/2}$ and $2^2p_{3/2} \to 1^2s_{1/2}$?

2.44 Obtain an expression for the Bohr magneton.

2.2.5 *Spectroscopy*

2.45 (a) Given the allowed values of the quantum numbers n, l, m and m_s of an electron in a hydrogen atom (b) What are the allowed numerical values of l and m for the $n = 3$ level? (c) Hence show that this level can accept 18 electrons.

[Adopted from University of London 2006]

2.46 State, with reasons, which of the following transitions are forbidden for electric dipole transitions

$^3D_1 \to {}^2F_3$

$^2P_{3/2} \to {}^2S_{1/2}$

$^2P_{1/2} \to {}^2S_{1/2}$

$^3D_2 \to {}^3S_1$

2.47 The $^9Be^+$ ion has a nucleus with spin $I = 3/2$. What values are possible for the hyperfine quantum number F for the $^2S_{1/2}$ electronic level?

[Aligarh University]

2.48 Obtain an expression for the Doppler linewidth for a spectral line of wavelength λ emitted by an atom of mass m at a temperature T

2.49 For the $2P_{3/2} \rightarrow 2S_{1/2}$ transition of an alkali atom, sketch the splitting of the energy levels and the resulting Zeeman spectrum for atoms in a weak external magnetic field (Express your results in terms of the frequency v_0 of the transition, in the absence of an applied magnetic field)

$$\left[\text{The Lande g-factor is given by } g = 1 + \frac{j(j+1) + s(s+1) - l(l+1)}{2j(j+1)}\right]$$

[Adapted from the University of London Holloway 2002]

2.50 The spacings of adjacent energy levels of increasing energy in a calcium triplet are 30×10^{-4} and 60×10^{-4} eV. What are the quantum numbers of the three levels? Write down the levels using the appropriate spectroscopic notation.

[Adapted from the University of London, Royal Holloway 2003]

2.51 An atomic transition line with wavelength 350 nm is observed to be split into three components, in a spectrum of light from a sun spot. Adjacent components are separated by 1.7 pm. Determine the strength of the magnetic field in the sun spot. $\mu_B = 9.17 \times 10^{-24}$ J T^{-1}

[Adapted from the University of London, Royal Holloway 2003]

2.52 Calculate the energy spacing between the components of the ground state energy level of hydrogen when split by a magnetic field of 1.0 T. What frequency of electromagnetic radiation could cause a transition between these levels? What is the specific name given to this effect.

[Adapted from the University of London, Royal Holloway 2003]

2.53 Consider the transition $2P_{1/2} \rightarrow 2S_{1/2}$, for sodium in the magnetic field of 1.0 T, given that the energy splitting $\Delta E = g\mu_B B m_j$, where μ_B is the Bohr magneton. Draw the sketch.

[Adapted from the University of London, Royal Holloway 2004]

2.54 To excite the mercury line 5,461 Å an excitation potential of 7.69 V is required. If the deepest term in the mercury spectrum lies at 84,181 cm^{-1}, calculate the numerical values of the two energy levels involved in the emission of 5,461 Å.

[The University of Durham 1963]

2.55 The mean time for a spontaneous $2p \rightarrow 1s$ transition is $1.6 \times 10^{-9} s$ while the mean time for a spontaneous $2s \rightarrow 1s$ transition is as long as 0.14 s. Explain.

2.56 In the Helium-Neon laser (three-level laser), the energy spacing between the upper and lower levels $E_2 - E_1 = 2.26$ in the neon atom. If the optical pumping operation stops, at what temperature would the ratio of the population of upper level E_2 and the lower level E_1, be 1/10?

2.2.6 Molecules

2.57 What are the two modes of motion of a diatomic molecule about its centre of mass? Explain briefly the origin of the discrete energy level spectrum associated with one of these modes.

[University of London 2003]

2.58 Rotational spectral lines are examined in the HD (hydrogen–deuterium) molecule. If the internuclear distance is 0.075 nm, estimate the wavelength of radiation arising from the lowest levels.

2.59 Historically, the study of alternate intensities of spectral lines in the rotational spectra of homonuclear molecules such as N_2 was crucial in deciding the correct model for the atom (neutrons and protons constituting the nucleus surrounded by electrons outside, rather than the proton–electron hypothesis for the Thomas model). Explain.

2.60 The force constant for the carbon monoxide molecule is 1,908 N m^{-1}. At 1,000 K what is the probability that the molecule will be found in the lowest excited state?

2.61 At a given temperature the rotational states of molecules are distributed according to the Boltzmann distribution. Of the hydrogen molecules in the ground state estimate the ratio of the number in the ground rotational state to the number in the first excited rotational state at 300 K. Take the interatomic distance as 1.06 Å.

2.62 Estimate the wavelength of radiation emitted from adjacent vibration energy levels of NO molecule. Assume the force constant $k = 1{,}550$ N m^{-1}. In which region of electromagnetic spectrum does the radiation fall?

2.63 Carbon monoxide (CO) absorbs energy at $1.153 \times 1{,}011$ Hz, due to a transition between the $l = 0$ and $l = 1$ rotational states.

(i) What is the corresponding wavelength? In which part of the electromagnetic spectrum does this lie?
(ii) What is the energy (in eV)?
(iii) Calculate the reduced mass μ. ($C = 12$ times, and $O = 16$ times the unified atomic mass constant.)
(iv) Given that the rotational energy $E = \frac{l(l+1)\hbar^2}{2\mu r^2}$, find the interatomic distance r for this molecule.

2.64 Consider the hydrogen molecule H_2 as a rigid rotor with distance of separation of H-atoms $r = 1.0$ Å. Compute the energy of $J = 2$ rotational level.

2.65 The $J = 0 \rightarrow J = 1$ rotational absorption line occurs at wavelength 0.0026 in $C^{12}O^{16}$ and at 0.00272 m in C^xO^{16}. Find the mass number of the unknown Carbon isotope.

2.66 Assuming that the H^2 molecule behaves like a harmonic oscillator with force constant of 573 N/m. Calculate the vibrational quantum number for which the molecule would dissociate at 4.5 eV.

2.2.7 Commutators

2.67 (a) Show that $e^{ip\alpha/\hbar}x\, e^{-ip\alpha/\hbar} = x + \alpha$
(b) If A and B are Hermitian, find the condition that the product AB will be Hermitian

2.68 (a) If A is Hermitian, show that e^{iA} is unitary
(b) What operator may be used to distinguish between
(a) e^{ikx} and e^{-ikx} (b) $\sin ax$ and $\cos ax$?

2.69 (a) Show that $\exp(i\sigma x\theta) = \cos\theta + i\sigma x \sin\theta$
(b) Show that $\left(\frac{\mathrm{d}}{\mathrm{d}x}\right)^{\dagger} = -\frac{\mathrm{d}}{\mathrm{d}x}$

2.70 Show that
(a) $[x, p_x] = [y, p_y] = [z, p_z] = i\hbar$
(b) $[x^2, p_x] = 2i\hbar x$

2.71 Show that a hermitian operator is always linear.

2.72 Show that the momentum operator is hermitian

2.73 The operators **P** and **Q** commute and they are represented by the matrices $\begin{pmatrix} 1 & 2 \\ 2 & 1 \end{pmatrix}$ and $\begin{pmatrix} 3 & 2 \\ 2 & 3 \end{pmatrix}$. Find the eigen vectors of **P** and **Q**. What do you notice about these eigen vectors, which verify a necessary condition for commuting operators?

2.74 An operator $\hat{A}$ is defined as $\hat{A} = \alpha\hat{x} + i\beta\hat{p}$, where α , β are real numbers
(a) Find the Hermitian adjoint operator $\hat{A}^{\dagger}$
(b) Calculate the commutators $[\hat{A}, \hat{x}]$, $[\hat{A}, \hat{A}]$ and $[\hat{A}, \hat{P}]$

2.75 A real operator A satisfies the lowest order equation.

$A^2 - 4A + 3 = 0$

(a) Find the eigen values of A (b) Find the eigen states of A (c) Show that A is an observable.

2.76 Show that (a) $[x, H] = \frac{i\hbar p}{\mu}$ (b) $[[x, H], x] = \frac{\hbar^2}{\mu}$ where H is the Hamiltonian.

2.77 Show that for any two operators A and B,

$[A^2, B] = A[A, B] + [A, B]A$

2.78 Show that $(\sigma.\mathbf{A})(\sigma.\mathbf{B}) = \mathbf{A}.\mathbf{B} + i\sigma.(\mathbf{A}\times\mathbf{B})$ where **A** and **B** are vectors and σ's are Pauli matrices.

2.79 The Pauli matrix $\sigma_y = \begin{pmatrix} 0 & -i \\ i & 0 \end{pmatrix}$

(a) Show that the matrix is real whose eigen values are real.
(b) Find the eigen values of σ_y and construct the eigen vectors.
(c) Form the projector operators P_1 and P_2 and show that

$$P_1^\dagger P_2 = \begin{pmatrix} 0\,0 \\ 0\,0 \end{pmatrix},\ P_1 P_1^\dagger + P_2 P_2^\dagger = I$$

2.80 The Pauli spin matrices are $\sigma_x = \begin{pmatrix} 0 & 1 \\ 1 & 0 \end{pmatrix}, \sigma_y = \begin{pmatrix} 0 & -i \\ i & 0 \end{pmatrix}, \sigma_z = \begin{pmatrix} 1 & 0 \\ 0 & -1 \end{pmatrix}$
Show that (i) $\sigma_x^2 = 1$ and (ii) the commutator $[\sigma_x, \sigma_y] = 2i\sigma_z$.

2.81 The condition that must be satisfied by two operators $\hat{A}$ and $\hat{B}$ if they are to share the same eigen states is that they should commute. Prove the statement.

2.2.8 Uncertainty Principle

2.82 Use the uncertainty principle to obtain the ground state energy of a linear oscillator

2.83 Is it possible to measure the energy and the momentum of a particle simultaneously with arbitrary precision?

2.84 Obtain Heisenberg's restricted uncertainty relation for the position and momentum.

2.85 Use the uncertainty principle to make an order of magnitude estimate for the kinetic energy (in eV) of an electron in a hydrogen atom.

[University of London 2003]

2.86 Write down the two Heisenberg uncertainty relations, one involving energy and one involving momentum. Explain the meaning of each term. Estimate the kinetic energy (in MeV) of a neutron confined to a nucleus of diameter 10 fm.

[University of London 2006]

2.3 Solutions

2.3.1 de Broglie Waves

2.1 (a) $E = hf = h\lambda/c$
(b) Each photon carries an energy

$$E_\gamma = \frac{m_\pi c^2}{2} = \frac{135}{2} = 67.5\,\text{MeV}$$

$$cP_\gamma = E_\gamma = 67.5\,\text{MeV}$$

$$\lambda = \frac{h}{p} = 2\pi\frac{\hbar c}{cp} = \frac{(2\pi)(197.3\,\text{MeV.fm})}{67.5\,\text{MeV}}$$

$$= 18.36\,\text{fm} = 1.836 \times 10^{-14}\,\text{m}.$$

2.2 $\lambda = \left(\frac{150}{V}\right)^{1/2} = \left(\frac{150}{54}\right)^{1/2} = 1.667\,\text{Å}$

2.3 $\lambda = \frac{h}{p} = \frac{2\pi\hbar c}{(2mc^2T)^{1/2}} = (2\pi) \times \frac{197.3\,\text{MeV} - \text{fm}}{(2 \times 939\,\text{T} - \text{MeV})^{1/2}}$

$$= 28.6 \times 10^{-5}\,\text{Å}/T^{1/2}$$

where T is in MeV. If T is in eV, $\lambda = 0.286\,\text{Å}/\text{T}^{1/2}$

2.4 $\lambda = \frac{h}{p} = \frac{6.63 \times 10^{-34}\,J\text{–}s}{(2 \times 9.1 \times 10^{-31} \times 1.6 \times 10^{-19})^{1/2}}$

$$= 12.286 \times 10^{-10}\text{m/V}^{1/2} = \left(\frac{151}{V}\right)^{1/2}.$$

2.5 (a) $\frac{m_n v^2}{2} = \frac{p^2}{2m_n} = \frac{3kT}{2} = \left(\frac{3}{2}\right) \times 1.38 \times 10^{-23} \times \frac{300}{1.6 \times 10^{-19}}$

$$= 0.0388\,\text{eV}$$

$$cCp = \left(\sqrt{2m_n c^2 \cdot E_n}\right)^{1/2} = (2 \times 940 \times 106 \times 0.0388)^{1/2} = 8{,}541\,\text{eV}$$

$$\lambda = \frac{h}{p} = \frac{hc}{cp} = 3.9 \times 10^{-15}(\text{eV} - \text{s}) \times 3 \times 10^8\,\text{m} - \text{s}^{-1}/8{,}541\,\text{eV}$$

$$= 1.37 \times 10^{-10}\,\text{m} = 1.37\,\text{Å}$$

Such neutrons can be diffracted by crystals as their deBroglie wavelength is comparable with the interatomic distance in the crystal.

(b) $\Delta p_x.\Delta x = \hbar$

Put the uncertainty in momentum equal to the momentum itself, $\Delta p_x = p$

$$cp = \frac{\hbar c}{\Delta x} = \frac{197.3\,\text{MeV} - \text{fm}}{1.0\,\text{fm}} = 197.3\,\text{MeV}$$

$$E = (c^2p^2 + m^2c^4)^{1/2} = [(197.3)^2 + (940)^2]^{1/2} = 960.48\,\text{MeV}$$

Kinetic energy $T = E - mc^2 = 960.5 - 940 = 20.5\,\text{MeV}$

2.6 $E^2 = c^2p^2 + m^2c^4$

$$\hbar^2\omega^2 = c^2\hbar^2k^2 + m^2c^4$$

$$\omega = \left(c^2k^2 + \frac{m^2c^4}{\hbar^2}\right)^{1/2}$$

$$v_p = \frac{\omega}{k} = \left(c^2k^2 + \frac{m^2c^4}{\hbar^2}\right)^{1/2} / k$$

$$v_g = \frac{d\omega}{dk} = kc^2\left(c^2k^2 + \frac{m^2c^4}{\hbar^2}\right)^{-1/2}$$

$$\therefore v_p v_g = c^2$$

2.3.2 Hydrogen Atom

2.7 Apart from the principle quantum number n, three other quantum numbers are required to specify fully an atomic quantum state viz l, the orbital angular quantum number, m_l the magnetic orbital angular quantum number, and m_s the magnetic spin quantum number.

For $n = 1, l = 0$, if there is only one electron as in H-atom, then it will be in $1s$ orbit. The total angular momentum $J = l \pm 1/2$, so that $J = 1/2$. In the spectroscopic notation, $^{2s+1}L_J$, the ground state is therefore a $^2S_{1/2}$ state. For $n = 2$, the possible states are ^2S and ^2P. if there are two electrons as in helium atom, both the electrons can go into the K-shell ($n = 1$) only when they have antiparallel spin direction ($\uparrow\downarrow$) on account of Pauli's principle. This is because if the spins were parallel, all the four quantum numbers would be the same for both the electrons ($n = 1, l = 0, m_l = 0, m_s = +1/2$). Therefore in the ground state $S = 0$, and since both electrons are $1s$ electrons, $L = 0$. Thus the ground state is a S state (closed shell). A triplet state is not given by this electron configuration. An excited state results when an electron goes to a higher orbit. Then both electrons can have, in addition, the same spin direction, that is we can have $S = 1$ as well as $S = 0$ Excited triplet and singlet spin states are possible (orthohelium and parahelium). The lowest triplet has the electron configuration $1s$ $2s$, it is a $^3s_1$, state. It is a metastable state. The corresponding singlet state is 2^1S_0, and lies somewhat higher.

Carbon has six electrons. The Pauli principle requires the ground state configuration $1S^2$ $2S^2$ $2P^2$. The superscripts indicate the number of electrons in a given state.

2.8 A carbon atom has 6 electrons. If all these electrons are replaced by π^- mesons then two differences would arise (i) As π^- mesons are bosons (spin 0) Pauli's principle does not operate so that all of them can be in the K-shell ($n = 1$) (ii) The total energy is enhanced because of the reduced mass μ.

$$\mu = \frac{m_c m_\pi}{m_c + m_\pi} = \frac{(12 \times 1{,}840)(270)}{[(12 \times 1{,}840) + (270)]} = 266.7\, m_e$$

For each π^-, $E = -13.6 \times 266.7 = 3{,}628\,\text{eV}$
For the 6 pions, $E = 3{,}628 \times 6 = 21{,}766\,\text{eV}$

2.9 In atomic physics the atomic units are as follows:

(i) (a) The Bohr radius $\hbar^2/m_e e^2$ is used as the unit of length. (b) The energy is measured in multiples of the ionization energy of hydrogen $m_e e^2/2\hbar^2$ (c) $\hbar^2 = 1$ (d) $e^2 = 2$

(ii) In atomic units the Schrodinger equation

$$-\frac{\hbar^2}{2m_e}\nabla^2 u - \frac{e^2 u}{r} = Eu$$

would read as

$$-\nabla^2 u - \frac{2u}{r} = Eu$$

2.10 (a) Let the separation between the two nuclei each of charge q be $2d$, then the negative charge Q on the electrons is at a distance d from either nuclei Total potential energy due to electrostatic interaction between three objects is

$$\frac{qQ}{d} + \frac{qQ}{d} + \frac{q^2}{2d} \leq 0$$

Taking the equality sign and cancelling q

$$Q = -\frac{q}{4}$$

(b) Let T_0 be the initial kinetic energy and p_0 the momentum of the ion and T the kinetic energy and p the momentum of the composite molecule and Q the excitation energy.

$$T_0 = T + Q \qquad \text{Energy conservation} \tag{1}$$

$$p_0 = p \qquad \text{Momentum conservation} \tag{2}$$

$$\therefore (2mT_0)^{1/2} = (2.2mT)^{1/2} \tag{3}$$

The mass of the composite being $2m$ as the excitation energy is expected to be negligible in comparison with the mass of the molecule. From (3) we get

$$T = \frac{T_0}{2} \tag{4}$$

Using (4) in (1), we find $Q = \frac{T_0}{2} = \frac{10^{-19}}{2} = 5 \times 10^{-20}$ J

2.11(a) Stationary orbits will be such that the circumference of a circular orbit is equal to an integral number of deBroglie wavelength so that constructive interference may take place i.e. $2\pi r = n\lambda$
But $\lambda = h/p$

$$\therefore L = rp = nh/2\pi \quad \text{(Bohr's quantization condition)} \tag{1}$$

(b) Equating the coulomb force to the centripetal force

$$Ze^2/4\pi\varepsilon_o r^2 = mv^2/r \tag{2}$$

Solving (1) and (2)

$$v = Ze^2/2\varepsilon_o nh \tag{3}$$

$$r = \varepsilon_o n^2 h^2/\pi m Z e^2 \tag{4}$$

$$\text{Total energy} E = K + U = \frac{1}{2}mv^2 - Ze^2/4\pi\varepsilon_o r \tag{5}$$

Substituting (3) and (4) in (5)

$$E = -me^4 Z^2/8\varepsilon_o^2 n^2 h^2 \tag{6}$$

(c) $E = -mZ^2e^4/8\varepsilon_o^2 n^2h^2 = -9me^4/8\varepsilon_o^2 n^2 h^2$ (for $Z = 3$)

(d) Ionization energy for $Be^{+3} = 13.6 \times 3^2 = 122.4\,eV$

2.12 For a hydrogen-like atom the energy in the nth orbit is $E_n = 13.6\,\mu Z^2/n^2$

For hydrogen atom the reduced mass $\mu \approx m_e$, while for muon mesic atom it is of the order of $200\,m_e$. Consequently, the transition energies are enhanced by a factor of about 200, so that the emitted radiation falls in the x-ray region instead of U.V., I.R. or visible part of electromagnetic spectrum. The radius is given by

$r_n = \varepsilon_0 n^2 h^2/\pi \mu e^2$

Here, because of inverse dependence on μ, the corresponding radii are reduced by a factor of about 200.

2.13 $r_1 = \dfrac{a_0}{\mu Z} = R = r_0 A^{1/3} = r_0(2Z)^{1/3}$

$$Z^{4/3} \approx \frac{0.529 \times 10^{-10}}{207 \times 1.3 \times 2^{1/3} \times 10^{-15}} = 156$$

Therefore $Z = 44.14$ or 44

The first orbit of mu mesic atom will be just grazing the nuclear surface in the atom of Ruthenium. Actually, in this region $A \approx 2.2\,Z$ so that the answer would be $Z \approx 43$

2.14 The first four lines of the Lyman series are obtained from the transition energies between $n = 2 \to 1, 3 \to 1, 4 \to 1, 5 \to 1$

Now $E_n = -\dfrac{\alpha^2 m_e c^2}{4n^2}$

$$\Delta E_{n,1} = \frac{\alpha^2 m_e c^2}{4}\left(\frac{1}{1^2} - \frac{1}{n^2}\right), n = 2, 3, 4, 5$$

Thus,

$$\Delta E_{21} = \left(\frac{1}{137}\right)^2 \left(\frac{0.511 \times 10^6}{4}\right)\left(1 - \frac{1}{4}\right) = 5.1 \text{ eV}$$

$$\lambda_{21} = \frac{1{,}241}{5.1} = 243.3 \text{ nm} = 2{,}433 \text{ Å}$$

The wavelengths of the other three lines can be similarly computed. They are 2,053, 1,946 and 1,901 Å.

2.15 (a) Using Bohr's theory of hydrogen atom

$$E_n = -\frac{\mu e^4}{8\varepsilon_0^2 h^2 n^2} \tag{1}$$

where μ is the reduced mass. But the fine structure constant

$$\alpha = \frac{e^2}{4\pi\varepsilon_0 \hbar c} \tag{2}$$

Combining (1) and (2)

$$E_n = -\frac{\alpha^2 \mu c^2}{2n^2} \tag{3}$$

For positronium, $= m_e/2$. Therefore for positron

$$E_n = -\frac{\alpha^2 m_e c^2}{4n^2} \tag{4}$$

(b) $$r_n = \varepsilon_0 \frac{n^2 h^2}{\pi \mu c^2} \tag{5}$$

$$r_n \propto \frac{1}{\mu} = \frac{2}{m_e}$$

Therefore the radii are doubled.

(c) $$E_n \propto \mu = \frac{m_e}{2}$$

Therefore the transition energies are halved.

2.16 (a) $U(r) = -\int f(r)\mathrm{d}r = \int kr\, \mathrm{d}r + C$

$= \frac{1}{2}kr^2 + C$

$U(0) = 0 \rightarrow C = 0$

$U(r) = \frac{1}{2}kr^2$

(b) Bohr's assumption of quantization of angular momentum gives

$$mvr = n\hbar \tag{1}$$

Equating the attracting force to the centripetal force.

$$mv^2/r = kr \tag{2}$$

solving (1) and (2)

$$v^2 = \left(\frac{n\hbar}{m}\right)\left(\frac{k}{m}\right)^{\frac{1}{2}} \tag{3}$$

$$r^2 = \frac{n\hbar}{\sqrt{km}} \tag{4}$$

(c) $E = U + T = \frac{1}{2}kr^2 + \frac{1}{2}mv^2$ (5)

Substituting (3) and (4) on (5) and simplifying

$E = n\hbar(k/m)^{1/2}$

(d) $\Delta E = E_n - E_{n-1} = \hbar\left(\frac{k}{m}\right)^{\frac{1}{2}} = 1.05 \times 10^{-34}\left(\frac{1,180}{3 \times 10^{-26}}\right)^{\frac{1}{2}}$ J

$= 0.13$ eV

$\lambda = \frac{1,241}{0.13} = 9,546$ nm

2.17 $E_n = -\frac{13.6}{n^2}$

$\Delta E = E_{n+1} - E_n = 13.6\left(\frac{1}{n^2} - \frac{1}{(n+1)^2}\right) = \frac{13.6(2n+1)}{n^2(n+1)^2}$

In the limit $n \to \infty$, $\Delta E \propto \frac{n}{n^4} = \frac{1}{n^3}$

2.18 The wavelength $\lambda = 486.1$ nm corresponds to the transition energy of $E = 1241/486.1 = 2.55$ eV Looking up Fig. 2.1, for the energy level diagram for hydrogen atom, the transition $n = 4 \to 2$ gives the energy difference $-0.85 - (-3.4) = 2.55$ eV

The line belongs to the Balmer series.

6
- 0.38
5
- 0.54
4
- 0.85
3
- 1.51
2
- 3.40
1
- 13.60

Fig. 2.1 Energy level diagram for hydrogen atom

2.19 Orbital velocity, $v = \frac{e^2}{2nh\varepsilon_0}$, $a_0 = n^2h^2\varepsilon_0/\pi e^2 m$

Orbital frequency $f = v/2\pi a_0$

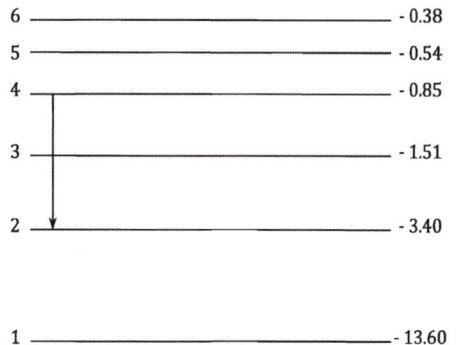

$$f = me^4/4n^3h^3\varepsilon_0^2$$

$$\nu = \frac{me^4}{8\varepsilon_0^2 h^3}\left(\frac{1}{n_f^2} - \frac{1}{n_i^2}\right) = \frac{me^4}{8\varepsilon_0^2 h^3}\left\{\frac{(n_i - n_f)(n_i + n_f)}{n_i^2 n_f^2}\right\}$$

If both n_i and n_f are large, and if we let

$$n_i = n_f + 1,\ \nu \approx \frac{me^4}{8\varepsilon_0^2 h^3}\left(\frac{2}{n_i^3}\right) = f$$

2.20 Energy difference for the transitions in the two series $\Delta E_{11} - \Delta E_{32} = 1{,}241/16.58 = 74.85$ eV

$$13.6Z^2\left\{\left(\frac{1}{1^2} - \frac{1}{2^2}\right) - \left(\frac{1}{2^2} - \frac{1}{3^2}\right)\right\} = 74.85$$

Solving for Z, we get $Z = 3$.
The ion is Li^{++}

2.21 Note that wave number is proportional to energy. The wavelength 486.1 nm in the Balmer series to the energy difference of 2.55 eV, and is due to the transition between $n = 4(E_4 = -0.85\,\text{eV})$ and $n = 2(E_2 = -3.4\,\text{eV})$. $\Delta E_{42} = -0.85 - (-3.4) = 2.55\,\text{eV}$. The wavelength 410.2 nm in the Balmer series corresponds to the energy difference of 3.0 eV and is due to the transition between $n = 6(E_6 = -0.38\,\text{eV})$ and $n = 2(E_2 = -3.4\,\text{eV})$. $\Delta E_{62} = -0.38 - (-3.4) = 3.02\,\text{eV}$
Thus $\Delta E_{62} - \Delta E_{42} = 3.02 - 2.55 = 0.47\,\text{eV}$

The difference of 0.47 eV is also equal to difference in $E_6 = -0.38\,\text{eV}$ $(n = 6)$ and $E_4 = -0.85$ eV $(n = 4)$. Thus the line arising from the transition $n = 6 \to n = 4$, must belong to Bracket series.

Note that in the above analysis we have used the well known law of spectroscopy, $\tilde{\upsilon}_{mn} - \tilde{\upsilon}_{kn} = \tilde{\upsilon}_{mk}$

2.3.3 X-rays

2.22 The wavelength $\lambda_{LK} = 0.0724$ nm corresponds to the energy
$E_\gamma = 1{,}241/\lambda_{LK} = 1{,}241/0.0724 = 17{,}141$ eV
Now $17{,}141 = 13.6 \times \frac{3}{4}(Z - \sigma)^2$
The factor 3/4 is due to the $L \to K$ transition. Substituting $Z = 42$, and solving for σ we obtain $\sigma = 1.0$

2.23 $E_\gamma = 13.6 \times 3(Z - \sigma)^2/4 = 13.6 \times 3(42 - 1)^2/4 = 17{,}146.2$ eV.
$\lambda_{LK} = 1{,}241/17{,}146.2 = 0.07238$ nm
$= 0.7238$ Å

2.24 Cobalt: $E_K = 13.6(Z - \sigma)^2 = 13.6(27 - 1)^2 = 9{,}193.6$ eV

$$\lambda_K = \frac{1{,}241}{9{,}193.6}\ \text{nm} = 0.135\ \text{nm} = 1.35\ \text{Å}$$

Iron: $E_K = 13.6(26-1)^2 = 8{,}500\,\text{eV}$

$$\lambda_K = \frac{1{,}241}{8{,}500} = 0.146\,\text{nm} = 1.46\,\text{Å}$$

2.25 The minimum wavelength of the photon will correspond to maximum frequency which will be determined by $E = h\upsilon_{\max}$

$$\lambda_{\min} = \frac{c}{\upsilon_{\max}} = \frac{hc}{h\upsilon_{\max}} = \frac{hc}{E} = \frac{2\pi\hbar c}{E}$$

$$= \frac{2\pi \times 197.3\,\text{fm} - \text{MeV}}{30 \times 10^{-3}\,\text{MeV}} = 4.13 \times 10^5\,\text{fm} = 4.13\,\text{Å}$$

2.26 $\lambda_C = \dfrac{hc}{eV}$

$$h = \frac{eV\lambda_C}{c} = \frac{1.6 \times 10^{-19} \times 80 \times 10^3 \times 0.15 \times 10^{-10}}{3 \times 10^8}$$

$$= 6.4 \times 10^{-34}\,\text{J} - \text{s}$$

2.27 $\lambda_c = \dfrac{hc}{eV}$

$$h = \frac{\lambda_c eV}{c} = \frac{0.247 \times 10^{-10} \times 1.6 \times 10^{-19} \times 50{,}000}{3 \times 10^8}$$

$$= 6.59 \times 10^{-34}\,\text{J} - \text{s}$$

2.28 According to Mosley's law

$$\frac{1}{\lambda} = A(Z-1)^2$$

$$\frac{1}{\lambda_I} = A(26-1)^2$$

$$\frac{1}{\lambda_{\text{Cu}}} = A(29-1)^2$$

$$\frac{\lambda_{\text{Cu}}}{\lambda_I} = \frac{25^2}{28^2} = 0.797 \rightarrow \lambda_{\text{Cu}} = 193 \times 0.797 = 153.8\,\text{pm}$$

2.29 $\lambda_K - \lambda_C = 84\text{ pm} = 0.84\,\text{Å}$ (1)

$$\frac{1{,}200}{(28-1)^2} - \frac{12.4}{V} = 0.84 \quad (2)$$

where $\lambda_C = \dfrac{hc}{eV} = \dfrac{12.4}{V}$ (V is in kV) (3)

Solving for V in (2), $V = 15.4\,\text{kV}$

2.30 The L_α line is produced due to transition $n = 3 \rightarrow n = 2$. For the $n = 2$ shell the quantum numbers are $l = 0$ or $l = 1$ and $j = l \pm \frac{1}{2}$, the energy states being $^2S_{1/2}$, $^2P_{1/2}$, $^2P_{3/2}$. For $n = 3$ shell the energy states are $^3S_{1/2}$, $^3P_{1/2}$, $^3P_{3/2}$, $^3d_{3/2}$, $^3d_{5/2}$

The allowed transitions are

$^3S_{1/2} \rightarrow {}^2P_{1/2}$, $^3S_{1/2} \rightarrow {}^2P_{3/2}$

$^3P_{1/2} \rightarrow {}^2S_{1/2}$, $^3P_{3/2} \rightarrow {}^2S_{1/2}$

$^3d \to {}^2P_{1/2}, {}^3d_{3/2} \to {}^2P_{3/2}$

$^3d_{5/2} \to {}^2P_{3/2}$

In all there are seven allowed transitions.

2.31 Let the wavelength difference be $\Delta\lambda$ when voltage V is applied.

$$\frac{1{,}200}{(Z-1)^2} - \frac{12.4}{V} = \Delta\lambda \tag{1}$$

$$\frac{1{,}200}{(Z-1)^2} - \frac{12.4}{10} = \Delta\lambda \tag{2}$$

$$\frac{1{,}200}{(Z-1)^2} - \frac{12.4}{20} = 3\Delta\lambda \tag{3}$$

Note that the first term on the LHS of (1) (Corresponding to the characteristic X-rays) is unaffected due to the application of voltage. Eliminating $\Delta\lambda$ between (2) and (3), we find $Z = 28.82$ or 29. The target material is Copper.

2.32 $\lambda_k = \dfrac{1{,}200}{(Z-1)^2}$ Å

$$Z_1 = 1 + \left(\frac{1{,}200}{2.4}\right)^{1/2} = 23.36$$

$$Z_2 = 1 + \left(\frac{1{,}200}{1.8}\right)^{1/2} = 26.82$$

The required elements have $Z = 23, 24, 25$ and 26

2.33 Bohr's theory gives

$$\nu = \frac{m_e e^4}{8\varepsilon_0^2 h^3}\left(\frac{1}{1^2} - \frac{1}{n^2}\right)(z-b)^2 \tag{1}$$

or

$$\sqrt{\upsilon} = \left[\frac{m_e e^4}{8\varepsilon_0^2 h^3}\left(\frac{1}{1^2} - \frac{1}{n^2}\right)\right]^{\frac{1}{2}}(z-b) \tag{2}$$

The factor within the square brackets is identified as a. Substitute $m_e = 9.11 \times 10^{-31}$ kg,

$e = 1.6 \times 10^{-19} C$,

$\varepsilon_0 = 8.85 \times 10^{-12}$ F/m and $h = 6.626 \times 10^{-34}$ J – s and put $n = 2$ to find a.

We get

$a = 4.956 \times 10^7\ \text{Hz}^{1/2}$

2.34 By Moseley's law

$$\frac{1}{\lambda} = A(Z_1 - 1)^2 = A(19 - 1)^2 \quad (1)$$

$$\frac{1}{\lambda/4} = A(Z - 1)^2 \quad (2)$$

Dividing (2) by (1) and solving for Z, we get $Z = 37$

2.3.4 *Spin and μ and Quantum Numbers – Stern–Gerlah's Experiment*

2.35 The existence of electron spin and its value was provided by the Stern–Gerlah experiment in which a beam of atoms is sent through an inhomogeneous magnetic field.

Schematic representation of the Stern–Gerlah experiment. to a force moment tending to align the magnetic moment along the field direction, but also to a deflecting force due to the difference in field strength at the two poles of the particle. Depending on its orientation, the particle will be driven in the direction of increasing or decreasing field strength. If atoms with all possible orientations in the field are present, a sharp beam should be split up into $2J+1$ components. In Fig. 2.2 the beam is shown to be split up into two components corresponding to $J = 1/2$

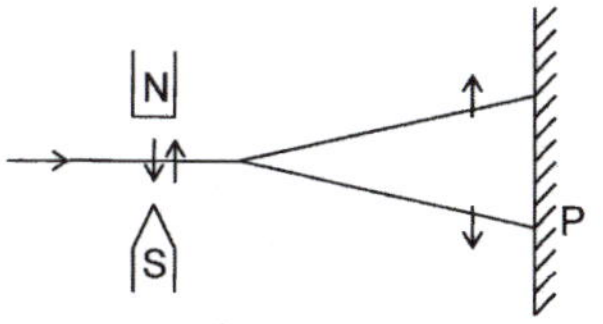

Fig. 2.2 Schematic drawing of Stern-Gerlah's apparatus

2.36 (i) If $l > s$, then there will be $2s + 1$ values of j; $j = l+s, l+s-1 \ldots l-s$
If $l < s$, then there will be $2l + 1$ values of j; $j = s+l, s+l-1 \ldots s-l$
(ii) The spectroscopic notation for a term is $^{2S+1}L_J$, $s, p, d, f \ldots$ refer to $l = 0, 1, 2, 3 \ldots$ respectively.

Term	L	S	J	Possible values of J
$^2S_{1/2}$	0	1/2	1/2	1/2
$^3D_2$	2	1	2	3, 2, 1
$^5P_3$	1	2	3	2, 1

(iii) Obviously the term $^5P_3$ cannot exist.

2.37 (a) By definition the magnetic moment of electron is given by the product of the charge and the area A contained by the circular orbit.

$$\mu = iA = -\frac{e\pi r^2}{T} = -\frac{\omega e\pi r^2}{2\pi}$$
$$= \frac{-em_e\omega r^2}{2m_e} = -\frac{eL}{2m_e}$$

(b) $\mu_l = -\frac{e}{2mc}(L(L+1))^{1/2}\hbar$

$\mu_s = -(2e/2mc)(S(S+1))^{1/2}\hbar$

2.38 The principle of the Stern–Gerlah experiment is described in Problem 2.35. While the atom is under the influence of inhomogeneous magnetic field the constant force acting on the atom along y-direction perpendicular to the straight line path OAF in the absence of the field, is a parabola (just like an object thrown horizontally in a gravitational field). The equation to the parabola is

$$y = kx^2 \tag{1}$$

where k is a constant, Fig 2.3. Let us focus on the atom which deviates upward. After leaving the field at D, its path along DE is a staright line. It hits the plate at E so that $EF = s$. When ED is extrapolated back, let it cut the line OAF in C.

Taking the origin at O, Eq. (1) satisfies the relation at D,

$$h = kl^2 \tag{2}$$

Furthermore at D,

$$\left(\frac{\mathrm{d}y}{\mathrm{d}x}\right)_D = 2Kx|_D = 2K.OA = 2K.l \tag{3}$$

$$\left(\frac{\mathrm{D}y}{\mathrm{d}x}\right)_D = \frac{AD}{CA} = \frac{h}{CA} \tag{4}$$

Combining (2), (3) and (4), we get

$$CA = \frac{l}{2} \tag{5}$$

Now the time taken for the atom along the x-component is the same as for along the y-component. Therefore

$$t = \frac{l}{v} = \left(\frac{2h}{a}\right)^{\frac{1}{2}} \tag{6}$$

or

$$h = \frac{l^2 a}{2v^2} \tag{7}$$

From the geometry of the figure,

$$\frac{EF}{CF} = \frac{DA}{CA}$$

or

$$\frac{s}{L + \frac{l}{2}} = \frac{h}{l/2} \rightarrow h = \frac{s.l}{2L + l} \qquad (8)$$

Eliminating h between (7) and (8), we find

$$a = \frac{2sv^2}{l(2L + l)} \qquad (9)$$

Now the acceleration,

$$a = \frac{F}{m} = \left(\frac{\mu}{m}\right)\left(\frac{\partial B}{\partial y}\right) \qquad (10)$$

Finally the separation between the images on the plate,

$$2s = \frac{l(2L + l)}{mv^2}\,\mu\left(\frac{\partial B}{\partial y}\right) \qquad (11)$$

$$1/2\,mv^2 = 2kT$$

$$2s = \frac{l(2L + l)\mu_B(\partial B/\partial y)}{4kT}$$

$$= \frac{[0.6(2 \times 1 + 0.6) \times 9.27 \times 10^{-24} \times 20]}{4 \times 1.38 \times 10^{-23} \times 600}$$

$$= 0.873 \times 10^{-2}\,\text{m} = 8.73\,\text{mm}$$

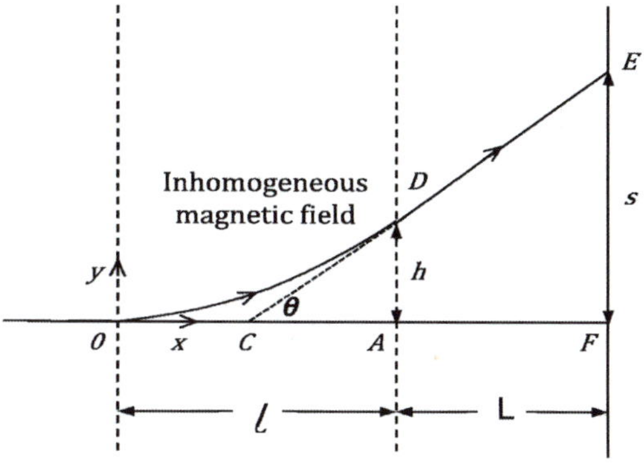

Fig. 2.3 Stern–Gerlah experiment

2.39 H Na
$1s$ $1s^2 2s^2 2p^6 3s$

Magnetic moment for both hydrogen and sodium is

1 Bohr magneton, $\mu_B = \frac{e\hbar}{2m_e} = 9.27 \times 10^{-24}\text{JT}^{-1}$

2.40 From Problem 2.38, the distance of separation on the plate

$$2s = \frac{l(2L+l)}{mv^2}\mu_B\left(\frac{\partial B}{\partial y}\right)$$

Therefore, $\tan\theta = \frac{s}{L+l/2} = \frac{2s}{2L+l} = \frac{l\mu_B\left(\frac{\partial B}{\partial y}\right)}{2E} = \frac{l\mu_B\left(\frac{\partial B}{\partial y}\right)}{2\times 2kT}$

Substituting $\theta = 0.14°$, $l = 1.0\,\text{m}$, $\frac{\partial B}{\partial y} = 6\,\text{Tm}^{-1}$, $k = 1.38 \times 10^{-23}\,\text{JK}^{-1}$

And $T = 400\,\text{K}$, we find $\mu_B = 8.99 \times 10^{-24}\,\text{J T}^{-1}$.

2.41 The total number of electrons is given by adding the numbers as superscripts for each term. This number which is equal to the atomic number Z is found to be 35. The transition elements have $Z = 21-30, 39-48, 72-80, 104-112$, while the rare earths comprising the Lanthanide series have $Z = 57 - 71$ and actinides have $Z = 89 - 100$. Thus the element with $Z = 35$ does not correspond to either a transition element or a rare earth element.

2.42 From Fig. 2.3 of Problem 2.38 the separation of the beams as they emerge from the magnetic field is given by

$2h = l^2a/v^2 = (l^2\mu/mv^2)(\partial B/\partial y)$

$= (l^2/4kT)\mu(\partial B/\partial y)$

Substituting $l = 0.07\,\text{m}$, $\mu = 9.27 \times 10^{-24}\,\text{J T}^{-1}$

$(\partial B/\partial y) = 5\,\text{Tmm}^{-1} = 5{,}000\,\text{Tm}^{-1}$, $k = 1.38 \times 10^{-23}\,\text{JK}^{-1}$, $T = 1{,}250\,\text{K}$.

we find $2l = 3.29 \times 10^{-3}$ m or 3.29 mm

2.43 (a) The magnetic moment for the silver atom is due to one unpaired electron

(b) In the $^3P_0$ state the atom has $J = 0$, therefore the magnetic moment is also zero.

(c) The beam of neutral atoms with total angular momentum J is split into $2J + 1$ components. $2J + 1 = 7$, so $J = 3$

(d) Ratio of intensities,

$$\frac{I_1}{I_2} = (2J_1 + 1)/(2J_2 + 1) = \frac{2 \times \frac{1}{2} + 1}{2 \times \frac{3}{2} + 1} = \frac{1}{2}$$

2.44 Let an electron move in a circular orbit of radius $r = \hbar^2/me^2$ around a proton. Assume that the z-component of the angular momentum is $L_z = \hbar$. Equating L_z to the classical angular momentum

$$L_z = \hbar = m_e r^2\omega \tag{1}$$

An electron orbiting the proton with frequency $v = \frac{\omega}{2\pi}$ constitutes a current $i = \frac{\omega e}{2\pi}$

The magnetic moment μ_0 produced is equal to this current multiplied by the area enclosed.

$$\mu_B = iA = \omega e.\frac{\pi r^2}{2\pi} \tag{2}$$

Using (1) in (2)

$$\mu_B = \frac{e\hbar}{2m_e} \tag{3}$$

μ_B is known as Bohr magneton.

Electron with total angular momentum $[j(j+1)]^{1/2}\hbar$ has a magnetic moment $\mu = [j(j+1)]^{1/2}\mu_B$

The z-component of the magnetic moment is $\mu_J = m_J \mu_B$

where $m_J\hbar$ is the z-component of the angular momentum.

2.3.5 Spectroscopy

2.45 The principle quantum number n denotes the number of stationary states in Bohr's atom model. $n = 1, 2, 3\ldots$

l is called azimuthal or orbital angular quantum number. For a given value of n, l takes the values $0, 1, 2 \ldots n-1$

The quantum number m_l, called the magnetic quantum number, takes the values $-l, -l+1, -l+2, \ldots, +l$ for a given pair of n and l values. This gives the following scheme:

n	1	2		3		
l	0	0	1	0	1	2
m_l	0	0	$-1\ 0\ +1$	0	$-1\ 0\ +1$	$-2\ -1\ 0\ +1\ +2$

n	4			
l	0	1	2	3
m_l	0	$-1\ 0\ +1$	$-2\ -1\ 0\ +1\ +2$	$-3\ -2\ -1\ 0\ +1\ +2\ +3$

m_s, the projection of electron spin along a specified axis can take on two values $\pm 1/2$.

Hence the total degeneracy is $2n^2$. For $n = 3$, $2 \times 3^2 = 18$ electrons can be accommodated.

2.46 According to Laporte rule, transitions via dipole radiation are forbidden between atomic states with the same parity. This is because dipole moment has odd parity and the integral $\int_{-\infty}^{\infty} \psi_f^*$ (dipole moment) $\psi_i d\tau$ will vanish between symmetric limits because the integrand will be odd when ψ_i and ψ_f have the same parity. Now the parity of the state is determined by the factor $(-1)^l$. Thus for the given terms the l-values and the parity are as below

Term	S	P	D	F
l	0	1	2	3
Parity $= (-1)^l$	+1	−1	+1	−1

2.47 $J = l + s = 0 + 1/2 = 1/2$
$F = I + J, I + J - 1, \ldots I - J$
$= 2, 1, 0$

2.48 The observed frequency (ω) of radiation from an atom that moves with the velocity v at an angle θ to the line of sight is given by

$$\omega = \omega_0(1 + (v/c)\cos\theta) \tag{1}$$

where ω_0 is the frequency that the atom radiates in its own frame of referenece. The Doppler shift is then

$$\frac{\Delta\omega}{\omega_0} = \frac{\omega - \omega_0}{a_0} = \left(\frac{v}{c}\right)\cos\theta \tag{2}$$

As the radiating atoms are subject to random thermal motion, a variety of Doppler shifts will be displayed. In equilibrium the Maxwellian distribution gives the fraction $\frac{dN}{N}$ of atoms with x-component of velocity lying between v_x and $v_x + dv_x$

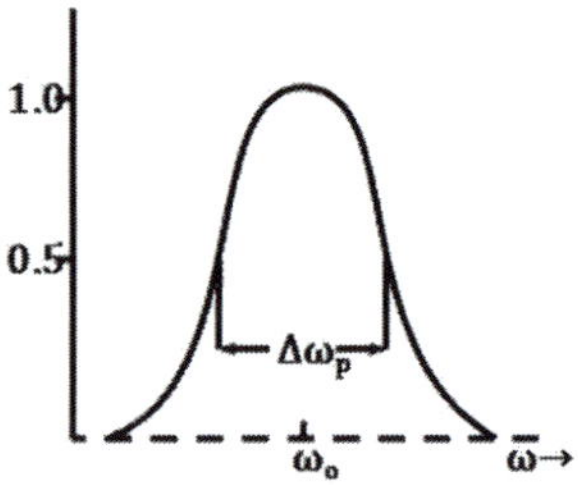

Fig. 2.4 Thermal broadening due to random thermal motion

$$\frac{dN}{N} = \frac{\exp\left[-\left(\frac{v_x}{U}\right)^2\right]}{\sqrt{\pi}}\frac{dv_x}{U} \tag{3}$$

where $u/\sqrt{2}$ is the root-mean-square velocity for particles of mass M at temperature T. Now

$$u = \left(\frac{2kT}{M}\right)^{1/2} \tag{4}$$

where $k = 1.38 \times 10^{-23}$ J/K is Boltzmann's constant.

Introducing the Doppler widths $\Delta\omega_D$ and $\Delta\lambda_D$ in frequency and wavelength

$$\frac{\Delta\omega_D}{\omega_0} = \frac{\Delta\lambda_D}{\lambda_0} = \frac{U}{c} = \left(\frac{2kT}{Mc^2}\right)^{1/2} \tag{5}$$

Further from (2) $d(\Delta\omega) = d\omega$. The relative distribution of Doppler shift is

$$\frac{dN}{N} = \frac{\exp\left[-\left(\frac{\Delta\omega}{\Delta\omega_D}\right)^2\right]}{\sqrt{\pi}} \frac{d\omega}{\Delta\omega_D} \tag{6}$$

Thus a Gaussian distribution is produced in the Doppler shift due to the random thermal motion of the source (Fig. 2.4).

The intensity of radiation is

$$I(\omega) = \frac{\exp\left[-\left(\frac{\Delta\omega}{\Delta\omega_D}\right)^2\right]}{\Delta\omega_D\sqrt{\pi}} \tag{7}$$

centered around the unshifted frequency ω_0. The width of the distribution at the frequencies where $I(\omega)$ falls to half the central intensity $I(\omega_0)$ is known as the half width

Doppler half width $= 2(ln2)^{1/2}\Delta\omega_D$

$$= 2\omega_0\left(\frac{2kT\ ln2}{Mc^2}\right)^{1/2} \tag{8}$$

Thermal broadening is most pronounced for light atoms such as hydrogen and high temperatures, for example the H_α line (6,563 Å) has a Doppler width of 0.6 Å at 400 K.

2.49 Lande' g-factor is

$g = 1 + j(j+1) + s(s+1) - l(l+1)/2j(j+1)$

For the term $^2P_{3/2}$, $l = 1$, $J = \frac{3}{2}$, $s = \frac{1}{2}$ and $g = \frac{4}{3}$

For $^2S_{1/2}$, $l = 0$, $j = \frac{1}{2}$, $s = \frac{1}{2}$ and $g = 2$

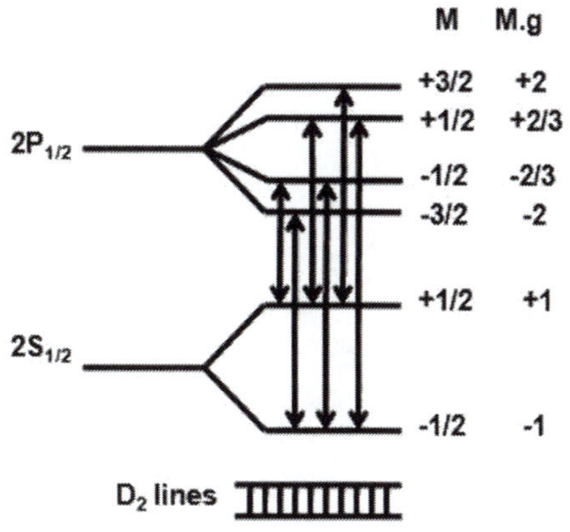

Fig. 2.5 Anamolous Zeeman effect in an alkali atom. The lines are not equidistant

The splitting of levels as in sodium is shown in Fig. 2.5. Transitions take place with the selection rule

$\Delta M = 0, \pm 1.$

2.50 Under the assumption of Russel–Saunders coupling, the ratios of the intervals in a multiplet can be easily calculated as follows. The magnetic field produced by L is proportional to $[L(L+1)]^{1/2}$, and the component of S in the direction of this field is $[S(S+1)]^{1/2}\cos(L, S)$. The energy in the magnetic field is

$$W = W_0 - B\mu_B \quad (1)$$

where μ_B is the component of the magnetic moment in the field direction and W_0 is the energy in the field-free case. From (1) the interaction energy is

$$\mu_B B = A[L(L+1)]^{1/2}[S(S+1)]^{1/2}\cos(L, S) \quad (2)$$

where A is a constant. From Fig. 2.6 It follows that

$$\cos(L, S) = \frac{J(J+1) - L(L+1) - S(S+1)}{2\sqrt{L(L+1)}\sqrt{S(S+1)}}$$

Consequently the interaction energy is $A[J(J+1) - L(L+1) - S(S+1)]/2$ As L and S are constant for a given multiplet term, the intervals between successive multiplet components are in the ratio of the differences of the corresponding $J(J+1)$ values. Now the difference between two successive $J(J+1)$ values is

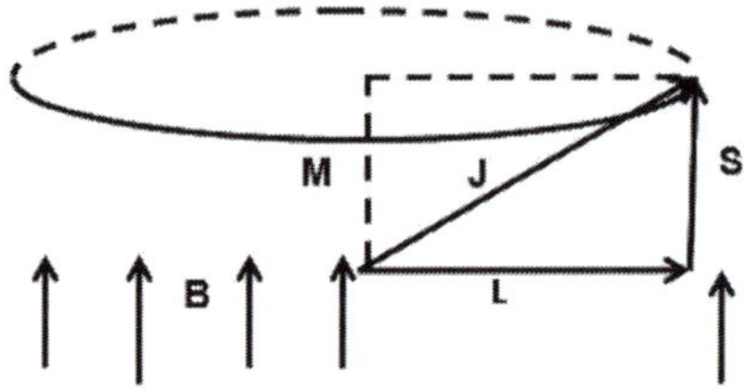

Fig. 2.6 Russel-Saunders coupling

$(J+1)(J+2) - J(J+1)$ or $2(J+1)$
and therefore proportional to $J+1$. This is known as Lande's interval rule.

For the calcium triplet

$(J+2)/(J+1) = 60 \times 10^{-4}/30 \times 10^{-4} = 2$
whence $J = 0$. The three levels of increasing energy have $J = 0$, 1 and 2.

Now $J = 0$, 1 and 2 are produced from the combination of L and S. With the spectroscopic notation $^{2S+1}L_J$ the terms for the three levels are $^3P_0$, $^3P_1$ and $^3P_2$.

2.51 $\Delta E = \mu_B B$

$h\Delta v = hc\Delta\lambda/\lambda^2 = \mu_B B$

$$B = \frac{hc\Delta\lambda}{\mu_B\lambda^2} = \frac{(6.63 \times 10^{-34})(3 \times 10^8)(1.7 \times 10^{-12})}{(9.17 \times 10^{-24})(350 \times 10^{-9})^2}$$
$= 0.3$ T

2.52 $\Delta E = \mu_B B \Delta m = \mu_B B$ (because $\Delta m = \pm 1$)
$= (9.27 \times 10^{-24})(1.0) = 9.27 \times 10^{-24}$ J $= 5.79 \times 10^{-5}$ eV.

The splitting of levels by equal amount in the presence of magnetic field is called normal Zeeman effect.

$f = \Delta E/h = 5.79 \times 10^{-5} \times 1.6 \times 10^{-19}/6.625 \times 10^{-34} = 1.398 \times 10^{10}$ c/s

2.53 For the term $^2P_{1/2}$, $l = 1$, $j = \frac{1}{2}$, $s = \frac{1}{2}$ and $g = \frac{2}{3}$. For the term $^2S_{1/2}$, $l = 0$, $j = \frac{1}{2}$, $s = \frac{1}{2}$ and $g = 2$. The energy levels and splitting of lines in sodium are shown in Fig. 2.7.

2.54 The ground state energy is

$$E_0 = h\nu = \frac{hc}{\lambda} = 6.63 \times 10^{-34} \times 3 \times 10^{10} \times 84{,}181/1.6 \times 10^{-19}$$
$$= 10.46 \text{ eV}$$

The excitation lines $E_2 = 10.46 + 7.69 = 18.15$ eV

The line 5461 Å is emitted when E_2 is deexcited to a lower level E_1 such that

$$E_2 - E_1 = \frac{1{,}241}{\lambda(\text{nm})} = \frac{1{,}241}{546.1} = 2.27 \text{ eV}$$

Thus $E_1 = 18.15 - 2.27 = 15.88$ eV

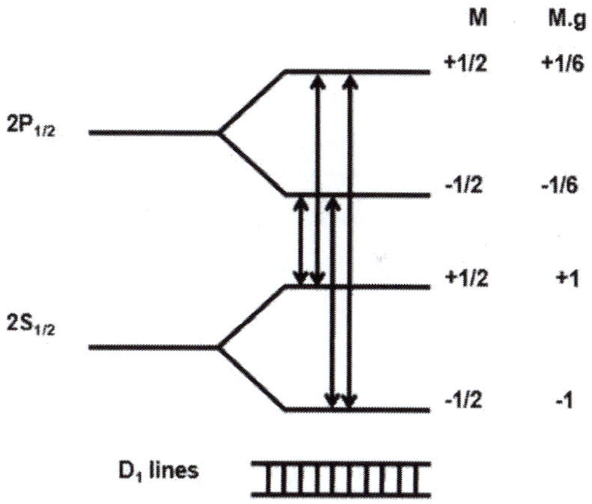

Fig. 2.7 Splitting of D_1 lines in magnetic field

Therefore the two levels involved in the emission of the 5,461 Å line are 18.15 eV and 15.88 eV

2.55 The 2*s* state of the hydrogen atom cannot decay by electric dipole radiation because a $2s \rightarrow 1s$ transition would violate the $\Delta l = \pm 1$ rule (Laporte rule). In point of fact the 2*s* state is a metastable state with a long life time which eventually decays to the 1*s* state by a mechanism, such as collision with other gas molecules, which is much less probable than an electric dipole transition.

2.56 $\frac{n(E_2)}{n(E_1)} = e^{-(E_2-E_1)/kT} = \frac{1}{10}$

$T = \frac{E_2 - E_1}{k\ln 10} = \frac{2.26}{8.625 \times 10^{-5} \times 2.3} = 1.14 \times 10^4\ \text{K}$

2.3.6 Molecules

2.57 The two modes of motion of a diatomic molecule are (i) rotation and (ii) vibration.

The first order rotational energy is $\hbar^2 J(J+1)/2I_0$, where $I_0 = MR_0^2$ is the moment of inertia of the molecule about an axis perpendicular to the line joining the nuclei; the energy being the same as for the rigid rotator. Clearly the spacing between successive levels is unequal; it progressively increases with the increasing value of J, where $J = 0, 1, 2\ldots$ The spectrum called band spectrum arises due to optical transitions between rotational levels. The band spectrum is actually a line spectrum, but is thus called because the lines are so closely spaced and unresolved with an ordinary spectrograph, and give the appearance of a band.

The second mode consists of to and fro vibrations of the atoms about the equilibrium position. The motion is described as simple harmonic motion. The energy levels are given by $E_n = \hbar\omega(n + 1/2)$, where $n = 0, 1, 2\ldots$ and are equally spaced. However as J or n increases, the spacing between levels becomes smaller than that predicted from the simple rigid rotator and harmonic oscillator.

2.58 The rotational energy levels are given by

$E_J = \hbar^2 J(J+1)/2I_o$

where I_o is the rotational inertia

$\Delta E = E_1 - E_0 = \hbar^2/I_o$

If μ is the reduced mass,

$$I = \mu r^2 = \frac{m_H m_D r^2}{m_H + m_D} = \frac{m_H \, . \, 2m_H r^2}{m_H + 2m_H} = \left(\frac{2}{3}\right) m_H r^2$$

(because $m_D \approx 2m_H$)

$$\Delta E = \frac{3\hbar^2}{2m_H r^2} = \frac{3}{2} \cdot \frac{(\hbar c\, \text{MeV} - \text{fm})^2}{m_H c^2 (0.075 \times 10^{-9}\text{m})^2}$$

$$= \frac{3}{2} \frac{(197.3 \times 10^{-15}\, \text{MeV} - \text{m})^2}{938(0.075 \times 10^{-9}\text{m})^2} = 0.011 \times 10^{-6}\ \text{MeV}$$

$= 0.011$ eV.

$\lambda = 1{,}241/0.011 = 1.128 \times 10^5$ nm

$= 0.113$ mm

2.59 All nuclei of even A, with zero or non-zero spin obey Bose statistics and all those of odd A obey Fermi statistics. The result has been crucial in deciding the model of the nucleus, that is discarding the electron–proton hypothesis. Consider the nitrogen nucleus. The electron–proton hypothesis implies 14 prorons+7 electrons. This means that it must have odd spin because the total number of particles is odd (21) and Fermi statistics must be obeyed. In the neutron–proton model the nitrogen nucleus has $7n + 7p = 14$ particles (even). Therefore Bose statistics must be obeyed. If the electronic wave function for the molecules is symmetric it was shown that the interchange of nuclei produces a factor $(-1)^J$ (J = rotational quantum number) in the total wave function of the molecule. Thus, if the nuclei obey Bose statistics symmetric nuclear spin function must be combined with even J rotational states and antisymmetric with odd J. Because of the statistical weight attached to spin states, the intensity of even rotational lines will be $(I + 1)/I$ as great as that of neighboring odd rotational lines where I is the nuclear spin. For Fermi statistics of the nuclei the spin and rotational states combine in a manner opposite to that stated previously, the odd rotational lines being more intense in the ratio $(I + 1)/I$. The experimental ratio $(I + 1)/I = 2$ for even to odd lines, giving $I = 1$, is consistent with the neutron–proton model.

2.60 The vibrational energy level is

$$E_n = \left(n + \frac{1}{2}\right)\hbar\omega, n = 0, 1, 2\ldots$$

with $\omega = \sqrt{(k/\mu)}$, k being the force constant and μ the reduced mass of the oscillating atoms.

$$\mu = \frac{m_c m_0}{m_c + m_0} = \frac{12 \times 16}{12 + 16} = 6.857\,\text{amu}$$

$$\omega = \left(\frac{1908}{6.857 \times 1.67 \times 10^{-27}}\right)^{1/2} = 4.082 \times 10^{14} S^{-1}$$

Number of molecules in state E_n is proportional to $\exp(-n\hbar\omega/kT)$, k being the Boltzmann constant and T the Kelvin temperature. The probability that the molecule is in the first excited state is

$$P_1 = \frac{\exp(-\hbar\omega/kT)}{\sum_0^\infty \exp(-n\hbar\omega/kT)} = \exp(-\hbar\omega/kT)[1 - \exp(-\hbar\omega/kT)]$$

$$\frac{\hbar\omega}{kT} = \frac{1.054 \times 10^{-34} \times 4.082 \times 10^{14}}{1.38 \times 10^{-23} \times 1,000} = 3.1177$$

Therefore, $P_1 = \exp(-3.117)[1 - \exp(-3.1177)]$

$= 0.042$

2.61 The rotational energy state is given by

$$E_J = J\frac{(J + 1)\hbar^2}{2I}, \quad J = 0, 1, 2\ldots$$

The state with quantum number J is proportional to $(2J + 1)\exp(-E_J/kT)$ The factor $(2J + 1)$ arises from the J state.

$N_0/N_1 = (1/3)\exp\left(\hbar^2/I_o kT\right)$

$$\mu = m_1/2 = 0.5\,m_p$$

$$N_0/N_1 = (1/3)\exp\left(\hbar^2/\mu r^2 kT\right) = (1/3)\exp\left(\hbar^2c^2/\mu c^2 r^2 kT\right)$$

$$= \frac{1}{3}\exp\left[\frac{\left(197.3 \times 10^{-15}\right)^2}{0.5 \times 938 \times \left(1.06 \times 10^{-10}\right)^2 \times \left(1.38 \times 10^{-23}/1.6 \times 10^{-13}\right) \times 300}\right]$$

$$= 0.445$$

2.62 $\omega = \sqrt{\dfrac{k}{\mu}}$

$$\mu = \frac{m_N m_0}{m_N + m_0} = 7.466\,\text{amu}$$

$$\omega = \left(\frac{1550}{7.466 \times 1.67 \times 10^{-27}}\right)^{1/2} = 3.526 \times 10^{14}\text{s}^{-1}$$

$$E = \hbar\omega = \frac{1.055 \times 10^{-34} \times 3.526 \times 10^{14}}{1.6 \times 10^{-19}} = 0.2325$$

$$\lambda = \frac{1{,}241}{0.2325}\,\text{nm} = 5{,}337\,\text{nm} = 5.34\,\mu\text{m}$$

This wavelength corresponds to Infrared region.

2.63 (i) $\lambda = \dfrac{c}{\nu} = \dfrac{3 \times 10^8}{1.153 \times 10^{11}} = 2.6 \times 10^{-3}\,\text{m} = 2.6\,\text{mm}$

It lies in the microwave part of electromagnetic spectrum.

(ii) $E(eV) = \dfrac{1241}{\lambda(\text{nm})} = \dfrac{1241}{2.6 \times 10^6(\text{nm})} = 0.000477\,\text{eV}$

(iii) $\mu = \dfrac{m_c m_0}{m_c + m_0} = \dfrac{12 \times 16}{12 + 16} = 6.857\,u$

(iv) $E_1 = \dfrac{1(1+1)\hbar^2}{2\mu r^2} = \dfrac{\hbar^2}{\mu r^2}$

$$E_0 = 0$$

$$\Delta E = E_1 - E_0 = \frac{\hbar^2}{\mu r^2}$$

$$r = \left(\frac{\hbar^2 c^2}{\mu c^2 \Delta E}\right)^{1/2} = \left(\frac{197.3^2(\text{MeV} - \text{fm})^2}{6.857 \times 931.5 \times 477 \times 10^{-12}}\right)^{1/2}$$

$$= 0.113 \times 10^{-9}\,\text{m}$$

$$= 1.13\,\text{Å}$$

2.64 $E_J = \dfrac{J(J+1)\hbar^2}{2I_o}$

$$I_o = \mu r^2 = \left(\frac{M_p}{2}\right) r^2$$

$$E_J = [J(J+1)\hbar^2 c^2]/(2)(0.5)(M_p c^2)r^2$$
$$E_2 = (2)(3)(197.3)^2(10^{-15})^2/(940) \times (10^{-10})^2$$
$$= 0.264 \times 10^{-10}\,\text{MeV}$$
$$= 2.64 \times 10^{-5}\,\text{eV}$$

2.65 $\Delta E_J = \dfrac{J\hbar^2}{I_o} = \dfrac{J\hbar^2}{\mu r^2}$

$$\Delta E_J = hc/\lambda$$

$$\therefore \lambda \propto \mu$$

$$\frac{\lambda_1}{\lambda_2} = \frac{0.00260}{0.00272} = \frac{\mu_1}{\mu_2} \tag{1}$$

$$\mu_1 = \frac{16 \times 12}{16+12}; \mu_2 = \frac{16x}{16+x} \tag{2}$$

Using (2) in (1) and solving for x, we get $x = 13.004$. Hence the mass number is 13.

2.66 $E_n = \hbar\omega\left(n + \dfrac{1}{2}\right) = \hbar\sqrt{\dfrac{k}{\mu}}\left(n + \dfrac{1}{2}\right)$

$$4.5 = \frac{1.054 \times 10^{-34}}{1.6 \times 10^{-19}}\left(\frac{573}{0.5 \times 1.67 \times 10^{-27}}\right)^{1/2}\left(n + \frac{1}{2}\right)$$

whence n = 7.75

Therefore the molecule would dissociate for $n = 8$.

2.3.7 Commutators

2.67 (a) Writing $x = i\hbar \dfrac{\partial}{\partial p}$

$$\left(e^{ip\alpha/\hbar}\left(i\hbar\frac{\partial}{\partial p}\right)e^{-ip\alpha/\hbar}\right)\psi|p|$$
$$= i\hbar\, e^{ip\alpha/\hbar}\left[-\frac{i\alpha}{\hbar}e^{\frac{ip\alpha}{\hbar}}\psi(p) + e^{-ip\alpha/\hbar}\frac{\partial\psi(p)}{\partial p}\right]$$
$$= \alpha + \frac{i\hbar\,\partial\psi(p)}{\partial p} = \alpha + x$$

(b) If A and B are Hermitian

$(AB)^\dagger = B^\dagger A^\dagger = BA$

If the product is to be Hermitian then $(AB)^\dagger = AB$ i.e. $AB = BA$. Thus, A and B must commute with each other.

2.68 (a) Let $f = e^{iA}$; then $f^{\dagger} = \left(e^{iA}\right)^{\dagger} = e^{-iA}$
Therefore $f^{\dagger} f = e^{-iA} e^{iA} = 1$
Thus e^{iA} is unitary.
(b) (a) Momentum (b) Parity

2.69 (a) $\exp(i\sigma_x\theta) = 1 + i\sigma_x\theta + (i\sigma_x\theta)^2/2! + (i\sigma_x\theta)^3/3! + \cdots$
$= (1 - \theta^2/2! + \theta^4/4! \ldots) + i\sigma_x(\theta - \theta^3/3! + \theta^5/5! \ldots)$
$= \cos\theta + i\sigma_x \sin\theta$
(where we have used the identity $\sigma_x^2 = 1$)

(b) $$\int \varphi^*(\mathrm{d}\psi/\mathrm{d}x)\mathrm{d}x = \varphi^*\psi - \int (\mathrm{d}\varphi^*/\mathrm{d}x)\psi \mathrm{d}x$$

But $\varphi^*\psi = 0$
Hence $\int \varphi^* \left(\frac{\mathrm{d}\psi}{\mathrm{d}x}\right) \mathrm{d}x = \int \left(-\frac{\mathrm{d}\varphi^*}{\mathrm{d}x}\right) \psi \mathrm{d}x = \int -(\mathrm{d}\varphi/\mathrm{d}x)^{\dagger} \psi \mathrm{d}x$
Therefore $\left(\frac{\mathrm{d}}{\mathrm{d}x}\right)^{\dagger} = -\mathrm{d}/\mathrm{d}x$

2.70 (a) $[x, P_x]\psi = xP_x\psi - P_x x\psi$
$$= x\left(-i\hbar\frac{\partial}{\partial x}\right)\psi + i\hbar\frac{\partial}{\partial x}(x\psi)$$
$$= -i\hbar x\frac{\partial\psi}{\partial x} + i\hbar x\frac{\partial\psi}{\partial x} + i\hbar\psi$$
$= i\hbar\psi$
$\therefore [x, P_x] = i\hbar$

(b) $$[x^2, P_x]\psi = x^2(-i\hbar\partial\,\psi/\partial x + i\hbar\frac{\partial}{\partial x}(x^2\psi)$$
$= -i\hbar x^2 \partial\,\psi/\partial x + i\hbar x^2 \partial\psi/\partial x + i\hbar(2x)\psi$
$= 2i\hbar x\psi$
Therefore $[x^2, p_x] = 2i\hbar x$

2.71 By definition a transformation A is said to be linear if for any constant (possibly complex) λ
$A(\lambda X) = \lambda\, A\, X$
And if for any two vectors x and y
$A(x + y) = Ax + Ay$
If H is a hermitian operator
$(x, H\lambda y) = (Hx, \lambda y) = \lambda(Hx, y) = \lambda(x, Hy) = (x, \lambda Hy)$
Or $H\lambda y = \lambda Hy$
for any y. Furthermore
$(z, H(x + y)) = (Hz, x + y) = (Hz, x) + (Hz, y)$
$= (z, Hx) + (z, Hy) = (z, Hx + Hy)$
$\therefore\ H(x + y) = Hx + Hy$

2.72 Consider the equation
$$\frac{\partial}{\partial x}(\psi^*\psi) = \psi^*\frac{\partial\psi}{\partial x} + \psi\frac{d\psi^*}{dx} \tag{1}$$

Integration of (1) yields

$$\int \frac{\partial}{\partial x}(\psi^*\psi)\mathrm{d}\tau + \int \psi^* \frac{\partial \psi}{\partial x}\mathrm{d}\tau + \int \psi \frac{d\psi^*}{dx}\mathrm{d}\tau \tag{2}$$

The integral on the LHS vanishes because $\psi^*\psi$ vanishes for large values of $|x|$ if the particle is confined to some finite region. Thus

$$\int \frac{\partial}{\partial x}(\psi^*\psi)\mathrm{d}x\mathrm{d}y\mathrm{d}z = \int |\psi^*\psi|_{-\infty}^{\infty}\mathrm{d}y\mathrm{d}z = 0$$

Therefore (2) becomes

$$\int \psi^* \left(\frac{\partial \psi}{\partial x}\right)\mathrm{d}\tau = -\int \psi \left(\frac{\partial \psi^*}{\partial x}\right)\mathrm{d}\tau \tag{3}$$

Generalizing to all the three coordinates

$$(\psi, \nabla\psi) = -(\nabla\psi, \psi) \tag{4}$$

Hence

$$(\psi, i\hbar\nabla\psi) = (i\hbar\nabla\psi, \psi) \tag{5}$$

where we have used, $i\hbar\nabla\psi, \psi = -\int i\hbar\nabla\ \psi^*\psi\ \mathrm{d}\tau$.

This completes the proof that the momentum operator is hermitian.

2.73 Using the standard method explained in Chap. 1, define the eigen values $\lambda_1 = 3$ and $\lambda_2 = -1$ for the matrix P and the eigen vectors $\frac{1}{\sqrt{2}}\begin{pmatrix}1\\1\end{pmatrix}$ and $\frac{1}{\sqrt{2}}\begin{pmatrix}1\\-1\end{pmatrix}$. For the matrix Q, the eigen values are $\lambda_1 = 5$ and $\lambda_2 = 1$, the eigen vectors being $\frac{1}{\sqrt{2}}\begin{pmatrix}1\\1\end{pmatrix}$ and $\frac{1}{\sqrt{2}}\begin{pmatrix}1\\-1\end{pmatrix}$. Thus the eigen vectors for the commutating matrices are identical.

2.74 (a) $A = \alpha x + i\beta p$

$A^\dagger = ax^\dagger - i\beta p^\dagger$

(b) $[A, x] = \alpha[x, x] + i\beta[p, x]$

$= 0 + i\beta(-i\hbar) = \beta\hbar$

$[A, A] = AA - AA = 0$

$[A, p] = \alpha[x, p] + i\beta[p, p]$

$= i\alpha\hbar + 0 = i\alpha\hbar$

2.75 (a) As A satisfies a quadratic equation it can be represented by a 2×2 matrix. Its eigen values are the roots of the quadratic equation

$\lambda^2 - 4\lambda + 3 = 0, \lambda_1 = 1, \lambda_2 = 3$

(b) A is represented by the matrix

$$A = \begin{pmatrix}1 & 0\\0 & 3\end{pmatrix}$$

The eigen value equation is

$$\begin{pmatrix}1 & 0\\0 & 3\end{pmatrix}\begin{pmatrix}a\\b\end{pmatrix} = \lambda\begin{pmatrix}a\\b\end{pmatrix}$$

This gives $a = 1, b = 0$ for $\lambda = 1$ and $a = 0, b = 1$ for $\lambda = 3$

Hence the eigen states of A are $\begin{pmatrix} 1 \\ 0 \end{pmatrix}$ and $\begin{pmatrix} 0 \\ 1 \end{pmatrix}$

(c) As $A = A^\dagger$, A is Hermitian and hence an observable.

2.76 (a) A general rule for commutators is

$$[A^2, B] = A[A, B] + [A, B]A$$

Here $H = P^2/2\mu$

$$[H, X] = (1/2\mu)[P^2, x] = (1/2\mu)(P[P, x] + [P, x]P)$$
$$= (1/2\mu)2p\hbar/i = p\hbar/i\mu$$

Therefore $[x, H] = i\hbar p/\mu$

(b) $[[x, H], x] = \left[\dfrac{i\hbar P_x}{\mu}, x\right] = \left(\dfrac{i\hbar}{\mu}\right)[P_x, x] = \left(\dfrac{i\hbar}{\mu}\right)(-i\hbar) = \dfrac{\hbar^2}{\mu}$

2.77 $[A^2, B] = A\ A\ B - B\ A\ A = A\ A\ B - A\ B\ A + A\ B\ A - B\ A\ A$
$= A[A, B] + [A, B]A$

2.78 $(\boldsymbol{\sigma}.\boldsymbol{A})(\boldsymbol{\sigma}.\boldsymbol{B}) = (\sigma_x A_x + \sigma_y A_y + \sigma_z A_z)(\sigma_x B_x + \sigma_y B_y + \sigma_z B_z)$

$$= A_x B_x \sigma_x^2 + A_y B_y \sigma_y^2 + A_z B_z \sigma_z^2 + \sigma_x\sigma_y A_x B_y + \sigma_x\sigma_z A_x B_z$$
$$+ \sigma_y\sigma_x A_y B_x + \sigma_y\sigma_z A_y B_z + \sigma_z\sigma_x A_z B_x + \sigma_z\sigma_y A_z B_y$$
$$= \boldsymbol{A}.\boldsymbol{B} + i\sigma_z(A_x\ B_y - A_y\ B_x) + i\sigma_x(A_y\ B_z - A_z\ B_y)$$
$$+ i\sigma_y(A_z\ B_x - B_z\ A_x)$$
$$= \boldsymbol{A}.\boldsymbol{B} + i[\boldsymbol{\sigma}.(\boldsymbol{A} \times \boldsymbol{B})]_z + i[\boldsymbol{\sigma}.(\boldsymbol{A} \times \boldsymbol{B})]_x + i[\boldsymbol{\sigma}.(\boldsymbol{A} \times \boldsymbol{B})]_y$$
$$= \boldsymbol{A}.\boldsymbol{B} + i\ \sigma.(\boldsymbol{A} \times \boldsymbol{B})$$

where we have used the identities in simplifying:

$$\sigma_x^2 = \sigma_y^2 = \sigma_z^2 = 1$$

and $\sigma_y\sigma_x = -\sigma_x\sigma_y$ etc.

2.79 (a) $\sigma_{y\mu} = \sigma_{\mu y}\dagger$ as can be seen from the matrix elements of σ_y. Therefore σ_y is Hermitian. It is the matrix of a Hermitian operator whose eigen values are real.

(b) The eigen values λ are found by setting

$$\begin{vmatrix} \sigma y_{11}\ -\lambda & \sigma y_{11} \\ \sigma y_{21} & \sigma y_{22}\ \ -\lambda \end{vmatrix} = \begin{vmatrix} -\lambda & -i \\ i & -\lambda \end{vmatrix} = 0$$

$\lambda^2 - 1 = 0,\ \lambda = \pm 1$

The eigen vector associated with $\lambda_1 = 1$ is

$$|\psi_1> = \sum_{n=1}^{2} C_n|n> \text{ with } -C_1 - iC_2 = 0, C_2 = iC_1,$$

$$|\psi_1> = \frac{1}{\sqrt{2}}|1> + \frac{i}{\sqrt{2}}|2>$$

The eigen vector associated with $\lambda_2 = -1$ is

$$|\psi_2> = \sum_{n=1}^{2} C_n|n> \text{ with } C_1 - iC_2 = 0, C_2 = -iC_1,$$

$$|\psi_2> = \frac{1}{\sqrt{2}}|1> - \frac{i}{\sqrt{2}}|2>$$

(c) The projector onto $|\psi_i>$ is $P_i = |\psi_i><\psi_i|$.

$$\text{Matrix of } P_1 = \begin{bmatrix} \frac{1}{2} & -\frac{i}{2} \\ \frac{i}{2} & \frac{1}{2} \end{bmatrix}, \text{ matrix of } P_2 = \begin{bmatrix} \frac{1}{2} & \frac{i}{2} \\ -\frac{i}{2} & \frac{1}{2} \end{bmatrix}$$

$$P_1{}^{\dagger} P_2 \begin{bmatrix} 0 & 0 \\ 0 & 0 \end{bmatrix} = 0,\ P_1 P_1{}^{\dagger} + P_2 P_2{}^{\dagger} = I$$

2.80 (i) $\sigma_x{}^2 = \begin{pmatrix} 0 & 1 \\ 1 & 0 \end{pmatrix}\begin{pmatrix} 0 & 1 \\ 1 & 0 \end{pmatrix} = \begin{pmatrix} 1 & 0 \\ 0 & 1 \end{pmatrix}$

(ii) $[\sigma_x, \sigma_y] = \begin{pmatrix} 0 & 1 \\ 1 & 0 \end{pmatrix}\begin{pmatrix} 0 & -i \\ i & 0 \end{pmatrix} - \begin{pmatrix} 0 & -i \\ i & 0 \end{pmatrix}\begin{pmatrix} 0 & 1 \\ 1 & 0 \end{pmatrix}$

$$= \begin{pmatrix} i & 0 \\ 0 & -i \end{pmatrix} - \begin{pmatrix} -i & 0 \\ 0 & i \end{pmatrix}$$

$$= \begin{pmatrix} 2i & 0 \\ 0 & -2i \end{pmatrix} = 2i\begin{pmatrix} 1 & 0 \\ 0 & -1 \end{pmatrix} = 2i\sigma_z$$

2.81 Proof : $AX = \lambda_1 X$ (1)

$BX = \lambda_2 X$ (2)

where λ_1 and λ_2 are the eigen values belonging to the same state λ.

$BAX = B\lambda_1 X = \lambda_1 BX = \lambda_1\lambda_2 X$ (3)

$ABX = A\lambda_2 X = \lambda_2 AX = \lambda_2\lambda_1 X = \lambda_1\lambda_2 X$ (4)

Subtracting (3) from (4)

$(AB - BA)X = 0$

Therefore $AB - BA = 0$, because $X \neq 0$
Operate with B on A in (1) and with A and B in (2)
Or $[A, B] = 0$

2.3.8 Uncertainty Principle

2.82 $\Delta x \Delta p_x \sim \hbar/2$

$$P = \frac{\hbar}{2x}$$

$$E = \frac{p^2}{2m} + 1/2\, m\omega^2 x^2$$

$$= \frac{\hbar^2}{8mx^2} + 1/2m\omega^2 x^2$$

The ground state energy is obtained by setting $\frac{\partial E}{\partial x} = 0$

$$\frac{\partial E}{\partial x} = -\frac{\hbar^2}{4mx^3} + m\omega^2 x = 0$$

whence $x^2 = \frac{\hbar}{2m\omega}$

$\therefore\ E = 1/4\hbar\omega + 1/4\hbar\omega = \frac{1}{2}\hbar\omega$

2.83 If E and p are to be measured simultaneously their operators must commute. Now

$H = -\hbar^2\nabla^2/2m + V$ and $p = -i\hbar\nabla$

$[H, p] = [-\hbar^2\nabla^2/2m + V, -i\hbar\nabla]$

$= i\hbar^3\nabla^2\nabla/2m - i\hbar V\nabla - i\hbar^3\nabla\nabla^2/2m + i\hbar\nabla V$

The first and the third term on the RHS get cancelled because $\nabla^2\nabla = \nabla\nabla^2$. Therefore

$$[H, P] = -i\hbar(V\nabla - \nabla V)$$

If V = constant, the commutator vanishes. To put it differently energy and momentum can be measured with arbitrary precision only for unbound particles.

2.84 Consider the motion of a particle along x-direction.
The uncertainty Δx is defined as

$$(\Delta x)^2 =< (x- < x >)^2 >=< x^2 > -2 < x >< x > + < x >^2$$
$$=< x^2 > - < x >^2 \quad (1)$$

Similarly

$$(\Delta P_x)^2 =< P_x^2 > - < P_x >^2 \quad (2)$$

The precise statement of the Heisenberg uncertainty principle is

$$\begin{aligned} \Delta P_x \Delta x &\geq \hbar/2 \\ \Delta P_y \Delta y &\geq \hbar/2 \\ \Delta P_z \Delta z &\geq \hbar/2 \end{aligned} \tag{3}$$

Consider the integral, a function of a real parameter λ

$$I(\lambda) = \int_{-\infty}^{\infty} \mathrm{d}x |(x- < x >)\psi + i\lambda(-i\hbar\partial\psi/\partial x- < P_x > \psi|^2 \tag{4}$$

By definition, $I(\lambda \geq 0)$. Expanding (4)

$$I(\lambda) = \int_{-\infty}^{\infty} \mathrm{d}x\psi^*(x- < x >)^2\psi + \lambda\hbar \int_{-\infty}^{\infty} \mathrm{d}x \left(\psi^* \frac{\partial\psi}{\partial x} + \psi \frac{\partial\psi^*}{\partial x}\right)(x- < x >)$$

$$+ \lambda^2\hbar^2 \int_{-\infty}^{\infty} \left(\frac{\partial\psi^*}{\partial x}\right) \frac{\partial\psi}{\partial x} - i\lambda^2 < P_x > \hbar \int_{-\infty}^{\infty} \mathrm{d}x[\psi \frac{\partial\psi^*}{\partial x} + \lambda^2 < P_x >^2 \int_{-\infty}^{\infty} \mathrm{d}x\psi^*\psi \tag{5}$$

The term in the second line can be written as

$$\int_{-\infty}^{\infty} \mathrm{d}x \frac{\partial}{\partial x}(\psi^*\psi)(x- < x >) = [(x- < x >)\psi^*\psi]_{-\infty}^{\infty} - \int_{-\infty}^{\infty} \mathrm{d}x\psi^*\psi = -1$$

because it is expected that $\psi \to 0$. Sufficiently fast as $x \to \pm\infty$ so that the integrated term is zero. Similarly the third term can be re-written as

$$\hbar^2 \int_{-\infty}^{\infty} \mathrm{d}x \left(\frac{\partial\psi^*}{\partial x}\right)\left(\frac{\partial\psi}{\partial x}\right) = \hbar^2 \left[\psi^* \frac{\partial\psi}{\partial x}\right]_{-\infty}^{\infty} + \int_{-\infty}^{\infty} \mathrm{d}x\psi^*(-\hbar^2\partial^2\psi/\partial x^2) =< P_x^2 >$$

In term (4) rewrite

$$-i\,\hbar \int_{-\infty}^{\infty} \mathrm{d}x \frac{\partial\psi^*}{\partial x}\psi = -i\hbar\left[\psi^*\psi\right]_{-\infty}^{\infty} + \int_{-\infty}^{\infty} \mathrm{d}x\psi^* i\hbar \frac{\partial\psi}{\partial x} = - < P_x >$$

So, the full term (4) becomes $-2 < P_x >^2$.

Collecting all the terms

$$I(\lambda) = (\Delta x)^2 - \hbar\lambda + (\Delta P_x)^2\lambda^2 \geq 0$$

Denoting $I(\lambda) = a\lambda^2 + b\lambda + c$, the condition $I(\lambda \geq 0)$ is satisfied if $b^2 - 4ac \geq 0$.

Thus, $\hbar^2 - 4(\Delta x)^2(\Delta P_x)^2 \leq 0$, and therefore

$$\Delta P_x \Delta x \geq \hbar/2$$

2.85 $\Delta x \Delta p = \hbar$

$$c\Delta P \approx cp = \frac{\hbar c}{\Delta x} = \frac{197.3\,\text{MeV} - \text{fm}}{0.529 \times 10^{-10}\,\text{m}}$$

$$= 372.97 \times 10^{-5}\,\text{MeV} = 3{,}730\,\text{eV}$$

$$T = c^2p^2/mc^2 = (3{,}730)^2/0.511 \times 10^6 = 13.61\,\text{eV}$$

This value is in agreement with 13.60 obtained from Bohr's theory of hydrogen atom.

2.86 (i) $\Delta x \Delta P_x = \hbar$ (ii) $\Delta E \Delta t = \hbar$

$$\Delta P_x = \frac{\hbar}{\Delta x}; c\Delta P_x = cP_x = \frac{\hbar c}{\Delta x} = \frac{197.3\,\text{MeV} - \text{fm}}{10\,\text{fm}} \approx 20\,\text{MeV/c}$$

$$T = \frac{P^2}{2M} = \frac{c^2 p^2}{2Mc^2} = \frac{20^2}{(2)(940)} = 0.21\,\text{MeV}$$

Chapter 3
Quantum Mechanics – II

3.1 Basic Concepts and Formulae

Schrodinger's equation

$$i\hbar\frac{\partial\psi}{\partial t} = -\frac{\hbar^2}{2\mu}\nabla^2\psi + V(\mathbf{r})\psi \tag{3.1}$$

$$-\frac{\hbar^2}{2\mu}\nabla^2\psi + (V - E)\,\psi = 0 \text{ (Time independent equation)} \tag{3.2}$$

$$\text{Probability density } \rho = \psi^*\psi \tag{3.3}$$

Continuity equation

$$\frac{\partial\rho}{\partial t} + \nabla.j = 0 \tag{3.4}$$

where j is the probability current density.

Normalization of the wave function

$$\int_{\text{all space}} \rho\,\mathrm{d}\tau = \int_{\text{all space}} \psi_n^*\psi_n\,\mathrm{d}\tau = 1 \tag{3.5}$$

$I = \int \psi_m\psi_n\mathrm{d}\tau$ is called the overlapping integral.

Orthogonality

$$\int_{-\infty}^{+\infty} \psi_n^*(x)\psi_m(x)\mathrm{d}x = 0, \text{ if } m \neq n \tag{3.6}$$

Table 3.1 Dynamic quantities and operators

Physical Quantity		Operator
Position	r	R
Momentum	P	$-i\hbar\nabla$
Kinetic energy	T	$-\dfrac{\hbar^2}{2\mu}\nabla^2$
Potential energy	V	$V(\mathrm{r})$
Angular momentum square	L^2	$l(l+1)\hbar^2$
z-component of angular momentum	L_z	$-i\hbar\dfrac{\partial}{\partial\phi}$

Expectation values of dynamical variables and operators

An arbitrary function of r has the expectation value

$$< f(\mathrm{r}) >= \int \psi^* f(\mathrm{r})\psi \, \mathrm{d}\boldsymbol{\tau} \tag{3.7}$$

The expectation value of P

$$< \mathbf{P} >= \int \psi^* \left(\frac{\hbar}{i}\nabla\psi\right) \mathrm{d}\boldsymbol{\tau} \tag{3.8}$$

The expectation value of the kinetic energy

$$< T >= \int \psi^* \left(-\frac{\hbar^2}{2\mu}\nabla^2\psi\right) \mathrm{d}\tau \tag{3.9}$$

Pauli spin matrices

$$\sigma_x = \begin{pmatrix} 0 & 1 \\ 1 & 0 \end{pmatrix}, \sigma_y = \begin{pmatrix} 0 & -i \\ i & 0 \end{pmatrix}, \sigma_z = \begin{pmatrix} 1 & 0 \\ 0 & -1 \end{pmatrix} \tag{3.10}$$

$$\sigma_x^2 = \sigma_y^2 = \sigma_z^2 = 1 \tag{3.11a}$$

$$\sigma_x\sigma_y = i\sigma_z, \sigma_y\sigma_z = i\sigma_x, \sigma_z\sigma_x = i\sigma_y \tag{3.11b}$$

These matrices are both Hermetian and unitary. Further, any two Pauli matrices anticommute

$$\sigma_x\sigma_y + \sigma_y\sigma_x = 0, \text{ etc.} \tag{3.11c}$$

Commutators

$$AB - BA = [A, B] \tag{3.12}$$

by definition.

Dirac's Bra and Ket notation

A ket vector, or simply ket, is analogous to the wave function for a state. The symbol $|m>$ denotes the ket vector that corresponds to the state m of the system. A bra vector, or bra, is analogous to the complex conjugate of the wave function for a state. The symbol $<n|$ denotes the bra vector that corresponds to the state n of the system. Then

$$\int \psi_n^* \psi_m \, d\tau = \langle n|m\rangle \tag{3.13}$$

And $$\int \psi_n^* \, \mathrm{H} \, \psi_m \, d\tau = \langle n|H|m\rangle \tag{3.14}$$

Parity

Parity of a function can be positive or negative, and some functions may not have any parity.

$$\begin{aligned}&\text{If } \psi(-x, -y, -z) = +\psi(x, y, z) \text{ then } \psi \text{ has positive or even parity.}\\&\text{If } \psi(-x, -y, -z) = -\psi(x, y, z) \text{ then } \psi \text{ has negative or odd parity.}\end{aligned} \tag{3.15}$$

Example of even parity is cos x. Example of odd parity is sin x.

For a function like e^x, parity cannot be defined. The parity due to orbital angular momentum is determined by the function $(-1)^l$.

Laporte rule

An integral vanishes between, symmetric limits if the integrand has odd parity. Considering that the operator of the electric dipole moment has odd parity, the expectation value of the electric dipole moment has odd parity, the expectation value of the electric dipole moment as well as the transition probability vanishes unless initial and final state have different parity.

Even a more restrictive selection rule is

$$\Delta l = \pm 1 \tag{3.16}$$

Table 3.2 Some selected eigen functions of hydrogen atom

State	N	L	m	u
1S	1	0	0	$A_n\, e^{-x}$
2S	2	0	0	$A_n\, e^{-x}(1-x)$
2P	2	1	0	$A_n e^{-x} x \cos\theta$
2P	2	1	± 1	$A_n \dfrac{e^{-x} x \sin\theta\, e^{\pm i\varphi}}{\sqrt{2}}$
3S	3	0	0	$A_n e^{-x}\left(1-2x+\frac{2x^2}{3}\right)$
3P	3	1	0	$A_n e^{-x}\sqrt{\frac{2}{3}}x(2-x)\cos\theta$
3P	3	1	± 1	$A_n e^{-x}\frac{1}{\sqrt{3}}x(2-x)\sin\theta e^{\pm i\Phi}$
3d	3	2	0	$A_n e^{-x}\frac{1}{2\sqrt{3}}x^2(3\cos^2\theta-1)$
3d	3	2	± 1	$A_n e^{-x}\frac{x^2}{\sqrt{3}}\sin\theta\cos\theta e^{\pm i\varphi}$
3d	3	2	± 2	$A_n e^{-x}\frac{1}{2\sqrt{3}}x^2\sin^2\theta e^{\pm i\varphi}$

where $x = r/n\,a_0$; $A_n = (1/\sqrt{\pi})(1/na_0)^{3/2}$; $a_0 = \hbar^2/me^2$ is the Bohr radius

Molecular spectra

Three types:

i. Electronic (Visible and ultraviolet)
ii. Vibrational (Near infrared)
iii. Rotational (Far infrared)

Because electron mass is much smaller than the nuclear mass, the three types of motion can be treated separately. This is the Born–Oppenheimer approximation, in which the complete ψ – function appears as the product of the wave functions of the three types of motion, and the total energy as the sum of the energies of electronic motion, of vibration, and of rotation.

$$\psi = \psi_{\text{el}} \cdot \psi_v \cdot \psi_{\text{rot}}$$

$$E = E_{\text{el}} + E_{\text{vibr}} + E_{\text{rot}} \tag{3.17}$$

$$E_{\text{el}} : E_{\text{vibr}} : E_{\text{rot}} = 1 : \sqrt{m/M} : m/M \tag{3.18}$$

where m and M are the mass of electron and nucleus.

Thus $E_{\text{el}} \gg E_{\text{vibr}} \gg E_{\text{rot}}$.

The rotational energy

$$E_R = \frac{\hbar^2}{2I_0} \cdot J(J+1) \tag{3.19}$$

Permanent dipole moment is necessary, molecules with center of symmetry such as C_2H_2 or O_2 have no dipole moment and do not exhibit rotational spectrum.

Hetero-nuclear diatomic molecules such as CO and linear polyatomic molecule such as HCN, do possess dipolemoment and so also rotational spectrum.

The selection rule is

$$\Delta J = 0, \pm 1. \tag{3.20}$$

The occurrence of a rotational Raman spectrum depends on the change in polarizibility. Such a change can occur in symmetric top molecules. The selection rule is

$$\Delta J = 0, \pm 2. \tag{3.21}$$

Required properties of the wave function

Ψ should be finite, single-valued and continuous. These requirements follow from the interpretation of $|\psi|^2$ as the probability density.

Ladder operators

Raising operator:

$$J_+|jm>=\sqrt{(j+m+1)(j-m)}|j,m+1>$$

Lowering operator:

$$J_-|jm>=\sqrt{(j-m+1)(j+m)}|j,m-1> \tag{3.23}$$

The Klein–Gordon equation for the relativistic free motion of a spinless particle of rest mass m.

$$\nabla^2\phi - \frac{1}{c^2}\frac{\partial^2\varphi}{\partial t^2} = \frac{m^2c^2\varphi}{\hbar^2} \tag{3.24}$$

If the quantum of nuclear field is assumed to have mass m, then its range

$$\mathrm{R} \sim \frac{\hbar}{mc} \tag{3.25}$$

The variation method

It consists of evaluating the integrals on the RHS of the inequality

$$E_0 \leqq \frac{\int \psi^* H \psi \, \mathrm{d}\tau}{\int |\psi|^2 \, \mathrm{d}\tau} \tag{3.26}$$

with a trial function ψ that depends on a number of parameters, and varying these parmeters until the expectation value of the energy is a minimum. This results in an upper limit for the ground state energy of the system, which will be close if the form of the trial function resembles that of the eigen function.

Table 3.3 Clebsch–Gordan coefficients (C.G.C)

$\frac{1}{2} \times \frac{1}{2}$

		$J = 1$	1	0	1
m_1	m_2	$M = +1$	0	0	-1
$+1/2$	$+1/2$	1			
$+1/2$	$-1/2$		$\sqrt{1/2}$	$\sqrt{1/2}$	
$-1/2$	$+1/2$		$\sqrt{1/2}$	$-\sqrt{1/2}$	
$-1/2$	$-1/2$				1

$1 \times \frac{1}{2}$

		$J = 3/2$	3/2	1/2	3/2	1/2	3/2
m_1	m_2	$M = +3/2$	$+1/2$	$+1/2$	$-1/2$	$-1/2$	$-3/2$
$+1$	$+1/2$	1					
$+1$	$-1/2$		$\sqrt{1/3}$	$\sqrt{2/3}$			
0	$+1/2$		$\sqrt{2/3}$	$-\sqrt{1/3}$			
0	$-1/2$				$\sqrt{2/3}$	$\sqrt{1/3}$	
-1	$+1/2$				$\sqrt{1/3}$	$-\sqrt{2/3}$	
-1	$-1/2$						1

1×1

		$J = 2$	2	1	2	1	0	2	1	2
m_1	m_2	$M = +2$	$+1$	$+1$	0	0	0	-1	-1	-2
$+1$	$+1$	1								
$+1$	0		$\sqrt{\frac{1}{2}}$	$\sqrt{\frac{1}{2}}$						
0	$+1$		$\sqrt{\frac{1}{2}}$	$-\sqrt{\frac{1}{2}}$						
$+1$	-1				$\sqrt{\frac{1}{6}}$	$\sqrt{\frac{1}{2}}$	$\sqrt{\frac{1}{3}}$			
0	0				$\sqrt{\frac{2}{3}}$	0	$-\sqrt{\frac{1}{3}}$			
-1	$+1$				$\sqrt{\frac{1}{6}}$	$-\sqrt{\frac{1}{2}}$	$\sqrt{\frac{1}{3}}$			
0	-1							$\sqrt{\frac{1}{2}}$	$\sqrt{\frac{1}{2}}$	
-1	0							$\sqrt{\frac{1}{2}}$	$-\sqrt{\frac{1}{2}}$	
-1	-1									1

The Born approximation

Here the entire potential energy of interaction between the colliding particles is regarded as a perturbation. The approximation works well when the kinetic energy of the colliding particles is large in comparision with the interaction energy. It therefore supplements the method of partial waves.

$$\sigma(\theta) = |f(\theta)|^2 \tag{3.27}$$

where

$$f(\theta) = -K^{-1} \int_0^{\infty} r \sin Kr \; V(r) \mathrm{d}r \tag{3.28}$$

and

$$K = 2k \sin \frac{\theta}{2}, \; k\hbar = p. \tag{3.29}$$

3.2 Problems

3.2.1 Wave Function

3.1 An electron is trapped in an infinitely deep potential well of width $L = 10^6$ fm. Calculate the wavelength of photon emitted from the transition $E_4 \rightarrow E_3$. (See Problem 3.18).

3.2 Given $\psi(x) = \left(\frac{\pi}{\alpha}\right)^{-\frac{1}{4}} \exp\left(-\frac{\alpha^2 x^2}{2}\right)$, calculate Var x

3.3 If $\psi(x) = \frac{N}{x^2+a^2}$, calculate the normalization constant N.

3.4 Find the flux of particles represented by the wave function $\psi(x) = A\, e^{ikx} + Be^{-ikx}$

3.5 For Klein – Gordon equation obtain expressions for probability density and current. Explain the significance of the result.

3.6 (a) Find the normalized wave functions for a particle of mass m and energy E trapped in a square well of width $2a$ and depth $V_0 > E$.
(b) Sketch the first two wave functions in all the three regions. In what respect do they differ from those for the infinite well depth.

3.7 The Thomas-Reich-Kuhn sum rule connects the complete set of eigen functions and energies of a particle of mass m. Show that

$$\left(\frac{2\mu}{\hbar^2}\right) \sum_k (E_k - E_s)|x_{sk}|^2 = 1$$

3.8 (a) State and explain Laporte rule for light emission.
(b) What are metastable states?

3.9 Show that the eigen values of a hermitian operator Q are real

3.10 The state of a free particle is described by the following wave function (Fig. 3.1)

$$\psi(x) = 0 \text{ for } x < -3a$$
$$= c \text{ for } -3a < x < a$$
$$= 0 \text{ for } x > a$$

(a) Determine c using the normalization condition
(b) Find the probability of finding the particle in the interval $[0, a]$

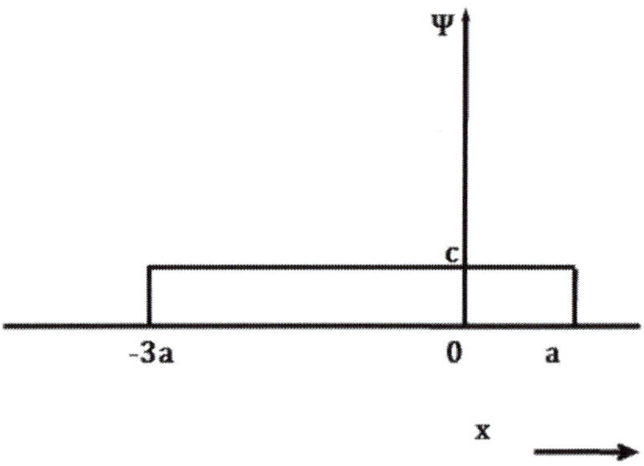

Fig. 3.1 Uniform distribution of ψ

3.11 In Problem 3.10,
(a) Compute $< x >$ and σ^2
(b) Calculate the momentum probability density.

3.12 Particle is described by the wavefunction

$$\psi = 0 \; x < 0$$
$$= \sqrt{2}e^{-x/L} x \geq 0$$

where $L = 1$ nm. Calculate the probability of finding the particle in the region $x \geq 1$ nm.

3.2.2 Schrodinger Equation

3.13 The radial Schrodinger equation, in atomic units, for an electron in a hydrogen atom for which the orbital angular momentum quantum number, $l = 0$, is

$$((\mathrm{d}^2/\mathrm{d}r^2) + (2/r) + (2E))F(r) = 0,$$

where E is the total energy.

(a) Put $F(r) = \exp(-r/\nu)\, y(r)$, where $E = -1/(2\nu^2)$, and show that
$$\frac{d^2y}{dx^2} = \frac{2}{\nu}\frac{d}{dr} - \frac{\nu}{r}y$$

(b) Assuming that $y(r)$ can be expanded as the series

$$y(r) = \sum_{p=0}^{\infty} a_p r^{p+1},$$

Show that the coefficients a_p in the series satisfy the recurrence relation,

$$p(p+1)a_p = \left(\frac{2}{\nu}\right)(p-\nu)a_{p-1}$$

(c) Solutions of the radial Schrodinger equation exist which are bounded for all r provided that $\nu = n$, where n is a positive integer. Show that the un-normalized radial function for the $n = 2$ state is

$$F(r) = a_0 e^{-r/2} r(1 - r/2)$$

3.14 State Ehrenfest's theorem. Show that

(a) $$\frac{d < x >}{dt} = \frac{< p_x >}{m}$$

(b) $$\frac{d < p_x >}{dt} = < -\partial V/\partial x >$$

3.15 Consider the time-independent Schrodinger equation in three dimensions

$$\left[\left(-\frac{\hbar^2}{2m}\right)\nabla^2 + V(r)\right]\psi = E\psi$$

In spherical coordinates

$$\nabla^2 = \left(\frac{1}{r^2}\right)\frac{\partial}{\partial r}\left(r^2\frac{\partial}{\partial r}\right) + \left(\frac{1}{r^2\sin\theta}\right)\frac{\partial}{\partial\theta}\left(\sin\theta\frac{\partial}{\partial\theta}\right) + \left(\frac{1}{r^2\sin^2\theta}\right)\frac{\partial^2}{\partial\varphi^2}$$

(a) Write $\psi(r,\theta,\varphi) = \psi_r(r)Y(\theta,\varphi)$ as a separable solution and split Schrodinger's equation into two independent differential equations, one depending on r and the other depending on θ and φ.
(b) Further separate the angular equation into θ and φ parts
(c) Combine the angular part and the potential part of the radial equation and write them as an effective potential V_e. Then make the substitution $\chi(r) = r\psi_r(r)$ and transform the radial equation into a form that resembles the one-dimensional Schrodinger equation.

3.16 Consider a three-dimensional spherically symmetrical system. In this case, Schrodinger's equation can be decomposed into a radial equation and an angular equation. The angular equation is given by

$$-\left[\left(\frac{1}{\sin\theta}\right)\left(\frac{\partial}{\partial\theta}\right)\left(\sin\theta\frac{\partial}{\partial\theta}\right) + \left(\frac{1}{\sin^2\theta}\right)\left(\frac{\partial^2}{\partial\varphi^2}\right)\right]Y(\theta,\varphi) = \lambda Y(\theta,\varphi)$$

Solve the equation and, in the process, derive the quantum number m.

3.17 In Problem 3.16,
(a) Consider the case where $m = 0$. Make the change of variable $\mu \cos\theta$ and consider a series solution to the equation for (μ). Derive a recurrence relation for the coefficients of the series solution.
(b) Explain why the series solution should be cut off at some finite term, give a mechanism for doing this and hence derive another quantum number l.

3.2.3 Potential Wells and Barriers

3.18 (a) The one-dimensional time-independent Schrodinger equation is

$$\left(-\frac{\hbar^2}{2m}\right)\frac{\mathrm{d}^2\psi(x)}{\mathrm{d}x^2} + U(x)\psi(x) = E\psi(x)$$

Give the meanings of the symbols in this equation.
(b) A particle of mass m is contained in a one-dimensional box of width a. The potential energy $U(x)$ is infinite at the walls of the box ($x = 0$ and $x = a$) and zero in between ($0 < x < a$).
Solve the Schrodinger equation for this particle and hence show that the normalized solutions have the form $\psi_n(x) = \left(\frac{2}{a}\right)^{\frac{1}{2}} \sin\left(\frac{n\pi x}{a}\right)$, with energy $E_n = h^2n^2/8ma^2$, where n is an integer ($n > 0$).
(c) For the case $n = 3$, find the probability that the particle will be located in the region $\frac{a}{3} < x < \frac{2a}{3}$.
(d) Sketch the wave-functions and the corresponding probability density distributions for the cases $n = 1, 2$ and 3.

3.19 Deuteron is a loose system of neutron and proton each of mass M. Assuming that the system can be described by a square well of depth V_0 and width R, show that to a good approximation

$$V_0 R^2 = \left(\frac{\pi}{2}\right)^2 \left(\frac{\hbar^2}{M}\right)$$

3.20 Show that the expectation value of the potential energy of deuteron described by a square well of depth V_0 and width R is given by

$$< V >= -V_0 A^2 \left[\frac{R}{2} - \frac{\sin 2kR}{4k}\right]$$

where A is a constant.

3.21 Assuming that the radial wave function

$U(r) = r\psi(r) = C\exp(-kr)$

is valid for the deuteron from $r = 0$ to $r = \infty$ find the normalization constant C.
Hence if $k = 0.232\,\mathrm{fm}^{-1}$ find the probability that the neutron – proton separation in the deuteron exceeds 2 fm. Find also the average distance of interaction for this wave function.

[Royal Holloway University of London 1999]

3.22 The small binding energy of the deuteron (2.2 MeV) implies that the maximum of $U(r)$ lies just inside the range R of the well. From this knowledge deduce the value of V_0 if R is approximately 1.5 fm.

[Osmania University]

3.23 Given that the normalized wave function

$$\psi = \left(\frac{1}{r}\right)\left(\frac{\alpha}{2\pi}\right)^{1/2} e^{-\alpha r}$$

$(1/\alpha = 4.3\,\text{fm})$ is a useful approximation to describe the ground state of the deuteron, find the root mean square separation of the neutron and proton in this nucleus.

[University of Durham 1972]

3.24 A particle of mass m_e trapped in an infinite depth well of width $L = 1\,\text{nm}$. Consider the transition from the excited state $n = 2$ to the ground state $n = 1$. Calculate the wavelength of light emitted. In which region of electromagnetic spectrum does it fall?

3.25 Consider a particle of mass m trapped in a potential well of finite depth V_0

$$V(x) = V_0, \; |x| > a$$
$$= 0; \; |x| < a$$

Discuss the solutions and eigen values for the class I and II solutions graphically.

3.26 Show that for deuteron, neutron and proton stay outside the range of nuclear forces for 70% of the time. Take the binding energy of deuteron as 2.2 MeV.

3.27 Show that the results of the energy levels for infinite well follow from those for the finite well.

3.28 Show that for deuteron excited states are not possible.

3.29 The small binding energy of the deuteron indicates that the maximum of $U(r)$ lies only just inside the range R of the square well potential. Use this information to estimate the value of V_0 if R is approximately 1.5 fm.

3.30 Consider a stream of particles with energy E travelling in one dimension from $x = -\infty$ to ∞. The particles have an average spacing of distance L. The particle stream encounters a potential barrier at $x = 0$. The potential can be written as

$$V(x) = 0 \text{ if } x < 0$$
$$= V \text{ if } 0 < x < a$$
$$= 0 \text{ if } x > a$$

Suppose the particle energy is smaller than the potential barrier, i.e., $< V_b$.

(a) For each of the three regions, write down Schrodinger's equation and calculate the wave-function ψ and its derivative $\mathrm{d}\psi/\mathrm{d}x$.

Use the constants to represent the amplitude of the reflected and transmitted particle streams respectively and take

$$k_1^2 = \frac{2mE}{\hbar^2} \text{ and } k_2^2 = \frac{2m(V_b - E)}{\hbar^2}$$

(b) At the boundaries to the potential barrier, ψ and $\mathrm{d}\psi/\mathrm{d}x$ must be continuous. Equate the solutions that you have at $x = 0$ and $x = a$ and manipulate these equations to derive the following expression for the transmission amplitude

$$\tau = \frac{4ik_1k_2e^{-ik_1a}}{[(ik_1 + k_2)^2e^{-k_2a}] - [(ik_1 - k_2)^2e^{k_2a}]}$$

3.31 In Problem 3.30,
(a) Show that the fraction of transmitted particles is given by $F_{\text{trans}} = \tau^*\tau$, which when calculated evaluates to

$$F_{\text{trans}} = \left[1 + \frac{V_b^2 \sinh^2(k_2a)}{4E(V_b - E)}\right]^{-1}$$

(b) How would F_{trans} vary if E $> V_b$.

3.32 A particle is trapped in a one dimensional potential given by $= kx^2/2$. At a time $t = 0$ the state of the particle is described by the wave function $\psi = C_1\psi_1 + C_2\psi_2$, where ψ is the eigen function belonging to the eigen value E_n. What is the expected value of the energy?

3.33 A particle is trapped in an infinitely deep square well of width a. Suddenly the walls are separated by infinite distance so that the particle becomes free. What is the probability that the particle has momentum between p and $p + \mathrm{d}p$?

3.34 The alpha decay is explained as a quantum mechanical tunneling. Assuming that the alpha particle energy is much smaller than the potential barrier the alpha particle has to penetrate, the transmission coefficient is given by

$$T \approx \exp\left\{\left(-\frac{2}{\hbar}\right)\int_a^b [2m(U(r) - E)]^{1/2}\mathrm{d}r\right\}$$

The integration limits a and b are determined as solutions to the equation $U(r) = E$, where $U(r)$ is the non-constant Coulomb's potential energy. Calculate the alpha transmission coefficient and the decay constant λ.

3.35 The one-dimensional square well shown in Fig. 3.2 rises to infinity at $x = 0$ and has range a and depth V_1. Derive the condition for a spinless particle of mass m to have (a) barely one bound state (b) two and only two bound states in the well. Sketch the wave function of these two states inside and outside the well and give their analytic expressions.

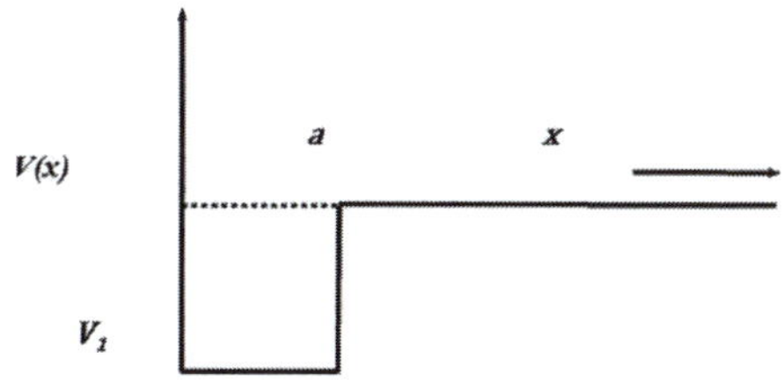

Fig. 3.2 Bound states in a square well potential

3.36 In Problem 3.25 express the normalization constant A in terms of α, β and a.

3.37 A particle of mass m is trapped in a square well of width L and infinitely deep. Its normalized wave function within the well for the nth state is

$$\psi_n = \left(\frac{2}{L}\right)^{1/2} \sin\left(\frac{n\pi x}{L}\right)$$

(a) Show that its mean position is $L/2$ and the variance is

$$\left(\frac{L^2}{12}\right)\left(1 - \frac{6}{n^2\pi^2}\right)$$

(b) Show that these expectations are in agreement with the classical values when $n \to \infty$.

3.38 The quantum mechanical Hamiltonian of a system has the form

$$H = (-\hbar^2/2m)\nabla^2 + ar^2\left(1 - \frac{5}{6}\sin^2\theta\cos^2\varphi\right)$$

Find the energy eigen value of the two lowest lying stationary states.

3.39 (a) Write down the three-dimensional time-independent Scrodinger equation in Cartesian co-ordinates. By separating the variables, $\psi\ (x, y, z) = X(x)Y(y)Z(z)$, solve this equation for a particle of mass m confined to a rectangular box of sides *a, b* and *c*, with zero potential inside.

(b) Show that the particle has energy given by

$$E = \left(\frac{\hbar^2}{8m}\right)\left[\frac{n_x^2}{a^2} + \frac{n_y^2}{b^2} + \frac{n_z^2}{c^2}\right]$$

3.40 In Problem 3.39, consider the special case of a cube $a = b = c$. Draw up a table listing the first six energy levels, stating the degeneracy for each level.

3.41 A particle of mass m is moving in a region where there is a potential step at $x = 0:\ V(x) = 0$ for $x < 0$ and $V(x) = U_0$ (a positive constant) for $x \geq 0$

(a) Determine $\psi(x)$ separately for the regions $x \ll 0$ and $x \gg 0$ for the cases:

(i) $U_0 < E$
(ii) $U_0 > E$.

(b) Write down and justify briefly the boundary conditions that $\psi(x)$ must satisfy at the boundary between the two adjacent regions. Use these conditions to sketch the form of $\psi(x)$ in the region around $x = 0$ for the cases (i) and (ii).

3.42 A steady stream of particles with energy $E(> V_0)$ is incident on a potential step of height V_0 as shown in Fig. 3.3.

The wave functions in the two regions are given by

$\psi_1(x) = A_0 \exp(ik_1 x) + A \exp(-ik_1 x)$

$\psi_2(x) = B \exp(ik_2 x)$

Write down expressions for the quantities k_1 and k_2 in terms of E and V_0. Show that

$$A = \left[\frac{k_1 - k_2}{k_1 + k_2}\right] A_0 \text{ and } B = \left[\frac{2k_1}{k_1 + k_2}\right] A_0$$

and determine the reflection and transmission coefficients in terms of k_1 and k_2.

If $E = 4\,V_0/3$ show that the reflection and transmission coefficients are 1/9 and 8/9 respectively.

Comment on why $A^2 + B^2$ is not equal to 1.

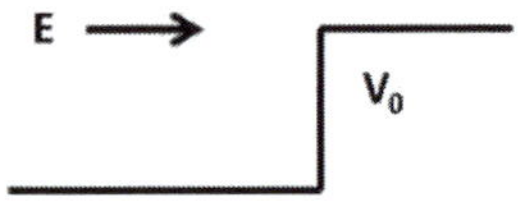

Fig. 3.3 Potential step

3.43 (a) What boundary conditions do wave-functions obey?

A particle confined to a one-dimensional potential well has a wave-function given by

$\psi(x) = 0$ for $x < -L/2$;

$\psi(x) = A \cos\left(\dfrac{3\pi x}{L}\right)$ for $-\dfrac{L}{2} \le x \le \dfrac{L}{2}$;

$\psi(x) = 0$ for $x > \dfrac{L}{2}$

(b) Sketch the wave-function $\psi(x)$.

(c) Calculate the normalization constant A.

(d) Calculate the probability of finding the particle in the interval $-\frac{L}{4} < x < \frac{L}{4}$.

(e) By calculating $d^2\psi/dx^2$ and writing the Schrodinger equation as

$$\left(-\frac{\hbar^2}{2m}\right)\left(\frac{d^2\psi}{dx^2}\right) = E\psi.$$

show that the energy E corresponding to this wave-function is $\frac{9\pi^2\hbar^2}{2mL^2}$.

3.44 (a) Sketch the one-dimensional "top hat" potential (1) $V = 0$ for $x < 0$; (2) $V = W =$ constant for $0 \le x \le L$; (3) $V = 0$ for $x > L$.

(b) Consider particles, of mass m and energy $E < W$ incident on this potential barrier from the left ($x < 0$). Including possible reflections from the barrier boundaries, write down general expressions for the wavefunctions in these regions and the form the time-independent Schrodinger equation takes in each region. What ratio of wavefunction amplitudes is needed to determine the transmission coefficient?

(c) Write down the boundary conditions for ψ and $d\psi/dx$ at $x = 0$ and $x = L$.

(d) A full algebraic solution for these boundary conditions is time consuming. In the approximation for a tall or wide barrier, the transmission coefficient T is given by

$$T = 16\left(\frac{E}{W}\right)\left(1 - \frac{E}{W}\right)e^{-2\alpha L}, \text{ where } \alpha^2 = 2m\left(\tfrac{W-E}{\hbar^2}\right)$$

Determine T for electrons of energy $E = 2\,\text{eV}$, striking a potential of value $W = 5\,\text{eV}$ and width $L = 0.3\,\text{nm}$.

(e) Describe four examples where quantum mechanical tunneling is observed.

3.45 A particle of mass m moves in a 2-D potential well, $V(x, y) = 0$ for $0 < x < a$ and $0 < y < a$, with walls at $x = 0,\ a$ and $y = 0,\ a$. Obtain the energy eigen functions and eigen values.

3.46 A particle of mass m is trapped in a 3-D infinite potential well with sides of length a each parallel to the x-, y-, z-axes. Obtain an expression for the number of states $N(N \gg 1)$ with energy, say less than E.

3.47 A particle of mass m is trapped in a hollow sphere of radius R with impenetrable walls. Obtain an expression for the force exerted on the walls of the sphere by the particle in the ground state.

3.48 Starting from Schrodinger's equation find the number of bound states for a particle of mass 2,200 electron mass in a square well potential of depth 70 MeV and radius 1.42×10^{-13} cm.

[University of Glasgow 1959]

3.49 A beam of particles of momentum $\hbar k_1$ are incident on a rectangular potential well of depth V_0 and width a. Show that the transmission amplitude is given by

$$\tau = \frac{4k_1k_2e^{-ik_1a}}{(k_2 + k_1)^2\, e^{-ik_2a} - (k_2 - k_1)^2\, e^{ik_2a}}$$

where $k_2 = \left[\dfrac{2m(E - V_0)}{\hbar^2}\right]^{1/2}$

Show that $\tau\tau^* = 1$ when $k_2a = n\pi$. Further, show graphically the variation of T, the transmission coefficient as a function of E/V_0, where E is the incident particle energy.

3.50 (a) What are virtual particles? What are space-like and Time-like four momentum vectors for real and virtual particles?

(b) Derive Klein – Gorden equation and deduce Yukawa's potential.

3.2.4 Simple Harmonic Oscillator

3.51 Show that the wavefunction $\psi_0(x) = A\exp(-x^2/2a^2)$ is a solution to the time- independent Schrodinger equation for a simple harmonic oscillator (SHO) potential.

$$\left(-\frac{\hbar^2}{2m}\right) \mathrm{d}^2\psi/\mathrm{d}x^2 + \left(\frac{1}{2}\right) m\omega_0 x^2\psi = E\psi$$

with energy $E_0 = \left(\frac{1}{2}\right)\hbar\omega_0$, and determine a in terms of m and ω_0.

The corresponding dimensionless form of this equation is

$$-\mathrm{d}^2\psi/\mathrm{d}R^2 + R^2\psi = \varepsilon\psi$$

where $R = x/a$ and $\varepsilon = E/E_0$.

Show that putting $\psi(R) = AH(R)\exp(-R^2/2)$ into this equation leads to Hermite's equation

$$\frac{\mathrm{d}^2 H}{\mathrm{d}R^2} - 2R\left(\frac{\mathrm{d}H}{\mathrm{d}R}\right) + (\varepsilon - 1)\,H = 0$$

$H(R)$ is a polynomial of order n of the form $a_n R^n + a_{n-2}R^{n-2} + a_{n-4}R^{n-4} + \ldots$ Deduce that ε is a simple function of n and that the energy levels are equally spaced.

[Adapted from the University of London, Royal Holloway and Bedford New College 2005]

3.52 Show that for a simple harmonic oscillator in the ground state the probability for finding the particle in the classical forbidden region is approximately 16%

3.53 Determine the energy of a three dimensional harmonic oscillator.

3.54 Show that the zero point energy of a simple harmonic oscillator could not be lower than $\hbar\omega/2$ without violating the uncertainty principle.

3.55 Show that when $n \to \infty$ the quantum mechanical simple harmonic oscillator gives the same probability distribution as the classical one.

3.56 Derive the probability distribution for a classical simple harmonic oscillator

3.57 The wave function (unnormalized) for a particle moving in a one dimensional potential well $V(x)$ is given by $\psi(x) = \exp(-ax^2/2)$. If the potential is to have minimum value at $x = 0$, determine (a) the eigen value (b) the potential $V(x)$.

3.58 Show that for simple harmonic oscillator $\Delta x.\Delta p_x = \hbar(n + 1/2)$, and that this is in agreement with the uncertainty principle.

3.59 In HCl gas, a number of absorption lines have been observed with the following wave numbers (in cm^{-1}): 83.03, 103.73, 124.30, 145.03, 165.51, and 185.86.

Are these vibrational or rotational transitions? (You may assume that transitions involve quantum numbers that change by only one unit). Explain your reasoning briefly. (a) If the transitions are vibrational, estimate the spring constant (in dyne/cm) (b) If the transitions are rotational, estimate the separation between H and Cl nuclei. What J values do they correspond to, and what is the moment of inertia of HCl (in g-cm^2)?

[Arizona State University 1996]

3.60 Determine the degeneracy of the energy levels of an isotropic harmonic oscillator.

3.61 At time $t = 0$, particle in a harmonic oscillator potential $V(x) = \frac{m\omega^2 x^2}{2}$ has a wavefunction

$$\psi(x, 0) = \left(\frac{1}{\sqrt{2}}\right)[\psi_0(x) + \psi_1(x)]$$

where $\psi_0(x)$ and $\psi_1(x)$ are real ortho-normal eigen functions for the ground and first-excited states of the oscillator. Show that the probability density $|\psi(x,t)|^2$ oscillates with angular frequency ω.

3.62 The quantum state of a harmonic oscillator has the eigen-function $\psi(x,t) =$

$$\left(\frac{1}{\sqrt{2}}\right)\psi_0(x)\exp\left(-\frac{iE_0t}{\hbar}\right) + \left(\frac{1}{\sqrt{3}}\right)\psi_1(x)\exp\left(-\frac{iE_1t}{\hbar}\right) + \left(\frac{1}{\sqrt{6}}\right)\psi_2(x)\exp\left(-\frac{iE_2t}{\hbar}\right)$$

where $\psi_0(x)$, $\psi_1(x)$ and $\psi_2(x)$ are real normalized eigen functions of the harmonic oscillator with energy E_0, E_1 and E_2 respectively. Find the expectation value of the energy.

3.63 (a) Show that the wave-function $\psi_0(x) = A\exp(-x^2/2a^2)$ with energy $E = \hbar\omega/2$ (where A and a are constants) is a solution for all values of x to the one-dimensional time-independent Schrodinger equation (TISE) for the simple harmonic oscillator (SHO) potential $V(x) = m\omega^2x^2/2$

(b) Sketch the function $\psi_1(x) = Bx\exp(-x^2/2a^2)$ (where $B =$ constant), and show that it too is a solution of the TISE for all values of x.

(c) Show that the corresponding energy $E = (3/2)\hbar\omega$

(d) Determine the expectation value $< p_x >$ of the momentum in state ψ_1

(e) Briefly discuss the relevance of the SHO in describing the behavior of diatomic molecules.

3.2.5 Hydrogen Atom

3.64 Find the expectation value of kinetic energy, potential energy, and total energy of hydrogen atom in the ground state. Take $\psi_0 = \frac{e^{-r/a_0}}{(\pi a_0^3)^{1/2}}$, where $a_0 =$ Bohr's radius

3.65 Show that (a) the electron density in the hydrogen atom is maximum at $r = a_0$, where a_0 is the Bohr radius (b) the mean radius is $3a_0/2$

3.66 Refer to the hydrogen wave functions given in Table 3.2. Show that the functions for $2p$ are normalized.

3.67 Show that the three $3d$ functions for H-atom are orthogonal to each other. (Refer to Table 3.2)

3.68 What is the degree of degeneracy for $n = 1, 2, 3$ and 4 in hydrogen atom?

3.69 What is the parity of the $1s$, $2p$ and $3d$ states of hydrogen atom.

3.70 Show that the $3d$ functions of hydrogen atom are spherically symmetric

3.71 In the ground state of hydrogen atom show that the probability (p) for the electron to lie within a sphere of radius R is
$P = 1 - \exp(-2R/a_0)\left(1 + 2R/a_0 + 2R^2/a_0^2\right)$

3.72 Locate the position of maximum and minimum electron density in the $2S$ orbit ($n = 2$ and $l = 0$) of hydrogen atom

3.73 When a negative muon is captured by an atom of phosphorous ($Z = 15$) in a high principal quantum number, it cascades down to lower state. When it reaches inside the electron cloud it forms a hydrogen-like mesic atom with the phosphorous nucleus.
(a) Calculate the wavelength of the photon for the transition $3d \to 2p$ state.
(b) Calculate the mean lifetimes of this mesic atoms in the $3d$-state, considering that the mean life of a hydrogen atom in the $3d$ state is 1.6×10^{-8} s. (mass of muon = 106 MeV)

3.74 The momentum distribution of a particle in three dimensions is given by $\psi(p) = [1/(2\pi\hbar)^{3/2}] \int e^{-p.r/\hbar}\psi(r)d\tau$. Take the ground state eigen function

$$\psi(r) = \left(\sqrt{\pi a_0^3}\right)^{-\frac{1}{2}} \exp(-r/a_0)$$

Show that for an electron in the ground state of the hydrogen atom the momentum probability distribution is given by

$$\psi|(p)|^2 = \frac{8}{\pi^2} \frac{\left(\frac{\hbar}{a_0}\right)^5}{\left[p^2 + \left(\frac{\hbar}{a_0}\right)^2\right]^4}$$

3.75 In Problem 3.74 (a) show that the most probable magnitude of the momentum of the electron is $\hbar/(\sqrt{3}\, a_0)$ and (b) its mean value is $8\hbar/3\pi\, a_0$, where a_0 is the Bohr radius.

3.76 Calculate the radius R inside which the probability for finding the electron in the ground state of hydrogen atom is 50%.

3.2.6 Angular Momentum

3.77 Given that $L = r \times p$, show that $[L_x, L_y] = i\hbar L_z$

3.78 The spin wave function of two electrons is $(x \uparrow x \downarrow - x \downarrow x \uparrow)/\sqrt{2}$. What is the eigen value of $S_1.S_2$? S_1 and S_2 are spin operators of 1 and 2 electrons

3.79 Show that for proton – neutron system

$$\begin{aligned}\sigma_p.\sigma_n &= -3 \text{ for singlet state}\\ &= 1 \text{ for triplet state}\end{aligned}$$

3.80 Write down an expression for the z-component of angular momentum, L_z, of a particle moving in the (x, y) plane in terms of its linear momentum components p_x and p_y.

Using the operator correspondence $p_x = -i\hbar\frac{\partial}{\partial x}$ etc., show that

$$L_z = -i\hbar\left(x\frac{\partial}{\partial y} - y\frac{\partial}{\partial x}\right)$$

Hence show that $L_z = -i\hbar\frac{\partial}{\partial \varphi}$, where the coordinates (x,y) and (r, φ) are related in the usual way.

Assuming that the wavefunction for this particle can be written in the form $\psi(r, \varphi) = R(r)\Phi(\varphi)$ show that the z-component of angular momentum is quantized with eigen value $\hbar$, where m is an integer.

3.81 Show that the operators L_x and L_y in the spherical polar coordinates are given by

$$\frac{L_x}{i\hbar} = \sin\varphi\frac{\partial}{\partial\theta} + \cot\theta\cos\varphi\frac{\partial}{\partial\varphi}$$

$$\frac{L_y}{i\hbar} = -\cos\varphi\frac{\partial}{\partial\theta} + \cot\theta\sin\varphi\frac{\partial}{\partial\varphi}$$

3.82 Using the commutator $[L_x, L_y] = i\,\hbar L_z$, and its cyclic variants, prove that total angular momentum squared and the individual components of angular momentum commute, i.e $[L^2, L_x] = 0$ etc.

3.83 Show that in the spherical polar coordinates

$$\frac{L^2}{(i\hbar)^2} = \frac{\partial^2}{\partial\theta^2} + \left(\frac{1}{\sin^2\theta}\right)\frac{\partial^2}{\partial\varphi^2} + \cot\theta\frac{\partial}{\partial\theta}$$

And show that in the expression for ∇^2 in spherical polar coordinates the angular terms are proportional to L^2.

3.84 (a) Obtain the angular momentum matrices for $j = 1/2$ particles
(b) Hence Obtain the matrix for J^2.

3.85 (a) Obtain the angular momentum matrices for $j = 1$
(b) Hence obtain the matrix for J^2.

3.86 Two angular momenta with $j_1 = 1$ and $j_2 = 1/2$ are vectorially added, obtain the Clebsch – Gordan coefficients.

3.87 The wave function of a particle in a spherically symmetric potential is

$$\Psi(x, y, z) = C\,(xy + yz + zx)e^{-\alpha r^2}$$

Show that the probability is zero for the angular momentum $l = 0$ and $l = 1$ and that it is unity for $l = 2$

3.88 Show that the states specified by the wave-functions

$$\psi_1 = (x + iy)f(r)$$
$$\psi_2 = zf(r)$$
$$\psi_3 = (x - iy)f(r)$$

are eigen states of the z-component of angular momentum and obtain the corresponding eigen values.

[Adapted from the University of Manchester 1959]

3.89 The Schrodinger wave function for a stationary state of an atom is

$$\psi = Af(r)\sin\theta\cos\theta e^{i\varphi}$$

where (r, θ, φ) are spherical polar coordinates. Find (a) the z component of the angular momentum of the atom (b) the square of the total angular momentum of the atom. (You may use the following transformations from Cartesian to spherical polar coordinates

$$x\frac{\partial}{\partial y} - y\frac{\partial}{\partial x} = \sin\varphi\frac{\partial}{\partial\theta} + \cot\theta\cos\varphi\frac{\partial}{\partial\varphi}$$

$$x\frac{\partial}{\partial z} - z\frac{\partial}{\partial x} = -\cos\varphi\frac{\partial}{\partial\theta} + \cot\theta\sin\varphi\frac{\partial}{\partial\varphi}$$

$$y\frac{\partial}{\partial x} - x\frac{\partial}{\partial y} = -\frac{\partial}{\partial\varphi}\Bigg)$$

[Adapted from the University of Durham 1963]

3.90 The normalized Schrodinger wavefunctions for one of the stationary states of the hydrogen atom is given in spherical polar coordinates, by

$$\psi(r, \theta, \varphi) = \left(\frac{1}{2a_0}\right)^{3/2}\frac{1}{\sqrt{3}}\frac{r}{a_0}\exp\left(-\frac{r}{2a_0}\right)\left(\frac{3}{8\pi}\right)^{1/2}\sin\theta\exp(-i\varphi)$$

(a) Find the value of the component of angular momentum along the z axis $(\theta = 0)$
(b) What is the parity of this wavefunction?
[a_0 is the radius of the first Bohr orbit]

[Adapted from the University of New Castle 1964]

3.91 The normalized $2p$ eigen functions of hydrogen atom are

$$\frac{1}{\sqrt{\pi}}\frac{1}{(2a_0)^{3/2}}e^{-r/2a_0}\frac{r}{2a_0}\sin\theta\, e^{i\Phi},\ \frac{1}{\sqrt{\pi}}\frac{1}{(2a_0)^{3/2}}e^{-r/2a_0}\frac{r}{2a_0}\cos\theta,$$
$$\frac{1}{\sqrt{\pi}}\frac{1}{(2a_0)^{3/2}}e^{-r/2a_0}\frac{r}{2a_0}\sin\theta e^{-i\Phi},\ \text{for } m = +1, 0, -1 \text{ respectively.}$$

Apply the raising operator $L_+ = L_x + iL_y$ and lowering operator to show that the states with $m = \pm 2$ do not exist.

3.92 How can nuclear spin be measured from the rotational spectra of diatomic molecules?

3.93 An electron is described by the following angular wave function

$$u(\theta, \varphi) = \frac{1}{4}\sqrt{\frac{15}{\pi}}\sin^2\theta\cos 2\varphi$$

Re-express u in terms of spherical harmonics given below. Hence give the probability that a measurement will yield the eigen value of L^2 equal to $6\hbar^2$ You may use the following:

$$Y_{20}(\theta, \varphi) = \sqrt{\frac{5}{16\pi}}\left(3\cos^2\theta - 1\right)$$

$$Y_{2\pm 1}(\theta, \varphi) = \sqrt{\frac{15}{8\pi}}\sin\theta\cos\theta\exp(\pm i\varphi)$$

$$Y_{2\pm 2}(\theta, \varphi) = \sqrt{\frac{15}{32\pi}}\sin^2\theta\exp(\pm 2i\varphi)$$

[University College, London]

3.94 Given that the complete wave function of a hydrogen-like atom in a particular state is $\psi(r, \theta, \varphi) = Nr^2\exp\left(-\frac{Zr}{3a_0}\right)\sin^2\theta\, e^{2i\varphi}$ determine the eigen value of L_z, the third component of the angular momentum operator.

3.95 Consider an electron in a state described by the wave function

$$\psi = \frac{1}{\sqrt{4\pi}}(\cos\theta + \sin\theta e^{i\varphi})f(r)$$

where $\int_0^\infty |f(r)|^2 r^2 \mathrm{d}r = 1$

(a) Show that the possible values of L_z are $+\hbar$ and zero
(b) Show that the probability for the occurrence of the L_z values in (a) is 2/3 and 1/3, respectively.

3.96 Show that (a) $[J_z, J_+] = J_+$ (b) $J_+|jm> = C_{jm} + \hbar|j, m+1>$ (c) $[J_x, J_y] = iJ_z$

3.2.7 Approximate Methods

3.97 Consider hydrogen atom with proton of finite size sphere with uniform charge distribution and radius R. The potential is

$$\begin{aligned} V(r) &= -\frac{3e^2}{2R^3}(R^2 - r^2/3) \text{ for } r < R \\ &= -e^2/r \qquad\qquad \text{for } r > R \end{aligned}$$

Calculate correction to first order for $n = 1$ and $n = 2$ with $l = 0$ states

[Adapted from University of Durham 1963]

3.98 A particle of mass m and charge q oscillating with frequency ω is subjected to a uniform electric field E parallel to the direction of oscillation. Determine the stationary energy levels.

3.99 Consider the Hermitian Hamiltonian $H = H_0 + H'$, where H' is a small perturbation. Assume that exact solutions $H_0|\psi >= E_0|\psi >$ are known, two of them, and that they are orthogonal and degenerate in energy. Work out to first order in H', the energies of the perturbed levels in terms of the matrix elements of H'.

3.100 The helium atom has nuclear charge $+2e$ surrounded by two electrons. The Hamiltonian is

$$H = \left(-\frac{\hbar^2}{2m}\right)(\nabla_1^2 + \nabla_2^2) - 2e^2\left(\frac{1}{r_1} + \frac{1}{r_2}\right) + \frac{e^2}{r_{12}}$$

where r_1 and r_2 are the position vectors of the two electrons with nucleus as the origin, and $r_{12} = |\mathrm{r}_1 - \mathrm{r}_2|$ is the distance between the two electrons. The expectation value for the first two terms are evaluated in a straight forward manner, the third term which is the interaction energy of the two electrons is evaluated by taking the trial function as the product of two hydrogenic wave functions for the ground state. The result is

$$< H >= \frac{e^2Z^2}{a_0} - \frac{4e^2Z}{a_0} + \frac{5e^2Z}{8a_0} = \left(\frac{e^2}{a_0}\right)\left(Z^2 - \frac{27Z}{8}\right)$$

Thus, the energy obtained by the trial function is

$$E(Z) = \left(-\frac{e^2}{2a_0}\right)\left(\frac{27Z}{4} - 2Z^2\right)$$

Determine the ionization energy of the helium atom.

3.101 Consider the first-order change in the energy levels of a hydrogen atom due to an external electric field of strength E directed along the z-axis. This phenomenon is known as Stark effect.

(a) Show that the ground state ($n = 1$) of hydrogen atom has no first-order effect.

(b) Show that two of the four degenerate levels for $n = 2$ are unaffected and the other two are split up by an energy difference of $3eEa_0$.

3.102 A particle of mass m is trapped in a potential well which has the form, $V = \frac{1}{2} m\omega^2 x^2$. Use the variation method with the normalized trial function $\left(1/\sqrt{a}\right)\cos(\pi x/2a)$ in the limits $-a < x < a$, to find the best value of a.

3.103 In Problem 3.45, consider the perturbation $W(x, y) = W_0$ for $0 < x < a/2$ and $0 < y < a/2$, and 0 elsewhere. Calculate the first order perturbation energy.

3.2.8 Scattering (Phase-Shift Analysis)

3.104 A beam of particles of energy $\hbar^2 k^2/2m$, moving in the $+z$ direction, is scattered by a short-range central potential $V(r)$. One looks for the stationary solution of the Schrodinger equation which is of the asymptotic form,

$\psi \approx e^{ikz} + f(\theta)e^{ikr}/r$

Derive the partial-wave decomposition

$$f(\theta) = (2ik)^{-1}\sum_{l=0}^{\infty}(2l+1)\,(\exp(2i\delta_l) - 1)P_l(\cos\theta)$$

[Adapted from the University College, Dublin, Ireland, 1967]

3.105 In the case of $\alpha - He$ scattering the measured scattered intensity at 45° (laboratory coordinates) is twice the classical result. Indicate how the wave-mechanical theory of collisions explains this experimental result.

[Adapted from the University of New Castle 1964]

3.106 In the analysis of scattering of particles of mass m and energy E from a fixed centre with range a, the phase shift for the lth partial wave is given by

$$\delta_l = \sin^{-1}\left[\frac{(iak)^l}{[(2l+1)!(l!)]^{1/2}}\right]$$

Show that the total cross-section at a given energy is approximately given by

$$\sigma = \left(\frac{2\pi\hbar^2}{mE}\right)\exp\left(-\frac{2mEa^2}{\hbar^2}\right)$$

[University of Cambridge, Tripos]

3.107 At what neutron lab energy will p-wave be important in n–p scattering?

3.108 1 MeV neutrons are scattered on a target. The angular distribution of the neutrons in the centre-of-mass is found to be isotropic and the total cross-section is measured to be 0.1 b. Using the partial wave representation, calculate the phase shifts of the partial waves involved.

3.109 Considering the scattering from a hard sphere of radius a such that only s- and p-waves are involved, the potential being

$$V(r) = \infty \text{ for } r < a$$
$$= 0 \text{ for } r > a.$$

show that $\sigma(\theta) = a^2\left[1 - \frac{(ka)^2}{3} + 2(ka)^2\cos\theta + \cdots\right]$
and $\sigma = 4\pi a^2[1 - (ka)^2/3]$

3.110 Find the elastic and total cross-sections for a black sphere of radius R.

3.111 Ramsauer (1921) observed that monatomic gases such as argon is almost completely transparent to electrons of 0.4 eV energy, although it strongly scatters electrons which are slower as well as those which are faster. How is this quantum mechanical peculiarity explained?

3.112 What conditions are necessary before the Schrodinger equation for the interaction of two nucleons can be reduced to the form

$$\frac{d^2U}{dr^2} + \left(\frac{2m}{\hbar^2}\right)[E - V(r)]U = 0$$

where $U(r) = r\psi(r)$ and the other symbols have their usual meanings?

By solving this equation for a square-well potential $V(r)$ for a neutron – proton collision show that the neutron – proton scattering cross-section, as calculated for high energies is about 3 barns compared with the experimental value of 20 barns. What is the explanation of this discrepeancy and how has this explanation been verified experimentally?

[Adapted from the University of Durham 1963]

3.2.9 Scattering (Born Approximation)

3.113 In the case of scattering from a spherically symmetric charge distribution, the form factor is given by

$$F(q^2) = \int_0^\infty \rho(r)\frac{\sin\left(\frac{qr}{\hbar}\right)}{qr/\hbar}4\pi\, r^2 dr$$

where $\rho(r)$ is the normalized charge distribution.

(a) If the charge distribution of proton is approximated by $\rho(r) = A\exp(-r/a)$, where A is a constant and a is some characteristic "radius" of the proton. Show that the form factor is proportional to $\left(1 + \frac{q^2}{q_0^2}\right)^{-2}$ where q_0 is $\hbar/a$.

(b) If $q_0^2 = 0.71\left(\frac{GeV}{c}\right)^2$, determine the characteristic radius of the proton.

3.114 The first Born approximation for the elastic scattering amplitude is

$$f = -\left(\frac{2\mu}{q\hbar^2}\right)\int V(r)e^{iq.r}d^3r$$

Show that for $V(r)$ spherically symmetric it reduces to

$$f = -\left(2\mu/q\hbar^2\right)\int r\sin(qr)\,V(r)dr$$

3.115 Given the scattering amplitude

$f(\theta) = (1/2ik) \sum (2l+1) \left[e^{2i\delta_l} - 1 \right] P_l(\cos\theta)$

Show that

$Im\ f(0) = k\sigma_t/4\pi$

3.116 Obtain the form factor $F(q)$ for electron scattering from an extended nucleus of radius R and charge Ze with constant charge density. Show that the minima occur when the condition

$\tan\ qR = qR$, is satisfied

3.117 In the Born's approximation the scattering amplitude is given by

$$f(\theta) = (-2\mu/q\hbar^2) \int_0^\infty V(r)\sin(qr)\, r\, \mathrm{d}r$$

where μ is the reduced mass of the target-projectile system, and $q\hbar$ is the momentum transfer. Show that the form factor is given by the expression

$$F(q) = (4\pi/q) \int_0^\infty \rho(r)\sin(qr) r \mathrm{d}r$$

where $\rho(r)$ is the charge density

3.118 Obtain the differential cross-section for scattering from the shielded Coulomb potential for a point charge nucleus of the form

$V = z_1 z_2 e^2 \exp(-r/r_0)/r$

where r_0 is the shielding radius of the order of atomic dimension. Thence deduce Rutherford's scattering law.

3.119 Electrons with momentum 300 MeV/c are elastically scattered through an angle of 12° by a nucleus of ^{64}Cu. If the charge distribution on the nucleus is assumed to be that of a hard sphere, by what factor would the Mott scattering be reduced?

3.120 An electron beam of momentum 200 MeV$/c$ is elastically scattered through an angle of 14° by a nucleus. It is observed that the differential cross-section is reduced by 60% compared to that expected from a point charge nucleus. Calculate the root mean square radius of the nucleus.

3.121 Assuming that the charge distribution in a nucleus is Gaussian, $\frac{e^{-(r^2/b^2)}}{\pi^{3/2}b^3}$ then show that the form factor is also Gaussian and that the mean square radius is $3b^2/2$

3.122 In the Born approximation the scattering amplitude is given by

$$f(\theta) = \left(-\frac{\mu}{2\pi\hbar^2}\right) \int V(r) e^{iq.r} \mathrm{d}^3 r$$

Show that for spherically symmetric potential it reduces to

$$f(\theta) = \left(-\frac{2\mu}{q\hbar^2}\right) \int r\sin(qr) V(r) \mathrm{d}r$$

3.123 Using the Born approximation, the amplitude of scattering by a spherically symmetric potential $V(r)$ with a momentum transfer q is given by

$$A = \int_0^\infty \left[\frac{\sin\left(\frac{qr}{\hbar}\right)}{\frac{qr}{\hbar}}\right] V(r) 4\pi r^2 \mathrm{d}r$$

Show that in the case of a Yukawa-type potential, this leads to an amplitude proportional to $(q^2 + m^2c^2)^{-1}$.

3.3 Solutions

3.3.1 Wave Function

3.1 $E_n = \dfrac{n^2h^2}{8mL^2} = \dfrac{\pi^2 n^2 \hbar^2 c^2}{2mc^2L^2}$

$$= \frac{\pi^2 \times (197.3\ \text{MeV} - \text{fm})^2 n^2}{2x0.511(\text{MeV}) \times (10^6\text{fm})^2} = 0.038\, n^2\ \text{eV}$$

$E_1 = 0.038$ eV, $E_2 = 0.152$ eV, $E_3 = 0.342$ eV, $E_4 = 0.608$ eV

$\Delta\, E_{43} = E_4 - E_3 = 0.608 - 0.342 = 0.266$ eV

$$\lambda = \frac{1,241}{0.266} = 4{,}665\ \text{nm}$$

3.2 $\psi(x) = (\pi/\alpha)^{-1/4} \exp\left(-\dfrac{\alpha^2}{2}x^2\right)$

Var $x = < x^2 > - < x >^2$

The expectation value

$$< x > = \int_{-\infty}^{\infty} \psi^* x\, \psi\, \mathrm{d}x = 0$$

because ψ and also ψ^* are even functions while x is an odd function. Therefore the integrand is an odd function

$$< x^2 > = \left(\frac{\pi}{\alpha}\right)^{-1/2} \int_{-\infty}^{\infty} x^2 \exp(-\alpha^2 x^2)\mathrm{d}x$$

Put $\alpha^2 x^2 = y$; $\mathrm{d}x = {}^1/_2\, \alpha\sqrt{y}$

$$< x^2 > = \left(\pi\alpha^5\right)^{-1/2} \int_0^\infty y^{1/2} e^{-y}\, \mathrm{d}y$$

But $\int_0^\infty y^{1/2} e^{-y} \mathrm{d}y = \Gamma(3/2) = \sqrt{\pi}/2$

Var $x = < x^2 > = (4\,\alpha^5)^{-1/2}$

3.3 Normalization condition is

$$\int_{-\infty}^{\infty} |\psi|^2 \mathrm{d}x = 1$$

$$N^2 \int_{-\infty}^{\infty} (x^2 + a^2)^{-2} \mathrm{d}x = 1$$

Put $x = a\ \tan\theta;\ \mathrm{d}x = \sec^2\theta\ \mathrm{d}\theta$

$$\left(\frac{2N^2}{\alpha^3}\right) \int_0^{\pi/2} \cos^2\ \theta\ \mathrm{d}\theta = N^2\pi/2a^3 = 1$$

Therefore $N = \left(\frac{2a^3}{\pi}\right)^{1/2}$

3.4 $\psi = Ae^{ikx} + Be^{-ikx}$

The flux $J_x = \left(\frac{\hbar}{2im}\right)\left[\psi^* \frac{\mathrm{d}\psi}{\mathrm{d}x} - \left(\frac{\mathrm{d}\psi^*}{\mathrm{d}x}\right)\psi\right]$

$$= \left(\frac{\hbar}{2im}\right)\left[\left(Ae^{-ikx} + Be^{ikx}\right) ik \left(Ae^{ikx} - Be^{-ikx}\right)\right.$$
$$\left. + ik\left(Ae^{-ikx} - Be^{ikx}\right)\left(Ae^{ikx} + Be^{-ikx}\right)\right]$$

$$= \left(\frac{\hbar k}{2m}\right)\left[A^2 - B^2 - ABe^{-2ikx} + ABe^{2ikx} + A^2 - B^2 + ABe^{-2ikx} - ABe^{2ikx}\right]$$

$$= \left(\frac{\hbar k}{m}\right)\left(A^2 - B^2\right)$$

3.5 In natural units ($\hbar = c = 1$) Klein – Gordon equation is

$$\nabla^2\varphi - \frac{\partial^2\varphi}{dt^2} - m^2\varphi = 0 \qquad (1)$$

The complex conjugate equation is

$$\nabla^2\varphi^* - \frac{\partial^2\varphi^*}{\partial t^2} - m^2\varphi^* = 0 \qquad (2)$$

Multiplying (1) from left by φ^* and (2) by φ and subtracting (1) from (2)

$$\varphi\nabla^2\varphi^* - \varphi^*\nabla^2\varphi - \varphi\frac{\partial^2\varphi^*}{\partial t^2} - \varphi\frac{\partial^2\varphi^*}{\partial t^2} + \varphi^*\frac{\partial^2\varphi}{\partial t^2} = 0$$

$$\nabla.\left(\varphi\nabla\varphi^*\nabla\varphi\right) - \frac{\partial}{\partial t}\left(\varphi\frac{\partial\varphi^*}{\partial t} - \varphi^*\frac{\partial\varphi}{\partial t}\right) = 0$$

Changing the sign through out and multiplying by $1/2im$

$$\frac{1}{2im}\nabla\cdot\left(\varphi^*\nabla\varphi - \varphi\nabla\varphi^*\right) - \frac{1}{2im}\frac{\partial}{\partial t}\left(\varphi * \frac{\partial\varphi}{\partial t} - \varphi\frac{\partial\varphi^*}{\partial t}\right) = 0$$

$$\nabla\cdot\left[\frac{1}{2im}\left(\varphi^*\nabla\varphi - \varphi\nabla\varphi^*\right)\right] + \frac{\partial}{\partial t}\left[\frac{i}{2m}\left(\varphi^*\frac{\partial\varphi}{\partial t} - \varphi\frac{\partial\varphi^*}{\partial t}\right)\right] = 0$$

$$Or\ \nabla\cdot J + \frac{\partial\rho}{\partial t} = 0$$

This is the continuity equation where the probability current $J = \frac{1}{2im}(\varphi^* \nabla\varphi - \varphi\nabla\varphi^*)$

And probability density

$$\rho = \frac{i}{2m}\left(\varphi^*\frac{\partial\varphi}{\partial t} - \varphi\frac{\partial\varphi^*}{\partial t}\right)$$

For a force free particle the solution of the Klein – Gordan equation is $\varphi = A\, e^{i(p.x - Et)}$

The probability density is

$$\begin{aligned}\rho &= \frac{i}{2m}\left[\left(A^* e^{-i(p.x-Et)}\right)(iAE)\, e^{-i(p.x-Et)} - \left(Ae^{-i(p.x-Et)}\right)\left(iA^*E\right) e^{-i(p.x-Et)}\right]\\ &= \frac{i}{2m}\left[A^*\, A(-iE) - A\, A^*\,(iE)\right]\\ &= \frac{|A|^2}{2m}[E + E] = E\frac{|A|^2}{m}\end{aligned}$$

As E can have positive and negative values, the probability density could then be negative

3.6 (a) Class I: Refer to Problem 3.25

$$\psi_1 = Ae^{\beta x}(-\infty < x < -a)$$
$$\psi_2 = D\,\cos\,\alpha x\;(-a < x < +a)$$
$$\psi_3 = Ae^{-\beta x}(a < x < \infty)$$

Normalization implies that

$$\int_{-\infty}^{-a}|\psi_1|^2\,\mathrm{dx} + \int_{-a}^{a}|\psi_2|^2\,\mathrm{dx} + \int_{a}^{\infty}|\psi_3|^2\,\mathrm{dx} = 1$$

$$\int_{-\infty}^{-a}A^2e^{2\beta x}\mathrm{dx} + \int_{-a}^{a}D^2\cos^2\alpha x\mathrm{dx} + \int_{a}^{\infty}A^2e^{-2\beta x}\mathrm{dx} = 1$$

$$A^2e^{-2\beta a}/2\beta + D^2\,[a + \sin(2\alpha a)/2\alpha] + A^2\,e^{-2\beta a}/2\beta = 1$$

Or

$$A^2\,e^{-2\beta a}/\beta + D^2(a + \sin(2\alpha a)/2\alpha) = 1 \tag{1}$$

Boundary condition at $x = a$ gives

$$D\,\cos\,\alpha\, a = a\, e^{-\beta a} \tag{2}$$

Combining (1) and (2) gives

$$D = \left(a + \frac{1}{\beta}\right)^{-1}$$

$$A = e^{\beta a}\,\cos\alpha a\,\left(a + \frac{1}{\beta}\right)^{-1}$$

(b) The difference between the wave functions in the infinite and finite potential wells is that in the former the wave function within the well terminates at the potential well, while in the latter it penetrates the well.

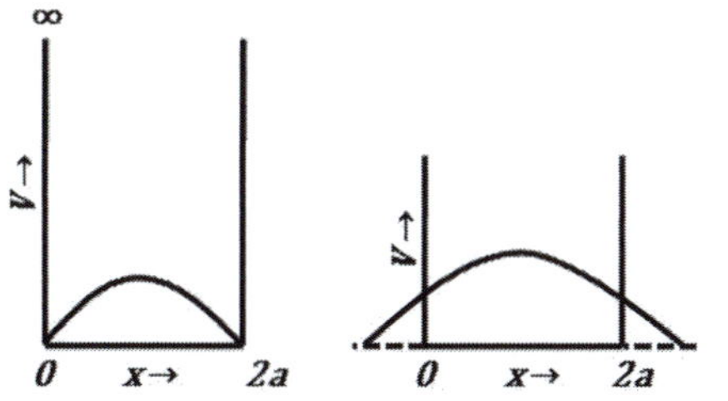

Fig. 3.4 Wave functions in potential wells of infinite and finite depths

3.7 Consider the expression

$$[x, [H, x]] = \frac{\hbar^2}{\mu} \quad (1)$$

The expectation value in the initial state s is

$$< s \left| \frac{\hbar^2}{\mu} \right| s >= \frac{\hbar^2}{\mu} < s\,[x, Hx - xH]\,|s > \quad (2)$$

Using the wave function ψ_s and expanding the commutator

$$\frac{\hbar^2}{\mu} =< s \left|2xHx - x^2H - Hx^2\right| s > \quad (3)$$

Further $< s|xHx|s >= \sum_k < s|x|k >< k|x|s > E_k = \sum |x_{ks}|^2 E_k \quad (4)$

where the summation is to be taken over all the excited states of the atom.

Also $< s|Hx^2|s >=< s|x^2H|s >$

$$= \sum_k < s|x|k >< k|x|s > \ E_s = \sum |x_{ks}|^2 E_s \quad (5)$$

Using (4) and (5) in (3) we get

$$\hbar^2/2\mu = \sum_k |x_{ks}|^2 (E_k - E_s)$$

3.8 (a) The wave functions of the hydrogen atom, or for that matter of any atom, with a central potential, are of the form

$u(r) = \lambda(r) P_l^m(\cos\theta) e^{im\varphi}$

Exchange of $r \to -r$ implies $r \to r, \theta \to \pi - \theta$ and $\varphi \to \pi + \varphi$ (parity operation)

where $P_l^m(\cos\theta)$ are the associated Legendre functions

Now, $P_l^m(\cos\theta) = (1-\cos^2\theta)^{m/2} d^m P_l(\cos\theta)/d\cos^m\theta$

$\therefore\ P_l^m(\cos(\pi-\theta)) = (1-\cos^2\theta)^{m/2} d^m P_l(-\cos\theta)/d(-\cos\theta)^m$

$= (1-\cos^2\theta)^{m/2}(-1)^l d^m P_l(\cos\theta)/d(-\cos\theta)^m$

$= (1-\cos^2\theta)^{m/2}(-1)^{l+m} d^m P_l(\cos\theta)/d\cos^m\theta = (-1)^{l+m} P_l{}^m(\cos\theta)$

Thus $P_l{}^m(\cos(\pi-\theta)) \to (-1)^{l+m} P_l{}^m(\cos\theta)$

Further $e^{im(\varphi+\pi)} = e^{im\varphi}.e^{im\pi} = e^{im\varphi}.(\cos m\pi + i\sin m\pi)$

$= (-1)^m e^{im\varphi}$

where m is an integer, positive or negative. So, under parity (p) operation, the function overall $F(r,\theta,\varphi)$, goes as

$PF(r,\theta,\varphi) = Pf(r)P_l{}^m(\cos\theta)e^{im\varphi} = f(r)P_l{}^m(\cos\theta)e^{im\varphi}(-1)^{l+m}(-1)^m$

$= (-1)^{l+2m}F(r,\theta,\varphi)$

$= (-1)^l F(r,\theta,\varphi)$

All the atomic functions with even values of l have even parity while those with odd values of l have odd parity. Considering that an integral vanishes between symmetrical limits if the integrand has odd parity, and that the operator of the electric dipole moment has odd parity, the following selection rule may be stated:- The expectation value of the electric dipole moment, as well as the transition probability vanishes unless initial and final state have different parity, that is $l_{\text{initial}} - l_{\text{final}} = \Delta l \neq 0, 2, 4\ldots$

This condition for the dipole radiation emission is known as Laportes's rule. Actually a more restrictive rule applies

$\Delta l = \pm 1$

Note that even if the matrix element of electric dipole moment vanishes, an atom will eventually go to the ground state by an alternative mechanism such as magnetic dipole or electric quadrupole etc for which the transition probability is much smaller than the dipole radiation.

(b) The $2s$ state of hydrogen can not decay to the $1s$ state via dipole radiation because that would imply $\Delta l = 0$. Furthermore, there are no other electric or magnetic moments to facilitate the transition. However, de-excitation may occur in collision processes with other atoms. Even in perfect vacuum transition may take place via two-photon emission, probability for which is again very small compared to one-photon emission. The result is that such a state is allowed to live for considerable time. Such states are known as metastable states.

3.9 $(\psi, Q\psi) = (\psi, q\psi) = q(\psi, \psi)$
$(Q\psi, \psi) = (q\psi, \psi) = q^*(\psi, \psi)$
since Q is hermitian,
$(\psi, Q\psi) = (Q\psi, \psi)$ and that $q = q^*$
That is, the eigen values are real. The converse of this theorem is also true, namely, an operator whose eigen values are real, is hermitian.

3.10 (a) The normalization condition requires

$$\int_{-\infty}^{\infty} |\psi|^2 \mathrm{d}x = \int_{-3a}^{a} |c|^2 \mathrm{d}x = 1 = 4a|c|^2$$

Therefore $c = 1/2\sqrt{a}$

(b) $\int_0^a |\psi|^2 \mathrm{d}x = \int_0^\alpha c^2 \, \mathrm{d}x = 1/4$

3.11 (a) The expectation values are

$$< x >= \int_{-\infty}^{\infty} \psi^* x \, \psi \, \mathrm{d}x = \int_{-3a}^{a} x \frac{\mathrm{d}x}{4a} = -a$$

$$< x^2 >= \int_{-\infty}^{\infty} \psi^* x^2 \, \psi \, \mathrm{d}x = \int_{-3a}^{a} (1/4a) \, x^2 \mathrm{d}x = \left(\frac{7}{3}\right) a^2$$

$$x\sigma^2 =< x^2 > - < x >^2 = \left(\frac{7}{3}\right) a^2 - (-a)^2 = \frac{4}{3} a^2$$

(b) Momentum probability density is $|\varphi(p)|^2$

$$\varphi(p) = (2\pi\hbar)^{-1/2} \int_{-\infty}^{\infty} \mathrm{d}x \, \psi\,(x) e^{-ipx/\hbar}$$

$$= (2\pi\hbar)^{-1/2} \int_{-3a}^{a} \mathrm{d}x c e^{-ipx/\hbar}$$

$$= \left(\frac{ic}{p}\right) \left(\frac{\hbar}{2\pi}\right)^{1/2} \left[e^{-\frac{ipa}{\hbar}} - e^{\frac{3ipa}{\hbar}} \right]$$

$$= \left(-\frac{ic}{p}\right) \left(\frac{\hbar}{2\pi}\right)^{1/2} e^{ipa/\hbar} \left[e^{\frac{2ipa}{\hbar}} - e^{\frac{-2ipa}{\hbar}} \right]$$

$$= \left(\frac{2c}{p}\right) \left(\frac{\hbar}{2\pi}\right)^{\frac{1}{2}} e^{\frac{ipa}{\hbar}} \sin\left(\frac{2pa}{\hbar}\right)$$

Therefore $|\varphi(p)|^2 = \frac{\hbar}{2\pi\, ap^2} \sin^2\left(\frac{2pa}{\hbar}\right)$

3.12 First the wave function is normalized

$$N^2 \int_0^\infty \psi^* \psi \, \mathrm{d}x = 1$$

$$N^2 \int_0^\infty \left(\sqrt{2}e^{-\frac{x}{L}}\right)^2 \mathrm{d}x = 1$$

$$N = 1/\sqrt{L}$$

The probability of finding the particle in the region $x \geq 1$ nm is

$$\left(\frac{1}{L}\right) \int_1^\infty \psi^* \psi \, \mathrm{d}x = \int_1^\infty \left(\left(\frac{1}{L}\right)^{\frac{1}{2}} e^{-\frac{x}{L}}\right)^2 \mathrm{d}x = \left(\frac{2}{L}\right) \int_1^\infty e^{-2x/L} \mathrm{d}x$$

$$= -e^{-2x/L}\Big|_1^\infty = e^{-2} = 0.135$$

3.3.2 Schrodinger Equation

3.13
$$\left(\frac{\mathrm{d}^2}{\mathrm{d}r^2} + \frac{2}{r} + 2E\right) F(r) = 0 \qquad (1)$$

(a) By using $F(r) = \exp(-r/v)\, y(r)$, and $E = -\frac{1}{2v^2}$, it is easily verified that

$$\frac{\mathrm{d}^2 y}{\mathrm{d}r^2} = \frac{2}{v}\left(\frac{\mathrm{d}}{\mathrm{d}r} - \frac{v}{r}\right) y \qquad (2)$$

(b) $y(r) = \sum_{p=0}^{\infty} a_p r^{p+1}$ (3)

$$\frac{\mathrm{d}y}{\mathrm{d}r} = \sum a_p (p+1) r^p \qquad (4)$$

$$\frac{\mathrm{d}^2 y}{\mathrm{d}r^2} = \sum a_p \, p(p+1) r^{p-1} \qquad (5)$$

Substitute (3), (4) and (5) in (2)

$$\Sigma a_p \, p(p+1) r^{p-1} = \frac{2}{v} \Sigma a_p (p+1) r^p - 2\Sigma a_p r^p$$

Replace p by $p - 1$ in the RHS and simplify

$$\Sigma a_p \, p(p+1) r^{p-1} = \frac{2}{v} \Sigma \, a_{p-1} (p - v) r^{p-1}$$

Comparing the coefficients of r^{p-1} on both sides

$$p(p+1) a_p = \frac{2}{v}(p - v) a_{p-1} \qquad (6)$$

(c) The series in (3) will terminate when $v = n$ where n is a positive integer.
Here $n = 2$
Using (3)

$$y(r) = \sum_0^1 ap \, r^{p+1} = a_0 r + a_1 r^2$$

From the recurrence relation (6)

$$a_1 = -\frac{a_0}{2}$$

Therefore, $y(r) = a_0 r\left(1 - \frac{r}{2}\right)$

$$F(r) = a_0 e^{-\frac{r}{2}} r\left(1 - \frac{r}{2}\right)$$

The normalization constant is given to be $1/\sqrt{2}$.

3.14 The statement that the variables in classical equations of motion can be replaced by quantum mechanical expectation values is known as Ehrenfest's theorem. For simplicity we shall prove the theorem in one-dimension although it can be adopted to three dimensions.

(a) $$\frac{\mathrm{d} <x>}{\mathrm{d}t} = \frac{\mathrm{d}}{\mathrm{d}t}\int \psi^* x\, \psi \mathrm{d}x = \int\left(\frac{\partial \psi^*}{\partial t} x\psi + \psi^* x\frac{\partial \psi}{\partial t}\right)\mathrm{d}x \quad (1)$$

Now ψ satisfies Schrodinger's one dimensional equation

$$i\hbar\frac{\partial \psi}{\partial t} = -\frac{\hbar^2}{2m}\frac{\partial^2 \psi}{\partial x^2} + V(x)\psi \quad (2)$$

$$-i\hbar\frac{\partial \psi^*}{\partial t} = -\frac{\hbar^2}{2m}\frac{\partial^2 \psi^*}{\partial x^2} + V(x)\psi^* \quad (3)$$

Premultiply (2) by $\psi^* x$ and post multiply (3) by $x\psi$ and subtract the resulting equations

$$i\hbar\left(\psi^* x\frac{\partial \psi}{\partial t} + \frac{\partial \psi^*}{\partial t}x\psi\right) = -\left(\frac{\hbar^2}{2m}\right)\left(\psi^* x\frac{\partial^2 \psi}{\partial x^2} - \frac{\partial^2 \psi^*}{\partial x^2}x\psi\right) \quad (4)$$

Using (4) in (1)

$$\frac{\mathrm{d} <x>}{\mathrm{d}t} = \frac{i\hbar}{2m}\int\left(\psi^* x\frac{\mathrm{d}^2 \psi}{\mathrm{d}x^2} - \frac{\partial^2 \psi^*}{\partial x^2}x\psi\right)\mathrm{d}x \quad (5)$$

The first integral can be evaluated by parts

$$\int \psi^* x\frac{\partial^2 \psi}{\partial x^2}\mathrm{d}x = \psi^* x\left.\frac{\mathrm{d}\psi}{\mathrm{d}x}\right|_0^\infty - \int(\mathrm{d}\psi/\mathrm{d}x)\left(\psi^* + x\frac{\mathrm{d}\psi^*}{\mathrm{d}x}\right)\mathrm{d}x \quad (6)$$

The first term on RHS is zero at both the limits.

$$\psi^* x\frac{\mathrm{d}^2 \psi}{\mathrm{d}x^2}\mathrm{d}x = -\int\frac{\mathrm{d}\psi}{\mathrm{d}x}\left(\psi^* + \frac{x\mathrm{d}\psi^*}{\mathrm{d}x}\right)\mathrm{d}x$$
$$= -\int \psi^*\frac{\mathrm{d}\psi}{\mathrm{d}x}\mathrm{d}x - \int x\left(\frac{\mathrm{d}}{\mathrm{d}x}\right)\left(\frac{\mathrm{d}}{\mathrm{d}x}\right)\mathrm{d}x \quad (7)$$

Furthermore

$$-\int\frac{\partial^2 \psi^*}{\partial x^2}x\psi\, \mathrm{d}x = -\psi x\left.\frac{\mathrm{d}\psi^*}{\mathrm{d}x}\right|_0^\infty - \frac{\mathrm{d}\psi^*}{\mathrm{d}x}\left(\psi + x\frac{\mathrm{d}\psi}{\mathrm{d}x}\right)\mathrm{d}x$$

The first term on the RHS is zero at both the limits.

$$-\int \frac{\partial^2\psi^*}{\partial x^2} x\psi \mathrm{d}x = -\int \frac{\mathrm{d}\psi^*}{\mathrm{d}x}\left(\psi + x\frac{\mathrm{d}\psi}{\mathrm{d}x}\right)\mathrm{d}x$$

$$= \psi\frac{\mathrm{d}\psi^*}{\mathrm{d}x}\mathrm{d}x + \int x\left(\frac{\mathrm{d}\psi^*}{\mathrm{d}x}\right)\left(\frac{\mathrm{d}\psi}{\mathrm{d}x}\right)\mathrm{d}x \tag{8}$$

Substituting (7) and (8) in (5), the terms underlined vanish together.

$$\frac{\mathrm{d} < x >}{\mathrm{d}t} = \left(\frac{i\hbar}{2m}\right)\left(-\int \psi^*\frac{\partial}{\partial x}\mathrm{d}x + \int \psi\frac{\mathrm{d}\psi^*}{\mathrm{d}x}\mathrm{d}x\right)$$

$$= \frac{1}{2m}\left(\int \psi^*\left(-i\hbar\frac{\partial}{\partial x}\right)\psi\,\mathrm{d}x + \int \psi\left(i\hbar\frac{\partial}{\partial x}\right)\psi^*\mathrm{d}x\right) \tag{9}$$

Now the operator for P_x is $-i\hbar\frac{\partial}{\partial x}$. The first term on RHS of (9) is the average value of the momentum P_x, the second term must represent the average value of P_x^*. But p_x being real, $P_x^* = P_x$. Therefore

$$\frac{\mathrm{d} < x >}{\mathrm{d}t} = \frac{1}{m} < P_x > \tag{10}$$

Thus (10) is similar to classical equation $x = p/m$
Equation (10) can be interpreted by saying that if the "position" and "momentum" vectors of a wave packet are regarded as the average or expectation values of these quantities, then the classical and quantum motions will agree.

(b) $$\frac{\mathrm{d} < P_x >}{\mathrm{d}t} = \frac{\mathrm{d}}{\mathrm{d}t}\int \psi^*\left(-i\hbar\frac{\partial}{\partial x}\right)\psi\mathrm{d}\tau = -i\hbar\frac{\mathrm{d}}{\mathrm{d}t}\int \psi^*\frac{\partial\psi}{\partial x}\mathrm{d}\tau$$

$$= -i\,\hbar\int \frac{\mathrm{d}\psi^*}{\mathrm{d}t}\frac{\partial\psi}{\partial x}\mathrm{d}\tau + \int \psi^*\frac{\partial}{\partial x}\frac{\partial\psi}{\partial t}\mathrm{d}\tau \tag{11}$$

Now $$i\hbar\frac{\partial\psi}{\partial t} = -\left(\frac{\hbar^2}{2m}\right)\nabla^2\psi + V\psi$$

$$-\frac{i\hbar\partial\psi^*}{\partial t} = -\left(\frac{\hbar^2}{2m}\right)\nabla^2\psi^* + V\psi^* \tag{12}$$

Using (12) in (11)

$$\frac{\mathrm{d}}{\mathrm{d}t} < P_x >= -\int \psi^*\frac{\partial}{\partial x}\left(-\left(\frac{\hbar^2}{2m}\right)\nabla^2\psi + V\psi\right)\mathrm{d}\tau$$

$$+\int\left(\left(-\frac{\hbar^2}{2m}\right)\nabla^2\psi + V\psi\right)\frac{\partial\psi}{\partial x}\mathrm{d}\,\tau$$

Integrating by parts twice

$$\frac{\mathrm{d}}{\mathrm{d}t} < P_x >= -\int \psi^* \left[\frac{\partial}{\partial x}(V\psi) - V\frac{\partial \psi}{\partial x}\right] \mathrm{d}\tau$$

$$= -\int \psi^* \frac{\partial V}{\partial x} \psi \mathrm{d}\tau =< \frac{-\partial V}{\partial x} >$$

These two examples support the correspondence principle as they show that the wave packet moves like a classical particle provided the expectation value gives a good representation of the classical variable.

3.15 (a) Using the Laplacian in the time-independent Schrodinger equation

$$-\frac{\hbar^2}{2m}\left[\frac{1}{r^2}\frac{\partial}{\partial r}\left(r^2\frac{\partial}{\partial r}\right) + \frac{1}{r^2 \sin\theta}\left(\frac{\partial}{\partial \theta}\left(\sin\theta\frac{\partial}{\partial \theta}\right) + \frac{1}{r^2\sin^2\theta}\frac{\partial^2}{\partial \varphi^2}\right)\right]$$

$$\psi(r,\theta,\varphi) + V(r)\psi(r,\theta,\varphi) = E\psi(r,\theta,\varphi) \tag{1}$$

We solve this equation by method of separation of variables

$$\text{Let } \psi\,(r,\theta,\varphi) = \psi_r\,(r)\,Y(\theta,\varphi) \tag{2}$$

Use (2) in (1) and multiply by $\left(-\frac{2m}{\hbar^2}.r^2\right)/\psi_r(r)Y(\theta,\varphi)$ and rearrange

$$\frac{1}{\psi_r}(r)\frac{\mathrm{d}}{\mathrm{d}r}\left(r^2 \mathrm{d}\psi_r(r)/\mathrm{d}r\right) + \frac{2mr^2}{\hbar^2}\left[E - V(r)\right]$$

$$= -\frac{1}{Y}(\theta,\varphi)\left[\frac{1}{\sin\theta}\frac{\partial}{\partial\theta}\left(\sin\theta\frac{\partial}{\partial\theta}\left(\sin\theta\partial Y\left(\theta,\varphi\right)/\partial\theta\right) + \right.\right.$$

$$\left.\left.\frac{1}{\sin^2\theta}\partial^2 Y\left(\theta,\varphi\right)/\partial\varphi^2\right)\right] \tag{3}$$

It is assumed that $V(r)$ depends only on r.

L.H.S. is a function of r only and R.H.S is a function of θ and φ only. Then each side must be equal to a constant, say λ.

$$\frac{1}{\sin\theta}\frac{\partial}{\partial\theta}\left(\sin\theta\frac{\partial Y}{\partial\theta}(\theta,\varphi)\right) + \frac{1}{\sin^2\theta}\frac{\partial^2 Y}{\partial\varphi^2}(\theta,\varphi) + \lambda Y\left(\theta,\varphi\right) = 0 \tag{4}$$

The radial equation is

$$\frac{\mathrm{d}}{\mathrm{d}r}r^2\frac{\mathrm{d}\psi_r(r)}{\mathrm{d}r} + \frac{2mr^2}{\hbar^2}\left[E - V(r) - \lambda\right]\psi_r(r) = 0 \tag{5}$$

(b) The angular equation (4) can be further separated by substituting

$$Y(\theta,\varphi) = f(\theta)g(\theta) \tag{6}$$

Following the same procedure

$$\frac{g(\varphi)}{\sin\theta}\frac{\mathrm{d}}{\mathrm{d}\theta}\sin\theta\frac{\mathrm{d}f(\theta)}{\mathrm{d}\theta}+\frac{f(\theta)}{\sin^2\theta}\frac{\mathrm{d}^2g(\varphi)}{\mathrm{d}\varphi^2}+\lambda f(\theta)g(\varphi)=0$$

$$\frac{\sin\theta}{f(\theta)}\frac{\mathrm{d}}{\mathrm{d}\theta}\left(\sin\theta\frac{\mathrm{d}f(\theta)}{\mathrm{d}\theta}\right)+\lambda\sin^2\theta=\frac{-1}{g(\varphi)}\frac{\mathrm{d}^2g(\varphi)}{\mathrm{d}\varphi^2}=m^2 \tag{7}$$

where m^2 is a positive constant

$$\frac{\mathrm{d}^2g}{\mathrm{d}\varphi^2}=-m^2\varphi \tag{8}$$

gives the normalized function

$$g=(1/\sqrt{2\pi})\,e^{im\varphi} \tag{9}$$

m is an integer since $g(\varphi+2\pi)=g(\varphi)$

Dividing (6) by $\sin^2\theta$ and multiplying by f, and rearranging

$$\frac{1}{\sin\theta}\frac{\mathrm{d}}{\mathrm{d}\theta}\left(\sin\theta\frac{df}{d\theta}\right)+\left(\lambda-\frac{m^2}{\sin^2\theta}\right)f=0 \tag{10}$$

(c) The physically accepted solution of (10) is Legendre polynomials when

$$\lambda=l(l+1) \tag{11}$$

and l is an integer.

With the change of variable $\psi_r(r)=\chi(r)/r$

The first term in (5) becomes

$$\frac{\mathrm{d}}{\mathrm{d}r}\left(r^2\frac{\psi_r}{dr}\right)=\frac{\mathrm{d}}{\mathrm{d}r}\left[r^2\left(-\frac{\chi}{r^2}+\frac{1}{r}\frac{\mathrm{d}\chi}{\mathrm{d}r}\right)\right]=\frac{\mathrm{d}}{\mathrm{d}r}\left[r\frac{\mathrm{d}\chi}{\mathrm{d}r}-\chi\right]$$

$$=r\frac{\mathrm{d}^2\chi}{\mathrm{d}r^2}+\frac{\mathrm{d}\chi}{\mathrm{d}r}-\frac{\mathrm{d}\chi}{\mathrm{d}r}=r\frac{\mathrm{d}^2\chi}{\mathrm{d}r^2}$$

With the substitution of λ from (11), (5) becomes upon rearrangement

$$\left(-\frac{\hbar^2}{2m}\right)\frac{\mathrm{d}^2\chi}{\mathrm{d}r^2}+\left[V(r)+\frac{l(l+1)\hbar^2}{2mr^2}\right]\chi=E\chi \tag{12}$$

Thus, the radial motion is similar to one dimensional motion of a particle in a potential

$$V_e=V(r)+\frac{l(l+1)\hbar^2}{2mr^2} \tag{13}$$

where V_e is the effective potential. The additional "potential energy" is interpreted to arise physically from the angular momentum. A classical particle that has angular momentum L about the axis through the origin perpendicular to the plane of its path has the angular velocity $\omega=L/mr^2$ where its radial distance from the origin is r. An inward force $m\omega^2r=mL^2/\omega r^3$ is required to keep the particle in the path. This "centripetal force" is supplied by the potential energy, and hence adds to the $V(r)$ which appears in (13) for the radial motion. This will have exactly the form indicated in (13) if we put $L = \sqrt{l(l+1)}\hbar$

3.16 $-\left[\frac{1}{\sin\theta}\frac{\partial}{\partial\theta}\left(\sin\theta\frac{\partial}{\partial\theta}\right)+\frac{1}{\sin^2\theta}\frac{\partial^2}{\partial\varphi^2}\right]Y(\theta,\varphi) = \lambda Y(\theta,\varphi)$

We solve the equation by the method of separation of variables.
Let $Y(\theta,\varphi) = f(\theta)g(\varphi)$ and multiply by $\sin^2\theta$

$$-g(\varphi)\sin\theta\frac{\partial}{\partial\theta}\left(\sin\theta\frac{\partial f}{\partial\theta}\right) + f(\theta)\frac{\partial^2 g}{\partial\varphi^2} = \lambda\sin^2\theta\ fg$$

Divide through out by $f(\theta)g(\varphi)$ and separate the θ and φ variables.

$$\frac{1}{f(\theta)}\sin\theta\frac{\partial}{\partial\theta}\left(\sin\theta\frac{\partial f}{\partial\theta} + \lambda\sin^2\theta\right) = -\frac{1}{g(\varphi)}\frac{\partial^2 g}{\partial\varphi^2} = m^2 \qquad (1)$$

LHS is a function of θ only and RHS function of φ only. The only way the above equation can be satisfied is to equate each side to a constant, say $-m^2$, where m^2 is positive.

$$\frac{1}{g(\varphi)}\frac{d^2 g(\varphi)}{d\varphi^2} = -m^2$$

Therefore $g(\varphi) = A\,e^{im\varphi}$
We can now normalize $g(\varphi)$ by requiring

$$\int g^*(\varphi)g(\varphi)d\varphi = 1$$

$$A^2\int_0^{2\pi}\left|e^{im\varphi}\right|^2 d\varphi = 2\pi A^2 = 1$$

Or $A = (2\pi)^{-1/2}$
We shall now show that m is an integer

$$g(\varphi + 2\pi) = g(\varphi)$$

$$g(\varphi + 2\pi) = (2\pi)^{-\frac{1}{2}}e^{im(\varphi+2\pi)} = (2\pi)^{-\frac{1}{2}}e^{im\varphi}.e^{2\pi mi}$$

$$= g(\varphi)e^{2\pi mi}$$

$$\therefore\ e^{2\pi mi} = \cos(2\pi m) + i\sin(2\pi m)$$

$$= \cos(2\pi m) = 1$$

Thus m is any integer, $m = 0, \pm1, \pm2\ \ldots$

3.17 (a) In Problem 3.16, going back to Eq. (1) and multiplying by $f(\theta)$ and dividing by $\sin^2\theta$ and putting $\mu = \cos\theta$.

$$\left(\frac{1}{\sin\theta}\right)\frac{\partial}{\partial\theta}\left(\sin\theta\frac{\mathrm{d}f}{\mathrm{d}\theta}\right)+\lambda\, f(\theta)=0 \tag{2}$$

$$\left(\frac{\mathrm{d}}{\mathrm{d}\theta}\right)=\left(\frac{\mathrm{d}}{\mathrm{d}\mu}\right)\cdot\left(\frac{\mathrm{d}\mu}{\mathrm{d}\theta}\right)=-\sin\theta\frac{\mathrm{d}}{\mathrm{d}\mu}$$

Writing $f(\theta)=P(\mu)$, Eq. (2) becomes

$$\frac{\mathrm{d}}{\mathrm{d}\mu}\left[\left(1-\mu^2\right)\frac{\mathrm{d}p}{\mathrm{d}\mu}\right]+\lambda P=0$$

$$\text{or }\left(1-\mu^2\right)\frac{\mathrm{d}^2p}{\mathrm{d}\mu^2}-2\mu\frac{\mathrm{d}p}{\mathrm{d}\mu}+\lambda p=0 \tag{3}$$

One can solve Eq. (3) by series method

$$\text{Let } P=\Sigma_{k=1}^{\infty}a_k\mu^k \tag{4}$$

$$\frac{\mathrm{d}p}{\mathrm{d}\mu}=\sum_k a_k k\mu^{k-1} \tag{5}$$

$$\frac{\mathrm{d}^2P}{\mathrm{d}\mu^2}=\sum a_k k(k-1)\mu^{k-2} \tag{6}$$

Using (4), (5) and (6) in (3)

$$\sum k(k-1)a_k\mu^{k-2}-\sum k(k-1)a_k\mu^k-2\sum ka_k\mu^k+\lambda\Sigma a_k\mu^k=0$$

Equating equal powers of k

$$(k+2)(k+1)a_{k+2}-[k(k-1)+2k-\lambda]\,a_k=0$$

$$\text{Or } a_{k+2}/a_k=[k\,(k+1)-\lambda]\,/\,(k+1)\,(k+2)$$

(b) If the infinite series is not terminated, it will diverge at $\mu=\pm 1$, i.e. at $\theta=0$ or $\theta=\pi$. Because this should not happen the series needs to be terminated which is possible only if $\lambda=k(k+1)$
i.e. $l(l+1)$; $l=0,1,2\ldots$ Here l is known as the orbital angular momentum quantum number. The resulting series $P(\mu)$ is then called Legendre polynomial.

3.3.3 Potential Wells and Barriers

3.18 (a) The term $-\frac{\hbar^2 d^2}{2mdx^2}$ is the kinetic energy operator, $U(x)$ is the potential energy operator, $\psi(x)$ is the eigen function and E is the eigen value.

(b) Put $U(x)=0$ in the region $0<x<a$ in the Schrodinger equation to obtain

$$\left(-\frac{\hbar^2}{2m}\right)\frac{\mathrm{d}^2\psi(x)}{\mathrm{d}x^2}=E\psi(x) \tag{1}$$

Or

$$\frac{d^2\psi(x)}{dx^2} + \left(\frac{2mE}{\hbar^2}\right)\psi(x) = 0 \tag{2}$$

Writing

$$\alpha^2 = \frac{2mE}{\hbar^2} \tag{3}$$

Equation (2) becomes

$$\frac{\mathrm{d}^2\psi}{\mathrm{d}x^2} + \alpha^2\psi = 0 \tag{4}$$

which has the solution

$$\psi(x) = A\sin\alpha x + B\cos\alpha x \tag{5}$$

where A and B are constants of integration. Take the origin at the left corner, Fig 3.5.

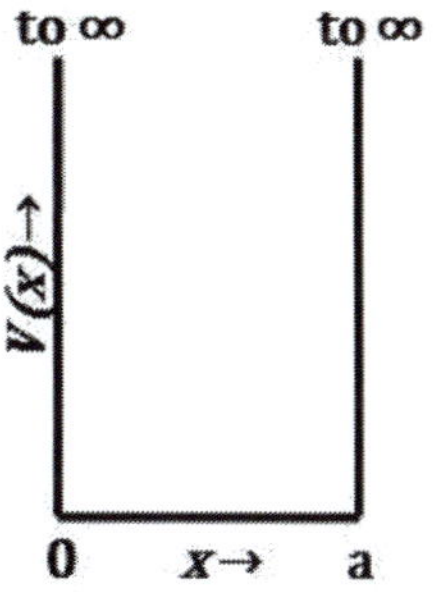

Fig. 3.5 Square potential well of infinite depth

Boundary condition:

$\psi(0) = 0;\ \psi(a) = 0$

The first one gives $B = 0$. We are left with

$$\psi = A\sin\alpha x \tag{6}$$

The second one gives

$$\alpha a = n\pi, n = 1, 2, 3\ldots \tag{7}$$

$n = 0$ is excluded as it would give a trivial solution.
Using the value of α in (6)

$$\psi_n(x) = A\sin\left(\frac{n\pi x}{a}\right) \tag{8}$$

This is an unnormalized solution. The constant A is determined from normalization condition.

$$\int_0^a \psi_n^*(x)\psi_n(x)\mathrm{d}x = 1$$

$$A^2 \int_0^\alpha \sin^2\left(\frac{n\pi x}{a}\right)\mathrm{d}x = 1$$

$$\left(\frac{A^2}{2}\right)\left(x - \cos\left(\frac{2n\pi x}{a}\right)\right)\Big|_0^a = A^2 a = 1$$

$$\text{Therefore, } A = \left(\frac{2}{a}\right)^{1/2} \tag{9}$$

The normalized wave function is

$$\psi_n(x) = \left(\frac{2}{a}\right)^{\frac{1}{2}} \sin\left(\frac{n\pi x}{a}\right) \tag{10}$$

Using the value of α from (7) in (3), the energy is

$$E_n = \frac{n^2 h^2}{8ma^2} \tag{11}$$

(c) probability $p = \int_0^a |\psi_3(x)|^2 \mathrm{d}x$

$$= \int_{\frac{a}{3}}^{\frac{2a}{3}} \left(\frac{2}{a}\right)\sin^2\left(\frac{3\pi x}{a}\right)\mathrm{d}x = \frac{1}{3}$$

(d) $\psi(n)$ and probability density $P(x)$ distributions for $n = 1, 2$ and 3 are sketched in Fig 3.6

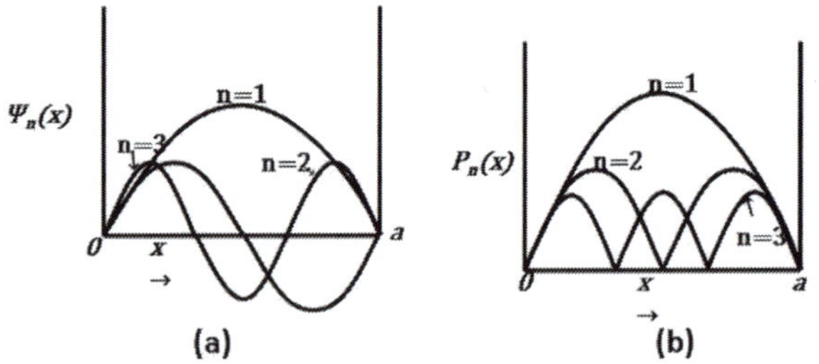

Fig. 3.6

3.19 The Schrodinger equation for the $n - p$ system in the CMS is

$$\nabla^2\psi(r,\theta,\varphi) + \left(\frac{2\mu}{\hbar^2}\right)[E - V(r)]\psi(r,\theta,\varphi) = 0 \tag{1}$$

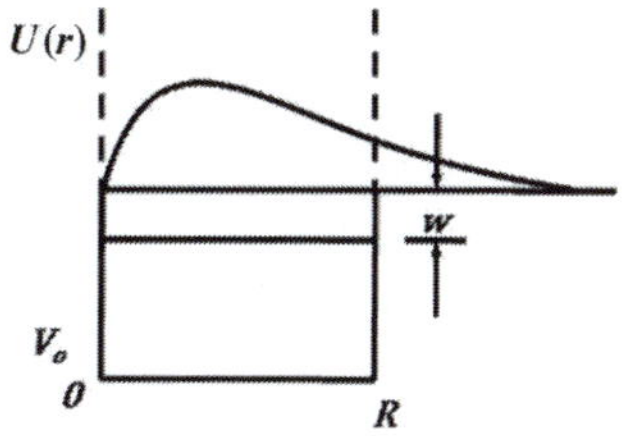

Fig. 3.7 Deuteron wave function and energy

where μ is the reduced mass $= M/2,\ M$, being neutron of proton mass. With the assumption of spherical symmetry, the angular derivatives in the Laplacian vanish and the radial equation is

$$\frac{1}{r^2}\frac{\mathrm{d}}{\mathrm{d}r}\left(r^2\frac{\mathrm{d}}{\mathrm{d}r}\right)\psi(r) + (M/\hbar^2)[E - V(r)]\psi(r) = 0 \tag{2}$$

With the change of variable

$$\psi(r) = \frac{u(r)}{r} \tag{3}$$

Equation (2) becomes

$$\frac{\mathrm{d}^2u}{\mathrm{d}r^2} + \left(\frac{M}{\hbar^2}\right)[E - V(r)]u = 0 \tag{4}$$

The total energy $= -W$, where $W =$ binding energy, is positive as the potential is positive

$V_0 = -V$, where V_0 is positive

Equation (4) then becomes

$$\frac{\mathrm{d}^2u}{\mathrm{d}r^2} + \left(\frac{M}{\hbar^2}\right)(V_0 - W)\,u = 0;\ r < R \tag{5}$$

$$\frac{\mathrm{d}^2u}{\mathrm{d}r^2} - \frac{MWu}{\hbar^2} = 0;\ r > R \tag{6}$$

where R is the range of nuclear forces, Fig. 3.7.

Calling

$$\frac{M(V_0 - W)}{\hbar^2} = k^2 \tag{7}$$

and

$$\frac{MW}{\hbar^2} = \gamma^2 \tag{8}$$

(5) and (6) become

$$\frac{\mathrm{d}^2u}{\mathrm{d}r^2} + k^2u = 0 \tag{9}$$

$$\frac{\mathrm{d}^2u}{\mathrm{d}r^2} - \gamma^2u = 0 \tag{10}$$

The solutions are

$$u_1(r) = A\sin kr + B\cos kr;\ r < R \tag{11}$$

$$u_2(r) = Ce^{-\gamma r} + De^{\gamma r};\ r > R \tag{12}$$

Boundary conditions: as $r \to 0,\ u_1 \to 0$
and as $r \to \infty,\ u_2$ must be finite. This means that $B = D = 0$.
Therefore the physically accepted solutions are

$$u_1 = A\sin kr \tag{13}$$

$$u_2 = Ce^{-\gamma r} \tag{14}$$

At the boundary, $r = R, u_1 = u_2$ and their first derivatives

$$\left(\frac{du_1}{dr}\right)_{r=R} = \left(\frac{du_2}{dr}\right)_{r=R}$$

These lead to

$$A\sin kR = Ce^{-\gamma r} \tag{15}$$

$$Ak\cos kR = -\gamma Ce^{-\gamma r} \tag{16}$$

Dividing the two equations

$$k\cot kR = -\gamma \tag{17}$$

Or

$$\cot kR = -\frac{\gamma}{k} \tag{18}$$

Now $V_0 \gg W$, so cot kR is a small negative quantity. Therefore $kR \approx \pi/2\ k^2R^2 = \left(\frac{\pi}{2}\right)^2$

$$Or\ \frac{M(V_0 - W)R^2}{\hbar^2} = \frac{\pi^2}{4}$$

Again neglecting W compared to V_0

$$V_0R^2 \approx \frac{\pi^2\hbar^2}{4M}$$

3.20 The inside wave function is of the form $u = A\ \sin\ kr$. Because $V(r) = 0$ for $r > R$, we need to consider contribution to $< V >$ from within the well alone.

$$< V > = \int_0^R u^*(-V_0)u\ dr = -V_0A^2\int_0^R \sin^2 kr dr$$

$$= \left(-\frac{V_0A^2}{2}\right)\int_0^R (1 - \cos 2kr)dr$$

$$= -V_0A^2\left[\frac{R}{2} - \frac{\sin 2kR}{4k}\right]$$

3.21 $u = C\,e^{-kr}$

$\int_0^\infty |u|^2 \mathrm{d}r = c^2 \int_0^\infty e^{-2kr} = \frac{c^2}{2k} = 1$

$C = \sqrt{2k}$

The probability that the neutron – proton separation in the deuteron exceeds R is

$$P = \int_R^\infty |u|^2 \mathrm{d}r = 2k \int_R^\infty e^{-2kr} \mathrm{d}r$$
$$= e^{-2kR} = e^{-(2\times 0.232\times 2)} \approx 0.4$$

Average distance of interaction

$$< r >= \int_0^\infty r|u^2|\mathrm{d}r = 2k \int_0^\infty r e^{-2kr} \mathrm{d}r$$
$$= \frac{1}{2k} = \frac{1}{2\times 0.232} = 2.16\ \mathrm{fm}$$

3.22 The inside wave function $u_1 = A\ \sin\ kr$ is maximum at $r \approx R$. Therefore $kR = \pi/2$

or $k^2R^2 = M(V_0 - W)R^2/\hbar^2 = \pi^2/4$

$$V_0 = \frac{\pi^2\hbar^2c^2}{4Mc^2R^2} + W$$

Substituting $\hbar c = 197.3\,\mathrm{MeV} - \mathrm{fm}$, $Mc^2 = 940\,\mathrm{MeV}$, $R = 1.5\,\mathrm{fm}$ and $W = 2.2\,\mathrm{MeV}$, we find $V_0 \approx 47\,\mathrm{MeV}$

3.23 $$< r^2 >= \int \psi^* r^2 \psi \mathrm{d}\tau$$
$$= \int_0^\infty \frac{r^2}{r^2}\left(\frac{\alpha}{2\pi}\right) e^{-2\alpha r} 4\pi r^2 \mathrm{d}r$$
$$= \frac{1}{2\alpha^2}$$
$$\sqrt{< r^2 >} = \frac{1}{\sqrt{2}\alpha} = \frac{4.3\times 10^{-15}\mathrm{m}}{\sqrt{2}} = 3.0\times 10^{-15}\mathrm{m} = 3.0\ \mathrm{fm}$$

3.24 Referring to Problem 3.18, the energy of the nth level is

$$E_n = \frac{n^2h^2}{8mL^2} \tag{1}$$

and

$$E_{n+1} = \frac{(n+1)^2h^2}{8ML^2} \tag{2}$$

Therefore $$E_{n+1} - E_n = \frac{(2n+1)h^2}{8mL^2} \tag{3}$$

The ground state corresponds to $n = 1$ and the first excited state to $n = 2$, $m = 8m_e$ and $L = 1$ nm $= 10^6$ fm. Putting $n = 1$ in (3)

$$h\nu = E_2 - E_1 = \frac{3h^2}{8mL^2} = 3\pi^2\hbar^2c^2/16m_ec^2L^2$$

$$= 3\pi^2(197.3)^2\ \text{MeV}^2 - \text{fm}^2/(16 \times 0.511\ \text{MeV})(10^6)^2\text{fm}^2$$

$$= 0.14 \times 10^{-6}\text{MeV} = 0.14\ \text{eV}$$

$$\lambda(\text{nm}) = \frac{1{,}241}{E(\text{eV})} = \frac{1{,}241}{0.14} = 8864\ \text{nm}$$

This corresponds to the microwave region of the electro-magnetic spectrum.

3.25 Consider a finite potential well. Take the origin at the centre of the well.

$$V(x) = V_0;\ |x| > a$$
$$= 0;\ |x| < a$$

$$\frac{d^2\psi}{dx^2} + \left(\frac{2m}{\hbar^2}\right)[E - V(r)]\,\psi = 0$$

Region 1 ($E < V_0$)

$$\frac{d^2\psi}{dx^2} - \left(\frac{2m}{\hbar^2}\right)(V_0 - E)\psi = 0 \quad (1)$$

$$\frac{d^2\psi}{dx^2} - \beta^2\psi = 0 \quad (2)$$

$$\text{where } \beta^2 = \left(\frac{2m}{\hbar^2}\right)(V_0 - E) \quad (3)$$

$$\psi_1 = Ae^{\beta x} + Be^{-\beta x} \quad (4)$$

where A and B are constants of integration.
Since x is negative in region 1, and ψ_1 has to remain finite we must set $B = 0$, otherwise the wave function grows exponentially. The physically accepted solution is

$$\psi_1 = Ae^{\beta x} \quad (5)$$

Region 2; ($V = 0$)

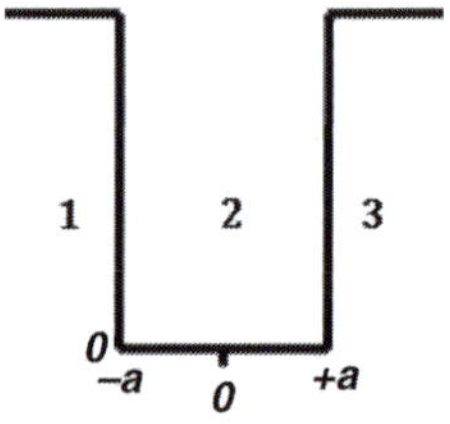

Fig. 3.8 Square potential well of finite depth

$$\frac{d^2\psi}{dx^2} + \left(\frac{2mE}{\hbar^2}\right)\psi = 0$$

$$\frac{d^2\psi}{dx^2} + \alpha^2\psi = 0 \tag{6}$$

$$\text{with } \alpha^2 = \frac{2mE}{\hbar^2} \tag{7}$$

$$\psi_2 = C \underset{\text{Odd}}{\sin \alpha x} + D \underset{\text{even}}{\cos \alpha x} \tag{8}$$

In this region either odd function must belong to a given value E or even function, but not both,

Region 3; ($E < V_0$)

Solution will be identical to (4)

$$\psi_3 = Ae^{\beta x} + Be^{-\beta x}$$

But physically accepted solution will be

$$\psi_3 = Be^{-\beta x} \tag{9}$$

because we must put $A = 0$ in this region where x takes positive values if the wave function has to remain finite.

Class I ($C = 0$)

$$\psi_2 = D \cos \alpha x \tag{10}$$

Boundary conditions

$$\psi_2(a) = \psi_3(a) \tag{11}$$

$$d\psi_2/dx|_{x=a} = d\psi_3/dx|_{x=a} \tag{11a}$$

These lead to

$$D \cos(\alpha a) = B\, e^{-\beta a} \tag{12}$$

$$-D\,\alpha \sin(\alpha a) = -B\,\beta\, e^{-\beta a} \tag{13}$$

Dividing (13) by (12)

$$\alpha \tan \alpha a = \beta \tag{14}$$

Class II ($D = 0$)

$$\psi_2 = C \sin(\alpha x) \tag{15}$$

Boundary conditions:

$$\psi_2(-a) = \psi_1(-a) \tag{16}$$

$$d\psi_2/dx|_{x=-a} = d\psi_1/dx|_{x=-a} \tag{17}$$

These lead to

$$C \sin(-\alpha a) = -C \sin(\alpha a) = Ae^{+\beta a} \tag{18}$$

$$C\alpha \cos(\alpha a) = A\beta e^{\beta a} \tag{19}$$

Dividing (19) by (18)

$$\alpha \cot(\alpha a) = -\beta \tag{20}$$

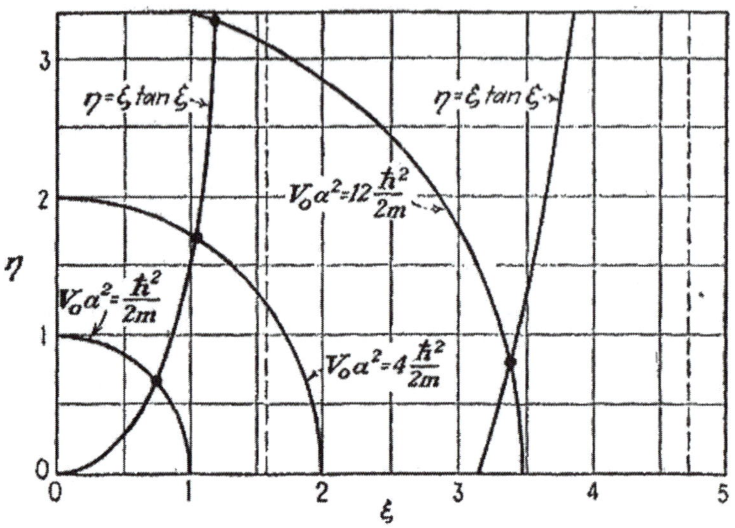

Fig. 3.9 $\eta - \xi$ curves for class I solutions. For explanation see the text

Note that from (15) and (2), $\alpha^2 = -\beta^2$, which is absurd because this implies that $\alpha^2 + \beta^2 = 0$, that is $2mV_0/\hbar^2 = 0$, but $V_0 \neq 0$. This simply means that class I and class II solutions cannot coexist

Energy levels:

Class I: set $\xi = \alpha a$; $\eta = \beta a$

where α and β are positive.

Equation (15) then becomes

$$\xi \tan \xi = \eta \tag{21}$$

$$\text{with } \xi^2 + \eta^2 = a^2(\alpha^2 + \beta^2) = 2mV_0a^2/\hbar^2 = \text{constant} \tag{22}$$

The energy levels are determined from the intersection of the curve $\xi \tan \xi$ plotted against η with the circle of known radius $\left(\frac{2mV_0a^2}{\hbar^2}\right)^{1/2}$, in the first quadrant since ξ and η are restricted to positive values.

The circles, Eq. (22), are drawn for $V_0a^2 = \hbar^2/2m,\ 4\hbar^2/2m$ and $9\hbar^2/2m$ for curves 1, 2 and 3 respectively Fig 3.9. For the first two values there is only one solution while for the third one there are two solutions.

For class II, energy levels are obtained from intersection of the same circles with the curves of $-\xi \cot \xi$ in the first quadrant, Fig 3.10.

Curve (1) gives no solution while the other two yield one solution each. Thus the three values of V_0a^2 in the increasing order give, one, two and three energy levels, respectively. Note that for a given particle mass the energy levels depend on the combination V_0a^2. With the increasing depth and/or width of the potential well, greater number of energy levels can be accommodated.

For $\xi = 0$ to $\pi/2$, that is V_0a^2 between 0 and $\pi^2\hbar^2/8m$ there is just one level of class I

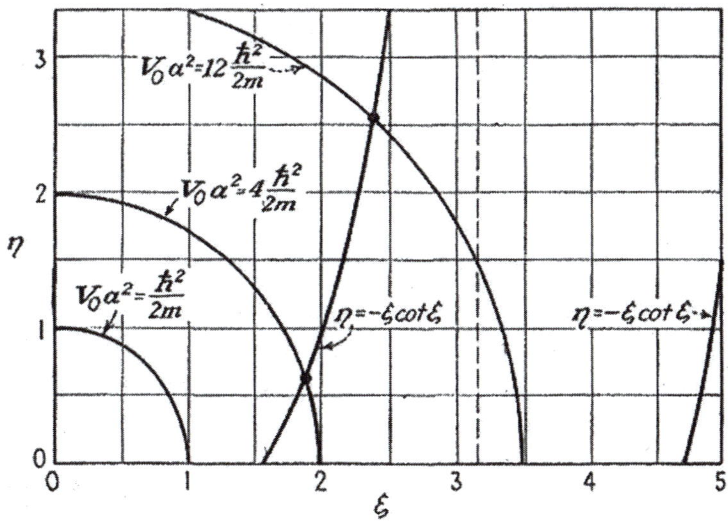

Fig. 3.10 $\eta - \xi$ curves for class II solutions. For explanation see the text (After Leonard I. Schiff, Quantum mechanics, McGraw-Hill 1955)

For V_0a^2 between $2\pi^2\hbar^2/8m$ and $4\pi^2\hbar^2/8m$ there is one energy level of each class or two altogether. As V_0a^2 increases, energy levels appear successively first of one class and next of the other.

3.26 The probability for neutron and proton to be found outside the range of nuclear forces (refer to Problem 3.21)

$$P = \int_R^\infty |\psi_2|^2 \mathrm{d}\tau = \int (|u_2(r)|/r)^2 4\pi r^2 \mathrm{d}r$$

$$= 4\pi C^2 \int_R^\infty e^{-2\gamma r} \mathrm{d}r$$

$$P = 2\pi C^2 e^{-2\gamma R/\gamma} \quad (1)$$

By Eq. (15) of solution 3.19

$$A \sin kR = Ce^{-\gamma R}$$

or

$$C^2 e^{-2\gamma R} = A^2 \sin^2 kR \approx A^2 \quad (2)$$

because, $kR \approx \pi/2$
Therefore,

$$p = \frac{2\pi A^2}{\gamma} \quad (3)$$

We can now find the constant A from the normalization condition

$$\int_0^R |\psi_1|^2 d\tau + \int_R^\infty |\psi_2|^2 d\tau = 1$$

$$\int_0^R u_1^2 .4\pi r^2 dr/r^2 + \int_R^\infty u_2^2 4\pi r^2 dr/r^2 = 1$$

$$A^2 \int_0^R \sin^2 kr dr + C^2 \int_R^\infty e^{-2\gamma r} dr = 1/4\pi$$

Integrating and using (2), we find

$$A^2 = \frac{\gamma}{2\pi(\gamma R + 1)} \tag{4}$$

Using (4) in (3)

$$P = \frac{1}{\gamma R + 1} \tag{5}$$

$$\text{Now } \gamma R = \left(\frac{MW}{\hbar^2}\right)^{1/2} R = \left(\frac{Mc^2 W}{\hbar^2 c^2}\right)^{1/2} R$$

$$= \left[\frac{940 \times 2.2}{(197.3)^2}\right]^{1/2} \times 2.1 = 0.48$$

where we have inserted $Mc^2 = 940\ MeV/c^2$,

$W = 2.2MeV$ and $R = 2.1\ fm$

Therefore $p = \frac{1}{0.48+1} = 0.67$

Thus neutron and proton stay outside the range of nuclear forces approximately 70% of time.

3.27 By Problem 3.25, for the finite well, for class I

$\alpha \tan \alpha a = \beta$

$$\text{with } \alpha = \frac{(2mE)^{\frac{1}{2}}}{\hbar}; \beta = \frac{[2m(V_0 - E)]^{\frac{1}{2}}}{\hbar}$$

As $V_0 \to \infty,\ \beta \to \infty$ and $\alpha a = n\pi/2$ (n odd)

Therefore, $\alpha^2 a^2 = \frac{2mEa^2}{\hbar^2} = \frac{n^2\pi^2}{4}$

Or $E = n^2\pi^2\hbar^2/8ma^2 \quad$ (n odd)

For class II

$\alpha \cot \alpha a = -\beta$

As $V_0 \to \infty,\ \beta \to \infty$ and $\alpha a = n\pi/2$ (n even)

$$\alpha^2 a^2 = \frac{2mEa^2}{\hbar^2} = \frac{n^2\pi^2}{4}$$

$$E = \frac{n^2\pi^2\hbar^2}{8ma^2} \ (n \text{ even})$$

$$\text{Thus } E = \frac{n^2\pi^2\hbar^2}{8ma^2}, n = 1, 2, 3 \ldots$$

3.28 From problem 3.27,

$$\alpha \cot \alpha a = -\beta = 0$$

The first solution is $\alpha a = \pi/2$, for the ground state.
The second solution, $\alpha a = 3\pi/2$, will correspond to the first excited state (with $l = 0$). This will give

$$\alpha^2 a^2 = 9\pi^2/4$$

Let the excited states be barely bound so that $W = 0$. Then,

$$\alpha^2 = 2mE/\hbar^2 = 9\pi^2/4a^2$$

$$E = V_1 = 9V_0$$

a value which is not possible. Thus, the physical reason why bound excited states are not possible is that deuteron is a loose structure as the binding energy (2.225 MeV) is small. The same conclusion is reached for higher excited states including $l = 1, 2 \ldots$

3.29 The inside wave function $u_1 = A \sin kr$ is maximum at $r \approx R$. Therefore

$$kR = \frac{\pi}{2}$$

$$\text{or } k^2 = \frac{M(V_0 - W)R^2}{\hbar^2} = \frac{\pi^2}{4}$$

$$\text{or } V_0 = \left(\frac{\pi^2\hbar^2c^2}{4Mc^2R^2}\right) + W$$

Substituting $\hbar c = 197.3\,\text{MeV} - \text{fm}$, $Mc^2 = 940\,\text{MeV}$, $R = 1.5\,\text{fm}$ and $W =$ 2.2 Mev, we find $V_0 \approx 47\,\text{MeV}$

3.30 Schrodinger's equation in one dimension

(a) $\frac{d^2\psi}{dx^2} + \left(\frac{2m}{\hbar^2}\right)(E - V)\psi = 0$
Region 1: $(x < 0) V = 0$; $\frac{d^2\psi}{dx^2} + k_1^2\psi = 0$
where $k_1^2 = \frac{2mE}{\hbar^2}$ Solution: $\psi_1 = \exp(ik_1x) + A \exp(-ik_1x)$
Region 2: $(0 < x < a) V = V_b$; $\frac{d^2\psi}{dx^2} - k_2^2\psi = 0$
where $k_2^2 = \left(\frac{2m}{\hbar^2}\right)(V_b - E)$
Solution: $\psi_2 = B \exp(k_2x) + C \exp(-k_2x)$

Region 3: $(x > a)\ V = 0$
Solution: $\psi_3 = D\exp(ik_1x)$

(b) Boundary conditions:

$$\psi_1(0) = \psi_2(0) \rightarrow 1 + A = B + C \quad (1)$$

$$\left.\frac{d\psi_1}{dx}\right|_{x=0} = \left.\frac{d\psi_2}{dx}\right|_{x=0} \rightarrow ik_1(1 - A) = k_2(B - C) \quad (2)$$

$$\psi_2(a) = \psi_3(a) \rightarrow B\exp(k_2a) + C\exp(-k_2a) = D\exp(ik_1a) \quad (3)$$

$$\left.\frac{d\psi_2}{dx}\right|_{x=a} = \left.\frac{d\psi_3}{dx}\right|_{x=a} \rightarrow k_2\,(B\exp(k_2a) - Ck_2\exp(-k_2a)) \quad (4)$$

$$= ik_1D\exp(ik_1a)$$

Eliminate A between (1) and (2) to get

$$B(k_2 + ik_1) - C(k_2 - ik_1) = 2ik_1 \quad (5)$$

Eliminate D between (3) and (4) to get

$$k_2(B\exp(k_2a) - Ck_2\exp(-k_2a)) = ik_1(B\exp(k_2a) + C\exp(-k_2a)) \quad (6)$$

Solve (5) and (6) to get

$$B = \frac{2ik_1(k_2 + ik_1)}{\left[(k_2 + ik_1)^2 - \exp(2k_2a)(k_2 - ik_1)^2\right]} \quad (7)$$

$$C = \frac{2ik_1(k_2 - ik_1)e^{2k_2a}}{\left[(k_2 + ik_1)^2 - e^{2k_2a}\,(k_2 - ik_1)^2\right]} \quad (8)$$

Using the values of B and C in (3),

$$\tau = D = \frac{4ik_1k_2\exp(-ik_1a)}{(k_2 + ik_1)^2\exp(-k_2a) - (ik_1 - k_2)^2\exp(k_2a)} \quad (9)$$

3.31 (a) $$F_{\text{trans}} = \tau^*\tau = |D|^2 = 16\frac{k_1^2k_2^2}{(k_1^2 + k_2^2)^2(e^{2k_2a} + e^{-2k_2a}) - 2(k_2^4 - 6k_2^2k_1^2 + k_1^4)}$$

This expression simplifies to

$$F_{\text{trans}} = T = \frac{4k_1^2\,k_2^2}{(k_1^2 + k_2^2)^2\sinh^2(k_2a) + 4k_1^2k_2^2} \quad (10)$$

use $k_1^2 = 2mE/\hbar^2$ and $k_2^2 = 2m(V_b - E)/\hbar^2$

The reflection coefficient R is obtained by substituting (7) and (8) in (1) to find the value of A. After similar algebraic manipulations we find

$$R = |A|^2 = \frac{(k_1^2 + k_2^2)^2\sinh^2(k_2a)}{(k_1^2 + k_2^2)^2\sinh^2(k_2a) + 4k_1^2k_2^2} \quad (11)$$

Note that $R + T = 1$

(b) When $E > V_b$, k_2 becomes imaginary and

$$\sinh(k_2a) = i\ \sin(k_2a) \quad (12)$$

Using (11) in (9) and noting $k_1^2 + k_2^2 = \frac{2mV_b}{\hbar^2}$

$$k_2^2 = \frac{2mV_b}{\hbar^2}$$

$$\text{and } k_1^2 k_2^2 = \left(\frac{2m}{\hbar^2}\right)^2 E(V_b - E)$$

we find

$$T = \frac{1}{1 + \frac{V_b^2 \sin^2 k_2 a}{4E(E - V_b)}} \tag{13}$$

and

$$R = \frac{1}{1 + \frac{4E(E - V_0)}{V_0^2 \sin^2 k_2 a}} \tag{14}$$

A typical graph for T versus $\frac{E}{V_b}$ is shown in Fig. 3.12

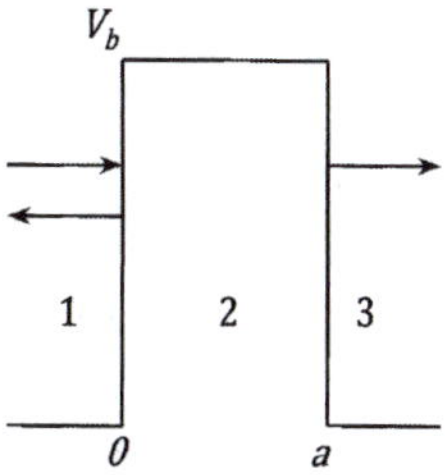

Fig. 3.11 Transmission through a rectangular potential barrier

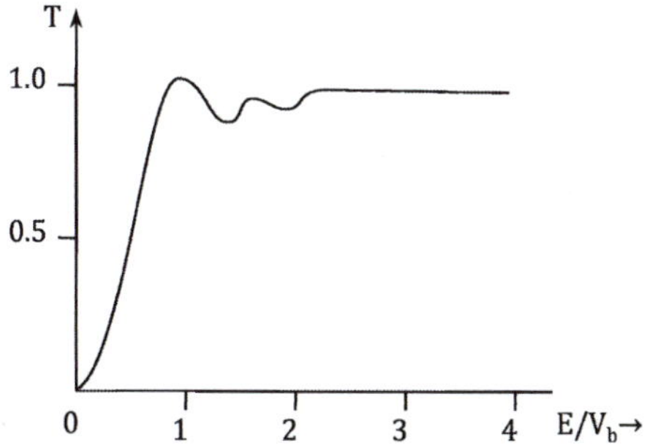

Fig. 3.12 Transmission as a function of E/V_b

3.32 The form of potential corresponds to that of a linear Simple harmonic Oscillator. The energy of the oscillator will be $E_1 = \frac{\hbar\omega}{2}$ and $E_2 = \frac{3\hbar\omega}{2}$.

$$
\begin{aligned}
&< E >=< \psi|H|\psi > \\
&=< (C_1\psi_1 + C_2\psi_2)|H|(C_1\psi_1 + C_2\psi_2) > \\
&=< C_1\psi_1 + C_2\psi_2|C_1 H\psi_1 + C_2 H\psi_2| > \\
&=< C_1\psi_1 + C_2\psi_2|C_1 E_1\psi_1 + C_2 E_2\psi_2| > \\
&= C_1^2 E_1 + C_2^2 E_2 \\
&= \frac{C_1^2\hbar\omega}{2} + \frac{C_2^2 3\hbar\omega}{2} \\
&= \frac{1}{2}\hbar\omega(C_1^2 + 3C_2^2)
\end{aligned}
$$

where $\omega = \left(\frac{k}{m}\right)^{1/2}$

3.33 The ground state is

$$\psi = \left(\frac{2}{a}\right)^{1/2} \sin(\pi x/a)$$

The wave function corresponding to momentum p is

$$\psi_i = (2\pi)^{-1/2} \sum_k C_k e^{ikx}$$

The probability that the particle has momentum between p and $p + dp$ is given by the value of
$|C_k|^2$, where C_k is the overlap integral

$$C_k = (2\pi)^{-\frac{1}{2}} \left(\frac{2}{a}\right)^{1/2} \int_0^a e^{ikx} \sin\left(\frac{\pi x}{a}\right) \mathrm{d}x$$

Itegrating by parts twice,

$$C_k = (\pi a)^{\frac{1}{2}} \left(e^{ika} + 1\right) \left(\pi^2 - k^2 a^2\right)^{-1}$$

The required probability is

$$
\begin{aligned}
|C_k|^2 &= \pi a \left(e^{ika} + 1\right) \left(e^{-ika} + 1\right) \left(\pi^2 - k^2 a^2\right)^{-2} \\
&= 4\pi \mathrm{a} \cos^2\left(\frac{ka}{2}\right) \left(\pi^2 - k^2 a^2\right)^{-2}
\end{aligned}
$$

3.34 The transmission coefficient is given by

$$T = e^{-G} \tag{1}$$

$$G = \frac{2}{\hbar} \int_a^b [2m(U(r) - E)^{1/2} \mathrm{d}r \tag{2}$$

Put

$$U(r) = \frac{zZe^2}{r} \tag{3}$$

for the Coulomb potential energy between the alpha particle and the residual nucleus at distance of separation r.

$$G = \frac{2}{\hbar}(2m)^{\frac{1}{2}} \int_a^b \left(\frac{zZe^2}{r} - E\right)^{1/2} dr$$

where $z = 2$

Now at distance b where the alpha energy with kinetic energy E, potential energy = kinetic energy

$$E = \frac{1}{2}mv^2 = zZe^2/b$$

$$G = \left(\frac{2}{\hbar}\right)(2mzZe^2)^{1/2} \int_a^b \left(\frac{1}{r} - \frac{1}{b}\right)^{1/2} dr$$

The integral is easily evaluated by the change of variable $r = b \cos^2 \theta$

$$I = \sqrt{b}\left\{\cos^{-1}\left(\frac{a}{b}\right) - \sqrt{\frac{a}{b} - \frac{a^2}{b^2}}\right\}$$

Finally

$$G = \frac{2}{\hbar}(2mz\,Z\,e^2 b)^{1/2}\left[\cos^{-1}\left(\frac{R}{b}\right) - \sqrt{\frac{R}{b} - \frac{R^2}{b^2}}\right]$$

where $a = R$, the nuclear radius.

If v_{in} is the velocity of the alpha particle inside the nucleus and $R = a$ is the nuclear radius then the decay constant $\lambda = 1/\tau \sim (v_{\text{in}}/R).e^{-G}$

3.35 (a) In Problem 3.19 the condition that a bound state be formed was obtained as

$$\cot kR = -\frac{\gamma}{k} = -\left[\frac{W}{V_0 - W}\right]^{1/2}$$

where V_0 is the potential depth and a is the width. Here the condition would read

$$\cot ka = -\left[\frac{W}{V_0 - W}\right]^{1/2}$$

where $k^2 = 2m(V_0 - W)a^2/\hbar^2$

If we now make $W = 0$, the condition that only one bound is formed is

$$ka = \frac{\pi}{2} \text{ or } V_0 = \frac{h^2}{32ma^2}$$

(b) The next solution is

$$ka = \frac{3\pi}{2}$$

Here $W_1 = 0$ for the first excited state

With the second solution we get

$$V_1 = \frac{9h^2}{32ma^2}$$

Note that in Problem 3.23 the reduced mass $\mu = M/2$ while here $\mu = m$. The graphs are shown in Fig. 3.13.

Fig. 3.13

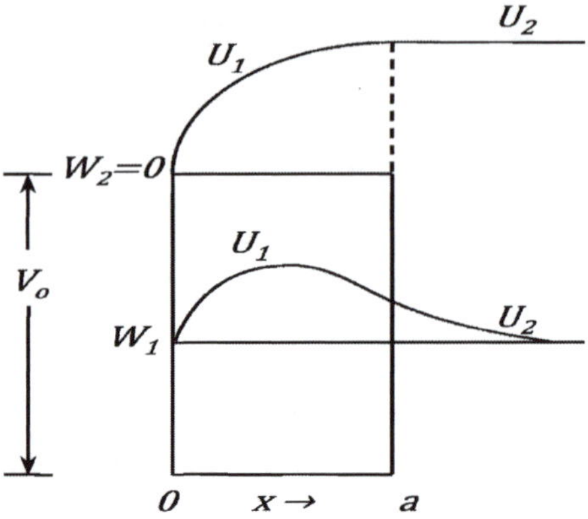

For (a) the inside and outside wave functions are as in the deuteron Problem 3.19. For (*b*) the inside wave function is similar but the outside function becomes constant ($W_1 = 0$) and is a horizontal line.

3.36 (a) Class I: Refer to Problem 3.25

$$\psi_1 = Ae^{\beta x}\,(-\infty < x < -a)$$

$$\psi_2 = D\cos ax(-a < x < +a)$$

$$\psi_3 = A\,e^{-\beta x}(a < x < \infty)$$

Normalization implies that

$$\int_{-\infty}^{-a} |\psi_1|^2\,dx + \int_{-a}^{a} |\psi_2|^2\,dx + \int_{a}^{\infty} |\psi_3|^2 dx = 1$$

$$\int_{-\infty}^{-a} A^2 e^{2\beta x} dx + \int_{-a}^{a} D^2\cos^2\alpha x\,dx + \int_{a}^{\infty} A^2 e^{-2\beta x} dx = 1$$

$$\frac{A^2 e^{-2\beta a}}{2\beta} + D^2\left[a + \frac{\sin(2\alpha a)}{2\alpha}\right] + \frac{A^{2e^{-2\beta a}}}{2\beta} = 1$$

Or

$$A^2 e^{-2\beta a}/\beta + D^2(a + \sin(2\alpha a)/2\alpha) = 1 \tag{1}$$

Boundary condition at $x = a$ gives

$$D_{\cos\alpha a} = ae^{-\beta a} \tag{2}$$

Combining (1) and (2) gives

$$D = \left(a + \frac{1}{\beta}\right)^{-1}$$

$$A = e^{\beta a}\cos\alpha a\left(a + \frac{1}{\beta}\right)^{-1}$$

3.37 (a) $u_n = \left(\frac{2}{L}\right)^{1/2} \sin\left(\frac{n\pi x}{L}\right)$

$$< x >= \int_0^L u_n^* x u_n \mathrm{d}x = \left(\frac{2}{L}\right) \int_0^L x \, \sin^2\left(\frac{n\pi x}{a}\right) \mathrm{d}x$$

$$= \frac{L}{2} + \left(\frac{L}{4n^2\pi^2}\right)(\cos(2n\pi) - 1)$$

The second term on the RHS vanishes for any integral value of n. Thus

$$< x >= \frac{L}{2}$$

$$\mathrm{Var}\, x = \sigma^2 =< (x- < x >)^2 >=< x^2 > - < x >^2 =< x^2 > - \frac{L^2}{4}$$

$$\text{Now } < x^2 >= \int_0^L u_n^* x^2 u_n \mathrm{d}x = \left(\frac{2}{L}\right) \int_0^L x^2 \sin^2\left(\frac{n\pi x}{L}\right) \mathrm{d}x$$

$$= \frac{L^2}{3} - \frac{L^2}{2n^2\pi^2}$$

$$\sigma^2 =< x^2 > - < x >^2 = \frac{L^2}{3} - \frac{L^2}{2n^2\pi^2} - \frac{L^2}{4} = \left(\frac{L^2}{12}\right)\left(1 - \frac{6}{n^2\pi^2}\right)$$

For $n \to \infty$, $< x >= \frac{L}{2}; \sigma^2 \to L^2/12$

(b) Classically the expected distribution is rectangular, that is flat.
The normalized function

$$f(x) = \frac{1}{L}$$

$$< x >= \int x f(x) \mathrm{d}x = \int_0^L \frac{x \mathrm{d}x}{L} = \frac{L}{2}$$

$$\sigma^2 =< x^2 > - < x >^2$$

$$< x^2 >= \int_0^L x^2 f(x) \mathrm{d}x = L^2/3$$

$$\therefore \sigma^2 = \frac{L^2}{3} - \frac{L^2}{4} = \frac{L^2}{12}$$

Fig. 3.14

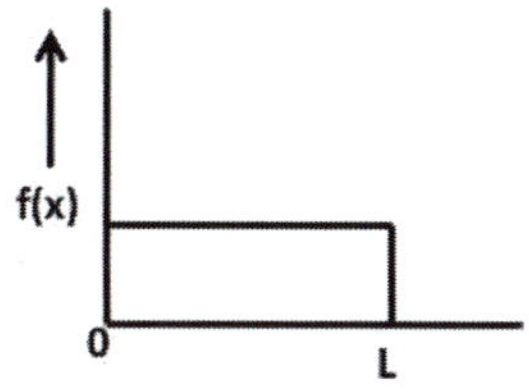

3.38 $$H = \left(-\frac{\hbar^2}{2m}\right) \nabla^2 + ar^2 \left[1 - \frac{5}{6} \sin^2\theta \cos^2\varphi\right] \quad (1)$$

In spherical coordinates $x = r \, \sin\theta \, \cos\,\varphi$. Therefore

$$H = \left(-\frac{\hbar^2}{2m}\right)\nabla^2 + a\left[x^2 + y^2 + z^2 - \frac{5}{6}x^2\right]$$
$$= \left(-\frac{\hbar^2}{2m}\right)\nabla^2 + a\left[\frac{x^2}{6} + y^2 + z^2\right] \quad (2)$$

The Schrodinger's equation is

$$H\psi(x, y, z) = E\psi(x, y, z) \quad (3)$$

This equation can be solved by the method of separation of variables. Let

$$\psi(x, y, z) = \psi_x\psi_y\psi_z \quad (4)$$

$$H\psi(x, y, z) = -\frac{\hbar^2}{2m}\left(\frac{\partial^2}{\partial x^2} + \frac{\partial^2}{\partial y^2} + \frac{\partial^2}{\partial z^2}\right)\psi_x\psi_y\psi_z + a\left[\frac{x^2}{6} + y^2 + z^2\right]$$

$$\psi_x\psi_y\psi_z = E\psi_x\psi_y\psi_z$$

$$-\frac{\hbar^2}{2m}\psi_y\psi_z\frac{\partial^2\psi_x}{\partial x^2} - \frac{\hbar^2}{2m}\psi_x\psi_z\frac{\partial^2\psi_y}{\partial x^2} - \frac{\hbar^2}{2m}\psi_x\psi_y\frac{\partial^2\psi_z}{\partial X^2}$$
$$+ \frac{ax^2}{6}\psi_x\psi_y\psi_z + ay^2\psi_x\psi_y\psi_z + az^2\psi_x\psi_y\psi_z = E\psi_x\psi_y\psi_z$$

Dividing throughout by $\psi_x\psi_y\psi_z$

$$-\frac{\hbar^2}{2m}\frac{1}{\psi_x}\frac{\partial^2\psi_x}{\partial x^2} - \frac{\hbar^2}{2m}\frac{1}{\psi_y}\frac{\partial^2\psi_y}{\partial y^2} - \frac{\hbar^2}{2m}\frac{1}{\psi_z}\frac{\partial^2\psi_z}{\partial z^2} + \frac{ax^2}{6} + ay^2 + az^2 = E \quad (5)$$

$$-\frac{\hbar^2}{2m}\frac{1}{\psi_x}\frac{\partial^2\psi_x}{\partial x^2} + \frac{ax^2}{6} = E_1 \qquad \frac{a}{6} = {}^1\!/\!_2 k_1 \quad (6)$$

$$-\frac{\hbar^2}{2m}\frac{1}{\psi_y}\frac{\partial^2\psi_y}{\partial y^2} + ay^2 = E_2 \qquad a = {}^1\!/\!_2 k_2 \quad (7)$$

$$-\frac{\hbar^2}{2m}\frac{1}{\psi_z}\frac{\partial^2\psi_z}{\partial z^2} + az^2 = E_3 \qquad a = {}^1\!/\!_2 k_2 \quad (8)$$

$$E_1 = (n_1 + {}^1\!/\!_2)\hbar\omega_1;\ E_2 = (n_1 + {}^1\!/\!_2)\hbar\omega_2;\ E_3 = (n_3 + {}^1\!/\!_2)\hbar\omega_3$$

$$\omega_1 = \sqrt{\frac{k_1}{m}} = \sqrt{\frac{a}{3m}};\ \omega_2 = \omega_3 = \sqrt{\frac{2a}{m}}$$

$$E = E_1 + E_2 + E_3 = \left(n_1 + \frac{1}{2}\right)\hbar\omega_1 + (n_2 + n_3 + 1)\hbar\omega_2$$

The lowest energy level corresponds to $n_1 = n_2 = n_3 = 0$, with

$$E = \frac{\hbar\omega_1}{2} + \hbar\omega_2 = \left(\sqrt{\frac{1}{12}} + \sqrt{2}\right)\hbar\sqrt{\left(\frac{a}{m}\right)}$$

It is non-degenerate.

The next higher state is degenerate with $n_1 = 1,\ n_2 = 0,\ n_3 = 0$;

$$E = \frac{3}{2}\hbar\sqrt{\frac{a}{3m}} + \hbar\sqrt{\frac{2a}{m}} = \left(\sqrt{\frac{3}{4}} + \sqrt{2}\right)\hbar\sqrt{\frac{a}{m}}$$

This is also non-degenerate.

3.39 (a) $$\left[\left(-\frac{\hbar^2}{2m}\right)\nabla^2 + V\right]\psi(x, y, z) = E\psi(x, y, z) \quad (1)$$

Put $V = 0$

$$\left(-\frac{\hbar^2}{2m}\right)\left[\frac{\partial^2}{\partial x^2} + \frac{\partial^2}{\partial y^2} + \frac{\partial^2}{\partial z^2}\right]\psi(x, y, z) = E\psi(x, y, z)$$

Let

$$\psi\,(x, y, z) = X\,(x)\,Y\,(y)\,Z\,(x) \quad (2)$$

$$YZ\frac{\partial^2 X}{\partial x^2} + ZX\frac{\partial^2 Y}{\partial y^2} + XY\frac{\partial Z}{\partial z^2} = -\left(\frac{2mE}{\hbar^2}\right)XYZ$$

Dividing throughout by *XYZ*

$$\frac{1}{Y}\frac{\partial^2 Y}{\partial y^2} + \frac{1}{Z}\frac{\partial^2 Z}{\partial z^2} = -\frac{1}{X}\frac{\partial^2 X}{\partial x^2} - \frac{2mE}{\hbar^2} \quad (3)$$

LHS is a function of y and z only while the RHS is a function of x only. The only way (3) can be satisfied is that each side is equal to a constant, say $-\alpha^2$.

$$\frac{1}{X}\frac{d^2 X}{dx^2} + \frac{2mE}{\hbar^2} - \alpha^2 = 0$$

$$\frac{\partial^2 X}{dx^2} + \left(\frac{2mE}{\hbar^2} - \alpha^2\right)X = 0$$

Or

$$\frac{\partial^2 X}{\partial x^2} + \beta^2 X = 0$$

where

$$\beta^2 = \left(\frac{2mE}{\hbar^2}\right) - \alpha^2 \quad (4)$$

$$X = A\sin\beta x + B\cos\beta x$$

Take the origin at the corner
Boundary condition: $X = 0$ when $x = 0$. This gives $B = 0$.

$$X = A\sin\beta x$$

Further, $X = 0$ when $x = a$

$$\mathrm{Sin}\beta a = 0 \rightarrow \beta a = n_x\pi$$

Or

$$\beta = \frac{n_x\pi}{a} \quad (n_x = \text{integer}) \quad (5)$$

Going back to (3)

$$\frac{1}{Y}\frac{\partial^2 Y}{\partial y^2} + \frac{1}{Z}\frac{\partial^2 Z}{\partial z^2} = -\alpha^2$$

$$\frac{1}{Z}\frac{\partial^2 Z}{\partial z^2} = -\frac{1}{Y}\frac{\partial^2 Y}{\partial y^2} - \alpha^2 \tag{6}$$

Each side must be equal to a constant, say $-\gamma^2$ for the same argument as before.

$$-\frac{1}{Y}\frac{\partial^2 Y}{\partial y^2} - \alpha^2 = -\gamma^2$$

Or

$$-\frac{1}{Y}\frac{\partial^2 Y}{\partial y^2} + \left(\alpha^2 - \gamma^2\right) = 0$$

Or

$$\frac{\partial^2 Y}{\partial y^2} + \mu^2 Y = 0$$

where $\mu^2 = \alpha^2 - \gamma^2$ (7)

$Y = D \sin \mu y$
$Y = 0\ at\ y = b$

This gives $\mu = \dfrac{n_y \pi}{b}$ (8)

Going back to (6)

$$\left(\frac{1}{Z}\right)\frac{\mathrm{d}^2 Z}{\mathrm{d}z^2} = -\gamma^2$$

This gives $Z = F \sin \gamma_z$

where $\gamma = \dfrac{n_z \pi}{c}$ (9)

$$\therefore \psi \sim \sin\left(\frac{n_x \pi x}{a}\right) \sin\left(\frac{n_y \pi y}{b}\right) \sin\left(\frac{n_z \pi z}{c}\right)$$

(b) Combining (4), (5), (7), (8) and (9)

$$\mu^2 = \alpha^2 - \gamma^2 = (2mE/\hbar^2) - \beta^2 - \gamma^2$$

Or

$$\frac{2mE}{\hbar^2} = \mu^2 + \beta^2 + \gamma^2 = \left(\frac{n_y \pi}{b}\right)^2 + \left(\frac{n_x \pi}{a}\right)^2 + \left(\frac{n_z \pi}{c}\right)^2$$

Or

$$E = \left(\frac{h^2}{8m}\right)\left(\frac{n_x^2}{a^2} + \frac{n_y^2}{b^2} + \frac{n_z^2}{c^2}\right) \tag{10}$$

3.40 For $a = b = c$
$E = (h/8ma^2)\left(n_x^2 + n_y^2 + n_z^2\right)$ (Equation 10 of Prob 3.39)

None of the numbers n_x, n_y, or n_z can be zero, otherwise $\psi(x, y, z)$ itself will vanish.

For an infinitely deep potential well $E_n = \frac{h^2}{8ma^2}\left[n_x^2 + n_y^2 + n_z^2\right]$. The combination $n_x = n_y = n_z = 0$ is ruled out because the wave function will be zero. Table 3.4 gives various energy levels along with the value of g, the degeneracy. The values of n_x, n_y *and* n_z are such that $n_x^2 + n_y^2 + n_z^2 = 8ma^2 E_n/h^2 =$ constant for a given energy E_n. The energies of the excited states are expressed in terms of the ground state energy $E_0 = h^2/8ma^2$

Table 3.4

n_x	n_y	n_z	g	E_n
0	0	1	3-fold	$E_0 = h^2/8ma^2$
0	1	0		
1	0	0		
0	1	1	3-fold	$2E_0$
1	0	1		
1	1	0		
1	1	1	Non-degenerate	$3E_0$
0	0	2	3-fold	$4E_0$
0	2	0		
2	0	0		
0	1	2	6-fold	$5E_0$
1	0	2		
1	2	0		
0	2	1		
2	0	1		
2	1	0		
1	1	2	3-fold	$6E_0$
1	2	1		
2	1	1		

3.41 (a) Case (i) $U_0 < E$, Region $x \ll 0$

Putting $V(x) = 0$, Schrodinger's equation is reduced to

$$\frac{d^2\psi}{dx^2} + \left(\frac{2mE}{\hbar^2}\right)\psi = 0 \tag{1}$$

which has the solution

$$\psi_1 = A\exp(ik_1x) + B\exp(-ik_1x) \tag{2}$$

$$\text{where } k_1^2 = \frac{2mE}{\hbar^2} \tag{3}$$

ψ_1 represents the incident wave moving from left to right (first term in (2)) plus the reflected wave (second term in (2)) moving from right to left

$$\text{Region } x \gg 0: \frac{d^2\psi}{dx^2} + \left[\frac{2m(E - U_0)}{\hbar^2}\right]\psi = 0 \tag{4}$$

which has the physical solution

$$\psi_2 = C\ \exp(ik_2 x) \tag{5}$$

where

$$k_2^2 = \frac{2m(E - U_0)}{\hbar^2} \tag{6}$$

It represents the transmitted wave to the right with reduced amplitude.

Note that the second term is absent in (5) as there is no reflected wave in the region $x > 0$.

Case (ii), $U_0 > E$

Region $x < 0$

$$\psi_3 = A\exp(ik_1 x) + B\exp(-ik_1 x) \tag{7}$$

Region $x > 0$

$$\frac{d^2\psi}{dx^2} - \frac{2m\psi(U_0 - E)}{\hbar^2} = 0$$

$$\frac{d^2\psi}{dx^2} - \alpha^2\psi = 0$$

$$\psi_4 = Ce^{-\alpha x} + De^{\alpha x}$$

where $\alpha^2 = \frac{2m(U_0 - E)}{\hbar^2}$

ψ must be finite everywhere including at $x = -\infty$. We therefore set $D = 0$. The physically accepted solution is then

$$\psi_4 = Ce^{-\alpha x} \tag{8}$$

(b) The continuity condition on the function and its derivative at $x = 0$ leads to Eqs. (9) and (10).

$$\psi_3(0) = \psi_4(0)$$

$$A + B = C \tag{9}$$

$$\left.\frac{d\psi_3}{dx}\right|_{x=0} = \left.\frac{d\psi_4}{dx}\right|_{x=0}$$

$$ik_1(A - B) = -C\alpha \tag{10}$$

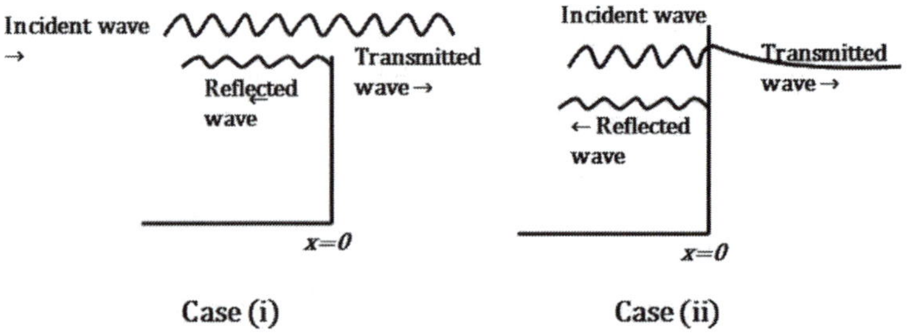

Fig. 3.15 Case(i)

Dividing (10) by (9) gives

$$\frac{ik(A-B)}{A+B} = -\alpha \tag{11}$$

Diagrams for ψ at around $x = 0$

3.42 $k_1 = \left(\frac{2mE}{\hbar^2}\right)^{1/2}$; $k_2 = \left(\frac{2m(E-V_0)}{\hbar^2}\right)^{1/2}$ (1)

Boundary condition at $x = 0$:

$$\psi_1(0) = \psi_2(0)$$

$$\left.\frac{\mathrm{d}\psi_1}{\mathrm{d}x}\right|_{x=0} = \left.\frac{\mathrm{d}\psi_2(x)}{\mathrm{d}x}\right|_{x=0} \tag{2}$$

These lead to

$$A_0 + A = B \tag{3}$$

$$ik_1(A_0 - A) = ik_2 B$$

Or

$$k_1(A_0 - A) = k_2 B \tag{4}$$

Solving (3) and (4)

$$A = \left[\frac{k_1 - k_2}{k_1 + k_2}\right] A_0 \tag{5}$$

$$B = \frac{2k_1 A_0}{k_1 + k_2} \tag{6}$$

Reflection coefficient,

$$R = \frac{|A|^2}{|A_0|^2} = \frac{(k_1 - k_2)^2}{(k_1 + k_2)^2} \tag{7}$$

Transmission coefficient,

$$T = \left(\frac{k_2}{k_1}\right)\frac{|B|^2}{|A|^2} = \frac{4k_1 k_2}{(k_1 + k_2)^2} \tag{8}$$

Substituting the expressions for k_1 and k_2 from (1) and putting $E = 4V_0/3$ we find that $R = 1/9$ and $T = 8/9$.

From (7) and (8) it is easily verified that

$$R + T = 1 \tag{9}$$

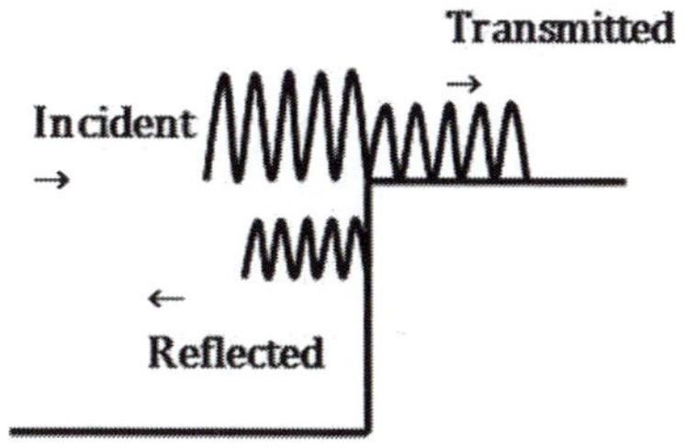

Fig. 3.16 Graphs for probability density

This is a direct result of the fact that the current density is constant for a steady state.

Thus $|A_0|^2 v_1 = |A|^2 v_1 + |B|^2 v_2$
where $v_1 = \frac{k_1 \hbar}{m}$ and $v_2 = \frac{k_2 \hbar}{m}$
$A^2 + B^2 \neq 1$ because the sum of the intensities of the reflected intensity and transmitted intensities does not add up to unity. What is true is relation (9) which is relevant to current densities.

3.43 (a) The wave function must be finite, single-valued and continous. At the boundary this is ensured by requiring the magnitude and the first derivative be equal.

(b)

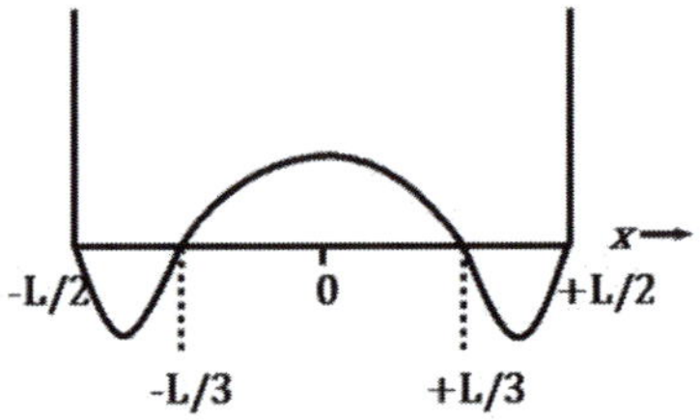

Fig. 3.17 Sketch of $\psi \sim \cos\left(\frac{3\pi x}{L}\right)$

(c) $$\int_{-\frac{L}{2}}^{\frac{L}{2}} |\psi|^2 \mathrm{d}x = A^2 \int_{-\frac{L}{2}}^{\frac{L}{2}} \cos^2 \frac{3\pi x}{L} \mathrm{d}x = 1$$

or

$$A^2 \int_{-\frac{L}{2}}^{\frac{L}{2}} (1 + \cos 6\pi x/L)\mathrm{d}x = A^2 L = 1$$

Therefore $A = 1/\sqrt{L}$

(d) $$P\left(-\frac{L}{4} < x < \frac{L}{4}\right) = A^2 \int_{\frac{L}{4}}^{\frac{L}{4}} \cos^2\left(\frac{3\pi x}{L}\right) \mathrm{d}x$$

$$= \left(\frac{1}{L}\right) \int_{-L/4}^{L/4} {}^1/_2 (1 + \cos\left(\frac{6\pi x}{L}\right) \mathrm{d}x = \frac{1}{4} + \frac{1}{6\pi} = 0.303$$

(e) $$\frac{\mathrm{d}^2\psi}{\mathrm{d}x^2} = \frac{\mathrm{d}^2}{\mathrm{d}x^2} A \cos\left(\frac{3\pi x}{L}\right) = -9\pi^2 \left(\frac{A}{L}\right) \cos\left(\frac{3\pi x}{L}\right) = -\left(\frac{9\pi^2}{L}\right)\psi$$

Therefore, $$\left(-\frac{\hbar^2}{2m}\right) \frac{d^2\psi}{dx^2} = \left(\frac{9\pi^2\hbar^2}{2mL}\right)\psi - E\psi$$

Or

$$E = \frac{9\pi^2\hbar^2}{2mL}$$

3.44 (a)

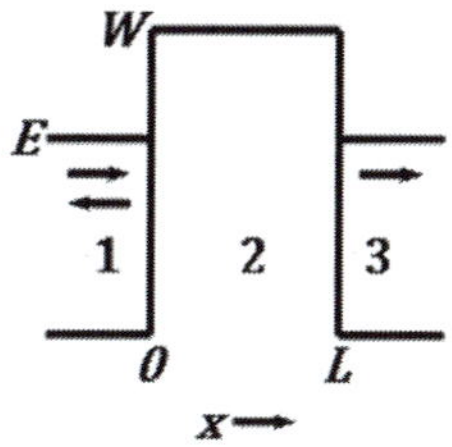

Fig. 3.18 Penetration of a rectangular barrier

(b) Region 1, $x < 0$

$$\frac{d^2\psi}{dx^2} + k^2\psi = 0$$

with $k^2 = \frac{2mE}{\hbar^2}$

$\psi_1 = Ae^{ikx} + Be^{-ikx}$

Incident reflected at $x = 0$

Region 2, $0 < x < L$

$$\frac{d^2\psi}{dx^2} - \alpha^2\psi = 0$$

with $\alpha^2 = \frac{2m(W-E)}{\hbar^2}$

$\psi_2 = Ce^{-\alpha x} + De^{\alpha x}$

Region 3, $x > L$

$$\frac{d^2\psi}{dx^2} + k^2\psi = 0$$

with $k^2 = 2mE/\hbar^2$

$\psi_3 = Fe^{ikx}$

The second term is absent as there is no reflected wave coming from right to left

The transmission coefficient $T = \frac{|F|^2}{|A|^2}$

(c) Boundary conditions

$\psi_1(0) = \psi_2(0)$

$$\left.\frac{d\psi_1}{dx}\right|_{x=0} = \left.\frac{d\psi_2}{dx}\right|_{x=0}$$

$\psi_2(L) = \psi_3(L)$

$$\left.\frac{d\psi_2}{dx}\right|_{x=L} = \left.\frac{d\psi_3}{dx}\right|_{x=L}$$

(d) $T = 16\left(\frac{E}{W}\right)\left(1 - \frac{E}{W}\right)e^{-2\alpha L}$

$$\alpha^2 = 2m\left(\frac{W-E}{\hbar^2}\right) \to \alpha = \frac{\sqrt{2mc^2(W-E)}}{\hbar c}$$

$$= \frac{\sqrt{(2 \times 0.511 \times (5-2) \times 10^{-6}}}{197.3 \times 10^{-15}} = 8.8748 \times 10^9 \text{ m}^{-1}$$

Therefore $2\alpha L = 2 \times 8.8748 \times 10^9 \times 0.3 \times 10^{-9} = 5.3249$

$$T = 16\left(\frac{2}{5}\right)\left(1 - \frac{2}{5}\right)e^{-5.3249} = 0.0187$$

(e) Examples of quantum mechanical tunneling
- (i) *α-decay* Observed α-energy may be ~ 5 MeV although the Coulomb barrier height is 20 or 30 MeV
- (ii) *Tunnel diode*
- (iii) *Josephson effect* In superconductivity electron emission in pairs through insulator is possible via tunneling mechanism
- (iv) *Inversion spectral line* in ammonia molecule. This arises due to tunneling through the potential barrier between two equilibrium positions of the nitrogen atom along the axis of the pyramid molecule which is perpendicular to the plane of the hydrogen atoms. The oscillation between the two equilibrium positions causes an intense spectral line in the microwave region.

3.45 The wave function to the zeroeth order in infinitely deep 2-D potential well is obtained by the method of separation of variables; the Schrodinger equation is

$$-\frac{\hbar^2}{2m}\frac{\partial^2\psi(x,y)}{\partial x^2} - \frac{\hbar^2}{2m}\frac{\partial^2\psi(x,y)}{\partial y^2} = E\psi(x,y)$$

Let $\psi(x, y) = \psi_x\psi_y$

$$-\frac{\hbar^2}{2m}\psi_y\frac{\partial^2\psi_x}{\partial x^2} - \frac{\hbar^2}{2m}\psi_x\frac{\partial^2\psi_y}{\partial y^2} = E\psi_x\psi_y$$

Divide through by $\psi_x\psi_y$

$$-\frac{\hbar^2}{2m}\frac{1}{\psi_x}\frac{\partial^2\psi_x}{\partial x^2} - \frac{\hbar^2}{2m}\frac{1}{\psi_y}\frac{\partial^2\psi_y}{\partial y^2} = E$$

$$-\frac{\hbar^2}{2m}\frac{1}{\psi_x}\frac{\partial^2\psi_x}{\partial x^2} - E = \frac{\hbar^2}{2m}\frac{1}{\psi_y}\frac{\partial^2\psi_y}{\partial y^2} = A = \text{constant}$$

$$\frac{\partial^2\psi_x}{\partial x^2} + \alpha^2\psi_x = 0$$

where $\alpha^2 = \left(\frac{2m}{\hbar^2}\right)(E + A)$

$\psi_x = C\sin\alpha x + D\cos\alpha x$

$\psi_x = 0\ at\ x = 0$

This gives $D = 0$

$\psi_x = C\sin\alpha x$

$\psi_x = 0\ at\ x = a$

This gives $\alpha a = n_1\pi$ or $\alpha = \frac{n_1\pi}{a}$

Thus $\psi_x = C\sin(n_1\pi x/a)$

Further $\frac{\partial^2\psi_y}{\partial y^2} = \frac{2mA\psi_y}{\hbar^2} = -\beta^2\psi_y$

The negative sign on the RHS is necessary, otherwise the ψ_y will have an exponential form which will be unphysical.

$\psi_y = G\sin\beta y$

When the boundary conditions are imposed,

$$\beta = \frac{n_2 \pi y}{a}$$

$$\psi_y = G \sin\left(\frac{n_2 \pi z}{a}\right)$$

Thus $\psi(x,\ y) = \psi_x \psi_y = K \sin\left(\frac{n_1 \pi x}{a}\right) \sin\left(\frac{n_2 \pi y}{a}\right)$ (K = constant)
and $\alpha^2 + \beta^2 = 2mE/\hbar^2 = \left(\frac{n_1 \pi}{a}\right)^2 + \left(\frac{n_2 \pi}{a}\right)^2$
or

$$E = \left(\frac{\hbar^2 \pi^2}{2ma^2}\right)\left(n_1^2 + n_2^2\right)$$

3.46 By Problem 3.39, $E = \left(\frac{h^2}{8ma^2}\right)\left(n_x^2 + n_y^2, n_z^2\right)$

Therefore the number N of states whose energy is equal to or less than E is given by the condition

$$n_x^2 + n_y^2 + n_z^2 \leq \frac{8ma^2 E}{h^2}$$

The required number, $N = \left(n_x^2 + n_y^2 + n_z^2\right)^{1/2}$, is numerically equal to the volume in the first quadrant of a sphere of radius $\left(8\, m\, a^2 \frac{E}{h^2}\right)^{1/2}$. Therefore

$$N = \left(\frac{1}{8}\right) \cdot \left(\frac{4\pi}{3}\right)\left(\frac{8ma^2 E}{\hbar^2}\right)^{3/2} = \frac{2\pi}{3}\left(\frac{ma^2 E}{2\hbar^2 \pi^2}\right)^{3/2}$$

3.47 Schrodinger's radial equation for spherical symmetry and $V = 0$ is

$$\frac{\mathrm{d}^2\psi(r)}{\mathrm{d}r^2} + \frac{2}{r}\frac{\mathrm{d}\psi(r)}{\mathrm{d}r} + \frac{2mE}{\hbar^2}\psi(r) = 0$$

Take the origin at the centre of the sphere. With the change of variable,

$$\psi = \frac{u(r)}{r} \tag{1}$$

The above equation simplifies to

$$\frac{\mathrm{d}^2 u}{\mathrm{d}r^2} + \frac{2mEu}{\hbar^2} = 0$$

The solution is

$$u(r) = A \sin kr + B \cos kr$$

$$\text{where } k^2 = \frac{2mE}{\hbar^2} \tag{2}$$

Boundary condition is: $u(0) = 0$, because $\psi(r)$ must be finite at $r = 0$. This gives $B = 0$
Therefore,

$$u(r) = A \sin kr \tag{3}$$

Further $\psi(R) = \frac{u(r)}{R} = 0$
Sin $kR = 0$

or

$$kR = n\pi \rightarrow k = \frac{n\pi}{R} \tag{4}$$

Complete unnormalized solution is

$$u(r) = A \sin\left(\frac{n\pi r}{R}\right) \tag{5}$$

The normalization constant A is obtained from

$$\int_0^R |\psi(r)|^2 \cdot 4\pi r^2 \mathrm{d}r = 1 \tag{6}$$

Using (1) and (5), we get

$$A = \frac{1}{\sqrt{2\pi R}} \tag{7}$$

The normalized solution is then

$$u(r) = (2\pi R)^{-\frac{1}{2}} \sin\left(\frac{n\pi r}{R}\right) \tag{8}$$

From (2) and (4)

$$E_n = \frac{\pi^2 n^2 \hbar^2}{2mR^2} \tag{9}$$

For ground state $n = 1$. Hence

$$E_1 = \frac{\pi^2 \hbar^2}{2mR^2} \tag{10}$$

The force exerted by the particle on the walls is

$$F = -\frac{\partial V}{\partial R} = -\frac{\partial H}{\partial R} = -\frac{\partial E_1}{\partial R} = \frac{\pi^2 \hbar^2}{mR^3}$$

The pressure exerted on the walls is

$$P = \frac{F}{4\pi R^2} = \frac{\pi \hbar^2}{4mR^5}$$

3.48 The quantity $\frac{\pi^2\hbar^2}{8m} = \frac{\pi^2\hbar^2 c^2}{8mc^2} = \frac{\pi^2(197.3)^2}{8\times 2{,}200\, m_e c^2}$

$$= \frac{\pi^2(197.3)^2}{8 \times 2,200 \times 0.511} = 42.719\,\text{MeV} - \text{fm}^2$$

Now $V_0 a^2 = 70 \times (1.42)^2 = 141.148\,\text{MeV} - \text{fm}^2$

It is seen that

$$\frac{\pi^2\hbar^2}{8m} < V_0 a^2 < \frac{4\pi^2\hbar^2}{8m}$$

$(42.7 < 141 < 169)$

From the results of Problem (3.25) there will be two energy levels, one belonging to class I function and the other to class II function.

The particle of mass $2{,}200\, m_e$ or $1{,}124\,\text{Mev}/c^2$ is probably Λ-hyperon (mass $1{,}116\,\text{MeV}/c^2$) which is sometimes trapped in a nucleus, to form a hypernucleus before it decays(Chap. 10).

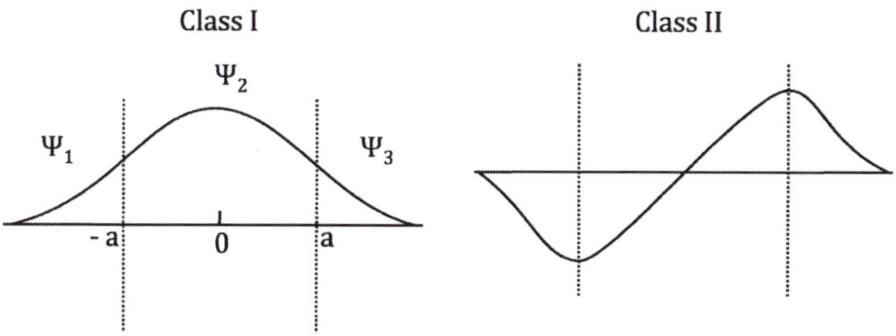

Fig. 3.19 Class I and Class II wave functions

3.49 The analysis for the reflection and transmission of stream of particles from the square well potential is similar to that for a barrier (Problem 3.30) except that the potential V_b must be replaced by $-V_0$ and in the region 2, k_2 must be replaced by ik_2. Thus, from Eq. (9) of Problem 3.30, we get

$$\tau = \frac{4k_1k_2\exp(-ik_1a)}{(k_2+k_1)^2\exp(-ik_2a)-(k_2-k_1)^2\exp(ik_2a)}$$

The fraction of transmitted particles when $k_2a = n\pi$ is determined by the imaginary exponential terms in the denominator.

$e^{+in\pi} = \cos n\pi \pm \text{i} \sin(n\pi) = \cos n\pi$
$= 1; (n = 0, 2, 4\cdot)$
$= -1; (n = 1, 3, 5\ldots)$

Fig. 3.20 Transmission coefficient T as a function of the ratio E/V_o for attractive square well potential

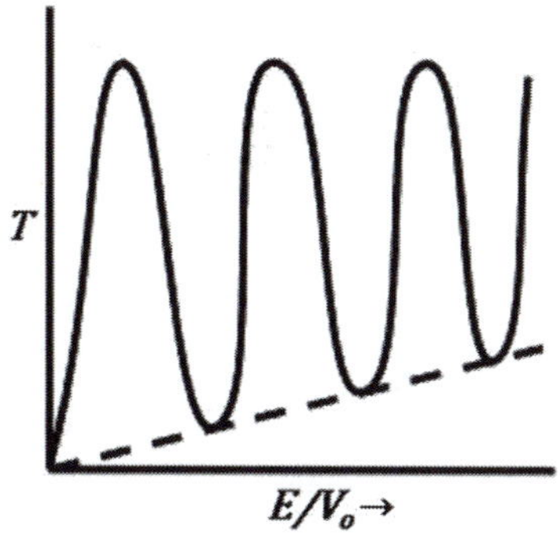

Therefore $\tau\tau^* = 1$

A typical graph for T as a function of E/V_0 is shown in Fig. 3.20.
In general we get the transmission coefficient

$$T = \left|1 + \frac{V_0^2\sin^2 k_2a}{4E(E+V_0)}\right|^{-1}$$

The transmission coefficient goes to zero at $E = 0$ because of the $1/E$ term in the denominator. For $E/V_0 \ll 1$, narrow transmission bands occur whenever the condition $k_2a = n\pi$ is satisfied, the width of the maxima becomes broader as the electron's incident energy increases. The occurrence of nearly complete transparency to incident electrons in the atoms of noble gases is known as the Ramsuer-Townsend effect.

The condition $k_2a = n\pi$ implies that $a = n\lambda/2$, $n = 1, 2\ldots$, that is whenever the barrier contains an integral number of half wavelengths leading to complete transparency. Interference phenomenon of this type is analogous to the transmission of light in optical layers.

3.50 (a) particles which can not be observed are called virtual particles

A 4-vector momemtum is $\mathbf{p} = (\mathbf{p}, iE)$ so that

$$(4 - \text{momentum})^2 = (3\ \underset{\text{Space}}{\text{momentum}})^2 - \underset{\text{time}}{(\text{energy})^2}$$

The components $p_{1,2,3}$ are said to be spacelike and the energy component E, timelike. If $\mathbf{q}$ denotes the 4-momentum transfer in a reaction i.e. $\mathbf{q} = \mathbf{p} - \mathbf{p}'$, where $\mathbf{p}, \mathbf{p}'$ are initial and 4-momenta, then

$q^2 > 0$ is spacelike as in the scattering process

$q^2 < 0$ is timelike as for (mass)^2 of free particle

$q^2 = 0$ is lightlike

(b) The relativistic relationship between total energy, momentum and mass for the field quantum is

$$E^2 - p^2c^2 - m^2c^4 = 0 \tag{1}$$

We can now use the quantum mechanical operators

$$E \to i\hbar\frac{\partial}{\partial t} \text{ and } p \to -i\hbar\nabla$$

To transform (1) into an operator equation

$$-\hbar^2\frac{\partial^2}{\partial t^2} + \hbar^2\nabla^2 - m^2c^4 = 0$$

representing the force between nucleons by a potential $\varphi(r, t)$ which may be regarded as a field variable, we can write

$$\left[\nabla^2 - \frac{1}{c^2}\frac{\partial^2}{\partial t^2} - \frac{m^2c^2}{\hbar^2}\right]\varphi = 0$$

as the wave equation describes the propagation of spinless particles in free space.

The time independent part of the equation is

$$\left[\nabla^2 - \frac{m^2c^2}{\hbar^2}\right]\varphi = 0 \tag{2}$$

For $m = 0$, this equation is the same as that obeyed in electromagnetism, for a point charge at the origin, the appropriate solution being

$$\varphi(r) = \left(\frac{e}{4\pi\varepsilon_0}\right)\frac{1}{r} \tag{3}$$

where ε_0 is the permittivity.

When $m \neq 0$, (2) can be written as

$$\left[\frac{\partial^2}{\partial r^2} + \left(\frac{2}{r}\frac{\partial}{\partial r} - \frac{m^2c^2}{\hbar^2}\right)\right]\varphi(r) = 0$$

or

$$\frac{1}{r^2}\frac{\partial}{\partial r}\left(r^2\frac{\partial\varphi}{\partial r}\right) = \frac{m^2c^2\varphi}{\hbar^2} \tag{4}$$

For values of $r > 0$ from a point source at the origin, $r = 0$. Integration gives

$$\varphi(r) = \frac{ge^{\frac{r}{R}}}{4\pi r} \tag{5}$$

where $R = \hbar/mc$ (6)

The quantity g plays the same role as charge in electrostatistics and measures the "strong nuclear charge".

3.3.4 Simple Harmonic Oscillator

3.51 By substituting $\psi(R) = AH(R)\exp(-R^2/2)$ in the dimensionless form of the equation and simplifying we easily get the Hermite's equation
The problem is solved by the series method

$$H = \Sigma H_n(R) = \Sigma_{n=0,2,4} a_n R^n$$

$$\frac{\mathrm{d}H}{\mathrm{d}R} = a_n n R^{n-1}$$

$$\frac{\mathrm{d}^2H}{\mathrm{d}R^2} = \Sigma n(n-1)a_n R^{n-2}$$

$$\Sigma n(n-1)a_n R^{n-2} - 2\Sigma a_n n R^n + (\varepsilon - 1)\Sigma a_n R^n = 0$$

Equating equal power of R^n

$$a_{n+2} = \frac{[2n - (\varepsilon - 1)]\,a_n}{(n+1)(n+2)}$$

If the series is to terminate for some value of n then

$2n - (\varepsilon - 1) = 0$ becuase $a_n \neq 0$. This gives $\varepsilon = 2n + 1$

Thus ε is a simple function of n

$$E = \varepsilon E_0 = (2n+1)^1\!/_2\hbar\omega,\ n = 0, 2, 4, \ldots$$
$$= {}^1\!/_2\hbar\omega,\ 3\hbar\omega/2,\ 5\hbar\omega/2, \ldots$$

Thus energy levels are equally spaced.

3.52 $u_0 = \left[\frac{\alpha}{\sqrt{\pi}}\right] e^{-\xi^2/2} H_0(\xi); \xi = \alpha x$

$$P = 1 - \int_{-a}^{a} |u_0|^2 \mathrm{d}x = 1 - 2\int_0^a (\alpha/\sqrt{\pi}) e^{-\xi^2/2} \mathrm{d}x$$

$$= 1 - \frac{2}{\sqrt{\pi}} \int_0^{\alpha a} e^{-\xi^2} \mathrm{d}\xi$$

$$E_0 = {}^1/_2 ka^2 = \frac{\hbar\omega}{2} (n = 0)$$

Therefore $a^2 = \frac{\hbar\omega}{k} = \left(\frac{\hbar}{k}\right)\left(\frac{k}{m}\right)^{1/2} = \frac{\hbar}{\sqrt{km}} = \frac{1}{\alpha^2}$

Therefore $\alpha^2 a^2 = 1$ or $\alpha a = 1$

$$P = 1 - \frac{2}{\sqrt{\pi}} \int_0^1 e^{-\xi 2} \mathrm{d}\xi$$

$$= 1 - \frac{2}{\sqrt{\pi}} \left[1 - \xi^2 + \frac{\xi^4}{2!} - \frac{\xi^6}{3!} + \frac{\xi^8}{4!}\right] \mathrm{d}\xi$$

$$= 1 - \frac{2}{\sqrt{\pi}} \left[1 - \frac{1}{3} + \frac{1}{10} - \frac{1}{42} + \frac{1}{216} \cdots\right]$$

≈ 0.16

Therefore, $p \approx 16\%$

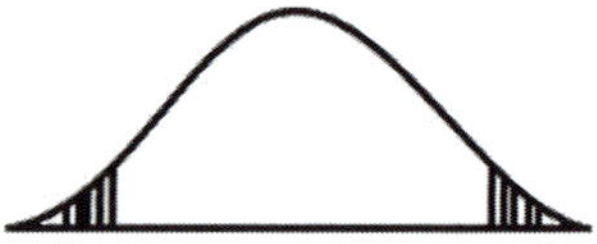

Fig. 3.21 Probability of the particle found outside the classical limits is shown shaded

3.53 The potential is of the form $V(r) = -V_0 + \gamma^2 r^2$ (1)

Schrodinger's radial equation is given by,

$$\frac{d^2u}{\mathrm{d}r^2} = \left[\frac{l(l+1)}{r^2} + \frac{2\mu}{\hbar^2(V(r) - E)}\right] u \quad (2)$$

Upon substituting (1) in (2), we obtain

$$\frac{d^2u}{\mathrm{d}r^2} + \left[\frac{2\mu}{\hbar^2(V_0 + E - \gamma^2 r^2)} - \frac{l(l+1)}{r^2}\right] u = 0 \quad (3)$$

The quantity γ^2 can be expressed in terms of the classical oscillator frequency

$$\gamma^2 = \frac{\mu\omega^2}{2} \quad (4)$$

For $r \to 0$, (3) may be approximated to

$$\frac{\mathrm{d}^2u}{\mathrm{d}r^2} - \frac{l(l+1)u}{r^2} = 0$$

The solution of which is,

$$u(r) = a\, r^{l+1} + \frac{b}{r}$$

with a and b as constants.

The boundary condition that u/r be finite at $r = 0$ demands that $b = 0$. Thus, ψ is proportional to r^l. The probability that a particle be in a spherical shell of radii r and $r + dr$ for small r, is proportional to $r^{2l+2}dr$. The larger l is, the smaller is the probability that the particle be in the vicinity of the origin. For the case of collision problems, there is a classical analogy: the larger the orbital angular momentum the larger the impact parameter.
Thus $u(r) \sim r^{l+1}(r \to 0)$

For $\to \infty$, we obtain, as an approximation to differential equation (3), as

$$\frac{d^2u}{dr^2} - \frac{2\mu\gamma^2r^2u}{\hbar^2} = 0$$

If we try a solution of the form,

$$u(r) = u_0 e^{-Br^2/2}$$

the asymptotically valid solution is satisfied provided we change

$$B = \frac{\gamma(2\mu)^{\frac{1}{2}}}{\hbar} = \frac{\mu\omega}{\hbar}$$

Inorder to solve (3) for all r, we may first separate the asymptotic behaviour by writing

$$u(r) = r^{l+1}e^{Br^2/2}V(r) \tag{5}$$

Insert (5) in (3), and dividing by $r^{l+1}e^{-Bv^2/2}$
We get

$$\frac{d^2v}{dr^2} + \frac{2dv}{dr}\left[\left(\frac{l+1}{r}\right) - Br\right] - Bv\left[2l + 3 - \frac{2}{\hbar\omega}(V_0 + E)\right]$$

Define $C = l + \frac{3}{2}$

$$4A = 2l + 3 - \frac{2}{\hbar\omega}(V_0 + E) \tag{6}$$

$$\frac{d^2v}{dr^2} + \frac{dv}{dr}\left[\left(\frac{2C-1}{r}\right) - 2Br\right] - 4ABv = 0 \tag{7}$$

Set

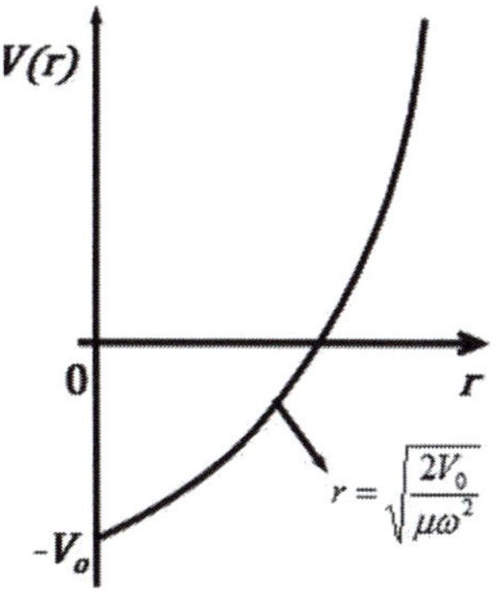

Fig. 3.22 The parabolic potential of the three dimensional harmonic oscillator

$$Br^2 = x \tag{8}$$

Substitute (8) in (7) and simplify, to obtain

$$\frac{x d^2 v}{dx^2} + \frac{dv}{dx}(C - x) - Av = 0 \tag{9}$$

which is the familiar confluent hyper geometric equation whose solution which is regular at $x = 0$ is;

$$V(A, C, x) = \alpha_0 \left[1 + \frac{Ax}{C} + \frac{A(A+1)x^2}{C(C+1)2!} + \frac{A(A+1)(A+2)x^3}{C(C+1)(C+2)3!} + \cdots . \right]$$

$$V(A, C, Br^2) = \alpha_0 \left[1 + \frac{ABr^2}{C} + \frac{A(A+1)B^2r^4}{C(C+1)2!} \right.$$

$$\left. + \frac{A(A+1)(A+2)B^3r^6}{C(C+1)(C+2)3!} + \cdots \right]$$

The asymptotic solution $V(r) \to 0$, while $r \to \infty$ implies that the series must break off for finite powers of Br^2 since $\alpha_0 \neq 0$. This means that A must equal a negative integer $-p$; where $p = 0, 1, 2, 3 \ldots$

Therefore $-4p = 2l + 3 - \frac{2}{\hbar\omega}(V_0 + E)$

Where we have used the definition of A (Eq. 6) from this we find, the energy eigen values,

$$E_{p,l} = -V_0 + \hbar\omega \left(2p + l + \frac{3}{2} \right)$$

$(p = 0, 1, 2, \ldots .)$

Setting $n = 2p + l$

$E_n = -V_0 + \hbar\omega \left(n + \frac{3}{2} \right)$ (which is different from one-dimensional harmonic oscillator)

$E_0 = -V_0 + \frac{3\hbar\omega}{2}$ corresponds to ground state.

It is a single state (not degenerate)

since $n = 0$ can be formed only by the combination $l = 0$, $p = 0$.

3.54 When the oscillator is in the lowest energy state

$$< H > = < V + T > = \left(\frac{m\omega^2}{2} \right) < x^2 > + \left(\frac{1}{2m} \right) < P^2 >$$

Now, if a, b and c are three real numbers such that $a + b = c$, then

$$ab = \frac{c^2}{4} - \left(\frac{a-b}{2} \right)^2$$

or

$$ab \leq \frac{c^2}{4}$$

Apply this inequality to $\left(\frac{m\omega^2}{2} \right) < x^2 >$, $\left(\frac{1}{2m} \right) < p^2 >$ and $\frac{\hbar\omega}{2}$

$$< \Delta x >^2 = < x^2 > - < x >^2 = < x^2 > \ and \ < x > = 0$$

Similarly $<\Delta p_x>^2=<p^2>$

$$\tfrac{1}{2}m\omega^2 <x^2> \cdot \left(\frac{1}{2m}\right) <p_x>^2 \leq \frac{1}{4}\left(\frac{\hbar\omega}{2}\right)^2$$

$$\left[<x^2><p_x^2>\right]^{1/2} \leq \frac{\hbar}{2}$$

or $\Delta x.\Delta\, p_x \leq \frac{\hbar}{2}$

Now compare this result with the uncertainty principle

$$\Delta x \cdot \Delta p_x \leq \frac{\hbar}{2}$$

We conclude that $\Delta x.\Delta p_x \geq \frac{\hbar}{2}$. Obviously the zero point energy could not have been lower than $\frac{\hbar\omega}{2}$ without violating the uncertainty principle.

3.55 The probability distribution for the quantum mechanical simple harmonic oscillator (S.H.O) is

$$P(x) = |\psi_n|^2 = \frac{\alpha \exp(-\xi^2) H_n^2(\xi)}{\sqrt{\pi}2^n n!} \quad (1)$$

$\xi = \alpha x; \alpha^4 = mk/\hbar^2$

Stirling approximation gives

$$n! \to (2n\pi)^{1/2} n^n e^{-n} \quad (2)$$

Furthermore the asymptotic expression for Hermite function is

$$H_n(\xi)(\text{for } n \to \infty) \to 2^{n+1}\frac{(n/2e)^{\frac{n}{2}}}{\sqrt{2}\,\cos\beta}\exp(n\beta^2)\cos\left[(2n+{}^1/_2)\beta - \frac{n\pi}{2}\right] \quad (3)$$

$$\text{where } \sin\beta = \xi/\sqrt{2n} \quad (4)$$

Using (2) and (3) in (1)

$$P(x) \to 2\alpha\exp(-\xi^2)\exp(2n\beta^2)\frac{\cos^2\left[\left(2n+\frac{1}{2}\right)\beta - \frac{n\pi}{2}\right]}{\pi\sqrt{2n}\cos\beta}$$

But $<\cos^2\left[\left(2n+\frac{1}{2}\right)\beta - \frac{n\pi}{2}\right]>=\frac{1}{2}$

$$\text{Therefore } P(x) = \frac{\alpha\exp(-\xi^2)\exp(2n\beta^2)}{\pi\sqrt{2n}\cos\beta} \quad (5)$$

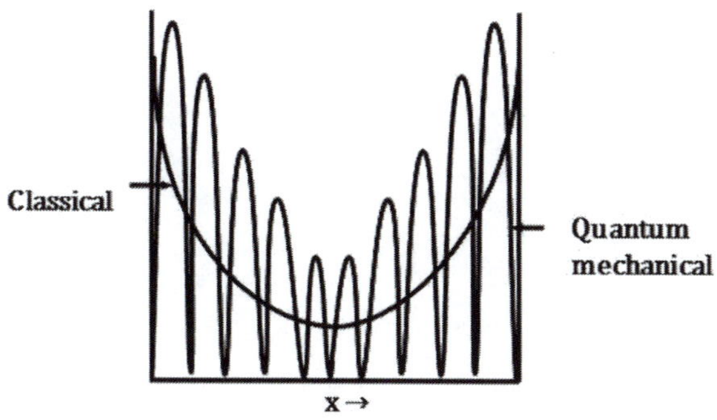

Fig. 3.23 Probability distribution of quantum mechanical oscillator and classical oscillator

Classically, $E = \frac{ka^2}{2} = \left(n + \frac{1}{2}\right)\hbar\omega$ (quantum mechanically) $\approx n\hbar\omega(n \to \infty)$
Therefore $a^2 = \frac{2n\hbar\omega}{k} = \left(\frac{2n\hbar}{k}\right)\left(\frac{k}{m}\right)^{1/2} = \frac{2n\hbar}{\sqrt{km}} = \frac{2n}{\alpha^2}$

$$\omega = \sqrt{\frac{k}{m}}$$

or

$$a = \frac{\sqrt{2n}}{\alpha} \tag{6}$$

$$\text{Sin}\,\beta = \frac{\xi}{\sqrt{2n}} = \frac{\alpha x}{\sqrt{2n}} = \frac{x}{a}$$

Therefore

$$\cos\beta = \frac{(a^2 - x^2)^{\frac{1}{2}}}{a} \tag{7}$$

Using (6) and (7) in (5)

$$P(x) = \frac{\exp(-\xi^2 \exp(2n\beta^2))}{\pi(a^2 - x^2)^{1/2}} \tag{8}$$

Now when $n \to \infty$, $\sin\beta \to \beta$ and
$\beta \to \xi/\sqrt{2n}$, and $exp(-\xi^2)\exp(2n\beta^2) \to 1)$
Therefore $P(x) = \frac{1}{\pi\sqrt{a^2 - x^2}}$ (classical)

3.56 One can expect the probability of finding the particle of mass m at distance x from the equilibrium position to be inversely proportional to the velocity

$$P(x) = \frac{A}{v} \tag{1}$$

where A =normalization constant. The equation for S.H.O. is

$$\frac{d^2x}{dt^2} + \omega^2 x = 0$$

which has the solution

$$x = a\sin\omega t; (\text{at } t = 0, x = 0)$$

where a is the amplitude.

$$v = \frac{dx}{dt} = \omega\sqrt{a^2 - x^2} \tag{2}$$

Using (2) in (1)

$$P(x) = A/\omega\sqrt{a^2 - x^2} \tag{3}$$

We can find the normalization constant A.

$$\int P(x)dx = \int_{-a}^{a} \frac{Adx}{\omega\sqrt{a^2 - x^2}} = \frac{A\pi}{\omega} = 1$$

Therefore,

$$A = \frac{\omega}{\pi} \tag{4}$$

Using (4) in (3), the normalized distribution is

$$P(x) = \frac{1}{\pi\sqrt{a^2 - x^2}} \tag{5}$$

3.57 Schrodinger's equation in one dimension is

$$\left(-\frac{\hbar^2}{2m}\right)\frac{d^2\psi}{dx^2} + V(x)\psi = E\psi \tag{1}$$

Given

$$\psi = \exp(-\frac{1}{2}ax^2) \tag{2}$$

Differentiating twice,
we get

$$\frac{d^2\psi}{dx^2} \exp(-\frac{1}{2}ax^2)(a^2x^2 - a) \tag{3}$$

Inserting (2) and (3) in (1), we get

$$V(x) = E + \left(\frac{\hbar^2}{2m}\right)(a^2x^2 - a) \tag{4}$$

Minimum value of $V(x)$ is determined from

$$\frac{dV}{dx} = \frac{\hbar^2 a^2 x}{m} = 0$$

Minimum of $V(x)$ occurs at $x = 0$
From (4) we find $0 = E - \frac{\hbar^2 a}{2m}$

(a) Or the eigen value $E = \frac{\hbar^2 a}{2m}$
(b) $V(x) = \frac{\hbar^2 a}{2m} + \left(\frac{\hbar^2}{2m}\right)(a^2x^2 - a) = \frac{\hbar^2 a^2 x^2}{2m}$

3.58 $$< V >_n = \frac{E_n}{2}$$

$$< \frac{P^2}{2m} >_n = < H >_n - < V >_n = {}^1/_2 E_n$$

$$< P^2 >_n = mE_n$$

Also $< x >_n = 0$; $< P >_n = 0$

$$< (\Delta x)^2 > = < x^2 >_n - < x^2 >_n$$

$$< \left(\Delta P^2\right) > = < P^2 >_n - < P_n >^2 = < P^2 >_n = mE_n$$

But $< x^2 >_n = \int_{-\infty}^{\infty} u_n^*(x)x^2u_n(x)dx$; $\xi = \alpha x$

$$= \frac{N_n^2}{\alpha^3} \int_{-\infty}^{\infty} H_n^2(\xi) \xi^2 e^{-\xi^2} d\xi$$

$$= \left(\frac{\alpha}{\sqrt{\pi} 2^n n!} \right) \left(\frac{1}{\alpha^3} \right) \frac{(2n+1)}{2} 2^n n! \sqrt{\pi}$$

$$= \frac{1}{\alpha^2} \left(n + \frac{1}{2} \right) = \frac{\hbar}{\sqrt{km}} \left(n + \frac{1}{2} \right)$$

$$\therefore \Delta x.\Delta P = \sqrt{m E_n \frac{\hbar}{\sqrt{km}} \left(n + \frac{1}{2} \right)}$$

Now, $\omega = \sqrt{\frac{k}{m}}$

$$\therefore \Delta x.\Delta P_x = \sqrt{\hbar\omega \left(n + \frac{1}{2} \right) \hbar \sqrt{\frac{m}{k}} \left(n + \frac{1}{2} \right)}$$

$$= \hbar \left(n + \frac{1}{2} \right)$$

Thus, $\Delta x.\Delta P \geq \frac{\hbar}{2}$ is in agreement with uncertainty principle.

3.59 The vibrational levels are equally spaced and so with the rule $\Delta n = 1$, the lines should coincide. The rotational levels are progressively further spaced such that the difference in the wave number of consecutive lines must be constant. This can be seen as follows:

$$E_J = \frac{J(J+1)\hbar^2}{2I_0}$$

$$\frac{\hbar c}{\lambda_J} = \Delta E = E_{J+1} - E_J = \frac{[(J+1)(J+2) - J(J+1)]\,\hbar^2}{2I_0} = \frac{(J+1)\hbar^2}{2I_0}$$

Therefore $\frac{1}{\lambda_{J+1}} - \frac{1}{\lambda_J} = \Delta \left(\frac{1}{\lambda_J} \right) \alpha[(J+2) - (J+1)] = \text{constant}$

$103.73 - 83.03 = 20.70\,\text{cm}^{-1}$; $124.30 - 103.73 = 20.57\,\text{cm}^{-1}$;

$145.03 - 124.30 = 20.73\,\text{cm}^{-1}$; $165.51 - 145.03$

$= 20.48\,\text{cm}^{-1}$; $185.86 - 165.51 = 20.35\,\text{cm}^{-1}$

The data are consistent with a constant difference, the mean value being, $20.556\,\text{cm}^{-1}$. Thus the transitions are rotational.

$$\frac{E_{J+1} E_J}{E_{J+2} - E_{J+1}} = \frac{2(J+1)}{2(J+2)} = \frac{83.03}{103.73} = 0.8$$

Therefore $J = 3$
The levels are characterized by $J = 3, 4, 5, 6, 7, 8, 9$

$$\hbar c\left(\frac{1}{\lambda_{j+1}}-\frac{1}{\lambda_j}\right)=\hbar c\times(20.556\ \mathrm{cm}^{-1})=\frac{\hbar^2}{I_0}$$

$$\text{Moment of inertia } \mathrm{I}_0=\frac{\hbar}{4\pi^2 c\times 20.556}$$

$$=\frac{6.63\times 10^{-27}\mathrm{erg-s}^{-1}}{4\pi^2\times 3\times 10^{10}\,\mathrm{cm-s}^{-1}\times 20.556\,\mathrm{cm}^{-1}}$$

$$=2.727\times 10^{-40}\mathrm{g-cm}^2$$

$$I_0=\mu r^2$$

$$\mu=\frac{m(\mathrm{H})m(\mathrm{Cl})}{[m(\mathrm{H})+m(\mathrm{Cl})]}=\frac{1\times 35\times 1.67}{1+35}\times 10^{-24}\ \mathrm{g}$$

$$=1.62\times 10^{-24}\ \mathrm{g}$$

$$r=\left(\frac{I_0}{\mu}\right)^{1/2}=\left(\frac{2.727\times 10^{-40}}{1.62\times 10^{-24}}\right)=1.3\times 10^{-8}\,\mathrm{cm}=1.3\,\text{Å}$$

3.60 For the 3-D isotropic oscillator the energy levels are given by

$$E_N=E_k+E_l+E_m=\left(\frac{3}{2}+n_k+n_l+n_m\right)\hbar\omega$$

where ω is the angular frequency
$N=n_k+n_l+n_m=0,1,2\ldots$

For a given value of N, various possible combinations of n_k, n_l and n_m are given in Table 3.5, and the degeneracy indicated.

Table 3.5 Possible combinations of n_k, n_l and n_m and degeneracy of energy levels

N	n_l	n_m	n_n	Degeneracy (D)
0	0	0	0	Non-degenerate
1	1	0	0	Three fold (1 + 2)
	0	1	0	
	0	0	1	
2	1	1	0	Sixfold (1 + 2 + 3)
	1	0	1	
	0	1	1	
	2	0	0	
	0	2	0	
	0	0	2	
3	1	1	1	Tenfold (1 + 2 + 3 + 4)
	1	2	0	
	1	0	2	
	2	1	0	
	2	0	1	
	0	2	1	
	0	1	2	
	3	0	0	
	0	3	0	
	0	0	3	

It is seen from the last column of the table that the degeneracy D is given by the sum of natural numbers, that is, $= n(n+1)/2$, if we replace n by $N+1$, $D = (N+1)(N+2)/2$.

3.61 As the time evolves, the eigen function would be

$\psi(x,t) = \sum_{n=0.1} C_n\psi_n(x) \exp(-iE_nt/\hbar)$
$= C_0\psi_0(x) \exp(-iE_0t/\hbar) + C_1\psi_1(x) \exp(-iE_1t/\hbar)$
The probability density
$|\psi(x,t)|^2 = C_0^2 + C_1^2 + C_0C_1\psi_0(x)\psi_1(x)[\exp(i(E_1 - E_0)t/\hbar)$
$- \exp(-i(E_1 - E_0)t/\hbar)]$
$= C_0^2 + C_1^2 + 2C_0C_1\psi_0(x)\psi_1(x) \cos \omega t$

where we have used the energy difference $E_1 - E_0 = \omega\hbar$. Thus the probability density varies with the angular frequency.

3.62 $< E >= \sum_{n=1,2,3} |C_n|^2 E_n = C_0^2 E_0 + C_1^2 E_1 + C_2^2 E_2$

$$= \left(\frac{1}{2}\right) \cdot \frac{\hbar\omega}{2} + \left(\frac{1}{3}\right) \cdot \frac{3\hbar\omega}{2} + \left(\frac{1}{6}\right) \cdot \frac{5\hbar\omega}{2} = \frac{7\hbar\omega}{6}.$$

3.63 (a) $\psi_0(x) = A \exp(-x^2/2a^2)$

Differentiate twice and multiply by $-\hbar^2/2m$

$$-\left(\frac{\hbar^2}{2m}\right)\frac{d^2\psi_0}{dx^2} = \left(\frac{A\hbar^2}{2ma^2}\right)\left(1 - \frac{x^2}{a^2}\right)\exp\left(-\frac{x^2}{2a^2}\right)$$
$$= \left(\frac{\hbar^2}{2ma^2}\right)\psi_0 - \left(\frac{\hbar^2x^2}{2ma^4}\right)\psi_0$$
$$\text{or } -\left(\frac{\hbar^2}{2m}\right)\frac{d^2\psi_0}{dx^2} + \left(\frac{\hbar^2x^2}{2ma^4}\right)\psi_0 = \left(\frac{\hbar^2}{2ma^2}\right)\psi_0$$

Compare the equation with the Schrodinger equation

$$E = \frac{\hbar^2}{2ma^2} = \frac{\hbar\omega}{2}$$
$$\omega = \frac{\hbar}{ma^2} \tag{1}$$
$$\text{or } a = \left(\frac{\hbar}{m\omega}\right)^{1/2}$$

Same relation is obtained by setting

$$V = \frac{\hbar^2x^2}{2ma^4} = \frac{m\omega^2x^2}{2}$$

(b) $\psi_1 = Bx \exp\left(-\frac{x^2}{2a^2}\right)$

Differentiate twice and multiply by $-\frac{\hbar^2}{2m}$

$$-\frac{\hbar^2}{2m}\frac{d^2\psi_1}{dx^2} = \frac{B\hbar^2x^3\exp\left(\frac{x^2}{a}\right)}{2ma^4} + \frac{3B\hbar^2\exp\left(-\frac{x^2}{2a^2}\right)}{2ma^2}$$

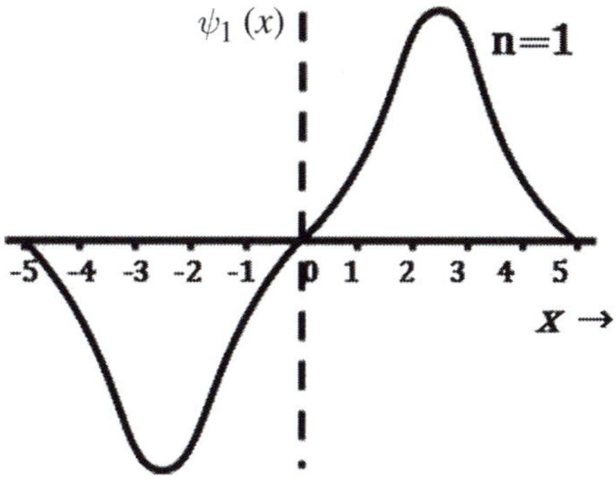

Fig. 3.24 $\psi_1(x)$ for SHO

Substitute
$Bx\ \exp(-x^2/2a) = \psi_1$ and $a = \left(\frac{\hbar}{m\omega}\right)^{1/2}$ and rearrange to get

$$-\left(\frac{\hbar^2}{2m}\right)\frac{d^2\psi_1}{dx^2} + \left(\frac{m\omega^2}{2}\right)\psi_1 = \left(\frac{3\hbar\omega}{2}\right)\psi_1$$

(c) $E = \frac{3\hbar\omega}{2}$

(d) $< p_x >= \int \psi_1^* \left(-\frac{i\hbar\partial\psi_1}{\partial x}\right) dx$

$= -i\hbar B^2 \int_{-\infty}^{\infty} x \exp\left(-\frac{x^2}{a^2}\right)\left(1 - \frac{x^2}{a^2}\right) dx$

= zero (because integration over an odd function between symmetrical limits is zero). This result is expected because half of the time the particle will be pointing along positive direction and for the half of time in the negative direction.

(e) The results of SHO are valuable for the analysis of vibrational spectra of diatomic molecules, identification of unknown molecules, estimation of force constants etc.

3.3.5 Hydrogen Atom

3.64 $< V >=< -\frac{e^2}{r} >= \int \psi^* V \psi d\tau = -\frac{e^2}{\pi a_0^3} \int_0^{\infty} (\exp\left(-\frac{2r}{a_0}\right)/r) 4\pi r^2 dr$

where e = charge of electron

$$= -\frac{4e^2}{a_0^3}\int_0^{\infty} \exp\left(-\frac{2r}{a_0}\right) r dr = -\frac{e^2}{a_0}$$

Therefore $< V >= -\frac{e^2}{a_0}$

$$< T >= \int \left[\psi^* \left(-\frac{\hbar^2}{2m}\right)\nabla^2\psi\right] d\tau \quad (1)$$

In polar coordinates (independent of θ and φ);

$$\nabla^2 = \frac{d^2}{dr^2} + \left(\frac{2}{r}\right)\frac{d}{dr} \tag{2}$$

Inserting (2) in (1) and performing the integration we get

$< T >= \dfrac{\hbar^2}{2ma_0^2}$

But $a_0 = \frac{\hbar^2}{me^2}$

Or $\frac{\hbar^2}{m} = a_0 e^2$

Therefore, $< T >= \frac{e^2}{2a_0}$

Also $< E >=< T > + < V >= \frac{e^2}{2a_0} - \frac{e^2}{a_0} = -e^2/2a_0$

3.65 The normalized eigen function for the ground state of hydrogen atom is

$$\psi_0 = 1/\left(\pi a_0^3\right)^{\frac{1}{2}} e^{-r/a_0}$$

where a_0 is the Bohr's radius

(a) The probability that the electron will be formed in the volume element $d\tau$ is

$$p(r)dr = |\psi_0|^2 d\tau = \left(\frac{e^{-\frac{2r}{a_0}}}{\pi a_0^3}\right).4\pi r^2 dr$$

$$= \left(\frac{4}{a_0^3}\right) r^2 e^{-2r/a_0} dr$$

Maximum probability is found by setting $\frac{dp}{dr} = 0$

$$\frac{d}{dr}\left(\frac{4r^2 e^{-\frac{2r}{a_0}}}{a_0^3}\right) = \left(\frac{8r}{a_0^3}\right) e^{-\frac{2r}{a_0}}\left(1 - \frac{r}{a_0}\right) = 0$$

Therefore $r = a_0$

(b) $$< r >= \int_0^\infty \psi^* r \psi d\tau = \frac{1}{\pi} a_0^3 \int_0^\infty r \exp\left(-\frac{2r}{a_0}\right).4\pi r^2 dr$$

$$= \left(\frac{4}{a_0^3}\right)\int_0^\infty r^3 \exp\left(-\frac{2r}{a_0}\right) dr$$

Let $\dfrac{2r}{a_0} = x;\ dr = \dfrac{a_0 dx}{2}$

$$< r >= \left(\frac{a_0}{4}\right)\int_0^\infty x^3 e^{-x} dx = \left(\frac{a_0}{4}\right) 3! = \frac{3a_0}{2}$$

3.66 $\int u^2{}_{210}\, d\tau = A_2{}^2 \int e^{-2x} x^2 \cos^2\theta r^2 \sin\theta\, d\theta\, dr\, d\varphi$

$$= \left(\frac{1}{\pi}\right)\left(\frac{1}{2a_0}\right)^3 \int_0^\infty \exp\left(-\frac{r}{a_0}\right)\left(\frac{r^2}{4a_0^2}\right) r^2 dr \int_{-1}^{+1} \cos^2\theta\, d(\cos\theta) \int_0^{2\pi} d\varphi$$

where we have put $x = r/2a_0$. Put $r/a_0 = y,\ dr = a_0 dy$

$$u_{210}^2\,\mathrm{d}\tau = \left(\frac{1}{8\pi a_0^3}\right)\left[\int_0^\infty y^4 e^{-y}\left(\frac{\mathrm{d}y}{4a_0^2}\right)a_0^5\right]\left[\frac{\cos^3\theta}{3}\right]_{-1}^{1}(2\pi)$$

$$= \left(\frac{1}{24}\right)\times 4! = 1$$

Similarly $u_{21\pm1}^2\mathrm{d}\tau = \frac{1}{\pi(2a_0)^3}\int \frac{e^{-2x}}{2}x^2\sin^2\theta\, e^{\pm i\varphi}\, e^{\mp i\varphi} r^2 \sin\theta \mathrm{d}\theta\mathrm{d}\varphi\mathrm{d}r$

$$= \left(\frac{1}{8\pi a_0^3}\right)\int_0^\infty\left(\frac{e^{-\frac{r}{a_0}}}{2}\right)\left(\frac{r^4\mathrm{d}r}{4a_0^2}\right)\int_{-1}^{+1}(1-\cos^2\theta)\,\mathrm{d}(\cos\theta)\int_0^{2\pi}\mathrm{d}\varphi$$

$$= \left(\frac{a_0^5}{64\pi a_0^5}\right)\left[\int_0^\infty y^4 e^{-y}\mathrm{d}y\right]\left(\frac{4}{3}\right)(2\,\pi)$$

$$= \left(\frac{1}{192}\right)(4!)(8) = 1$$

3.67 $\int u_{21\pm1}\, u_{210}\, \mathrm{d}r = \frac{A_2^2}{\sqrt{2}}\int e^{-x}\, x\cos\theta\, e^{-x}\, x\, \sin\theta\, e^{+\varphi} r^2 \sin\theta\, \mathrm{d}\theta\mathrm{d}\varphi\mathrm{d}r$
The integral $\int_0^{2\pi} \mathrm{e}^{\pm i\varphi}\mathrm{d}\varphi = 0$
Therefore $u_{21\pm1}$ and u_{210} are orthogonal.
Further the integral

$$\int u_{211}^* U_{21-1} d\tau \text{ involves the integral}$$

$$\int_0^{2\pi} e^{-i\varphi}\, e^{-i\varphi}\mathrm{d}\varphi \text{ or } \int_0^{2\pi} e^{-2i\varphi}\,\mathrm{d}\varphi = 0$$

So, the functions u_{211} and u_{21-1} are also orthogonal.

3.68 The degree of degenerating is given by $2n^2$. So for $n = 1$, degenerary is 2, for $n = 2$ it is 8, for $n = 3$, it is 18 and for $n = 4$, it is 32.

3.69 Parity of the state is determined by the factor $(-1)^l$. For $1s$, $l = 0$, parity= +1, for $2p$, $l = 1$, parity $= -1$ and for $3d$, $l = 2$, parity $= +1$.

3.70 To show that the probability density of the $3d$ state is independent of the polar angle θ. We form
$u^*\,u = u^*(3,\ 2,\ 0)u(3,\ 2,\ 0) + 2\,u^*\,(3,\ 2,\ 1)u(3,\ 2,\ 1) + 2\,u^*(3,\ 2,\ 2)\, u(3,\ 2,\ 2)$
The factor 2 takes care of m values ±1 and ±2, as in Table 3.2. Inserting the functions the azimuth part, $e^{i\varphi}$ or $e^{-i\varphi}$ drop off when we multiply with the complex conjugate,

i.e. $(e^{i\varphi})^* e^{i\varphi} = 1$ or $(e^{-i\varphi})^*(e^{-i\varphi}) = 1$

$$u^*u = A_3^2 e^{-2x}x^4\left[\left(\frac{1}{18}\right)(3\,\cos^2\theta - 1)^2 + \left(\frac{2}{3}\right)\sin^2\theta\cos^2\theta + \left(\frac{1}{6}\right)\sin^4\theta\right]$$

Writing $\sin^2\theta = 1 - \cos^2\theta$ and simplifying we get $u^*u = \frac{2}{9}A_3^2 e^{-2x}x^4$ which is independent of both θ and φ. Therefore the $3d$ functions are spherically symmetrical or isotropic.

3.71 $\psi_{100} = \left(\pi a_0^3\right)^{-\frac{1}{2}} \exp\left(-\frac{r}{a_0}\right)$

The probability p of finding the electron within a sphere of radius R is

$$P = \int_0^R |\psi_{100}|^2 . 4\pi r^2 \mathrm{d}r = \left(\frac{4\pi}{\pi a_0^3}\right) \int_0^R r^2 \exp\left(-\frac{2r}{a_0}\right) \mathrm{d}r$$

Set $\frac{2r}{a_0} = x$; $\mathrm{d}r = \left(\frac{a_0}{2}\right) \mathrm{d}x$

$$P = \left(\frac{4}{a_0^3}\right) \cdot \left(\frac{a_0^2}{4}\right) \cdot \left(\frac{a_0}{2}\right) \int x^2 e^{-x}\,\mathrm{d}x = {}^1/_2 \int x^2 e^{-x} \mathrm{d}x$$

Integrating by parts

$$P = {}^1/_2 [-x^2 e^{-x} + 2 \int x e^{-x} \mathrm{d}x]$$

$$= {}^1/_2 \left[-x^2 e^{-x} + 2\left\{-x\, e^{-x} + \int e^{-x} \mathrm{d}x\right\}\right]$$

$$= {}^1/_2 \left[-x^2 e^{-x} - 2x e^{-x} - 2e^{-x}\right]_0^{2R/a_0}$$

$$= {}^1/_2 \left[-\left(\frac{2R}{a_0}\right)^2 \exp\left(-\frac{2R}{a_0}\right) - 2\left(\frac{2R}{a_0}\right) \exp\left(-\frac{2R}{a_0}\right) - 2\exp\left(-\frac{2R}{a_0}\right) + 2\right]$$

$$P = 1 - e^{-\frac{2R}{a_0}} \left(1 + \frac{2R}{a_0} + \frac{2R^2}{a_0^2}\right)$$

3.72 The hydrogen wave function for $n = 2$ orbit is

$$\psi_{200} = \left(\frac{1}{4}\right) \left(2\pi a_0^3\right)^{-\frac{1}{2}} \left(2 - \frac{r}{a_0}\right) e^{-r/2a_0}$$

The probability of finding the electron at a distance r from the nucleus

$$P = |\psi_{200}|^2 \cdot 4\pi r^2 = \frac{1}{8}\frac{r^2}{a_0^3} \left(2 - \frac{r}{a_0}\right)^2 \exp\left(-\frac{r}{a_0}\right)$$

Maxima are obtained from the condition $\mathrm{d}p/\mathrm{d}r = 0$.
The maxima occur at $r = (3 - \sqrt{3})a_0$ and $r = (3 + \sqrt{3})a_0$ while minimum occur at $r = 0,\ 2a_0$ and ∞ (Fig. 3.25) (the minima are found from the condition $p = 0$).

3.73 (a) $h\nu = 13.6\, Z^2 \left(\frac{m_\mu}{m_e}\right) \left(\frac{1}{2^2} - \frac{1}{3^2}\right)$

Put $m_\mu = 106\,\mathrm{MeV}$ (instead of 106.7 MeV for muon)
$Z = 15$ and $m_e = 0.511\,\mathrm{MeV}$

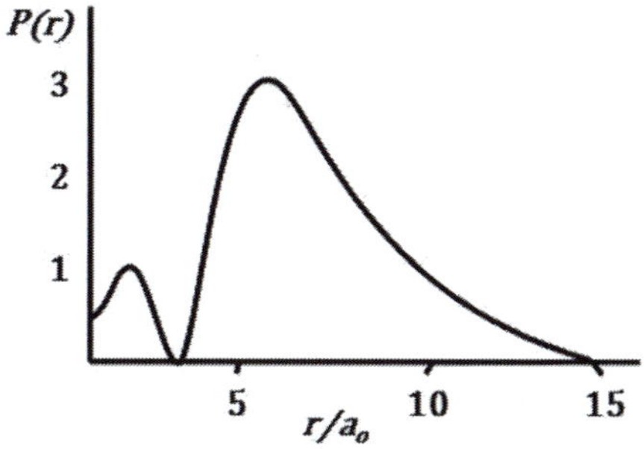

Fig. 3.25 Probability distribution of electron in $n = 2$ orbit of H-atom

$h\nu = 0.088\,\text{MeV}$
$\lambda = \frac{1241}{8.8\times10^4}\,\text{nm} = 0.0141\,\text{nm}$

(b) The transition probability per unit time is

$A \propto \omega^3 |r_{kk'}|^2$

For hydrogen-like atoms, such as the meisc atom

$$|r_{kk'}| \propto \frac{1}{z},\ \omega \propto m_\mu Z^2,\ \text{so that}$$

$A \propto m_\mu^3 Z^4$

The mean life time of the mesic atom in the 3d state is

$$\tau_\mu = \left(\frac{A_H}{A_\mu}\right)\tau_H = \left(\frac{m_e}{m_\mu}\right)^3 \tau_H / Z^4$$
$$= \left(\frac{0.511}{106}\right)^3 \left(1.6\times10^{-8}\right)\left(\frac{1}{15^4}\right) = 3.5\times10^{-20}\ \text{s}.$$

3.74 Take k along the z-axis so that $\boldsymbol{p}.r/\hbar = \boldsymbol{k}.r = kr\cos\theta$. Write

$$d\tau = r^2 dr d(\cos\theta) d\varphi$$

$$\psi(p) = \frac{1}{(2\pi\hbar)^{\frac{3}{2}}}\int_{r=0}^{\infty}\int_{\theta=0}^{\pi}\int_{\varphi=0}^{2\pi} e^{-ikr\cos\theta}\left(\pi a_0^3\right)^{-\frac{1}{2}}\exp\left(-\frac{r}{a_0}\right) r^2 dr\, d(\cos\theta)\, d\varphi \tag{1}$$

$$\int_0^{2\pi} d\varphi = 2\pi \tag{2}$$

$$\int_{-1}^{+1} e^{-ikr\cos\theta} d(\cos\theta) = \left(\frac{2}{kr}\right)\sin kr \tag{3}$$

With the aid of (2) and (3), (1) becomes

$$\psi(p) = \left(\frac{\sqrt{2}}{\pi k}\right)(\hbar a_0)^{-\frac{3}{2}}\int_0^\infty r\ \sin kr \exp\left(-\frac{r}{a_0}\right) dr \tag{4}$$

Now $\int_0^\infty r\ \sin kr\ e^{-br} = \frac{-\partial}{\partial b}\int_0^\infty \sin\ kr\ e^{-br}$

$$= -\frac{\partial}{\partial b}\left(\frac{k}{b^2+k^2}\right) = \frac{2kb}{(b^2+k^2)^2}$$

Therefore the integral in (4) is evaluated as

$$\left(\frac{2k}{a_0}\right) \Big/ \left[\left(\frac{1}{a_0^2}\right) + k^2\right] \tag{5}$$

Using the result (5) in (4), putting $k = p/\hbar$, and rearranging, we get

$$\psi(p) = \left(\frac{2\sqrt{2}}{\pi}\right)\left(\frac{\hbar}{a_0}\right)^{\frac{5}{2}} \Big/ \left[p^2 + \left(\frac{\hbar}{a_0}\right)^2\right]^2$$

or

$$|\psi(p)|^2 = \frac{8}{\pi^2}\frac{(\hbar/a_0)^5}{[p^2+(\hbar/a_0)^2]^4} \tag{6}$$

3.75 (a) $|\psi(p)|^2 = \frac{8}{\pi^2}\left(\frac{\hbar}{a_0}\right)^5 .4\,\pi p^2 \Big/ \left[p^2 + \left(\frac{\hbar}{a_0}\right)^2\right]^4$ (1)

Maximize (1)

$$\frac{\mathrm{d}}{\mathrm{d}p}|\psi(p)|^2 = 0$$

This gives $P_{\text{most probable}} = \hbar/\sqrt{3}\,a_0$

(b) $< p >= \int_0^\infty \psi_p^* p \psi_p .4\pi p^2 \mathrm{d}p$

$$= \left(\frac{32}{\pi}\right)\left(\frac{\hbar}{a_0}\right)^5 \int_0^\infty p^3 \mathrm{d}p \Big/ \left[p^2 + \left(\frac{\hbar}{a_0}\right)^2\right]^4$$

The integral I_1 is easily evaluated by the change of variable $p = \left(\frac{\hbar}{a_0}\right)\tan\theta$. Then

$$I_1 = \left(\frac{1}{8}\right)\left(\frac{a_0}{\hbar}\right)^4 \int_0^{\pi/2} \sin^3 2\theta \mathrm{d}\theta = \left(\frac{1}{12}\right)\left(\frac{a_0}{\hbar}\right)^4$$

Thus $< p >= \frac{8\hbar}{3\pi a_0}$

3.76 By Problem 3.71, the probability that

$$P\left(\frac{r}{a_0}\right) = 1 - \exp\left(-\frac{2r}{a_0}\right)\left(1 + \frac{2r}{a_0} + \frac{2r^2}{a_0^2}\right)$$

Put $p\left(\frac{r}{a_0}\right) = 0.5$ and solve the above equation numerically (see Chap. 1). We get $r = 1.337a_0$, with an error of 2 parts in 10^5.

3.3.6 Angular Momentum

3.77 $L = \begin{vmatrix} i & j & k \\ x & y & z \\ p_x & p_y & p_z \end{vmatrix}$

$= i(yp_z - zp_y) - j(xp_z - zp_x) + k(xp_y - yp_x)$

$= iL_x + jL_y + kL_z$

Thus, $L_x = yp_z - zp_y,\ L_y = zp_x - xp_z,\ L_z = xp_y - yp_x$ (1)

$[L_x, L_y] = L_xL_y - L_yL_x$

$= (yp_z - zp_y)(zp_x - xp_z) - (zp_x - xp_z)(yp_z - zp_y)$

$= yp_zzp_x - yp_zxp_z - zp_yzp_x + zp_yxp_z - zp_xyp_z + zp_xzp_y$
$+ xp_zyp_z - xp_zzp_y$ (2)

But $[p_x, p_y] = [x, p_y] = 0$, etc (3)

(2) becomes

$[L_x, L_y] = yp_xp_zz - yxp_z^2 - z^2p_yp_x + xp_yzp_y - yp_xzp_z + z^2p_xp_y$
$+ yxp_z^2 - xp_yp_zz = [z, p_z](xp_y - yp_z)$ (4)

But $[z, p_z] = [z, -i\hbar\frac{\partial}{\partial z}] = -i\hbar\left[z, \frac{\partial}{\partial z}\right]$ (5)

Further, $\left[z, \frac{\partial}{\partial z}\right] = -1$ (6)

So

$[z, p_z] = i\hbar$ (7)

Combining (1), (4) and (7) we get

$[L_x, L_y] = i\hbar L_z$ (8)

3.78 Given spin state is a singlet state, that is $S = 0$
$S_1 + S_2 = S$
Form scalar product by itself

$S_1 \cdot S_1 + S_1 \cdot S_2 + S_2 \cdot S_1 + S_2 \cdot S_2 = S \cdot S$

${S_1}^2 + 2\,S_1 \cdot S_2 + {S_2}^2 = S^2 = 0$

Now, ${S_1}^2 = {S_2}^2 = (1/2)(1/2 + 1) = 3/4$
Therefore $S_1 \cdot S_2 = -(3/4)\,\hbar^2$

3.79 For the n – p system

$\mathbf{S}_\mathrm{p} + \mathbf{S}_\mathrm{n} = \mathbf{S}$
and ${S_p}^2 = {S_n}^2 = s(s+1)$ with $s = 1/2$

(i) For singlet state, $S = 0$

$\therefore (S_p + S_n) \cdot (S_p + S_n) = S \cdot S$
$S_p{}^2 + S_n{}^2 + 2\,S_p \cdot S_n = S^2 = 0$
$½(1/2 + 1) + ½(1/2 + 1) + 2\,S_p \cdot S_n = 0$
Or $S_p \cdot S_n = -3/4$. Or $\sigma_p \cdot \sigma_n = -3$

(ii) For triplet state S = 1
$3/4 + 3/4 + 2\,S_p \cdot S_n = 1(1 + 1)$
$\therefore\ S_p \cdot S_n = 1/4$
But $S_p = ½\sigma_p$ and $S_n = ½\,\sigma_n$
$\therefore\ \sigma_p \cdot \sigma_n = 1$

3.80 From the definition of angular momentum
$L = r \times p$, we can write

$$L = \begin{vmatrix} i & j & k \\ x & y & z \\ p_x & p_y & p_z \end{vmatrix}$$

$$= i(yp_z - zp_y) + j(zp_x - xp_z)$$
$$+ k(xp_y - yp_x)$$
$$= iL_x + jL_y + kL_z$$

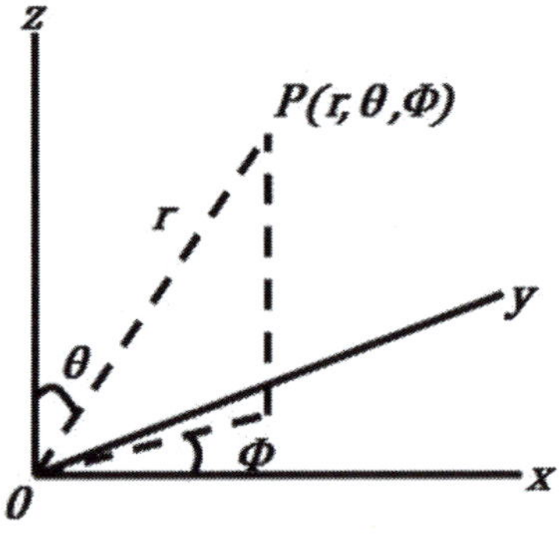

Fig. 3.26 Cartesian and polar coordinates

$$L_x = yp_z - zp_y = -i\hbar\left(y\frac{\partial}{\partial z} - z\frac{\partial}{\partial y}\right)$$
$$L_y = zp_x - xp_z = -i\hbar\left(z\frac{\partial}{\partial x} - x\frac{\partial}{\partial z}\right) \quad (1)$$
$$L_z = xp_y - yp_x = -i\hbar\left(x\frac{\partial}{\partial y} - y\frac{\partial}{\partial x}\right)$$

If θ is the polar angle, φ the azimuthal angle and r the radial distance, (Fig. 3.26). Then

$$x = r\sin\theta\cos\varphi$$

$$y = r\sin\theta\sin\varphi \tag{2}$$

$$z = r\cos\theta$$

$$r^2 = x^2 + y^2 + z^2 \tag{3}$$

$$\tan^2\theta = \frac{x^2 + y^2}{z^2} \tag{4}$$

$$\tan\varphi = \frac{y}{x} \tag{5}$$

Differentiating (3), (4) and (5) partially with respect to x

$$\frac{\partial r}{\partial x} = \sin\theta\cos\varphi;\ \frac{\partial r}{\partial y} = \sin\theta\ \sin\varphi;\ \frac{\partial r}{\partial z} = \cos\theta \tag{6}$$

$$\frac{\partial\theta}{\partial x} = \left(\frac{1}{r}\right)\cos\theta\cos\varphi;\ \frac{\partial\theta}{\partial y} = \left(\frac{1}{r}\right)\cos\theta\sin\varphi;\ \frac{\partial\theta}{\partial z} = -\frac{\sin\theta}{r} \tag{7}$$

$$\frac{\partial\varphi}{\partial x} = -\left(\frac{1}{r}\right)\mathrm{cosec}\theta\ \sin\varphi;\ \frac{\partial\varphi}{\partial y} = \frac{\cos\varphi}{r\sin\theta};\ \frac{\partial\varphi}{\partial z} = 0 \tag{8}$$

$$L_z\psi(r,\theta\varphi) = -i\hbar\left(\frac{x\partial\psi}{\partial y} - \frac{y\partial\psi}{\partial x}\right)$$

$$= -i\hbar\left[x\left(\frac{\partial\psi}{\partial r}\right)\cdot\left(\frac{\partial r}{\partial y}\right) + \left(\frac{\partial\psi}{\partial\theta}\right)\cdot\left(\frac{\partial\theta}{\partial y}\right) + \left(\frac{\partial\psi}{\partial\varphi}\right)\cdot\left(\frac{\partial\varphi}{\partial y}\right)\right.$$

$$\left. -y\left(\frac{\partial\psi}{\partial r}\right)\cdot\left(\frac{\partial r}{\partial x}\right) + \left(\frac{\partial\psi}{\partial\theta}\right)\cdot\left(\frac{\partial\theta}{\partial x}\right) + \left(\frac{\partial\psi}{\partial\varphi}\right)\cdot\left(\frac{\partial\varphi}{\partial x}\right)\right]$$

$$L_z\psi(r,\theta,\varphi)$$

$$= -i\hbar\left[\frac{\partial\psi}{\partial r}\left(x\frac{\partial r}{\partial y} - y\frac{\partial r}{\partial x}\right) + \frac{\partial\psi}{\partial\theta}\left(x\frac{\partial\theta}{\partial y} - y\frac{\partial\theta}{\partial x}\right)\right.$$

$$\left. + \frac{\partial\psi}{\partial\varphi}\left(x\frac{\partial\varphi}{\partial y} - y\frac{\partial\varphi}{\partial x}\right)\right] \tag{9}$$

Substituting (2), (6), (7) and (8) in (9) and simplifying, the first two terms drop off and the third one reduces to $\partial\psi/\partial\varphi$, yielding

$$L_z = -i\hbar\frac{\partial}{\partial\varphi} \tag{10}$$

In Problem 3.15 it was shown that the Schrodinger equation was separated into radial (r) and angular parts (θ and φ). The angular part was shown to be separated into θ and φ components. The solution to φ was shown to be

$$g(\varphi) = \frac{1}{\sqrt{2\pi}}e^{im\varphi}$$

where m is an integer.

Now $L_z g(\varphi) = -i\hbar\frac{\partial g}{\partial \varphi} = m\hbar g(\varphi)$

Thus the z-component of angular momentum is quantized with eigen value $\hbar$.

3.81 One can do similar calculations for L_x and L_y as in Problem 3.80 and obtain

$$\frac{L_x}{i\hbar} = \sin\varphi\frac{\partial}{\partial\theta} + \cot\theta\,\cos\varphi\frac{\partial}{\partial\varphi}$$

$$\frac{L_y}{i\hbar} = -\cos\varphi\frac{\partial}{\partial\theta} + \cot\theta\sin\varphi\frac{\partial}{\partial\varphi}$$

3.82 Using $L^2 = L_x^2 + L_y^2 + L_z^2$, the commutator with total angular momentum squared can be evaluated

$$[L^2, L_z] = \left[L_x^2 + L_y^2 + L_z^2, L_z\right]$$

$$= \left[L_x^2 + L_y^2, L_z\right]$$

$$= L_x[L_x, L_z] + [L_x, L_z]L_x + L_y\left[L_y, L_z\right] + \left[L_y, L_z\right]L_y \quad (1)$$

$$= -i\hbar L_xL_y - i\hbar L_yL_x + i\hbar L_yL_x + i\hbar L_xL_y = 0$$

Similarly $\left[L^2, L_x\right] = [L^2, L_y] = [L^2, L] = 0$

3.83 $L^2 = L_x^2 + L_y^2 + L_z^2$

Using the expressions for L_x, L_y and L_z from problem (3.81)

$$\frac{L^2}{(i\hbar)^2} = \left(\sin\,\varphi\frac{\partial}{\partial\theta} + \cot\,\theta\cos\,\varphi\frac{\partial}{\partial\varphi}\right)\left(\sin\,\varphi\frac{\partial}{\partial\theta} + \cot\,\theta\cos\,\varphi\frac{\partial}{\partial\varphi}\right)$$

$$+\left(-\cos\,\varphi\frac{\partial}{\partial\theta} + \cot\,\theta\sin\,\varphi\frac{\partial}{\partial\varphi}\right)\left(-\cos\,\varphi\frac{\partial}{\partial\theta} + \cot\,\theta\sin\varphi\frac{\partial}{\partial\varphi}\right)$$

$$+\left(-\frac{\partial}{\partial\varphi}\right)\left(-\frac{\partial}{\partial\varphi}\right)$$

$$= \sin^2\varphi\frac{\partial^2}{\partial\theta^2} + \cot^2\theta\cos^2\varphi\frac{\partial^2}{\partial\varphi^2} - \sin\varphi\cos\,\varphi\,\mathrm{cosec}^2\theta\frac{\partial}{\partial\varphi} + \cos^2\varphi\cot\,\theta\frac{\partial}{\partial\theta}$$

$$+\cos^2\varphi\frac{\partial^2}{\partial\theta^2} + \cot^2\theta\sin^2\varphi\frac{\partial^2}{\partial\varphi^2} + \sin\varphi\cos\varphi\,\mathrm{cosec}^2\theta\frac{\partial}{\partial\varphi}$$

$$+\sin^2\varphi\cot\theta\frac{\partial}{\partial\theta} + \frac{\partial^2}{\partial\varphi^2}$$

The cross terms get cancelled and the expression is reduced to

$$\frac{\partial^2}{\partial\theta^2} + \frac{1}{\sin^2\theta}\frac{\partial^2}{\partial\varphi^2} + \cot\theta\frac{\partial}{\partial\theta}$$

$$\nabla^2\psi = \frac{1}{r^2}\frac{\partial}{\partial r}\left(r^2\frac{\partial\psi}{\partial r}\right) + \frac{1}{r^2\sin\theta}\frac{\partial}{\partial\theta}\left(\sin\theta\frac{\partial\psi}{\partial\theta}\right) + \frac{1}{r^2\sin^2\theta}\frac{\partial^2\psi}{\partial\varphi^2}$$

Apart from the factor $1/r^2$, the angular part is seen to be

$$\frac{\partial^2 \psi}{\partial \theta^2} + \cot\theta \frac{\partial \psi}{\partial \theta} + \frac{1}{\sin^2} \frac{\partial^2 \psi}{\partial \varphi^2}$$

3.84 (a) The (i, j)th matrix element of an operator O is defined by

$$O_{ij} = < i|o|j > \quad (1)$$

For $j = 1/2, \; m = 1/2$ and $-1/2$. The two states are

$$|1 >= |\frac{1}{2}, \frac{1}{2} > \text{ and } |2 >= |\frac{1}{2}, -\frac{1}{2} > \quad (2)$$

With the notation $|j, m >$

Now

$$< j'm'|J_z|jm >= m\hbar\delta_{jj'}\delta_{mm'} \quad (3)$$

Thus

$$(J_z)_{11} = < 1|J_z|1 >= \frac{1}{2}\hbar \quad (4)$$

$$(J_z)_{22} = < 2|J_z|2 >= -\frac{1}{2}\hbar \quad (5)$$

$$(J_z)_{12} = < 1|J_z|2 >= \left\langle \frac{1}{2}, \frac{1}{2}|J_z|\frac{1}{2}, \frac{1}{2} \right\rangle = 0 \quad (6)$$

because of (3).

Similarly

$$(J_z)_{21} = 0 \quad (7)$$

Therefore

$$J_z = \frac{\hbar}{2}\begin{pmatrix} 1 & 0 \\ 0 & -1 \end{pmatrix} \quad (8)$$

For J_x and J_y, we use the relations

$$J_x = {}^1\!/_2(J_+ + J_-) \text{ and } J_y = -\left(\frac{1}{2i}\right)(J_+ - J_-)$$

$$< j, m|J_x|j, m >=< j, m|\frac{1}{2}(J_+ + J_-)|j, m >$$

$$= {}^1\!/_2\hbar[(j + m + 1)(j - m)]^{1/2} < j, m'|j, m + 1 >$$

$$+ {}^1\!/_2[(j - m + 1)(j + m)]^{1/2} < j, m'|j, m - 1 >$$

$$= {}^1\!/_2[(j + m + 1)(j - m)]^{1/2}\hbar\delta_{m',m+1}$$

$$+ {}^1\!/_2[(j - m + 1)(j + m)]^{1/2}\delta_{m',m-1}$$

That is the matrix element is zero unless $m' = m + 1$ or $m' = m - 1$. The first delta factor survives

$$(J_x)_{12} = < 1|J_x|2 > = \left\langle \frac{1}{2}, \frac{1}{2}\, |J_x|\, \frac{1}{2}, -\frac{1}{2} \right\rangle$$

$$= \frac{1}{2}\left[\left(\frac{1}{2} - \frac{1}{2} + 1\right)\left(\frac{1}{2} + \frac{1}{2}\right)\right]^{1/2} = {}^1\!/_2\, \hbar$$

$$(J_x)_{21} = \left\langle \frac{1}{2}, -\frac{1}{2}\, |J|\, \frac{1}{2}, \frac{1}{2} \right\rangle = {}^1\!/_2\, \hbar$$

because the second delta factor survives
$(J_x)_{11} = < 1|J_x|1 > = \langle \frac{1}{2}, \frac{1}{2}|J_x|\frac{1}{2}, \frac{1}{2}\rangle = 0$ because of delta factors.
Similarly, $(J_x)_{22} = 0$
Thus $J_x = \frac{\hbar}{2}\begin{pmatrix} 0 & 1 \\ 1 & 0 \end{pmatrix}$; $J_y = \frac{\hbar}{2}\begin{pmatrix} 0 & -i \\ i & 0 \end{pmatrix}$

$$J_z = \frac{\hbar}{2}\begin{pmatrix} 1 & 0 \\ 0 & -1 \end{pmatrix} \tag{9}$$

These three matrices are known as Pauli matrices.

(b) $J^2 = J_x^2 + J_y^2 + J_z^2$
Using the matrices given in (6), squaring them and adding we get

$$J^2 = {}^3\!/_4 \hbar^2 \begin{pmatrix} 1 & 0 \\ 0 & 1 \end{pmatrix}$$

3.85 (a) For $j = 1$, $m = 1, 0$ and -1, the three base states are denoted by $|1>$, $|2>$ and $|3>$. In the $|j, m>$ notation $|1> = |1, 1>$, $|2> = |1, 0>$, $|3> = |1, -1>$

$$J_z|1> = m\hbar|1> = \hbar|1>$$
$$J_z|2> = 0.\hbar|2> = 0$$
$$J_z|3> = -\hbar|3>$$
$$(J_z)_{11} = < 1|J_z|1 > = < 1, 1|J_z|1, 1 > = \hbar$$
$$(J_z)_{22} = < 2|J_z|2 > = < 1, 0|J_z|1, 0 > = 0$$
$$(J_z)_{33} = < 3|J_z|3 > = < 1, -1|J_z|1, -1 > = -\hbar$$
$$(J_z)_{12} = (J_z)_{21} = (J_z)_{13} = (J_z)_{31} = (J_z)_{23} = (J_z)_{32} = 0$$

because of δ – factor $\delta_{mm'}$

$$J_z = \hbar\begin{pmatrix} 1 & 0 & 0 \\ 0 & 0 & 0 \\ 0 & 0 & -1 \end{pmatrix}$$

For the calculation of J_x and J_y, we need to work out J_+ and J_-.

$$J_+|1> = 0$$
$$J_+|2> = [(j - m)(j + m + 1)]^{1/2}\hbar|1>$$

$= [(1-0)(1+0+1)]^{1/2}\hbar|1> = \sqrt{2}\hbar|1>$

$J_+ \left|3> = [[1-(-1)](1-1+1)]^{\frac{1}{2}}\hbar\right| 2> = \sqrt{2}\,\hbar|2>$

$J_-|1> = [(j+m)(j-m+1)]^{1/2}\hbar|2>$

$= [(1+1)(1-1+1)]^{1/2}\hbar\,|2> = \sqrt{2}\,\hbar|2>$

$J_-|2> = \sqrt{2}\,\hbar|3>$

$J_-|3> = 0$

Matrices for J_x and J_y:

$(J_x)_{11} = <1|J_x|1> = 1/2 <1|J_+ + J_-|1>$

$= 1/2 <1|J_+|1> + \frac{1}{2} <1|J_-|1>$

$= 1/2[(j-m)(j+m+1)]^{1/2}\hbar\delta_{m'm+1}$

$+ 1/2[(j+m)(j-m+1)]^{1/2}\delta_{m',m-1=0}$

Similarly; $(J_x)_{22} = (J_x)_{33} = 0$

$(J_x)_{12} = <1|J_x|2> = 1/2 <1|J_+ + J_-|2> = 1/2\ <1,1|J_+ + J_-|1,0>$

$= {}^1/_2[(j-m)(j+m+1)]^{1/2}\delta_{m'm+1} + 1/2[(j+m)(j-m+1)]^{1/2}\delta_{m'm-1}$

The second delta is zero

$\therefore \quad (J_x)_{12} = \frac{1}{2}[(1+0)(1+0+1)]^{1/2}\hbar = \frac{\hbar}{\sqrt{2}}$

Similarly, $(J_x)_{21} = (J_x)_{23} = (J_x)_{32} = \frac{\hbar}{\sqrt{2}}$

By a similar procedure the matrix elements of J_y can be found out. Thus

$$J_x = \frac{\hbar}{\sqrt{2}}\begin{pmatrix}0&1&0\\1&0&1\\0&1&0\end{pmatrix};\ J_y = \frac{\hbar}{\sqrt{2}}\begin{pmatrix}0&-i&0\\i&0&-i\\0&i&0\end{pmatrix};\ J_z = \hbar\begin{pmatrix}1&0&0\\0&0&0\\0&0&-1\end{pmatrix}$$

(b) For the matrix elements of J we can use the relation

$< j'm'|J^2|jm> = j(j+1)\hbar^2\delta_{j'j}\delta_{m'm}$

Thus $(J^2)_{11} = (J^2)_{22} = (J^2)_{33} = 1(1+1)\hbar^2 = 2\hbar^2$

$(J^2)_{12} = (J^2)_{21} = (J^2)_{13} = (J^2)_{31} = (J^2)_{23} = (J^2)_{32} = 0$

$$J^2 = 2\hbar^2\begin{pmatrix}1&0&0\\0&1&0\\0&0&1\end{pmatrix}$$

Alternatively, $J^2 = J_x^2 + J_y^2 + J_z^2$

Using the matrices which have been derived the same result is obtained.

3.86 With the addition of $j_1 = 1$ and $j_2 = 1/2$, one can get $J = 3/2$ or ½. In the $(J,\ M)$ notation in all one gets 6 states

$$\psi\left(\frac{3}{2},\frac{3}{2}\right),\psi\left(\frac{3}{2},\frac{1}{2}\right),\psi\left(\frac{3}{2},-\frac{1}{2}\right),\psi\left(\frac{3}{2},-\frac{3}{2}\right)\text{ and }\psi\left(\frac{1}{2},\frac{1}{2}\right),\psi\left(\frac{1}{2},-\frac{1}{2}\right)$$

Clearly the states $\psi\left(\frac{3}{2}, \frac{3}{2}\right)$ and $\psi\left(\frac{3}{2}, -\frac{3}{2}\right)$ can be formed in only one way

$$\psi\left(\frac{3}{2}, \frac{3}{2}\right) = \varphi(1, 1)\varphi\left(\frac{1}{2}, \frac{1}{2}\right) \tag{1}$$

$$\psi\left(\frac{3}{2}, -\frac{3}{2}\right) = \varphi(1, -1)\varphi\left(\frac{1}{2}, -\frac{1}{2}\right) \tag{2}$$

We now use the ladder operators J_+ and J_- to generate the second and the third states.

$$J_+\varphi(j, m) = [(j - m)(j + m + 1)]^{\frac{1}{2}}\varphi(j, m + 1) \tag{3}$$

$$J_-\varphi(j, m) = [(j + m)(j - m + 1)]^{\frac{1}{2}}\varphi(j, m - 1) \tag{4}$$

Applying (4) to (1) on both sides

$$J_-\psi\left(\frac{3}{2}, \frac{3}{2}\right) = \left[\left(\frac{3}{2} + \frac{3}{2}\right)\left(\frac{3}{2} - \frac{3}{2} + 1\right)\right]^{1/2} = \sqrt{3}\psi\left(\frac{3}{2}, \frac{1}{2}\right)$$

$$= J_-\varphi(1, 1)\varphi\left(\frac{1}{2}, \frac{1}{2}\right)$$

$$= \varphi(1, 1)J_-\left(\frac{1}{2}, \frac{1}{2}\right) + \varphi\left(\frac{1}{2}, \frac{1}{2}\right)J_-\varphi(1, 1)$$

$$= \varphi(1, 1)\varphi\left(\frac{1}{2}, -\frac{1}{2}\right) + \varphi\left(\frac{1}{2}, \frac{1}{2}\right)\sqrt{2}\varphi(1, 0)$$

Thus $$\psi\left(\frac{3}{2}, \frac{1}{2}\right) = \sqrt{\frac{2}{3}}\varphi(1, 0)\varphi\left(\frac{1}{2}, \frac{1}{2}\right) + \sqrt{\frac{1}{3}}\varphi(1, 1)\varphi\left(\frac{1}{2}, -\frac{1}{2}\right) \tag{5}$$

Similarly, applying J_+ operator given by (3) to the state $\psi\left(\frac{3}{2}, -\frac{3}{2}\right)$ we obtain

$$\psi\left(\frac{3}{2}, -\frac{1}{2}\right) = \sqrt{\frac{2}{3}}\varphi(1, 0)\varphi\left(\frac{1}{2}, -\frac{1}{2}\right) + \sqrt{\frac{1}{3}}\varphi(1, -1)\varphi\left(\frac{1}{2}, \frac{1}{2}\right) \tag{6}$$

The $J = 1/2$ state can be obtained by making it as a linear combination

$$\psi\left(\frac{1}{2}, \frac{1}{2}\right) = a\varphi(1, 1)\varphi\left(\frac{1}{2}, -\frac{1}{2}\right) + b\varphi(1, 0)\varphi\left(\frac{1}{2}, \frac{1}{2}\right) \tag{7}$$

For normalization reason,

$$a^2 + b^2 = 1 \tag{8}$$

We can obtain one other relation by making (7) orthogonal to (5)

$$a\sqrt{\frac{1}{3}} + b\sqrt{\frac{2}{3}} = 0$$

Or $a = -\sqrt{2}b$ (9)

Same result is obtained by applying J_+ operator to (7). $J_+\psi(1/2, 1/2) = 0$

Solving (8) and (9), $= \sqrt{\frac{2}{3}},\ b = -\sqrt{\frac{1}{3}}$ (10)

Thus

$$\psi\left(\frac{1}{2},\frac{1}{2}\right)=\sqrt{\frac{2}{3}}\varphi(1,1)\varphi\left(\frac{1}{2},-\frac{1}{2}\right)-\sqrt{\frac{1}{3}}\varphi(1,0)\varphi\left(\frac{1}{2},\frac{1}{2}\right) \tag{11}$$

Similarly

$$\psi\left(\frac{1}{2},-\frac{1}{2}\right)=\sqrt{\frac{1}{3}}\varphi(1,0)\varphi\left(\frac{1}{2},-\frac{1}{2}\right)-\sqrt{\frac{2}{3}}\varphi(1,-1)\varphi\left(\frac{1}{2},\frac{1}{2}\right) \tag{12}$$

The coefficients appearing in (1), (2), (5), (6), (11) and (12) are known as Clebsch – Gordon coefficients. These are displayed in Table 3.3.

3.87 In spherical coordinates

$$x = r\ \sin\theta\ \cos\varphi\ ;\ y = r\ \sin\theta\ \sin\varphi\ ;\ z = r\ \cos\theta \tag{1}$$

So that $xy + yz + zx = r^2 \sin^2\theta \sin\ \varphi\ \cos\ \varphi + r^2 \sin\ \theta\ \cos\ \theta\ \sin\ \varphi$

$$+ r^2 \sin\ \theta\ \cos\ \theta\ \cos\ \varphi \tag{2}$$

The spherical harmonics are

$$Y_{00} = \left(\frac{1}{4\pi}\right)^{1/2};\ Y_{10} = \left(\frac{3}{4\pi}\right)^{1/2} \cos\ \theta\ ;\ Y_{1\pm1} = \mp\left(\frac{3}{8\pi}\right)^{1/2} \sin\ \theta\ e^{\pm\varphi}$$

$$Y_{20} = \left(\frac{5}{16\pi}\right)^{\frac{1}{2}} (3\cos^2\ \theta - 1);\ Y_{2\pm1} = \mp\left(\frac{15}{8\pi}\right)^{\frac{1}{2}} \sin\ \theta\ \cos\ \theta\ e^{\pm i\varphi};$$

$$Y_{2\pm2} = \left(\frac{15}{32\pi}\right)^{1/2} \sin^2\theta\ e^{\pm 2i\varphi} \tag{3}$$

Expressing (2) in terms of (3),

$$\sin^2\theta\ \sin\ \varphi\ \cos\ \varphi = {}^1/_2\ \sin^2\theta \sin 2\varphi = \frac{Y_{22} - Y_{2-2}}{4i}\left(\frac{32\pi}{15}\right)^{1/2}$$

Similarly $\sin\ \theta\ \cos\ \theta\ \sin\ \varphi = \left(\frac{8\pi}{15}\right)^{1/2} \frac{(Y_{21} - Y_{2-1})}{2i}$

$$\sin\ \theta\ \cos\ \theta\ \cos\ \varphi = \left(\frac{8\pi}{15}\right)^{\frac{1}{2}} (Y_{21} - Y_{2-1})/2$$

Hence, $xy + yz + zx = r^2\left(\frac{8\pi}{15}\right)^{\frac{1}{2}} [(Y_{22} - Y_{2-1})/i + (Y_{21} + Y_{2-1})/2i + (Y_{21} - Y_{2-1})/2]$

The above expression does not contain Y_{00} corresponding to $l = 0$, nor Y_{10} and $Y_{1\pm1}$ corresponding to $l = 1$. All the terms belong to $l = 2$, and the probability for finding $l = 2$ and therefore $L^2 = l(l + 1)\hbar = 6\hbar^2$ is unity.

3.88 $L_z = -i\hbar\dfrac{\partial}{\partial\varphi}$

$x = r\ \sin\ \theta\ \cos\ \varphi\ ;\ y = r \sin\ \theta\ \sin\ \varphi\ ;\ z = r \cos\ \theta$

$\psi_1 = (x + iy)f(r) = r \sin\ \theta(\cos\ \varphi + i \sin\ \varphi)f(r)$

$= (\cos\ \varphi + i\ \sin\ \varphi) \sin\ \theta F(r)$

where $F(r) = r f(r)$

$$L_z\psi_1 = -i\hbar\frac{\partial}{\partial\varphi}(\cos\varphi + i\sin\varphi)\sin\theta\, F(r)$$
$$= -i\hbar(-\sin\varphi + i\cos\varphi)\sin\theta\, F(r)$$
$$= \hbar(\cos\varphi + i\sin\varphi)\sin\theta\, F(r)$$
$$= \hbar\psi_1$$

Thus ψ_1 is the eigen state and the eigen value is $\hbar$

$$\psi_2 = zf(r) = r\cos\theta f(r)$$
$$L_z\psi_2 = -i\hbar\frac{\partial}{\partial\varphi}(r\cos\theta f(r)) = 0$$

The eigen value is zero.

$$\psi_3 = (x - iy)f(r) = r\sin\theta(\cos\varphi - i\sin\varphi)f(r)$$
$$= (\cos\varphi - i\sin\varphi)\sin\theta\, F(r)$$
$$L_z\psi_3 = -i\hbar\frac{\partial}{\partial\varphi}(\cos\varphi - i\sin\varphi)\sin\theta\, F(r)$$
$$= -i\hbar(-\sin\varphi - i\cos\varphi)\sin\theta\, F(r)$$
$$= i\hbar(\sin\varphi + i\cos\varphi)\sin\theta\, F(r)$$
$$= -\hbar(\cos\varphi - i\sin\varphi)\sin\theta\, F(r) = -\hbar\psi_3$$

Thus ψ_3 is an eigen state and the eigen value is $-\hbar$.

3.89 (a) $L_z = -i\hbar\dfrac{\partial}{\partial\varphi}$

$$L_z\psi = -i\hbar\frac{\partial}{\partial\varphi}Af(r)\sin\theta\cos\theta\, e^{i\varphi}$$
$$= (i)(-i\hbar A f(r)\sin\theta\cos\theta\, e^{i\varphi})$$
$$= \hbar A f(r)\sin\theta\cos\theta\, e^{i\varphi}$$
$$= \hbar\psi$$

Therefore, the z-component of the angular momentum is $\hbar$.

(b) $L^2 = -\hbar^2\left\{\frac{\partial^2}{\partial\theta^2} + \cot\theta\frac{\partial}{\partial\theta} + \left(\frac{1}{\sin^2\theta}\right)\frac{\partial^2}{\partial\varphi^2}\right\}$

Expressions for L_z and L^2 are derived in Problems 3.80 and 3.83.

$$L^2\psi = -\hbar^2\left\{\frac{\partial^2}{\partial\theta^2} + \cot\theta\frac{\partial}{\partial\theta} + \left(\frac{1}{\sin^2\theta}\right)\cdot\frac{\partial^2}{\partial\varphi^2}\right\}Af(r)\sin\theta\cos\theta\, e^{i\varphi}$$
$$= -\hbar^2 Af(r)e^{i\varphi}\left\{-4\sin\theta\cos\theta + \cot\theta\,(\cos^2\theta - \sin^2\theta) - \frac{\sin\theta\cos\theta}{\sin^2\theta}\right\}$$
$$= 6\hbar^2 Af(r)\sin\theta\cos\theta\, e^{i\varphi}$$
$$= 6\,\hbar^2\psi$$

Thus $L^2 = 6\,\hbar^2$

But $L^2 = l(l+1)$. Therefore $l = 2$

3.90 With reference to the Table 3.2, the function $\psi(r, \theta, \varphi)$ is the $2p$ function for the hydrogen atom.

(a) $L_z = -i\hbar\frac{\partial}{\partial\varphi}$
Applying L_z to the wavefunction

$$L_z\psi(r,\theta,\varphi) = -i\hbar\frac{\partial\psi(r,\theta,\varphi)}{\partial\varphi}$$

$$= (-i\hbar)(-i)\psi(r,\theta,\varphi)$$

$$= -\hbar\psi(r,\theta,\varphi)$$

Therefore, the value of L_z is $-\hbar$

(b) As it is a p-state, $l = 1$ and the parity is $(-1)^l = (-1)^l = -1$, that is an odd parity.

3.91 $$L_x = i\,\hbar\left(\sin\,\varphi\,\frac{\partial}{\partial\theta} + \cot\,\theta\cos\,\varphi\frac{\partial}{\partial\varphi}\right) \tag{1}$$

$$L_y = i\hbar\left(-\cos\,\varphi\frac{\partial}{\partial\theta} + \cot\,\theta\,\sin\,\varphi\frac{\partial}{\partial\varphi}\right) \tag{2}$$

$$L_+ = L_x + iL_y = i\hbar\left(\sin\,\varphi\frac{\partial}{\partial\theta} + \cot\,\theta\,\cos\,\varphi\frac{\partial}{\partial\varphi}\right)$$

$$-\hbar\left(-\cos\,\varphi\frac{\partial}{\partial\theta} + \cot\,\theta\,\sin\,\varphi\frac{\partial}{\partial\varphi}\right) \tag{3}$$

Apply (3) to the $m = +1$ state which is proportional to $\sin\theta\, e^{i\varphi}$

$$L_+(\sin\,\theta\, e^{i\varphi}) = i\hbar(\sin\,\varphi\,\cos\,\theta + i\cot\,\theta\,\cos\,\varphi)e^{i\varphi}$$

$$-\hbar(-\cos\,\varphi\cos\,\theta i + \cot\,\theta\sin\,\varphi\sin\,\theta)e^{i\varphi}$$

$$= i\hbar(\sin\,\varphi\cos\,\theta - \cot\,\theta\sin\,\varphi\sin\,\theta)e^{i\varphi}$$

$$-\hbar(\cos\,\theta\cos\,\varphi - \cos\,\varphi\cos\,\theta)e^{i\varphi} = 0$$

Thus the state with $m = 2$ does not exist. Similarly by applying L_- to the state with $m = -1$, it can be shown that the $m = -2$ state does not exist.

3.92 Particles with even spin (0, 2, 4 ...) obey Bose statistics and those with odd spin (1/2, 3/2, 5/2 ...) obey Fermi – Dirac statistics.

Consider a diatomic molecule with identical nuclei. The total wave function may be written as

$\psi = \psi_{\text{elec}}\zeta_{\text{vib}}\rho_{\text{rot}}\sigma_{\text{nuc}}$

Let p be an operator which exchanges the space and spin coordinates.

Now $p\psi_{\text{elec}} = \pm\psi_{\text{elec}}$

It is known from molecular spectroscopy, that for the ground state it is positive. Furthermore, $P\zeta_{\text{vib}} = +\zeta_{\text{vib}}$, because ζ_{vib} depends only on the distance of separation of nuclei.

Now $\rho = P_l^{\text{m}}\,(\cos\theta)e^{im\varphi}$, where θ is the polar angle and φ the azimuth angle; ρ is represented by the associated Legendre function.

Exchange of $x \rightarrow -x$, $y \rightarrow -y$, $z \rightarrow -z$ implies $\theta \rightarrow \pi - \theta$ and $\varphi \rightarrow \pi + \varphi$, so that

$P_1{}^m (\cos\theta) \rightarrow (-1)^{l+m}$

and $e^{im\varphi} \rightarrow (-1)^m e^{im\varphi}$, Thus $P\rho = (-1)^{2m}(-1)^l = (-1)^l \rho$

where m is an integer. Thus ρ is symmetrical for even l and antisymmetrical for odd l.

First consider zero nuclear spin. The total wave function ψ is antisymmetrical for odd l and symmetrical for even l. As the nuclei must obey either Fermi or Bose statistics, either only the $l =$ odd states must exist or only the $l =$ even states must exist. It turns out that for nuclei with zero spin only the even rotational states exist and odd rotational states are missing.

Next consider the case of non-zero spin. A nucleus of total angular momentum I can have a component M in any prescribed direction taking $2I+1$ values in all $(I, I-1, \ldots -I)$, that is $2I+1$ states exist. For the two identical nuclei $(2I+1)^2$ wave functions of the form $\psi_{M1}(A)\psi_{M2}(B)$ can be constructed. If the two nuclei are identical, these simple products must be replaced by linear combination of those products which are symmetric or antisymmetric for interchange of nuclei. If $M_1 = M_2$, the products themselves are $(2I+1)$ symmetric wave functions, the remaining $(2I+1)^2 - (2I+1) = 2I(2I+1)$ functions with $M_1 \neq M_2$ have the form $\psi_{M1}(A)\psi_{M2}(B)$ *and* $\psi_{M2}(A)\psi_{M1}(B)$. Each such pair can be replaced by one symmetric and one antisymmetric wave function of the form $\psi_{M1}(A)\psi_{M2}(B) \pm \psi_{M2}(A)\psi_{M1}(B)$. Thus, half of $2I(2I+1)$ functions, that is $I(2I+1)$ are symmetric and an equal number antisymmetric. Therefore, total number of symmetric wave functions $= (2I+1) + I(2I+1) = (2I+1)(I+1)$. Total number of antisymmetric wave functions $= I(2I+1)$. Therefore, the ratio of the number of symmetric and antisymmetric functions is $(I+1)/I$.

From the previous discussion it was shown that for the symmetric electronic wave function of the molecule the interchange of nuclei produces a factor $(-1)^l$ in the molecular wave function. Thus, for nuclei obeying Bose statistics symmetric nuclear spin functions must combine with even l. Because of the statistical weight attached to spin states, the intensity of even rotational lines will be $(I+1)/I$ as great as that of neighboring odd rotational lines.

For nuclei obeying Fermi statistics, the spin and rotational states combine in a manner opposite to the previously described and the odd rotational lines are more intense in the ratio $(I+1)/I$.

Thus, by determining which lines are more intense, even or odd, the nuclear statistics is determined and by measuring the ratio of intensities of adjacent lines the nuclear spin is obtained. The reason for comparing the intensity of neighboring lines is that the intensity of rotational lines varies according to the occupation number of rotational states governed by the Boltzmann distribution.

3.93
$$u(\theta,\varphi) = {}^1\!/_4\sqrt{\frac{15}{\pi}}\sin 2\theta\cos 2\varphi$$
$$= {}^1\!/_4\sqrt{\frac{15}{\pi}}\sin^2\theta\left[\frac{(e^{2i\varphi}+e^{-2i\varphi})}{2}\right] \tag{1}$$
$$\text{But } Y_{2+2}(\theta,\varphi) = \sqrt{\frac{15}{32\pi}}\sin^2\theta e^{2i\varphi} \tag{2}$$
$$Y_{2-2}(\theta,\varphi) = \sqrt{\frac{15}{32\pi}}\sin^2\theta e^{-2i\varphi} \tag{3}$$

Adding (2) and (3)

$$Y_{2+2}(\theta,\varphi) + Y_{2-2}(\theta,\varphi) = \sqrt{\frac{15}{32\pi}}\sin^2\theta(e^{2i\varphi}+e^{-2i\varphi}) \tag{4}$$

Dividing (1) by (4) and simplifying we get

$$u(\theta,\varphi) = \frac{1}{\sqrt{2}}(Y_{2+2}(\theta,\varphi) + Y_{2-2}(\theta,\varphi))$$

We know that
$L^2 Y_{lm} = \hbar^2 l(l+1)Y_{lm}$
So $L^2 u(\theta,\varphi) = \frac{L^2[Y_{22}(\theta,\varphi)+Y_{2-2}(\theta,\varphi)]}{\sqrt{2}}$
$= 2(2+1)\hbar^2\frac{[Y_{22}(\theta,\varphi)+Y_{2-2}(\theta,\varphi)]}{\sqrt{2}}$
$= 6\hbar^2 u(\theta,\varphi)$
Thus the eigen value of L^2 is $6\hbar^2$

3.94 The wave function is identified as ψ_{322}

$$L_Z\psi = -i\hbar\frac{\partial\psi}{\partial\varphi} = 2\hbar\psi$$

Thus the eigen value of L_Z is $2\hbar$.

3.95 (a) Using the values, $Y_{10} = \sqrt{\frac{3}{4\pi}}\cos\theta$ and $Y_{1,\pm 1} = \mp\sqrt{\frac{3}{8\pi}}\sin\theta\exp(\pm i\varphi)$, we can write

$$\psi = \sqrt{\frac{1}{3}}\left(-\sqrt{2}Y_{11} + Y_{10}\right) f(r)$$

Hence the possible values of L_Z are $+\hbar$, 0

(b) First we show that the wavefunction is normalized.

$$\int |\psi|^2 \mathrm{d}r = \frac{1}{4\pi}\int_0^\infty |g(r)|^2 r^2 \mathrm{d}r \int_0^\pi \mathrm{d}\theta \int_0^{2\pi} (1+\cos\varphi \sin 2\theta)\sin\theta \mathrm{d}\varphi$$

$$= \frac{1}{2}\int_0^\pi \sin\theta \mathrm{d}\theta = 1$$

The probability for the occurrence of $L_z = \hbar$ is

$$\int \left[\sqrt{\frac{1}{3}}\left(-\sqrt{2}\right) Y_{11}\right]^2 \mathrm{d}\Omega = (2/3)\int (3/8\pi)\sin^2\theta .2\pi \sin\theta \mathrm{d}\theta$$

$$= (1/2)\int_{-1}^{+1} (1-\cos^2\theta) d\cos\theta = 2/3$$

The probability for the occurrence of $L_z = 0$ is

$$\int \left[\sqrt{\frac{1}{3}} \mathrm{Y}_{10}\right]^2 \mathrm{d}\Omega = \int_{-1}^{+1} \left(\frac{3}{4\pi}\right) .2\pi \cos^2\theta d\cos\theta = 1/3$$

3.96 (a) $[J_z, J_+] = J_z J_+ - J_+ J_z = J_z(J_x + iJ_y) - (J_x + iJ_y)J_z$
$= J_z J_x - J_x J_z + i(J_z J_y - J_y J_z)$
$= [J_z, J_x] + i[J_z, J_y] = i\hbar J_y - ii\hbar J_x$
$= i\hbar J_y + \hbar J_x = \hbar(J_x + iJ_y) = \hbar J_+$
$= J_+$, in units of $\hbar$.

(b) From (a), $J_z J_+ = J_+ J_z + \hbar J_+$
$J_z J_+ |j\ m >= J_+ J_z |jm > + \hbar J_+ |jm >$
$= J_+ m\hbar |jm > + \hbar J_+ |jm >$
$= (m+1)\hbar J_+ |jm >$
$J_+|jm >$ is nothing but $|j, m+1 >$ apart from a possible normalization constant. Thus
$J_+|jm >= C_{jm^+}\hbar |j, m+1 >$

Given a state $|jm >$, the state $|j, m+1 >$ must exist unless C_{jm^+} vanishes for that particular m. Since j is the maximum value of m by definition. There can not be a state $|j, j+1 >$, i.e. C_{jj^+} must vanish. J_+ is known as the ladder operator. Similarly, J_- lowers m by one unit.

(c) $J_+ = J_x + iJ_y$
$J_- = J_x - iJ_y$

Therefore, $J_x = \frac{1}{2}(J_+ + J_-) = \begin{pmatrix} 0 & 1/\sqrt{2} & 0 \\ 1/\sqrt{2} & 0 & 1/\sqrt{2} \\ 0 & 1/\sqrt{2} & 0 \end{pmatrix}$

$$J_y = \frac{1}{2i}(J_+ - J_-) = \begin{pmatrix} 0 & -i/\sqrt{2} & 0 \\ i/\sqrt{2} & 0 & -i/\sqrt{2} \\ 0 & i/\sqrt{2} & 0 \end{pmatrix}$$

$$[J_x, J_y] = J_x J_y - J_y J_x = i\begin{pmatrix} 1 & 0 & 0 \\ 0 & 0 & 0 \\ 0 & 0 & -1 \end{pmatrix} = iJ_z$$

3.3.7 Approximate Methods

3.97 $\Delta E = <\psi|\delta U|\psi>$

$\delta U = U$ (interaction energy of electron with point charge nucleus)

$$\delta U = e^2/r - \frac{3e^2}{2R}\left(R^2 - \frac{r^2}{3}\right) \quad \text{for } r \le R$$

$$= 0 \quad \text{for } r \ge R$$

(a) First we consider $n = 1$ state

$$\Delta E = \left(\frac{e^2}{\pi a_0^3}\right)\int_0^R \exp\left(-\frac{2r}{a_0}\right)\left[\frac{1}{r} - \frac{3}{2R} + {}^1/_2\frac{r^2}{R^3}\right]4\pi r^2 dr$$

$$= \frac{2e^2}{a_0^3}\int_0^R e^{-2r/a_0}\left[2r - \frac{3r^2}{R} + \frac{r^4}{R^3}\right]dr$$

Now $R = 10^{-13}$ cm $\ll 10^{-8}$ cm $= a_0$, the factor $\exp\left(-\frac{2r}{a_0}\right) \approx 1$

$$\Delta E = \left(\frac{2e^2}{a_0^3}\right)\left(R^2 - R^2 + \frac{R^2}{5}\right) = \frac{4}{5}\left(\frac{e^2}{2a_0}\right)\left(\frac{R}{a_0}\right)^2$$

$$= (0.8)(13.6)\left(\frac{10^{-13}}{0.53 \times 10^{-8}}\right)^2$$

$$= 3.87 \times 10^{-9}\,\text{eV}$$

(b) $n = 2$

$$\psi_{200} = \left(\frac{1}{8\pi a_0^3}\right)^{\frac{1}{2}}\left(2 - \frac{r}{a_0}\right)\exp\left(-\frac{r}{2a_0}\right)$$

$$\Delta E = \left(\frac{e^2}{8\pi a_0^3}\right)\int_0^R \exp\left(-\frac{r}{a_0}\right)\left(2 - \frac{r}{a_0}\right)^2\left[\frac{1}{r} - \frac{3}{2R} + {}^1/_2\frac{r^2}{R^3}\right]4\pi r^2 dr$$

Here also $\exp\left(-\frac{r}{a_0}\right) \sim 1$, for reasons indicated in (a)

When the remaining factors are integrated we get

$$\Delta E = \left(\frac{e^2}{2a_0}\right)\left(\frac{R^2}{a_0^2}\right)\left[\frac{2}{5} - \frac{1}{6}\frac{R}{a_0} + \frac{3}{140}\frac{R^2}{a_0^2}\right]$$

$$\approx \frac{2}{5}\cdot\left(\frac{e^2}{2a_0}\right)\left(\frac{R}{a_0}\right)^2 \quad \left(\text{as}\frac{R}{a_0} << 1\right)$$

$$= {}^1/_2 \times 3.87 \times 10^{-9} = 1.93 \times 10^{-9}\,\text{eV}$$

where we used the result of (a)

3.98 Schrodinger's equation in the presence of electric field is

$$\left[\left(-\frac{\hbar^2}{2m}\right)\frac{d^2}{dx^2} + {}^1/_2 m\,\omega^2 x^2 + qEx\right]\psi_n = E_n x \qquad (1)$$

Now, ${}^1/_2 m\omega^2 x^2 + qEx = {}^1/_2 m\omega^2\left[x^2 + \frac{2qEx}{m\omega^2}\right]$

$$= \frac{1}{2}\, m\,\omega^2[(x + qE/m\omega^2)^2 - q^2E^2/m^2\omega^4]$$

$$= \frac{1}{2}\, m\omega^2\left(x + \frac{qE}{m\omega^2}\right)^2 - \frac{q^2E^2}{2m\omega^2} \tag{2}$$

Put $X = x + \frac{qE}{m\omega^2}$; then we can write

$$\frac{\mathrm{d}}{\mathrm{d}x} = \frac{\mathrm{d}}{\mathrm{d}X} \text{ and } \frac{\mathrm{d}^2}{\mathrm{d}x^2} = \frac{\mathrm{d}^2}{\mathrm{d}X^2}$$

Equation (1) becomes

$$\left(-\frac{\hbar^2}{2m}\frac{\mathrm{d}^2}{\mathrm{d}X^2} + \frac{1}{2}m\omega^2X^2\right)\psi_n(X) = \left(E_n + \frac{q^2E^2}{2m\omega^2}\right)\psi_n(X) \tag{3}$$

Left hand side of (3) is the familiar Hamiltonian for the Simple harmonic oscillator. The modified eigen values are then given by the right hand side

$$E_n + \frac{q^2E^2}{2m\omega^2} = \left(n + \frac{1}{2}\right)\hbar\omega$$

or

$$E_n = \left(n + \frac{1}{2}\right)\hbar\omega - q^2E^2/2m\omega^2 \tag{4}$$

3.99 The matrix of H' is

$$(H') = \begin{pmatrix} H'_{11} & H'_{12} \\ H'_{21} & H'_{22} \end{pmatrix}, \text{ with } H'_{12} = H'^{*}_{21}$$

The matrix of H is

$$< H >= \begin{bmatrix} E_0 + H'_{11} & H'_{12} \\ H'_{21} & E_0 + H'_{22} \end{bmatrix}$$

$$\begin{vmatrix} E_0 + H'_{11} - E & H'_{12} \\ H'_{21} & E_0 + H'_{22} - E \end{vmatrix} = 0$$

$$E_1 = \frac{1}{2}\left(2E_0 + H'_{11} + H'_{22}\right) + \frac{1}{2}\left[\left(H'_{11} - H'_{22}\right)^2 + 4\left|H'_{12}\right|^2\right]^{1/2}$$

$$E_2 = \frac{1}{2}\left(2E_0 + H'_{11} + H'_{22}\right) - \frac{1}{2}\left[\left(H'_{11} - H'_{22}\right)^2 + 4\left|H'_{12}\right|^2\right]^{1/2}$$

These are the energy levels of a two state system with Hamiltonian $H = H_0 + H'$. The perturbation theory requires finding the eigen values of H' and adding them to E_0, which gives an exact result.

3.100 The nuclear charge seen by the electron is Z and not 2. This is because of screening the effective charge is reduced.

The smallest value of $E(Z) = -\frac{e^2}{2a_0}\left(\frac{27}{4}Z - 2Z^2\right)$ must be determined which is done by setting

$$\frac{\partial E}{\partial Z} = 0; \text{ this yields } Z = \frac{27}{16}$$

$$E\left(\frac{27}{16}\right) = -\left(\frac{2e^2}{2a_0}\right)\left(\frac{27}{16}\right)^2 = -(2)(13.6)\left(\frac{27}{16}\right)^2 = -77.45\,\text{eV}$$

The estimated value of ionization energy 77.4 eV may be compared to the 74.4 eV derived by perturbation theory and an experimental value of 78.6 eV. In general the variation method gives better accuracy then the perturbation theory.

3.101 (a) The unperturbed Hamiltonian for a hydrogen atom is

$$H_0 = -\frac{\hbar^2}{2\mu}\nabla^2 - \frac{e^2}{r} \tag{1}$$

where μ is the reduced mass.
H' is the extra energy of the nucleus and electron due to external field and is

$$H' = -eEz = eEr\cos\theta \tag{2}$$

where the polar axis is in the direction of positive z.

Now, the perturbation in (2) is an odd operator since it changes sign when the coordinates are reflected through the origin. Thus, the only non-vanishing matrix elements are those for unperturbed states that have opposite parities. In particular all diagonal elements of H' of hydrogen atom wave functions are zero. This shows that a non-degenerate state, like the ground state ($n = 1$) has no first-order Stark effect.

(b) The first excited state ($n = 2$) of hydrogen atom is fourfold degenerate, the quantum numbers *n, l* and *m* have the values (2,0,0), (2,1,1), (2,1,0), (2,1,−1). The first one (2S) has even parity while the remaining three(2P) states have odd parity of the degenerate states only $|2,0,0>$ and $|2,1,0>$ are mixed by the perturbation, but $|2,1,1>$ and $(2,1,-1>$ are not and do not exhibit the Stark effect. It remains to solve the secular equation,

$$\begin{vmatrix} e|E| < 2,0,0|z|2,0,0> -\lambda & e|E| < 2,0,0|z|2,1,0> \\ e|E| < 2,1,0|z|2,0,0> & e|E| < 2,1,0|z|2,1,0> -\lambda \end{vmatrix} = 0$$

Because the conservation of parity the diagonal elements $< 2,0,0|z|2,0,0>$ and $< 2,1,0|z|2,1,0>$ vanish. Hence the first-order change in energy $\lambda = \pm e|E < 2,0,0|z|2,1,0>$

Thus, only one matrix element needs to be evaluated, using the unperturbed eigen functions explicitly. (see Table 3.2)

$$2S(m=0);\ \psi_{2s}(0) = (4\pi)^{-\frac{1}{2}}\left(\frac{1}{2a}\right)^{\frac{3}{2}}\left(2-\frac{r}{a}\right)\exp\left(-\frac{r}{a}\right)$$

$$2P(m=0);\ \psi_{2p}(0) = (4\pi)^{-\frac{1}{2}}\left(\frac{1}{2a}\right)^{\frac{3}{2}}\left(\frac{r}{a}\right)\exp\left(-\frac{r}{2a}\right)\cos\theta$$

We can calculate

$$< 2,0,0|z|2,1,0 >=< 2,0,0|r\cos\theta|2,1,0 >$$

$$= \left(\frac{1}{4\pi}\right)\left(\frac{1}{2a}\right)^{3}\left(\frac{1}{a}\right)\int_0^\infty r^4\left(2-\frac{r}{a}\right)\exp\left(-\frac{r}{a}\right)\mathrm{d}r$$

$$\int_0^\pi \cos^2\theta\ \sin\theta \mathrm{d}\theta \int_0^{2\pi} \mathrm{d}\varphi$$

$$= -3a$$

Thus, the linear Stark effect splits the degenerate $m = 0$ level into two components, with the shift

$$\Delta E = \pm 3\ ae\ |E|$$

The corresponding eigen functions are $\frac{1}{\sqrt{2}}(\psi_s(0) \mp \psi_p(0))$
The two components being mixed in equal proportion (Fig. 3.27).

Fig. 3.27 Stark effect in Hydrogen

3.102 $$E = \int_{-a}^{+a}\left(\frac{1}{\sqrt{a}}\right)\cos\left(\frac{\pi x}{2a}\right)\left[\left(-\frac{\hbar^2}{2m}\right)\frac{d^2}{dx^2} + 1/2m\omega^2x^2\right]\frac{1}{\sqrt{a}}\cos\left(\frac{\pi x}{2a}\right)\mathrm{d}x$$

$$= \frac{\pi^2\hbar^2}{8ma^2} + \frac{m\omega^2 a^4}{10} + \frac{8a^5}{\pi^2}\left(1-\frac{6}{\pi^2}\right)$$

The best approximation to the ground-state wave function is obtained by setting $\frac{\partial E}{\partial \alpha} = 0$. This gives

$$a = \left[\frac{3\pi^2\hbar^2}{5m^2\,\omega^2(\pi^2-3)}\right]^{1/4}$$

3.103 The unperturbed wave function is

$$\psi^0 = k\sin\left(\frac{n_1\pi x}{a}\right)\sin(n_2\pi\ y/a);\ H' = W_0$$

$$E = \left(\frac{\pi\hbar^2}{2ma^2}\right)\left(n_1^2+n_2^2\right)$$

First order correction is

$$\Delta E = \frac{\int \psi_0^* H' \psi \mathrm{d}\tau}{\int \psi_0^* \psi \mathrm{d}\tau}$$

$$\int \psi_0^* H' \psi \mathrm{d}\tau = K^2 W \int_0^{\frac{a}{2}} \sin^2\left(\frac{n_1 \pi x}{a}\right) \mathrm{d}x \int_0^{a/2} \sin^2\left(\frac{n_2 \pi y}{a}\right) \mathrm{d}y$$

$$= K^2 W \left[\frac{x}{2} - \left(\frac{a}{4n_1\pi}\right)\sin\left(\frac{2n_1\pi x}{a}\right)\right]_0^{a/2} \left[\frac{y}{2} - \left(\frac{a}{4n_2\pi}\right)\sin\left(\frac{2n_2\pi y}{a}\right)\right]_0^{a/2}$$

$$= \frac{K^2 W a^2}{16}$$

$$\int_0^a \psi_0^* \psi_0 \mathrm{d}\tau = K^2 \int_0^a \int_0^a \frac{\sin^2 n_1 \pi x}{a} \frac{\sin^2 n_2 \pi y}{a} \mathrm{d}x\, \mathrm{d}y = \frac{k^2 a^2}{4}$$

Therefore $\Delta E = \frac{K^2 W_0 a^2}{16} / \frac{K^2 a^2}{4} = \frac{W_0}{4}$

3.3.8 Scattering (Phase Shift Analysis)

3.104 Let the total wave function be

$$\psi = \psi_i + \psi_s \tag{1}$$

where ψ_i represents the incident wave and ψ_s the scattered wave.

In the absence of potential, the incident plane wave

$$\psi_i = Ae^{ikz} = e^{ikz} \tag{2}$$

where we have dropped off A to choose unit amplitude.

Assume

$$\psi_s = \frac{f(\theta)e^{ikr}}{r} \tag{3}$$

which ensures inverse square r dependence of the scattered wave from the scattering centre.

$$\sigma(\theta) = |f(\theta)|^2 \tag{4}$$

f(θ) being the scattering amplitude.

We can write (1)

$$\psi = e^{ikrcos\theta} + \frac{f(\theta)e^{ikr}}{r} \tag{5}$$

or

$$f\theta = re^{-ikr}(\psi - e^{ikrcos\theta}) \tag{6}$$

$Lt\ r \to \infty$

The azimuth angle φ has been omitted in $f(\theta)$ as the scattering is assumed to have azimuthal symmetry. In the absence of potential ψ_i is the most general solution of the wave equation.

$$\nabla^2 \psi_i + k^2 \psi_i = 0 \tag{7}$$

Now ψ_{i} can be expanded as a sum of partial waves

$$\psi_i = e^{ikr\cos\theta} = \sum_{l=0}^{\infty} A_l\, j_l(kr) p_l(\cos\theta) \tag{8}$$

where $j_l(kr)$ are the spherical Bessel functions and $p_l(\cos\theta)$ are the Legendre polynomials of degree l. For $r \to \infty$, $j_l(kr) \approx \frac{1}{kr}\sin\left(kr - \frac{\pi l}{2}\right)$. The A_l are some constants which can be evaluated as follows.

Multiply both sides of (8) by $P_l(\cos\theta)\sin\theta\mathrm{d}\theta$ and integrate. Put $\cos\theta = t$

$$A_l\, j_l(kr) 2/(2l+1) = \int_{-1}^{+1} e^{ikrt}\, p_l(t)(\mathrm{d})t$$

where we have used the orthonormal property of Legendre polynomials.

Integrating the RHS by parts

$$(1/ikr)\left[e^{ikrt}\, p_l(t)\right]_{-1}^{+1} - (1/ikr)\int e^{ikrt} p_l{}'(t)\mathrm{d}t$$

where prime ($'$) means differentiation with respect to t. The second term is of the order of $1/r^2$ which can be neglected. Therefore

$$\left[\frac{2}{2l+1}\right] A_l\, j_l(kr) \approx \left(\frac{1}{ikr}\right)\left[e^{ikr} - e^{-ikr}(-1)^l\right] \tag{9}$$

where we have used $p_l(1) = 1$ and $p_l(-1) = (-1)^l$

Also, using the identity

$$e^{i\pi l/2} = i^l \tag{10}$$

(9) becomes

$$\left[\frac{2}{2l+1}\right] A_l\, j_l(kr) \approx \frac{\left[\frac{2i^l}{kr}\right]\left[e^{i\left(kr-\frac{\pi l}{2}\right)} - e^{-i\left(kr-\frac{\pi l}{2}\right)}\right]}{2i}$$

$$= \frac{2i^l\sin\left(kr - \frac{\pi l}{2}\right)}{kr}$$

Thus

$$A_l\, j_l(kr) = \frac{(2l+1)i^l\sin\left(kr - \frac{\pi l}{2}\right)}{kr} \tag{11}$$

Similarly, we can expand the total wave function into components

$$\psi(r,\theta) = \sum_{l=0}^{\infty} B_l\, R_l\,(r) p_l(\cos\theta)$$

$$= \sum_{r\to\infty}\left(\frac{Bk}{kr}\right)\sin\left(kr - \frac{\pi l}{2} + \delta_l\right) p_l(\cos\theta)$$

where B_l are arbitrary coefficients and δ_l is the phase-shift of the lth wave. From (6)

$$f(\theta) = re^{-ikr}\left[\sum B_l\left(\frac{1}{kr}\sin\left(kr - \frac{\pi l}{2} + \delta_l\right) p_l(\cos\theta)\right)\right.$$

$$\left. - \frac{\Sigma\, i^l(2l+1)}{kr}\sin\left(kr - \frac{\pi l}{2}\right) p_l(\cos\,\theta)\right]$$

$$or\ e^{ikr} f(\theta) = \frac{\sum B_l p_l(\cos\theta)\left[e^{i(kr-\frac{\pi l}{2}+\delta_l)} - e^{-i(kr-\frac{\pi l}{2}+\delta_l)}\right]}{2ik}$$

$$- \frac{\sum i^l(2l+1)p_l(\cos\theta)\left[e^{i(kr-\frac{\pi l}{2})} - e^{-i(kr-\frac{\pi l}{2})}\right]}{2ik}$$

Equating coefficients of e^{-ikr}

$$0 = -\sum\left(\frac{1}{2ik}\right) B_l\, p_l(\cos\theta)\left[e^{-i(-\pi l/2+\delta_l)} + \frac{\sum i^l(2l+1)P_l(\cos\theta)}{2ik} e^{i\pi l/2}\right]$$

Therefore

$$B_l = i^l(2l+1)e^{i\delta_l} \tag{12}$$

Equating coefficients of e^{ikr}, and using the value of B_l

$$f(\theta) = \left(\frac{1}{2ik}\right)\Big[\sum i^l(2l+1)e^{i\delta_l}\, p_l(\cos\theta)e^{i(-\frac{\pi l}{2}+\delta_l)}$$

$$-\sum i^l(2l+1)p_l(\cos\theta)e^{-\frac{\pi l}{2}}\Big]$$

$$= \left(\frac{1}{2ik}\right)\sum i^l\,(2l+1)P_l(\cos\theta)e^{-i\pi l/2}\left[e^{2i\delta_l} - 1\right] \tag{13}$$

Using (10), formula (12) becomes

$$f(\theta) = \frac{1}{2ik}\sum (2l+1)(e^{2i\delta_l} - 1)P_l(\cos\theta)$$

The above method is called the method of partial wave analysis. The summation over various integral values of l means physically summing over various values of angular momenta associated with various partial waves. The quantity δ_l is understood to be the phase shift when the potential is present. At low energies only a few l values would be adequate to describe the scattering.

3.105 The differential cross-section for the scattering of identical particles of spin s is given by

$$\sigma(\theta)^* = |f(\theta^*)|^2 + |f(\pi-\theta^*)|^2 + \left[\frac{(-1)^{2s}}{2s+1}\right] 2R_e[f(\theta^*)f^*(\pi-\theta^*)] \tag{1}$$

where f is assumed to be independent of the azimuth angle φ. The angles refer to the CM-system. The first two terms on RHS are given by the Rutherford scattering, one for the scattered particle and the other for the target particle. In the CMS the identical particles are oppositely directed and the detector cannot tell one from the other. The third term on the RHS is due to quantum mechanical interference and does not occur in the classical formula. Now for alpha-alpha scattering $s = 0$ and (1) reduces to

$$\sigma(\theta^*) = |f(\theta^*)|^2 + |f(\pi-\theta^*)|^2 + 2R_e[f(\theta^*)f^*(\pi-\theta^*)].$$

Furthermore if the scattering at $\theta^* = 90°$ is considered then obviously $f(\pi - \theta^*) = f(\theta^*)$ and the lab angle $\theta = 45°$. In that case classically $\sigma_L(45°) = 2|f(90°)|^2_{CM}$ while quantum mechanically

$\sigma_L(45°) = 4|f(90°)|_{CM}$.

Thus quantum mechanics explains the experimental result

3.106 $\sigma = \left(\frac{4\pi}{k^2}\right)\sum_{l=0}(2l+1)\sin^2\delta_l$ (1)

By problem

$$\sin\delta_l = \frac{(iak)^l}{\sqrt{(2l+1)l!}} \quad (2)$$

Therefore,

$$\sin^2\delta_l = \frac{(-a^2k^2)^l}{(2l+1)l!} \quad (3)$$

Using (3) in (1)

$$\sigma = \left(\frac{4\pi}{k^2}\right)\sum_{l=0}\frac{(-a^2k^2)l}{l!}$$

Summing over infinite number of terms for the summation and writing

$$k^2 = \frac{2mE}{\hbar^2},$$

$$\sigma = \left(\frac{2\pi\hbar^2}{mE}\right)\exp(-a^2k^2)$$

$$= \frac{2\pi\hbar^2}{mE}\exp\left(-\frac{2mEa^2}{\hbar^2}\right)$$

3.107 Let b be the impact parameter. In the c-system

$$bP_{cm} = l\hbar = \hbar$$

where we have set $l = 1$ for the p-wave scattering

$E_{CM} = \frac{P^2_{CM}}{2\mu} = \frac{P^2_{CM}}{M} = \hbar^2/Mb^2$ (Since the reduced mass $\mu = M/2$, where M is the mass of neutron or proton)

$$E_{\text{Lab}} = 2E_{CM} = \frac{2\hbar^2}{Mb^2} = \frac{2\hbar^2c^2}{Mc^2b^2}$$

Inserting $\hbar c = 197.3$ MeV.fm, $Mc^2 = 940$ MeV and $b = 2$ fm, we find $E_{\text{Lab}} = 20.6$ MeV. Thus up to 20 MeV Lab energy, s-waves ($l = 0$) alone are important

3.108 Only s-waves ($l = 0$) are expected to be involved as the scattering is isotropic.

$$\sigma = \frac{4\pi\sin^2\delta_0}{k^2}$$

Now $k^2\hbar^2 = p^2 = 2mE$

$\sin^2\delta_0 = \frac{2mE\sigma}{4\pi\hbar^2} = \frac{2mc^2E\sigma}{4\pi\hbar^2c^2}$

Inserting $mc^2 = 940$ MeV; $E = 1.0$ MeV,

$\sigma = 0.1\,b = 10^{-25}\,\text{cm}^2 = 10\,\text{fm}^2,\ \hbar c = 197.3$ MeV – fm

$\sin^2\delta_0 = 0.03845$

$\delta_0 = \pm 11.3°$

3.109 By Problem 3.104

$$\sigma(\theta) = \frac{1}{k^2}\left[\sin^2\delta_0 + 6\sin\delta_0\sin\delta_1\cos(\delta_1 - \delta_0)\cos\theta + 9\sin^2\delta_1\cos^2\theta\right] \quad (1)$$

We assume that at low energies $\delta_1 \ll \delta_0$. Now in the scattering with a hard sphere

$$\tan\,\delta_l = -\frac{(ka)^{2l+1}}{(2l+1)(1.1.3.5\ldots 2l-1)^2}$$

δ_0 (H.sphere) $= -ka$, for all ka

And δ_1 (H.sphere) $= -\frac{(ka)^3}{3}$, for $ka \ll 1$

Neglecting higher powers of $\delta's$, we can write (1)

$$\sigma(\theta) = \frac{1}{k^2}\left[\left(\delta_0 - \frac{\delta_0^3}{3!}\right)^2 + 6\delta_0\delta_1\cos\delta\right]$$

$$= \frac{1}{k^2}\left[\delta_0^2 - \frac{\delta_0^4}{3} + 6\delta_0\delta_1\cos\delta\right]$$

$$= \frac{1}{k^2}\left[k^2a^2 - \frac{k^4a^4}{3} + 6(-ka)\left(-\frac{k^3a^3}{3}\right)\cos\theta\right]$$

$$= a^2\left[1 - \frac{k^2a^2}{3} + 2k^2a^2\cos\theta\right]$$

$$\sigma = \int\left(\frac{d\sigma}{d\Omega}\right).2\pi\,\sin\theta\,d\theta = 2\pi\int_{-1}^{+1} a^2\left(1 - \frac{k^2a^2}{3} + 2k^2a^2\cos\theta\right)d\cos\theta$$

$$= 4\pi a^2\left[1 - \frac{(ka)^2}{3}\right]$$

3.110 A spherical nucleus of radius R will be totally absorbing, or appear "black" when the angular momentum $l < R/\lambda$. In that case $\eta_l = 0$ in the reaction and scattering formulae.

$$\sigma_r = \pi\lambda^2\sum_l(2l+1)(1 - |\eta_l|^2)$$

$$\sigma_s = \pi\lambda^2\sum_l(2l+1)|1 - \eta_l|^2$$

$(| > \eta_l > 0)$

Putting $\eta_l = 0$

$$\sigma_r = \sigma_s = \pi\lambda^2\sum_{l=0}^{R/\lambda}(2l+1)$$

The summation can be carried out by using the formula for arithmetic progression

$$S = na + \frac{n(n-1)d}{2}$$

Here $a = 1, d = 2, n = R/\lambda$

$$\sum_{l=0}^{R/\lambda}(2l+1) = (R/\lambda)^2$$

$$\therefore\ \sigma_r = \sigma_s = \pi\lambda^2\left(\frac{R}{\lambda^2}\right) = \pi R^2$$

The total cross-section

$\sigma_t = \sigma_r + \sigma_s = 2\pi R^2$

which is twice the geometrical cross-section

3.111 The potential which an electron sees as it approaches an atom of a monatomic gas can be qualitatively represented by a square well. Slow particles are considered.

$$V(r) = -V_0;\ r \le R$$
$$= 0;\qquad r > R$$

corresponding to an attractive potential. Scattering of slow particles for which $kR << 1$, is determined by the equation

$$\left(\nabla^2 + k^2 - \frac{2\mu V}{\hbar^2}\right)\psi_2 = 0 \text{ (inside the well)} \tag{1}$$

with $k^2 = 2\mu E/\hbar^2$, and the wave number $k = p/\hbar$

Outside the well the equation is

$$(\nabla^2 + k^2)\psi_1 = 0 \tag{2}$$

Further writing

$k_1^2 = k^2 + k_0^2$

where $k_0^2 = \frac{2\mu V}{\hbar^2}$

and $V = -V_0$

The solutions are found to be

$$\psi_2 = A \sin k_1 r \tag{3}$$

$$\psi_1 = B \sin(kr + \delta_0) \tag{4}$$

$\psi_1(r)$ is the asymptotic solution at large distances with the boundary condition

$\psi_1(0) = 0$

Matching the solutions (3) and (4) at $r = R$ both in amplitude and first derivative,

$$A \sin k_1 R = B \sin(kR + \delta_0) \tag{5}$$

$$Ak_1 \cos k_1 R = Bk \cos(kR + \delta_0) \tag{6}$$

Dividing one equation by the other, and setting $k_1 \cot k_1 R = \frac{1}{D}$, and with simple algebraic manipulations we get

$$\tan\delta_0 = (kD - \tan kR)(1 + kD \tan kR)^{-1} \tag{7}$$

The phase shift δ_0 determined from (7) is a multivalued function but we are only interested in the principle value lying within the interval $-\frac{\pi}{2} \le \delta_0 \le \frac{\pi}{2}$. For small values of the energy of the relative motion

$$\tan\, kR \approx kR + \frac{(kR)^3}{3} + \cdots$$

We therefore have

$$\tan\delta_0 \approx \frac{k\left[D - R - \left(\frac{(kR)^3}{3k}\right)\right]}{1 + k^2 DR}$$

If the inequalities $kR << 1$ and $k^2DR << 1$ are satisfied we can still further simplify the expression for tan δ_0.

$$\tan\ \delta_0 \approx k(D - R) = kR\left(\frac{\tan k_1 R}{k_1 R} - 1\right) \tag{8}$$

The total cross-section is then

$$\sigma = \frac{4\pi}{k^2}\sin^2\delta_0 \approx 4\pi(D - R)^2 = 4\pi R^2\left(1 - \frac{\tan k_1 R}{k_1 R}\right)^2 \tag{9}$$

It follows that if the condition

$$\tan k_1 R = k_1 R \tag{10}$$

is satisfied, the phase shift and the scattering cross-section both vanish. This phenomenon is known as the Ramsauer-Townsend effect. The field of the inert gas atoms decreases appreciably faster with distance than the field of any other atom, so that to a first approximation, we can replace this field by a rectangular spherical well with sharply defined range and use Equation (10) to evaluate the cross-section for slow electrons.

Physically, the Ramasuer – Townsend effect is explained as the diffraction of the electron around the rare-gas atom, in which the wave function inside the atom is distorted in such a way that it fits on smoothly to an undistorted function outside.

Here the partial wave wave with $l = 0$ has exactly a half cycle more of oscillation inside the atomic potential then the wave in the force-free field, and the wavelength of the electron is large enough in comparision with R so that higher l phase-shifts are negligible.

3.112 In order that the Schrodinger equation is reduced to the given form is that the potential $V(r)$ does not depend on time. From Problem 3.104 the total wave function

$$\psi = \Sigma i^l(2l + 1)\left[\frac{e^{i\delta l}}{kr}\right]\sin\left(kr - \frac{1}{2}\pi l + \delta_l\right) p_l(\cos\,\theta)$$

For slow neutrons only the first term ($l = 0$) in the summation is important. As $p_0(\cos\,\theta) = 1$

$$\psi = \frac{\exp(i\delta_0)}{kr}\sin(kr + \delta_0)$$

$$u = \psi r = \frac{\exp(i\delta_0}{k}\sin(kr + \delta_0) = \text{const.}\ \sin(kr + \delta_0)$$

We assume that the wave function inside the well is identical with that in the deuteron problem. This is justifiable since the total energy inside the potential well is raised by little over 2 MeV corresponding to the binding energy W of the deuteron, which is much smaller than the well depth (~25 MeV). In the deuteron problem the outside function $Ce^{-\gamma r}$, where $\gamma = \sqrt{MW/\hbar^2}$, is matched with the inside function $A \sin kr$. Here we match the functions $\sin(kr + \delta_0)$ and $Ce^{-\gamma r}$ at $r = R$, both in magnitude and first derivative.

This gives us $k \cot(kR+\delta_0) = -\gamma$. Further, $R \approx 0$. This is also reasonable since for the square well the main features of the deuteron problem remain unaltered by narrowing the well width and deepening the well. It follows that
$\sin^2 \delta_0 = k^2/k^2 + \gamma^2)$

But the s-wave cross-section is given by

$\sigma = 4\pi \sin^2 \delta_0/k^2 = 4\pi/(k^2 + \gamma^2)$

Substituting $k^2 = ME/\hbar^2$ and $\gamma^2 = MW/\hbar^2$

$$\sigma = \frac{4\pi\hbar^2}{M}\frac{1}{W+E} \tag{1}$$

where M is proton or neutron mass, W is the deuteron binding energy (2.225 MeV), and E is the lab kinetic energy.

Formula (1) agrees well with experiment at relatively higher energies (say 5–10 MeV) but fails badly at very low energies. For $E \ll W$, for example, (1) predicts $\sigma = 2$ barns which is far from the experimental value of 20 *barns*. Wigner pointed out that in $n-p$ scattering the spins of the colliding nucleons could be either parallel or antiparallel. Formula (1) holds for the parallel case because the analogy is made with the deuteron problem which has parallel spins. Now for random orientations of spins:

$$\sigma = \frac{3}{4}\sigma_t + \frac{1}{4}\sigma_s \tag{2}$$

where σ_t and σ_s are the cross-sections for the triplet and singlet scattering, the factors $\frac{3}{4}$ and $\frac{1}{4}$ being the statistical weights. In (1), W is the binding energy of the $n-p$ system for the triplet state. Corresponding to the singlet state the quality W_s is introduced, although it is a virtual state.

Combining (1) and (2)

$$\sigma = \frac{3\pi\hbar^2}{M(E+W)} + \frac{\pi\hbar^2}{M(E+W_s)} \tag{3}$$

W_s takes a value of 70 keV if agreement is to reach with the experiments. Agreement at higher energies is preserved because for $E \gg W$ or W_s, (3) reduces to (1).

3.3.9 Scattering (Born Approximation)

3.113 (a) $F(q) \approx \int_0^\infty \frac{\rho(r)\sin(qr/\hbar)4\pi r^2}{qr/\hbar}\mathrm{d}r$

$\rho(r) = A\exp(-r/a)$

$$F(q) \approx 4\pi A \int_0^\infty r\exp(-r/a)[\sin(qr/\hbar)/(q/\hbar)]\mathrm{d}r$$

Put $\alpha = 1/a$ and $\beta = q/\hbar$

$$F(q) \approx 4\pi A \int_0^\infty r\,\mathrm{e}^{-\alpha r} \frac{\sin \beta r}{\beta} \mathrm{d}r$$

$$I = -\frac{\partial}{\partial \alpha} \int_0^\infty e^{-\alpha r} \sin \beta r \mathrm{d}r$$

$$= \frac{-1}{\beta} \frac{\partial}{\partial \alpha} \frac{\beta}{\alpha^2 + \beta^2} = \frac{1}{\beta} \frac{2\alpha\beta}{(\alpha^2 + \beta^2)^2}$$

$$= \frac{2\alpha}{(\alpha^2 + \beta^2)^2} = \frac{2}{\alpha^3} \left(1 + \frac{\beta^2}{\alpha^2}\right)^{-2}$$

$$= 2a^3/(1 + q^2 a^2/\hbar^2)$$

$F(q) = 8\pi A\, a^3 / \left(1 + q^2/q_o^2\right)$, where $q_o = \hbar/a$

thus $F(q) \approx 1/\left(1 + \frac{q^2}{q_0^2}\right)^2$

(b) The characteristic radius

$$a = \frac{\hbar}{q_o} = \frac{\hbar c}{q_o c} = \frac{197.3\,\mathrm{MeV} - \mathrm{fm}}{0.71 \times 1{,}000\,\mathrm{MeV}} = 0.278\,\mathrm{fm}$$

3.114 $f(\theta) = -\frac{\mu}{2\pi\hbar^2} \int V(r) e^{iq.r} \mathrm{d}^3 r$

$$= -\frac{\mu}{2\pi\hbar^2} \int_{r=0}^\infty \int_{\theta=0}^\pi \int_{\varphi=0}^{2\pi} V(r) e^{iqr\cos\theta} r^2 \sin\theta \mathrm{d}\theta \mathrm{d}\varphi \mathrm{d}r$$

$$= -\frac{\mu}{2\pi\hbar^2} \int_0^\infty V(r) r^2 \mathrm{d}r \int_{-1}^{+1} e^{iqr\cos\theta} d(\cos\theta \int_0^{2\pi} \mathrm{d}\varphi$$

$$= -\frac{2\mu}{\hbar^2} \int \frac{V(r) r^2 \mathrm{d}r}{qr} \left[\frac{e^{iqr} - e^{-iqr}}{2i}\right]$$

$$= -\frac{2\mu}{q\hbar^2} \int r \sin(qr) V(r) \mathrm{d}r$$

3.115 From the partial wave analysis of scattering the scattering amplitude

$$f(\theta) = \frac{1}{k} \Sigma_l (2l + 1)(\eta_l \exp(2i\delta_l) - 1)/2i) p_l(\cos\theta).$$

For elastic scattering without absorption $\eta_l = 1$, and

$$f(\theta) = \frac{1}{k} \Sigma_l (2l + 1) \left[\exp(2i\delta_l) - 1)/2i\right] p_l(\cos\theta)$$

$$= \frac{1}{k} \Sigma_l (2l + 1) \exp(i\delta_l) \sin\delta_l\, p_l(\cos\theta).$$

Now for $\theta = 0$, $p_l(\cos\,\theta) = p_l(1) = 1$ for any value of l, and $\exp(i\delta_l) = \cos\,\delta_l + i\,\sin\,\delta_l$. Therefore the imaginary part of the forward scattering amplitude

$$Im\, f(0) = \frac{1}{k} \sum_l (2l + 1) \sin^2 \delta_l.$$

But the total cross-section is given by

$$\sigma_t = \frac{4\pi}{k^2}(2l+1)\sin^2\delta_l.$$

It follows that $Im\, f(0) = k\sigma_t/4\pi$. The last equation is known as the optical theorem.

3.116 $$V(r) = \left(-\frac{Ze^2}{2R}\right)\left(3 - \frac{r^2}{R^2}\right);\ 0 < r < R \quad (1)$$

$$= -\frac{Ze^2 e^{-ar}}{r};\ R < r < \infty \quad (2)$$

Inside the nucleus the electron sees the potential as given by (1) corresponding to constant charge distribution, while outside it sees the shielded potential given by (2). The scattering amplitude is given by

$$f(\theta) = -(2\mu/q\hbar^2)\int_0^\infty V(r) r \sin(qr)\mathrm{d}r$$

$$= 2\mu\frac{Ze^2}{q\hbar^2}\left[\left(\frac{1}{2R}\right)\int_0^R \left(3 - \frac{r^2}{R^2}\right) r\sin(qr)\mathrm{d}r + \int_R^\infty \sin(qr)e^{-ar}\mathrm{d}r\right] \quad (3)$$

The first integral is easily evaluated and the second integral can be written as

$$\int_R^\infty \sin(qr)e^{-ar}\mathrm{d}r = \int_0^\infty \sin(qr)e^{-ar}\mathrm{d}r - \int_0^R \sin(qr)e^{-ar}\mathrm{d}r \quad (4)$$

$$= \frac{q}{q^2+a^2} - \int_0^R \sin(qr)e^{-ar}\mathrm{d}r \quad (5)$$

$$(\text{Lim } a \to 0) = \frac{1}{q} - \int_0^R \sin(qr)\mathrm{d}r = \frac{1}{q}\cos(qr)$$

We finally obtain

$$f(\theta) = \left(-\frac{2\mu Ze^2}{q^2\hbar^2}\right)\left(\frac{3}{q^2R^2}\right)\left(\frac{\sin(qR)}{qR} - \cos qR\right)$$

$$\sigma(\theta)\text{finite size} = \sigma(\theta)_{\text{point charge}}|F(q)|^2$$

where the form factor is identified as

$$F(q) = \left(\frac{3}{q^2R^2}\right)\left(\frac{\sin(qR)}{qR} - \cos(qR)\right)$$

The angular distribution no longer decreases smoothly but exhibits sharp maxima and minima reminiscent of optical diffraction pattern from objects with sharp edges. The minima occur whenever the condition $\tan\ qR = qR$, is satisfied. This feature is in contrst with the angular distribution from a smoothly varying charge distribution, such as Gaussian, Yakawa, Wood-Saxon or exponential, wherein the charge varies smoothly and the maxima and minima are smeared out, just as in the case of optical diffraction from a diffuse boundary of objects characterized by a slow varying refractive index.

3.117 $f(\theta) = -(2\mu/q\hbar^2)\int_0^\infty V(r)\sin(qr)\, r\, dr$

Integrate by parts

$$\int_0^\infty V(r)\sin(qr) r\, dr = V(r)\left[\frac{1}{q^2} sin\ qr - \frac{r}{q} cos\ qr\right]_0^\infty$$

$$-\int_0^\infty \frac{dV}{dr}\left(\frac{1}{q^2}\sin qr - \frac{r}{q}\cos qr\right) dr$$

The first term on the right hand side vanishes at both limits because $V(\infty) = 0$, Therefore:

$$\int_0^\infty V(r)\sin(qr) r\, dr = -\frac{1}{q^2}\int_0^\infty \frac{dV}{dr}\sin qr dr + \frac{1}{q}\int_0^\infty \frac{dV}{dr} r\cos qr dr$$

Evaluate the second integral by parts

$$\frac{1}{q}\int_0^\infty \frac{dV}{dr} r\cos(qr) dr = \frac{1}{q}\left[\frac{dV}{dr}\left(\frac{r}{q}\sin qr + \frac{\cos qr}{q^2}\right)\right]_0^\infty$$

$$-\frac{1}{q}\int_0^\infty \left(\frac{r}{q}\sin qr + \frac{\cos qr}{q^2}\right)\frac{d^2V}{dr^2} dr$$

Now the term $\left(\frac{1}{q^2}\right) r \left(\frac{dV}{dr}\right)\sin qr\Big|_0^\infty$ vanishes at both the limits because it is expected that $(dV/dr)_{r=\infty} = 0$.

Integrating by parts again

$$\left(\frac{1}{q^3}\right)\int \cos qr \frac{d^2V}{dr^2} dr = \left(\frac{1}{q^3}\right)\cos qr \left.\frac{dV}{dr}\right|_0^\infty + \left(\frac{1}{q^2}\right)\int_0^\infty \left(\frac{dV}{dr}\right)\sin qr dr$$

$$\left(\frac{1}{q}\right)\int_0^\infty \left(\frac{dV}{dr}\right) r\cos qr\, dr = \left(\frac{1}{q^3}\right)\left(\frac{dV}{dr}\right)\cos qr\Big|_0^\infty$$

$$-\left(\frac{1}{q^2}\right)\int_0^\infty \left(\frac{d^2V}{dr^2}\right) r\sin qr\, dr - \left(\frac{1}{q^3}\right)\left(\frac{dV}{dr}\right)\cos qr\Big|_0^\infty -$$

$$\left(\frac{1}{q^2}\right)\int_0^\infty \left(\frac{dV}{dr}\right)\sin qr\, dr.$$

The first and third terms on the right hand side get cancelled

$$\int_0^\infty V(r)\sin(qr) r\, dr = -\left(\frac{1}{q^2}\right)\int\left(\frac{d^2V}{dr^2} + \frac{2}{r}\frac{dV}{dr}\right)\sin(qr) r\, dr.$$

Now for spherically symmetric potential

$$\nabla^2 V = \frac{d^2V}{dr^2} + \left(\frac{2}{r}\right)\frac{dV}{dr}.$$

Furthermore by Poisson's equation:

$$\nabla^2 V = -4\pi Ze^2 \rho$$

where *Ze* is the nuclear charge and ρ is the charge density.

$$\therefore\ f(\theta) = -8\pi\mu\left(\frac{Ze^2}{q^3\hbar^2}\right)\int_0^\infty \rho(r)\sin(qr)\, r\ \mathrm{d}r$$

$$= \left(\frac{2\mu Ze^2}{q^2\hbar^2}\right)\left(\frac{4\pi}{q}\right)\int_0^\infty \rho(r) sin\ (qr)\, r\ \mathrm{d}r$$

The quantity $\left(\frac{4\pi}{q}\right)\int_0^\infty \rho(r)\ \sin(qr)\, r\ dr$ is known as the form factor.

3.118 $$f(\theta) = \left(-\frac{2\mu}{q\hbar^2}\right)\int_0^\infty V(r)\sin(qr) r\ \mathrm{d}r \qquad (1)$$

Substituting,

$$V(r) = \frac{z_1 z_2 e^2}{r} e^{-\mathrm{ar}} \qquad (2)$$

Where $a = 1/r_o$, (1) becomes

$$f(\theta) = -\left(\frac{2\mu z_1 z_2 e^2}{q\hbar^2}\right)\int_0^\infty e^{-ar}\ \sin(qr)\mathrm{d}r$$

$$= \frac{-2\mu z_1 z_2 e^2}{q\hbar^2}\frac{q}{q^2 + a^2} = \frac{-2\mu z_1 z_2 e^2}{\hbar^2\left(q^2 + 1/r_0^2\right)} \qquad (3)$$

But the momentum transfer

$$q\hbar = 2k\hbar \sin\left(\frac{\theta}{2}\right) \qquad (4)$$

The differential cross-section

$$\sigma(\theta) = |f(\theta)|^2 = \frac{4\mu^2 z_1^2 z_2^2 e^4}{\hbar^4\left(4k^2\sin^2(\theta/2) + 1/r_0^2\right)^2} \qquad (5)$$

The general angular distribution of scattered particles is reminiscent of Rutherford scattering. However for $\theta < \theta_0$, where

$$\sin(\theta_o/2) \approx 1/2kr_o \qquad (6)$$

the curve does not rise indefinitely but tends to flatten out because when $qr_o \ll 1$, the angular dependence of $\sigma(\theta)$ is damped out resulting in the flattening of the curve. The angle θ_o may be considered as the limiting angle below which the Rutherford scattering is inoperative because of the shielding of the atomic nucleus by the electron cloud.

Rutherford scattering is derived from (5) by letting $r_o \to \infty$, in which case the scattering would occur from a bare nucleus. The screening potential (2) now reduces to Coulomb potential. Furthermore, writing $k\hbar = p = \mu v$, (5) becomes

$$\sigma(\theta) = \frac{1}{4}\left(\frac{z_1 z_2 e^2}{\mu v^2}\right)^2 \frac{1}{\sin^4\left(\frac{\theta}{2}\right)} \quad \text{(Rutherford scattering formula)}$$

3.119 By Problem 3.116

$$F(q^2) = \frac{3}{q^2R^2}\left(\frac{\sin qR}{qR} - \cos qR\right) \tag{1}$$

$$R = r_o A^{1/3} = 1.3 \times (64)^{1/3} = 5.2\,\text{fm}$$

$$q\hbar = 2p_o \sin(\theta/2)$$

$$qR = 2cp_o R \sin(\theta/2)/\hbar c = 2 \times 300 \times \frac{5.2 \times \sin 6^\circ}{197.3}$$

$$= 1.653\,\text{radians} \tag{2}$$

$$\sin qR = 0.9966, \;\; \cos qR = -0.0819 \tag{3}$$

Inserting (2) and (3) in (1), we find $F(q) = 0.75$, $F^2 \approx 0.57$. Thus Mott's scattering is reduced by 57%.

3.120 $F(q^2) = 1 - \frac{q^2}{6\hbar^2} < r^2 > + \cdots$

$$q\hbar = 2p_o \sin(\theta/2) = 2 \times 200 \times (\sin 7^\circ)\,\text{MeV}/c = 48.75\,\text{MeV}/c$$

$$< r^2 >= \frac{6\hbar^2}{q^2}[1 - F(q^2)]$$

$$= 6 \times \frac{(197.3)^2}{(48.75)^2}(1 - 0.6)\,\text{fm}^2$$

$$= 39.3$$

$\therefore$ Root mean square radius = 6.27 fm

3.121 $F(q) = (4\pi/q)\int_0^\infty \rho(r)\sin(qr)r\,\text{d}r$

$$= (4\pi/\pi^{3/2}b^3q)\int_0^\infty e^{-r^2/b^2}\sin(qr)r\,\text{d}r$$

$$= (-4/\pi^{1/2}b^3q)\frac{\partial}{\partial_q}\int_0^\infty e^{-r^2/b^2}\cos(qr)\,\text{d}r$$

$$= (-4/\pi^{1/2}b^3q)\frac{\partial}{\partial_q}\left[\frac{1}{2}(\pi b^2)^{1/2}e^{-b^2q^2/4}\right]$$

$$F(q) = \exp(-b^2q^2/4)$$

$$< r^2 >= \frac{\int_0^\infty r^2\rho(r)4\pi r^2\text{d}r}{\int \rho(r)4\pi r^2\text{d}r} = \frac{\int_0^\infty r^4 e^{r^2/b^2}\text{d}r}{\int_0^\infty r^2 e^{-r^2/b^2}\text{d}r}$$

where we have put $\rho(r) = (1/\pi^{3/2}b^3)e^{-r^2/b^2}\text{d}r$

With the change of variable $r^2/b^2 = x$, we get

$$< r^2 >= \frac{b^2\int_0^\infty x^{3/2}e^{-x}\text{d}x}{\int_0^\infty x^{1/2}e^{-x}\text{d}x} = \frac{\Gamma\left(\frac{5}{2}\right)b^2}{\Gamma\left(\frac{3}{2}\right)} = \frac{3b^2}{2}$$

3.122 $f(\theta) = -\left(\frac{\mu}{2\pi\hbar^2}\right)\int V(r)e^{iqr}\mathrm{d}^3r$

$$= -\left(\frac{\mu}{2\pi\hbar^2}\right)\int_0^\infty V(r)r^2\mathrm{d}r\int_{-1}^{+1} e^{iqr\cos\theta}\ \mathrm{d}(\cos\theta)\int_0^{2\pi}\mathrm{d}\varphi$$

$$= -\left(\frac{\mu}{2\pi\hbar^2}\right)\int V(r)r^2\left[\frac{(e^{iqr}-e^{-iqr})}{iqr}\right]2\pi\,\mathrm{d}r$$

$$= -\left(\frac{2\mu}{q\hbar^2}\right)\int_0^\infty V(r)r\ \sin(qr)\mathrm{d}r$$

3.123 $A = \int_0^\infty \frac{\sin(qr/\hbar)}{(qr/\hbar)}V(r)4\pi r^2\mathrm{d}r$
Substitute $V(r) \sim \frac{e^{-r/R}}{r}$
where $R = \hbar/mc$

$$A \sim \int_0^\infty e^{-r/R}\frac{\sin(qr/\hbar)}{q}\mathrm{d}r$$

Put $1/R = a$ and $q/\hbar = b$

$$A \sim \int_0^\infty e^{-ar}\frac{\sin(br)}{b}\mathrm{d}r$$

$$= \frac{1}{b}\left[\frac{b}{a^2+b^2}\right] = \frac{1}{a^2+b^2}$$

$$= \frac{1}{\frac{1}{R^2}+\frac{q^2}{\hbar^2}} \sim \frac{1}{\frac{\hbar^2}{R^2}+q^2} = \frac{1}{q^2+m^2c^2}$$

Chapter 4
Thermodynamics and Statistical Physics

4.1 Basic Concepts and Formulae

Kinetic theory of gases

Pressure

$$p = \frac{1}{3}\rho < v^2 > \tag{4.1}$$

Root-mean-square velocity

$$v_{\text{rms}} = \sqrt{3P/\rho} \tag{4.2}$$

$$v_{\text{rms}} = \sqrt{3kT/m} = \sqrt{3RT/M} \tag{4.3}$$

Average speed

$$< v >= \sqrt{\frac{8kT}{\pi m}} = \sqrt{\frac{8RT}{\pi M}} \tag{4.4}$$

Most probable speed

$$v_p = \sqrt{\frac{2kT}{m}} = \sqrt{\frac{2RT}{M}} \tag{4.5}$$

where m is the mass of the molecule, M is the molar weight, ρ the gas density, $k = 1.38 \times 10^{-23}$/K, the Boltzmann constant, $R = 8.31$ J/mol-K, is the universal gas constant, and K is the Kelvin (absolute) temperature.

$$v_p :< v >: v_{\text{rms}} :: \sqrt{2} : \sqrt{8/\pi} : \sqrt{3} \tag{4.6}$$

The Maxwell distribution

$$N(v)\mathrm{d}v = 4\pi\left(\frac{m}{2\pi kT}\right)^{3/2} v^2 e^{-mv^2/2kT}\,\mathrm{d}v \tag{4.7}$$

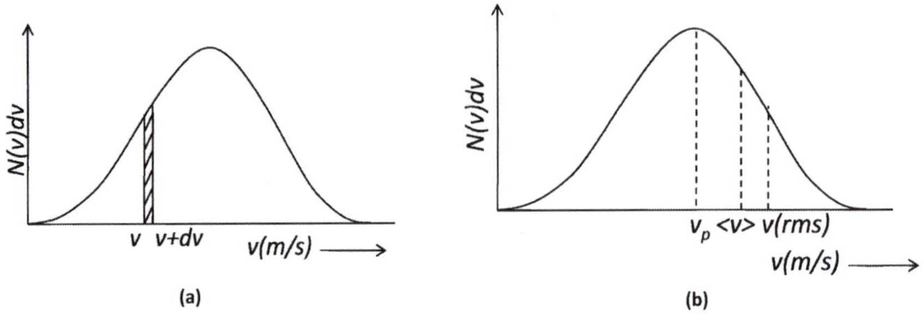

Fig. 4.1 The Maxwell distribution

Flux

$$\emptyset = \frac{1}{4}n < v > \quad \text{(number of molecules striking unit area per second)} \tag{4.8}$$

where n is the number of molecules per unit volume.

Mean free path (M.F.P)

$$\lambda = \frac{1}{\sqrt{2}\pi n\sigma^2} \tag{4.9}$$

where n is the number of molecules per unit volume and σ is the diameter of the molecule.

Collision frequency

$$f = \frac{< v >}{\lambda} \tag{4.10}$$

Viscosity of gas (η)

$$\eta = \frac{1}{3}\rho\lambda < v > \tag{4.11}$$

Thermal conductivity (K)

$$K = \eta C_v \tag{4.12}$$

where C_v is the specific heat at constant volume.

Coefficient of diffusion (D)

$$D = \frac{\eta}{\rho} \tag{4.13}$$

Clausius Clepeyron equation

$$\frac{\mathrm{d}P}{\mathrm{d}T} = \frac{L}{T(\nu_2 - \nu_1)} \tag{4.14}$$

where ν_1 and ν_2 are the initial and final specific volumes (volume per unit mass) and L is the latent heat.

Vander Waal's equation

$$\left(P + \frac{a}{V^2}\right)(V - b) = RT \quad \text{(for one mole of gas)} \tag{4.15}$$

The Stefan-Boltzmann law

$$E = \sigma T^4 \tag{4.16}$$

If a blackbody at absolute temperature T be surrounded by another blackbody at absolute temperature T_0, the amount of energy E lost per second per square metre of the former is

$$E = \sigma(T^4 - T_0^4) \tag{4.17}$$

where $\sigma = 5.67 \times 10^{-8}\ \mathrm{W/m^2.K^4}$ is known as Stefan-Boltzmann constant.

Maxwell's thermodynamic relations

$$\text{First relation: } \left(\frac{\partial S}{\partial V}\right)_T = \left(\frac{\partial P}{\partial T}\right)_V \tag{4.18}$$

$$\text{Second relation: } \left(\frac{\partial S}{\partial P}\right)_T = -\left(\frac{\partial V}{\partial T}\right)_P \tag{4.19}$$

$$\text{Third relation: } \left(\frac{\partial T}{\partial V}\right)_S = -\left(\frac{\partial P}{\partial S}\right)_V \tag{4.20}$$

$$\text{Fourth relation: } \left(\frac{\partial T}{\partial P}\right)_S = \left(\frac{\partial V}{\partial S}\right)_P \tag{4.21}$$

Thermodynamical potentials

(i) Internal energy (U)
(ii) Free energy (F)
(iii) Gibb's function (G)
(iv) Enthalpy (H)

$$H = U + PV \tag{4.22}$$

$$\left(\frac{\partial U}{\partial T}\right)_V = C_V \tag{4.23}$$

$$\left(\frac{\partial H}{\partial S}\right)_P = T \tag{4.24}$$

$$\left(\frac{\partial H}{\partial P}\right)_S = V \tag{4.25}$$

$$\left(\frac{\partial H}{\partial T}\right)_P = C_P \tag{4.26}$$

The Joule-Kelvin effect

$$\Delta T = \frac{\left[T\left(\frac{\partial V}{\partial T}\right)_P - V\right]\Delta P}{C_P} \tag{4.27}$$

Black body radiation

$$P_{\text{radiatio}} = \frac{u}{3} \tag{4.28}$$

where u is the radiation density

Wein's displacement law

$$\lambda_m T = 0.29 \text{ cm - K} \tag{4.29}$$

Planck's radiation law

$$u_\nu d\nu = \frac{8\pi h\nu^3 d\nu}{c^3(e^{h\nu/kT} - 1)} \tag{4.30}$$

$$u_\lambda d\lambda = \frac{8\pi hc}{\lambda^5} \cdot \frac{1}{(e^{hc/\lambda kT} - 1)} \tag{4.31}$$

$$\sigma = \frac{2}{15} \cdot \frac{\pi^5 k^4}{h^3 c^2} \tag{4.32}$$

$$\Delta S = \frac{\Delta Q}{T} \tag{4.33}$$

$$\Delta S = k \ln(\Delta W) \tag{4.34}$$

where W is the number of accessible states.
Probability for finding a particle in the nth state at temperature T

$$P(n, T) = \frac{e^{-E_n/kT}}{\Sigma_{n=0}^{\infty} e^{-E_n/kT}} \tag{4.35}$$

Stirling's approximation

$$n! = \sqrt{2\pi n}\, n^n e^{-n} \tag{4.36}$$

4.2 Problems

4.2.1 Kinetic Theory of Gases

4.1 Derive the formula for the velocity distribution of gas molecules of mass m at Kelvin temperature T.

4.2 Assuming that low energy neutrons are in thermal equilibrium with the surroundings without absorption and that the Maxwellian distribution for velocities is valid, deduce their energy distribution.

4.3 In Problem 4.1 show that the average speed of gas molecule $< v >= \sqrt{8kT/\pi m}$.

4.4 Show that for Maxwellian distribution of velocities of gas molecules, the root mean square of speed $< v^2 >^{1/2}= (3kT/m)^{1/2}$

4.5 (a) Show that in Problem 4.1 the most probable speed of the gas molecules $v_p = (2kT/m)^{1/2}$
(b) Show that the ratio $v_p :< v >:< v^2 >^{1/2}:: \sqrt{2} : \sqrt{8/\pi} : \sqrt{3}$

4.6 Estimate the rms velocity of hydrogen molecules at NTP and at 127°C

[Sri Venkateswara University 2001]

4.7 Find the rms speed for molecules of a gas with density of 0.3 g/l of a pressure of 300 mm of mercury.

[Nagarjuna University 2004]

4.8 The Maxwell's distribution for velocities of molecules is given by $N(v)\mathrm{d}v = 2\pi N(m/2\pi kT)^{3/2} v^2 \exp(-mv^2/2kT)\mathrm{d}v$

Calculate the value of $< 1/v >$

4.9 The Maxwell's distribution of velocities is given in Problem 4.8. Show that the probability distribution of molecular velocities in terms of the most probable velocity between α and $\alpha + \mathrm{d}\alpha$ is given by

$$N(\alpha)\mathrm{d}\alpha = \frac{4N}{\sqrt{\pi}}\alpha^2 e^{-\alpha^2}\mathrm{d}\alpha$$

where, $\alpha = \nu/\nu_p$ and $\nu_p = (2kT/m)^{1/2}$.

4.10 Calculate the fraction of the oxygen molecule with velocities between 199 m/s and 201 m/s at 27°C

4.11 Assuming that the hydrogen molecules have a root-mean-square speed of 1,270 m/s at 300 K, calculate the rms at 600 K.

4.12 Clausius had assumed that all molecules move with velocity v with respect to the container. Under this assumption show that the mean relative velocity $< \nu_{\text{rel}} >$ of one molecule with another is given by $< \nu_{\text{rel}} >= 4\nu/3$.

4.13 Estimate the temperature at which the root-mean-square of nitrogen molecule in earth's atmosphere equals the escape velocity from earth's gravitational field. Take the mass of nitrogen molecule = 23.24 amu, and radius of earth = 6,400 km.

4.14 Calculate the fraction of gas molecules which have the mean-free-path in the range λ to 2λ.

4.15 If ρ is the density, $< \nu >$ the mean speed and λ the mean free path of the gas molecules, then show that the coefficient of viscosity is given by $\eta = \frac{1}{3}\rho < \nu > \lambda$.

4.16 At STP, the rms velocity of the molecules of a gas is 10^5 cm/s. The molecular density is $3 \times 10^{25}\ \text{m}^{-3}$ and the diameter (σ) of the molecule 2.5×10^{-10} m. Find the mean-free-path and the collision frequency.

[Nagarjuna University 2000]

4.17 When a gas expands adiabatically its volume is doubled while its Kelvin temperature is decreased by a factor of 1.32. Calculate the number of degrees of freedom for the gas molecules.

4.18 What is the temperature at which an ideal gas whose molecules have an average kinetic energy of 1 eV?

4.19 (a) If γ is the ratio of the specific heats and n is the degrees of freedom then show that for a perfect gas

$$\gamma = 1 + 2/n$$

(b) Calculate γ for monatomic and diatomic molecules without vibration.

4.20 If K is the thermal conductivity, η the coefficient of viscosity, C_ν the specific heat at constant volume and γ the ratio of specific heats then show that for the general case of any molecule

$$\frac{K}{\eta C_\nu} = \frac{1}{4}(9\gamma - 5)$$

4.2.2 Maxwell's Thermodynamic Relations

4.21 Obtain Maxwell's Thermodynamic Relations

(a) $\left(\frac{\partial s}{\partial V}\right)_T = \left(\frac{\partial P}{\partial T}\right)_V$

(b) $\left(\frac{\partial s}{\partial P}\right)_T = -\left(\frac{\partial V}{\partial T}\right)_P$

4.22 Obtain Maxwell's thermodynamic relation.

$$\left(\frac{\partial T}{\partial V}\right)_S = -\left(\frac{\partial p}{\partial S}\right)_V$$

4.23 Obtain Maxwell's thermodynamic relation.

$$\left(\frac{\partial T}{\partial P}\right)_S = \left(\frac{\partial V}{\partial S}\right)_P$$

4.24 Using Maxwell's thermodynamic relations deduce Clausius Clapeyron equation

$$\left(\frac{\partial p}{\partial T}\right)_{\text{saturation}} = \frac{L}{T(\nu_2 - \nu_1)}$$

where p refers to the saturation vapor pressure, L is the latent heat, T the temperature, ν_1 and ν_2 are the specific volumes (volume per unit mass) of the liquid and vapor, respectively.

4.25 Calculate the latent heat of vaporization of water from the following data: $T = 373.2\,\text{K}$, $\nu_1 = 1\,\text{cm}^3$, $\nu_2 = 1,674\,\text{cm}^3$, $\mathrm{d}p/\mathrm{d}T = 2.71\,\text{cm}$ of mercury K^{-1}

4.26 Using the thermodynamic relation

$$\left(\frac{\partial s}{\partial V}\right)_T = \left(\frac{\partial p}{\partial T}\right)_V,$$

derive the Stefan-Boltzmann law of radiation.

4.27 Use the thermodynamic relations to show that for an ideal gas $C_P - C_V = R.$

4.28 For an imperfect gas, Vander Waal's equation is obeyed

$$\left(p + \frac{a}{V^2}\right)(V - b) = RT$$

with the approximation $b/V \ll 1$, show that

$$C_P - C_V \cong R\left(1 + \frac{2a}{RTV}\right)$$

4.29 If E is the isothermal bulk modulus, α the coefficient of volume expansion then show that

$$C_P - C_V = TE\alpha^2 V$$

4.30 Obtain the following $T\mathrm{d}s$ equation
$T\mathrm{d}s = C_V \mathrm{d}T + T\alpha E_T \mathrm{d}V$

where $E_T = -V\left(\frac{\partial P}{\partial V}\right)_T$ is the isothermal elasticity and $\alpha = \frac{1}{V}\left(\frac{\partial V}{\partial T}\right)_P$ is the volume coefficient of expansion, S is the entropy and T the Kelvin temperature.

4.31 Obtain the equation
$T\mathrm{d}s = C_p \mathrm{d}T - TV\alpha \mathrm{d}p$

4.32 Obtain the equation

$$T\mathrm{d}s = C_V \left(\frac{\partial T}{\partial P}\right)_V \mathrm{d}P + C_P \left(\frac{\partial T}{\partial V}\right)_P \mathrm{d}V$$

4.33 Obtain the formula for the Joule–Thompson effect

$$\Delta T = \frac{[T(\partial V/\partial T)_P - V]\Delta P}{C_P}$$

4.34 (a) Show that for a perfect gas governed by the equation of state $PV = RT$ the Joule-Thompson effect does not take place.
(b) Show that for an imperfect gas governed by the equation of state $\left(P + \frac{a}{V^2}\right)(V - b) = RT$, the Joule-Thompson effect is given by

$$\Delta T = \frac{1}{C_P}\left(\frac{2a}{RT} - b\right)\Delta P.$$

4.35 Explain graphically the condition for realizing cooling in the Joule-Thompson effect using the concept of the inversion temperature.

4.36 Prove that for any substance the ratio of the adiabatic and isothermal elasticities is equal to the ratio of the two specific heats.

4.37 Prove that the ratio of the adiabatic to the isobaric pressure coefficient of expansion is $1/(1-\gamma)$.

4.38 Show that the ratio of the adiabatic to the isochoric pressure coefficient is $\gamma/(\gamma - 1)$.

4.39 If U is the internal energy then show that for an ideal gas $(\partial U/\partial V)_T = 0$.
[Nagarjuna University 2004]

4.40 Find the change in boiling point when the pressure on water at 100°C is increased by 2 atmospheres. ($L = 540\,\mathrm{Cal\,g^{-1}}$, volume of 1 g of steam = 1,677 cc)
[Nagarjuna University 2000]

4.41 If 1 g of water freezes into ice, the change in its specific volume is 0.091 cc Calculate the pressure required to be applied to freeze 10 g of water at −1°C.
[Sri Venkateswara University 1999]

4.42 Calculate the change of melting point of naphthalene per atmospheric change of pressure, given melting point = 80°C, latent heat = 35.5 cal/g, density of solid = 1.145 g/cc and density of liquid = 0.981 g/cc

[University of Calcutta]

4.43 The total energy of blackbody radiation in a cavity of volume V at temperature T is given by $u = aVT^4$, where $a = 4\sigma/c$ is a constant.

(a) Obtain an expression for the entropy S in terms of T, V and a.
(b) Using the expression for the free energy F, show that the pressure $P = \frac{1}{3}u$.

4.44 Given that the specific heat of Copper is 387 J/kg K^{-1}, calculate the atomic mass of Copper in amu using Dulong Petit law.

4.2.3 Statistical Distributions

4.45 Calculate the ratio of the number of molecules in the lowest two rotational states in a gas of H_2 at 50 K (take inter atomic distance = 1.05 A°)

[University of Cambridge, Tripos 2004]

4.46 Consider a photon gas in equilibrium contained in a cubical box of volume $V = a^3$. Calculate the number of allowed normal modes of frequency ω in the interval dω.

4.47 Show that for very large numbers, the Stirling's approximation gives
$n! \cong \sqrt{2\pi n}\, n^n e^{-n}$

4.48 Show that the rotational level with the highest population is given by

$$\text{J max(pop)} = \frac{\sqrt{I_0 kT}}{\hbar} - \frac{1}{2}$$

4.49 Assuming that the moment of inertia of the H_2 molecule is 4.64×10^{-48} kg-m^2, find the relative population of the $J = 0, 1, 2$ and 3 rotational states at 400 K.

4.50 In Problem 4.49, at what temperature would the population for the rotational states $J = 2$ and $J = 3$ be equal.

4.51 Calculate the relative numbers of hydrogen atoms in the chromosphere with the principal quantum numbers $n = 1, 2, 3$ and 4 at temperature 6,000 K.

4.52 Calculate the probability that an allowed state is occupied if it lies above the Fermi level by kT, by $5kT$, by $10\,kT$.

4.53 If n is the number of conduction electrons per unit volume and m the electron mass then show that the Fermi energy is given by the expression

$$E_{\mathrm{F}} = \frac{h^2}{8m}\left(\frac{3n}{\pi}\right)^{2/3}$$

4.54 The probability for occupying the Fermi level $P_F = 1/2$. If the probability for occupying a level ΔE above E_F is P_+ and that for a level ΔE below E_F is P_-, then show that for $\frac{\Delta E}{kT} \ll 1$, P_F is the mean of P_+ and P_-

4.55 Find the number of ways in which two particles can be distributed in six states if
(a) the particles are distinguishable
(b) the particles are indistinguishable and obey Bose-Einstein statistics
(c) the particles are indistinguishable and only one particle can occupy any one state.

4.56 From observations on the intensities of lines in the optical spectrum of nitrogen in a flame the population of various vibrationally excited molecules relative to the ground state is found as follows:

v	0	1	2	3
N_v/N_0	1.000	0.210	0.043	0.009

Show that the gas is in thermodynamic equilibrium in the flame and calculate the temperature of the gas ($\theta_v = 3{,}350$ K)

4.57 How much heat (in eV) must be added to a system at 27°C for the number of accessible states to increase by a factor of 10^8?

4.58 The counting rate of Alpha particles from a certain radioactive source shows a normal distribution with a mean value of 10^4 per second and a standard deviation of 100 per second. What percentage of counts will have values
(a) between 9,900 and 10,100
(b) between 9,800 and 10,200
(c) between 9,700 and 10,300

4.59 A system has non-degenerate energy levels with energy $E = \left(n + \frac{1}{2}\right)\hbar\omega$, where $\hbar\omega = 8.625 \times 10^{-5}$ eV, and $n = 0, 1, 2, 3 \ldots$ Calculate the probability that the system is in the $n = 10$ state if it is in contact with a heat bath at room temperature ($T = 300$ K). What will be the probability for the limiting cases of very low temperature and very high temperature?

4.60 Derive Boltzmann's formula for the probability of atoms in thermal equilibrium occupying a state E at absolute temperature T.

4.2.4 Blackbody Radiation

4.61 A wire of length 1 m and radius 1 mm is heated via an electric current to produce 1 kW of radiant power. Treating the wire as a perfect blackbody and ignoring any end effects, calculate the temperature of the wire.

[University of London]

4.62 When the sun is directly overhead, the thermal energy incident on the earth is $1.4\ \text{kWm}^{-2}$. Assuming that the sun behaves like a perfect blackbody of radius 7×10^5 km, which is 1.5×10^8 km from the earth show that the total intensity of radiation emitted from the sun is $6.4 \times 10^7\ \text{Wm}^{-2}$ and hence estimate the sun's temperature.

[University of London]

4.63 If u is the energy density of radiation then show that the radiation pressure is given by $P_{\text{rad}} = u/3$.

4.64 If the temperature difference between the source and surroundings is small then show that the Stefan's law reduces to Newton's law of cooling.

4.65 The pressure inside the sun is estimated to be of the order of 400 million atmospheres. Estimate the temperature corresponding to such a pressure assuming it to result from the radiation.

4.66 The mass of the sun is 2×10^{30} Kg, its radius 7×10^8 m and its effective surface temperature 5,700 K.
(a) Calculate the mass of the sun lost per second by radiation.
(b) Calculate the time necessary for the mass of the sun to diminish by 1%.

4.67 Compare the rate of fall of temperature of two solid spheres of the same material and similar surfaces, where the radius of one surface is four times of the other and when the Kelvin temperature of the large sphere is twice that of the small one (Assume that the temperature of the spheres is so high that absorption from the surroundings may be ignored).

[University of London]

4.68 A cavity radiator has its maximum spectral radiance at a wavelength of 1.0 μm in the infrared region of the spectrum. The temperature of the body is now increased so that the radiant intensity of the body is doubled.
(a) What is the new temperature?
(b) At what wavelength will the spectral radiance have its maximum value? (Wien's constant $b = 2.897 \times 10^{-3}$ m-K)

4.69 In the quantum theory of blackbody radiation Planck assumed that the oscillators are allowed to have energy, 0, ε, $2\varepsilon \ldots$ Show that the mean energy of the oscillator is $\bar{\varepsilon} = \varepsilon/[\exp(\varepsilon/kT) - 1]$ where $\varepsilon = h\nu$

4.70 Planck's formula for the blackbody radiation is

$$u_\lambda \mathrm{d}\lambda = \frac{8\pi hc}{\lambda^5} \frac{1}{e^{hc/\lambda kT} - 1} \mathrm{d}\lambda$$

(a) Show that for long wavelengths and high temperatures it reduces to Rayleigh-Jeans law.
(b) Show that for short wavelengths it reduces to Wien's distribution law

4.71 Starting from Planck's formula for blackbody radiation deduce Wien's displacement law and calculate Wien's constant b, assuming the values of h, c and k.

4.72 Using Planck's formula for blackbody radiation show that Stefan's constant

$$\sigma = \frac{2}{15}\frac{\pi^5 k^4}{h^3 c^2} = 5.67 \times 10^{-8}\ \mathrm{W.m^{-2}.K^{-4}}$$

4.73 A blackbody has its cavity of cubical shape. Determine the number of modes of vibration per unit volume in the wavelength region 4,990–5,010 A°.

[Osmania University 2004]

4.74 A cavity kept at 4,000 K has a circular aperture 5.0 mm diameter. Calculate (a) the power radiated in the visible region (0.4–0.7 μm) from the aperture (b) the number of photons emitted per second in the visible region

4.75 Planck's formula for the black body radiation is

$$u_\lambda \mathrm{d}\lambda = \frac{8\pi hc}{\lambda^5}\frac{1}{e^{hc/\lambda kT} - 1}\mathrm{d}\lambda$$

Express this formula in terms of frequency.

4.76 Estimate the temperature T_E of the earth, assuming that it is in radiation equilibrium with the sun (assume the radius of sun $R_s = 7 \times 10^8$ m, the earth-sun distance $r = 1.5 \times 10^{11}$ m, the temperature of solar surface $T_s =$ 5,800 K)

4.77 Calculate the solar constant, that is the radiation power received by 1 m^2 of earth's surface. (Assume the sun's radius $R_s = 7 \times 10^8$ m, the earth-sun distance $r = 1.5 \times 10^{11}$ m, the earth's radius $R_E = 6.4 \times 10^6$ m, sun's surface temperature, $T_s = 5{,}800$ K and Stefan-Boltzmann constant $\sigma = 5.7 \times 10^{-8}\frac{\mathrm{W}}{\mathrm{m}^2} - \mathrm{K}^4$).

4.78 A nuclear bomb at the instant of explosion may be approximated to a blackbody of radius 0.3 m with a surface temperature of 10^7 K. Show that the bomb emits a power of 6.4×10^{20} W.

4.3 Solutions

4.3.1 Kinetic Theory of Gases

4.1 Consider a two-body collision between two similar gas molecules of initial velocity v_1 and v_2. After the collision, let the final velocities be v_3 and v_4. The probability for the occurrence of such a collision will be proportional to the number of molecules per unit volume having these velocities, that is to the product $f(v_1)f(v_2)$. Thus the number of each collisions per unit volume per unit time is $c\ f(v_1)f(v_2)$ where c is a constant. Similarly, the number of inverse collisions per unit volume per unit time is $c'\ f(v_3)f(v_4)$ where c' is also a constant. Since the gas is in equilibrium and the velocity distribution is unchanged by collisions, these two rates must be equal. Further in the centre of mass these two collisions appear to be equivalent so that $c' = c$. We can then write

$$f(v_1)f(v_2) = f(v_3)f(v_4)$$
$$\text{or} \ln f(v_1) + \ln f(v_2) = \ln f(v_3) + \ln f(v_4) \quad (1)$$

Since kinetic energy is conserved

$$v_1^2 + v_2^2 = v_3^2 + v_4^2 \quad (2)$$

Equations (1) and (2) are satisfied if

$$\ln f(v) \propto v^2 \quad (3)$$
$$\text{or} f(v) = A\exp(-\alpha v^2) \quad (4)$$

where A and α are constants. The negative sign is essential to ensure that no molecule can have infinite energy.

Let $N(v)\mathrm{d}v$ be the number of molecules per unit volume with speeds v to $+\mathrm{d}v$, irrespective of direction. As the velocity distribution is assumed to be spherically symmetrical, $N(v)\mathrm{d}v$ is equal to the number of velocity vectors whose tips end up in the volume of the shell defined by the radii v and $+\mathrm{d}v$, so that

$$N(v)\mathrm{d}v = 4\pi v^2 f(v)\mathrm{d}v \quad (5)$$

Using (4) in (5)

$$N(v)\mathrm{d}v = 4\pi Av^2 exp(-\alpha v^2) \quad (6)$$

We can now determine A and α. If N is the total number of molecules per unit volume,

$$N = \int_0^\infty N(v)\mathrm{d}v \quad (7)$$

Using (6) in (7)

$$N = 4\pi A \int_0^\infty v^2 \exp(-\alpha v^2)\,\mathrm{d}v = 4\pi A(1/4)(\pi/\alpha^3)^{1/2}$$
$$\text{or } N = A(\pi/\alpha)^{3/2} \quad (8)$$

If E is the total kinetic energy of the molecules per unit volume

$$E = \frac{1}{2}m\int_0^\infty v^2 N(v)\mathrm{d}v = \frac{4\pi Am}{2}\int_0^\infty v^4 \exp(-\alpha v^2)\mathrm{d}v$$
$$\text{or } E = (3mA/4)(\pi^3/\alpha^5)^{1/2} \quad (9)$$

where gamma functions have been used for the evaluation of the two integrals. Further,

$$E = 3NkT/2 \tag{10}$$

Combining (8), (9) and (10)

$$\alpha = \frac{m}{2kT} \tag{11}$$

$$\text{and} \quad A = N(\alpha/\pi)^{3/2} = N(m/2\pi kT)^{3/2} \tag{12}$$

Using (11) and (12) in (5)

$$N(v)\mathrm{d}v = 4\pi N(m/2\pi kT)^{3/2} v^2 exp(-mv^2/2kT)\mathrm{d}v$$

4.2 $N(v)\mathrm{d}v = 4\pi N(m/2\pi kT)^{3/2} v^2 exp(-mv^2/2kT)\mathrm{d}v$ (1)

Put $E = \frac{1}{2}mv^2$, $\mathrm{d}E = mv\mathrm{d}v$ (2)

Use (2) in (1) and simplify to obtain

$$N(E)\mathrm{d}E = \frac{2\pi N E^{1/2}}{(\pi kT)^{3/2}} \exp\left(-\frac{E}{kT}\right)\mathrm{d}E$$

4.3 The average speed

$$< v >= \frac{\int_0^\infty vN(v)\mathrm{d}v}{N} = 4\pi\left(\frac{m}{2\pi kT}\right)^{3/2}\int_0^\infty v^3 \exp(-mv^2/2kT)\mathrm{d}v \tag{1}$$

where we have used the Maxwellian distribution

$$\text{Put } \alpha = \frac{m}{2kT} \tag{2}$$

$$\text{so that } \int_0^\infty v^3 e^{-\alpha v^2}\mathrm{d}v = \frac{1}{2\alpha^2} \tag{3}$$

Combining (1), (2) and (3)

$$< v >= \left(\frac{8kT}{\pi m}\right)^{1/2} = \sqrt{\frac{8RT}{M}} \tag{4}$$

where m is the mass of the molecule, M is the molecular weight and R the gas constant.

4.4 $< v^2 >= \dfrac{\int_0^\infty v^2 N(v)\mathrm{d}v}{N} = 4\pi\left(\dfrac{m}{2\pi kT}\right)^{3/2}\displaystyle\int_0^\infty v^4 \exp(-mv^2/2kT)\mathrm{d}v$

with $\alpha = \dfrac{m}{2kT}$ and $x = \alpha v^2$; $\mathrm{d}x = 2\alpha v\mathrm{d}v$

The integral, $I = \displaystyle\int_0^\infty v^4 e^{-\alpha v^2}\mathrm{d}v = \frac{1}{2\alpha^{5/2}}\int_0^\infty x^{3/2}e^{-x}\mathrm{d}x = \frac{3\sqrt{\pi}}{8\alpha^{5/2}}$

Therefore, $< v^2 >= 4\pi\left(\dfrac{m}{2\pi kT}\right)^{3/2}\dfrac{3\sqrt{\pi}}{8\left(\frac{m}{2kT}\right)^{5/2}} = \dfrac{3kT}{m}$

$$< v^2 >^{1/2} = (3kT/m)^{1/2}$$

4.5 (a) v_p is found by maximizing the Maxwellian distribution.

$$\frac{d}{dv}[v^2 \exp(-mv^2/2kT)] = 0$$
$$\exp(-mv^2/kT)[2v - mv^3/kT] = 0$$
whence $v = v_p = (2kT/m)^{1/2}$

(b) $v_p :< v >:< v^2 >^{1/2} :: (2kT/m)^{1/2} : (8kT/\pi m)^{1/2} : (3kT/m)^{1/2}$
$= \sqrt{2} : \sqrt{8/\pi} : \sqrt{3}$

4.6 $$< v^2 >^{1/2} = \left(\frac{3kT}{m}\right)^{1/2} = \left(\frac{3 \times 1.38 \times 10^{-23} \times 273}{1.67 \times 10^{-27}}\right)^{1/2}$$

$= 2{,}601$ m/s at N.T.P

$$< v^2 >^{1/2} = \left(\frac{3 \times 1.38 \times 10^{-23} \times 400}{1.67 \times 10^{-27}}\right)^{1/2} = 3{,}149 \text{ m/s at } 127^\circ\text{C}.$$

4.7 $$< v^2 >^{1/2} = \left(\frac{3p}{\rho}\right)^{1/2} = \left(\frac{3 \times (300/760) \times 1.013 \times 10^5}{0.3}\right)^{1/2} = 632 \text{ m/s}$$

4.8 $$< \frac{1}{v} > = \frac{1}{N}\int_0^\infty \frac{1}{v} N(v)\, dv$$

$$= \frac{1}{N}\int_0^\infty \frac{1}{v} .4\pi N \left(\frac{m}{2\pi kT}\right)^{3/2} v^2 \exp(-mv^2/2kT)\, dv$$

Set $mv^2/2kT = x; v dv = kT\, dx/m$

$$< \frac{1}{v} > = (2m/\pi kT)^{1/2} \int_0^\infty \exp(-x)\, dx = (2m/\pi kT)^{1/2}$$

4.9 $N(v)dv = 4\pi N(m/2\pi kT)^{3/2} v^2 \exp(-mv^2/2kT)dv$ (1)

$v_p = (2kT/m)^{1/2}$ (2)

Let $v/v_p = \alpha; dv = v_p d\alpha$ (3)

Use (2) and (3) in (1)

$$N(\alpha)d\alpha = \frac{4N}{\sqrt{\pi}} \alpha^2 \exp(-\alpha^2)d\alpha$$

4.10 Fraction

$$f = \frac{N(v)dv}{N} = 4\pi \left[\frac{m}{2\pi kT}\right]^{3/2} v^2 \exp(-mv^2/2kT)dv$$

$$v = \frac{199 + 201}{2} = 200 \text{ m/s}$$

$dv = 201 - 199 = 2$ m/s

$$f = 4\pi \left(\frac{32 \times 1.67 \times 10^{-27}}{2\pi \times 1.38 \times 10^{-23} \times 300} \right)^{3/2} (200)^2$$
$$\exp\left(-\frac{32 \times 1.67 \times 10^{-27} \times 200^2}{2 \times 1.38 \times 10^{-23} \times 300} \right) \times (2)$$
$$= 2.29 \times 10^{-3}$$

4.11 $< v^2 >^{1/2} = \left(\frac{3kT}{m} \right)^{1/2}$

$$v_{\text{rms}}(600\text{ K}) = [v_{\text{rms}}(300\text{ K})](600/300)^{1/2}$$
$$= 1270 \times \sqrt{2} = 1{,}796\,\text{m/s}$$

4.12 Relative velocity v_{rel} of one molecule and another making an angle θ is
$v_{\text{rel}} = (v^2 + v^2 - 2(v)(v)\cos\theta)^{1/2} = 2v\sin(\theta/2)$

Now, all the direction of velocities v are equally probable. The probability $f(\theta)$ that v lies within an element of solid angle between θ and $\theta + \mathrm{d}\theta$ is given by

$$f(\theta) = 2\pi \sin\theta \mathrm{d}\theta / 4\pi = \frac{1}{2}\sin\theta \mathrm{d}\theta$$

v_{rel} is obtained by integrating over $f(\theta)$ in the angular interval 0 to π.

$$< v_{\text{rel}} > = \int_0^\pi v_{\text{rel}} f(\theta) = \int_0^\pi 2v \sin\left(\frac{\theta}{2}\right)\left(\frac{1}{2}\sin\theta \mathrm{d}\theta\right)$$

$$= 2v \int_0^\pi \sin^2\left(\frac{\theta}{2}\right)\cos\frac{\theta}{2}\mathrm{d}\theta = 4v \int_0^\pi \sin^2\frac{\theta}{2}\mathrm{d}\left(\sin\frac{\theta}{2}\right) = 4v/3$$

4.13 $v_e = (2gR)^{1/2}$; $v_{\text{rms}} = (3kT/m)^{1/2}$

$$v_{\text{rms}} = v_e$$

$$T = \frac{2mgR}{3k} = \frac{2 \times (2 \times 23.24 \times 10^{-27})(9.8)(6.37 \times 10^6)}{3 \times 1.38 \times 10^{-23}}$$
$$= 1.4 \times 10^5\text{ K}$$

4.14 Fraction of gas molecules that do not undergo collisions after path length x is $\exp(-x/\lambda)$. Therefore the fraction of molecules that has free path values between λ to 2λ is

$$f = \exp(-\lambda/\lambda) - \exp(-2\lambda/\lambda)$$
$$= \exp(-1) - \exp(-2)$$
$$= 0.37 - 0.14 = 0.23$$

4.15 Consider a volume element $\mathrm{d}V = 2\pi r^2 \sin\theta \mathrm{d}\theta \mathrm{d}r$ located on a layer at a height $z = r\cos\theta$. If mu is the momentum of a molecule at the XY-plane at $z = 0$, then its value at $\mathrm{d}V$ will be $mu + \left(\frac{\mathrm{d}}{\mathrm{d}z}mu\right) r\cos\theta$ (Fig. 4.2). At an identical layer below the reference plane $\mathrm{d}A$, the momentum would be

$$mu - \left(\frac{\mathrm{d}}{\mathrm{d}z}mu\right) r\cos\theta$$

Let dn be the number of molecules with velocity between ν *and* $\nu + \mathrm{d}\nu$ per unit volume. The number of molecules with velocity ν *and* $\nu + \mathrm{d}\nu$ in the volume element dν is dndν. Molecules within the volume element undergo collisions and are scattered in various directions.

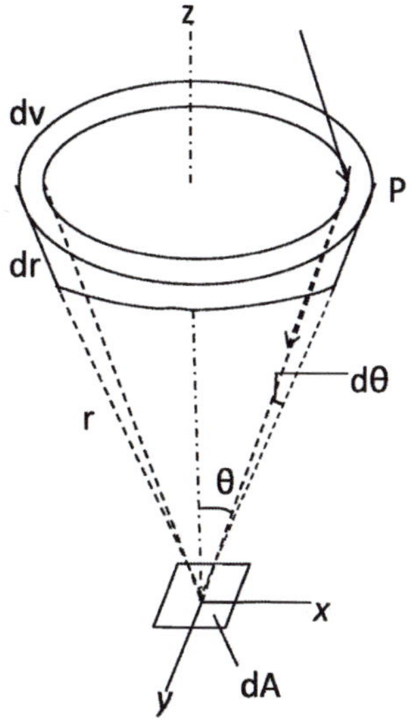

Fig. 4.2 Transport of momentum of gas molecules

Number of collisions that occur in dV in time dt will be $\frac{1}{2}\frac{\nu}{\lambda}\mathrm{d}t$. The factor $\frac{1}{2}$ is introduced to avoid counting each collision twice, since the collision between molecules 1 and 2 and that between 2 and 1 is same.

Each collision results in two new paths for the scattered molecules. Hence the number of molecules that are scattered in various directions from this volume element dV in time dt will be $2 \times \frac{1}{2}\frac{\nu}{\lambda}\mathrm{d}t \times \mathrm{d}n\mathrm{d}V$ or $\frac{\nu}{\lambda}\,\mathrm{d}t\mathrm{d}n\mathrm{d}V$.

Now the solid angle subtended by dA of the reference plane at dV is $\mathrm{d}A\cos\theta/r^2$.

Assuming the scattering to be isotropic the number of molecules moving downward toward dA is

$$\frac{\nu}{\lambda}\frac{\mathrm{d}t\mathrm{d}n\mathrm{d}V\mathrm{d}A\cos\theta}{4\pi r^2}$$

$$\text{or } \frac{\nu\mathrm{d}t\mathrm{d}n(2\pi r^2\sin\theta\mathrm{d}\theta\mathrm{d}r)\mathrm{d}A\cos\theta}{\lambda.4\pi r^2}$$

$$\text{or } \frac{\nu\mathrm{d}t\mathrm{d}n\mathrm{d}A\sin\theta\cos\theta}{2\lambda}$$

Transport of momentum downward from molecules in the upper hemisphere through dA in time dt is

$$P_- = \frac{\mathrm{d}A\mathrm{d}t}{2\lambda}\int_0^\infty v\mathrm{d}n \int_0^\infty e^{-r/\lambda}\mathrm{d}r \int_0^{\pi/2} \sin\theta\cos\theta\left(mu + r\cos\theta\frac{\mathrm{d}mu}{\mathrm{d}z}\right)\mathrm{d}\theta$$

The factor $e^{-r/\lambda}$ is included to ensure that the molecule in traversing the distance r toward $\mathrm{d}A$ does not get scattered and prevented from reaching $\mathrm{d}A$.

Similarly, transport of momentum upward, from molecules in the lower hemisphere through $\mathrm{d}A$ in time $\mathrm{d}t$ is

$$P_+ = \frac{\mathrm{d}A\mathrm{d}t}{2\lambda}\int_0^\infty v\mathrm{d}n \int_0^\infty e^{-r/\lambda}\mathrm{d}r \int_0^{\pi/2} \sin\theta\cos\theta\left(mu - r\cos\theta\frac{\mathrm{d}mu}{\mathrm{d}z}\right)\mathrm{d}\theta$$

Hence net momentum transfer to the reference plane through an area $\mathrm{d}A$ in time $\mathrm{d}t$ is

$$P = P_- - P_+ = \frac{\mathrm{d}A\mathrm{d}t}{\lambda}\frac{m\mathrm{d}u}{\mathrm{d}z}\int_0^\infty v\mathrm{d}n \int_0^\infty re^{-r/\lambda}\mathrm{d}r \int_0^{\pi/2} \cos^2\theta\sin\theta\mathrm{d}\theta$$

$$= \frac{\mathrm{d}A\mathrm{d}t}{\lambda}m\frac{\mathrm{d}u}{\mathrm{d}z}n < v > \frac{\lambda^2}{3}$$

$$= \frac{m}{3}\mathrm{d}A\mathrm{d}t\frac{\mathrm{d}u}{\mathrm{d}z}\lambda n < v >$$

(the first integral gives $n < v >$, the second one λ^2 and the third one a factor 1/3)

Momentum transported per second is force

$$F = \frac{\lambda}{3}\mathrm{d}An < v > m\frac{\mathrm{d}u}{\mathrm{d}z}$$

The viscous force is

$$\eta\mathrm{d}A\frac{\mathrm{d}u}{\mathrm{d}z} = \frac{\lambda}{3}\mathrm{d}An < v > m\frac{\mathrm{d}u}{\mathrm{d}z}$$

or $\eta = \frac{1}{3}mn < v > \lambda = \frac{1}{3}\rho < v > \lambda$

where $mn = \rho =$ density of molecules.

4.16 $\lambda = \frac{1}{\sqrt{2}\pi n\sigma^2} = \frac{1}{\sqrt{2}\pi \times 3 \times 10^{25} \times (2.5 \times 10^{-10})^2}$

$= 1.2 \times 10^{-7}$ m

$$f = \frac{v}{\lambda} = \frac{1,000}{1.2 \times 10^{-7}} = 8.33 \times 10^9\ \mathrm{s}^{-1}$$

4.17 $T_1V_1^{\gamma-1} = T_2V_2^{\gamma-1}, \quad \left(\frac{V_2}{V_1}\right)^{\gamma-1} = \frac{T_1}{T_2}$

or $2^{\gamma-1} = 1.32, \quad \gamma = 1.4$

Number of degrees of freedom,

$$f = \frac{2}{\gamma - 1} = \frac{2}{1.4 - 1} = 5$$

4.18 $1\text{eV} = kT$

$$T = \frac{1eV}{k} = \frac{1.6 \times 10^{-19} J}{1.38 \times 10^{-23} J/K} = 11{,}594\,\text{K}$$

4.19 (a) For a perfect gas at temperature T, the kinetic energy from translation motion

$$\frac{1}{2}m < v_x^2 > +\frac{1}{2}m < v_y^2 > +\frac{1}{2}m < v_z^2 >= \frac{3}{2}\frac{RT}{N_0} \qquad (1)$$

where R is the gas constant and N_0 is Avagadro's number. The energy of the 3 degrees of freedom of translation is therefore on the average equal to $\frac{3}{2}RT/N_0$ for each molecule. Using this result together with the principle of the equipartition of energy, it is concluded that in a system at temperature T each degree of freedom contributes, $\frac{1}{2}\frac{R}{N_0}T$ to the total energy.

If each molecule has n degrees of freedom, the total internal energy U of a gram-molecule of a perfect gas at temperature T,

$$U = \frac{1}{2}nRT \qquad (2)$$

The molecular heat at constant volume C is equal to $\left(\frac{\partial U}{\partial T}\right)_v$, and is therefore given by

$$C_v = \frac{1}{2}nR \qquad (3)$$

For a perfect gas

$$C_p - C_v = R \qquad (4)$$

Therefore $C_p = C_v + R = \dfrac{(n+2)R}{2}$ (5)

and $\gamma = \dfrac{C_p}{C_v} = 1 + \dfrac{2}{n}$ (6)

(b) For monatomic molecule $n = 3$, for translation (rotation and vibration are absent), $\gamma = 1.667$.

For diatomic molecule $n = 5$ (3 from translation and only 2 from rotation as the rotation about an axis joining the centres of atoms does not contribute) and $\gamma = 1.4$

If vibration is included then $n = 7$ and $\gamma = 1.286$

4.20 According to Chapman and Enskog

$$K = \frac{\eta}{m}\left[\frac{5}{2}\frac{d\overline{E}_t}{dT} + \frac{d\overline{E}'}{dT}\right] \qquad (1)$$

where $\overline{E}_t$ is the translational energy and $\overline{E}'$ the energy of other types.

If β denotes the number of degrees of freedom of the molecule due to causes other than translation, the total number of degrees of freedom of the molecule will be $3 + \beta$.

From the law of equipartition of energy we have

$$\frac{\mathrm{d}\overline{E}_t}{\mathrm{d}T} = \frac{3}{2}k; \frac{\mathrm{d}\overline{E}'}{\mathrm{d}T} = \frac{\beta}{2}k \tag{2}$$

Hence,

$$\frac{K}{\eta} = \left[\frac{5}{2}.\frac{3}{2} + \frac{\beta}{2}\right]\frac{k}{m} \tag{3}$$

We can express the result in terms of C_v and γ. From the law of equipartition of energy

$$C_v = \frac{(3+\beta)}{2}.\frac{k}{m}; \quad C_p = \frac{(5+\beta)}{2}.\frac{k}{m}$$

whence $\gamma = \dfrac{C_p}{C_v} = 1 + \dfrac{2}{3+\beta}$

$$\text{or } \beta = \frac{5-3\gamma}{\gamma-1} \tag{4}$$

Furthermore

$$C_v = \frac{k}{m(\gamma-1)} \tag{5}$$

Combining (3), (4) and (5)

$$\frac{K}{\eta C_v} = \frac{1}{4}(9\gamma - 5)$$

4.3.2 Maxwell's Thermodynamic Relations

4.21 Let $f(x, y) = 0$ (1)

$$\mathrm{d}f = \left(\frac{\partial f}{\partial x}\right)_y \mathrm{d}x + \left(\frac{\partial f}{\partial y}\right)_x \mathrm{d}y = 0 \tag{2}$$

Equation of state can be written as $f(P, V, T) = 0$. By first law of thermodynamics

$$\mathrm{d}Q = \mathrm{d}U + \mathrm{d}W \tag{3}$$

By second law of thermodynamics

$$\mathrm{d}Q = T\mathrm{d}s \tag{4}$$

for infinitesimal reversible process

$$\mathrm{d}W = p\mathrm{d}V \tag{5}$$

Therefore,

$$dU = T\mathrm{d}s - P\mathrm{d}V \tag{6}$$

where U is the internal energy, Q the heat absorbed, W the work done by the system, S the entropy, P the pressure and T the Kelvin temperature.
Let the independent variables be called x and y. Then

$$U = U(x, y); V = V(x, y); S = S(x, y) \tag{7}$$

Now,

$$\mathrm{d}f = \left(\frac{\partial f}{\partial x}\right)_y \mathrm{d}x + \left(\frac{\partial f}{\partial y}\right)_x \mathrm{d}y \tag{8}$$

Therefore

$$\mathrm{d}U = \left(\frac{\partial U}{\partial x}\right)_y \mathrm{d}x + \left(\frac{\partial U}{\partial y}\right)_x \mathrm{d}y \tag{9}$$

$$\mathrm{d}V = \left(\frac{\partial V}{\partial x}\right)_y \mathrm{d}x + \left(\frac{\partial V}{\partial y}\right)_x \mathrm{d}y \tag{10}$$

$$\mathrm{d}S = \left(\frac{\partial S}{\partial x}\right)_y \mathrm{d}x + \left(\frac{\partial S}{\partial y}\right)_x \mathrm{d}y \tag{11}$$

Eliminating internal energy U and substituting (9), (10) and (11) in (6)

$$\left(\frac{\partial U}{\partial x}\right)_y \mathrm{d}x + \left(\frac{\partial U}{\partial y}\right)_x \mathrm{d}y = T\left[\left(\frac{\partial S}{\partial x}\right)_y \mathrm{d}x + \left(\frac{\partial S}{\partial y}\right)_x \mathrm{d}y\right]$$
$$-P\left[\left(\frac{\partial V}{\partial x}\right)_y \mathrm{d}x + \left(\frac{\partial V}{\partial y}\right)_x \mathrm{d}y\right] \tag{12}$$

Equating the coefficients of $\mathrm{d}x$ and $\mathrm{d}y$

$$\left(\frac{\partial U}{\partial x}\right)_y = T\left(\frac{\partial S}{\partial x}\right)_y - P\left(\frac{\partial V}{\partial x}\right)_y \tag{13}$$

$$\left(\frac{\partial U}{\partial y}\right)_x = T\left(\frac{\partial S}{\partial y}\right)_x - P\left(\frac{\partial V}{\partial y}\right)_x \tag{14}$$

Differentiating (13) with respect to y with x fixed, and differentiating (14) with respect to x with y fixed

$$\left\{\frac{\partial}{\partial y}\left(\frac{\partial U}{\partial x}\right)_y\right\}_x = \left(\frac{\partial T}{\partial y}\right)_x\left(\frac{\partial S}{\partial x}\right)_y + T\left\{\frac{\partial}{\partial y}\left(\frac{\partial S}{\partial x}\right)_y\right\}_x - \left(\frac{\partial P}{\partial y}\right)_x\left(\frac{\partial V}{\partial x}\right)_y - P\left\{\frac{\partial}{\partial y}\left(\frac{\partial V}{\partial x}\right)_y\right\}_x \tag{15}$$

$$\left\{\frac{\partial}{\partial x}\left(\frac{\partial U}{\partial y}\right)_x\right\}_y = \left(\frac{\partial T}{\partial x}\right)_y\left(\frac{\partial S}{\partial y}\right)_x + T\left\{\frac{\partial}{\partial x}\left(\frac{\partial S}{\partial y}\right)_x\right\}_y - \left(\frac{\partial P}{\partial x}\right)_y\left(\frac{\partial V}{\partial y}\right)_x - P\left\{\frac{\partial}{\partial x}\left(\frac{\partial V}{\partial y}\right)_x\right\}_y \tag{16}$$

Since the order of differentiation is immaterial, dU being a perfect differential, the left hand sides of (15) and (16) are equal. Further, since dS and dV are perfect differentials.

$$\left\{\frac{\partial}{\partial y}\left(\frac{\partial S}{\partial x}\right)_y\right\}_x = \left\{\frac{\partial}{\partial x}\left(\frac{\partial S}{\partial y}\right)_x\right\}_y \tag{17}$$

and

$$\left\{\frac{\partial}{\partial y}\left(\frac{\partial V}{\partial x}\right)_y\right\}_x = \left\{\frac{\partial}{\partial x}\left(\frac{\partial V}{\partial y}\right)_x\right\}_y \tag{18}$$

Using (15), (16), (17), and (18),

$$\left(\frac{\partial P}{\partial x}\right)_y\left(\frac{\partial V}{\partial y}\right)_x - \left(\frac{\partial P}{\partial y}\right)_x\left(\frac{\partial V}{\partial x}\right)_y = \left(\frac{\partial T}{\partial x}\right)_y\left(\frac{\partial S}{\partial y}\right)_x - \left(\frac{\partial T}{\partial y}\right)_x\left(\frac{\partial S}{\partial x}\right)_y \tag{19}$$

Equation (19) can be written in the form of determinants

$$\begin{vmatrix} \left(\frac{\partial P}{\partial x}\right)_y & \left(\frac{\partial P}{\partial y}\right)_x \\ \left(\frac{\partial V}{\partial x}\right)_y & \left(\frac{\partial V}{\partial y}\right)_x \end{vmatrix} = \begin{vmatrix} \left(\frac{\partial T}{\partial x}\right)_y & \left(\frac{\partial T}{\partial y}\right)_x \\ \left(\frac{\partial S}{\partial x}\right)_y & \left(\frac{\partial S}{\partial y}\right)_x \end{vmatrix} \tag{20}$$

(a) Let the temperature and volume be independent variables. Put $x = T$ and $y = V$ in (20). Then

$$\left(\frac{\partial T}{\partial x}\right)_y = \left(\frac{\partial V}{\partial y}\right)_x = 1; \quad \left(\frac{\partial T}{\partial y}\right)_x = \left(\frac{\partial V}{\partial x}\right)_y = 0$$

Since T and V are independent, we find

$$\left(\frac{\partial S}{\partial V}\right)_T = \left(\frac{\partial P}{\partial T}\right)_V \tag{21}$$

(b) Let the temperature and pressure be independent variables. Put $x = T$ and $y = P$ in (20).

$$\left(\frac{\partial T}{\partial x}\right)_y = \left(\frac{\partial P}{\partial y}\right)_x = 1; \quad \left(\frac{\partial T}{\partial y}\right)_x = \left(\frac{\partial P}{\partial x}\right)_y = 0$$

$$\left(\frac{\partial S}{\partial P}\right)_T = -\left(\frac{\partial V}{\partial T}\right)_P \tag{22}$$

4.22 In Problem 4.21 let the entropy and volume be independent variables. Put $x = s$ and $y = V$ in Eq. (19)

$$\left(\frac{\partial S}{\partial x}\right)_y = \left(\frac{\partial V}{\partial y}\right)_x = 1; \quad \left(\frac{\partial S}{\partial y}\right)_x = \left(\frac{\partial V}{\partial x}\right)_y = 0 \quad \left(\frac{\partial T}{\partial V}\right)_T = -\left(\frac{\partial P}{\partial S}\right)_V \tag{23}$$

4.23 Let the entropy and pressure be independent variables. Put $x = s$ and $y = p$ in Eq. (19) of Problem 4.21.

$$\left(\frac{\partial S}{\partial x}\right)_y = \left(\frac{\partial P}{\partial y}\right)_x = 1$$

Therefore,

$$\left(\frac{\partial T}{\partial p}\right)_s = \left(\frac{\partial V}{\partial S}\right)_p \tag{24}$$

4.24 Consider Maxwell's relation (21) of Problem 4.21

$$\left(\frac{\partial S}{\partial V}\right)_T = \left(\frac{\partial P}{\partial T}\right)_V \tag{1}$$

Multiply both sides by T,

$$T\left(\frac{\partial S}{\partial V}\right)_T = T\left(\frac{\partial P}{\partial T}\right)_V \tag{2}$$

or

$$\left(\frac{\partial Q}{\partial V}\right)_T = T\left(\frac{\partial P}{\partial T}\right)_V \tag{3}$$

which means that the latent heat of isothermal expansion is equal to the product of the absolute temperature and the rate of increase of pressure with temperature at constant volume. Apply (3) to the phase transition of a substance. Consider a vessel containing a liquid in equilibrium with its vapor. The pressure is due to the saturated vapor pressure which is a function of temperature only and is independent of the volume of liquid and vapor present. If the vessel is allowed to expand at constant temperature the vapor pressure would remain constant. However, some liquid of mass δm would evaporate to fill the extra space with vapor. If L is the latent heat absorbed per unit mass,

$$\delta Q = L\mathrm{d}m \tag{4}$$

If v_1 and v_2 are the specific volumes (volumes per unit mass) of the liquid and vapor respectively

$$\delta v = (v_2 - v_1)\mathrm{d}m \tag{5}$$

Using (4) and (5) in (3)

$$\frac{L}{v_2 - v_1} = T\left(\frac{\partial P}{\partial T}\right)_V \tag{6}$$

Here, various thermodynamic quantities refer to a mixture of the liquid and vapor in equilibrium. In this case

$$\left(\frac{\partial P}{\partial T}\right)_V = \left(\frac{\partial V}{\partial T}\right)_{\text{sat}}$$

since the pressure is due to the saturated vapor and is therefore independent of V, being only a function of T. Thus (6) can be written as

$$\left(\frac{\partial P}{\partial T}\right)_{\text{sat}} = \frac{L}{T(v_2 - v_1)} \quad \text{(Clapeyron's equation)} \tag{7}$$

4.25 $L = T(v_2 - v_1)\dfrac{\mathrm{d}P}{\mathrm{d}T}$

$$= 373.2(1{,}674 - 1) \times \left(\frac{2.71}{76}\right) \times 1.013 \times 10^6$$
$$= 2.255 \times 10^{10}\,\text{erg g}^{-1}$$
$$= 2.255\,\text{J/g}$$
$$= \frac{2.255}{4.18} = 539.5\,\text{cal/g}$$

4.26 $$\left(\frac{\partial S}{\partial V}\right)_T = \left(\frac{\partial P}{\partial T}\right)_V \tag{1}$$

Substitute

$$\mathrm{d}S = \frac{\mathrm{d}U + P\mathrm{d}V}{T} \tag{2}$$

in (1)

$$\left(\frac{\partial U}{\partial V}\right)_T = T\left(\frac{\partial P}{\partial T}\right)_V - P \tag{3}$$

If u is the energy density and P the total pressure, $\left(\frac{\partial U}{\partial V}\right) = u$ and the total pressure $P = u/3$, since the radiation is diffuse. Hence (3) reduces to

$$u = \frac{T}{3}\frac{\partial u}{\partial T} - \frac{u}{3} \quad \text{or} \quad \frac{\mathrm{d}u}{u} + 4\frac{\mathrm{d}T}{T} = 0$$

Integrating,

$$\ln\, u = 4\ln T + \ln a = \ln aT^4$$

where $\ln a$ is the constant of integration. Thus,

$$u = aT^4$$

4.27 $S = f(T, V)$
where T and V are independent variables.

$$\mathrm{d}S = \left(\frac{\partial S}{\partial T}\right)_V \mathrm{d}T + \left(\frac{\partial S}{\partial V}\right)_T \mathrm{d}V$$

$$\left(\frac{\partial S}{\partial T}\right)_p = \left(\frac{\partial S}{\partial T}\right)_V + \left(\frac{\partial S}{\partial V}\right)_T \left(\frac{\partial V}{\partial T}\right)_P$$

Multiplying out by T and re-arranging

$$T\left(\frac{\partial S}{\partial T}\right)_p - T\left(\frac{\partial S}{\partial T}\right)_V = T\left(\frac{\partial S}{\partial V}\right)_T \left(\frac{\partial V}{\partial T}\right)_P$$

Now,

$$T\left(\frac{\partial S}{\partial T}\right)_p = C_p; \quad T\left(\frac{\partial S}{\partial T}\right)_v = C_v$$

and from Maxwell's relation,

$$\left(\frac{\partial S}{\partial V}\right)_T = \left(\frac{\partial P}{\partial T}\right)_v$$

Therefore,

$$C_p - C_v = T\left(\frac{\partial P}{\partial T}\right)_V \left(\frac{\partial V}{\partial T}\right)_P \tag{1}$$

For one mole of a perfect gas, $PV = RT$. Therefore

$$\left(\frac{\partial P}{\partial T}\right)_V = \frac{R}{V} \text{ and } \left(\frac{\partial V}{\partial T}\right)_P = \frac{R}{P}$$

It follows that

$$C_p - C_v = RT$$

4.28 $$\left(P + \frac{a}{V^2}\right)(V - b) = RT \tag{1}$$

Neglecting b in comparison with V,

$$P = \frac{RT}{V} - \frac{a}{V^2} \tag{2}$$

$$\left(\frac{\partial P}{\partial T}\right)_V = \frac{R}{V} \tag{3}$$

Re-writing (1)

$$PV + \frac{a}{V} = RT$$

Differentiating V with respect to T, keeping P fixed

$$P\left(\frac{\partial V}{\partial T}\right)_P - \frac{a}{V^2}\left(\frac{\partial V}{\partial T}\right)_P = R$$

or

$$\left(\frac{\partial V}{\partial T}\right)_P = \frac{R}{P - a/V^2} \tag{4}$$

Now,

$$C_p - C_v = T\left(\frac{\partial P}{\partial T}\right)_V \left(\frac{\partial V}{\partial T}\right)_P \tag{5}$$

(By Problem 4.27)

Using (3) and (4) in (5)

$$C_p - C_v = \frac{R^2 T}{V(P - a/V^2)} = R\frac{(P + a/V^2)}{(P - a/V^2)} \approx R(1 + 2a/PV^2)$$

$$= R\left(1 + \frac{2a}{RTV}\right)$$

4.29 If $f(x, y, z) = 0$, then it can be shown that

$$\left(\frac{\partial x}{\partial y}\right)_z \left(\frac{\partial y}{\partial z}\right)_x \left(\frac{\partial z}{\partial x}\right)_y = -1 \tag{1}$$

Thus, if $f(P, V, T) = 0$

$$\left(\frac{\partial P}{\partial V}\right)_T \left(\frac{\partial V}{\partial T}\right)_P \left(\frac{\partial T}{\partial P}\right)_V = -1 \tag{1}$$

or $\left(\frac{\partial P}{\partial T}\right)_V = -\left(\frac{\partial P}{\partial V}\right)_T \left(\frac{\partial V}{\partial T}\right)_P$ (2)

and $\left(\frac{\partial V}{\partial T}\right)_P = -\left(\frac{\partial P}{\partial T}\right)_V \left(\frac{\partial V}{\partial P}\right)_T$ (3)

But

$$C_P - C_V = T\left(\frac{\partial P}{\partial T}\right)_V \left(\frac{\partial V}{\partial T}\right)_P \tag{4}$$

Use (2) and (3) in (4)

$$C_P - C_V = -T\left(\frac{\partial P}{\partial V}\right)_T\left(\frac{\partial V}{\partial T}\right)_P^2 \tag{5}$$

$$C_P - C_V = -T\left(\frac{\partial V}{\partial P}\right)_T\left(\frac{\partial P}{\partial T}\right)_V^2 \tag{6}$$

Equation (5) can be written in terms of the bulk modulus E at constant temperature and the coefficient of volume expansion α.

$$E = -\left(\frac{\partial P}{\partial V/V}\right); \quad \alpha = \frac{1}{V}\left(\frac{\partial V}{\partial T}\right) \tag{7}$$

$$C_p - C_v = TE\alpha^2 V \tag{8}$$

4.30 Taking T and V as independent variables

$$S = f(T, V)$$

$$\mathrm{d}S = \left(\frac{\partial S}{\partial T}\right)_V \mathrm{d}T + T\left(\frac{\partial S}{\partial V}\right)_T \mathrm{d}V$$

Multiplying by T,

$$T\mathrm{d}S = T\left(\frac{\partial S}{\partial T}\right)_V \mathrm{d}T + T\left(\frac{\partial S}{\partial V}\right)_T \mathrm{d}V$$

$$= C_V\mathrm{d}T + T\left(\frac{\partial S}{\partial V}\right)_T \mathrm{d}V$$

But $\left(\frac{\partial S}{\partial V}\right)_T = \left(\frac{\partial P}{\partial T}\right)_v$

$$\therefore T\mathrm{d}S = C_V\mathrm{d}T + T\left(\frac{\partial P}{\partial T}\right)_V \mathrm{d}V$$

Also,

$$\left(\frac{\partial P}{\partial T}\right)_V = -\left(\frac{\partial P}{\partial V}\right)\left(\frac{\partial V}{\partial T}\right)_P$$

$$\therefore T\mathrm{d}S = C_V\mathrm{d}T - T\left(\frac{\partial P}{\partial V}\right)\left(\frac{\partial V}{\partial T}\right)_P \mathrm{d}V$$

Introducing relations $\alpha = \frac{1}{V}(\partial V/\partial T)_P$ and $E_T = -V(\partial P/\partial V)_T$ for volume coefficient of expansion and isothermal elasticity

$$T\mathrm{d}S = C_V\mathrm{d}T + T\alpha E_T\mathrm{d}V$$

4.31 Taking T and P as independent variables

$$S = f(T, P)$$

$$dS = \left(\frac{\partial S}{\partial T}\right)_P dT + \left(\frac{\partial S}{\partial P}\right)_T dP$$

$$\text{or } TdS = T\left(\frac{\partial S}{\partial T}\right)_P dT + T\left(\frac{\partial S}{\partial P}\right)_T dP$$

$$= C_P dT + T\left(\frac{\partial S}{\partial P}\right)_T dP$$

$$\text{or } TdS = C_P dT - T\left(\frac{\partial V}{\partial T}\right)_P dP$$

$$= C_P dT - TV\alpha dP$$

4.32 Taking P and V as independent variables,

$$S = f(P, V)$$

$$dS = \left(\frac{\partial S}{\partial P}\right)_V dP + \left(\frac{\partial S}{\partial V}\right)_P dV$$

$$TdS = T\left(\frac{\partial S}{\partial P}\right)_V dP + T\left(\frac{\partial S}{\partial V}\right)_P dV$$

$$= T\left(\frac{\partial S}{\partial T}\right)_V \left(\frac{\partial T}{\partial P}\right)_V dP + T\left(\frac{\partial S}{\partial T}\right)_P \left(\frac{\partial T}{\partial V}\right)_P dV$$

$$= C_V\left(\frac{\partial T}{\partial P}\right)_V dP + C_P\left(\frac{\partial T}{\partial V}\right)_P dV$$

4.33 In the Joule–Thompson effect heat does not enter the expanding gas, that is $\Delta Q = 0$. The net work done by the external forces on a unit mass of the gas is $(P_1V_1 - P_2V_2)$, where P_1 and P_2 refer to higher and lower pressure across the plug respectively.

$$\Delta W = P_1V_1 - P_2V_2$$

If the internal energy of unit mass is U_1 and U_2 before and after the gas passes through the plug

$$\Delta U = U_1 - U_2$$

By the first law of Thermodynamics

$$\Delta Q = 0 = \Delta W + \Delta U$$

$$\text{or} \quad U_2 - U_1 = P_1V_1 - P_2V_2$$

$$\text{or} \quad \Delta(U + PV) = 0$$

$$\text{or} \quad \Delta H = 0$$

where H is the enthalpy

$$\therefore T\Delta S + V\Delta P = 0$$

But by Problem 4.31

$$T\Delta S = C_P\Delta T - T\left(\frac{\partial V}{\partial T}\right)_P \Delta P$$

$$\therefore C_P\Delta T + \left[V - T\left(\frac{\partial V}{\partial T}\right)_P\right]\Delta P = 0$$

or $$\Delta T = \frac{\left[T\left(\frac{\partial V}{\partial T}\right)_P - V\right]\Delta P}{C_P}$$

4.34 (a) For perfect gas

$$PV = RT$$

$$P\left(\frac{\partial V}{\partial T}\right)_P = R$$

$$T\left(\frac{\partial V}{\partial T}\right)_P = \frac{TR}{P} = V$$

or $$T\left(\frac{\partial V}{\partial T}\right)_P - V = 0$$

$\therefore \Delta T = 0$ by Problem 4.31

(b) For imperfect gas

$$\left(P + \frac{a}{V^2}\right)(V - b) = RT$$

or $$PV = RT - \frac{a}{V} + bP + \frac{ab}{V^2}$$

$$P\left(\frac{\partial V}{\partial T}\right)_P = R + \frac{a}{V^2}\left(\frac{\partial V}{\partial T}\right)_P - \frac{2ab}{V^3}\left(\frac{\partial V}{\partial T}\right)_P$$

Re-arranging

$$\left(\frac{\partial V}{\partial T}\right)_P = \frac{R}{P - \frac{a}{V^2} + \frac{2ab}{V^3}} = \frac{R}{\frac{RT}{V-b} - \frac{2a}{V^2}\left(1 - \frac{b}{V}\right)}$$

Multiplying both numerator and denominator of RHS by $(V - b)/R$

$$T\left(\frac{\partial V}{\partial T}\right)_P = (V - b)\left[1 - \frac{2a}{RTV^3}(V - b)^2\right]^{-1}$$

$$= (V - b)\left[1 + \frac{2a}{RTV^3}(V - b)^2\right]$$

$$= (V - b) + \frac{2a}{RTV^3}(V - b)^3$$

$$T\left(\frac{\partial V}{\partial T}\right)_P - V = \frac{2a}{RT} - b \qquad (\because b \ll V)$$

Using this in the expression for Joule–Thompson effect (Problem 4.31),

$$\Delta T = \frac{1}{C_p}\left(\frac{2a}{RT} - b\right)\Delta P$$

4.35 The equation of state for an imperfect gas is

$$\left(p + \frac{a}{V^2}\right)(V - b) = RT$$

It can be shown that

$$\Delta T = \frac{1}{C_p}\left(\frac{2a}{RT} - b\right)\Delta p$$

If $T < 2a/bR$, $\Delta T/\Delta p$ is positive and there will be cooling.

If $T > 2a/bR$, $\Delta T/\Delta p$ will be negative and the gas is heated on undergoing Joule–Kelvin expansion.

If $T = 2a/bR$, $\Delta T/\Delta p = 0$, there is neither heating nor cooling.

The temperature given by $T_i = \frac{2a}{bR}$ is called the temperature of inversion since on passing through this temperature the Joule–Kelvin effect changes its sign. Figure 4.3 shows the required curve.

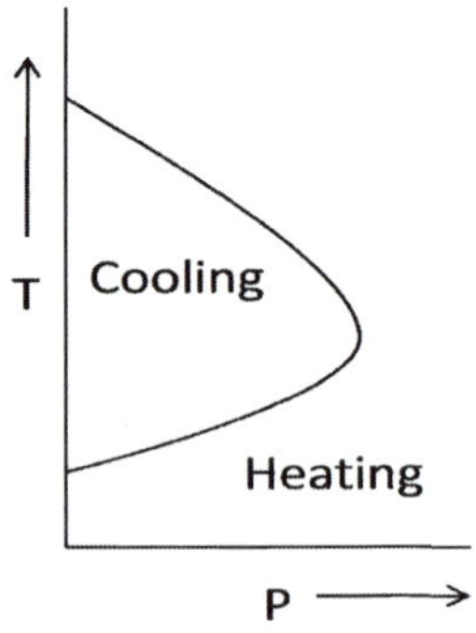

Fig. 4.3 Joule-Thompson effect

4.36 By definition

$$E_T = -V\left(\frac{\partial P}{\partial V}\right)_T ; E_S = -V\left(\frac{\partial P}{\partial V}\right)_S$$

$$\frac{E_S}{E_T} = \frac{(\partial P/\partial V)_S}{(\partial P/\partial V)_T} = \frac{(\partial P/\partial V)_S \left(\dfrac{\partial T}{\partial V}\right)_S}{(\partial P/\partial S)_T \left(\dfrac{\partial S}{\partial V}\right)_T}$$

$$= \frac{(\partial T/\partial V)_S \left(\dfrac{\partial S}{\partial P}\right)_T}{(\partial T/\partial P)_S \left(\dfrac{\partial S}{\partial V}\right)_T} = \frac{(\partial P/\partial S)_V(\partial V/\partial T)_P}{(\partial V/\partial S)_P(\partial P/\partial T)_V}$$

from the relations given in Problems 4.21 and 4.22

$$\therefore \frac{E_S}{E_T} = \frac{(\partial S/\partial T)_P}{(\partial S/\partial T)_V} = \frac{(\partial Q/\partial T)_P}{(\partial Q/\partial T)_V} = \frac{C_P}{C_V} = \gamma$$

4.37 $$\frac{(\partial V/\partial T)_S}{(\partial V/\partial T)_P} = \frac{1}{(\partial T/\partial V)_S(\partial V/\partial T)_P} = \frac{1}{-\left(\dfrac{\partial P}{\partial S}\right)_V \left(\dfrac{\partial V}{\partial T}\right)_P}$$

where we have used Eq. (23) of Problem 4.22.
Writing

$$\left(\frac{\partial P}{\partial S}\right)_V = \left(\frac{\partial P}{\partial T}\right)_V \left(\frac{\partial T}{\partial S}\right)_V = \frac{(\partial P/\partial T)_V}{(\partial S/\partial T)_V}$$

$$\frac{(\partial V/\partial T)_S}{(\partial V/\partial T)_P} = \frac{(\partial S/\partial T)_V}{(\partial P/\partial T)_V(\partial V/\partial T)_P} = \frac{(\partial S/\partial T)_V}{-(C_P - C_V)/T}$$

(by Eq. (4.1) of Problem 4.27

$$\frac{T(\partial S/\partial T)_V}{-(C_P - C_V)} = \frac{C_V}{-(C_P - C_V)} = \frac{1}{1-\gamma}$$

4.38 $$\frac{(\partial P/\partial T)_S}{(\partial P/\partial T)_V} = \frac{1}{(\partial T/\partial P)_S(\partial P/\partial T)_V} = \frac{1}{\left(\dfrac{\partial V}{\partial S}\right)_P \left(\dfrac{\partial P}{\partial T}\right)_V}$$

$$= \frac{1}{\left(\dfrac{\partial V}{\partial T}\right)_P \left(\dfrac{\partial T}{\partial S}\right)_P \left(\dfrac{\partial P}{\partial T}\right)_V} = \frac{(\partial S/\partial T)_P}{\left(\dfrac{\partial V}{\partial T}\right)_P (\partial P/\partial T)_V}$$

$$= \frac{T(\partial S/\partial T)_P}{(C_P - C_V)} = \frac{C_P}{(C_P - C_V)} = \frac{\gamma}{\gamma - 1}$$

where we have used Eq. (4.24) of Problem 4.22 and the relation

$$C_P = T\left(\frac{\partial S}{\partial T}\right)_P$$

4.39 By Maxwell's first equation

$$\left(\frac{\partial S}{\partial V}\right)_T = \left(\frac{\partial P}{\partial T}\right)_V \tag{1}$$

$$\mathrm{d}S = \frac{\mathrm{d}U + P\mathrm{d}V}{T} \tag{2}$$

using (2) in (1)

$$\left(\frac{\partial U}{\partial V}\right)_T = T\left(\frac{\partial P}{\partial T}\right) - P$$

For perfect gases,

$$P = \frac{RT}{V}$$

$$\left(\frac{\partial U}{\partial V}\right)_T = \frac{RT}{V} - P = 0$$

Thus, temperature remaining constant, the internal energy of an ideal gas is independent of the volume.

4.40 $$\frac{\mathrm{d}P}{\mathrm{d}T} = \frac{L}{T(v_2 - v_1)}$$

$$\mathrm{d}T = \frac{T}{L}(v_2 - v_1)\mathrm{d}P$$

$$= \frac{373(1677 - 1)(2 \times 10^6)}{546 \times 4.2 \times 10^7} = 55.1^\circ\mathrm{C}$$

4.41 $$v_1 = 1\ \mathrm{cm}^3;\ v_2 = \frac{1}{0.091} = 10.981\ \mathrm{cm}^3$$

$$\mathrm{d}P = \frac{L\mathrm{d}T}{T(v_2 - v_1)} = \frac{80 \times 4.2 \times 10^7 \times 1}{(-1 + 273)(10.981 - 1.0)}$$

$$= 1.238 \times \frac{10^6\mathrm{dynes}}{\mathrm{cm}^2} = 1.24\ \mathrm{atm}$$

$$P_2 = P_1 + \mathrm{d}P = 1.0 + 1.24 = 2.24\ \mathrm{atm}$$

4.42 $$v_1 = \frac{1}{\rho_1} = \frac{1}{1.145} = 0.873\ \mathrm{cm}^3/\mathrm{g}$$

$$v_2 = \frac{1}{\rho_2} = \frac{1}{0.981} = 1.019\ \mathrm{cm}^3/\mathrm{g}$$

$$\mathrm{d}T = \frac{T(v_2 - v_1)\mathrm{d}P}{L}$$

$$= \frac{(80 + 273)(1.019 - 0.873)(1.0 \times 10^6)}{35.5 \times 4.2 \times 10^7}$$

$$= 0.0346^\circ\mathrm{C}$$

4.43 (a) Use the relation

$$dU = T\,ds - P\,dV \tag{1}$$

Here,

$dV = 0(\because V = \text{constant})$ and

$$U = aVT^4 \tag{2}$$

$$dU = 4aVT^3 dT = T\,ds$$

$$\left(\frac{ds}{dT}\right)_V = 4aVT^2$$

Integrating $S = \frac{4}{3}aT^3V$

(b) $F = U - TS = aVT^4 - \dfrac{4}{3}aT^4V = -\dfrac{1}{3}aVT^4$

$$p = -\left(\frac{\partial F}{\partial V}\right)_T = \frac{1}{3}aT^4 = \frac{1}{3}u$$

4.44 According to Dulong-Petit's law the molar specific heats of all substances, with a few exceptions like carbon, have values close to 6 cal/mol°C^{-1}. The specific heat of Cu is $\frac{387}{\text{kgK}^{-1}} = \frac{0.387\text{J}}{\text{gK}^{-1}} = 0.0926\text{cal/gK}^{-1}$. Therefore, the atomic mass of Cu $= \frac{6}{0.0926} = 64.79$ amu.

4.3.3 Statistical Distributions

4.45 Probability for the rotational state to be found with quantum number J is given by the Boltzmann's law.

$P(E) \propto (2J+1)\exp[-J(J+1)\hbar^2/2I_0kT$

where I_0 is the moment of inertia of the molecule, k is Boltzmann's constant, and T the Kelvin temperature. The two lowest states have $J = 0$ and $J = 1$

$I_0 = M(r/2)^2 + M(r/2)^2 = \dfrac{1}{2}Mr^2$, where $M = 938$ MeV/c^2

$2I_0 = Mr^2 = 938 \times (1.05 \times 10^{-10})^2/c^2$

$\hbar c = 197.3\,\text{MeV} - 10^{-15}\text{m}$

$$kT = 1.38 \times 10^{-23} \times \frac{50}{1.6 \times 10^{-13}} = 43.125 \times 10^{-10}$$

$$\frac{\hbar^2}{2I_0kT} = \frac{\hbar^2c^2}{Mc^2r^2kT} = \frac{(197.3)^2 \times 10^{-30}}{938 \times (1.05 \times 10^{-10})^2 \times 43.125 \times 10^{-10}} = 0.8728$$

For $J = 1$, $\dfrac{J(J+1)\hbar^2}{2kT} = 1 \times (1+1) \times 0.8728 = 1.7457$

For $J = 0$, $P(E_0) \propto 1.0$

For $J = 1$, $P(E_1) \propto (2 \times 1 + 1)\exp(-1.7457) = 0.52$

$\therefore P(E_0) : P(E_1) :: 1 : 0.52$

4.46 For stationary waves, in the x-direction

$k_x a = n_x \pi$

or $n_x = k_x a/\pi$

$dn_x = (a/\pi)dk_x$

Similar expressions are obtained for y and z directions.

$$dn = dn_x dn_y dn_z$$
$$= (a/\pi)^3 d^3k$$

However only the first octant of number space is physically meaningful. Therefore

$$dn = (1/8)(a/\pi)^3 d^3k$$

Taking into account the two possible polarizations

$$dn = \frac{2V}{(2\pi)^3} d^3k = \frac{2V}{8\pi^3} .4\pi k^2 dk$$

But $k = \dfrac{\omega}{c}$; $dk = d\omega/c$

$$\therefore dn = \frac{V\omega^2 d\omega}{\pi^2 c^3}$$

4.47 $n! = n(n-1)(n-2)\ldots(4)(3)(2)$

Take the natural logarithm of n!

$$\ln n! = \ln 2 + \ln 3 + \ln 4 + \cdots + \ln(n-2) + \ln(n-1) + \ln n$$
$$= \Sigma_{n=1}^{n} \ln n$$
$$= \int_1^n \ln n \, dn$$
$$= n \ln n - n + 1$$
$$\approx n \ln n - n$$

where we have neglected 1 for $n \gg 1$

4.48 $p(E) = (2J+1)e^{-J(J+1)\hbar^2/2ikT}$

The maximum value of $p(E)$ is found by setting $dp(E)/dJ = 0$

$$\left[2 - \frac{(2J+1)^2\hbar^2}{2I_0kT}\right] e^{-J(J+1)\hbar^2/2I_0kT} = 0$$

Since the exponential factor will be zero only for $J = \infty$,

$$\left[2 - \frac{(2J+1)^2\hbar^2}{2I_0kT}\right] = 0$$

Solving for J, we get

$$J_{\max} = \frac{\sqrt{I_0kT}}{\hbar} - \frac{1}{2}$$

4.49 $p(E_J) = (2j+1)e^{\frac{-J(J+1)\hbar^2}{2I_0kT}}$

The factor $\dfrac{\hbar^2}{2I_0k} = \dfrac{(1.055\times10^{-34})^2}{2\times4.64\times10^{-48}\times1.38\times10^{-23}\ \text{J}} = 86.9$

$p(E_0) = 1$

$p(E_1) = 3e^{-2\times86.9/400} = 1.942$

$p(E_2) = 5e^{-6\times86.9/400} = 1.358$

$p(E_3) = 7e^{-12\times86.9/400} = 0.516$

4.50 $p(E_2) = 5e^{-6\times86.9/T} = 5e^{-521.4/T}$ (1)

$p(E_3) = 7e^{-12\times86.9/T} = 7e^{-1042.8/T}$

Equating $p(E_2)$ and $p(E_3)$ and solving for T, we find $T = 1{,}549$ K

4.51 For Boltzmann statistics $p(E)\propto e^{-E/kT}$ Therefore,

$$\frac{p(E_n)}{p(E_1)} = e^{-(E_n-E_1)/kT}$$

In hydrogen atom, if the ground state energy $E_1 = 0$, then $E_2 = 10.2$, $E_3 = 12.09$ and $E_4 = 12.75$ eV

The factor $kT = 8.625\times10^{-5}\times6{,}000 = 0.5175$

$P(E_2)/P(E_1) = e^{-10.2/0.5175} = 2.75\times10^{-9}$

$P(E_3)/P(E_1) = e^{-12.09/0.5175} = 1.4\times10^{-10}$

$P(E_4)/P(E_1) = e^{-12.75/0.5175} = 1.99\times10^{-11}$

Thus $P(E_1) : P(E_2) : P(E_3) :: 1 : 2.8\times10^{-9} : 1.4\times10^{-10} : 2.0\times10^{-11}$

This then means that the hydrogen atoms in the chromospheres are predominantly in the ground state.

4.52 $p(E) = \dfrac{1}{e^{(E-E_F)/kT}+1}$

For $E - E_F = kT$, $p(E) = \dfrac{1}{e+1} = 0.269$

For $E - E_F = 5kT$, $p(E) = \dfrac{1}{e^5+1} = 6.69\times10^{-3}$

For $E - E_F = 10kT$, $p(E) = \dfrac{1}{e^{10}+1} = 4.54\times10^{-5}$

4.53 For the conduction electrons, the number of states per unit volume with energy in the range E and $E+\mathrm{d}E$, can be written as $n(E)\mathrm{d}E$ where $n(E)$ is the density of states. Now, for a free electron gas

$$n(E) = \frac{8\sqrt{2}\pi m^{3/2}}{h^3}E^{1/2}$$

Let $P(E)$ be the probability function which gives the probability of the state at the energy E to be occupied. At $T = 0$ all states below a certain energy are filled ($P = 1$) and all states above that energy are vacant ($P = 0$). The highest occupied state under the given conditions is called the Fermi energy.
The product of the density $n(E)$ of available states and the probability $P(E)$ that those states are occupied, gives the density of occupied states $n_0(E)$; that is

$$n_0(E) = n(E)P(E)$$

The total number of occupied states per unit volume is given by

$$n = \int_0^{E_F} n_0(E)\mathrm{d}E$$

$$= \frac{8\sqrt{2}\pi m^{3/2}}{h^3}\int_0^{E_F} E^{1/2}\mathrm{d}(E)$$

$$= \frac{8\sqrt{2}\pi m^{3/2}}{h^3}\cdot\frac{2}{3}E_F^{3/2}$$

$$\text{or}\quad E_F = \frac{h^2}{8m}\left(\frac{3n}{\pi}\right)^{2/3}$$

4.54 $P_+ = \dfrac{1}{e^{(E-E_F)/kT}+1} = \dfrac{1}{e^{\Delta/kT}+1} \approx \dfrac{1}{2+\Delta/kT} = \dfrac{1}{2}(1-\Delta/2kT)$

$$P- = \frac{1}{2}(1+\Delta/2kT)$$

$$\therefore \frac{P_+ + P_-}{2} = 1/2 = P_F$$

4.55 (a) For n states, the number of ways is $N = n^2$. Therefore, for $n = 6$ states $N = 36$
(b) For n states the number of ways is $N = n^2-(n-1) or n^2-n+1$. Therefore, for $n = 6$, $N = 31$
(c) For n states, $N = n^2 - n + 1 - n$ *or* $n^2 - 2n + 1$. Therefore for $n = 6$, $N = 25$

4.56 If the gas is in equilibrium, the number of particles in a vibrational state is

$$N_\nu = N_0 \exp\left(-\frac{h\nu}{kT}\right) = N_0 \exp\left(-\frac{\theta}{T}\right).$$

The ratios, $N_0/N_1 = 4.7619$, $N_1/N_2 = 4.8837$, $N_2/N_3 = 4.7778$, are seen to be constant at 4.8078. Thus the ratio $N_\nu/N_{\nu+1}$ is constant equal to 4.81, showing the gas to be in equilibrium at a temperature
$T = 3,350/(\ln 4.81) \approx 2,130$ K

4.57 $\Delta S = k\ln(\Delta W)$
But $\Delta S = \Delta Q/T$

or $\Delta Q = T\Delta S = kT \ln \Delta W$
$= (1.38 \times 10^{-23})(300) \ln 10^8$
$= 7.626 \times 10^{-20}\text{J} = 0.477\text{ eV}$

4.58 The Gaussian (normal) distribution is

$$f(x) = \frac{1}{\sigma\sqrt{2\pi}} e^{-(x-\mu)^2/2\sigma^2}$$

where μ is the mean and σ is the standard deviation. The probability is found from

(a) $P(\mu - \sigma < x < \mu + \sigma) = \int_{\mu-\sigma}^{\mu+\sigma} f(x)\mathrm{d}x$

Letting $z = \frac{x-\mu}{\sigma}$

$$P(-1 < z < 1) = \int_{-1}^{1} \phi(z)\mathrm{d}z$$

$= 2\int_0^1 \phi(z)\mathrm{d}z$ (from symmetry)
$= 2 \times 0.3413 = 0.6826$ (from tables)
or 68.26%(shown shaded under the curve, Fig 4.4)

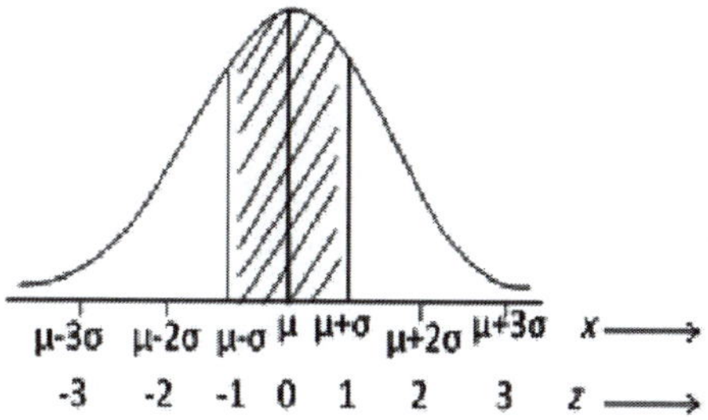

Fig. 4.4

(b) Similarly

$P(\mu - 2\sigma) < x < \mu + 2\sigma) = 0.9544$ or 95.44%

(c) $P(\mu - 3\sigma) < x < (\mu + 3\sigma) = 0.9973$ or 99.73%

4.59 $P(n, T) = \dfrac{e^{-\frac{\left(n+\frac{1}{2}\right)\hbar\omega}{kT}}}{\Sigma_{n=0}^{\infty} e^{-\frac{\left(n+\frac{1}{2}\right)\hbar\omega}{kT}}}$

$$= \frac{e^{-(n+\frac{1}{2})\hbar\omega/kT}}{e^{-\frac{1}{2}\hbar\omega/kT}\,\Sigma_{n=1}^{\infty} e^{n\hbar\omega/kT}}$$

$$= \frac{e^{-n\hbar\omega/kT}}{\dfrac{e^{-\hbar\omega/kT}}{1 - e^{-\hbar\omega/kT}}} = \frac{e^{-n\hbar\omega/kT}}{\frac{1}{e^{\hbar\omega/kT} - 1}}$$

$$= e^{-n\hbar\omega/kT}\left(e^{\hbar\omega/kT} - 1\right)$$

Substitute $n = 10$, $\dfrac{\hbar\omega}{k} = \dfrac{8.625 \times 10^{-5}}{(1.38 \times 10^{-23}/1.6 \times 10^{-19})} = 1.0$

$P(10, 300) = 3.2 \times 10^{-3}$

In the limit $T \to 0$, the state $n = 0$ alone is populated so that $n = 10$ state is unpopulated.

In the limit $T \to \infty$, probability for $n = 10$ again goes to zero, as higher states which are numerous, are likely to be populated.

4.60 Consider a collection of N molecules of a large number of energy states, E_1, E_2, E_3 etc such that there are N_1 molecules in state E_1, N_2 in E_2 and so on. The nature of energy is immaterial. The number of ways in which N molecules can be accommodated in various states is given by

$$W = \frac{N!}{N_1!N_2!\ldots} \tag{1}$$

The underlying idea is that the state of the system would be state if W is a maximum.

Taking logs on both sides and applying Stirling's approximation $\ln W = N \ln N - N - \Sigma N_i \ln N_i + \Sigma N_i$

$$= N \ln N - \Sigma N_i \ln N_i \tag{2}$$

$$\text{because } \Sigma N_i = N \tag{3}$$

$$\Sigma N_i E_i = E \tag{4}$$

If the system is in a state of maximum thermodynamic probability, the variation of W with respect to change in N_i is zero, that is

$$\Sigma \delta N_i = 0 \tag{5}$$

$$\Sigma E_i \delta N_i = 0 \tag{6}$$

$$\Sigma(1 + \ln N_i)\delta N_i = 0 \tag{7}$$

We now use the Lagrange method of undetermined multipliers. Multiplying (5) by α and (6) by β and adding to (7), we get

$$\Sigma\{(1 + \ln N_i) + \alpha + \beta E_i\}\delta N_i = 0 \tag{8}$$

Therefore

$$\ln N_i + 1 + \alpha + \beta E_i = 0 \tag{9}$$

$$\text{or } N_i = Ce^{-\beta E_i} \tag{10}$$

Where C = constant which can be determined as follows.

$$\Sigma N_i = N = C\Sigma e^{-\beta E_i} \tag{11}$$

$$\text{or } C = \frac{N}{\Sigma e^{-\beta E_i}} \tag{12}$$

Equation (10) then becomes

$$N_i = \frac{Ne^{-\beta E_i}}{\Sigma e^{-\beta E_i}} \tag{13}$$

The denominator in (13)

$$Z = \Sigma e^{-\beta E_i} \tag{14}$$

Is known as the partition function. It can be shown that the quantity

$$\beta = \frac{1}{kT} \tag{15}$$

where k is the Boltzmann constant and T is the absolute temperature.

$$\alpha = \frac{N}{Z} \tag{16}$$

4.3.4 Blackbody Radiation

4.61 Electric power = power radiated

$$W = \sigma T^4 A$$

$$A = 2\pi rl = 2\pi \times 10^{-3} \times 1.0 = 6.283 \times 10^{-3}\text{m}^2$$

$$T = \left[\frac{W}{\sigma A}\right]^{1/4} = \left[\frac{1{,}000}{5.67 \times 10^{-8} \times 6.283 \times 10^{-3}}\right]^{1/4} = 1{,}294\ K$$

4.62 The Solar constant S is the heat energy received by 1 m^2 of earth's surface per second. If R is the radius of the sun and r the earth-sun distance, then the total intensity of radiation emitted from the sun will be σT^4 W m^{-2} and from the sun's surface $\sigma T^4.4\pi R^2$. The radiation received per second per m^2 of earth's surface will be

$$S = \sigma T^4.\frac{4\pi R^2}{4\pi r^2}$$

Solving,

$$\sigma T^4 = S.\frac{r^2}{R^2} = 1{,}400\left(\frac{1.5 \times 10^8}{7 \times 10^5}\right)^2 = 6.43 \times 10^7\ \text{W m}^{-2}$$

$$T = \left(\frac{6.43 \times 10^7}{\sigma}\right)^{1/4} = \left(\frac{6.43 \times 10^7}{5.67 \times 10^{-8}}\right)^{1/4} = 5{,}800\ \text{K}$$

4.63 Using the analogy between radiation (photon gas) and gas molecules, the photons move in a cavity at random in all directions, rebounding elastically from the walls of the cavity. The pressure exerted by an ideal photon gas is

$$p = \frac{1}{3}\rho < v^2 >$$

where ρ is the mass density. In the case of photon gas, the speed of all photons is identical being equal to c. Furthermore, from Einstein's relation

$$u = \rho c^2$$

where u is the energy density. Replacing $< v^2 >$ by c^2

$$p_{\text{rad}} = \frac{1}{3}\rho c^2 = \frac{u}{3}$$

4.64 Let T and T_0 be the Kelvin temperatures of the body and the surroundings. Then, by Stefan–Boltzmann law, the rate of loss of heat per unit area of the body is

$$\frac{\mathrm{d}Q}{\mathrm{d}t} = \sigma(T^4 - T_0^4)$$

$$= \sigma(T - T_0)(T + T_0)(T^2 + T_0^2)$$

If $(T - T_0)$ be small, $(T \approx T_0)$, and

$$\frac{\mathrm{d}Q}{\mathrm{d}t} = \sigma(T - T_0) \times 4T_0^3$$

Since T_0 is constant,

$$\frac{\mathrm{d}T}{\mathrm{d}t} \propto (T - T_0); \qquad \text{(Newton's law of cooling).}$$

4.65 The energy density u and pressure p of radiation are related by

$$p = \frac{u}{3}$$

Furthermore, $u = 4\sigma T^4/c$
Eliminating u,

$$T = \left(\frac{3cp}{4\sigma}\right)^{1/4} = \left(\frac{3 \times 3 \times 10^8 \times 4 \times 10^8 \times 1.013 \times 10^5}{4 \times 5.67 \times 10^{-8}}\right)^{1/4} = 2 \times 10^7\ \text{K}$$

4.66 (a) Power, $P = \sigma A T^4 = 4\pi R^2 \sigma T^4$

$$= 4\pi(7 \times 10^8)^2(5.67 \times 10^{-8})(5{,}700)^4$$

$$= 3.68 \times 10^{26}\ \text{W}$$

Mass lost per second, $m = P/c^2 = \dfrac{3.68 \times 10^{26}}{(3 \times 10^8)^2} = 4.1 \times 10^9\,\text{kg/s}$

(b) Time taken for the mass of sun (M) to decrease by 1% is

$$t = \frac{M}{100} \times \frac{1}{m} = \frac{2 \times 10^{30}}{100} \times \frac{1}{4.1 \times 10^9} = 4.88 \times 10^{18}\ \text{s}$$

$$= \frac{4.88 \times 10^{18}}{3.15 \times 10^7} = 1.55 \times 10^{11}\ \text{years}$$

4.67 Power radiated, $P = \sigma A T^4 = 4\pi R^2 \sigma T^4$

$$\frac{P_2}{P_1} = \frac{R_2^2}{R_1^2}.\frac{T_2^4}{T_1^4} = \frac{(4R_1)^2}{R_1^2}.\frac{(2T_1)^4}{T_1^4} = 256$$

Furthermore, $\dfrac{P_2}{P_1} = \dfrac{\mathrm{d}Q_2/\mathrm{d}t}{\mathrm{d}Q_1/\mathrm{d}t} = \dfrac{m_2 s(\mathrm{d}T/\mathrm{d}t)_2}{m_1 s(\mathrm{d}T/\mathrm{d}t)_1}$

where s is the specific heat
But $m_2 \propto R_2^3$ and $m_1 \propto R_1^3$

$$\therefore \frac{(\mathrm{d}T/\mathrm{d}t)_2}{(\mathrm{d}T/\mathrm{d}t)_1} = \frac{P_2}{P_1}.\frac{R_1^3}{R_2^3} = \frac{256}{4^3} = 4$$

4.68 (a) $\lambda_m.T = b$

$$T = \frac{b}{\lambda_m} = \frac{2.897 \times 10^{-3}}{1 \times 10^{-6}} = 2{,}897\,\mathrm{K}$$

$$\frac{P_2}{P_1} = \frac{T_2^4}{T_1^4} = 2$$

New temperature, $T_2 = T_1 \times 2^{1/4} = 2{,}897 \times 1.189 = 3{,}445\mathrm{K}$

(b) The wavelength at which the radiation has maximum intensity

$$\lambda_m = \frac{2.897 \times 10^{-3}}{3445} = 0.84 \times 10^{-6} m = 0.84\,\mu\mathrm{m}$$

4.69 The mean value $\overline{\in}$ is determined from;

$$\overline{\in} = \frac{\Sigma_{n=0}^{\infty} n \in e^{-\beta n \in}}{\Sigma_{n=0}^{\infty} e^{-\beta n \in}} = -\frac{\mathrm{d}}{\mathrm{d}\beta} \ln \sum_{n=0}^{\infty} e^{-\beta n \in}$$

$$= -\frac{\mathrm{d}}{\mathrm{d}\beta} \ln \left(1 + e^{-\beta \in} + e^{-2\beta \in} + \cdots\right)$$

$$= -\frac{\mathrm{d}}{\mathrm{d}\beta} \ln \frac{1}{1 - e^{-\beta \in}}$$

where we have used the formula for the sum of terms of an infinite geometric series.

$$\overline{\in} = \frac{\in e^{-\beta \in}}{1 - e^{-\beta \in}} = \frac{\in}{e^{\beta \in} - 1} \quad (\beta = 1/kT)$$

4.70 (a) $$u_\lambda \mathrm{d}\lambda = \frac{8\pi hc}{\lambda^5}.\frac{1}{e^{hc/\lambda kT} - 1}\mathrm{d}\lambda \quad \text{(Planck's formula)} \qquad (1)$$

For long wavelengths (low frequencies) and high temperatures the ratio $\frac{hc}{\lambda kT} \ll 1$ so that we can expand the exponential in (1) and retain only the first two terms

$$u_\lambda \mathrm{d}\lambda = \frac{8\pi hc}{\lambda^5[(1 + hc/\lambda kT + \ldots) - 1]} = \frac{8\pi kT}{\lambda^4}\mathrm{d}\lambda$$

writing $\lambda = \frac{c}{\upsilon}$; $\mathrm{d}\lambda = -\frac{c}{\upsilon^2}\mathrm{d}\upsilon$

$$u_\nu = \frac{8\pi\nu^2}{c^3}kT \qquad \text{(Rayleigh-Jeans law)}$$

(b) If $h\nu/kT \gg 1$ i.e $hc/\lambda kT \gg 1$ then we can ignore 1 in the denominator in comparison with the exponential term in Planck's formula
$u_\lambda d\lambda = c_1 e^{-c_2/\lambda kT} d\lambda$ (Wien's distribution law)
where the constants, $c_1 = 8\pi hc$ and $c_2 = hc$

4.71 $$u_\lambda d\lambda = \frac{8\pi hc}{\lambda^5}.\frac{1}{e^{hc/\lambda kT}-1}d\lambda \qquad \text{(Planck's formula)}$$

The wavelength λ_m corresponding to the maximum of the distribution curve is obtained from the condition

$$\left(\frac{du_\lambda}{d\lambda}\right)_{\lambda=\lambda_m} = 0$$

Differentiating and writing $hc/kT\lambda_m = \beta$, gives

$$e^{-\beta} + \frac{\beta}{5} - 1 = 0$$

This is a transcental equation and has the solution

$\beta = 4.9651$, so that

$$\lambda_m T = \frac{hc}{4.9651k} = b = \text{constant.}$$

Thus, the constant

$$b = \frac{6.626068 \times 10^{-34} \times 2.99792 \times 10^8}{4.9651 \times 1.38065 \times 10^{-23}} = 2.8978 \times 10^{-3}\ \text{m-K}$$

a value which is in excellent agreement with the experiment.

4.72 By definition

$$u = \int u_\nu d\nu = aT^4 \tag{1}$$

Inserting Planck's formula in (1)

$$u = aT^4 = \frac{8\pi h}{c^3}\int_0^\infty \frac{\nu^3 d\nu}{e^{h\nu/kT}-1} = \frac{8\pi k^4 T^4}{h^3c^3}\int_0^\infty \frac{x^3 dx}{e^x - 1}$$

where $x = h\nu/kT$

$$a = \frac{8\pi k^4}{h^3c^3}\int_0^\infty x^3(e^{-x} + e^{-2x} + \ldots e^{-rx} + \ldots)$$

Now, $\int_0^\infty x^3 e^{-rx} dx = \frac{6}{r^4}$, and $\Sigma_{r=1}^{\infty}\frac{1}{r^4} = \frac{\pi^2}{90}$

$$a = \frac{48\pi k^4}{h^3c^3}.\frac{\pi^4}{90} = \frac{8}{15}\frac{\pi^5 k^4}{h^3c^3}$$

$$\therefore \sigma = \frac{ac}{4} = \frac{2}{15}\frac{\pi^5 k^4}{h^3c^2}$$

$$= \frac{2}{15}\frac{(3.14159)^5(1.38065)^4 \times 10^{-92}}{(6.626068 \times 10^{-34})^3(2.99792 \times 10^8)^2}$$

$$= 5.67 \times 10^{-8}\text{W-m}^{-2}\text{-K}^{-4}$$

a value which is in excellent agreement with the experiment.

4.73 Number of modes per m^3 in the frequency interval $d\nu$ is

$$N = \frac{8\pi\nu^2 d\nu}{c^3}$$

But,

$$\nu = \frac{c}{\lambda}; d\nu = -\frac{d\lambda}{\lambda^2}; \lambda = \frac{4{,}990 + 5{,}010}{2} = 5{,}000\ A^0$$

$$d\lambda = 5{,}010 - 4{,}990 = 20\ A^0$$

$$\therefore N = \frac{8\pi d\lambda}{\lambda^4} = \frac{8\pi \times 20 \times 10^{-10}}{(5 \times 10^{-7})^4} = 8.038 \times 10^{17}/\text{m}^3$$

4.74 (a) $P = AE_\lambda d\lambda = \dfrac{8\pi hcAd\lambda}{\lambda^5(e^{hc/\lambda kT} - 1)}$ (2)

Mean wavelength $\lambda = 0.55\ \mu\text{m} = 5.5 \times 10^{-7}\text{m}$. (1)

$$d\lambda = (0.7 - 0.4)\ \mu\text{m} = 3 \times 10^{-7}\text{m}$$

$$A = \pi r^2 = \pi(2.5 \times 10^{-3})^2 = 1.96 \times 10^{-5}\text{m}^2$$

$$\frac{hc}{\lambda kT} = \frac{(6.63 \times 10^{-34})(3 \times 10^8)}{(5.5 \times 10^{-7})(1.38 \times 10^{-23})(4{,}000)} = 6.55$$

Using the above values in (2) we find

$$P = AE_\lambda d\lambda = 0.84 \times 10^{-6}\text{W} = 0.84\ \mu\text{W}.$$

(b) $h\bar{\nu} = \dfrac{hc}{\bar{\lambda}} = \dfrac{6.63 \times 10^{-34} \times 3 \times 10^8}{5.5 \times 10^{-7}} = 3.616 \times 10^{-19}$

Number of photons emitted per second

$$n = \frac{P}{h\bar{\nu}} = 0.84 \times 10^{-6}/3.616 \times 10^{-19} = 2.32 \times 10^{12}/\text{s}$$

4.75 $u_\lambda d\lambda = \dfrac{8\pi hc}{\lambda^5}\dfrac{1}{e^{hc/\lambda kT} - 1}d\lambda$ (1)

Put $\lambda = c/\nu$ (2)

and $d\lambda = -\left(\dfrac{c}{\nu^2}\right)d\nu$ (3)

in the RHS of (1) and simplify

$$u_\nu d\nu = \frac{8\pi h\nu^3}{c^3(e^{h\nu/kT} - 1)}d\nu \quad (4)$$

The negative sign in (3) is omitted because as λ increases ν decreases.

4.76 Power radiated from the sun $= \sigma \times$ (surface area) $\times T_s^4$

$P_s = \sigma 4\pi R_s^2 T_s^4$
Power received by the earth,

$$P_E = \frac{\pi R_e^2}{4\pi r^2} . P_s$$

The factor πR_e^2 represents the effective (projected) area of the earth on which the sun's radiation is incident at a distance r from the sun. The factor $4\pi r^2$ is the surface area of a sphere scooped with the centre on the sun. Thus $\pi R_e^2 / 4\pi r^2$ is the fraction of the radiation intercepted by the earth's surface area.

Now power radiated by earth,

$P_E = \sigma 4\pi R_E^2 T_E^4$

For radiation equilibrium, power radiated by the earth=power received by the earth.

$$\sigma 4\pi R_E^2 T_E^4 = \sigma 4\pi R_s^2 T_s^4 . \frac{\pi R_E^2}{4\pi r^2}$$

$$\text{or } T_E = T_s \left(\frac{R_s}{2r} \right)^{1/2} = 5{,}800 \left[\frac{7 \times 10^8}{2 \times 1.5 \times 10^{11}} \right]^{1/2}$$

$= 280\ \text{K} = 7^\circ\text{C}$

Note that the calculations are approximate in that the earth and sun are not black bodies and that the contribution of heat from the interior of the earth has not been taken into account.

4.77 Power radiated by the sun, $P_s = \sigma 4\pi R_s^2 T_s^4$
Power received by 1 m^2 of earth's surface,

$$S = \frac{\sigma 4\pi R_s^2 T_s^4}{4\pi r^2}$$
$$= \frac{(5.7 \times 10^{-8})(7 \times 10^8)^2 (5{,}800)^4}{(1.5 \times 10^{11})^2}$$
$$= 1{,}400\ \text{W/m}^2$$

4.78 $P = 4\pi r^2 \sigma T^4$
$= 4\pi (0.3)^2 (5.67 \times 10^{-8})(10^7)^4$
$= 6.4 \times 10^{20}$ W

Chapter 5
Solid State Physics

5.1 Basic Concepts and Formulae

Crystal Structure

There are seven crystal systems Cubic, Tetragonal, Orhtorhombic, monoclinic, triclinic, Rhombohedral, Hexagonal. They are distinguished by the axial lengths and axial angles. The lengths are taken as *a, b* and *c*. In the cubic system $a = b = c$ and the angle between any two axes is a right angle.

Bragg's equation

$$2d \sin\theta = n\lambda \tag{5.1}$$

where d is the distance between parallel diffraction planes, θ is the angle between the incident beam and the diffraction plane and n is the order of diffraction. The distance d can be related to the lattice parameters of the crystal cell. In the simple cubic cell the distance between (100) planes is "a", the lattice parameter. The distance between parallel (110) planes passing through lattice points is $a/\sqrt{2}$; for (111), it is $a/\sqrt{3}$. In general distance between parallel planes of indices (hkl) in terms of the parameter "a" for the cubic system is

$$d^2 = \frac{a^2}{h^2 + k^2 + l^2} \tag{5.2}$$

Electrical properties of crystals

The principal attractive force between ions of opposite sign is an electrostatic force. The repulsive force arises from the interaction of the electron clouds surrounding an atom. This force arises because of the exclusion principle and is not electrostatic in nature. Empirically this is represented by b/r^n, where b and n are constants, and r is the anion-cation distance.

The total energy of the lattice is the sum of the attractive and repulsive energies

$$U = -NA\frac{e^2}{r} + \frac{b}{r^n} \tag{5.3}$$

where N is the number of molecules and A is known as the Madelung constant.

For equilibrium

$$\frac{\mathrm{d}U}{\mathrm{d}r} = 0 \tag{5.4}$$

The force is

$$-\frac{\mathrm{d}U}{\mathrm{d}r} = 0 \tag{5.5}$$

The current density

$$J = i/A \tag{5.6}$$

where A is the cross-section of the conductor.

The drift speed

$$v_d = j/ne \tag{5.7}$$

where n is the number of conduction electrons per unit volume. The resistivity is given by

$$R = \rho L/A \tag{5.8}$$

The conductivity is given by

$$\sigma = 1/\rho \tag{5.9}$$

$$v_d = eE\tau/m \tag{5.10}$$

where E is the electric field and τ is the mean time between collisions.

$$\rho = m_\mathrm{e}/ne^2\tau \tag{5.11}$$

$$\tau = m_\mathrm{e}\sigma/ne^2 \tag{5.12}$$

The mean free path

$$\lambda = \tau <v> \tag{5.13}$$

Hall effect

If a thin strip of material carrying a constant current is placed in a magnetic field B perpendicular to the strip a potential difference appears across the strip. This is known as Hall effect.

$$E = jB/qn \tag{5.14}$$

Metals, insulators and semiconductors

Materials are distinguished by the extent to which the valence and conduction bands are filled by electrons. The bands in solids may be filled, partially or empty. A good conductor has a conduction band that is approximately half filled or the conduction band overlaps the next higher band. In this case it is very easy for the valence electron to be raised to a higher energy level under the application of electric field and provide electrical conduction.

In an insulator the valency band is completely filled and the energy gap (E_g) with the conduction gap is large ($\sim 5\,\text{eV}$).

In the case of semiconductors the valence band is completely filled, like an insulator. However, the conduction band is empty, so that at room temperature some of the electrons acquire sufficient energy to be found in the conduction band. Furthermore, the electrons leave behind unfilled "holes" into which other electrons in the valence band can move in the electrical conduction regime. The excitation of electron into these holes has the net effect of positive charge carriers aiding the electrical conduction. Such semiconductors are known as intrinsic semiconductors. However, with the introduction of certain impurities into a material in a controlled way, a procedure known as doping conduction is dramatically increased. Such doped semiconductors are known as extrinsic semiconductors, on which are based numerous semiconductor devices. If the majority charge carriers are electrons, the material is called an n-type semiconductor and if the holes are the majority charge carrier the material is called a p-type semiconductor.

The Fermi energy E_F lies in the middle of the energy gap.

The mobility of charge carriers is defined as

$$\mu = v_d/E \tag{5.15}$$

The conductivity σ has two contributions, one from the electrons and the other from the holes.

$$\sigma = n_n e\mu_n + n_p e\mu_p \tag{5.16}$$

$$n = \sigma/e\mu \tag{5.17}$$

$$\tau = \mu m/e \tag{5.18}$$

Superconductivity

Some materials when cooled below a certain temperature, called critical temperature (T_c), have zero resistance. The material is said to be a superconductor. T_c varies from one superconductor to another.

When a superconductor is placed in a magnetic field, T_c decreases with the increasing B. When B is increased beyond a critical magnetic field B_c, the superconductivity will not take place no matter how low the temperature.

$$T_c(B) = T_{c0}(1 - B/B_c)^{1/2} \tag{5.19}$$

where T_{c0} is the critical temperature with zero magnetic field, and B is the applied field.

According to the BCS theory superconductivity is due to a weak binding of two electrons of equal and opposite momenta and spin to form the so-called a Cooper pair which behaves as a single particle, a Boson.

An energy E_g, called the superconducting energy gap, is required to break the Cooper pair. At $T = 0$,

$$E_g = 3.53\,kT_c \tag{5.20}$$

where k is the Boltzmann constant.

5.2 Problems

5.2.1 Crystal Structure

5.1 Show that $\pi/6$ of the available volume is occupied by hard spheres in contact in a simple cubic arrangement.

5.2 Show that $\sqrt{3}\pi/8$ of the available volume is occupied by hard spheres in contact in a body-centered cubic arrangement.

5.3 Calculate the separations of the sets of planes which produce strong x-ray diffractions beams at angles 4° and 8° in the first order, given that the x-ray wavelength is 0.1 nm.

5.4 At what angle will a diffracted beam emerge from the (111) planes of a face centered cubic crystal of unit cell length 0.4 nm? Assume diffraction occurs in the first order and that the x-ray wavelength is 0.3 nm.

5.5 An x-ray beam of wavelength 0.16 nm is incident on a set of planes of a certain crystal. The first Bragg reflection is observed for an incidence angle of 36°. What is the plane separation? Will there be any higher order reflections?

5.6 In the historical experiment of Davisson and Germer electrons of 54 eV at normal incidence on a crystal showed a peak at reflection angle $\theta_r = 40^0$. At what energy neutrons would also show a peak at $\theta_r = 40^0$ for the same order.

5.7 Write down the atomic radii r in terms of the lattice constant a, for (a) Simple cubic structure (b) FCC structure (c) BCC structure (d) Diamond structure.

5.2.2 Crystal Properties

5.8 Show that the Madelung constant for a one-dimensional array of ions of alternating sign with equal distance between successive ions is equal to 2 ln 2.

5.9 Write down the first five terms for the Madelung constant corresponding to the NaCl crystal.

5.10 The energy of interaction of two atoms a distance r apart can be written as:

$$E(r) = -\frac{a}{r} + \frac{b}{r^7}$$

where a and b are constants.

(a) Show that for the particles to be in equilibrium, $r = r_o = (7\,b/a)^{1/6}$
(b) In stable equilibrium, show that the energy of attraction is seven times that of the repulsion in contrast to the forces of attraction and repulsion being equal.

5.11 In Problem 5.10, if the two atoms are pulled apart, show that they will separate most easily when $r = (28\,b/a)^{1/6}$.

5.12 Let the interaction energy between two atoms be given by:

$$E(r) = -\frac{A}{r^2} + \frac{B}{r^8}$$

If the atoms form a stable molecule with an inter-nuclear distance of 0.4 nm and a dissociation energy of 3 eV, calculate A and B.

5.13 Lead is a fcc with lattice constant 4.94 Å. Lead melts when the average amplitude of its atomic vibrations is 0.46 Å. Assuming that for lead the Young's modulus is $1.6 \times 10^{10}\,\mathrm{N/m^2}$, find the melting point of lead.

5.2.3 Metals

5.14 Take the Fermi energy of silver to be 5.52 eV.

(a) Find the corresponding velocity of conduction electron.
(b) If the resistivity of silver at room temperature is $1.62 \times 10^{-8}\,\Omega\mathrm{m}$ estimate the average time between collisions.
(c) Determine the mean free path. Assume the number of conduction electrons as $5.86 \times 10^{28}\,\mathrm{m^{-3}}$.

5.15 Find the drift velocity of electron subjected to an electric field of $20\,\mathrm{Vm^{-1}}$, given that the inter-collision time is 10^{-14} s.

5.16 Aluminum is trivalent with atomic weight 27 and density $2.7\,\mathrm{g/cm^3}$, while the mean collision time between electrons is 4×10^{-14} s. Calculate the current flowing through an aluminum wire 20 m long and $2\,\mathrm{mm^2}$ cross-sectional area when a potential of 3 V is applied to its ends.

5.17 Given that the conductivity of sodium is $2.17 \times 10^7\,\Omega^{-1}\mathrm{m}^{-1}$, calculate:

(a) The inter-collision time at 300 K, and
(b) The drift velocity in a field of $200\,\mathrm{Vm^{-1}}$.

5.18 Given that the inter-collision time in copper is 2.3×10^{-14} s, calculate its thermal conductivity at 300 K. Assume the Wiedemann-Franz constant is $C_{\mathrm{WF}} = 2.31 \times 10^{-8}\,\mathrm{W}\,\Omega\mathrm{K}^{-2}$.

5.19 The resistivity of a certain material is $1.72 \times 10^{-8}\,\Omega$m whilst the Hall coefficient is $-0.55 \times 10^{-10}\,\mathrm{m^3C^{-1}}$. Deduce:

(a) The electrical conductivity (σ)
(b) Mobility (μ)
(c) The inter-collision time (τ)
(d) Electron density (n)

5.20 In a Hall effect experiment on zinc, a potential of 4.5 μV is developed across a foil of thickness 0.02 mm when a current of 1.5 A is passed in a direction perpendicular to a magnetic field of 2.0 T. Calculate:

(a) The Hall coefficient for zinc
(b) The electron density

5.21 The density of states function for electrons in a metal is given by:
$Z(E)\mathrm{d}E = 13.6 \times 10^{27} E^{1/2}\mathrm{d}E$
Calculate the Fermi level at a temperature few degrees above absolute zero for copper which has 8.5×10^{28} electrons per cubic metre.

5.22 Using the results of Problem 5.21, calculate the velocity of electrons at the Fermi level in copper.

5.23 For silver ($A = 108$), the resistivity is $1.5 \times 10^{-8}\,\Omega$m at 0 °C density is 10.5×10^3 kg/m^3 and Fermi energy $E_F = 5.5$ eV. Assuming that each atom contributes one electron for conduction, find the ratio of the mean free path λ to the interatomic spacing d.

5.24 Calculate the average amplitude of the vibrations of aluminum atoms at 500 K, given that the force constant $K = 20$ N/m.

5.25 The Fermi energy in gold is 5.54 eV (a) calculate the average energy of the free electrons in gold at 0 °K. (b) Find the corresponding speed of free electrons (c) What temperature is necessary for the average kinetic energy of gas molecules to posses this value?

5.26 The density of copper is 8.94 g/cm^3 and its atomic weight is 63.5 per mole, the effective mass of electron being 1.01. Calculate the Fermi energy in copper assuming that each atom gives one electron.

5.27 Find the probability of occupancy of a state of energy (a) 0.05 eV above the Fermi energy (b) 0.05 eV below the Fermi energy (c) equal to the Fermi energy. Assume a temperature of 300 K.

5.28 What is the probability at 400 K that a state at the bottom of the conduction band is occupied in silicon. Given that $E_g = 1.1$ eV

5.29 The Debye temperature θ for iron is known to be 360 K. Calculate ν_m, the maximum frequency.

5.30 Einstein's model of solids gives the expression for the specific heat

$$C_v = 3N_0k\left(\frac{\theta_E}{T}\right)^2 \frac{e^{\theta_E/T}}{(e^{\theta_E/T}-1)^2}$$

where $\theta_E = h\nu_E/k$.
The factor θ_E is called the characteristic temperature. Show that (a) at high temperatures Dulong Petit law is reproduced. (b) But at very low temperatures the T^3 law is not given.

5.31 Debye's model of solids gives the expression for specific heat

$$C_v = 9N_0k\frac{1}{x^3}\int_0^x \frac{\xi^4 e^\xi}{(e^\xi - 1)^2}\mathrm{d}\xi$$

where $\xi = h\nu/kT$, $x = h\nu_\mathrm{m}/kT$ and $\theta_\mathrm{D} = h\nu_\mathrm{m}/k$ is the Debye's characteristic temperature. Show that (a) at high temperatures Debye's model gives Dulong Petit law (b) at low temperatures it gives $C_\mathrm{v} \propto T^3$ in agreement with the experiment.

5.32 For a free electron gas in a metal, the number of states per unit volume with energies from E to $E + \mathrm{d}E$ is given by

$$n(E)\mathrm{d}E = \frac{2\pi}{h^3}(2m)^{3/2}E^{1/2}\mathrm{d}E$$

Show that the total energy $= 3NE_{\mathrm{max}}/5$.

5.33 Assuming that the conduction electrons in a cube of a metal on edge 1 cm behave as a free quantized gas, calculate the number of states that are available in the energy interval 4.00–4.01 eV, per unit volume.

5.34 Calculate the Fermi energy for silver given that the number of conduction electrons per unit volume is $5.86 \times 10^{28}\,\mathrm{m}^{-3}$.

5.35 Calculate for silver the energy at which the probability that a conduction electron state will be occupied is 90%. Assume $E_\mathrm{F} = 5.52\,\mathrm{eV}$ for silver and temperature $T = 800\,\mathrm{K}$.

5.2.4 Semiconductors

5.36 An LED is constructed from a *Pn* junction based on a certain semi-conducting material with energy gap of 1.55 eV. What is the wavelength of the emitted light?

5.37 Suppose that the Fermi level in a semiconductor lies more than a few kT below the bottom of the conduction band and more than a few kT above the top of the valence band, then show that the product of the number of free electrons and the number of free holes per cm^3 is given by

$n_e n_h = 2.33 \times 10^{31} T^3 e^{-E_g/kT}$

where E_g is the gap width

5.38 The effective mass m^* of an electron or hole in a band is defined by

$$\frac{1}{m^*} = \frac{1}{\hbar^2} \cdot \frac{d^2E}{dk^2}$$

where k is the wave number ($k = 2\pi/\lambda$). For a free electron show that $m^* = m$.

5.39 After adding an impurity atom that donates an extra electron to the conduction band of silicon ($\mu_n = 0.13\,\text{m}^2/\text{Vs}$), the conductivity of the doped silicon is measured as 1.08 (Ωm^{-1}). Determine the doped ratio (density of silicon is 2,420 kg/m^3).

5.40 Estimate the ratio of the electron densities in the conduction bands of silicon ($E_g = 1.14\,\text{eV}$) and germanium ($E_g = 0.7\,\text{eV}$) at 400 K.

5.41 Show that at the room temperature (300 K) the electron densities in the conduction bands of the insulator carbon ($E_g = 5.33\,\text{eV}$) and the semiconductor like germanium ($E_g = 0.7\,\text{eV}$) is extremely small.

5.42 A current of 8×10^{-11} A flows through a silicon $p-n$ junction at temperature 27 °C. Calculate the current for a forward bias of 0.5 V.

5.43 Calculate the depletion layer width for a *pn* junction with zero bias in germanium, given that the impurity concentrations are $N_a = 1 \times 10^{23}\,\text{m}^{-3}$ and $N_d = 2 \times 10^{22}\,\text{m}^{-3}$, respectively at $T = 300\,\text{K}$, $\epsilon_r = 16$ and contact potential difference $V_0 = 0.8\,\text{V}$.

5.44 Consider the Shockley equation for the diode

$I = I_0 \exp[(eV/kT) - 1]$

Show that the slope resistance r_e of the $I - V$ curve at a particular d.c bias is given to a good approximation, at room temperature ($T = 300\,\text{K}$) by the expression $r_e = \frac{26}{I}\,\Omega$ (forward bias) where I is in milliampere, and that for the reverse bias r_e tends to infinity.

5.45 Given that a piece of n-type silicon contains $8 \times 10^{21}\,\text{m}^{-3}$ phosphorus impurity atoms, calculate the carrier concentrations at room temperature. It may be assumed that the intrinsic electron concentration in silicon at room temperature is $1.6 \times 10^{16}\,\text{m}^{-3}$.

5.2.5 Superconductor

5.46 It is required to break up a Cooper pair in lead which has the energy gap of 2.73 eV. What is the maximum wavelength of photon which will accomplish the job?

5.47 Given that the maximum wavelength of photon to break up Cooper pair in tin is 1.08×10^{-3} m, calculate the energy gap.

5.48 A Josephson junction consists of two super conductors separated by a very thin insulating layer. When a DC voltage is applied across the junction an AC current is produced, a phenomenon called Josephson effect. Calculate the frequency of the AC current produced when a DC voltage of 1.5 μV is applied.

5.49 Use the BCS theory to calculate the energy gap for indium whose critical temperature $T_c = 3.4$ K.

5.50 For lead superconductivity ensues at 7.19 K, when there is a zero applied magnetic field. When the magnetic field of 0.074 T is applied at temperature 2.0 K superconductivity will stop. Find the magnetic field that should be applied so that superconductivity will not occur at any temperature?

5.51 An ac current of frequency 1 GHz is observed through a Josephson junction. Calculate the applied dc voltage.

5.52 If a superconducting Quantum Interference Device which consists of a 3 mm ring can measure 1/5,000 of a fluxon, calculate the magnetic field that can be detected (1 fluxon, $\phi_0 = h/2e = 2.0678 \times 10^{-15}$ T m^2, is the smallest unit of flux).

5.3 Solutions

5.3.1 Crystal Structure

5.1 The cross-section of the portions of four spheres each of radius r touching each other and lying in a cell of edge a is shown in Fig. 5.1. The volume of each sphere lying within the cell is $\frac{1}{4} \times \frac{4}{3}\pi r^3$ or $\frac{\pi}{3}r^3$. Volume of four spheres lying within the cell is $\frac{4\pi}{3}r^3$. Volume of the cell is a^3 or $(2r)^3$. Therefore, the available volume occupied by hard spheres in the simple cubic structure is $\frac{4\pi r^3/3}{(2r)^3}$ or $\frac{\pi}{6}$.

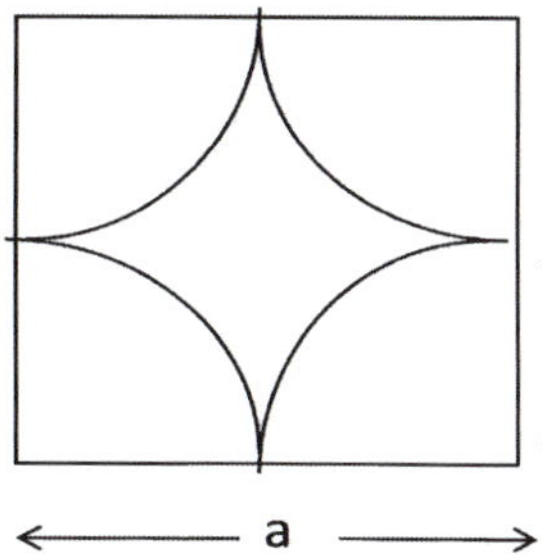

Fig. 5.1 Hard spheres in the simple structure

5.2 Volume of the unit cell $= a^3$. Since there are two atoms per unit cell, $8 \times 1/8$ for the corner atoms and 1×1 for the centre atom,

Volume $= 2 \times \frac{4}{3}\pi r^3$

Since the body diagonal atoms touch one another,

$4r = a\sqrt{3}$

Volume of atoms in terms of a is

$$2 \times \frac{4}{3}\pi r^3 = 2 \times \frac{4}{3}\pi [a\sqrt{3}/4]^3$$
$$= \sqrt{3}\pi a^3/8$$

Or the fraction of the volume occupied by the body-centred cubic structure is $\sqrt{3}\pi/8$.

5.3

$$2d \sin\theta = n\lambda$$
$$d_1 = \frac{1.\lambda}{2\sin\theta} = \frac{0.1}{2\sin 4^\circ} = 0.717\,\text{nm}$$
$$d_2 = \frac{0.1}{2\sin 8^\circ} = 0.359\,\text{nm}$$

5.4 $$n\lambda = \frac{2a}{(h^2+k^2+l^2)^{1/2}}\sin\theta = \frac{2\times 0.4}{(1^2+1^2+1^2)^{1/2}}\sin\theta$$
$$\sin\theta = \frac{0.3\sqrt{3}}{0.8} = 0.6495$$
$$\theta = 40.5^\circ$$

5.5 $$2d\sin\theta = n\lambda$$
$$d = \frac{1.\lambda}{2\sin\theta} = \frac{0.16}{2\sin 30^\circ} = 0.136\,\text{nm}$$

For $n = 2$,

$$\sin\theta = \frac{2\times 0.16}{2\times 0.136} = 1.176$$

a value which is not possible. Thus higher order reflections are not possible.

5.6 The de Broglie wavelength for electrons is calculated from

$$\lambda = \sqrt{\frac{150}{V}} = \sqrt{\frac{150}{54}} = 1.66\,\text{Å}$$

Bragg's equation will be satisfied for neutrons of the same wavelength.

$$\lambda = \frac{0.286}{\sqrt{E}}\,\text{Å}$$

where E is in eV,

$$E = \left(\frac{0.286}{\lambda}\right)^2 = \left(\frac{0.286}{1.66}\right)^2 = 0.0297\,\text{eV}.$$

5.7 (*a*) $r = a/2$

(*b*) $r = \sqrt{2}\,a/4$

(*c*) $r = \sqrt{3}\,a/4$

(*d*) $\sqrt{3}\,a/8$

5.3.2 Crystal Properties

5.8 Consider an infinite line of ions of alternating sign, as in Fig. 5.2. Let a negative ion be a reference ion and let a be the distance between adjacent ions. By definition the Madelung Constant $\propto$ is given by:

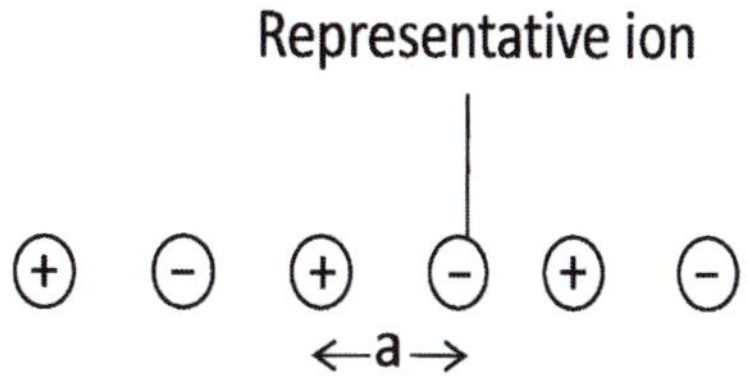

Fig. 5.2 Infinite line of ions of alternating sign

$$\frac{\alpha}{a} = \sum_j \frac{(\pm)}{r_j} \tag{1}$$

where r_j is the distance of the jth ion from the reference ion and a is the nearest neighbor distance. Thus:

$$\frac{\alpha}{a} = 2\left[\frac{1}{a} - \frac{1}{2a} + \frac{1}{3a} - \frac{1}{4a} + \cdots\right]$$

$$\text{Or, } \alpha = 2\left[1 - \frac{1}{2} + \frac{1}{3} - \frac{1}{4} + \cdots\right] \tag{2}$$

The factor 2 occurs because there are two ions, one to the right and one to the left, at equal distances r_j. We sum the series by the expansion:

$$\ln(1+x) = x - \frac{x^2}{2} + \frac{x^3}{3} - \frac{x^4}{4} + \cdots \tag{3}$$

Putting $x = 1$, the RHS in (3) is identified as In 2. Thus $\propto = 2$ ln 2.

5.9 The lattice of the NaCl structure which is face-centered is represented in Fig. 5.3. The shortest inter ionic distance is represented by r. A given sodium ion is surrounded by $6Cl^-$ ions at a distance r, $12Na^+$ ions at a distance $r\sqrt{2}$, $8Cl^-$ ions at a distance $r\sqrt{3}$, $6Na^+$ ions at a distance $r\sqrt{4}$, $24Cl^-$ ions at a distance $r\sqrt{5}$, etc. The coulomb energy of of this ion in the field of all other ions is therefore

$$E_c = -\frac{e^2}{r}\left(\frac{6}{\sqrt{1}} - \frac{12}{\sqrt{2}} + \frac{8}{\sqrt{3}} - \frac{6}{\sqrt{4}} + \frac{24}{\sqrt{5}} - \cdots\right)$$

where e is the charge per ion. Series of this sort which consists of pure numbers depends on the crystal structure and is known as the Madelung constant.

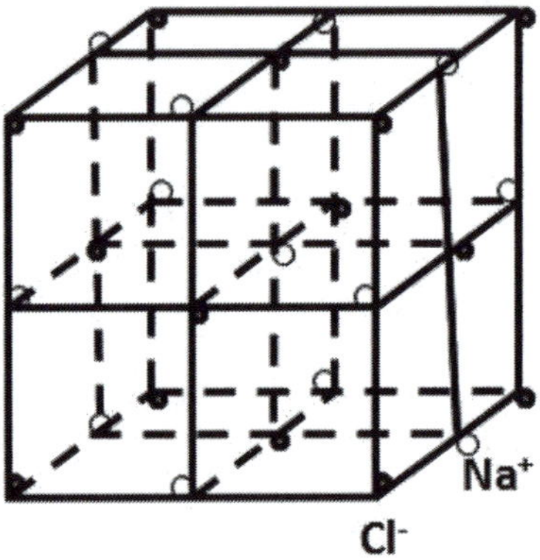

Fig. 5.3 The lattice of the NaCl structure

5.10 $E_r = -\frac{a}{r} + \frac{b}{r^7}$

(a) For equilibrium, E_r must be minimum, so that $\frac{dE}{dr} = 0$

$$\frac{a}{r^2} - \frac{7b}{r^8} = 0$$

Or, $r = r_o = (7b/a)^{1/6}$

(b) Energy of attraction, $E_A = \frac{-a}{r_o}$

Energy of repulsion, $E_R = \frac{b}{r_0^7}$

$$\frac{|E_A|}{|E_R|} = \frac{a}{b} r_o^6 = \frac{a}{b} \cdot \frac{7b}{a} = 7$$

Force, $F = -\frac{dV}{dr}$

Attractive force $F_A = \frac{a}{r_o^2}$ $(r = r_o)$

Repulsive force $F_R = \frac{-7b}{r_o^8} = \frac{-7b}{r_o^2} \cdot \frac{a}{7b} = -\frac{a}{r_o^2}$

Thus the two forces are equal in magnitude.

5.11 Force, $F = -\dfrac{\mathrm{d}V}{\mathrm{d}r} = \dfrac{a}{r^2} - \dfrac{7b}{r^8}$

The particles will separate most easily when the force between them is a minimum, that is when $\dfrac{\mathrm{d}F}{\mathrm{d}r} = 0$. This gives:

$$\frac{\mathrm{d}F}{\mathrm{d}r} = -\frac{2a}{r^3} + \frac{56b}{r^9} = 0$$

$$r = \left(\frac{28b}{a}\right)^{1/6}$$

5.12 The inter-nuclear distance is found from $\dfrac{\mathrm{d}E}{\mathrm{d}r} = 0$

$$\frac{2A}{r_o^3} - \frac{8B}{r_o^9} = 0 \rightarrow r_o^6 = \frac{4B}{A} \qquad (1)$$

The dissociation energy D is formed from $-D = E(r_o)$

$$-D = -\frac{A}{r_o^2} + \frac{B}{r_o^8} = -\frac{A}{r_o^2} + \frac{A}{4r_o^2} = -\frac{3A}{4r_o^2} \text{ where we have used (1).}$$

$$A = \frac{4Dr_o^2}{3}$$

$$= \frac{4}{3} \times 3 \times 1.6 \times 10^{-19} \times (0.4 \times 10^{-9})^2 = 1.02 \times 10^{-37}$$

$$B = \frac{A}{4} r_0^6 = \frac{1.02 \times 10^{-37}}{4} \times (0.4 \times 10^{-9})^6$$

$$= 1.04 \times 10^{-91}$$

5.13 $<A>= \left(\dfrac{2kT}{K}\right)^{1/2}$

The force constant $K = Ya_0 = 1.6 \times 10^{10} \times 4.94 \times 10^{-10} = 7.9\,\mathrm{N/m^2}$

$$T = \frac{K}{2k}\langle A\rangle^2 = \frac{7.9 \times (0.46 \times 10^{-10})^2}{2 \times 1.38 \times 10^{-23}} = 606\,\mathrm{K} = 333^\circ\mathrm{C}$$

5.3.3 Metals

5.14 (a) $v_F = \left(\dfrac{2E_F}{m}\right)^{1/2} = \left[\dfrac{2 \times 5.52 \times 1.6 \times 10^{-19}}{9.11 \times 10^{-31}}\right]^{1/2}$

$$= 1.39 \times 10^6\,\mathrm{m/s}$$

(b) $\tau = \dfrac{m}{ne^2\rho} = \dfrac{9.11 \times 10^{-31}}{(5.86 \times 10^{28})(1.6 \times 10^{-19})^2(1.62 \times 10^{-8})}$

$$= 3.7 \times 10^{-14}\,\mathrm{s}$$

(c) $\lambda = v_F \tau = (1.39 \times 10^6)(3.7 \times 10^{-14})$

$$= 5.14 \times 10^{-8}\ \text{m}$$

5.15 $\langle v_D \rangle = \dfrac{e}{m}\varepsilon\tau$

$$= \frac{(1.6 \times 10^{-19})(20)(10^{-14})}{9.11 \times 10^{-31}} = 0.0351\ \text{m/s}$$

$= 3.51\ \text{cm/s}$

Note that the drift velocities are much smaller than the average thermal velocities which are of the order of 10^5 m/s. $\left[\langle v_T \rangle = (3kT/m_e)^{1/2}\right]$

5.16 Current,

$$i = \frac{V}{R} \quad (1)$$

$$R = \frac{\rho l}{A} \quad (2)$$

where the resistivity,

$$\rho = \frac{m_e}{ne^2\tau} \quad (3)$$

$$n = \frac{N_o d}{A} \times 3 \times 10^4 (4) \quad (4)$$

where n is the number of electrons per m^3, N_o being Avagardro's number, A the atomic weight and d the density, the factor 3 is for the trivalency.

$$n = 6.02 \times 10^{23} \times \frac{2.7}{27} \times 3 \times 10^4 = 1.806 \times 10^{27}$$

$$\rho = \frac{9.11 \times 10^{-31}}{1.806 \times 10^{27} \times (1.6 \times 10^{-19})^2 \times 4 \times 10^{-14}} = 4.92 \times 10^{-9}$$

$$R = \frac{4.92 \times 10^{-9} \times 20}{2 \times 10^{-6}} = 0.0492\ \Omega$$

$$i = \frac{3}{0.0492} = 61\ \text{Å}$$

5.17 (a) $\tau = \dfrac{m\sigma}{ne^2}$

Assuming that one conduction electron will be available for each sodium atom,

$$n = \frac{N_o \rho}{A} = \frac{6.02 \times 10^{23} \times 0.97}{23}\text{cm}^{-3} = 2.539 \times 10^{28}\ \text{m}^{-3}$$

$$\tau = \frac{9.11 \times 10^{-31} \times 2.17 \times 10^7}{2.539 \times 10^{28} \times (1.6 \times 10^{-19})^2} = 3.04 \times 10^{-14}\text{s}$$

(b) $\langle v_D \rangle = \dfrac{e}{m}\varepsilon\tau$

$$= \frac{(1.6 \times 10^{-19})(200)(3.04 \times 10^{-14})}{9.11 \times 10^{-31}} = 1.07\ \text{m/s}$$

5.18 $\sigma = \dfrac{ne^2\tau}{m}$

$$n = \frac{6.02 \times 10^{23} \times 8.88 \times 10^6}{63.57} = 8.38 \times 10^{28}\,\text{m}^{-3}$$

$$\sigma = \frac{8.38 \times 10^{28} \times (1.6 \times 10^{-19})^2 \times 2.3 \times 10^{-14}}{9.11 \times 10^{-31}} = 5.422 \times 10^7\ \Omega^{-1}\text{m}^{-1}$$

$$K = \sigma C_{\text{WF}} T = (5.422 \times 10^7)(2.31 \times 10^{-8})(300) = 376\,\text{Wm}^{-1}\text{K}^{-2}$$

5.19 (*a*) $\sigma = \dfrac{1}{\rho} = \dfrac{1}{1.72 \times 10^{-8}} = 5.8 \times 10^7\Omega^{-1}\text{m}^{-1}$

(*b*) $\mu = R_H\sigma = (0.55 \times 10^{-10})(5.8 \times 10^7) = 0.0032\,\text{m}^2\text{V}^{-1}\text{s}^{-1}$

(*c*) $\tau = \dfrac{\mu m}{e} = \dfrac{(0.0032)(9.11 \times 10^{-31})}{1.6 \times 10^{-19}} = 1.82 \times 10^{-14}\,\text{s}$

(*d*) $n = \dfrac{\sigma}{e\mu} = \dfrac{5.8 \times 10^7}{(1.6 \times 10^{-19})(0.0032)} = 1.13 \times 10^{29}\,\text{m}^{-3}$

5.20 (*a*) $R_{\text{H}} = \dfrac{V_{\text{H}}t}{iB} = \dfrac{(4.5 \times 10^{-6})(2 \times 10^{-5})}{(1.5)(2)} = 0.3 \times 10^{-10}\,\text{m}^3\text{C}^{-1}$

(*b*) $n = \dfrac{1}{R_{\text{H}}e} = \dfrac{1}{(0.3 \times 10^{-10})(1.6 \times 10^{-19})} = 2.08 \times 10^{29}\,\text{m}^{-3}$

5.21 Integrating $n(E)\,\text{d}E$ from zero to E_{F}:

$$13.6 \times 10^{27} \int_0^{E_{\text{F}}} E^{1/2}\,\text{d}E = 8.5 \times 10^{28}$$

$E_{\text{F}}^{3/2} = 9.375$

Or $E_{\text{F}} = 4.445\,\text{eV}$

5.22 $v = \sqrt{\dfrac{2E}{m}} = \sqrt{\dfrac{2 \times 4.445 \times 1.6 \times 10^{-19}}{9.11 \times 10^{-31}}} = 1.25 \times 10^6\,\text{m/s}$

5.23 $v_{\text{F}} = \left(\dfrac{2E_{\text{F}}}{mc^2}\right)^{1/2} c = \left(\dfrac{2 \times 5.5}{0.511 \times 10^6}\right)^{1/2} (3 \times 10^8) = 1.39 \times 10^6\,\text{m/s}$

Since each atom contributes one electron, the density of electrons is equal to that of atoms.

$$n = \frac{6.02 \times 10^{26} \times 10.5 \times 10^3}{108} = 5.85 \times 10^{28}\,\text{e/m}^3$$

Each atom occupies approximately a volume d^3. Therefore

$$d = \left(\frac{1}{5.85 \times 10^{28}}\right)^{1/3} = 2.576 \times 10^{-10}\,\text{m}$$

Now, $\rho = \dfrac{m_e V_F}{e^2 n\lambda}$

$$\text{or } \lambda = \frac{m_e V_F}{\rho e^2 n} = \frac{(9.11\times10^{-31})(1.39\times10^{6})}{(1.5\times10^{-8})(1.6\times10^{-19})^2(5.85\times10^{28})} = 9.02\times10^{-8}\,\text{m}$$

$$\therefore \frac{\lambda}{d} = \frac{9.02\times10^{-8}}{2.58\times10^{-10}} = 350$$

5.24 $$<A> = \left(\frac{2kT}{K}\right)^{1/2} = \left(\frac{2\times1.38\times10^{-23}\times500}{20}\right)^{1/2} = 2.63\times10^{-11}\,\text{m} = 0.26\,\text{Å}$$

5.25 (a) $<E> = \dfrac{3}{5}E_F = \dfrac{3\times5.54}{5} = 3.32\,\text{eV}$

(b) $v = c\left(\dfrac{2E}{Mc^2}\right)^{1/2} = 3\times10^8\left(\dfrac{2\times3.32}{0.511\times10^6}\right)^{1/2} = 1.08\times10^6\text{m/s}$

(c) $\dfrac{3}{2}kT = \dfrac{3}{5}E_F = 3.32\,\text{eV} = 3.32\times1.6\times10^{-19}\text{J}$

$$T = \frac{2}{3}\times\frac{3.32\times1.6\times10^{-19}}{1.38\times10^{-23}} = 2.56\times10^4\text{K}$$

5.26 $$E_F = \frac{h^2}{2m^*}\left(\frac{3N}{8\pi V}\right)^{2/3}$$

where m^* is the effective mass.

Density of Cu atoms $= \dfrac{N_0\rho}{A} = \dfrac{6.02\times10^{23}\times8.94}{63.5}$

$= 8.475\times10^{22}\text{atoms/cm}^3$

$= 8.475\times10^{28}\text{atoms/m}^3$

$= 8.475\times10^{28}\text{e/m}^3$

($\because$ each atom gives one electron)

$$E_F = \frac{(6.625\times10^{-34})^2}{2\times1.01\times9.11\times10^{-31}}\left(\frac{3}{8\pi}\times8.475\times10^{28}\right)^{2/3}$$

$= 11.157\times10^{-19}J = 6.97\,\text{eV}$

5.27 The probability is given by Fermi-Dirac distribution $p(E) = \dfrac{1}{e^{(E-E_F)/kT}+1}$

(a) $K = 1.38\times10^{-23}J = 8.625\times10^{-5}\,\text{eV K}^{-1}$

$$\frac{E-E_F}{kT} = \frac{0.05}{(8.625\times10^{-5})(300)} = 1.932$$

$$p(E) = \frac{1}{e^{1.932}+1} = 0.126$$

(b) $p(E) = \frac{1}{e^{-1.932} + 1} = 0.873$

(c) $p(E) = \frac{1}{e^{0} + 1} = 0.5$

5.28 Assuming that the Fermi energy is to be at the middle of the gap between the conduction and valence bands, $E - E_F = ½\, E_g$

$$p(E) = \frac{1}{e^{(E-E_F)/kT+1}} = \frac{1}{e^{E_g/2kT} + 1}$$

The factor $E_g/2kT = \frac{1.1}{2 \times 8.625 \times 10^{-5} \times 400} = 15.942$

$$p(E) \approx e^{-15.942} = 8.4 \times 10^{-6}$$

5.29 The Debye temperature θ is

$$\theta = \frac{h\nu_m}{k}$$

$$\nu_m = \frac{k}{h}\theta = \frac{1.38 \times 10^{-23} \times 360}{6.625 \times 10^{-34}} = 7.5 \times 10^{12}\,\text{Hz}$$

5.30 (a) At high temperatures $T >> \theta_E$, in the denominator $(e^{\theta_E/T} - 1)^2 \approx \theta_E^2/T^2$, and in the numerator $e^{\theta_E/T} \to 1$, so that $C_v \to 3N_0k = 3R$, the Dulong – Petit's value

(b) When the temperature is very low $T << \theta_E$, and in the bracket of the denominator, 1 is negligible in comparison with the exponential term. Therefore, $C_v \to 3R(\theta_E/T)^2 e^{-\theta_E/T}$. Thus the specific heat goes to zero as $T \to 0$. However, the experimentally observed specific heats at low temperatures decrease more gradually than the exponential decrease suggested by Einstein's formula.

5.31 $$C_v = \frac{9R}{x^3}\int_0^x \frac{\xi^4 e^\xi}{(e^\xi - 1)^2}\mathrm{d}\xi \quad (1)$$

This equation may be integrated by parts,

$$\int_0^x \frac{\xi^4 e^\xi}{(e^\xi - 1)^2}\mathrm{d}\xi = -\int \xi^4 \frac{\mathrm{d}}{\mathrm{d}\xi}\left(\frac{1}{e^\xi - 1}\right)\mathrm{d}\xi$$

$$= -\xi^4 \frac{1}{e^\xi - 1} + \int \frac{1}{e^\xi - 1}\frac{\mathrm{d}\xi^4}{\mathrm{d}\xi}\mathrm{d}\xi$$

$$= -\xi^4 \frac{1}{e^\xi - 1} + 4\int \frac{\xi^3}{e^\xi - 1}\mathrm{d}\xi$$

Thus (1) becomes

$$C_v = 9R\left[\frac{4}{x^3}\int_0^x \frac{\xi^3}{e^\xi - 1}\mathrm{d}\xi - \frac{x}{e^x - 1}\right] \quad (2)$$

(a) At high temperatures, $\theta_D >> T$, or $x << 1$, and the exponential can be expanded to give

$$C_v = 9R\left(\frac{4}{3} - 1\right) = 3R \quad \text{(Dulong Petit's law)}$$

(b) At very low temperatures $T << \theta_D$ $x >> 1$, (2) can be approximated to

$$C_v = 9R\frac{4}{x^3}\int_0^\infty \frac{\xi^3 d\xi}{e^\xi - 1} = \frac{12}{5}\pi^4\left(\frac{T}{\theta_D}\right)^3$$

where the value of the integral is $\pi^4/15$. Thus, $C_v \propto T^3$

5.32 If there are N free electrons in the metal there will be $N/2$ occupied quantum states at the absolute zero of temperature in accordance with the Fermi Dirac statistics. In Fermi-Dirac statistics at absolute zero, kinetic energy is not zero as would be required if the Boltzmann statistics were assumed.
As $N(E)dE$ gives the number of states per unit volume, in a crystal of volume V, the number of electrons in the range from E to $E + dE$ is

$$2V \cdot \frac{2\pi}{h^3}(2m)^{3/2}E^{1/2}dE \tag{1}$$

The total energy of these electrons would be

$$E_{\text{total}} = \int_0^{E_{\max}} \frac{4\pi V}{h^3}(2m)^{3/2}E^{3/2}dE = \frac{4\pi V(2m)^{\frac{3}{2}}}{h^3} \cdot \frac{2}{5}E_{\max}^{5/2} \tag{2}$$

But,

$$E_{\max} = \frac{h^2}{8m}\left(\frac{3N}{\pi V}\right)^{2/3} \tag{3}$$

Combining (2) and (3),

$$E_{\text{total}} = \frac{3}{5}NE_{\max} \tag{4}$$

or per electron $3E_{\max}/5$. The quantity $E_{\max} = E_F$, the Fermi energy

5.33 The density of states $n(E)$ (the number of states per unit volume of the solid in the unit energy interval) is given by

$$n(E) = \frac{8\sqrt{2}\pi m^{3/2}}{h^3}E^{1/2}$$
$$= \frac{(8\sqrt{2}\pi)(9.11 \times 10^{-31})^{3/2}}{(6.63 \times 10^{-34})^3}(4 \times 1.6 \times 10^{-19})^{1/2}$$
$$= 8.478 \times 10^{46}\text{m}^{-3}\text{J}^{-1} = 1.356 \times 10^{28}\text{m}^{-3}\text{eV}^{-1}$$

Number of states N that lie in the range $E = 4.00$eV to $E = 4.01$eV, for volume, $V = a^3$

$$N = n(E)\Delta E a^3$$
$$= 1.356 \times 10^{28} \times 0.01 \times (10^{-2})^3$$
$$= 1.356 \times 10^{20}$$

5.34 $$E_F = \frac{h^2}{8m}\left(\frac{3n}{\pi}\right)^{2/3}$$
$$= \frac{(6.63 \times 10^{-34})^2}{(8)(9.11 \times 10^{-31})}\left(\frac{3 \times 5.86 \times 10^{28}}{\pi}\right)^{2/3}$$
$$= 8.827 \times 10^{-19}\,\text{J} = 5.517\,\text{eV}$$

5.35 $$P(E) = \frac{1}{e^{\Delta E/kT} + 1} = 0.9$$
Substituting $kT = 5.52 \times 10^{-5} \times 800 = 0.04416\,\text{eV}$
Solving for ΔE, we get $\Delta E = E - E_F = -2.2 \times 0.04416 = -0.097$
Therefore, $E = 5.52 - 0.10 = 5.42\,\text{eV}$

5.3.4 Semiconductors

5.36 $$\lambda = \frac{1241}{1.55} = 800\,\text{nm}$$

5.37 The number of electrons and holes per unit volume are given by

$$n_e = 2\left(2\pi m\frac{kT}{h^2}\right)^{3/2} e^{(E_F - E_g)/kT} \quad (1)$$

$$\text{and } n_h = 2\left(2\pi m\frac{kT}{h^2}\right)^{3/2} e^{-E_F/kT} \quad (2)$$

Multiplying (1) and (2), one can write

$$n_e n_h = 4\left(\frac{mc^2 k}{2\pi \hbar^2 c^2}\right)^3 T^3 e^{-E_g/kT} \quad (3)$$
$$= 2.34 \times 10^{31} T^3 e^{-E_g/kT}\,\text{cm}^{-6}$$

where we have substituted the values of the constants.

5.38 $$p = k\hbar \quad (1)$$
$$E = p^2/2m = k^2\hbar^2/2m \quad (2)$$
$$\frac{1}{m^*} = \frac{1}{\hbar^2}\frac{d^2E}{dk^2} \quad (3)$$

Using (2) in Eq. (3)

$$\frac{1}{m^*} = \frac{1}{\hbar^2}\frac{d^2}{dk^2}\left(\frac{k^2\hbar^2}{2m}\right) = \frac{2\hbar^2}{2m\hbar^2} = \frac{1}{m}$$
$$\therefore m^* = m$$

5.39 The number of silicon atoms/m^3

$$n = \frac{N_0 d}{A} = \frac{6.02 \times 10^{26} \times 2{,}420}{28} = 0.52 \times 10^{29}$$

Let x be the fraction of impurity atom (donor). The general expression for the conductivity is

$$\sigma = n_\mathrm{n} e\mu_\mathrm{n} + n_\mathrm{p} e\mu_\mathrm{p}$$

where n_n and n_p are the densities of the negative and positive charge carriers. Because $n_\mathrm{n} \gg n_\mathrm{p}$,

$$\sigma \cong n_\mathrm{n} e\mu_\mathrm{n} = x n_\mathrm{p} e\mu_\mathrm{n}$$

$$x = \frac{\sigma}{n_\mathrm{p} e\mu_\mathrm{n}} = \frac{1.08}{0.52 \times 10^{29} \times 1.6 \times 10^{-19} \times 0.13} = \frac{9.985}{10^{10}}$$

or 1 part in 10^9.

Note that in normal silicon the conductivity is of the order of $10^{-4}(\Omega\text{– m})^{-1}$. A small fraction of doping (10^{-9}) has dramatically increased the value by four orders of magnitude.

5.40 $n_\mathrm{e} = (4.83 \times 10^{21}) T^{3/2} e^{-E_\mathrm{g}/2kT}\ \mathrm{e/m^3}$

$$\frac{n_\mathrm{Ge}}{n_\mathrm{Si}} = e^{(E_\mathrm{Si} - E_\mathrm{Ge})/2kT}$$

$$kT = \frac{1.38 \times 10^{-23} \times 400}{1.6 \times 10^{-19}} = 0.0345\,\mathrm{eV}$$

$$\frac{n_\mathrm{Ge}}{n_\mathrm{Si}} = e^{(1.14-0.7)/(2\times 0.0345)} = 588$$

5.41 $$kT = \frac{1.38 \times 10^{-23} \times 300}{1.6 \times 10^{-19}} = 0.0259\,\mathrm{eV}$$

$$\frac{n_\mathrm{C}}{n_\mathrm{Ge}} = e^{-(E_{Ge} - E_\mathrm{C})/2KT} = e^{-(5.33-07)/0.052} = e^{-89} \approx 2.2 \times 10^{-39}$$

5.42 $I = I_0[\exp(\mathrm{eV}/kT) - 1]$

where I_0 is the forward bias saturation current.

$$I = 8\times 10^{-11}\left[\exp\left(\frac{0.5 \times 1.6 \times 10^{-19}}{1.38 \times 10^{-23} \times 300}\right) - 1\right] = 19.7\times 10^{-3}\mathrm{A} = 19.7\,\mathrm{mA}$$

5.43 $$W = \left[\frac{2\epsilon_0\epsilon_r}{e}(V_0 - V_\mathrm{b})\left(\frac{1}{N_\mathrm{a}} + \frac{1}{N_\mathrm{d}}\right)\right]^{1/2}$$

where ϵ_r is the relative permittivity, V_b is the bias voltage applied to the junction (here $V_\mathrm{b} = 0$), N_a and N_d are carrier concentrations in n-type and p-type respectively.

$$W = \left[\frac{2 \times 8.85 \times 10^{-12} \times 16 \times 0.8}{1.6 \times 10^{-19}}\left(\frac{1}{1 \times 10^{23}} + \frac{1}{2 \times 10^{22}}\right)\right]^{1/2} = 0.29\,\mu\mathrm{m}$$

5.44 $I = I_0[\exp(\mathrm{eV}/kT) - 1]$ (1)

$$\frac{1}{r_e} = \frac{dI}{dV} = \frac{eI_0}{kT}\exp(eV/kT) \quad (2)$$

But exp $(\mathrm{eV}/kT) \gg 1$. Therefore

$$\frac{1}{r_e} = \frac{eI}{kT}$$

$$\text{or } r_e = \frac{kT}{eI} = \frac{1.38 \times 10^{-23} \times 300}{1.6 \times 10^{-19} I} = \frac{25.875 \times 10^{-3}}{I}$$

If I is in milliamp.

$$r_e \approx \frac{26}{I} \quad \text{(forward bias)} \quad (3)$$

For the reversed bias we note from (2)

$$\frac{1}{r_e} = \frac{dI}{dV} = \frac{e}{kT} I_0 \exp\,(\mathrm{eV}/kT) = 0$$

For V $\leq -4kT/e$, $r_e \to \infty$. (reverse bias) (4)

5.45 For a semiconductor in equilibrium the product of $n(= N_d)$ and $p(= N_a)$ is equal to n_i^2, the square of the intrinsic concentration.

$$n \times p = n_i^2$$

$$p = \frac{n_i^2}{n} = \frac{(1.6 \times 10^{16})^2}{8 \times 10^{21}} = 3.2 \times 10^{10}\,\mathrm{m}^{-3}$$

5.3.5 Superconductor

5.46 $\lambda = \dfrac{1241}{E(\mathrm{eV})}\mathrm{nm} = \dfrac{1,241}{2.73 \times 10^{-3}} = 4.546 \times 10^5\ \mathrm{nm}$

5.47 $E_g = \dfrac{1241}{\lambda(\mathrm{nm})} = \dfrac{1,241}{1.08 \times 10^6} = 1.15 \times 10^{-3}\,\mathrm{eV}$

5.48 $f = \dfrac{2\mathrm{eV}}{h} = \dfrac{2(1.602 \times 10^{-19})(1.5 \times 10^{-6})}{6.626 \times 10^{-34}} = 7.253 \times 10^8\ \mathrm{Hz} = 0.7253\,\mathrm{GHz}$

5.49 $E_g = 3.53kT_c = (3.53)\dfrac{(1.38 \times 10^{-23})}{1.6 \times 10^{-19}}(3.4) = 1.035 \times 10^{-3}\mathrm{eV}$

5.50 $T_c(B) = T_c\left(1 - \dfrac{B}{B_c}\right)^{1/2}$

$$2.0 = 7.19\left(1 - \frac{0.074}{B_c}\right)^{1/2}$$

Solving for B_c, we get $B_c = 0.079\,T$.

5.51 $f = 2\text{eV}/h \rightarrow V = hf/2e$
$V = (6.625 \times 10^{-34})(10^{9})/2 \times (1.6 \times 10^{-19})$
$= 2.07 \times 10^{-6}\ \text{V}$
$= 2.07\ \mu\text{V}$

5.52 $\Delta B = \dfrac{\varphi}{A} = \left(\dfrac{1}{5,000}\right)\varphi_0/\pi(3 \times 10^{-3})^2$
Substituting $\phi_0 = 2.0678 \times 10^{-15}\ \text{Tm}^2$, we get
$\Delta B = 2.93 \times 10^{-14}\ \text{T}$

Chapter 6
Special Theory of Relativity

6.1 Basic Concepts and Formulae

Inertial frame

Laws of mechanics take the same form (invariant) in all inertial frames. An inertial frame of reference is the one which moves with constant relative velocity in which Newton's laws of motion are valid. The principle that all inertial frames are equivalent for the description of nature is called the principle of relativity.

Galilean Transformations

Reference frame S' moves along x-axis with velocity v relative to S. Spatial coordinates x, y, z are measured in S and x', y', z' in S' and time t and t' in S and S' respectively. For simplicity, x and x' axes coincide. At the beginning ($t = 0$), S and S' coincide. After time t, S' would have moved through a distance vt. The Galilean transformations are given by the set of relations.

$$x' = x - vt \tag{6.1}$$

$$y' = y \tag{6.2}$$

$$z' = z \tag{6.3}$$

$$t' = t \tag{6.4}$$

In Galilean relativity time is absolute.

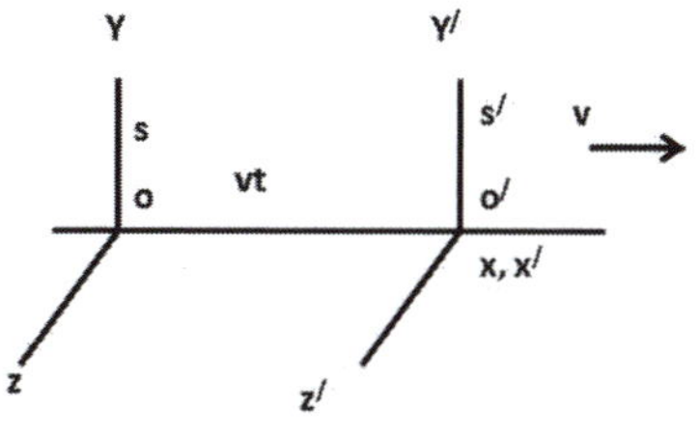

Fig. 6.1 Reference frames S and S'

Transformation of velocities
Differentiating (6.1) with respect to time and noting that $t' = t$ and v is constant

$$\frac{\mathrm{d}x'}{\mathrm{d}t} = \left(\frac{\mathrm{d}x}{\mathrm{d}t}\right) - v \tag{6.5}$$

$$U' = U - v$$

or

$$V = U' + v \tag{6.6}$$

Invariance of Newton's second law of motion

In S, force is given by

$$F = ma = \frac{m\mathrm{d}^2x}{\mathrm{d}t^2} = m\frac{\mathrm{d}}{\mathrm{d}t}\left(\frac{\mathrm{d}x}{\mathrm{d}t}\right) = m\frac{\mathrm{d}u}{\mathrm{d}t}$$

In S', force is given by

$$F' = \frac{m\mathrm{d}^2x}{\mathrm{d}t^2} = \frac{m\mathrm{d}^2x}{\mathrm{d}t^2} = \frac{m\mathrm{d}^2x}{\mathrm{d}t^2} - 0 = F \tag{6.7}$$

Galilean Relativity fails for Electromagnetism as evidenced by the negative result of Michelson–Morley experiment to measure earth's velocity in the hypothetical medium of ether.

Einstein took the view that the principle of Relative is correct but time is not absolute but only relative.

Postulates of special Theory of relativity

(1) The laws of physics apply equally well in all inertial frames of reference, that is no preferred system exists (the principle of relativity)
(2) The speed of light in free space has the same value $c(= 3 \times 10^8\ \mathrm{ms}^{-1})$ in all inertial frames (the principle of constancy of light)

Lorentz Transformations

$$x' = \gamma(x - vt) \tag{6.8}$$

$$y' = y \tag{6.9}$$

$$z' = z \tag{6.10}$$

$$t' = \gamma\left(t - \frac{vx}{c^2}\right) \tag{6.11}$$

Inverse transformations

$$x = \gamma(x' + vt') \tag{6.12}$$

$$y = y' \tag{6.13}$$

$$z = z' \tag{6.14}$$

$$t = \gamma\left(t' + \frac{vx'}{c^2}\right) \tag{6.15}$$

with

$$\gamma = \frac{1}{\sqrt{(1-\beta^2)}} = \frac{1}{\sqrt{(1-v^2/c^2)}} \tag{6.16}$$

and

$$\beta = \frac{v}{c} \tag{6.17}$$

Transformation matrix

The Lorentz transformations (6.8), (6.9), (6.10), and (6.11) can be condensed in the matrix form

$$X' = \Lambda X \tag{6.18}$$

$$\text{where } X = \begin{bmatrix} x_1 \\ x_2 \\ x_3 \\ x_4 \end{bmatrix} \quad \text{and} \quad X' = \begin{bmatrix} x_1' \\ x_2' \\ x_3' \\ x_4' \end{bmatrix} \tag{6.19}$$

are the column vectors with components

$$x_1 = x, x_2 = y, x_3 = z, x_4 = \tau = ict \tag{6.20}$$

$$x_1' = x', x_2' = y', x_3' = z', x_4' = \tau' = ict' \tag{6.21}$$

with $i = \sqrt{-1}$, and Λ is an orthogonal matrix

$$\Lambda = \begin{bmatrix} \gamma & 0 & 0 & i\beta\gamma \\ 0 & 1 & 0 & 0 \\ 0 & 0 & 1 & 0 \\ -i\beta\gamma & 0 & 0 & \gamma \end{bmatrix} \tag{6.22}$$

Inverse transformations

The inverse transformations (6.12), (6.13), (6.14), and (6.15) are immediately written with the aid of inverse matrix

$$\Lambda^{-1} = \tilde{\Lambda}$$

$$\Lambda^{-1} = \begin{bmatrix} \gamma & 0 & 0 & -i\beta\gamma \\ 0 & 1 & 0 & 0 \\ 0 & 0 & 1 & 0 \\ i\beta\gamma & 0 & 0 & \gamma \end{bmatrix} \tag{6.23}$$

Four vectors

$$\text{If } s = \sqrt{\Sigma_\mu x_\mu^2} = \text{invariant}, \;\; \mu = 1, 2, 3, 4 \tag{6.24}$$

under Lorentz transformation, then s is said to be a four vector, s can be positive or negative or zero. Examples of Four vectors are

$$\boldsymbol{X} = (x_1, x_2, x_3, ict) \tag{6.25}$$

$$c\boldsymbol{P} = (cP_x, cP_y, cP_z, iE) \tag{6.26}$$

In (6.25) the first three space components of $\boldsymbol{X}$ define the ordinary three-dimensional position vector $\boldsymbol{x}$ and the fourth, a time component ict. The four-momentum in (6.26) has the first three components of ordinary momentum and E is the total energy of the particle. The four-vectors have the properties which are similar to those of ordinary vectors. Thus, the scalar product of two four-vectors, $\boldsymbol{A.B} = A_1B_1 + A_2B_2 + A_3B_3 + A_4B_4$

Consequences of Lorentz transformations

Time Dilation

$$\Delta t' = \gamma \Delta t \tag{6.27}$$

Rule: Every clock appears to go at its fastest rate when it is at rest relative to the observer. If it moves relative to the observer with velocity ν, its rates appears slowed down by the factor $\sqrt{1 - \nu^2/c^2}$. No distinction need be made between the stationary and moving frame. Each observer will think that the other observer's clock has slowed down. What matters is the only the relative motion.

The Lorentz contraction

$$l' = l\sqrt{1 - \beta^2} \tag{6.28}$$

Rule: Every rigid body appears to be longest when at rest relative to the observer. When it is moving relative to the observer it appears contracted in the direction of its relative motion by the factor $\sqrt{1 - v^2/c^2}$, while its dimensions perpendicular to the direction of motion are unaffected.

Addition of velocities

$$\beta = \frac{\beta_1 + \beta_2}{1 + \beta_1 \beta_2} \tag{6.29}$$

Mass, energy and momentum

$$m = m_0 \gamma = \frac{m_0}{\sqrt{1 - v^2/c^2}} \tag{6.30}$$

where m is the effective mass and m_0 is the rest mass.

The rest mass energy

$$E_0 = m_0 c^2 \tag{6.31}$$

The total energy of a free particle is

$$E = T + m_0 c^2 = mc^2 = m_0 \gamma c^2 \tag{6.32}$$

where T is the kinetic energy.

The momentum p is given by

$$P = mv = m_0 \gamma \beta c \tag{6.33}$$

$$E^2 = c^2 p^2 + m_0^2 c^4 \tag{6.34}$$

$$cp = \beta E \tag{6.35}$$

$$c^2 p^2 = T^2 + 2T\, m_0 c^2 \tag{6.36}$$

Lorentz transformations of momentum and energy

$$cp_{x'} = \gamma (cp_x - \beta E) \tag{6.37}$$

$$cp_{y'} = cp_y \tag{6.38}$$

$$cp_z = cp_z \tag{6.39}$$

$$E' = \gamma (E - \beta cp_x) \tag{6.40}$$

Inverse transformations

$$cp_x = \gamma(cp'_x + \beta E') \tag{6.41}$$

$$cp_y = cp'_y \tag{6.42}$$

$$cp_z = cp'_z \tag{6.43}$$

$$E = \gamma(E' - \beta cp'_{x*}) \tag{6.44}$$

$$E'^2 - c^2 p'^2 = E^2 - c^2 p^2 = m_0^2 = \text{Invariant} \tag{6.45}$$

$$\gamma' = \gamma\gamma_0(1 - \beta\beta_0 \cos\theta_0) \tag{6.46}$$

$$\gamma_0 = \gamma\gamma'(1 + \beta\beta^* \cos\theta^*) \tag{6.47}$$

where zeros refer to the particle's velocity, Lorentz factor and the angle in the S-system while primes refer to the corresponding quantities in the S' system.

Transformation of angles

$$\tan\theta' = \frac{\sin\theta}{\gamma\left(\cos\theta - \frac{\beta}{\beta_0}\right)} \tag{6.48}$$

$$\tan\theta = \frac{\sin\theta'}{\gamma_c(\cos\theta' - \beta/\beta')} \tag{6.49}$$

Optical Doppler effect

$$\nu' = \gamma\nu(1 - \beta\cos\theta) \tag{6.50}$$

$$\nu = \gamma\nu'(1 + \beta\cos\theta^*) \tag{6.51}$$

where ν is the frequency in the S-$system$ and ν' is the frequency in the S'-$system$, θ and θ' are the corresponding angles, β is the source velocity and γ is the corresponding Lorentz factor.

Threshold for particle production

Consider the reaction

$$m_1 + m_2 \to m_3 + m_4 + M \tag{6.52}$$

$$T(\text{threshold}) = \frac{1}{2m_2}[(m_3 + m_4 + M)^2 - (m_1 + m_2)^2]$$

$$T(\text{threshold}) = \frac{(\text{Sum of final masses})^2 - (\text{sum of initial masses})^2}{2 \times \text{mass of target particle}} \tag{6.53}$$

6.2 Problems

6.2.1 Lorentz Transformations

6.1 In the inertial system S, an event is observed to take place at point A on the x-axis and 10^{-6} S later another event takes place at point B, 900 m further down. Find the magnitude and direction of the velocity of S' with respect to S in which these two events appear simultaneous.

6.2 Show that the Lorentz-transformations connecting the S' and S systems may be expressed as
$x_1{}' = x_1 \cosh\alpha - ct \sinh\alpha$
$x_2{}' = x_2$
$x_3{}' = x_3$
$t' = t \cosh\alpha - (x_1 t \sinh\alpha)/c$
where $\tanh\alpha = v/c$. Also show that the Lorentz transformations correspond to a rotation through an angle $i\alpha$ in four-dimensional space.

6.3 A pion moving along x-axis with $\beta = 0.8$ in the lab system decays by emitting a muon with $\beta' = 0.268$ along the incident direction (x'-*axis*) in the rest system of pion. Find the velocity of the muon (magnitude and direction) in the lab system.

6.4 In Problem 6.3, the muon is emitted along the y'-*axis*. Find the velocity of muon in the lab frame

6.5 In Problem 6.3, the muon is emitted along the positive y-axis (i.e. perpendicular to the incidental direction of pion in the lab frame). Find the speed of muon in the lab frame and the direction of emission in the rest frame of pion. Assume $\beta_c = 0.2$

6.6 Show that Maxwell's equations for the propagation of electromagnetic waves are Lorentz invariant.

6.7 A neutral K meson decays in flight via $K^0 \rightarrow \pi^+\pi^-$. If the negative pion is produced at rest, calculate the kinetic energy of the positive pion.
[Mass of K^0 is 498 MeV/c^2; that of $\pi^\pm$ is 140 MeV/c^2]

6.8 A pion travelling with speed $v = |\mathbf{v}|$ in the laboratory decays via $\pi \rightarrow \mu + \nu$. If the neutrino emerges at right angles to $\mathbf{v}$, find an expression for the angle θ at which the muon emerges.

6.9 Determine the speed of the Lorentz transformation in the x-direction for which the velocity in the frame S of a particle is $\mathbf{u} = (c/\sqrt{2}, c/\sqrt{2})$ and the velocity in frame S' is seen as
$\boldsymbol{u}' = (-c/\sqrt{2}, c/\sqrt{2})$.

6.10 A particle decays into two particles of mass m_1 and m_2 with a release of energy Q. Calculate relativistically the energy carried by the decay products in the rest frame of the decaying particle.

6.11 A π-meson with a kinetic energy of 140 MeV decays in flight into μ-meson and a neutrino. Calculate the maximum energy which **(a)** the μ-meson **(b)** the neutrino may have in the Laboratory system (Mass of π-meson = 140 MeV/c^2, mass of μ-meson = 106 MeV/c, mass of neutrino = 0)
[University of Bristol 1968]

6.12 A positron of energy E_+ , and momentum p_+ and an electron, energy E_-, momentum p_- are produced in a pair creation process
(a) What is the velocity of their CMS?
(b) What is the energy of either particle in the CMS?

6.13 A particle of mass m collides elastically with another identical particle at rest. Show that for a relativistic collision
$\tan\theta \tan\varphi = 2/(\gamma + 1)$
where θ, φ are the angles of the out-going particles with respect to the direction of the incident particle and γ is the Lorentz factor before the collision. Also, show that $\theta + \varphi \leq \pi/2$ where the equal sign is valid in the classical limit

6.14 A K^+ meson at rest decays into a π^+ meson and π^0 meson. The π^+ meson decays into a μ meson and a neutrino. What is the maximum energy of the final μ meson? What is its minimum energy?
($m_K = 493.5\,\text{MeV/c}^2$, $m_{\pi^+} = 139.5\,\text{MeV/c}^2$, $m_{\pi^0} = 135\,\text{MeV/c}^2$, $m_\mu = 106\,\text{MeV/c}^2$, $m_v = 0$)

6.15 An unstable particle decays in its flight into three charged pions (mass 140 MeV/c^2). The tracks recorded are shown in Fig. 6.2, the event being coplanar. The kinetic energies and the emission angles are
$T_1 = 190$ MeV, $T_2 = 321$ MeV, $T_3 = 58$ MeV
$\theta_1 = 22.4°$, $\theta_2 = 12.25°$
Estimate the mass of the primary particle and identify it. In what direction was it moving?

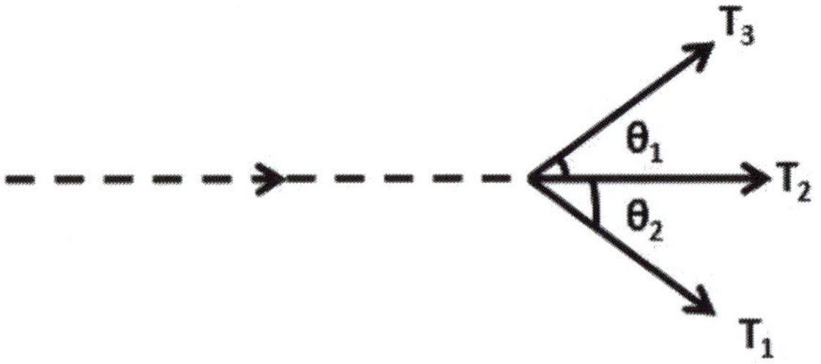

Fig. 6.2 Decay of a kaon into three poins

6.2.2 *Length, Time, Velocity*

6.16 If a rod travels with a speed $v = 0.8\,c$ along its length, how much does it shrink?

6.17 If a rod is to appear shrunk by half along its direction of motion, at what speed should it travel?

6.18 Assuming that the rest radius of earth is 6,400 km and its orbital speed about the sun is 30 km^{-1}, how much does earth's diameter appear to be shortened to an observer on the sun, due to earth's orbital motion?

6.19 The mean life-time of muons at rest is found to be about 2.2×10^{-6} s, while the mean life time in a burst of cosmic rays is found to be 1.5×10^{-5} s. What is the speed of these cosmic ray muons?

6.20 A beam of muons travels with a speed of $v = 0.6\,c$. Their mean life-time as observed in the laboratory is found to be 2.9×10^{-6} s. What is the mean life-time of muons when they decay at rest?

6.21 (a) If the mean proper life-time of muons is 2.2×10^{-6} s, what average distance would they travel in vacuum before decaying in the reference frame in which its velocity is measured as $0.6\,c$?
(b) Compare this distance with the distance the muon sees while travelling.

6.22 With what constant velocity must a person travel from the centre to the edge of our galaxy so that the trip may last 40 years (proper time)? Assume that the radius of the galaxy is 3×10^4 light years?

6.23 A pion is produced in a high energy collision of a primary cosmic ray particle in the earth's atmosphere 1 km above the sea level. It proceeds vertically down at a speed of $0.99\,c$ and decays in its rest frame 2.5×10^{-8} s after its production. At what altitude above the sea level is it observed from the ground to decay?

6.24 One cosmic particle approaches the earth along its axis with a velocity of $0.9\,c$ toward the North Pole and another one with a velocity of $0.5\,c$ toward the South Pole. Find the relative speed of approach of one particle with respect to another.

6.25 A 100 MeV electron moves along the axis of an evacuated tube of length 4 m fixed to the laboratory frame. What length of the tube would be measured by the observer moving with the electron?

6.26 A man has a mass of 100 kg on the earth. When he is in the space-craft, an observer from the earth registers his mass as 101 kg. Determine the speed of the space-craft.

6.27 At the time a space ship moving with speed $\nu = 0.5\,c$ passes a space station located near Mars, a radio signal is sent from the station to earth. This signal is received on earth 1,125 s later.
How long does the spaceship take to reach the earth according to the observers on earth?

6.28 In Problem 6.27, what is the duration according to the crew of the spaceship?

6.29 A spaceship is moving away from earth with speed $v = 0.6\,c$. When the ship is at a distance $d = 5 \times 10^8$ km from earth. A radio signal is sent to the ship by the observers on earth.
How long does the signal take to reach the ship as measured by the scientist on earth?

6.30 In Problem 6.29, how long does the signal take to reach the ship as measured by the crew of the spaceship?

6.31 If the mean track length of 100 MeV π mesons is 4.88 m up to the point of decay, calculate their mean lifetime.

[University of Durham 1962]

6.32 A beam of π^+ mesons of energy 1 GeV has an intensity of 10^6 particles per sec at the beginning of a 10 m flight path. Calculate the intensity of the neutrino flux at the end of the flight path (mass of π meson = 139 MeV/c^2, lifetime = 2.56×10^{-8} s)

[University of Durham 1961]

6.33 A π^+ meson at rest decays into a μ^+ meson and a neutrino in 2.5×10^{-8} s. Assuming that the π^+ meson has kinetic energy equal to its rest energy. What distance would the meson travel before decaying as seen by an observer at rest?

6.34 Beams of high-energy muon neutrinos can be obtained by generating intense beams of π^+ mesons and allowing them to decay while in flight. What fraction of the π^+ mesons in a beam of momentum 200 GeV/c will decay while travelling a distance of 300 m?
At the end of the decay path (an evacuated tunnel) the beam is a mixture of π^+-mesons, muons and neutrinos. What distinguishes these particles in their interactions with matter, and how is a neutrino beam free of contamination by π-mesons and muons obtained? [π^+-meson mean-life: 2.6×10^{-8} s]

[Osmania University]

6.35 A beam of 140 MeV kinetic energy π^+ mesons through three counters A, B, C spaced 10 m apart. If 1,000 pions pass through counter A and 470 in B. (a) how many pions are expected to be recorded in C? (b) Find the mean life time of pions (Take mass of pion as 140 MeV/c^2)

6.36 A moving object heads toward a stationary one with a velocity αc. At what velocity βc would an observer have to move so that in his frame of reference the objects would have equal and opposite velocities?

6.37 A beam of identical unstable particles flying at a speed βc is sent through two counters separated by a distance L. It is observed that N_1 particles are recorded at the first counter and N_2 at the second counter, the reduction being solely due to the decay of the particles in flight.
(i) Show that the lifetime of the particles at rest is given by

$$\tau = \frac{L}{\ln\left(\frac{N_1}{N_2}\right)(\sqrt{\gamma^2 - 1}\,c}$$

where the Lorentz factor γ is defined as usual
(ii) Hence determine the lifetime of muons at rest, knowing that when travelling at a speed $c\sqrt{8}/3$ through the apparatus described above (with $L = 200\,\text{m}$) N_1 and N_2 were measured to be 10,000 and 8,983, respectively. [adapted from University of London, Royal Holloway and Bedford New College]

6.38 A particle X at rest is a sphere of rest-mass m and radius r and has a proper lifetime τ. If the particle is moving with speed $\frac{\sqrt{3}}{2}c$ with respect to the lab frame (c is the speed of light):

(a) Determine the total energy of the particle in the lab frame
(b) The average distance the particle travels in the lab before decaying
(c) Sketch the shape and dimensions of the particle when viewed perpendicular to its motion in the lab frame, include an arrow to indicate its direction of motion on your sketch.

[adapted from the University of London Royal Holloway and Bedford New College 2006]

6.39 The Lorentz velocity transformation is $v' = \frac{v-u}{1-uv/c^2}$, where v' and v are the velocities of an object parallel to u as measured in two inertial frames with relative velocity u. Show that a photon moving at c, the speed of light will have the same speed in all frames of reference.

6.2.3 Mass, Momentum, Energy

6.40 The mean life-time of muons at rest is 2.2×10^{-6} s. The observed mean life-time of muons as measured in the laboratory is 6.6×10^{-6} s. Find

(a) The effective mass of a muon at this speed when its rest mass is $207\,m_e$
(b) its kinetic energy (c) its momentum

6.41 Calculate the energy that can be obtained from complete annihilation of 1 g of mass.

6.42 What is the speed of a proton whose kinetic energy equals its rest energy? Does the result depend on the mass of proton?

6.43 What is the speed of a particle when accelerated to 1.0 GeV when the particle is (a) proton (b) electron

6.44 (a) Calculate the energy needed to break up the ^{12}C nucleus into its constituents. The rest masses in amu are:
^{12}C 12.000000; p 1.007825; n 1.008665; α 4.002603
(b) If ^{12}C nucleus is to break up into 3 alphas. Calculate the energy that is released.

(c) If the alphas are to further break into neutrons and protons, then show that the overall energy needed is identical with the results in (a)

6.45 The kinetic energy and the momentum of a particle deduced from measurements on its track in nuclear photographic emulsions are 250 MeV and 368 MeV/c, respectively. Determine the mass of the particle in terms of electron mass and identify it.

[University of Durham]

6.46 What is the rest mass energy of an electron ($m_e = 9.1 \times 10^{-31}$ kg)?

6.47 What potential difference is required to accelerate an electron from rest to velocity $0.6\,c$?

6.48 At what velocity does the relativistic kinetic energy differ from the classical energy by
(a) 1% (b) 10%?

6.49 Prove that if $v/c \ll 1$, the kinetic energy of a particle will be much less than its rest energy. Further show that the relativistic expression reduces to the classical one for small velocities.

6.50 Find the effective mass of a photon for
(a) $\lambda = 5{,}000$Å (visible region) (b) $\lambda = 1$Å (X-ray region)

6.51 Show that 1 amu = 931.5 MeV/c^2

6.52 Estimate the energy that is released in the explosion of a fission bomb containing 5.0 kg of fissionable material

6.53 A proton moving with a velocity βc collides with a stationary electron of mass m and knocks it off at an angle θ with the incident direction. Show that the energy imparted to the electron is approximately
$T = 2mc^2\beta^2 \cos^2\theta/(1 - \beta^2 \cos^2\theta)$

6.54 A positive pion ($m_\pi = 273\,m_e$) decays into a muon ($m_\mu = 207\,m_e$) and a neutrino ($m_\nu = 0$) at rest. Calculate the energy carried by the muon and neutrino, given that $m_e c^2 = 0.511$ MeV

6.55 A body of rest mass m travelling initially at a speed of $\beta = 0.6$ makes a completely inelastic collision with an identical body initially at rest. Find (a) the speed of the resulting body (b) its rest mass in terms of m.

6.56 A neutral particle is observed to decay into a kaon and a pion. They are produced in the opposite direction, each of them with momentum 861 MeV/c. Calculate the mass of the neutral particle and identify it. (Mass of kaon is 494 MeV/c^2, Mass of pion is 140 MeV/c^2).

6.57 A pion at rest decays via $\pi \to \mu + \nu$. Find the speed of the muon in terms of the masses involved.

6.58 A particle A decays at rest via $A \rightarrow B + C$. Find the total energy of B in terms of the masses of A, B and C.

6.59 Calculate the maximum energy of the positron emitted in kaon decay at rest $K^+ \rightarrow e^+ + \pi^0 + \gamma_e$.

6.60 Consider a symmetric elastic collision between a particle of mass m and kinetic energy T and a particle of the same mass at rest. Relativistically, show that the cosine of the angle between the two particles after the collision is $T/(T + 4\,\text{mc}^2)$

6.61 An electron has kinetic energy equal to its rest energy. Show that the energy of a photon which has the same momentum as this electron is given by $E_\gamma = \sqrt{3}E_0$, where $E_0 = m_e c^2$

6.62 Consider the decay of muon at rest. If the energy released is divided equally among the final leptons, then show that the angle between paths of any two leptons is approximately 120° (neglect the mass of leptons compared to the mass of muon mass).

6.63 If a proton of 10^9 eV collides with a stationary electron and knock it off at 3° with respect to the incident direction, what is the energy acquired by the electron?

[Osmania University 1963]

6.64 Calculate (a) the mass of the pion in terms of the mass of the electron, given that the kinetic energy of the muon from the pion decay at rest is 4.12 MeV and (b) the maximum energy of electron (in MeV) from the decay of muon at rest (mass of muon is 206.9 m_e). The mass of the electron is equivalent to 0.511 MeV)

[University of Durham 1961]

6.65 Antiprotons are captured at rest in deuterium giving rise to the reaction.
$p^- + d \rightarrow n + \pi^0$
Find the total energy of the π^0. The rest energies for p^-, d, n, π^0 are 938.2, 1875.5, 939.5 and 135.0 MeV respectively.

6.66 As a result of a nuclear interaction a K^{*+} particle is created which decays to a K meson and a π^- meson with rest masses equal to 966 m_e and 273 m_e respectively. From the curvature of the resulting tracks in a magnetic field, it is concluded that the momentum of the secondary K and π mesons are 394 MeV/c and 254 MeV/c respectively, their initial directions of motion being inclined to one another at 154°. Calculate the rest mass of the K^* particle

[Bristol 1964, 1966]

6.67 A proton of kinetic energy 940 MeV makes an elastic collision with a stationary proton in such a way that after collision, the protons are travelling at equal angles on either side of the incident proton. Calculate the angle between the directions of motion of the protons.

[Liverpool 1963]

6.68 A proton of momentum p large compared with its rest mass M, collides with a proton inside a target nucleus with Fermi momentum p_f. Find the available kinetic energy in the collision, as compared with that for a free-nucleon target, when p and p_f are (a) parallel (b) anti parallel (c) orthogonal.

6.69 An antiproton of momentum 5 GeV/c suffers a scattering. The angles of the recoil proton and scattered antiproton are found to be 82° and 2°30′ with respect to the incident direction. Show that the event is consistent with an elastic scattering of an antiproton with a free proton.

6.70 Show that if E is the ultra-relativistic laboratory energy of electrons incident on a nucleus of mass M, the nucleus will acquire kinetic energy
$E_N = (E^2/Mc^2)(1 - \cos\theta)/(1 + E(1 - \cos\theta)/Mc^2)$
where θ is the scattering angle.

6.71 A particle of mass $M \gg m_e$ scatters elastically from an electron. If the incident particle's momentum is p and the scattered electron's relativistic energy is E and ϕ is the angle the electron makes with the incident particle, show that
$M = P[\{[E + m_e]/[E - m_e]\} \cos^2\phi - 1]]^{1/2}$

6.72 A neutrino of energy 2 GeV collides with an electron. Calculate the maximum momentum transfer to the electron.

6.73 A particle of mass m_1 collides elastically target particle of mass m_2 at relativistic energy. Show that the maximum angle at which m_1 is scattered in the lab system is dependent only on the masses of particles provided $m_1 > m_2$

6.74 Show that if energy $\nu(> m_e c^2)$ and momentum q are transferred to a free stationary electron the four-momentum transfer squared is given by $q^2 = -2m_e\nu$

6.75 A photon of energy E travelling in the $+x$ direction collides elastically with an electron of mass m moving in the opposite direction. After the collision, the photon travels back along the $-x$ direction with the same energy E.
(a) Use the conservation of energy and momentum to demonstrate that the initial and final electron momenta are equal and opposite and of magnitude E/c.
(b) Hence show that the electron speed is given by
$v/c = (1 + (m_c{}^2/E)^2)^{-1/2}$

[adapted from the University of Manchester 2008]

6.2.4 Invariance Principle

6.76 Use the invariance of scalar product of two four-vectors under Lorentz transformation to obtain the expression for Compton scattering wavelength shift.

6.77 Show that for a high energy electron scattering at an angle θ, the value of the squared four-momentum transfer is given approximately by $Q^2 = 2E^2$

$(1 - \cos\theta)/c^2$, where E is the total initial electron's energy in the lab system. State when this approximation is justified.

6.78 A neutral unstable particle decays into π^+ and π^-, each of which has a momentum 530 MeV/c. The angle between the two pions is 90°. Calculate the mass of the unstable particle.

6.79 If a particle of mass M decays in flight into m_1 and another m_2; m_1 has momentum p_1 and total energy E_1, where as m_2 has momentum p_2 and total energy E_2. p_1 and p_2 make an angle θ. Show that
$E_1E_2 - p_1p_2\cos\theta = \text{invariant} = \frac{1}{2}[M^2 - {m_1}^2 - {m_2}^2]$

6.80 The Mandelstam variables s, t, and u are defined for the reaction $A + B \rightarrow C + D$, by
$s = (P_{\mathrm{A}} + P_{\mathrm{B}})^2/c^2, t = (P_{\mathrm{A}} - P_{\mathrm{C}})^2/c^2, u = (P_{\mathrm{A}} - P_{\mathrm{D}})^2/c^2$
where P_{A}, P_{B}, P_{C}, P_{D} are the relevant energy-momentum four vectors. Show that
$s + t + u = \sum \mathrm{m}_j{}^2 (j = A, B, C, D)$

6.81 In Problem 6.80 show for the elastic scattering $t = -2p^2(1 - \cos\theta)/c^2$ where $p = |\mathbf{p}|.\mathbf{p}$ is the center of mass momentum of particle and θ is its scattering angle in the CMS.

6.82 A neutral pion undergoes radioactive decay into two γ-rays. Obtain the expression for the laboratory angle between the direction of the γ-rays, and find the minimum value for the angle when the pion energy is 10 GeV ($m_\pi = 0.14$ GeV)

[University of Bristol 1965]

6.83 A bubble chamber event was identified in the reaction
$p^- + p \rightarrow \pi^+ + \pi^- + \omega^0$
The total energy available was 2.29 GeV while the kinetic energy of the residual particles was 1.22 GeV. What is the rest energy of ω^0 in MeV?

6.84 A particle of rest mass m_1 and velocity v_1 collides with a particle of mass m_2 at rest after which the two particles coalesce. Show that the mass M and velocity v of the composite particle are related by $M^2 = {m_1}^2 + {m_2}^2 + 2m_1m_2/\sqrt{1 - v^2/c^2}$

6.85 Show that for the decay in flight of a Λ-hyperon into a proton and a pion with Laboratory momenta P_p and P_π respectively, the Q value can be calculated from
$Q = ({m_p}^2 + {m_\pi}^2 + 2E_pE_\pi - 2P_pP_\pi\cos\theta)^{1/2} - (m_p + m_\pi)$
where θ is the angle between P_p and P_π in the Laboratory system and E is the total relativistic energy

[University of Dublin 1967]

6.86 Two particles are moving with relativistic velocities in directions at right angles, they have momenta p_1 and p_2 and total energies E_1 and E_2. If they are the sole products of disintegration of a heavier object, what was the rest mass, velocity and direction of motion?

[University of Bristol 1965]

6.87 A V-type of event is observed in a bubble chamber. The curvature measurements on the two tracks show that their momenta are $p_+ = 1.670\,\text{GeV}/c$ and $p_- = 0.408\,\text{GeV/c}$. The angle contained between the two tracks is $\theta = 15°$. It is obviously due to the decay of a neutral unstable particle. It is suspected that it is due to the decay **(a)** $K^0 \rightarrow \pi^+ + \pi^-$ or **(b)** $\Lambda \rightarrow p + \pi^-$. Identify the neutral particle.

6.88 Derive the formula in Problem 6.103 using the invariance of $(\Sigma E)^2 - |\Sigma \mathbf{p}|^2$

6.89 Calculate the maximum four momentum transfer to proton in the decay of neutron at rest.

6.2.5 *Transformation of Angles and Doppler Effect*

6.90 Find the Doppler shift in wavelength of H line at 6,563 Å emitted by a star receding with a relative velocity of $3 \times 10^6\,\text{ms}^{-1}$.

6.91 Certain radiation of a distant nebula appears to have a wavelength 656 nm instead of 434 nm as observed in the laboratory. (a) if the nebula is moving in the line of sight of the observer, what is its speed? (b) Is the nebula approaching or receding?

6.92 Show that for slow speeds, the Doppler shift can be approximated as
$\Delta\lambda/\lambda = v/c$
where $\Delta\lambda$ is the change in wavelength.

6.93 A physicist was arrested for going over the railway level crossing on a motorcycle when the lights were red. When he was produced before the magistrate the physicist declared that he was not guilty as red lights ($\lambda = 670\,\text{nm}$) appeared green ($\lambda = 525\,\text{nm}$) due to Doppler Effect. At what speed he was travelling for the explanation to be valid? Do you think such a speed is feasible?

6.94 A spaceship is receding from earth at a speed of $0.21\,c$. A light from the spaceship appears as yellow ($\lambda = 589.3\,\text{nm}$) to an observer on earth. What would be its color as seen by the passenger of the spaceship?

6.95 Find the wavelength shift in the Doppler effect for the sodium line 589 nm emitted by a source moving in a circle with a constant speed $0.05\,c$ observed by a person fixed at the center of the circle.

6.96 A neutrino of energy E_0 and negligible mass collides with a stationary electron. Find an expression for the laboratory angle of emission of the electron in terms of its recoil energy E and calculate its value when $E_0 = 2\,\text{GeV}$ and $E = 0.5\,\text{GeV}$

6.97 Assume the decay $K^0 \rightarrow \pi^+ + \pi^-$. Calculate the mass of the primary particle if the momentum of each of the secondary particles is 3×10^8 eV and the angle between the tracks is 70°

[University of Durham 1960]

6.98 Neutral pions of fixed energy decay in flight into two γ-rays. Show that the velocity of pion is given by
$\beta = (E_{\text{max}} - E_{\text{min}})/(E_{\text{max}} + E_{\text{min}})$
where E is the γ-ray energy in the laboratory

6.99 In Problem 6.98 show that the rest mass energy of π^0 is given by $mc^2 = 2(E_{\text{max}} E_{\text{min}})^{1/2}$

6.100 In Problem 6.98 show that the energy distribution of γ-rays in the laboratory is uniform under the assumption that γ-rays are emitted isotropically in the rest system of π^0

6.101 In Problem 6.98 show that the angular distribution of γ-rays in the laboratory is given by
$I(\theta) = 1/4\pi\gamma^2(1 - \beta\cos\theta)^2$

6.102 In Problem 6.98 show that the locus of the tip of the momentum vector is an ellipse

6.103 In Problem 6.98 show that in a given decay the angle ϕ between two γ-rays is given by
$\sin(\phi/2) = mc^2/2(E_1 E_2)^{1/2}$

6.104 In Problem 6.98 show that the minimum angle between the two γ-rays is given by
$\phi_{\text{min}} = 2mc^2/E_\pi$

6.105 In Problem 6.98 find an expression for the disparity D (the ratio of energies) of the γ-rays and show that $D > 3$ in half the decays and $D > 7$ in one quarter of them

6.106 In the interaction $\pi^- + p \rightarrow K^*(890) + Y_0{}^*(1,800)$ at pion momentum 10 GeV/c , K^* is produced at an angle θ in the lab system. Calculate the maximum value θ_m, given $m_\pi = 0.140\,\text{GeV}/\text{c}^2$ and $m_p = 0.940\,\text{GeV}/\text{c}^2$

6.107 A particle of mass m_1 travelling with a velocity $v = \beta c$ collides elastically with the particle m_2 at rest. The scattering angles of m_1 in the LS and CMS are θ and θ^*. Show that
(a) $\gamma_c = (\gamma + \nu)/(1 + 2\gamma\nu + \nu^2)^{1/2}$
(b) $\gamma^* = (\gamma + 1/\gamma)/\sqrt{(1 + 2\gamma/\nu + 1/\nu^2)}$
(c) $\tan\theta = \sin\theta^*/\gamma_c(\cos\theta^* + \beta_c/\beta^*)$
(d) $\tan\theta^* = \sin\theta/\gamma_c(\cos\theta - \beta_c/\beta^*)$
where β_c is the CMS velocity, $\beta^* c$ is the velocity of m_1 in CMS, $\gamma_c = (1 - \beta_c{}^2)^{-1/2}$, $\gamma^* = (1 - \beta^{*2})^{-1/2}$, $\nu = m_2/m_1$

6.108 A linear accelerator produces a beam of excited carbon atoms of kinetic energy 120 MeV. Light emitted on de-excitation is viewed at right angles to the beam and has a wavelength λ'. If λ is the wavelength emitted by a stationary atom, what is the value of $(\lambda' - \lambda)/\lambda$? (Take the rest energies of both protons and neutrons to be 10^9 eV)

[University of Manchester 1970]

6.109 A certain spectral line of a star has natural frequency of 5×10^{17} c/s. If the star is approaching the earth at 300 km/s, what would be the fractional change of frequency?

6.110 A neutral pion (mass 135 MeV/c^2) travelling with speed $\beta = v/c = 0.8$ decays into two photons at right angle to the line of flight. Find the angle between the two photons as observed in the lab system.

6.111 An observer O sights light coming to him by an object X at 45° to its path as in the diagram. If the corresponding angle of emission of light in the frame of reference of the object is 60°, calculate the velocity of the object.

Fig. 6.3 Aberration of light

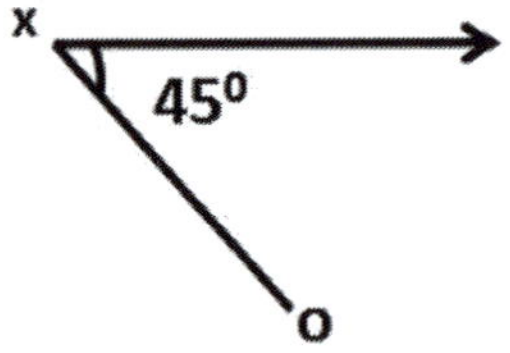

6.112 In an inertial frame S a rod of proper length L is at rest and at an angle θ with respect to the x-axis with relativistic speed relative to S

(a) Show that the product $\tan\theta.\ L_x$ is independent of which frame it is evaluated in:
$\tan\theta.L_x = \tan\theta'.L_x{}'$
where L_x and $L_x{}'$ are the projections of the rod length onto the x-axis in frames S and S', respectively.

(b) In frame S the rod is at an angle $\theta = 30°$ knowing that an observer in frame S' measures the rod to be at an angle $\theta' = 45°$ with respect to the x-axis, determine the speed at which the rod is moving with respect to the observer.

[adapted from University of London 2004 Royal Holloway]

6.2.6 *Threshold of Particle Production*

6.113 Show that the threshold energy for the production of a proton–antiproton pair in the collision of a proton with hydrogen target ($P + P \to P + P + P + P^-$) is 6 Mc^2, where M is the mass of a proton or antiproton.

6.114 A positron–electron pair production can occur in the interaction of a gamma ray with electron, via $\gamma + e^- \rightarrow e^- + e^+ + e^-$. Determine the threshold.

6.115 Find the threshold energy for the pion production in the reaction $N + N \rightarrow N + N + \pi$, given
$M_N c^2 = 940\,\text{MeV}$ and $m_\pi c^2 = 140\,\text{MeV}$

6.116 Show that a Fermi energy of 25 MeV lowers the threshold incident kinetic energy for antiproton production by proton incident on nucleus to 4.3 GeV.

6.117 Find the threshold energy of the reaction
$\gamma + p \rightarrow K^{+^*} + \Lambda$
The laboratory proton is at rest. The following rest energies may be assumed, for proton 940 MeV for K^* 890 MeV, for Λ 1,110 MeV

6.118 Find the threshold energy for the production of two pions by a pion incident on a hydrogen target. Assume the rest masses of the pion and proton are 273 m_e and 1,837 m_e, where the rest energy of the electron $m_e c^2 = 0.51\,\text{MeV}$

[University of Manchester 1959]

6.119 Calculate the threshold energy of the following reaction $\pi^- + p \rightarrow K^0 + \Lambda$
The masses for π^- and p, K^0 and Λ are, 140, 938, 498, 1,115 MeV respectively

6.120 Show that the threshold kinetic energy in the Laboratory for the production of n pions in the collision of protons with a hydrogen target is given by
$T = 2nm_\pi(1 + nm_\pi/4m_p)$
where m_π and m_p are respectively the pion and proton masses.

6.121 A gamma ray interacts with a stationary proton and produces a neutral pion according to the scheme
$\gamma + p \rightarrow p + \pi^0$
Calculate the threshold energy given $M_p = 940\,\text{MeV}$ and $M_\pi = 135\,\text{MeV}$

6.122 Calculate the threshold energy of the reaction
$\pi^- + p \rightarrow \Xi^- + K^+ + K^0$
The masses for π^-, p, Ξ^-, K^+, K^0 are respectively 140, 938, 1,321, 494, 498, MeV

[University of Durham 1970]

6.123 Attempts have been made to produce a hypothetical new particle, the W^+, using the reaction $\nu + p \rightarrow p + \mu^- + W^+$ where stationary protons are bombarded with neutrinos. If a neutrino energy of 5 GeV is not high enough for this reaction to proceed, estimate a lower limit for the mass of the W^+ (mass of proton 938 MeV/c^2, mass of muon 106 MeV/c^2, neutrino has zero rest mass)

[University of Manchester 1972]

6.124 For the reaction $p + p \rightarrow p + \Lambda + K^+$ calculate the threshold energy and the invariant mass of the system at threshold energy. The rest energies of the p, Λ and K^+ are respectively 938, 1,115 and 494 MeV

[University of London 1969]

6.125 The Ω^- has been produced in the reaction
$K^- + p \rightarrow K^0 + K^+ + \Omega^-$
What is the minimum momentum of the K^- in the Laboratory for this reaction to proceed assuming that the target proton is at rest in the laboratory?

Assume that a Ω^- is produced in the above reaction with this momentum of K^-. What is the probability that Ω^- will travel 3 cm in the Lab before decaying? You may ignore any likelihood of the Ω^- interacting. The rest energies of K^+, K^-, K^0, P, Ω^- are 494, 494, 498, 938, 1,675 MeV, respectively. Lifetime of $\Omega^- = 1.3 \times 10^{-10}$ s.

[University of Bristol 1967]

6.126 Assuming that the nucleons in the nucleus behave as particles moving independently and contained within a hard-walled box of volume $(4/3)\pi R^3$ where R is the nuclear radius, calculate the maximum Fermi momentum for a proton in ${}_{29}\mathrm{Cu}^{63}$ the nuclear radius being 5.17 fm.

A proton beam of kinetic energy 100 MeV (momentum 570.4 MeV/c) is incident on a target ${}_{29}\mathrm{Cu}^{63}$. Would you expect any pions to be produced by the reaction $p + p \rightarrow d + \pi$ from protons within the nucleus (neglect the binding energy of nucleons in the copper, i.e assume a head-on collision with a freely moving proton having maximum Fermi momentum and calculate the total energy in the CMS of the $p + p$ collision). The binding energy of the deuteron is 2.2 MeV

[University of Bristol 1969]

6.3 Solutions

6.3.1 Lorentz Transformations

6.1 $t_1 = \gamma(t_1{}' + vx_1{}'/c^2)$

$t_2 = \gamma(t_2{}' + vx_2{}'/c^2)$

$t_2 - t_1 = \gamma(t_2{}' - t_1{}') + \gamma v(x_2{}' - x_1{}')/c^2$

$= 0 + \gamma v(x_2{}' - x_1{}')/c^2$

$10^{-6} = \gamma\beta \times 900/c$

$\gamma\beta = 10^{-6} \times 3 \times 10^8/900 = 1/3$

$\beta = 0.316$. The velocity of S' is $0.316\,c$ with respect to S along the positive direction.

6.2 The transformation matrix is

$$\Lambda = \begin{bmatrix} \gamma & 0 & 0 & i\beta\gamma \\ 0 & 1 & 0 & 0 \\ 0 & 0 & 1 & 0 \\ -i\beta\gamma & 0 & 0 & \gamma \end{bmatrix}$$

Set $\gamma = \cosh\ \alpha$ and $\beta = \tanh\ \alpha$, so that $\gamma\beta = \sinh\ \alpha$, the transformation matrix becomes

$$\Lambda = \begin{bmatrix} \cosh\ \alpha & 0 & 0 & i \sinh\ \alpha \\ 0 & 1 & 0 & 0 \\ 0 & 0 & 1 & 0 \\ -i \sinh\ \alpha & 0 & 0 & \cosh\ \alpha \end{bmatrix}.$$

Since we can write $i\ \sinh\ \alpha\ =\ \sin i\alpha$ and $\cosh\ \alpha\ =\ \cos i\alpha$, the matrix Λ corresponds to a rotation through an angle $i\alpha$ in four-dimensional space, Further the transformation equations can be obtained from

$$\begin{bmatrix} x'_1 \\ x'_2 \\ x'_3 \\ ict' \end{bmatrix} = \begin{bmatrix} \cosh\ \alpha & 0 & 0 & i \sinh\ \alpha \\ 0 & 1 & 0 & 0 \\ 0 & 0 & 1 & 0 \\ -i \sinh\ \alpha & 0 & 0 & \cosh\ \alpha \end{bmatrix} \begin{bmatrix} x_1 \\ x_2 \\ x_3 \\ ict \end{bmatrix}.$$

and the inverse transformation equations from

$$\begin{bmatrix} x_1 \\ x_2 \\ x_3 \\ ict \end{bmatrix} = \begin{bmatrix} \cosh\ \alpha & 0 & 0 & -i \sinh\ \alpha \\ 0 & 1 & 0 & 0 \\ 0 & 0 & 1 & 0 \\ i \sinh\ \alpha & 0 & 0 & \cosh\ \alpha \end{bmatrix} \begin{bmatrix} x'_1 \\ x'_2 \\ x'_3 \\ ict' \end{bmatrix}.$$

6.3 $\beta^* = 0.268$ and $\beta_c = 0.8$

$\gamma_c = 1/(1 - {\beta_c}^2)^{1/2} = 1/(1 - 0.8^2)^{1/2} = 1/0.6 = 1.667$

$\gamma^* = 1/(1 - \beta^{*2})^{1/2} = 1/(1 - 0.268^2)^{1/2} = 1.038$

$\gamma = \gamma_c\gamma^*(1 + \beta_c\beta^* \cos\theta^*),\ \ \beta = (\gamma^2 - 1)^{1/2}/\gamma$

$\tan\theta = \sin\theta^*/\gamma_c(\cos\theta^* + \beta_c/\beta^*)$

$\theta^* = 0$

$\gamma = (1.667 \times 1.038)(1 + 0.8 \times 0.268) = 1.4248$

$\beta = 0.712$

$\tan\theta = 0$ (Because $\theta^* = 0$)

Therefore, $\theta = 0$

6.4 $\theta^* = 90°$
$\gamma = \gamma_c\gamma^* = 1.667 \times 1.038 = 1.73$
$\beta = (1.73^2 - 1)^{1/2}/1.73 = 0.816$
$\tan\theta = \sin\theta^*/(\gamma_c\beta_c/\beta^*) = 1/(1.667\times0.8/0.268) = 0.2$
$\theta = 11.3°$

6.5 $\theta = 90°$
$\gamma^* = \gamma_c\gamma(1 - \beta_c\beta\cos\theta) = \gamma_c\gamma$
$\gamma = \gamma^*/\gamma_c$
$\gamma_c = 1/[1 - (0.2)^2]^{1/2} = 1.02$
$\gamma^* = 1/[1 - (0.268)^2]^{1/2} = 1.038$
$\gamma = 1.038/1.02 = 1.0176$
$\beta = (\gamma^2 - 1)^{1/2}/\gamma = (1.0176^2 - 1)^{1/2}/1.0176 = 0.185$
$\tan\theta^* = \sin\theta/\gamma_c(\cos\theta - \beta_c/\beta^*)$
$= -\beta^*/\beta_c\gamma_c$ (Because $\theta = 90°$)
$\beta_c = \beta_\pi = 0.2$
$\tan\theta^* = -0.268/1.02 \times 0.2 = 1.3137$
$\theta^* = 127°$

6.6 The scalar wave equation for the propagation of electromagnetic waves derivable from Maxwell's equations is:

$$(\partial^2/\partial x^2 + \partial^2/\partial y^2 + \partial^2/\partial z^2 - (1/c^2)\partial^2/\partial t^2)\varphi(x, y, z, t) = 0 \quad (1)$$

for the S system. We are required to show that in S' system, the equation has the form:

$$\frac{\partial^2}{\partial x'^2} + \frac{\partial^2}{\partial y'^2} + \frac{\partial^2}{\partial z'^2} - \left(\frac{1}{c^2}\right)\left(\frac{\partial^2}{\partial t'^2}\right)\varphi(x', y', z', t') = 0$$

The Lorentz transformations are:

$$x' = \gamma(x - \beta ct) \quad (2)$$

$$y' = y \quad (3)$$

$$z' = z \quad (4)$$

$$t' = \gamma(t - \beta x'c) \quad (5)$$

Assume that we have propagation along x-axis so that the wave function will depend only on x and t. Now the function $\varphi(x', y', z', t') = 0$ is obtained from $\varphi(x, y, z, t) = 0$ by a substitution of variables. We have $\varphi(x, t) = \varphi(x', t')$. Then,

$$d\varphi = (\partial\varphi/\partial x)\,dx + (\partial\varphi/\partial t)\,dt = (\partial\varphi/\partial x')\,dx' + (\partial\varphi/\partial t')dt' \quad (6)$$

Differentiating (2) and (5),

$$dx' = \gamma(dx - \beta c dt) \quad (7)$$

$$dt' = \gamma(dt - \beta c\, dt) \quad (8)$$

Substituting (7) and (8) in (6) and equating the coefficients of dx and dt:
$(\partial\varphi/\partial x)\,dx + (\partial\varphi/\partial t)\,dt = (\gamma\partial\varphi/\partial x' - \gamma\beta c\partial\varphi/\partial t')\,dx +$
$(\gamma\partial\varphi/\partial t' - \gamma\beta c\partial\varphi/\partial x')\,dt$

$$\partial\varphi/\partial x = \gamma(\partial\varphi/\partial x' - \beta c\partial\varphi/\partial t') \tag{9}$$

$$\partial\varphi/\partial t = \gamma(\partial\varphi/\partial t' - \beta c\partial\varphi/\partial x') \tag{10}$$

$$\frac{\partial^2\varphi}{\partial x^2} = \gamma^2\frac{\partial^2\varphi}{\partial x'^2} + (\gamma^2\beta^2c^2)\frac{\partial^2 c\varphi}{\partial t'^2} - (2\gamma\beta c)\frac{\partial^2\varphi'}{\partial x'\partial t'} \tag{11}$$

$$\frac{\partial^2\varphi}{\partial t^2} = \frac{\gamma^2\partial^2\varphi}{\partial t'^2} + (\gamma^2\beta^2c^2)\frac{\partial^2\varphi}{\partial x'^2} - (2\gamma\beta c)\frac{\partial^2\varphi'}{\partial x'\partial t'} \tag{12}$$

Dividing (12) through c^2 and subtracting the resulting equation from (11)

$\frac{\partial^2\varphi}{\partial x^2} - \left(\frac{1}{c^2}\right)\frac{\partial^2\varphi}{\partial t^2}$

$$= (\gamma^2 - \gamma^2\beta^2)\frac{\partial^2\varphi}{\partial x'^2} - \left(\frac{1}{c^2}\right)(\gamma^2 - \gamma^2\beta^2)\frac{\partial^2\varphi}{\partial t'^2}$$

$$= \frac{\partial^2\varphi}{\partial x'^2} - \left(\frac{1}{c^2}\right)\frac{\partial^2\varphi}{\partial t'^2}$$

since $\gamma^2 - \gamma^2\beta^2 = \gamma^2(1-\beta^2) = 1$

Similarly, the Klein–Gordon equation

$(\nabla^2 - (1/c^2)\partial^2/\partial t^2 + m^2c^2/\hbar^2) = 0$ is Lorentz invariant.

6.7 The only way π^- is emitted at rest in the lab system is when it is emitted at $\theta_1{}^* = 180°$ in the CMS (rest frame of $K°$) with with the same speed as $K°$ in the lab system. In that case π^- will be emitted at $\theta^*{}_2 = 0°$ in the CMS.

The energy released $Q = 498 - 2 \times 140 = 218$ MeV

As the product particles are identical, each pion carries half of the enrgy, 109 MeV

$\gamma^* = 1 + T^*/m_\pi = 1 + 109/140 = 1.778$

From the above discussion

$\gamma_c = \gamma^*, \beta_c = \beta^*$

$\gamma = \gamma^*\gamma_c(1 + \beta^*\beta_c) = \gamma^{*2}(1 + \beta^{*2})$

$= \gamma^{*2}(1 + (\gamma^{*2} - 1)/\gamma^{*2}) = 2\gamma^{*2} - 1$

$= 2 \times 1.778^2 - 1 = 5.3266$

$T = (\gamma - 1)m_\pi = (5.3266 - 1) \times 140$

$= 605.7$ MeV

6.8 $\tan\theta_\nu{}^* = \sin\theta_\nu/\gamma_c(\cos\theta_\nu - \beta_c/\beta_\nu{}^*) = -1/\gamma_c\beta_c$ (1)

(Because $\theta_\nu = 90°$ and $\beta_\nu^* = 1$). Here β_c is the velocity of the pion.

It follows that

$\sin\theta_\nu{}^* = 1/\gamma_c$ and $\cos\theta_\nu^* = -\beta_c$ (2)

In the CMS (the system in which the pion is at rest)

$\theta_\mu{}^* = \pi - \theta_\nu{}^*$, because the muon and neutrino must fly in the opposite direction to conserve momentum.

$\tan\theta_\mu{}^* = \tan(\pi - \theta_\nu{}^*) = -\tan\theta_\nu{}^* = -1/\gamma_c\beta_c$

or

$\tan\theta_\mu{}^* = 1/\beta_c\gamma_c$ (3)

$\tan\theta_\mu = \sin\theta_\mu{}^*/\gamma_c(\cos\theta_\mu{}^* + \beta_c/\beta_\mu{}^*) = 1/\gamma_c{}^2\beta_c(1 + 1/\beta^*{}_\mu)$ (4)

(Since $\theta_\mu{}^* = \pi - \theta_\nu{}^*$) and we have used (2)

But $\beta^*{}_\mu = (m_\pi{}^2 - m_\mu{}^2)/(m_\pi{}^2 + m_\mu{}^2)$ (5)

Substituting (5) in (4) and simplifying

$\tan\theta_\mu = (m_\pi{}^2 - m^2{}_\mu)/2\gamma^2{}_c\beta_c m^2{}_\pi$

6.9 The z-component of velocity is zero; hence the particle must be moving in the xy =plane. Further, the y-component of velocity is unchanged. This implies that the Lorentz transformation is to be made along x-axis

$cP_x = \gamma(cP'_x + \beta E')$ (1)

$c^2\, m\beta_x\gamma_0 = \gamma(c^2\, m\beta'_x\gamma' + m\beta\gamma' c^2)$ (2)

$\beta_x = 1/2^{1/2}, \gamma_0 = 2^{1/2}, \beta'_x = -1/2^{1/2}, \gamma' = 2^{1/2}, \gamma = 1/(1-\beta^2)^{1/2}$ (3)

Using (4) in (1) and simplifying we get $\beta = 2 \times 2^{1/2}/3$

6.10 Energy conservation gives

$T_1 + T_2 = Q$ (1)

Momentum conservation gives

$P_1 + P_2 = 0$

or $p_1{}^2 = p_2{}^2$

$T_1{}^2 + 2T_1\, m_1 = T_2{}^2 + 2T_2\, m_2$ (2)

Solving (1) and (2)

$T_1 = Q(Q + 2m_2)/2(m_1 + m_2 + Q)$; $T_2 = Q(Q + 2m_1)/2(m_1 + m_2 + Q)$

6.11 $\gamma_\pi = 1 + T_\pi/m_\pi = 1 + 140/140 = 2$

$\beta_\pi = (\gamma_\pi{}^2 - 1)^{1/2}/\gamma_\pi = (2^2 - 1)^{1/2}/2 = 0.866$

By Problem 6.54 , $T_\mu{}^* = 4.0\,\text{MeV}$ therefore

$\gamma_\mu{}^* = 1 + T_\mu{}^*/m_\mu = 1 + (4/106) = 1.038$

$\beta_\mu{}^* = (1.03777^2 - 1)^{1/2}/1.0377 = 0.267$

$\gamma_\mu = \gamma\gamma_\mu{}^*(1 + \beta_\pi\beta_\mu{}^*\cos\theta^*)$

$\gamma_\mu(\max) = \gamma_\pi\gamma_\mu{}^*(1 + \beta_\pi\beta_\mu{}^*) = 2 \times 1.038(1 + 0.866 \times 0.267) = 2.556$

(Because $\theta^* = 0$)

$T_\mu(\max) = (\gamma_\mu(\max) - 1)m_\mu = 165\,\text{MeV}$

Using the formula for optical Doppler effect

$T_\nu(\max) = \gamma_\pi T_\nu{}^*(1 + \beta_\pi) = 2 \times 29.5(1 + 0.866) = 110\,\text{MeV}$

6.12 $\beta_c = |p_+ + p_-|/(E_+ + E_-)$

Using the invariance principle

(total energy)^2 – (total momentum)^2 = invariant

$(E_+ + E_-)^2 - |p_+ + p_-|^2 = (E_1{}^* + E_2{}^*)^2 - |p_1{}^* + p_2{}^*|^2$

But $E_1{}^* = E_2{}^*$ since the particles have equal masses. Also by definition of center of mass, $|p_1{}^* + p_2{}^*| = 0$

Therefore, $E_1{}^{*2} = E_2{}^{*2} = \frac{1}{4}[E_+ + E_-)^2 - (p_+{}^2 + p_-{}^2 + 2\, p_+p_-\cos\theta)]$

where θ is the angle between $e^+ - e^-$ pair

$E_1{}^* = E_2{}^* = (1/4)[(E_+{}^2 - p_+{}^2 + E_-{}^2 - p_-{}^2 + 2(E_+E_- - p_+p_- \cos\theta)]$
$E_1{}^* = E_2{}^* = (1/4)[m^2 + m^2 + 2(E_+E_- - p_+p_- \cos\theta)]$
$= (1/2)(m^2 + E_+E_- - p_+p_- \cos\theta)$
or $E_1{}^* = E_2{}^* = [1/2(m^2 + E_+E_- - p_+p_- \cos\theta)]^{1/2}$

6.13 $\tan\theta = \sin\theta^*/\gamma_c(\cos\theta^* + \beta_c/\beta_1{}^*) = (1/\gamma_c)\tan\theta^*/2$ (1)
(Because $\beta_c = \beta_1{}^*$)
Also
$\tan\varphi = (1/\gamma_c)\tan\theta^*/2$ (2)
Where θ^* and φ^* are the corresponding angles in the CM system
Multiply (1) and (2)
$\tan\theta \tan\varphi = (1/\gamma_c{}^2).\tan\theta^*/2.\tan\varphi^*/2$
But $\varphi^* = \pi - \theta^*$ and for $m_1 = m_2$, $\gamma_c = ((\gamma + 1)/2)^{1/2}$
$\tan\theta \tan\varphi = 2/(\gamma + 1)$
In the classical limit, $\gamma \to 1$ and $\tan\theta \tan\varphi = 1$
But since $\tan(\theta + \varphi) = (\tan\theta + \tan\varphi)/(1 - \tan\theta\tan\varphi)$
$\tan(\varphi + \theta) \to \infty$
i.e. $\theta + \varphi = \pi/2$
For $\gamma > 1$, $\tan\theta\tan\varphi < 1$. For $\tan(\theta + \varphi)$ to be finite $(\theta + \varphi) < \pi/2$
Hence, for $\gamma > 1; \theta + \varphi < \pi/2$

6.14 The total energy carried by π^+ in the rest frame of K^+ can be calculated from
$E_{\pi^+} = (m_K{}^2 + m_\pi{}^2 - m_\nu{}^2)/2m_k = 265$ MeV (1)
$\gamma_c = \gamma^*_{\pi^+} = 265/139.5 = 1.9$
and $\beta_c = \beta_\pi{}^* = (\gamma^{*2}_{\pi^+} - 1)^{1/2}/\gamma^*_{\pi^+}$

In the rest frame of pion, the total energy of muon is obtained again by (1)
$E_{\mu^+}{}^2 = (m_{\mu^+}{}^2 + m_{\pi^+}{}^2 - 0)/2m_{\pi^+} = 110$ MeV
$\gamma_\mu{}^* = 110/106 = 1.0377$
$\beta_\mu{}^* = (\gamma^{*2}_\mu - 1)^{1/2}/\gamma_\mu{}^* = 0.267$
The Lorentz factor γ_μ for the muon in the LS is obtained from $\gamma_\mu = \gamma_c\gamma^*_\mu(1 + \beta^*_\mu\beta_c\cos\theta^*_\mu)$ where θ^*_μ is the emission angle of the muon in the rest frame of pion. Put $\theta^*_\mu = 0$ to obtain γ_μ(max) and $\theta^*_\mu = 180°$ to obtain γ_μ(min). The maximum and minimum kinetic energies are 150 and 55.5 MeV.

6.15 First we find the momenta of three pions by using the formula $P_1 = (E_1^2 - m^2)^{1/2}$ etc, where the total energy, $E_1 = T_1 + m$, $m = 140$ MeV is the pion mass. We need to find the total momentum of the three particles. Take the direction of the middle particle as x-axis. Calculate these components as below:

$T_1(\text{MeV}) = 190$, $E_1 = 330(\text{MeV})$, $P_1 = 299$ MeV/c,
$p_1(x) = 276.4$ MeV/c, $p_1(y) = -114$ MeV/c
$T_2 = 321$, $E_2 = 461$, $P_2 = 439$, $p_2(x) = 439$, $p_2(y) = 0$
$T_3 = 58$, $E_3 = 198$, $P_3 = 140$, $p_3(x) = 137$, $p_3(y) = 227$
$\sum E = 989$ MeV, $\sum p(x) = 852.4$ MeV/c
$\sum P(y) = -84$ MeV/c

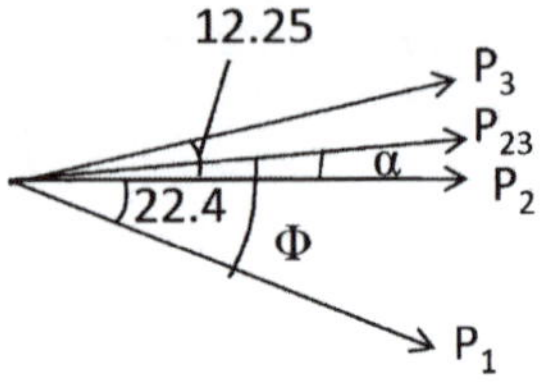

Fig. 6.4 Decay of a charged unstable particle into three pions

Therefore, $P = [\sum p(x)^2 + (\sum p(y)^2]^{1/2} = 856.4\ \text{MeV/c}$

The mass of the particle is given by

$$M = \left[\left(\sum E\right)^2 - \left|\sum p\right|^2\right]^{1/2}$$

$= [(989)^2 - (856.4)^2]^{1/2}$

$= 494.7$ MeV

It is a K meson

We can find the direction of K meson by calculating the resultant momenta of the three pions and its orientation with respect to one of the pions.

By the vector addition of p_2 and p_3 we find the resultant $P_{23} = 576.6\ \text{MeV/c}$ inclined at angle $\alpha = 2.84°$ above p_2 as in the Fig. 6.4. The angle inclined between P_{23} and p_1 is $\varphi = 2.84 + 22.4 = 25.24°$.

When p_{23} is combined with vector p we find that the resultant is inclined at an angle of 16.7° above p_1.

6.3.2 Length, Time, Velocity

6.16 $L = \gamma L_0 = (1 - \beta^2)^{1/2} L_0 = (1 - 0.8^2)^{1/2} L_0$
$= 0.6 L_0$
$\Delta L = L_0 - L = 0.4 L_0$

6.17 $L = L_0/\gamma = L_0/2$
$\gamma = 2 \rightarrow \beta = (\gamma^2 - 1)^{1/2}/\gamma = (2^2 - 1)^{1/2}/2 = 0.866$
$v = \beta c = 0.866 \times 3 \times 10^8 = 2.448 \times 10^8\ \text{ms}^{-1}$

6.18 $\beta = v/c = 30\ \text{km s}^{-1}/3 \times 10^5\ \text{km s}^{-1} = 10^{-4}$
$1/\gamma = (1 - \beta^2)^{1/2} = (1 - 1/2 \times \beta^2)$
$\Delta L = L_0 - L = L_0 - L_0/\gamma = L_0 - L_0(1 - \beta^2)^{1/2} = \frac{1}{2} L_0 \beta^2 = 1/2 \times 6{,}400 \times 10^{-8}\ \text{km} = 3.2\ \text{cm}$
Thus the earth appears to be shrunk by 3.2 cm.

6.19 $\tau = \gamma \tau_0$
$1.5 \times 10^{-5} = 2.2 \times 10^{-6} \gamma$
$\gamma = 6.818$
$\beta = (\gamma^2 - 1)^{1/2}/\gamma = 0.9892$
$v = 0.9890c$

6.20 $\tau_0 = \tau/\gamma = \tau(1-\beta^2)^{1/2}$
$= 2.9 \times 10^{-6} \times (1-0.6^2)^{1/2}$
$= 2.32 \times 10^{-6}$ s

6.21 (a) Time, $\tau = \tau_0/(1-\beta^2)^{1/2} = 2.2 \times 10^{-6}/(1-0.6^2)^{1/2}$
$= 2.75 \times 10^{-6}$ s
Distance $d = v\tau = 0.6 \times 3 \times 10^8 \times 2.75 \times 10^{-6} = 495$ m
(b) $d_0 = v\tau_0 = 0.6 \times 3 \times 10^8 \times 2.2 \times 10^{-6} = 396$ m.

6.22 $d = vt = v\gamma t_0 = 3 \times 10^4 c$
$\gamma\beta = 3 \times 10^4/40 = 750$
$\beta/(1-\beta^2)^{1/2} = 750$
$\beta = 0.99999956$

6.23 $d = v\gamma t_0 = \beta\gamma c t_0 = 0.99 \times 3 \times 10^8 \times 2.5 \times 10^{-8}/(1-0.99^2)^{1/2}$
$= 373$ m.
It is therefore observed at an altitude of 1,000 − 373 = 627 m above the sea level.

6.24 $\beta = (\beta_1 + \beta_2)/(1+\beta_1\beta_2) = (0.9+0.9)/(1+0.9 \times 0.9) = 0.994475$

6.25 $E = T + m_0c^2 = 100 + 0.51 = 100.51$ MeV
$\gamma = E/m_0c^2 = 100.51/0.51 = 197$
$L = L_0/\gamma = 4/197 = 0.02\text{ m} = 2\text{ cm}$

6.26 $m = \gamma\, m_0$
$\gamma = m/m_0 = 101/100 = 1.01$
$\beta = (\gamma^2 - 1)^{1/2}/\gamma = 0.14$

6.27 If the space station is located at a distance d from the earth then d is fixed by the time taken by the radio signal to reach the earth is
$d = ct$
As observed from the earth, at $t_1 = 0$ the spaceship was at a distance d approaching with speed 0.5 c. It will arrive at time
$t_1 = d/\beta c = ct/\beta c = 1{,}125/0.5 = 2{,}250$ s

6.28 The time t_2 recorded in the spaceship related to t_1 is shortened by γ, the Lorentz factor.
$t_2 = t_1/\gamma = t(1-\beta^2)^{1/2} = t(1-0.5^2)^{1/2} = 0.866\,t$
$= 0.866 \times 2{,}250 = 1{,}948$ s

6.29 Let system S be attached to the ground and S' to the spaceship.
Let t_1 be the time when the radio signal reaches the ship. In that time the signal traveled a distance
$d_1 = ct_1$
At time $t_1 = 0$, the ship was at a distance d.
At time t_1 it is now at a distance
$d_2 = d + vt_1 = d + \beta c t_1$
Now $d_1 = d_2$

So $ct_1 = d + \beta ct_1$
Solving for t_1,
$t_1 = d/c(1 - \beta)$
For $\beta = 0.6$, $\gamma = 1.25$
$t_1 = 5 \times 10^8 \times 10^3/3 \times 10^8 \times 0.4 = 4167$ s

6.30 From Lorentz transformations we get
$\Delta t' = \gamma \Delta t = \gamma d_1/c = 1.25 \times 5 \times 10^8 \times 10^3/3 \times 10^8$
$= 2{,}083$ s

6.31 $\tau = d/v\gamma = d/\gamma\beta c = d/c(\gamma^2 - 1)^{1/2}$ (1)
$\gamma = E/m = 1 + (T/m) = 1 + (100/140) = 1.714$
$D = 4.88\,\text{m},\ c = 3 \times 10^8\,\text{ms}^{-1}$
Using these values in (1), we get $\tau = 1.17 \times 10^{-8}$ s

6.32 $I = I_0\, e^{-t/\tau} = I_0\, e^{-\gamma d/c\beta\tau}$
$\gamma = 1/0.14 = 7.143,\ d = 10\,\text{m},\ c = 3 \times 10^8\,\text{ms}^{-1}$
$\beta = (1 - 1/7.143^2)^{1/2} = 0.99,\ \tau = 2.56 \times 10^{-8}$ s
$I_0 = 10^6$
Using the above values, we find $I = 83$

6.33 $(\gamma - 1)M = M$
Or $\gamma = 2$
$\beta = (\gamma^2 - 1)^{1/2}/\gamma = (2^2 - 1)^{1/2}/2 = \sqrt{3}/2$
The dilated time $T = \gamma\, T_0 = 2 \times 2.5 \times 10^{-8} = 5 \times 10^{-8}$ s
The distance traveled before decaying is
$d = vT = \beta\, cT = \sqrt{3}/2 \times 3 \times 10^8 \times 5 \times 10^{-8} = 13$ m

6.34 Time $t = d/v = 300/3 \times 10^8 = 1.0 \times 10^{-6}$ s
As $v \approx c$ at ultrarelativistic velocity
The proper lifetime is dilated
$\tau = \tau_0\gamma = \tau_0\, E/m = 2.6 \times 10^{-8} \times (200 \times 10^3 + 140)/140$
$= 3.71 \times 10^{-3}$ s
Fraction f of pions decaying is given by the radioactive law
$f = 1 - \exp(-T/\tau)$
$= 1 - \exp(-0.0269)$
$= 0.027$
The pions and muons are subsequently stopped in thick walls of steel and concrete, pions through their nuclear interactions and muons through absorption by ionization. The neutrinos being stable, neutral and weakly interacting will survive.

6.35 Assuming that the pions decay exponentially (the law of radioactivity), then after time t they travel a distance d, with velocity $v = \beta c$ so that $t = d/\beta c$ and their mean lifetime is lengthened by the Lorentz factor γ.
(a) The intensity at counter B will be

$$I_B = I_A \exp\,(-d/\gamma\beta c\tau) \qquad (1)$$

and at the counter C

$$I_c = I_A \exp(-2d/\gamma\beta c\tau) \tag{2}$$

$$\therefore I_c = \frac{I_B^2}{I_A} = \frac{(470)^2}{1000} = 221$$

(b) Take logarithm on both sides of (1) and simplify.

$$\tau = \frac{d}{\gamma\beta c \ln(I_A/I_B)} \tag{3}$$

$\gamma = 1 + T/mc^2 = 1 + 140/140 = 2.0$
$\beta = (\gamma^2 - 1)^{1/2}/\gamma = 0.866$
$d = 10$ m; $c = 3 \times 10^8$ m/s
$\ln(I_A/I_B) = \ln(1000/470) = 0.755$
Substituting the above values in (3),
$\tau = 2.55 \times 10^{-8} s$
The accepted value is $2.6 \times 10^{-8} s$

6.36 The stationary object will appear to move with velocity $-\beta c$ toward the observer. The object moving with velocity αc toward the stationary object would appear to have velocity
$(\alpha c - \beta c)/(1 - \alpha\beta)$, as seen by the observer. If these two velocities are to be equal then $(\alpha c - \beta c)/(1 - \alpha\beta) = \beta c$

Cross multiplying and simplifying we get the quadratic equation whose solution is $\beta = [1 - (1 - \alpha^2)^{1/2}]/\alpha$

6.37

(i) $$t = \frac{L}{\beta c} \tag{1}$$

$$N_2 = N_1 \exp[-t/\gamma\tau] = N_1 \exp\left[-\frac{L}{\gamma\beta c\tau}\right] \tag{2}$$

Therefore (2) becomes
$\exp\left[\frac{L}{\gamma\beta c\tau}\right] = N_1/N_2$
Take logarithm on both sides
$\frac{L}{\gamma\beta c\tau} = \ln\left(\frac{N_1}{N_2}\right)$
But $\gamma\beta = \sqrt{\gamma^2 - 1}$
Therefore $\tau = \frac{L}{\ln\left(\frac{N_1}{N_2}\right)\sqrt{\gamma^2 - 1c}}$

(ii) $\gamma = 1/(1 - \beta^2)^{1/2} = 1/(1 - 8/9)^{1/2} = 3$

$$\tau = \frac{200}{\ln\left(\frac{10{,}000}{8{,}983}\right)\sqrt{3^2 - 1} \times 3 \times 10^8}$$
$= 2.2 \times 10^{-6}$ s.

6.38 (a) $\beta = v/c = \frac{\sqrt{3}}{2}$
$\gamma = (1 - \beta^2)^{-1/2} = (1 - 3/4)^{-1/2} = 2$
Total energy of the particle
$E = \gamma M c^2 = 2Mc^2$
(b) Distance traveled on an average
$d = \gamma \beta c t_0 = 2 \times \frac{\sqrt{3}}{2} c\tau = \sqrt{3}\, c\tau$
(c) The sphere will shrink in the direction of motion but will not in the transverse direction. Consequently, its shape would appear as that of a spheroid as shown in Fig. 6.5

Fig. 6.5

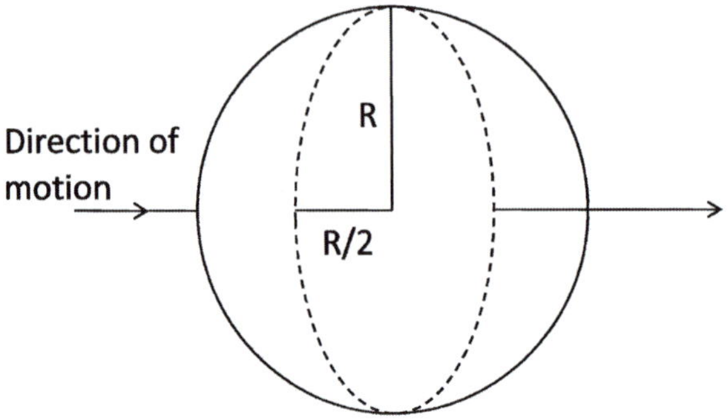

6.39 $v' = \frac{v-u}{1-uv/c^2}$
put $v = c$
$v' = \frac{c-u}{1-uc/c^2} = \frac{c-u}{1-u/c} = c$

6.3.3 Mass, Momentum, Energy

6.40 (a) $m = m_0 \gamma$
$\gamma = \tau/\tau_0 = 6.6 \times 10^{-6}/2.2 \times 10^{-6} = 3$
$m = 3 \times 207 = 621\ \mathrm{m_e}$
(b) $T = (\gamma - 1) m_0 c^2 = (3 - 1)(207 \times 0.51)$
$= 211\ \mathrm{MeV}$
(c) Total energy $E = mc^2 = 621 m_e c^2 = 621 \times 0.511 = 3.173\ \mathrm{MeV}$
$p = \beta E/c$
$\beta = (\gamma^2 - 1)^{1/2}/\gamma = (3^2 - 1)^{1/2}/3 = 0.9428$
$p = (0.9428)(317.3)/c$
$= 299\ \mathrm{MeV}/c$

6.41 $E = m_0\, c^2 = (1 \times 10^{-3}\ \mathrm{kg})(3 \times 10^8)^2 = 9 \times 10^{13}\ \mathrm{J}$

6.42 $T = (\gamma - 1)m = m$
$\gamma = 2$
$\beta = (\gamma^2 - 1)^{1/2}/\gamma = 0.866$
The result is independent of the mass of the particle.

6.43 (a) $\gamma = 1 + T/m = 1 + 1,000/940 = 2.064$
$\beta = (\gamma^2 - 1)^{1/2}/\gamma = [(2.064)^2 - 1]^{1/2}/2.064 = 0.87$
(b) $\gamma = 1 + 1,000/0.511 = 1,958$
$\beta = [1,958^2 - 1]^{1/2}/1,958 = 0.999973$

6.44 (a) There are 6 protons and 6 neutrons in ^{12}C nucleus.
$\Delta mc^2 = [(1.007825 \times 6 + 1.008665 \times 6 - 12.000] \times 931.5$
$= 92.16\,\text{MeV}$
(b) $\Delta mc^2 = [3 \times 4.002603 - 12.000] \times 931.5$
$= 7.27\,\text{MeV}$
(c) $\Delta mc^2 = [2 \times (m_n + m_p) - m_\alpha] \times 931.5$
$= [2 \times (1.008665 + 1.007825) - 4.002603] \times 931.5$
$= 28.296\,\text{MeV}$
Total energy required $= 3 \times 28.296 + 7.27 = 92.16\,\text{MeV}$
which is identical with that in (a)

6.45 $c^2p^2 = T^2 + 2Tmc^2$
$m = (c^2p^2 - T^2)/2T = (368^2 - 250^2)/2 \times 250 = 145.85\,\text{MeV}$
Or $m = 145.85/0.511 = 285.4\, m_e$
The particle is identified as the pion whose actual mass is 273 m_e

6.46 $E_0 = m_0c^2 = (9.1 \times 10^{-31}\,\text{kg})(3 \times 10^8\,\text{ms}^{-1})^2$
$= 8.19 \times 10^{-14}\,\text{J}$
$= (8.19 \times 10^{-14}\,\text{J})(1\,\text{MeV}/1.6 \times 10^{-13}\,\text{J})$
$= 0.51\,\text{MeV}$

6.47 $\gamma = 1/(1 - \beta^2)^{1/2} = 1/(1 - 0.6^2)^{1/2} = 1.25$
Energy acquired by electron
$T = (\gamma - 1)m_ec^2 = (1.25 - 1) \times 0.51 = 0.1275\,\text{MeV}$
1 eV energy is acquired when an electron (or any singly charged particle) is accelerated from rest through a P.D of 1 V. Hence the required P.D is 0.1275 Mega volt or 127.5 kV.

6.48 (a) $K(\text{relativistic}) = (\gamma - 1)m_0c^2$ (1)
$K(\text{classical}) = (1/2)m_0v^2 = (1/2)m_0c^2\beta^2 = (1/2)m_0c^2(\gamma^2 - 1)/\gamma^2$ (2)
$$\frac{\Delta K}{K} = \frac{K(\textit{relativistic}) - K(\textit{Classical})}{K(\textit{relativistic})} = \frac{\gamma - 1}{2\gamma} \quad (3)$$
where we have used (1) and (2)
Putting $\Delta K/K = 1/100$, we find $\gamma = 50/49$. Using
$\beta = (\gamma^2 - 1)/\gamma$ (4)
we obtain $\beta = 0.199$
Or $v = 0.199\,c$
(b) Putting $\Delta K/K = 10/100 = 1/10$ in (3), we find $\gamma = 5/4$, Using (4) we obtain $\beta = 0.6$
Or $v = 0.6\,c$

6.49 $T = (\gamma - 1)m_0c^2$
$= [1/(1 - \beta^2)^{1/2} - 1]m_0c^2 = [(1 - \beta^2)^{-1/2} - 1]m_0c^2$
$= [1 + \frac{1}{2}\beta^2 + \frac{3}{8}\beta^4 + \ldots - 1]m_0c^2$
$= [1 + \frac{3}{4}\beta^2 + \ldots](\frac{1}{2}m_0\beta^2c^2)$
where we have expanded the radical binomially
For $v/c \ll 1$ or $\beta \ll 1$, obviously $T \ll m_0c^2$
For small velocities $T = m_0\beta^2c^2/2 +$ small terms $= m_0v^2/2$

6.50 (a) The photon energy $E_\gamma = 1,240/500\,\text{nm} = 2.48\,\text{eV}$
$= 2.48 \times (1.6 \times 10^{-19})\,\text{J}$
$= 3.968 \times 10^{-19}\,\text{J}$
Effective mass, $m = E_\gamma/c^2 = 3.968 \times 10^{-19}/(3 \times 10^8)^2 = 4.4 \times 10^{-36}\,\text{kg}$
(b) $E = 1{,}240/0.1\,\text{nm} = 12{,}400\,\text{eV} = 1.984 \times 10^{-15}\,\text{J}$
$m = 1.984 \times 10^{-15}/(3 \times 10^8)^2\,\text{kg} = 2.2 \times 10^{-32}\,\text{kg}$.

6.51 $1\text{amu} = 1.66 \times 10^{-27}\,\text{kg}$
$$= \frac{(1.66 \times 10^{-27}\,\text{kg})(c^2)}{c^2}$$
$= 1.66 \times 10^{-27} \times 2.998 \times 10^8)^2\,\text{J}/\text{c}^2$
$= 1.492 \times 10^{-10}\,\text{J}/\text{c}^2$
$= 1.492 \times 10^{-10}\,\text{J}/\text{MeV.MeV}/\text{c}^2$
$= 1.492 \times 10^{-10}/1.602 \times 10^{-13}\quad \text{MeV}/\text{c}^2 = 931.3\,\text{MeV}/\text{c}^2$

6.52 Number of Uranium atoms in 1.0 g is
$N = N_0/A = 6.02 \times 10^{23}/235 = 2.56 \times 10^{21}$
Number in 5 kg $= 2.56 \times 10^{21} \times 5{,}000 = 1.28 \times 10^{25}$
In each fission ~ 200 MeV energy is released.
Therefore, total energy released
$= 1.28 \times 10^{25} \times 200 = 2.56 \times 10^{27}\,\text{MeV}$
$= (2.56 \times 10^{27}\,\text{MeV/J})(1.6 \times 10^{-13}\,\text{J})$
$= 4 \times 10^{14}\,\text{J}$

6.53 The analysis is similar to that for Compton scattering except for some approximations.
Energy conservation gives

$$E = E' + T \quad (1)$$

From momentum triangle (Fig 6.6)

$$p'^2 = p^2 + p_e{}^2 - 2pp_e\cos\theta \quad (2)$$
Using $c^2p'^2 = E'^2 - M^2c^4$ (3)
$$c^2p_e{}^2 = T^2 + 2Tmc^2 \quad (4)$$
$$p = \gamma M\beta c \quad (5)$$
$$\gamma = 1/(1 - \beta)^{1/2} \quad (6)$$

Combining (1) – (6) and simplifying and using the fact that $mc^2 \ll E$. we easily obtain the desired result.

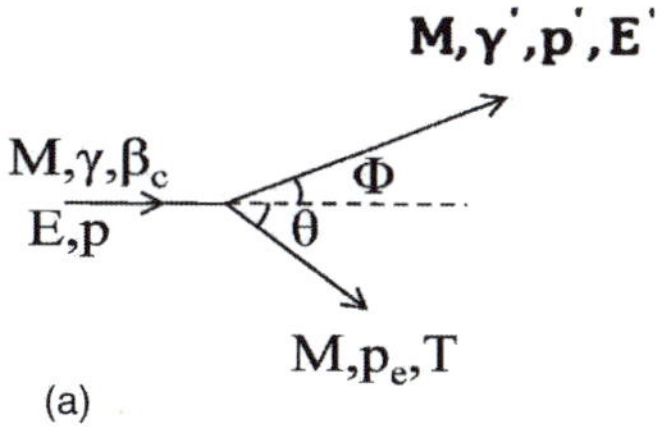

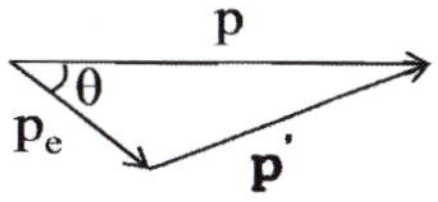

Fig. 6.6 **(a)** Scattering of a proton with a stationary electron **(b)** Momentum triangle

6.54 The energy released in the decay of pion is
$Q = m_\pi c^2 - (m_\mu c^2 + m_\nu c^2)$
$= (273 - 207 - 0)m_e c^2$
$= 66 \times 0.511 = 33.73\,\text{MeV}$
Energy conservation gives
$T_\mu + T_\nu = 33.73$ (1)
In order to conserve momentum, muon neutrino must move in opposite direction
$p_\nu = p_\mu$ (2)

Multiplying (2) by c and squaring

$c^2 p^2{}_\nu = T^2{}_\nu = K_\mu + 2K_\mu m_\mu c^2$ (3)

Solving (1) and (3) and using
$m_\mu c^2 = 207 m_e c^2 = 207 \times 0.511 = 105.77\,\text{MeV}$
$K_\mu = 4.08\,\text{MeV},\ K_\nu = 29.65\,\text{MeV}$
Observe that the lighter particle carries greater energy.

6.55 Let the mass of the final single body be M which moves with a velocity βc.
Momentum conservation gives
$m \times 0.6c/(1 - 0.6^2)^{1/2} = M\beta c/(1 - \beta^2)^{1/2}$

Or $3m/4 = M\beta/(1 - \beta^2)$ (1)
Since the total energy is conserved

$mc^2/(1 - 0.6^2)^{1/2} + mc^2 = Mc^2/(1 - \beta^2)^{1/2}$ (2)
(a) Using (2) in (1), $\beta = 1/3$
(b) Using $\beta = 1/3$ in (2), $M = 2.12\,\text{m}$

6.56 $T_K = [(M_D - M_K)^2 - M^2{}_\pi]/2M_D = [(1865 - 494)^2 - 140^2]/2 \times 1865$
$= 498.67\,\text{MeV}$
$P_K = (T_K{}^2 + 2M_K T_K)^{1/2} = 861\,\text{MeV/c}$
$E_K{}^2 = p_K{}^2 + m_K{}^2 = (0.861)^2 + (0.494)^2 = 0.9853\,\text{GeV}^2$
$E_K = 0.9926\,\text{GeV/c}$
$E_\pi{}^2 = p_\pi{}^2 + m_\pi{}^2 = (0.861)^2 + (0.140)^2 = 0.7609\,\text{GeV}^2$
$E_\pi = 0.8736\,\text{GeV}$
$M_x = E_K + E_\pi = 1.8662\,\text{GeV/c}^2$
It is a D^0 meson.

6.57 The mass of neutrino is zero. Applying conservation laws of energy and momentum

$$E_\mu + E_\nu = m_\pi c^2 \quad (1)$$

$$p_\mu = p_\nu \quad (2)$$

Multiplying (2) by c and squaring
$c^2 p_\mu{}^2 = c^2 p_\nu{}^2$
Or $E_\mu{}^2 - m_\mu{}^2 c^4 = E_\nu{}^2$
Or $E_\mu{}^2 - E_\nu{}^2 = m_\mu{}^2 c^4$ (3)
Solve (1) and (3)
$\gamma_\mu = (m^2{}_\pi + m^2{}_\mu)/2m_\pi m_\mu$
$\beta_\mu = (1 - 1/\gamma_\mu^2)^{1/2} = (m_\pi{}^2 - m_\mu{}^2)/(m_\mu{}^2 + m_\mu{}^2)$

6.58 $E_B + E_C = m_A c^2$ (energy conservation) (1)
$P_B = P_C$ (momentum conservation) (2)
Or $c^2 P_B{}^2 = c^2 P_C{}^2$ (3)
Using the relativistic equations $E^2 = c^2 p^2 + m^2 c^4$, (3) becomes
$E_B{}^2 - m_B{}^2 c^4 = E_C{}^2 - m_C{}^2 c^4$ (4)
Eliminating E_C between (1) and (4), and simplifying
$E_B = (m_A{}^2 + m_B{}^2 - m_C{}^2)c^2/2m_A$ (5)

6.59 $K^+ \to e^+ + \pi^\circ + \nu_e$
The maximum energy of positron will correspond to a situation in which the neutrino is at rest. In that case the total energy carried by electron will be
$E_e(\text{max}) = (m_K^2 + m_e^2 - m_{\pi^0}^2)/2m_K = (494^2 + 0.5^2 + 135^2)/2 \times 494 = 228.5\text{MeV}$
$\therefore T_e(\text{max}) = 228$ MeV

6.60 Let the incident particle carry momentum p_0. As the scattering is symmetrical, each particle carries kinetic energy $T/2$ and momentum P after scattering, and makes an angle $\theta/2$ with the incident direction.
Momentum conservation along the incident direction gives
$p_0 = p\cos\theta/2 + p\cos\theta/2 = 2p\cos\theta/2$ (1)
Or $(T^2 + 2Tmc^2)^{1/2} = 2(T^2/4 + 2T/2mc^2)^{1/2}\cos\theta/2$ (2)
Squaring (2), and using the identity, $\cos^2\theta/2 = (1 + \cos\theta)/2$
We get the result $\cos\theta = T/(T + 4mc^2)$

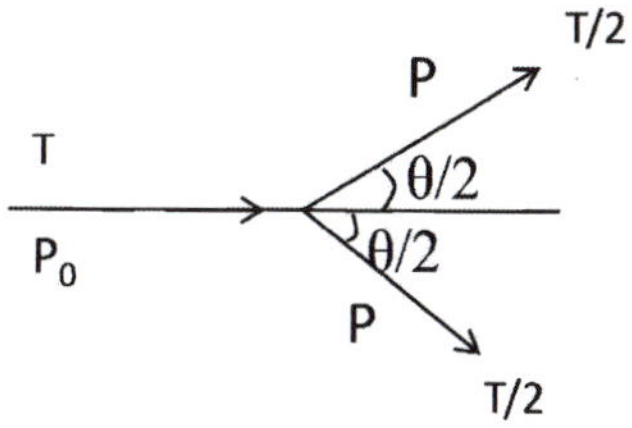

Fig. 6.7 Symmetrical elastic collision between identical particles

6.61 $(\gamma - 1)mc^2 = mc^2$
$\gamma = 2$
$\beta = (1 - 1/\gamma^2)^{1/2} = \sqrt{3}/2$
$p_e = m\gamma\beta c$
$cp_e = mc^2\gamma\beta = mc^2 \times 2 \times \sqrt{3}/2 = mc^2 \times \sqrt{3}\,\text{MeV}$
$p_\gamma = p_e = \sqrt{3}E_0\,\text{MeV/c}$

6.62 For the three particles the energies (total) are equal.
$E_1 = E_2 = E_3$ (3)
(masses are neglected)
The magnitude of momenta are also equal
$p_1 = p_2 = p_3$
The momenta represented by the three vectors AC, CB and BA form the closed Δ ABC.
$180° - \theta = 60°$
Therefore, $\theta = 120°$
Thus the paths of any two leptons are equally inclined to 120°

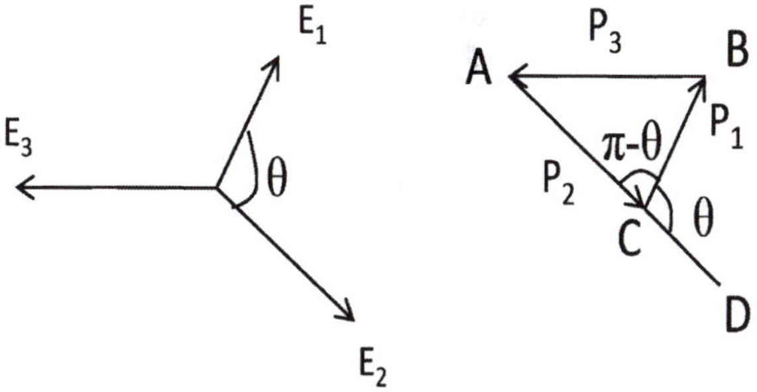

Fig. 6.8 Decay of a muon at rest into three leptons whose masses are neglected

6.63 By Problem 6.53, $T = 2mc^2\beta^2\cos^2\theta/[1 - \beta^2\cos^2\theta]$ (1)
$T = 10^9\,\text{eV} = 1,000\,\text{MeV}$
$\gamma = 1 + T/M = 1 + 1,000/940 = 2.0638$
$\beta = [1 - (1/\gamma^2)]^{1/2} = [1 - (1/2.0638)^2]^{1/2} = 0.875$
Using $mc^2 = 0.511\,\text{MeV}$, $\beta = 0.875$ and $\theta = 3°$ in (1),
We find $T = 3.3\,\text{MeV}$

6.64 (a) $\pi^+ \to \mu^+ + \nu_\mu$
$T = Q(Q + 2m_\nu)/2(m_\mu + m_\nu + Q) = (m_\pi - m_\mu)^2/2m_\pi$
$(Q = m_\pi - m_\mu - m_\nu = m_\pi - m_\mu$ and $m_\nu = 0)$
$m_\mu = 206.9m_e = 206.9 \times 0.511 = 105.7\,\text{MeV}$
Therefore,
$T_\mu = 4.12 = (m_\pi - 105.7)^2/2m_\pi$
Solving for m_π
$m_\pi = 141.39\,\text{MeV/c}^2 = 141.39/0.511 m_e$
$= 276.7m_e$
(b) $\mu^+ \to e^+ + \overline{\nu}_e + \nu_\mu$
T_{max} for electron is obtained when $\overline{\nu}_e$ and ν_μ fly together in opposite direction. Thus, the three-body problem is reduced to a two-body one.
$m_\mu c^2 = 206.9 \times 0.511 = 105.72\,\text{MeV}$
$Q = 105.72 - 0.51 = 105.21\,\text{MeV}$
$T_e(\text{max}) = Q^2/2(m_e + Q) = (105.72)^2/2(0.51 + 105.72) = 52.6\,\text{MeV}$

6.65 $Q = 938.2 + 1{,}875.5 - (939.5 + 135.0) - 2.2 = 1{,}737\,\text{MeV}$
$T_\pi = Q(Q + 2m_n)/2(Q + m_\pi + m_n)$
$= 1{,}737(1{,}737 + 2 \times 939.5)/2(1{,}737 + 135 + 939.5) = 1{,}117\,\text{MeV}$
Total energy $E_\pi = 1{,}117 + 135 = 1{,}252\,\text{MeV}$

6.66 $M^2 = {m_1}^2 + {m_2}^2 + 2(E_1E_2 - P_1P_2\cos\theta)$ (1)
$m_1 = 966\,m_e = 966 \times 0.511\,\text{MeV} = 493.6\,\text{MeV}$ (2)
$m_2 = 273\,m_\text{e} = 273 \times 0.511\,\text{MeV} = 139.5\,\text{MeV}$ (3)
$p_1 = 394\,\text{MeV},\ p_2 = 254\,\text{MeV}$ (4)
$E_1 = ({p_1}^2 + {m_1}^2)^{1/2} = [(394)^2 + (93.6)^2]^{1/2} = 631.6$ (5)
$E_2 = ({p_2}^2 + {m_2}^2)^{1/2} = [(254)^2 + (139.5)^2]^{1/2} = 289.5$ (6)
$\cos\theta = \cos 154^\circ = -0.898$ (7)
Using (2) to (7) in (1), $M = 899.4\,\text{MeV}$

6.67 Using the results of Problem 6.60, the angle between the outgoing particles after the collision is given by
$\cos\theta = T/(T + 4M)$
Here $T = 940\,\text{MeV} = M$
Therefore, $\cos\theta = 0.2 \to \theta = 78.46^\circ$

6.68 Using the invariance, $E^2 - |\sum p|^2 = E^{*2}$
$E^{*2} = (E + E_\text{f})^2 - (p^2 + p_\text{f}^2 + 2pp_\text{f}\cos\theta)$
$= (E + E_\text{f})^2 - (E^2 - M^2 + E_\text{f}^2 - M^2 + 2pp_\text{f}\cos\theta$
$= 2M^2 + 2(EE_\text{f} - \mathbf{p}.\mathbf{p}_\text{f})$
(a) For parallel momenta, $\theta = 0$, $\mathbf{p}.\mathbf{p}_\text{f} = +pp_\text{f}$
(b) For anti-parallel momenta $\theta = \pi$, $\mathbf{p}.\mathbf{p}_\text{f} = -pp_\text{f}$
(c) For orthogonal momenta $\theta = \pi/2$, $\mathbf{p}.\mathbf{p}_\text{f} = 0$

6.69 By problem 6.13 it is sufficient to show that $\tan\theta\tan\varphi = 2/(\gamma + 1)$
$p = \gamma\beta m = (\gamma^2 - 1)^{1/2}\,m$

Therefore, $\gamma = [1 + (p/m)^2]^{1/2} = [1 + (5/0.938)^2]^{1/2} = 5.41$
$\tan\theta \tan\varphi = \tan 82^\circ \tan 2^\circ 30' = 7.115 \times 0.04366 = 0.3106$
$2/(\gamma + 1) = 2/(5.41 + 1) = 0.3120$
Hence the event is consistent with elastic scattering.

6.70 Let $\mathbf{p}$, $\mathbf{p}_e$, $\mathbf{p}_N$ be the momentum of the incident electron, scattered electron and recoil nucleus, respectively. From the momentum triangle, Fig. 6.9.

$$p_N{}^2 = p_e{}^2 + p^2 - 2pp_e \cos\theta = E_N{}^2 + 2E_N M \quad (1)$$

where we have put $c = 1$
From energy conservation

$$E_N + E_e = E \quad (2)$$

As

$$E_e \approx p_e \quad (3)$$

$$E \approx P \quad (4)$$

(2) Can be written as

$$E_N + p_e = E \quad (5)$$

Combiining (1), (3), (4) and (5), we get
$E_N = E^2(1 - \cos\theta)/M[1 + E/M(1 - \cos\theta)]$
Restoring c^2, we get the desired result.

Fig. 6.9 Momentum triangle

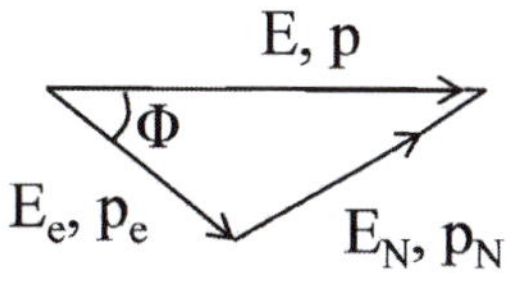

6.71 Use the result of Problem 6.53,

$$T = 2mc^2\beta^2 \cos^2\varphi/(1 - \beta^2 \cos^2\varphi) \quad (1)$$

Put $c = 1$, $T = E - m$ (2)

$$P = M\beta\gamma = M\beta/(1 - \beta^2)^{1/2}$$

whence $\beta^2 = P^2/(P^2 + M^2)$ (3)

Use (2) and (3) in (1) and simplify to get the desired result.

6.72 The formula for the recoil energy of electron in Compton scattering is
$T = (E^2/mc^2)(1 - \cos\theta)/[1 + \alpha(1 - \cos\theta)]$

Here neutrinos are assumed to be massless, so that the same formula which is based on relativistic kinematics can be used.

The maximum recoil energy will occur when the neutrino is scattered back, that is $\theta = 180^\circ$. Substituting $E = 2\,\text{GeV}$ for the incident neutrino energy, $mc^2 = 0.511\,\text{MeV} = 0.511 \times 10^{-3}\,\text{GeV}$, and $\alpha = E/mc^2 = 2/0.511 \times 10^{-3} = 3{,}914$, we find the maximum energy transferred to electron is 1.9997 GeV. The maximum momentum transfer
$p_{max} = (T_{max}^2 + 2\,m_e c^2.T_{max})^{1/2} = 2.0437\ \text{GeV/c}$

6.73 Expressing θ in terms of θ^*

$\tan\theta = \sin\theta^*/\gamma_c(\cos\theta^* + \beta_c/\beta^*)$ (1)

Differentiating with respect to θ^* and setting $\partial\tan\theta/\partial\theta^* = 0$, gives

$\cos\theta^* = -\beta^*/\beta_c$ (2)

And $\sin\theta^* = (\beta_c^2 - \beta^{*2})^{1/2}/\beta_c$ (3)

Using (2) and (3) in (1), and the equations

$\beta^{*2}\gamma^{*2} = \gamma^{*2} - 1$ (4)

${\beta_c}^2{\gamma_c}^2 = {\gamma_c}^2 - 1$ (5)

as well as

$m_1\beta^*\gamma^* = m_2\beta_c\gamma_c$ (momentum conservation in the CMS) (6)

and simplifying, we get

$\tan\theta_{\text{max}} = [{m_2}^2/({m_1}^2 - {m_2}^2)]^{1/2}$

6.74 Let P_1 be the electron four-momentum before the collision and P_2 the final four-momentum. If Q is the four-momrntum transfer then the conservation of energy and momentum requires that

$\mathbf{Q} = \mathbf{P_2} - \mathbf{P_1}$ (1)

and $Q^2 = Q.Q = {P_2}^2 + {P_1}^2 - 2\mathbf{P_1}\mathbf{P_2}$ (2)

But $P = (\mathbf{p}, E/c)$ (3)

$P^2c^2 = E^2 - \mathbf{p}.\mathbf{p}c^2 = {m_e}^2c^4$ (4)

For stationary electron, the initial four-momrntum is

$\mathbf{P_1} = (0, m_e c)$ (5)

and final four-momentum

$\mathbf{P_2} = (p_2, m_e c + \nu/c)$ (6)

Therefore

$\mathbf{P_1}.\mathbf{P_2} = m_e c(m_e c + \nu/c)$ (7)

Substituting (7) in (2) and using (4)

$Q^2 = 2{m_e}^2c^4 - 2m_e c(m_e c + \nu/c)$

$= -2\,m_e\nu$

6.75 (a) As the collision is elastic the total initial energy = total final energy.

$E + E_e = E' + {E_e}'$ (1)

But $p' = p$ (2)

$\therefore E' = E$ (3)

It follows that $E_e' = E_e$ (4)

Consequently ${p_e}' = p_e$ (5)

Momentum conservation gives

$p - p_e = -p' + {p_e}'$ (6)

Because of (2) and (5)

$p'_e = p = E/c$

(b) $\beta = v/c = (1 - 1/\gamma^2)^{1/2} = (1 - m^2/E_e{}^2)^{1/2}$
$= (1 - m^2/(p_e{}^2 + m^2))^{1/2} = (p_e{}^2/(p_e{}^2 + m^2))^{1/2}$
$= (1 + (m_c{}^2/E)^2)^{-1/2}$
where $P_e = E$ and $c = 1$.

6.3.4 Invariance Principle

6.76 Referring to Fig. 7.8, let $\mathbf{P}_\gamma$ and $\boldsymbol{P}_\gamma{}'$ be the four-momentum vectors of photon before and after the scattering, respectively, $\mathbf{P}_e$ and $\boldsymbol{P}_e{}'$ the four vectors of electron before and after scattering respectively. Form the scalar product of the four-vector of the photon and the four-vector of photon + electron. Since this scalar product is invariant.

$$\mathbf{P}_\gamma.(\mathbf{P}_\gamma + \mathbf{P}_e) = \boldsymbol{P}_\gamma{}'.(\boldsymbol{P}_\gamma{}' + \boldsymbol{P}_e{}') \tag{1}$$

Further, total four-momentum is conserved.

$$\mathbf{P}_\gamma + \mathbf{P}_e = \boldsymbol{P}_\gamma{}' + \boldsymbol{P}_e{}' \tag{2}$$

Now, $\mathbf{P}_\gamma = [\mathrm{h}\nu,\ 0,\ 0,\ \mathrm{ih}\nu]$ (3)

$\mathbf{P}_e = [0,\ 0,\ 0,\ \mathrm{imc}]$ (4)

$\boldsymbol{P}_\gamma{}' = [h\,\nu' \cos\theta,\ h\,\nu' \sin\theta,\ 0,\ ih\nu]$

where m is the rest mass of the electron.
Using (1) to (4)

$[h\nu, 0, 0, ih\nu].[h\nu, 0, 0, i(h\nu + mc^2)]$

$= [h\nu' \cos\theta, h\nu' \sin\theta, 0, ih\nu].[h\nu, 0, 0, i(h\nu + mc^2)]$

Therefore, $h^2\nu^2 - h\nu(h\nu + mc^2) = h^2\nu\nu' \cos\theta - h\nu'(h\nu + mc^2)$
Simplifying

$h\nu'\nu(1 - \cos\theta) = mc^2(\nu - \nu')$

Or $h/mc(1 - \cos\theta) = c(1/\nu' - 1/\nu) = \lambda' - \lambda$

Or $\Delta\lambda = \lambda' - \lambda = (h/mc)(1 - \cos\theta)$

This is the well known formula for Compton shift in wavelength (formula 7.37).

6.77 Let the initial and final four momenta of the electron be $\mathbf{P_i} = (E_i/c, \mathbf{p_i})$ and $\mathbf{P_f} = (E_f, \mathbf{p_f})$, respectively. The squared four-momentum transfer is defined by

$$Q^2 = (\mathbf{P_i} - \mathbf{P_f})^2 = -2m^2c^2 + \frac{2E_iE_f}{c^2} - 2\mathbf{P_i} \cdot \mathbf{P_f}$$

However, $E_i = E_f = E$ and $|\mathbf{p_i}| = |\mathbf{p_f}| = E/c$; so neglecting the electron mass

$$Q^2 = 2E^2(1 - \cos\theta)/c^2$$

6.78 $\frac{1}{2}(M^2 - m_1{}^2 - m_2{}^2) = E_1\, E_2 - p_1\, p_2 \cos\,\theta$
$\theta = 90^0$
$P_\pm = 530\,\text{MeV/c}$
$E_\pm = (P_\pm^2 + m_\pi{}^2)^{1/2} = 548$
$2 \times 548^2 + 2 \times 140^2 = M^2$
$M = 800\,\text{MeV/c}$
It is a ρ meson.

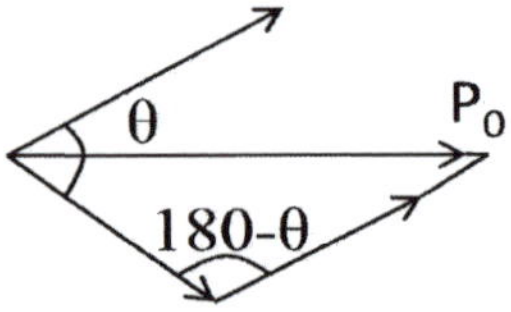

Fig. 6.10

6.79 Using the invariance of squared four-momentum before and after the decay

$E_i{}^2 - P_i{}^2 = E_f{}^2 - |\mathbf{p_f}|^2$
$M^2 = (E_1 + E_2)^2 - (p_1{}^2 + p_2{}^2 + 2p_1\, p_2 \cos\theta)$
$= (E_1{}^2 - p_1{}^2) + (E_2{}^2 - p_2{}^2) + 2(E_1\, E_2 - p_1\, p_2 \cos\,\theta)$
$= m_1{}^2 + m_2{}^2 + 2(E_1\, E_2 - p_1\, p_2 \cos\,\theta)$
Or $E_1\, E_2 - p_1\, p_2 \cos\,\theta = \frac{1}{2}(M^2 - m_1{}^2 - m_2{}^2) = \text{Invariant}$

6.80 $s + t + u = (1/c^2)[(P_A + P_B)^2 + (P_A - P_C)^2 + (P_A - P_D)^2]$
$= (1/c^2)[3P_A{}^2 + P_B{}^2 + P_C{}^2 + P_D{}^2 + 2P_A(P_B - P_C - P_D)]$ (1)
From four-momentum conservation, $P_A + P_B = P_C + P_D$ (2)
(1) becomes $(s + t + u)c^2 = (3m_A{}^2 + m_B{}^2 + m_C{}^2 + m_D{}^2) - 2P_A{}^2$
Using $P_A = m_A c$, $P_B = m_B c$, $P_C = m_C c$ and $P_D = m_D c$
$(s + t + u)c^2 = (m_A{}^2 + m_B{}^2 + m_C{}^2 + m_D{}^2)c^2$
Or $s + t + u = \sum_{i=A,B,C,D} m_i^2$

6.81 $t = (P_A - P_C)^2/c^2 = (1/c^2)(P_A{}^2 + P_C{}^2 - 2P_A P_C)$
$= (1/c^2)[m_A{}^2c^2 + m_C{}^2c^2 - 2(E_A E_C/c^2 - \mathbf{P}_A.\mathbf{P}_C)]$
For elastic scattering $A \equiv C$. Thus $E_A = E_C$ and $|\mathbf{P}_A| = |\mathbf{P}_C| = p$
So that $\mathbf{P}_A.\mathbf{P}_C = p^2 \cos\theta$.
$c^2 t = 2m_A{}^2c^2 - 2(E_A{}^2/c^2 - p^2 \cos\theta)$
But $E_A{}^2 = c^2 p^2 + m_A{}^2 c^4$
Therefore, $t = -2p^2(1 - \cos\theta)/c^2$

6.82 $\sin\varphi/2 = m_\pi c^2/2(E_1 E_2)^{1/2}$ (see Prob. 6.103 and 6.104)
Minimum angle is $\varphi_{\min} = 2/\gamma$

But $\gamma = 10/0.14$
Therefore, $\varphi_{\min} = 0.028$ rad or 1.6°

6.83 Rest mass energy of ω^0 = Total available energy – (total kinetic energy + mass energy of π^+ and π^-)

$$m_\omega c^2 = 2.29 - (1.22 + 0.14 + 0.14) = 0.79\,\text{GeV}$$

6.84 Energy conservation gives

$$m_1 \gamma_1 + m_2 = M \gamma \quad (1)$$

Momentum conservation gives

$$m_1 \gamma_1 \beta_1 = M \gamma \beta \quad (2)$$

Squaring (1)

$$m_1^2 \gamma_1^2 + m_2^2 + 2\gamma_1 m_1 m_2 = M^2 \gamma^2 \quad (3)$$

Squaring (2)

$$m_1^2 \gamma_1^2 \beta_1^2 = M^2 \gamma^2 \beta^2 \quad (4)$$

Using $\beta_1 = (1 - 1/\gamma_1^2)^{1/2}$ and $\beta = (1 - 1/\gamma^2)^{1/2}$
(4) becomes

$$m_1^2(\gamma_1^2 - 1) = M^2(\gamma^2 - 1) \quad (5)$$

Subtracting (5) from (3)

$$m_1^2 + m_2^2 + 2m_1 m_2/(1 - v^2/c^2)^{1/2} = M^2$$

6.85 $E_0 = E_p + E_\pi$ (energy conservation) (1)

$$Q = m_0 - (m_p + m_\pi) \quad (2)$$

$$P_0^2 = P_p^2 + P_\pi^2 + 2P_p P_\pi \cos\theta \quad (3)$$

Or $E_0^2 - m_0^2 = E_p^2 - m_p^2 + E_\pi^2 - m_\pi^2 + 2P_p P_\pi \cos\theta$ (4)

Using (1) in (4) and simplifying

$$2E_p E_\pi - 2P_p P_\pi \cos\theta + m_p^2 + m_\pi^2 = m_0^2 = (Q + m_p + m_\pi)^2 \quad (5)$$

Or $Q = (m_p^2 + m_\pi^2 + 2E_p E_\pi - 2P_p P_\pi \cos\theta)^{1/2} - (m_p + m_\pi)$ (6)

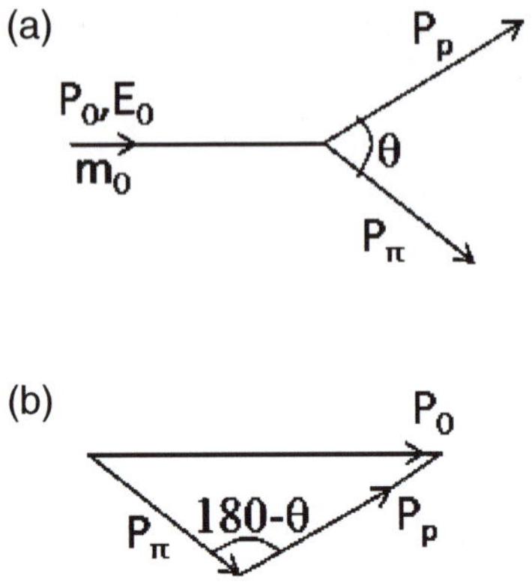

Fig. 6.11 **(a)** Decay $\wedge \to P + \pi^-$ in flight. **(b)** Momentum triangle

6.86 $M_0\beta\gamma = p_1 \sin\theta + p_2 \cos\theta$ (momentum conservation along x-axis) (1)

$p_1 \cos\theta = p_2 \sin\theta$ (momentum conservation along y-axis) (2)

$M_0\gamma = E_1 + E_2$ (Energy conservation) (3)

Solving (1), (2) and (3)

$M_0 = ({m_1}^2 + m_2^2 + 2E_1\ E_2)^{1/2}$

$\beta = ({p_1}^2 + {p_2}^2)^{1/2}/(E_1 + E_2)$

$\theta = \tan^{-1}(p_1/p_2)$

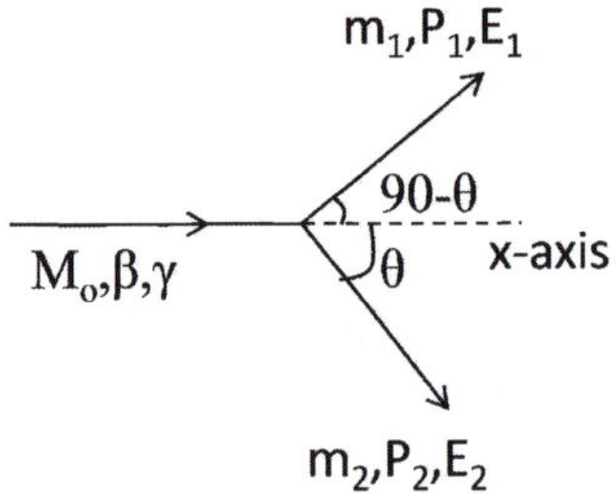

Fig. 6.12 Decay $M_0 \to m_1 + m_2$

6.87 Under the assumption (a)

$M^2 = 2(E_{\pi+} + E_{\pi-} - P_{\pi+}P_{\pi-}\cos\theta) + {m_{\pi+}}^2 + {m_{\pi-}}^2$ (1)

$E_{\pi+} = ({p_\pi}^2 + {m_{\pi+}}^2)^{1/2}$ (2)

$E_{\pi-} = ({P_{\pi-}}^2 + {m_{\pi-}}^2)^{1/2}$ (3)

$m_{\pi+} = m_{\pi-} = 0.14\,\mathrm{GeV/c^2}, \theta = 15^\circ$ (4)

$p_+ = 1.67,\ p_- = 0.408\,\mathrm{GeV/c}$ (5)

Using (2), (3), and (4) in (1) and solving for M, we find $M = 0.239\,\mathrm{GeV/c^2}$, a value quite different from the standard value, $m_{k0} = 0.498\,\mathrm{GeV/c^2}$

Under the assumption (b)

$$M^2 = 2(E_p E_{\pi -} - P_p\, P_{\pi -} \cos\theta) + m_{\pi +}{}^2 + m_{\pi -}{}^2 \quad (6)$$

$$E_p = (P_+{}^2 + m_p{}^2)^{1/2} \quad (7)$$

$$m_p = 0.938\,\text{GeV}/\text{c}^2 \quad (8)$$

Using (3), (4), (5), (7) and (8) in (6) and solving for M, we find $M = 1.109\,\text{GeV}/\text{c}^2$ which is in good agreement with the mass $m_\Lambda = 1.115\,\text{GeV}/\text{c}$ Thus, the neutral particle is Λ.

6.88 Let the momenta of photons in the LS be p_1 ad p_2 ad energies E_1 and E_2. The invariant mass W of the initial state is given by

$W^2 = E^2 - p^2 = m^2$

In the final state

$E^2 - p^2 = (E_1 + E_2)^2 - |(p_1 + p_2)|^2$

$= 2E_1\, E_2(1 - \cos\varphi) = 4E_1\, E_2 \sin^2(\varphi/2)$ (because $E_1 = p_1$ and $E_2 = p_2$ and $\mathbf{p}_1.\mathbf{p}_2 = E_1\, E_2 \cos\varphi$)

Invariance of $E^2 - p^2$ gives

Sin $(\varphi/2) = mc^2/2(E_1\, E_2)^{1/2}$

6.89 $n \to p + e^- + \overline{\nu}$

The proton will carry maximum energy when the neutrino with negligible mass is at rest.

$(q_n - q_p)^2 = (E_n - E_p) - P_p{}^2$

$= (m_n - E_p)^2 - (E_p{}^2 - m_p{}^2)$; (because neutron is at rest)

$= m_n{}^2 + m_p{}^2 - 2m_n\, E_p$

But $P_p = P_e \to P_p{}^2 = P_e{}^2$

Or $E_p{}^2 - m_p{}^2 = E_e{}^2 - m_e{}^2 = (m_n - E_p)^2 - m_e{}^2$

$= m_n{}^2 - 2m_n\, E_p + E_p{}^2 - m_e{}^2$

$\therefore m_n{}^2 + m_p{}^2 - 2m_n\, E_p = m_e{}^2$

Thus $(q_n - q_p)^2 = m_e^2$

Or $q_n - q_p = m_e c^2 = 0.511$ MeV/c

6.3.5 Transformation of Angles and Doppler Effect

6.90 $\lambda = \lambda' \sqrt{(1 + \beta)/(1 - \beta)}$

$\beta = v/c = 3 \times 10^6 / 3 \times 10^8 = 0.01$

$\lambda' = 6{,}563\,\text{Å}$

$\lambda = 6{,}629\,\text{Å}$

$\Delta\lambda = \lambda - \lambda' = 6{,}629 - 6{,}563 = 66\,\text{Å}$

6.91 (a) $\lambda/\lambda' = [(1+\beta)/(1-\beta)]^{1/2} = 656/434 = 1.5115$

$v = \beta c = 1.17 \times 10^8\ \text{ms}^{-1}$

(b) the nebula is receding

6.92 $\lambda/\lambda' = [(1+\beta)/(1-\beta)]^{1/2} = (1+\beta)^{1/2}(1-\beta)^{-1/2}$

$\approx (1+\beta/2+\ldots)(1+\beta/2-\ldots)$

$= 1+\beta+\ldots$ (neglecting higher order terms)

$\Delta\lambda/\lambda' = (\lambda/\lambda') - 1 = \beta = v/c$

6.93 $\lambda/\lambda' = [(1+\beta)/(1-\beta)]^{1/2} = 670/525$

$\beta = 0.239$

$v = \beta c = 0.239 \times 3 \times 10^8 = 7.17 \times 10^7\ \text{ms}^{-1}$

$= 7.17 \times 10^4\ \text{kms}^{-1}$

This speed exceeds the escape velocity. Hence the explanation is not valid.

6.94 $\lambda' = \lambda[(1-\beta)/(1+\beta)]^{1/2} = 589.3[(1-0.21)/(1+0.21)]^{1/2} = 476.2\,\text{nm}$

The color is blue

6.95 The source velocity is perpendicular to the line of sight. $\theta = 90°$,

$\nu' = \nu\gamma$

$\lambda = \gamma\lambda'$

$\gamma = 1/(1-\beta^2)^{1/2} = 1/(1-0.05^2)^{1/2} = 1.00125$

$\Delta\lambda = \lambda - \lambda' = \lambda'(\gamma - 1) = 589(1.00125 - 1)$

$= 0.736\,\text{nm} = 7.36\,\text{Å}$

6.96 Let the electron recoil at angle φ with momentum p, and neutrino get scattered with energy E' and momentum p'.

Energy conservation gives

$$E_0 + m = E + E' \tag{1}$$

From the momentum triangle

$$p'^2 = p_0{}^2 + p^2 - 2p_0\, p\cos\varphi \tag{2}$$

We can write $p' = E'$, $p = E$, $p_0 = E_0$, so that (2) becomes

$$E'^2 = E_0{}^2 + E^2 - 2E_0E\cos\varphi \tag{3}$$

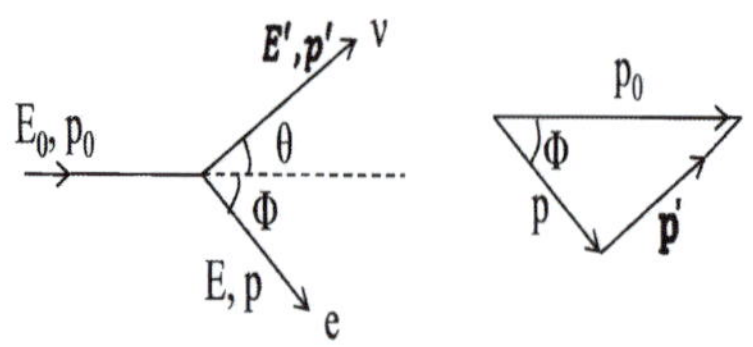

Fig. 6.13 Collision of an energetic neutrino with a stationary electron

From (1) we have

$E'^2 = (E_0 - E + m)^2$ (4)

Comparing (3) and (4) and simplifying

$\cos\varphi = 1 - m(E_0 - E - m)/E_0E \approx 1 - m(E_0 - E)/E_0E$ (5)

where we have neglected m in comparison with $E_0 - E$.

For small angle, (5) becomes $\varphi = [2m(E_0 - E)/E_0E]^{1/2}$ (6)

For $E_0 = 2\,\text{GeV}$, $E = 0.5\,\text{GeV}$, $m = 0.51 \times 10^{-3}\,\text{GeV}$
$\varphi = 0.039$ radians $= 2.24°$

6.97 Let the mass of the primary particle be M, and that of secondary particles m_1 and m_2. Let the total energy of the secondary particles in the LS be E_1 and E_2, and momenta p_1 and p_2. Using the invariance of (total energy)^2– (total momentum)^2

$M^2 = (E_1 + E_2)^2 - ({p_1}^2 + {p_2}^2 + 2p_1\, p_2 \cos\,\theta)$

$= {E_1}^2 - {p_1}^2 + {E_2}^2 - {p_2}^2 + 2(E_1^2 - {p_1}^2 \cos\,\theta)$

$= {m_1}^2 + {m_2}^2 + 2({E_1}^2 - {p_1}^2 \cos\,\theta)$ (Since $E_1 = E_2$, $p_1 = p_2$)

$= 2{m_1}^2 + 2({m_1}^2 + {p_1}^2 - {p_1}^2 \cos\,\theta)$

$= 4{m_1}^2 + 4{p_1}^2 \sin^2\theta/2$

$= 4(140)^2 + 4(300)^2 \sin^2 35°$

$= 196{,}836$

$M = 444\,\text{MeV/c}^2$

6.98 Consider one of the two γ-rays. From Lorentz transformation

$cp_x = \gamma_c(cp_x{}^* + \beta_c E^*)$

where the energy and momentum refer to one of the two γ-rays and the subscript C refers to π^0. Starred quantities refer to the rest system of π^0.

$cp\cos\theta = \gamma(cp^*\cos\theta^* + \beta E^*)$

where we have dropped off the subscript C. But for γ-rays $cp^* = E^*$ and $cp = E$, and because the two γ-rays share equal energy in the CMS, $E^* = mc^2/2$, where m is the rest mass of $\pi°$.

Therefore $cp\cos\theta = (\gamma mc^2/2)(\beta + \cos\theta^*)$ (1)

Also, $E = \gamma(E^* + \beta cp_x^*) = \gamma(E^* + \beta cp^*\cos\theta^*)$

or

$cp = E = (\gamma mc^2/2)(1 + \beta\cos\theta^*)$ (2)

When $\theta^* = 0$

$$E_{\max} = \frac{1}{2}E_{\pi°}(1 + \beta) \quad (3)$$

When $\theta^* = \pi$

$$E_{\min} = \frac{1}{2}E_{\pi°}(1 - \beta) \quad (4)$$

From (3) and (4),$(E_{\max} - E_{\min})/(E_{\max} + E_{\min}) = \beta$ (5a)

From the measurement of $E_{\max}$ and $E_{\min}$, the velocity of π^0 can be determined.

6.99 In the solution of Problem 6.98 multiply (3) and (4) and write $E = \gamma m_\pi c^2$

$m_\pi c^2 = 2(E_{\max} E_{\min})^{1/2}$ (5b)

From the measurement of $E_{\max}$ and $E_{\min}$, mass of π^0 can be determined.
It $E_{\max} = 75\,\mathrm{MeV}$ and $E_{\min} = 60\,\mathrm{MeV}$, then $mc^2 = 2 \times (75 \times 60)^{1/2} = 134.16\,\mathrm{MeV}$
Hence the mass of π^0 is $134.16/0.51 = 262.5m_\mathrm{e}$

6.100 $\mathrm{d}N/\mathrm{d}E = (\mathrm{d}N/\mathrm{d}\Omega^*).\mathrm{d}\Omega^*/\mathrm{d}E = (1/4\pi).2\sin\theta^*\mathrm{d}\theta^*/\mathrm{d}E$

$= (1/2)\mathrm{d}\cos\theta^*/\mathrm{d}E$ (6)

where we have put $\mathrm{d}\Omega^* = 2\pi \sin\theta^*\, \mathrm{d}\,\theta^*$ for the element of solid angle and $\mathrm{d}N/\mathrm{d}\Omega^* = 1/4\pi$ under the assumption of isotropy.
Differentiating (2) with respect to $\cos\theta^*$

$\mathrm{d}E/\mathrm{d}\cos\theta^* = \gamma\beta mc^2/2$

or $\mathrm{d}\cos\theta^*/2\mathrm{d}E = 1/\gamma\beta mc^2$ (7)

Combining (6) and (7), the normalized distribution is

$\mathrm{d}N/\mathrm{d}E = 1/\gamma\beta mc^2 = \text{constant}$ (8)

This implies that the energy spectrum is rectangular or uniform. It extends from a minimum to maximum, Fig. 6.14.
From (3) and (4),

$E_{\max} - E_{\min} = \beta E_\pi = \gamma\beta mc^2$ (9)

Note that the area of the rectangle is height × length

$(\mathrm{d}N/\mathrm{d}E) \times (E_{\max} - E_{\min}) = (1/\gamma\beta mc^2) \times \gamma\beta mc^2 = 1$

That is, the distribution is normalized as it should.

The higher the π^0 energy the larger is the spread in the γ-ray energy spectrum. For mono-energetic source of π^0 s, we will have a rectangular distribution of γ-ray energy as in Fig. 6.14. But if the γ-rays are observed from π^0 s, of varying energy, as in cosmic ray events the rectangular distributions may be superimposed so that the resultant distributions may look like the solid curve, shown in Fig. 6.15.

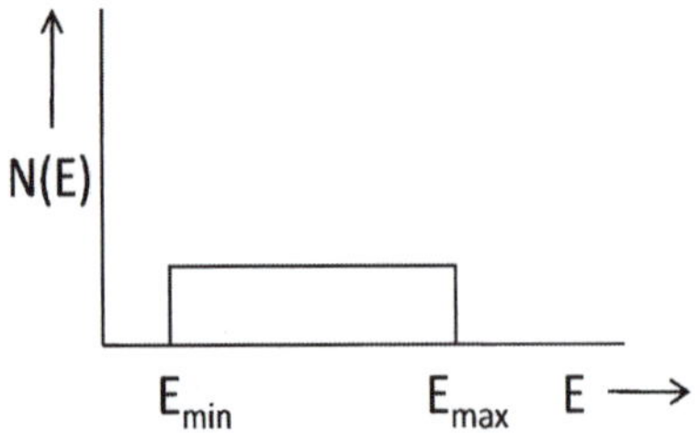

Fig. 6.14 γ-ray energy spectrum from π^0 decay at fixed energy

Note that if the π^0 s were to decay at rest ($\gamma = 1$) then the rectangle would have reduced to a spike at $E = 67.5$ MeV, half of rest energy of π^0.

6.101 The γ-rays of intensity $I(\theta^*)$ which are emitted in the solid angle d Ω^* in the CMS will appear in the solid angle dΩ^* in the LS with intensity $I(\theta)$. Therefore

$I(\theta)\mathrm{d}\,\Omega = I(\theta^*)\mathrm{d}\,\Omega^*$

or

$$I(\theta) = I(\theta^*)\sin\theta^*\mathrm{d}\theta^*/\sin\theta\mathrm{d}\theta \tag{10}$$

Fig. 6.15 γ-ray energy spectrum from π^0 decay in cosmic ray events

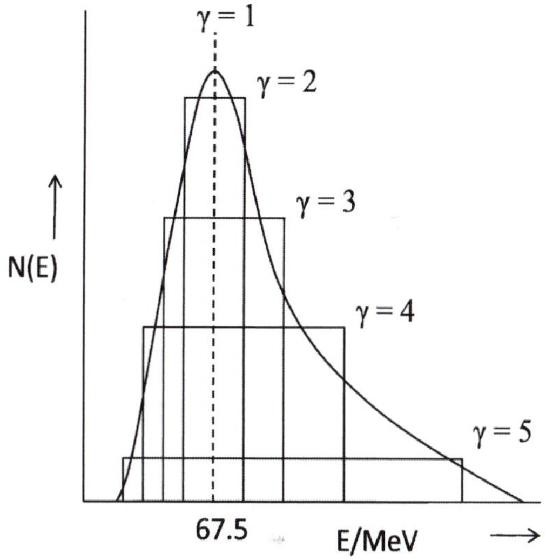

From the Lorentz transformation

$E^* = \gamma E(1 - \beta\cos\theta)$

$= \gamma E^*\gamma(1 + \cos\theta^*)(1 - \cos\theta)$

Where we have used (2)

$1/\gamma^2(1 - \beta\cos\theta) = 1 + \beta\cos\theta^*$

Differentiating

$-\beta\sin\theta\mathrm{d}\theta/\gamma^2(1 - \beta\cos\theta)^2 = -\beta\sin\theta^*\mathrm{d}\theta^*$

Therefore $\sin\theta^*\mathrm{d}\theta^*/\sin\theta\mathrm{d}\theta = 1/\gamma^2(1 - \beta\cos\theta)^2$ (11)

Also $I(\theta^*) = 1/4\pi$ (12)

because of assumption of isotropy of photons in the rest frame of π^0

Combining (10), (11) and (12)

$$I(\theta) = 1/4\pi\gamma^2(1 - \beta\cos\theta)^2 \tag{13}$$

This shows that small emission angles of photons in the lab system are favored.

6.102 In Fig. 6.16, AB and AD represent the momentum vectors of the two photons in the LS. BC is drawn parallel to AD so that ABC forms the momentum triangle, that is

AB + BC = AC

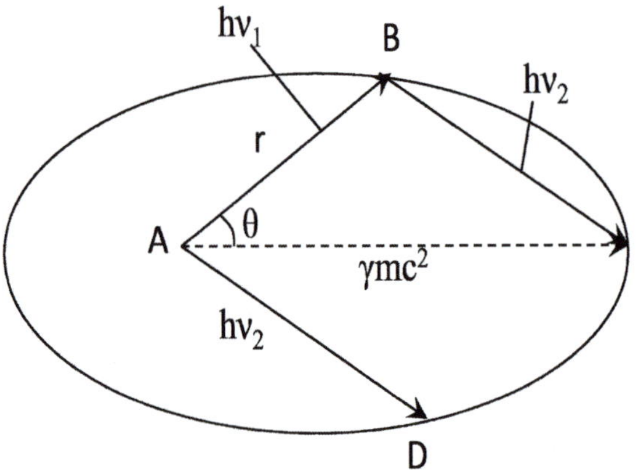

Fig. 6.16 Locus of the tip of the momentum vector of γ-rays from π^0 decay is an ellipse

From energy conservation

$h\,\nu_1 + h\,\nu_2 = \gamma\, mc^2/2(1+\beta\,\cos\,\theta^*) + (\gamma\, mc^2/2)(1+\beta\,\cos(\pi-\theta^*))$
$= \gamma\, mc^2 = \text{const}$

where we have used the fact that the angles of emission of the two photons in the rest frame of π^0 , are supplementary.
Since momentum is given by $p = h/c$, it follows that AB + BC = constant, which means that the locus of the tip of the momentum vector is an ellipse.

$$E = mc^2/2\gamma(1-\beta\cos\theta) \qquad (14)$$

Compare this with the standard equation for ellipse

$$r = a(1-\varepsilon^2)/(1-\varepsilon\cos\theta)$$

We find $\varepsilon = \beta$

$a(1-\beta^2) = mc^2/2\,\gamma$
Or $a = \gamma\, mc^2/2$

The larger the velocity of π^0, the greater will be the eccentricity, ε.

6.103 The angle between the two γ-rays in the LS can be found fom the formula

$$m^2 c^4 = m_1{}^2 c^4 + m_2{}^2 c^4 + 2(E_1 E_2 - c^2 p_1 p_2 \cos \varphi)$$

Putting $m_1 = m_2 = 0, cp_1 = E_1, cp_2 = E_2$

$$m^2 c^4 = 2E_1 E_2(1 - \cos \varphi) = 4E_1 E_2 \sin^2(\varphi/2)$$

$$\sin (\varphi/2) = mc^2/2(E_1 E_2)^{1/2} \quad (15)$$

6.104 For small angle φ,

$$\varphi = mc^2/(E_1 E_2)^{1/2} \quad (16)$$

Set $E_2 = \gamma mc^2 - E_1$

$$\varphi = mc^2/[E_1(\gamma mc^2 - E_1)]^{1/2} \quad (17)$$

For minimum angle d $\varphi/\mathrm{d}E_1 = 0$. This gives $E_1 = \gamma mc^2/2$.
Using this value of E_1 in (17), we obtain

$$\varphi_{\min} = 2/\gamma \quad (18)$$

Measurement of $\varphi_{\min}$ affords the determination of E_π via γ.

$$\varphi_{\min} = 2mc^2/E_\pi$$

6.105 $E_1 + E_2 = E$ (energy conservation) (1)

$\mathbf{p}_1 + \mathbf{p}_2 = \mathbf{p}$ (momentum conservation) (2)

Taking the scalar product

$$(\mathbf{p}_1 + \mathbf{p}_2).(\mathbf{p}_1 + \mathbf{p}_2) = \mathbf{p}.\mathbf{p}$$

or

$$p_1{}^2 + p_2{}^2 + 2p_1 p_2 \cos\theta = p^2 \quad (3)$$

Using $c = 1$, Eq. (3) becomes

$$E_1{}^2 + E_2{}^2 + 2E_1 E_2 \cos\theta = E^2 - m^2 \quad (4)$$

Let $E_2/E_1 = D$, (5)

the disparity factor. Then (1) becomes

$$E_1(D + 1) = E \quad (6)$$

Combining (4), (5) and (6)

$$2DE^2(1 - \cos\theta) = m^2$$

Or $\sin\theta/2 = [m/2E][\sqrt{}D + 1/\sqrt{}D]$
For small θ,

$$\theta = [m/E][\sqrt{}D + 1\sqrt{}D]$$

The minimum angle is found by setting $d\theta/dD = 0$
This gives us $D = 1$, that is $E_1 = E_2 = E/2$.

$\theta_{\text{min}} = 2mc^2/E$

The Lorentz transformation of angles gives us the relation

$E = [m\,\gamma/2][1 + \beta\ \cos\ \theta^*)$

We need to consider one of the photons in the forward hemisphere. The fraction of photons emitted in the CMS (rest frame of π^0) within the angle θ^* is $(1 - \cos\theta^*)$. This fraction is 1/2 for $\theta^* = 60^0$, that is $\cos\theta^* = 1/2$. When one photon goes at $\theta^* = 60^0$, the other photon will go at $\theta^* = 120^0$ with the direction of flight of π^0. Hence $\cos\theta^* = -1/2$ for the second photon. The disparity factor

$D = E_2/E_1 = (1 + \beta/2)/(1 - \beta/2) = (2 + \beta)/(2 - \beta)$.

For relativistic pions $\beta = 1$ Hence $D > 1$.

A quarter of pions will be emitted within an angle $\theta^* = 41.4^0$, that is $\cos\theta^* = 0.75$. In this case

$D = (1 + 3\beta/4)/(1 - 3\beta/4)$

And the previous argument gives us $D > 7$

6.106 First find E^* the total energy available in the CMS

$E^{*2} = (E_\pi + m_p)^2 - P_\pi^2 \approx (P_\pi + m_p)^2 - P_\pi{}^2$ (Because $E_\pi \gg m_\pi$)

$E^* = 4.436\,\text{GeV}$

Total energy carried by K^* in the CMS

$E_k{}^* = (E^{*2} + m_k{}^{*2} - m_{\gamma^0})/2E^* = 1.942\,\text{GeV}$

$\gamma_K{}^* = E_K{}^*/m_K{}^* = 1.942/0.89 = 2.18$

$\beta_K{}^* = 0.8888$

$\gamma_c = (\gamma + \nu)/(1 + 2\gamma\nu + \nu^2)^{1/2}$

$\gamma = 10/0.14 = 71.4$

$\nu = m_2/m_1 = 0.940/0.140 = 6.71$

$\gamma_c = 2.466$, $\beta_c = 0.9141$

$\tan\theta = \sin\theta^*/(\cos\theta^* + \beta_c/\beta^*)$ (1)

Differentiate with respect to θ^* and set

$\partial\tan\theta/\partial\theta^* = 0$.This gives $\cos\theta^* = -\beta^*/\beta_c$

$\cos\theta^* = -0.8888/0.9141 = -0.9723$

$\theta^* = 166.5^0$

Using the values of θ^*, γ_c, and the ratio β_c/β^* in (1) we find $\theta_m = 59.3^\circ$

6.107 (a), (b) In the CMS, m_2 will move with the velocity β_c in a direction opposite to that of m_1. By definition, the total momentum in the CMS before and after the collision is zero. In natural units $c = 1$.

$$m_1 \gamma^* \beta^* = m_2 \gamma_c \beta_c \quad (1)$$

Squaring (1) and expressing the velocities in terms of Lorentz factors

$$m_1{}^2(\gamma^{*2} - 1) = m_2{}^2(\gamma_c{}^2 - 1) \quad (2)$$

Using the invariance

$$(\Sigma E)^2 - |\Sigma \mathbf{P}|^2 = (\Sigma E^*)^2 - |\Sigma \mathbf{P}^*|^2 = (\Sigma E^*)^2 \quad (3)$$

(because $\sum \mathbf{P}^* = 0$, in the CMS)

$$(m_1 \gamma + m_2)^2 - m_1{}^2(\gamma^2 - 1) = (m_1 \gamma^* + m_2 \gamma_c)^2 \quad (4)$$

Combining (2) and (4) and calling $\upsilon = m_2/m_1$

$$\gamma_c = (\gamma + \nu)/(1 + 2\gamma\nu + \nu^2)^{1/2} \quad (5)$$

$$\gamma^* = (\gamma + 1/\nu)/(1 + 2\gamma/\nu + 1/\nu^2)^{1/2} \quad (6)$$

For the special case, $m_1 = m_2$, as in the P–P collision

$$\gamma_c = \gamma^* = [(\gamma + 1)/2]^{1/2} \quad (7)$$

In addition if $\gamma \gg 1$

$$\gamma_c \approx (\gamma/2)^{1/2} \quad (8)$$

(c), (d)
The Lorentz transformations are

$$P \cos\theta = \gamma_c(p^* \cos\theta^* + E^*) \quad (9)$$

$$P \sin\theta = p^* \sin\theta^* \quad (10)$$

Dividing (10) by (9)

$$\tan\theta = p^* \sin\theta^*/\gamma_c(p^* \cos\theta^* + \beta_c E^*) = \sin\theta^*/\gamma_c(\cos\theta^* + \beta_c/\beta^*) \quad (11)$$

(because $p^*/E^* = \beta^*$)
From the inverse transformation

$$P^* \cos\theta^* = \gamma_c(P \cos\theta - \beta_c E) \quad (12)$$

and (10) we get

$$\tan\theta^* = \sin\theta/\gamma_c(\cos\theta - \beta_c/\beta) \quad (13)$$

6.108 At the right angle to the direction of source velocity the Doppler shift in wavelength is calculated from

$$\nu' = \gamma\nu \text{ or } \lambda' = \lambda/\gamma$$

where γ is the Lorentz factor of the carbon atoms and T is the kinetic energy of carbon and Mc^2 is the approximate rest mass energy, the quantity

$(\lambda' - \lambda)/\lambda$ becomes $(1 - \gamma)/\gamma$

But $\gamma = 1 + T/Mc^2 = 1 + 120/12 \times 10 = 1.01$
Hence $(\lambda' - \lambda)/\lambda = -0.01/1.01 = -0.0099$

6.109 The observed frequency v due to Doppler effect is given by

$$\nu = \gamma \nu'(1 + \beta \cos\theta')$$

Where ν' is the natural frequency
When the star is moving toward the observer $\theta' = 0$

$$\beta = v/c = (300\,\text{km/s})/3 \times 10^5\,\text{km/s} = 10^{-3}$$

$$\gamma = 1/(1 - \beta^2)^{1/2} \approx 1 + (1/2)\beta^2 = 1 + 5 \times 10^{-7}$$

Neglecting small terms, $v = (1 + 10^{-3})\nu'$
Fractional change in frequency

$$(\nu - \nu')/\nu' = 10^{-3}$$

6.110 Use the formula for Lorentz transformation of angles from CMS to LS

$$\tan\theta = \sin\theta^*/\gamma_c(\cos\theta^* + \beta_c/\beta^*). \tag{1}$$

For one of the photons, in the rest system of $\pi^\circ.\theta^* = 90^\circ.\beta^* = 1$. From the given value $\beta_c = 0.8$ we find $\gamma_c = 1.6666$. Inserting these values in (1) $\tan\theta = 0.75$ or $\theta = 36.87^\circ$ in the LS. From symmetry the second photon will also be emitted at the same angle on the other side of the line of flight and be coplanar. Hence the angle between the two photons will be $2\theta = 73.75^\circ$

6.111 Use the formula for the transformation of angles.

$$\tan\theta = \sin\theta^*/\gamma_c(\cos\theta^* + \beta_c/\beta^*)$$

Use $\theta = 45^\circ$, $\theta^* = 60^\circ$, $\beta^* = 1$, $\gamma_c = 1/(1 - \beta_c{}^2)^{1/2}$ in the above formula, and simplify to obtain a quadratic equation in β_c. On solving this equation we find the velocity of the object

$$v = \beta_c c = 0.535\,c$$

6.112 (a) The y-component of the rod is unchanged that is $L_y = L_y{}'$

$$\text{or } L\sin\theta = L'\sin\theta' \tag{1}$$

$$\text{Also } L_x = L\cos\theta \tag{2}$$

$$L'_x = L'\cos\theta' \tag{3}$$

Eliminating L and L', (4)

$$L_x.\tan\theta = L_x{}'.\tan\theta' \tag{5}$$

(b) $L_x{}' = L_x/\gamma$
where γ is the Lorentz factor. Using (4) in (5)

$$\gamma = L_x/L_x{}' = \tan\theta/\tan\theta = \tan 45^\circ/\tan 30^\circ = \sqrt{3}$$

$$\beta = (\gamma^2 - 1)^{1/2}/\gamma = \sqrt{\frac{2}{3}} = 0.816$$

The speed at which the rod is moving is $v = \beta c = 0.816\,c$

6.3.6 Threshold of Particle Production

6.113 For the production reaction

$$m_1 + m_2 \to m_3 + m_4$$

The threshold energy for m_1 when m_2 is at rest is

$$T_1 = [(m_3 + m_4)^2 - (m_1 + m_2)^2]/2m_2$$

In the given reaction we can put

$$m_3 + m_4 = 4M, m_1 = m_2 = M$$

$$T_1 = 6M \text{ or } T_1 = 6Mc^2$$

6.114 Here $m_1 = 0$, $m_2 = m$, $(m_3 + m_4) = 3m$

$$T_1 = 4mc^2$$

6.115 If m_1 is the projectile mass, m_2 target mass, and $m_3 + m_4 + m_5$, the mass of product particles. The threshold is given by formula

$$T_1 = [(m_3 + m_4 + m_5)^2 - (m_1 + m_2)^2]/2m_2$$

$$= [(940 + 940 + 140)^2 - (940 + 940)^2]/2 \times 940$$

$$= 290.4\,\text{MeV}$$

The threshold energy is thus slightly greater than twice the rest-mass energy of pion (140 MeV). Non-relativistically, the result would be 280 MeV, that is double the rest mass energy of Pion. The extra energy of 10 MeV is to be regarded as relativistic correction

6.116 Use the invariance of $E^2 - P^2 = E^{*2} - P^{*2} = E^{*2} - 0 = E^{*2}$

$$E^{*2} = (4m)^2 = (T + m + m + 0.025)^2 - (P_1 - 0.218)^2$$

Putting $m = 0.938$, $P_1 = (T^2 + 2Tm)^{1/2}$ and
solving for T, we find that T (threshold) $= 4.3\,\text{GeV}$

6.117 $T_{\text{threshold}} = [(m_3 + m_4)^2 - (m_1 + m_2)^2]/2m_2$

$$= [(0.89 + 1.11)^2 - (0 + 0.94)^2]/2 \times 0.94$$

$$= 3.12\,\text{GeV}$$

6.118 $T_{\text{threshold}} = [(m_p + m_p + m_p)^2 - (m_p + m_p)^2]/2m_p$ by Eq. (6.53)

$m_p = 1{,}837\, m_e = 1{,}837 \times 0.00.00051\,\text{GeV} = 0.937\,\text{GeV}$

$M = 273 \times 0.00051\,\text{GeV} = 0.137\,\text{GeV}$

Using the above values we find $T_{\text{threshold}} = 0.167$ GeV

6.119 $T_{\text{threshold}} = (m_k + m_\Lambda)^2 - (m_p + m_p)^2/2m_p$

$m_k = 0.498\,\text{GeV}, m_\Lambda = 1.115\,\text{GeV}, m_\pi = 0.140\,\text{GeV}, m_p = 0.938\,\text{GeV}$

Using these values, we find $T_{\text{threshold}} = 0.767\,\text{GeV} = 767\,\text{MeV}$

Note that when pions are used as bombarding particles the threshold for strange particle production is lowered then in N–N collisions. However, first a beam of pions must be produced in N–N collisions.

6.120 Consider the reaction $P + P \rightarrow P + P + n\pi$

$T_{\text{threshold}} = [(m_p + m_p + nm_\pi)^2 - (m_p + m_p)^2]/2m_p$

Simplifying we get the desired result

6.121 $T_{\text{threshold}} = [(m_p + m_{\pi0})^2 - (m_p + 0)^2]/2m_p$
Using $m_p = 940\,\text{MeV}$ and $m_{\pi0} = 135\,\text{MeV}$, we
find $T_{\text{threshold}} = 145\,\text{MeV}$

Note that the threshold energy for pion production in collision with gamma rays is only half of that for N–N collisions. But the cross-section is down by two orders of magnitude as the interaction is electromagnetic.

6.122 $T_{\text{threshold}} = [(m_{\Xi-} + m_k + m_{k0})^2 - (m_{\pi-} + m_p)^2]/2m_p$

$= [(1{,}321 + 494 + 498)^2 - (140 + 938)^2]/2 \times 938$

$= 2{,}233\,\text{MeV}$

Note that for Ξ production, the threshold is much higher than that for $\sum^-$ production as it has to be produced in association with two other strange particles (see Chaps.9 and 10).

6.123 Using the invariance, $E^2 - |\sum p|^2 = E^{*2} - |\sum p^*|^2$
At threshold: $(E + M_p)^2 - E_\nu{}^2 = (M_p + M_\mu + M_\text{w})^2 - 0$

$(5 + 0.938)^2 - 5^2 = (0.938 + 0.106 + M_\text{w})^2$

$M_\text{w} = 2.16\,\text{GeV}$

Since the reaction does not proceed, $M_\text{w} > 2.16\,\text{GeV}$

6.124 $T_{\text{thr}} = [(m_p + m_\Lambda + m_k)^2 - (2m_p)^2]/2m_p$

$= [(0.938 + 1.115 + 0.494)^2 - (2 \times 0.938)^2]/2 \times 0.938$

$= 1.58\,\text{GeV}$

6.125 $T_{\text{thr}} = [(m_{k0} + m_k + m_{\Omega-})^2 - (m_{k-} + m_p)^2]/2m_{k-}$

$= [(0.498 + 0.494 + 1.675)^2 - (0.494 + 0.938)^2]/2 \times 0.938 = 2.7\,\text{GeV}$

Minimum momentum $p_{K-} = (T^2 + 2Tm)^{1/2} = (2.7^2 + 2 \times 2.7 \times 0.494)^{1/2} = 3.15\,\text{GeV/c}$

$P_{\text{thr}} = 3.15\,\text{GeV/c}$

$E_K = 2.7 + 0.494 = 3.194\,\text{GeV}$

$\gamma_K = E_k/m_k = 3.194/0.494 = 6.46$

$\gamma_c = (\gamma + m_2/m_1)/[1 + 2\gamma m_2/m_1 + (m_2/m_1)^2]^{1/2}$

$m_2/m_1 = 938/494 = 1.9$

$\gamma_c = (6.46 + 1.9)/(1 + 2 \times 6.46 \times 1.9 + 1.9^2)^{1/2} = 1.61$

$\gamma_\Omega = \gamma_c = 1.61; \beta_\Omega = (\gamma_\Omega^2 - 1)^{1/2}/\gamma_\Omega = 0.79$

Proper time $t_0 = d/v = d/\beta c$

Observed time $t = \gamma t_0 = \gamma d/\beta c = 1.61 \times 0.03/0.79 \times 3 \times 10^8 = 2 \times 10^{-10}\,\text{s}$

Probability that Ω^- will travel 3 cm before decay.

$= \exp(-t/\tau)$

$= \exp(-2 \times 10^{-10}/1.3 \times 10^{-10})$

$= 0.21$

6.126 $T_{\text{F}}(\text{max})_{\text{p}} = (9/32\pi^2)^{2/3} 4\pi^2(\hbar^2 c^2/2m_{\text{p}}c^2 r_0^2)(Z/A)^{2/3}$

$R = r_0 A^{1/3}$

$r_0 = R/A^{1/3} = 5.17/(63)^{1/3} = 1.3\,\text{fm}$

$T_{\text{F}}(\text{max})_{\text{p}} = (9/32\pi^2)^{2/3} \times 4\pi^2(197\,\text{MeV} - \text{fm})^2(29/63)^{2/3}/2$

$\times 938 \times (1.3)^2 = 26.886\,\text{MeV}$

$P_{\text{F}}(\text{max}) = (T^2 + 2Tm)^{1/2} = [(26.886)^2 + 2 \times 26.886 \times 938]^{1/2}$

$= 226.2\,\text{MeV/c}$

$E^2 - (p_1 + p_2)^2 = E^{*2}$ (maximum energy will be available when p_1 and p_2 are antiparallel)

$E^* = [(938 + 160 + 938 + 27)^2 - (570.4 - 226.2)^2]^{1/2}$

$= 2034\,\text{MeV}$

$m_d + m_\pi = 938 + 939 - 2.2 + 139.5 = 2,014\,\text{MeV}$

As the energy available in the CMS is in excess of the required energy, we do expect the pions to be produced.

Chapter 7
Nuclear Physics – I

7.1 Basic Concepts and Formulae

Solid angle

In two dimensions the angle in radians is defined as the ratio of the arc of the circle and the radius, that is $\theta = s/r$. In three dimensions, the element of solid angle $d\Omega$ is defined as the elementary area A at a distance d from a point, perpendicular to the line joining the point and the area, divided by the square of the distance, that is $d\,\Omega = \Delta A/d^2$. For a ring of radii r and $r + dr$, located on the surface of a sphere of radius R, the element of solid angle in polar coordinates is subtended at the centre O is given by (Fig. 7.1)

$$d\,\Omega = 2\pi\ \sin\theta\, d\theta \tag{7.1}$$

We assume an azimuthal symmetry, that is scattering is independent of the azimuthal angle β.

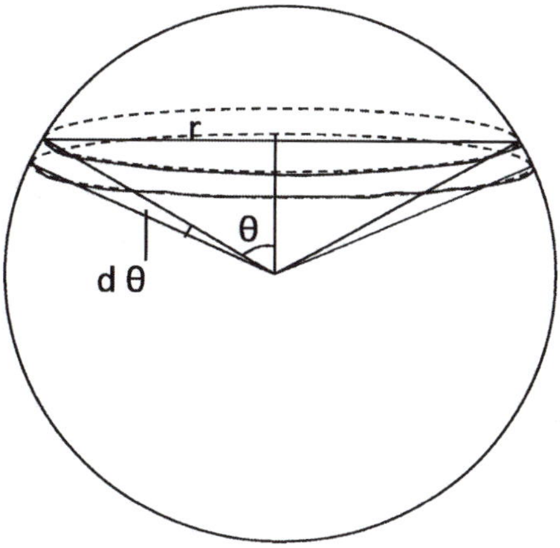

Fig. 7.1 Concept of solid angle

The solid angle for $\theta_1 = 0$ and $\theta_2 = \theta$ is

$$\Omega = \int \mathrm{d}\Omega = \int_0^{\theta} 2\pi \sin\theta \, \mathrm{d}\theta = 2\pi(1 - \cos\theta)$$

The maximum solid angle $\Omega = 4\pi$ (for $\theta = \pi$)

Kinematics of scattering

Relations between velocities, angles etc. in the Lab system (LS) and centre of mass system (CMS)

In the centre of mass system the total momentum of particles is zero. Let a particle of mass m_1 moving with velocity u_1 in the LS be scattered by the target particle of mass initially at rest and be scattered at angle θ with velocity v_1. The target particle recoils with velocity v_2 at angle φ the angles being measured with the incident direction (Fig. 7.2). The corresponding angles in the CMS will be denoted by θ^* and φ^*.

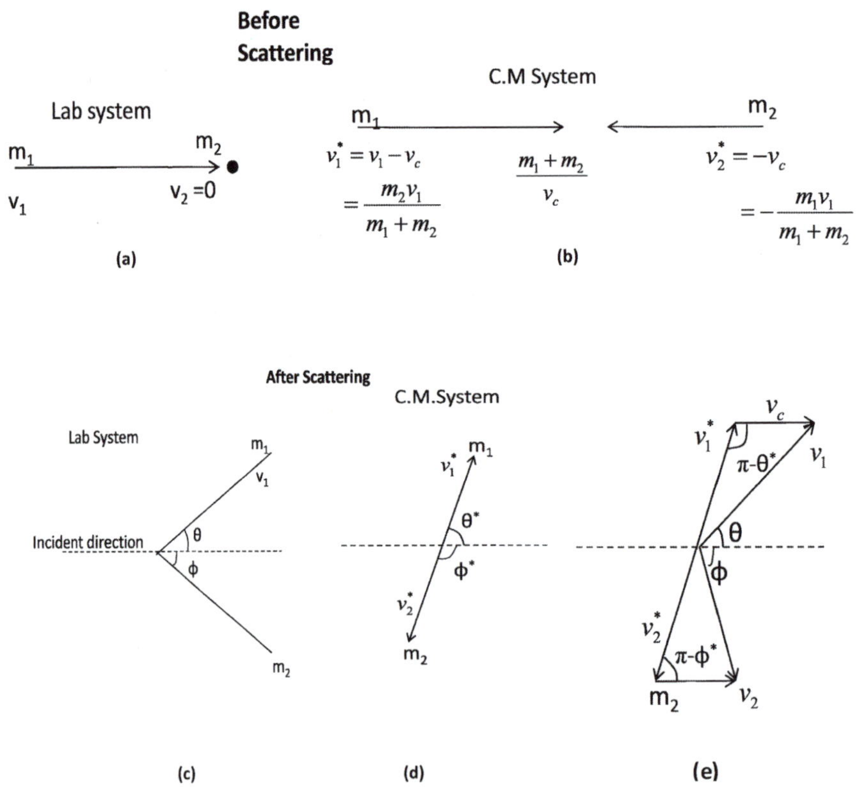

Fig. 7.2 Scattering angle and recoil angle in the LS and CMS

$$\tan\theta = \sin\theta^*/(\cos\theta^* + m_1/m_2) \tag{7.2}$$

$$\begin{aligned} &\text{If } m_2 > m_1;\ 0 < \theta < \pi \\ &m_2 = m_1;\ 0 < \theta < \pi/2 \\ &m_2 < m_1;\ 0 < \theta < \theta_{max} \\ &\text{where } \theta_{max} = \sin^{-1}(m_2/m_1) \end{aligned} \tag{7.3}$$

CM velocity combined with $v_1{}^*$ or $v_2{}^*$ gives v_1 or v_2, respectively

$$\varphi = \varphi^*/2 \text{ (regardless of the ratio } m_1/m_2) \tag{7.4}$$

$$\varphi_{max} = \pi/2 \tag{7.5}$$

Total kinetic energy available in the CMS

$$T^* = (1/2)\,\mu\, v_1{}^2 \tag{7.6}$$

where μ is the reduced mass given by

$$\mu = m_1 m_2/(m_1 + m_2) \tag{7.7}$$

Energy associated with the CMS is

$$(1/2)(m_1 + m_2)v_c^2 \tag{7.8}$$

Scattering cross-section

Let I_0 be the beam intensity of the projectiles, that is the number of incident particles crossing unit area per second, and I be the intensity of the scattered particles going into a solid angle dΩ per second, and n the number of target particles intercepting the beam, then

$$I = I_0\, n\, \sigma(\theta,\ \varphi)\, \mathrm{d}\,\Omega \tag{7.9}$$

If we assume here an azimuthal symmetry, then we can omit the azimuth angle and simply write $\sigma(\theta)$.

The constant of proportionality $\sigma(\theta)$, also written as $\mathrm{d}\sigma(\theta)/\mathrm{d}\Omega$, is known as the differential cross-section. It is a measure of the probability of scattering in a given direction $(\theta,\ \beta)$ per unit solid angle from the given target nucleus. The integral over the solid angle is known as total scattering cross-section (Fig. 7.3).

$$\sigma = \int \sigma\,(\theta,\ \varphi)\,\mathrm{d}\,\Omega \tag{7.10}$$

Fig. 7.3 Concept of differential cross-section

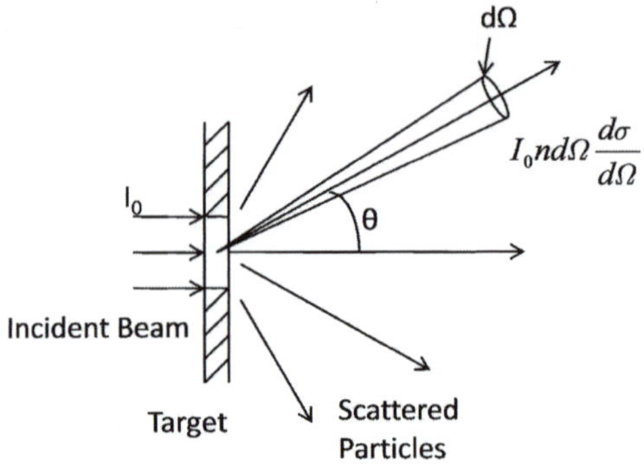

The unit of σ is a Barn (B). 1 Barn $= 10^{-24}\,\text{cm}^2$, $1\text{mB} = 10^{-27}\,\text{cm}^2$ and $1\,\mu\text{B} = 10^{-30}\,\text{cm}^2$. The unit of $\sigma(\theta,\ \varphi)$ is Barn/Steradian, where steradian (sr) is the unit of solid angle.

Relation between the differential cross-sections in the LS and CMS

$$\sigma(\theta) = \frac{(1+\gamma^2+2\gamma\cos\theta^*)^{\frac{3}{2}}}{|1+\gamma\cos\theta^*|}\sigma(\theta^*) \tag{7.11}$$

where $\gamma = m_1/m_2$

Note that the total cross-section is the same for both LS and CMS because the occurrence of the total number of collisions is independent of the description of the process.

Geometric cross-section

$$\sigma_g = \pi R^2 \tag{7.12}$$

This is the projected area of a sphere of radius R.

Rutherford Scattering

$$\sigma(\theta) = [1.295(zZ/T)^2/\sin^4(\theta/2)]\ \text{mb/sr} \tag{7.13}$$

$$\sigma(\theta', \theta'' = (\pi/4)\text{R}_0{}^2[\cot^2(\theta'/2) - \cot^2(\theta''/2)] \tag{7.14}$$

represents the cross-section for particles to be scattered between angles θ' and θ''

$$R_0\,(\text{fm}) = 1.44zZ/T_0(\text{MeV}) \tag{7.15}$$

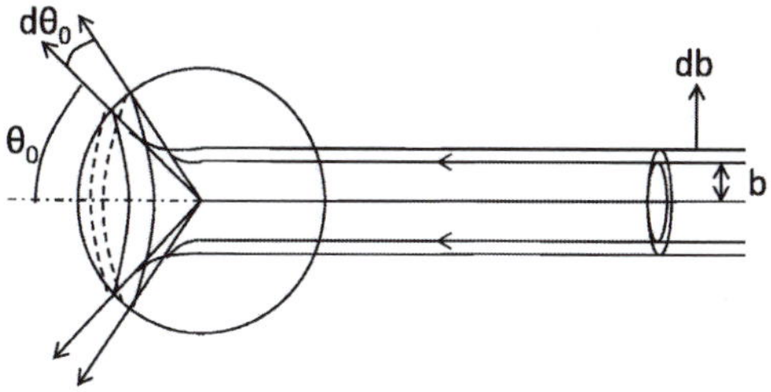

Fig. 7.4 Rutherford scattering

R_0 is also the minimum distance of approach in head-on collision for positively charged particles of energy below Coulomb barrier (Fig. 7.4).

Impact parameter (b) and scattering angle (θ)

$$\tan(\theta/2) = R_0/2b \tag{7.16}$$

Minimum distance of approach

$$\frac{R_0}{2}\left[1+\left(1+\frac{4b^2}{R_0^2}\right)^{1/2}\right] = (R_0/2)[1+\mathrm{cosec}\,(\theta/2)] \tag{7.17}$$

Multiple scattering angle

The root mean square angle of multiple scattering

$$\sqrt{<\Theta^2>} = k\sqrt{t}\, ze/pv \tag{7.18}$$

where k is the scattering constant

$$k = [8\pi\, N\, Z^2\, e^2 \ln(\mathrm{b}_{\max}/\mathrm{b}_{\min})]^{1/2}$$

t=thickness, N=number of atoms/cm^3, Ze and ze are the charge of nuclei of medium and projectile, pv is momentum times the velocity of the incident particle. For photographic emulsions, $\ln(\mathrm{b}_{\max}/\mathrm{b}_{\min})$ is of the order of 10.

Cross-section and mean free path

If n is the number of atoms/cm^3, then the macroscopic cross-section

$$\Sigma = n\,\sigma \quad (\mathrm{cm}^{-1}) \tag{7.19}$$

And mean free path

$$\lambda = 1/\Sigma \tag{7.20}$$

$$\text{Also, } n = N_0\,\rho/A \tag{7.21}$$

where N_0 =Avagadro's number, ρ =density, A the atomic weight.

Ionization

Bethe's quantum mechanical formula

$$-\,\mathrm{d}E/\mathrm{d}x = (4\pi\, z^2\, e^4\, n/mv^2)[\ln(2mv^2/I) - \ln(1-\beta^2) - \beta^2] \qquad (7.22)$$

where n =number of electrons/cm^3, I =ionization potential, $v = \beta\, c$ is the particle velocity and ze is its charge, m is the mass of electron

Note that $-\mathrm{d}E/\mathrm{d}x$ is independent of the mass of the incident particle (Fig. 7.5).

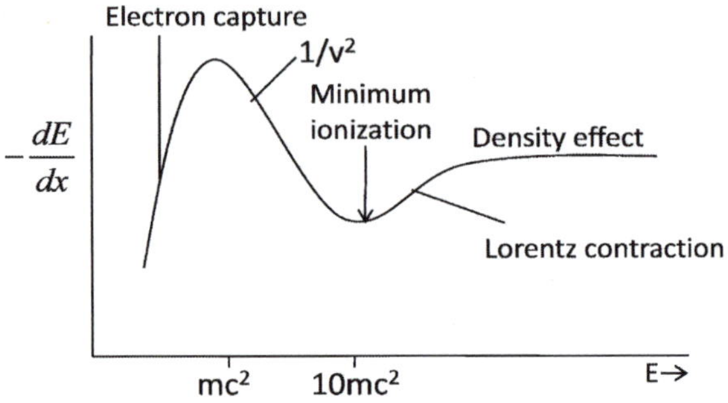

Fig. 7.5 Ionization ($-\mathrm{d}E/\mathrm{d}x$) versus particle energy

Range–Energy-relation

$$E = kz^{2n}\, M^{1-n}\, R^n \qquad (7.23)$$

where k and n are empirical constants which depend on the nature of the absorber, M is the mass of the particle in terms of proton mass.

If two particles of mass M_1 and M_2 and atomic number z_1 and z_2 enter the absorber with the same velocity then the ratio of their ranges

$$R_1/R_2 = (M_1/M_2)({z_2}^2/{z_1}^2) \qquad (7.24)$$

Range in air – Geiger's rule

$$R = \text{const. } v^3$$

$$R = 0.32\, E^{3/2} \qquad \text{(alphas in air)} \qquad (7.25)$$

Valid for 4–10 MeV α particles. R is in cm and E in MeV

The Bragg–Kleeman rule

If R_1, ρ_1 and A_1 are the range, density and atomic weight in medium 1, the corresponding quantities R_2, ρ_2 and A_2 in medium 2, then

$$R_2/R_1 = (\rho_1\,/\,\rho_2)\,(A_2/A_1)^{1/2} \qquad (7.26)$$

Straggling

When a mono-energetic beam of charged particles traverses a fixed absorber thickness, Δr, there will be fluctuations in the energy of the emerging beam about a mean value due to finite number of collisions with the atoms of the medium along the path. This phenomenon is known as Energy straggling, Fig. 7.6.

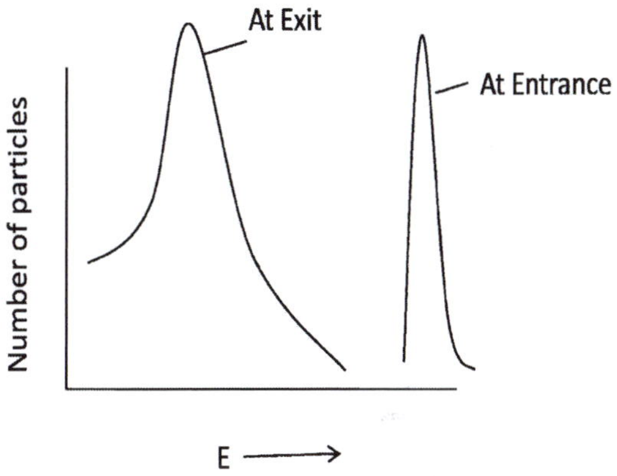

Fig. 7.6 Energy straggling

The variance of the energy distribution of the emerging particles is given by

$$\sigma^2 = 4\pi\, n\, z^2\, e^4 .\, \Delta\, r \tag{7.27}$$

where n is the number of electrons/cm^3, ze is the charge of beam particles. When a mono-energetic beam of particles is arrested in the absorber, there will be fluctuation in the ranges of the paths of the particles about a mean value. If σ_R is the standard deviation of the range distribution and $\bar{R}$ the mean range, then the ratio $\sigma/\bar{R}$ for the particle of mass number A is related to that for α-particle by

$$(\sigma/R)_A/(\sigma\,/R)_\alpha = (4/A)^{1/2} \tag{7.28}$$

Delta rays

In the collision of a charged particle with the atoms one or more electrons are ejected. The more energetic ones of these are called Delta rays are responsible for the secondary ionization, that is the production of further ions due to collisions with other atoms of the absorber.

The kinetic energy of the delta ray is given by

$$W = 2mv^2 \cos^2 \varphi \tag{7.29}$$

Motion of a charged particle in a magnetic field

Centripetal force = Magnetic force

$$mv^2/\rho = qvB \tag{7.30}$$

$$\text{Momentum } P(\text{GeV/c}) = 0.3\,B\rho \tag{7.31}$$

ρ is in meters and B is in Tesla

$1\,\text{T} = 10\,k\,\text{G}$

Cerenkov radiation

Electromagnetic radiation is emitted when a charged particle on passing through a medium with a velocity $v = \beta c$ which exceeds the phase velocity of light c/n, where n is the index of refraction, the radiation is instantaneous and possesses a sharply pronounced spatial symmetry.

The radiation is soft and is mostly emitted in the blue part of the spectrum. The radiation is emitted on a conical surface BDA, as in Fig. 7.7

$$\cos\theta = 1/\beta n \tag{7.32}$$

Fig. 7.7 Cerenkov radiation

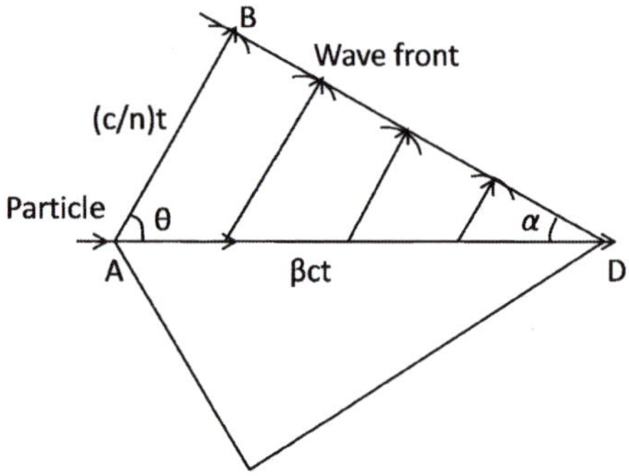

The threshold velocity

$$\beta\,(\text{thresh}) = 1/n \tag{7.33}$$

The number of photons radiated in the interval $\mathrm{d}E = h\,\mathrm{d}\nu$ by a particle of charge ze in track length $\mathrm{d}x$ is given by

$$\mathrm{d}^2 N_\gamma / \mathrm{d}x\,\mathrm{d}E = (\alpha z^2/\hbar c)[1 - (1/\beta^2 n^2)] \tag{7.34}$$

Threshold Cerenkov counters can be used to discriminate between two relativistic particles of the same momentum p and different masses m_1 and m_2, if the heavier particle (m_2) is just below the threshold. In that case

$$\sin^2\theta_1 \approx ({m_2}^2 - {m_1}^2)/p^2 \tag{7.35}$$

Bremsstrahlung

When a relativistic particle of mass m, charge ze moves close to a target nucleus of charge Ze, it undergoes acceleration which is proportional to zZ/m, and emits radiation known as bremsstrahlung. Thus the radiation losses for electrons under identical conditions, are 3×10^6 times greater than for protons. Since energy loss is a one-shot process, the law of energy degradation is exponential. If E_0 is the initial energy then at distance x

$$<E>= E_0 \exp(-x/X_0) \tag{7.36}$$

Radiation length X_0 is defined as that absorber thickness which reduces the particle energy by a factor e.

Passage of radiation through matter

When electromagnetic radiation passes through matter the type of interaction depends on (1) photon energy (2) Z of the material (3) particle or field with which the photon interacts. The important processes in Nuclear physics are

1. Compton scattering
2. Photoelectric effect
3. Electron–positron pair production
4. Nuclear resonance fluorescence

The compton effect

The process of elastic scattering of photon by a free electron with reduced frequency (or increased wavelength) is known as Compton scattering, Fig. 7.8

Fig. 7.8 Compton scattering

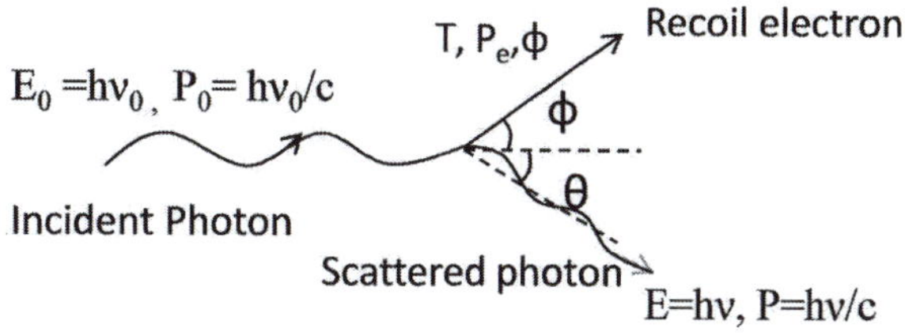

Shift in wavelength

$$\Delta\lambda = \lambda - \lambda_0 = (h/mc)(1 - \cos\theta) \tag{7.37}$$

The Compton wavelength

$$\lambda_c = h/mc = 2.43 \times 10^{-12}\,\text{m} \tag{7.38a}$$

Maximum shift in wavelength

$$(\Delta\,\lambda)_{\max} = 2\,\lambda_c \tag{7.38b}$$

Shift in frequency

$$\nu = \nu_0/(1+\alpha(1-\cos\,\theta)) \tag{7.39a}$$

$$\text{where } \alpha = h\,\nu_0\,/mc^2 \tag{7.39b}$$

$$\Delta\,\nu = \nu_0 - \nu = \alpha\,\nu_0(1-\cos\,\theta)/(1+\alpha(1-\cos\theta))$$

Energy of recoil electron

$$T = h\,\nu_0 - h\,\nu = \alpha\,h\,\nu_0\,(1-\cos\,\theta)/(1+\alpha\,(1-\cos\,\theta)) \tag{7.40}$$

Angular relation

$$\cot\varphi = (1+\alpha)\tan(\theta\,/2) \tag{7.41}$$

Compton attenuation coefficients

$$I = I_0\,exp\,(-\mu_c\,x) \tag{7.42}$$

If x is in cm then μ_c is in cm^{-1}; if x is in g/cm^2 which is obtained by multiplying x cm by ρ (the density), then μ is in cm^2/g

Photoelectric effect

If the incident photon is absorbed by an electron in the atom or metal, and the electron is ejected then the process is known as photoelectric effect. The kinetic energy of the photoelectron

$$T = h\,\nu - W \qquad (\text{Einstein's equation}) \tag{7.43}$$

where W is the ionization energy of the ejected electron

$$eV_o = \frac{1}{2}m\,v_{\max}^2 = h\nu - W \tag{7.43a}$$

$$\text{Threshold energy } h\,\nu_0 = W \tag{7.44}$$

At low Photon energy

$$\sigma_{\text{ph}} \propto Z^5/(h\nu)^{7/2} \tag{7.45}$$
$$\mu_{\text{ph}} = N\,\sigma_{\text{ph}} \tag{7.46}$$
$$I = I_0 \exp(-\mu_{\text{ph}}\, x) \tag{7.47}$$

Low photon energy can be measured from the observation of absorption edges.

$$\text{E(eV)} = 13.6\,(Z - \sigma)^2/n^2 \tag{7.48}$$

where $n = 1$ for the K-series and $n = 2$ for the L-series etc, σ is the screening constant and Z is the atomic number of the absorber.

The photoelectric absorption also follows the exponential law.

Pair production

At incident photon energies greater than $2mc^2$ (1.02 MeV), the electron–positron pair production becomes important.

$$h\,\nu = T_- + T_+ + 2mc^2 \tag{7.49}$$
$$\mu_{\text{p}} \propto Z^2 \tag{7.50}$$
$$\mu_{\text{total}} = \mu_{\text{c}} + \mu_{\text{ph}} + \mu_\beta \tag{7.51}$$

Importance of the three processes is shown in Fig. 7.9

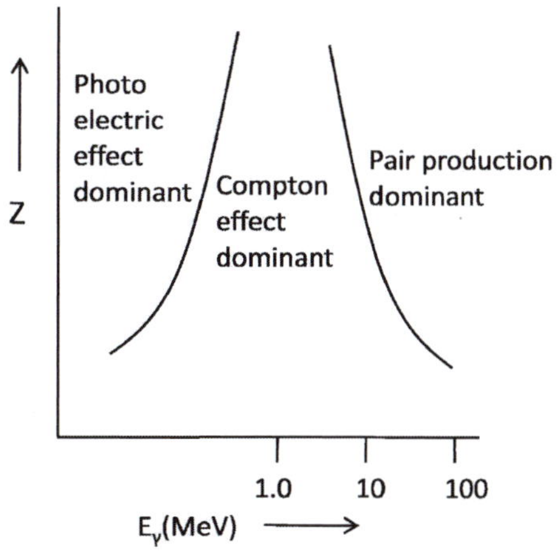

Fig. 7.9 Importance of the three processes at increasing photon energy

Nuclear resonance fluorescence

If E_0 is the transition energy in an atom of mass M, the resonance energy $E_\gamma = h\nu$ is given by

$$E_\gamma = E_0 - E^2{}_\gamma / 2Mc^2 \tag{7.52}$$

The recoil energy

$$E_\mathrm{R} = E^2{}_\gamma \; / 2Mc^2 \tag{7.53}$$

The width of the energy level Γ is the full width at half maximum, Fig. 7.10. Γ is calculated from the mean life time by the uncertainty principle

$$\Gamma . \tau = \hbar \tag{7.54}$$

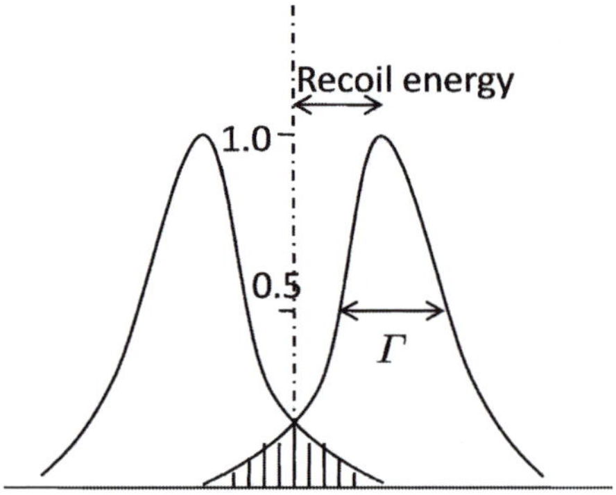

Fig. 7.10 Nuclear resonance fluorescence

Compensation for the recoil energy loss

$$\text{Source velocity } v = E_0 / Mc \tag{7.55}$$

Mosbauer Effect is the recoilless emission and absorption of nuclear radiation

$$\Delta E_\mathrm{r} / E_\mathrm{r} = v/c. \tag{7.56}$$

Radioactivity

Radioactivity is the spontaneous disintegration of an atomic nucleus. In natural radioactivity, the decay may occur via alpha, beta or gamma emission. In artificial radioactivity, neutron or proton may also be emitted.

Units

1 Curie (ci) = 3.7×10^{10} disintegrations/s
1 Rutherford = 10^6 disintegrations/s
1 Becquerel (Bq) = 1 disintegration/s

Radiation dose

1 Rad = 100 ergs/g absorbed

Radioactive law

$$\mathrm{d}N/\mathrm{d}t = -N\lambda \tag{7.57}$$

$$N = N_0 e^{-\lambda t} \tag{7.58}$$

$$T_{1/2} = 0.693\tau \tag{7.59}$$

$$\text{Activity} = |\mathrm{d}N/\mathrm{d}t| = n\lambda \tag{7.60}$$

Successive decays

$$A \rightarrow B \rightarrow C$$

$$\mathrm{d}N_B/\mathrm{d}t = \lambda_A N_A - \lambda_B N_B \tag{7.61}$$

$$N_B = \frac{\lambda_A N_A^0}{\lambda_B - \lambda_A}[\exp(-\lambda_A t) - \exp(-\lambda_B t)] \tag{7.62}$$

Transient equilibrium ($\lambda_A < \lambda_B$)

$$N_B/N_A = \lambda_A/(\lambda_B - \lambda_A) \tag{7.63}$$

Secular equilibrium ($\lambda_A << \lambda_B$)

$$N_B/N_A = \lambda_A/\lambda_B \tag{7.64}$$

Alpha-decay (Gamow's formula)

$$\lambda = (v/R)\exp(-2\pi\, zZ/137\beta) \tag{7.65}$$

Geiger–Nuttal law

$$\log\lambda = K\log x + C \tag{7.66}$$

Beta-decay

$$\lambda = \frac{G^2|M_{\mathrm{if}}|^2 E_0^5}{60\pi^3(\hbar c)^6\hbar}\,(E_0 \gg m_e c^2) \tag{7.67}$$

Selection rules

$$\textbf{Fermi rule}: \ \Delta I = 0,\ I_{\mathrm{i}} = 0 \to I_{\mathrm{f}} = 0 \text{ allowed},\ \Delta\pi = 0 \tag{7.68}$$

$$\textbf{GT rule}: \ \Delta I = 0, \pm 1,\ I_{\mathrm{i}} = 0 \to I_{\mathrm{f}} = 0 \text{ forbidden},\ \Delta\pi = 0 \tag{7.69}$$

7.2 Problems

7.2.1 Kinematics of Scattering

7.1 A particle of mass M is elastically scattered from a stationary proton of mass m. The proton is projected at an angle $\varphi = 22.1^\circ$ while the incident particle is scattered through an angle $\theta = 5.6^\circ$ with the incident direction. Calculate M in atomic mass units. (This event was recorded in photographic emulsions in the Wills Lab. Bristol).

7.2 A particle of mass M is elastically scattered through an angle θ from a target particle of mass m initially at rest ($M > m$). (a) Show that the largest possible scattering angle $\theta_{\max}$ in the Lab. System is given by $\sin\theta_{\max} = m/M$, the corresponding angle in the CMS being $\cos\theta^{*}_{\max} = -m/M$. (b) Further show that the maximum recoil angle for m is given by $\sin\varphi_{\max} = [(M - m)/2M]^{1/2}$. (c) Calculate the angle $\theta_{\max} + \varphi_{\max}$ for elastic collisions between the incident deuterons and target protons.

7.3 A deuteron of velocity u collides with another deuteron initially at rest. The collision results in the production of a proton and a triton (^{3}H), the former moving at an angle 45° with the direction of incidence. Assuming that this re-arrangement collision may be approximated to an elastic collision (quasi-scattering), calculate the speed and direction of triton in the Lab and CM system.

7.4 An α-particle from a radioactive source collides with a stationary proton and continues with a deflection of 10°. Find the direction in which the proton moves (α-mass = 4.004 amu; Proton mass = 1.008 amu).

[University of Durham]

7.5 When α-particles of kinetic energy 20 MeV pass through a gas, they are found to be elastically scattered at angles up to 30° but not beyond. Explain this, and identify the gas. In what way if any, does the limiting angle vary with energy?

[University of Bristol]

7.6 A perfectly smooth sphere of mass m_1 moving with velocity v collides elastically with a similar but initially stationary sphere of mass m_2 ($m_1 > m_2$) and is deflected through an angle θ_{L}. Describe how this collision would appear in the center of mass frame of reference and show that the relation

between θ_L and the angle of deflection θ_M, in the center of mass frame is $\tan\theta_L = \sin\theta_M/[m_1/m_2 + \cos\theta_M]$
Show also that θ_L can not be greater than about 15° if $m_1/m_2 = 4$.
[University of London]

7.7 Show that the maximum velocity that can be imparted to a proton at rest by non-relativistic alpha particle is 1.6 times the velocity of the incident alpha particle.

7.8 Show that the differential cross section $\sigma(\theta)$ for scattering of protons by protons in the Lab system is related to $\sigma(\theta^*)$ corresponding to the CMS by the formula $\sigma(\theta) = 4\ \cos(\theta^*/2)\,\sigma(\theta^*)$.

7.9 If E_0 is the neutron energy and σ the total cross-section for low energy n–p scattering assumed to be isotropic in the CMS, then show that in the LS, the proton energy distribution is given by $d\sigma_p/dE_p = \sigma/E_0 = \text{constant}$.

7.10 Particles of mass m are elastically scattered off target nuclei of mass M initially at rest. Assuming that the scattering in The CMS is isotropic show that the angular distribution of M in the LS has $\cos\ \varphi$ dependence.

7.11 A beam of particles of negligible size is elastically scattered from an infinitely heavy hard sphere of radius R. Assuming that the angle of reflection is equal to the angle of incidence in any encounter, show that $\sigma(\theta)$ is constant, that is scattering is isotropic and that the total cross-section is equal to the geometric cross-section, πR^2. (Osmania University)

7.2.2 Rutherford Scattering

7.12 Show that for the Rutherford scattering the differential cross section for the recoil nucleus in the Lab system is given by $\sigma(\varphi) = (zZe^2/2T)^2/\cos^3\varphi$

7.13 A beam of α-particles of kinetic energy 5 MeV passes through a thin foil of $_4Be^9$. The number of alphas scattered between 60° and 90° and between 90° and 120° is measured. What would be the ratio of these numbers?

7.14 If the probability of α-particles of energy 10 MeV to be scattered through an angle greater than θ on passing through a thin foil is 10^{-3}, what is it for 5 MeV protons passing through the same foil?
[University of Bristol]

7.15 What α-particle energy would be necessary in order to explore the field of force within a radius of 10^{-12} cm of the center of nucleus of atomic number 60, assuming classical mechanics to be adequate?
[University of London]

7.16 In an elastic collision with a heavy nucleus when the impact parameter b is just equal to the collision radius $R_0/2$, what is the value of the scattering angle θ^* in the CMS?

7.17 In the elastic scattering of deuterons of 11.8 MeV from ${}_{82}Pb^{208}$, the differential cross-section is observed to deviate from Rutherford's classical prediction at 52°. Use the simplest classical model to calculate the closest distance of approach d to which this angle of scattering corresponds. You are given that for an angle of scattering θ, d is given by $(d_0/2)\ [1 + \text{cosec}(\theta/2)]$, where d_0 is the value of d in a head-on collision.

[University of Manchester]

7.18 Given that the angle of scattering is $2\ \tan^{-1}\ (a/2b)$, where "a" is the least possible distance of approach, and b is the impact parameter. Calculate what fraction of a beam of 0.5 MeV deuterons will be scattered through more than 90° by a foil of thickness 10^{-5} cm of a metal of density $5\,\text{g cm}^{-3}$ atomic weight 100 and atomic number 50.

[University of Liverpool]

7.19 An electron of energy 1.0 keV approaches a bare nucleus ($Z = 50$) with an impact parameter corresponding to an orbital momentum $\hbar$. Calculate the distance from the nucleus at which this has a minimum (take $\hbar = 10^{-27}$ j-s, $e = 1.6 \times 10^{-19}\ C$ and $m = 10^{-30}$ kg)

[University of Manchester]

7.20 A beam of protons of 5 MeV kinetic energy traverses a gold foil, one particle in 5×10^6 is scattered so as to hit a surface $0.5\,\text{cm}^2$ in area at a distance 10 cm from the foil and in a direction making an angle of 60° with the initial direction of the beam. What is the thickness of the foil?

[Saha Institute]

7.21 A narrow beam of alpha particles falls normally on a silver foil behind which a counter is set to register the scattered particles. On substitution of platinum foil of the same mass thickness for the silver foil, the number of alpha particles registered per unit time increases 1.52 times. Find the atomic number of platinum, assuming the atomic number of silver and the atomic masses of both platinum and silver to be known.

7.22 Derive an expression for the differential cross-section for energy transfer in elastic collision between a heavy charged particle and an electron.

[University of London]

7.23 If σ_g is the geometrical cross-section (πR^2) for neutrons interaction with a nucleus of charge Z and radius R, then show that for positively charged particles(+ze) the cross-section will decrease by a factor $(1 - R_0/R)$, where $R_0 = zZe^2/4\pi\,\varepsilon_0\,E_0$, and E_0 is the neutron energy.

7.24 Alphas of 4.5 MeV bombarded an aluminum foil and undergo Rutherford scattering. Calculate the minimum distance of approach if the scattered alphas are observed at 60° with the beam direction.

7.25 If the radius of silver nucleus ($Z = 47$), is 7×10^{-15} m, what is the minimum energy that the particle should have to just reach it? Give your answer in MeV.

[University of Manchester]

7.26 The following counting rates (in arbitrary units) were obtained when α particles were scattered through 180° from a thin gold ($Z = 79$) target. Deduce a value for the radius of a gold nucleus from these results.

Energy of α particle (MeV)	8	12	18	22	26	27	30	34
Counting rate	91,000	40,300	18,000	12,000	8,400	100	12	1.1

[University of Manchester]

7.27 If a silver foil is bombarded by 5.0 MeV alpha particles, calculate the deflection of the alpha particles when the impact parameter is equal to the distance of closest approach.

7.28 Calculate the minimum distance of approach of an alpha particle of energy 0.5 MeV from stationary ^{7}Li nucleus in a head-on collision. Take the nuclear recoil into account.

7.29 A narrow beam of alpha particles with kinetic energy $T = 500$ keV falls normally on a golden foil incorporating 1.0×10^{19} nuclei cm^{-3}. Calculate the fraction of alpha particles scattered through the angles $\theta < \theta_0 = 30°$

7.30 A narrow beam of protons with kinetic energy $T = 1.5$ MeV falls normally on a brass foil whose mass thickness $\rho t = 2.0$ mg cm^{-2}. The weight ratio of copper and zinc in the foil is equal to 7:3. Find the fraction of the protons scattered through the angles exceeding $\theta = 45°$. For copper, $Z = 29$ and $A = 63.55$ and for zinc $Z = 30$ and $A = 65.38$

7.31 The effective cross-section of a gold nucleus corresponding to the scattering of monoergic alpha particles at angles exceeding 90° is equal to $\Delta\sigma = 0.6$ kb. Find (a) the energy of alpha particles (b) the differential cross-section $\sigma(\theta)$ at $\theta = 90°$

7.32 Derive Darwin's formula for scattering (modified Rutherford's formula which takes into account the recoil of the nucleus).

7.2.3 Ionization, Range and Straggling

7.33 Show that the order of magnitude of the ratio of the rate of loss of kinetic energy by radiation for a 10-MeV deuteron and a 10-MeV electron passing through lead is 10^{-7}.

7.34 Suppose at the sea level the central core of an extensive shower consists of a narrow vertical beam of muons of energy 60 GeV which penetrate the interior of the earth. Assuming that the ionization loss in rock is constant at 2 MeV g^{-1} cm^2, and the rock density is 3.0 g cm^{-3}, find the depth of the rock through which the muons can penetrate.

7.35 Show that deuteron of energy E has twice the range of proton of energy $E/2$.

7.36 If the mean range of 10 MeV protons in lead is 0.316 mm, calculate the mean range of 20 MeV deuterons and 40 MeV α-particles.

[University of Manchester]

7.37 Show that the range of α-particles and protons of energy 1–10 MeV in aluminium is 1/1,600 of the range in air at 15 °C, 760 mm of Hg.

7.38 Show that except for small ranges, the straggling of a beam of ^{3}He particles is greater than that of a beam of ^{4}He particles of equal range.

[University of Cambridge]

7.39 The range of a 15 MeV proton is 1,100 μm in nuclear emulsions. A second particle whose initial ionization is the same as the initial ionization of proton has a range of 165 μm. What is the mass of the particle? (The rate at which a singly ionized particle loses energy E, by ionization along its range is given by $\mathrm{d}E/\mathrm{d}R = K/(\beta c)^2$ MeV μm where βc is the velocity of the particle, and K is a constant depending only on emulsion; the mass of proton is 1,837 mass of electron)

[University of Durham]

7.40 α-particles and deuterons are accelerated in a cyclotron under identical conditions. The extracted beam of particles is passed through an absorber. Show that the range of deuteron will be approximately twice that of α-particles.

7.41 The α-particle from *Th C*′ have an initial energy of 8.8 MeV and a range in standard air of 8.6 cm. Find their energy loss per cm in standard air at a point 4 cm distance from a thin source.

[University of Liverpool]

7.42 Compare the stopping power of a 4 MeV proton and a 8 MeV deuteron in the same medium.

7.43 (a) Show that the specific ionization of 480 MeV α-particle is approximately equal to that of 30 MeV proton
(b) Show that the rate of change of ionization with distance is different for the two particles and indicate how this might be used to identify one particle, assuming the identity of the other is known.

[University of Bristol]

7.44 Calculate the thickness of aluminum in g cm^{-2} that is equivalent in stopping power of 2 cm of air. Given the relative stopping power for aluminum $S = 1{,}700$ and its density = 2.7 g cm^{-3}.

7.45 Calculate the minimum energy of an α-particle that can be counted with a GM counter if the counter window is made of stainless steel ($A \approx 56$) with 2.5 mg cm^{-2} thickness. Take the density of air as 1.226 g cm^{-3}, and atomic weight as 14.6.

7.46 Calculate the range in aluminum of a 5 MeV α-particle if the relative stopping power of aluminum is 1,700.

7.47 The range of 5 MeV α's in air at NTP is 3.8 cm. Calculate the range of 10 MeV α's using the Geiger–Nuttal law.

7.48 Mean range of α-particles in air under standard conditions is given by the formula $R(\text{cm}) = 0.98 \times 10^{-27}\, v_0{}^3$, where v_0 (cm s^{-1}) is the initial velocity of an α-particle of 5.0 MeV, find (a) its mean range (b) the average number of ion pairs formed by the α-particle over the whole path as well as over its first half, given that the ion pair formation energy of an ion pair is 34 eV.

7.49 Protons and deuterons are accelerated to the same energy and passed through a thin sheet of material. Compare their energy losses.

7.50 Protons and deuterons lose the same amount of energy in passing through a thin sheet of material. How are their energies related?

7.51 Determine the average radiative energy loss of electrons of $p = 2.7\,\text{GeV/c}$ crossing one radiation length of lead.

7.52 A beam of electrons of energy 500 MeV traverses normally a foil of lead 1/10th of a radiation length thick. Show that the angular distribution of bremsstrahlung photons of energy 400 MeV is determined more by multiple scattering of the electrons than by the angular distribution in the basic radiation process.

7.2.4 Compton Scattering

7.53 In the Compton scattering, the photon of energy $E_0 = h\nu_0$ and momentum $P_0 = h\nu_0/c$ is scattered from a free electron of rest mass m. Show that
(a) the scattered photon will have energy $E = E_0/[1 + \alpha(1 - \cos\theta)]$, where θ is the angle through which the photon is scattered and $\alpha = h\,\nu_0/mc^2$
(b) the kinetic energy acquired by the electron is
$T = \alpha\, E_0(1 - \cos\theta)/[1 + \alpha(1 - \cos\theta)]$
(c) $\tan(\theta/2) = (1 + \alpha)\, \tan\varphi$, where φ is the recoil angle of the electron.

7.54 Calculate the maximum fractional frequency shift for an incident photon of wavelength $\lambda = 1$ Å scattering off a proton initially at rest (Compton scattering analogue with proton instead of electron)
[adopted from the University College, Dublin, Ireland 1967]

7.55 A 30 keV x-ray photon strikes the electron initially at rest and the photon is scattered through an angle of 30°, what is the recoil velocity of electron?
[University of New Castle 1966]

7.56 A collimated beam of 1.5 MeV gamma rays strikes a thin tantalum foil. Electrons of 0.7 MeV energy are observed to emerge from the foil. Are these due to the photoelectric effect, Compton scattering or pair production? Assume that any electrons produced in the initial interaction with the material of the tantalum foil do not undergo a second interaction.
[University of Manchester 1972]

7.57 X-rays are Compton scattered at an angle of 60°. If the wavelength of the scattered radiation is 0.312 Å, find the wavelength of the incident radiation.

7.58 Compare the energy loss of a photon in the following situations.
(a) One single Compton scattering through 180°
(b) Two successive scatterings through 90° each
(c) Three successive scatterings through 60° each

7.59 For Aluminum, and a photon energy of 0.06 MeV, the atomic absorption cross-section due to the Compton effect is $8.1 \times 10^{-24}\,\text{cm}^2$ and due to the photo-effect is $4.0 \times 10^{-24}\,\text{cm}^2$. Calculate how much the intensity of a given beam is reduced by $3.7\,\text{g cm}^{-2}$ of aluminum and state the ratio of the intensities absorbed due to the Compton effect and due to the photo effect.

[University of Bristol 1963]

7.60 Calculate the maximum change in the wavelength of Compton scattered radiation.

7.61 A photon is Compton scattered off a stationary electron through an angle of 45° and its final energy is half its initial energy. Calculate the value of the initial energy in MeV

7.62 What is the range of energies of gamma rays from the annihilation radiation which are Compton scattered?

7.2.5 Photoelectric Effect

7.63 The wavelength of the photoelectric threshold for silver is $3{,}250 \times 10^{-10}\,\text{m}$. Determine the velocity of electron ejected from a silver surface by ultraviolet light of wavelength $2{,}536 \times 10^{-10}\,\text{m}$

[University of Durham 1960]

7.64 The gamma ray photon from ^{137}Cs when incident upon a piece of uranium ejects photo-electrons from its K-shell. The momentum measured with a magnetic beta ray spectrometer, yields a value of $Br = 3.08 \times 10^{-3}\,\text{Wb/m}$. The binding energy of a K-electron in Uranium is 115.59 keV. Determine
(a) the kinetic energy of the photoelectrons
(b) the energy of the gamma ray photons

[University of Durham 1962]

7.65 Ultraviolet light of wavelength 2,537 Å from a mercury arc falls upon a silver photocathode. If the photoelectric threshold wavelength for silver is 3,250 Å, calculate the least potential difference which must be applied between the anode and the photo-cathode to prevent electrons from the photo-cathode.

[University of Durham]

7.66 Show that photoelectric effect can not take place with a free electron.

7.67 Estimate the thickness of lead (density 11.3 g cm^{-3}) required to absorb 90% of gamma rays of energy 1 MeV. The absorption cross-section or gammas of 1 MeV in lead ($A = 207$) is 20 barns/atom.

7.68 An X-ray absorption survey of a specimen of silver shows a sharp absorption edge at the expected $\lambda_{k\alpha}$ value for silver of 0.0485 nm and a smaller edge at 0.0424 nm due to an impurity. If the atomic number of silver is 47, identify the impurity as being $_{44}$Ru, $_{45}$Rh, $_{46}$Pd, v$_{48}$Cd, V$_{49}$In or $_{50}$Sn.

7.69 A metal surface is illuminated with light of different wavelengths and the corresponding stopping potentials of the photoelectrons V, are found to be as follows.

λ (Å)	3,660	4,050	4,360	4,920	5,460	5,790
V (V)	1.48	1.15	0.93	0.62	0.36	0.24

Determine Planck's constant, the threshold wavelength and the work function.

[University of Durham 1970]

7.70 A 4 cm diameter and 1 cm thick NaI is used to detect the 660 keV gammas emitted by a 100 μCi point source of ^{137}Cs placed on its axis at a distance of 1 m from its surface. Calculate separately the number of photoelectrons and Compton electrons released in the crystal given that the linear absorption coefficients for photo and Compton processes are 0.03 and 0.24 per cm, respectively. What is the number of 660 keV gammas that pass through the crystal without interacting? (1 Curie $= 3.7 \times 10^{10}$ disintegration per second)

[Osmania University 1974]

7.71 A photon incident upon a hydrogen atom ejects an electron with a kinetic energy of 10.7 eV. If the ejected electron was in the first excited state, calculate the energy of the photon. What kinetic energy would have been imparted to an electron in the ground state?

7.72 Ultraviolet light of wavelengths, 800 Å and 700 Å, when allowed to fall on hydrogen atoms in their ground state, are found to liberate electrons with kinetic energy 1.8 and 4.0 eV, respectively. Find the value of Planck's constant.

[Indian Institute of Technology 1983]

7.73 What is the maximum wavelength (in nm) of light required to produce any current via the photoelectric effect if the anode is made of copper, which has a work function of 4.7 eV?

[University of London 2006]

7.74 Photons of energy 4.25 eV strike the surface of a metal A. The ejected photoelectrons have maximum kinetic energy T_A eV and de Broglie wavelength λ_A. The maximum kinetic energy of photoelectrons liberated from another metal by photons of energy 4.70 eV is $T_B = (T_A - 1.50)$ eV.

If the de Broglie wavelength of the photoelectrons is $\lambda_B = 2\lambda_A$, then calculate the kinetic energies T_A and T_B, and the work functions W_A and W_B.

[Indian Institute of Technology]

7.2.6 Pair Production

7.75 Calculate the maximum wavelength of γ-rays which in passing through matter, can lead to the creation of electrons.

[University of Bristol 1967]

7.76 A positron and an electron with negligible kinetic energy meet and annihilate one another, producing two γ-rays of equal energy. What is the wavelength of these γ-rays?

[University of Dublin 1969]

7.77 Show that electron–positron pair cannot be created by an isolated photon.

7.2.7 Cerenkov Radiation

7.78 Pions and muons each of 150 MeV/c momentum pass through a transparent material. Find the range of the index of refraction of this material over which the muons alone give Cerenkov light. Assume $m_\pi c^2 = 140\,\text{MeV}$; $m_\mu c^2 = 106\,\text{MeV}$.

7.79 A beam of protons moves through a material whose refractive index is 1.8. Cerenkov light is emitted at an angle of 11° to the beam. Find the kinetic energy of the proton in MeV.

[University of Manchester]

7.80 The rate of loss of energy by production of Cerenkov radiation is given by the relation $-\mathrm{d}W/\mathrm{d}l = (z^2e^2/c^2)\int\left(1 - \frac{1}{\beta^2\mu^2}\right)\omega\mathrm{d}\omega$ erg cm^{-1} where βc is the velocity, ze is the charge, μ is the refractive index of the medium and $\omega/2\pi$ is the frequency of radiation. Make an order of magnitude estimate of the number of photons emitted in the visible region, per cm of track, by a particle having $\beta = 0.9$ passing through water. The fine structure constant $\alpha = e^2/\hbar c = 1/137$

[University of Durham]

7.2.8 Nuclear Resonance

7.81 The 129 keV gamma ray transition in ^{191}Ir was used in a Mösbauer experiment in which a line shift equivalent to the full width at half maximum (Γ) was observed for a source speed of 1 cm s^{-1}. Estimate the value of Γ and the mean lifetime of the excited state in ^{191}Ir.

7.82 An excited atom of total mass M at rest with respect to a certain inertial system emits a photon, thus going over into a lower state with an energy smaller by Δw. Calculate the frequency of the photon emitted.

[University of Durham 1961]

7.83 Calculate the spread in energy of the 661 keV internal conversion line of ^{137}Cs due to the thermal motion of the source. Assume that all atoms move with the root mean square velocity for a temperature of 15 °C.

[Osmania University]

7.84 Pound and Rebeka at Harward performed an experiment to verify the Red shift predicted by general theory of Relativity. The experiment consisted of the use of 14 keV γ-ray ^{57}Fe source placed on the top of a tower 22.6 m high and the absorber at the bottom. The red shift was detected by the Mosbauer technique. What velocity of the absorber foil was required to compensate the red shift, and in which direction?

7.2.9 *Radioactivity (General)*

7.85 The disintegration rate of a radioactive source was measured at intervals of four minutes. The rate was found to be (in arbitrary units) 18.59, 13.27, 10.68, 9.34, 8.55, 8.03, 7.63, 7.30, 6.99, 6.71, and 6.44. Assuming that the source contained only one or two types of radio nucleus, calculate the half lives involved.

[University of Durham]

7.86 100 millicuries of radon which emits 5.5 MeV α-particles are contained in a glass capillary tube 5 cm long with internal and external diameters 2 and 6 mm respectively. Neglecting end effects and assuming that the inside of the tube is uniformly irradicated by the particles which are stopped at the surface, calculate the temperature difference between the walls of a tube when steady thermal conditions have been reached.
Thermal conductivity of glass = 0.025 Cal $cm^{-2}s^{-1}$ C^{-1}
Curie = 3.7×10^{10} disintegrations per second
J = 4.18 joule Cal^{-1}

[University of Durham]

7.87 Radium being a member of the uranium series occurs in uranium ores. If the half lives of uranium and radium are respectively 4.5×10^9 and 1,620 years, calculate the relative proportions of these elements in a uranium ore, which has attained equilibrium and from which none of the radioactive products have escaped.

[University of Durham]

7.88 A sealed box was found which stated to have contained an alloy composed of equal parts by weight of two metals A and B. These metals are radioactive, with half lives of 12 years and 18 years, respectively and when the container was opened it was found to contain 0.53 kg of A and 2.20 kg of B. Deduce the age of the alloy.

[University of New Castle]

7.89 Determine the amount of $^{210}_{84}$Po necessary to provide a source of α-particles of 10 millicuries strength. Half life of Polonium = 138 D.

7.90 A radioactive substance of half life 100 days which emits β-particles of average energy 5×10^{-7} ergs is used to drive a thermoelectric cell. Assuming the cell to have an efficiency 10%, calculate the amount (in gram-molecules) of radioactive substance required to generate 5 W of electricity.

7.91 The radioactive isotope, $^{14}_{6}$C does not occur naturally but it is found at constant rate by the action of cosmic rays on the atmosphere. It is taken up by plants and animals and deposited in the body structure along with natural carbon, but this process stops at death. The charcoal from the fire pit of an ancient camp has an activity due to $^{14}_{6}$C of 12.9 disintegrations per minute, per gram of carbon. If the percentage of $^{14}_{6}$C compared with normal Carbon in living trees is 1.35×10^{-10}%, the decay constant is $3.92 \times 10^{-10}\ \text{s}^{-1}$ and the atomic weight = 12.0, what is the age of the campsite?

[University of Liverpool]

7.92 Consider the decay scheme RaE $\xrightarrow{\beta}$ RaF $\xrightarrow{\beta}$ RaG (stable). A freshly purified sample of RaE weighs 5×10^{-10} g at time $t = 0$. If the sample is undisturbed, calculate the time at which the greatest number of atoms of RaF will be present and find this number. Derive any necessary formula [Half life of RaE $\left(^{210}_{83}\text{Bi}\right) = 5.0$ days; Half life of RaF $\left(^{210}_{84}\text{Po}\right) = 138$ days]

7.93 It is found that a solution containing 1 g of the α– emitter radium (^{226}Ra) never accumulates more than 6.4×10^{-6} g of its daughter element radon which has a half life of 3.825 days. Explain how the half life of radium may be deduced from this information and calculate its value.

[University of London]

7.94 Find the mean-life of ^{55}Co radionuclide if its activity is known to decrease 4.0% per hour. The decay product is non-radioactive.

7.95 What proportion of ^{235}U was present in a rock formed $3{,}000 \times 10^{6}$ years ago, given that the present proportion of ^{235}U to ^{238}U is 140?

[University of Liverpool]

7.96 A source consisting of 1 μg of ^{242}Pu is spread thinly over one plate of an ionization chamber. Alpha-particle pulses are observed at the rate of 80 per second, and spontaneous fission pulses at the rate of 3 per hour. Calculate the half life of ^{242}Pu and the partial decay constants for the two modes of decay.

[Osmania University]

7.97 ^{90}Sr decays to ^{90}Y by β decay with a half-life of 28 years. ^{90}Y decays by β decay to ^{90}Zr with a half-life of 64 h. A pure sample of ^{90}Sr is allowed to decay. What is the composition after (a) 1 h (b) after 10 years?

[University of Manchester]

7.98 Natural Uranium, as found on earth, consists of two isotopes in the ratio of ${}^{235}_{92}U/{}^{238}_{92}U = 0.7\%$. Assuming that these two isotopes existed in equal amounts at the time the earth was formed; calculate the age of the earth. [Mean life times: ${}^{238}_{92}U = 6.52 \times 10^9$ years, ${}^{235}_{92}U = 1.02 \times 10^9$ years]

[University of Cambridge, Tripos 2004]

7.99 Calculate the activity (in Ci) of 2.0 μg of ^{224}ThX. ThX ($T_{1/2} = 3.64$ D)

7.100 Calculate the energy in calories absorbed by a 20 kg boy who has received a whole body dose of 40 rad.

7.101 A small volume of solution, which contained a radioactive isotope of sodium had an activity of 16,000 disintegrations per minute/cm^3 when it was injected into the blood stream of a patient. After 30 h, the activity of 1.0 cm^3 of the blood was found to be 0.8 disintegrations per minute. If the half-life of the sodium isotope is taken as 15 h, estimate the volume of the blood in the patient.

7.2.10 Alpha-Decay

7.102 If two α-emitting nuclei, with the same mass number, one with $Z = 84$ and the other with $Z = 82$ had the same decay constant, and if the first emitted α-particles of energy 5.3 MeV, estimate the energy of α-particles emitted by the second.

[Osmania University]

7.103 Calculate the energy to be imparted to an α-particle to force it into the nucleus of ${}^{238}_{92}U$ ($r_0 = 1.2$ fm)

7.104 Radium, Polonium and RaC are all members of the same radioactive series. Given that the range in air at S.T.P of the α-particles from Radium (half-life time 1,622 Year) the range is 3.36 cm where as from polonium (half life time 138 D) the range is 3.85 cm. Calculate the half-life of RaC' for which the α-particle range at S.T.P is 6.97 cm assuming the Geiger Nuttal rule.

[Osmania University]

7.105 The α particles emitted in the decays of ${}_{88}Ra^{226}$ and ${}_{90}Th^{226}$ have energies 4.9 MeV and 6.5 MeV, respectively. Ignoring the difference in their nuclear radii, find the ratio of their half life times.

[Osmania University]

7.2.11 Beta-Decay

7.106 Classify the following transitions (the spin parity, J^P, of the nuclear states are given in brackets):-

${}^{14}O \rightarrow {}^{14}N^* + e^+ + \nu \quad (0^+ \rightarrow 0^+)$

${}^{6}He \rightarrow {}^{6}Li + e^- + \nu\ (0^+ \rightarrow 1^+)$

Why is the transition

$^{17}F \rightarrow {}^{17}O + e^- + \nu \; (5/2^+ \rightarrow 5/2^+)$

called a super-allowed transition?

7.107 Beta particles were counted from Mg nuclide. At time $t_1 = 2.0$ s, the counting rate was N_1 and at $t_2 = 6.0$ s, the counting rate was $N_2 = 2.66\ N_1$. Estimate the mean lifetime of the given nuclide.

7.108 Determine the half-life of β emitter ^{6}He whose end point energy is 3.5 MeV and $|M_{\text{if}}|^2 = 6$. Take $G/(\hbar c)^3 = 1.166 \times 10^{-5}\,\text{GeV}^{-2}$

7.109 The maximum energy $E_{\max}$ of the electrons emitted in the decay of the isotope ^{14}C is 0.156 MeV. If the number of electrons with energy between E and $E + \mathrm{d}E$ is assumed to have the approximate form

$N(E)\mathrm{d}E \propto \sqrt{E}\,(E_{\max} - E)^2\,\mathrm{d}E$

find the rate of evolution of heat by a source of ^{14}C emitting 3.7×10^7 electrons per sec.

[University of Cambridge]

7.110 In the Kurie plot of the decay of the neutron, the end point energy of the electron is 0.79 MeV. What is the threshold energy required by an antineutrino for the inverse reaction.

$\bar{\nu} + p \rightarrow n + e^+$

[University of Durham]

7.111 $_{36}\text{Kr}^{88}$ decays to $_{37}\text{Rb}^{88}$ with the emission of β-rays with a maximum energy of 2.4 MeV. The track of a particular electron from the nuclear process has a curvature in a field of 10^3 gauss of 6.1 cm. Determine

i. the energy of this electron in eV and that of the associated neutrino.
ii. the maximum possible kinetic energy of the recoiling nucleus

[University of Bristol]

7.112 If the β-ray spectrum is represented by $n(E)\mathrm{d}E \propto \sqrt{E}\,(E_{\max} - E)^2\,\mathrm{d}E$ Show that the most intense energy occurs at $E = E_{\max}/5$

7.3 Solutions

7.3.1 Kinematics of Scattering

7.1 We shall find an expression for the ratio of the masses M/m in terms of the angles θ and φ. To this end we start with the equation for the transformation of angle from CMS to LS.

$$\tan\theta = \sin\theta^*/(\cos\theta^* + M/m) \quad (1)$$

$$\theta^* = \pi - \varphi^* = \pi - 2\varphi$$

(because m and M are oppositely directed in the CMS, and the recoil angle of the proton in the CMS is always twice the angle in the LS)
Therefore $\sin\ \theta^* = \sin(\pi - 2\varphi) = \sin\ 2\varphi$
and $\cos\ \theta^* = \cos(\pi - 2\varphi) = -\cos\ 2\varphi$
Equation (1) then becomes
$\tan\ \theta = \sin\ \theta / \cos\ \theta = \sin\ 2\varphi/(M/m - \cos\ 2\varphi)$
Cross multiplying the second equation and re-arranging
$(M/m)\ \sin\ \theta = \sin\ \theta\ \cos\ 2\varphi + \cos\ \theta\ \sin\ 2\varphi = \sin(\theta + 2\varphi)$
$M/m = \sin(\theta + 2\varphi)/\sin\ \theta$
Using $\theta = 5.6^0$ and $\varphi = 22.1^0$, we find $M = 7.8\,\mathrm{m} \approx 7.8\,\mathrm{amu}$.

7.2 (a) We can work out this problem in the LS, but we prefer the CMS for convenience. Writing the equation for transformation of angles

$$\tan\ \theta = \sin\ \theta^*/(\cos\ \theta^* + M/m) \qquad (1)$$

The condition for the maximum angle of scattering, θ_{max} is $d\ \tan\ \theta/d\theta = 0$. This gives us

$$\cos\ \theta_{max} = -m/M \qquad (2)$$

$$\sin\ \theta^*_{max} = (M^2 - m^2)^{1/2}/M \qquad (3)$$

When (2) and (3) are used in (1) we find $\cot\ \theta_{max} = (M^2 - m^2)^{1/2}/m$, whence

$$\sin \theta_{max} = m/M \qquad (4)$$

(b) $\sin\varphi_{max} = \sin(\varphi^*{}_{max}/2) = \sin[(\pi - \theta^*{}_{max})/2] = \cos(\theta^*{}_{max}/2)$

$$= [(1 + \cos\ \theta^*{}_{max})/2]^{1/2} = [(1 - m/M)/2]^{1/2} = [(M - m)/2M]^{1/2} \qquad (5)$$

(c) Using $m = 1$ and $M = 2$ in (4) and (5)
we find $\theta_{max} + \varphi_{max} = 30^\circ + 30^\circ = 60^\circ$

7.3 We prefer to work in the CMS.
Let m be the proton mass.
Energy available in the CMS in the d-d collision is $E^* = \frac{1}{2}\,\mu\,u^2 = \frac{1}{2}\,mu^2$
The centre of mass velocity $v_c = u/2$

The energy E^* is partitioned between the product particles, proton and triton as follows.
$E_p{}^* = 3E^*/4$ and $E_\theta{}^* = E^*/4$. The corresponding velocities in the CMS will be, $v_p{}^* = \sqrt{3}u/2$ and $v_t{}^* = u/2\sqrt{3}$, respectively. Using the formula for the transformation of angles from CM to LS (see formula 7.2)

$$\tan\ \theta = \frac{\sin\theta^*}{\cos\theta^* + v_c/v^*} \qquad (1)$$

And using $\theta = \theta_p = 45^\circ$, $v_c = u/2$ and $v^* = v_p{}^* = \sqrt{3}u/2$ and solving for θ^*, we find $\theta_p{}^* = \theta^* = 69^\circ$.

As triton will be emitted in the opposite direction in the CMS
$\theta_t^* = 111°$
Using formula (1) again, with the substitution.
$\theta_t^* = 111°,\ v_c = u/2$ and $v_t^* = u/2\sqrt{3}$, we can solve for θ, and obtain $\theta_t = 34°$ in the LS.
Finally, we can use the inverse transformation
$\tan\theta^* = \frac{\sin\theta}{\cos\theta - v_c/v}$
And substitute $\theta^* = 111°,\ \theta = 34°$ and $v_c = u/2$ to find $v = 0.48\,u$.

7.4 Use the result of Problem 7.1,

$M/m = \sin(\theta + 2\varphi)/\sin\theta$
$M = 4.004,\ m = 1.008,\ \theta = 10^0.$

Substituting these values in the above equation and solving for φ, we find $\varphi = 16.8°$ which is the recoil angle of proton

7.5 Alphas of 20 MeV energy means that we are dealing with non-relativistic particles. From the results of Problem 7.2, the maximum scattering angle θ_{max} for $m_1 > m_2$, is given by

$\sin\theta_m = m_2/m_1$
$\sin 30^0 = 0.5 = m_2/4$ or $m_2 = 2$

The gas is deuterium. The limiting angle θ_{max} is independent of the incident energy.

7.6 The description of the scattering event in the CM system is shown in Fig. 7.11. From the velocity triangle

$v_L \sin\theta_L = v_M \sin\theta_M$
$v_L \cos\theta_L = v_M \cos\theta_M + v_C$
Dividing the two equations and simplifying

$$\tan\theta_L = \frac{\sin\theta_M}{\cos\theta_M + v_C/v_M} = \frac{\sin\theta_M}{\cos\theta_M + m_1/m_2}$$

$$\left(\because v_C = \frac{v m_1}{(m_1 + m_2)} \text{ and } v_m = \frac{v m_2}{(m_1 + m_2)}\right)$$

The maximum scattering angle θ_m is given by

$\sin\theta_M = m_2/m_1 = 1/4 = 0.25$
$\theta_M = 15°$

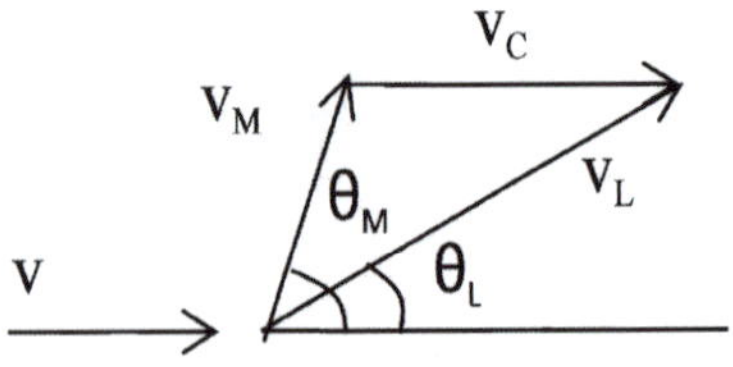

Fig. 7.11 Velocity triangle for elastic scattering

7.7 Let the alpha particle of mass m moving with velocity v_0 and momentum p_0 collide with proton of mass m. After the collision the maximum velocity v_2 and momentum p_2 will be acquired by proton when it is emitted in the incident direction. Since the alpha particle is heavier than proton, it must also proceed in the same direction as the proton with velocity v_1 and momentum p_1. Assuming that the collision is elastic,

$$P_0^2/2m_1 = p_1^2/2m_1 + p_2^2/2m_2 \quad \text{(energy conservation)} \quad (1)$$

$$P_0 = p_1 + p_2 \quad \text{(momentum conservation)} \quad (2)$$

Noting $m_1 = 4m_2$, p_1 can be eliminated between (1) and (2) to yield $p_2 = 0.4p_0$
or $m_2\, v_2 = 0.4m_1\, v_0 = 0.4 \times 4m_2 v_0$
$v_2 = 1.6v_0$

7.8 Use the transformation equation for the differential cross-sections in the CMS and LS.

$\sigma(\theta) = (1 + \gamma^2 + 2\gamma \cos\theta^*)^{3/2}\sigma(\theta^*)/|1 + \gamma\cos\theta^*|$
where $\gamma = m_1/m_2 = m/m = 1$. Then the equation simplifies to
$\sigma(\theta) = 4\cos(\theta^*/2)\,\sigma(\theta^*)$

7.9 We can write by chain rule

$$\mathrm{d}\sigma_p/\mathrm{d}E_p = (\mathrm{d}\sigma_p/\mathrm{d}\,\Omega^*).(\mathrm{d}\Omega^*/\mathrm{d}E_p) \quad (1)$$

Let the proton be scattered through an angle θ^* in the CMS with the direction of incidence of neutron (left to right). The CMS velocity will be $v_0/2$. $v_c = v_0\, m_1/(m_1 + m_2) = v_0/2$ as the masses of neutron and proton are approximately equal. The proton has velocity v_c in the CMS both before and after scattering. The velocity of the scattered proton is combined with the CMS velocity to give the velocity v_0 in the LS as shown in Fig. 7.12. From the velocity triangle we have

$v^2 = v_c^2 + v_c^2 + 2v_c^2 \cos\theta^* = 2\,v_c^2(1 + \cos\theta^*)$
$= v_0^2(1 + \cos\theta^*)/2$

Therefore, $E_p = E_0(1 + \cos\theta^*)/2$
Differentiating
$\mathrm{d}E_p = -E_0 \sin\theta^*\, \mathrm{d}\theta^*/2 = -E_0\, \mathrm{d}\Omega^*/4\pi$
The negative sign means that as θ^* increases E_p decreases
Thus,

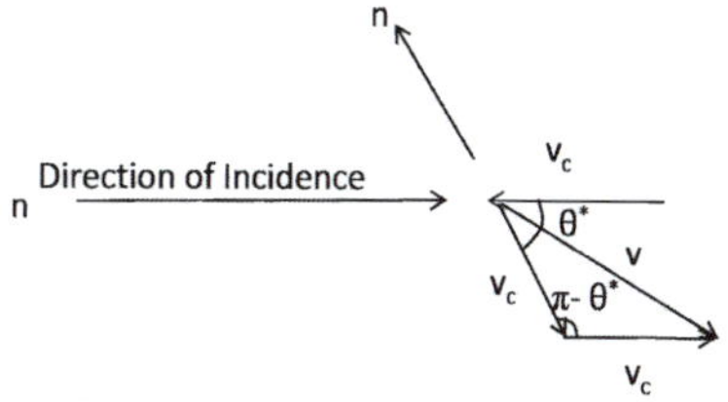

Fig. 7.12 n−p scattering in CMS

$$d\Omega^*/dE_p = 4\pi/E_0 \quad (2)$$

$$\text{And } d\sigma_p/d\Omega^* = \sigma/4\pi \quad (3)$$

as the scattering is isotropic in the CMS.
Using (2) and (3) in (1) we get $d\sigma_p/dE_p = \sigma/E_0$

7.10 As the scattering is isotropic in the CMS the differential cross-section of the recoiling nuclei is constant and is given by $\sigma(\varphi^*) = \sigma/4\pi$ = constant.
Now the differential cross-sections in the LS and CMS are related by
$\sigma(\varphi) = (\sin\varphi^* d\varphi^* / \sin\varphi\, d\varphi).\ \sigma(\varphi^*)$
But $\varphi^* = 2\varphi$ and $d\varphi^* = 2\, d\varphi$
$\sigma(\varphi) = (\sin 2\varphi.2\, d\varphi / \sin\varphi\, d\varphi)(\sigma/4\pi) = \frac{\sigma}{\pi}\cos\varphi$
Thus, $\sigma(\varphi)$ has $\cos\varphi$ dependence. It is of interest to note that
$\int \sigma(\varphi)\, d\Omega = \int_0^{\pi/2} \sigma\cos\varphi.2\pi\sin\varphi\, d\varphi/\pi = \sigma$
as it should. The upper limit for the integration is confined to 90° as the target nucleus can not recoil in the backward sphere in the LS.

7.11 In Fig. 7.13, b denotes the impact parameter. Consider particles I_0 going through a ring perpendicular to the central axis, its area being $2\pi b\, db$. On hitting the sphere, the same number of particles are scattered through a solid angle $d\Omega = 2\pi \sin\theta\, d\theta$

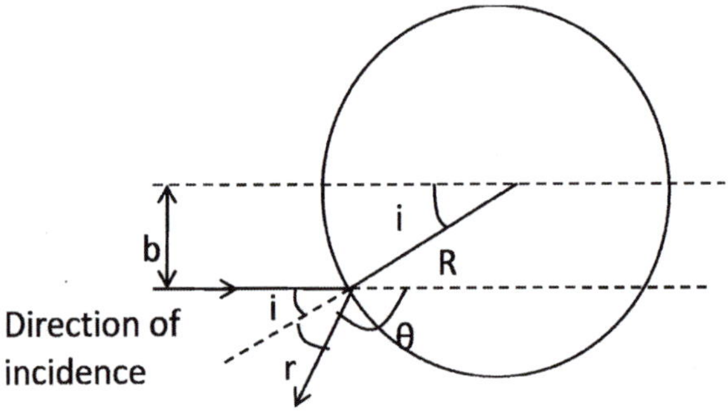

Fig. 7.13 Scattering of particles of negligible size from an infinitely heavy hard sphere of radius R

Therefore, $I_0\, \sigma(\theta).2\pi \sin\theta\, d\theta = -I_0\, 2\pi b\, db$

$$\text{Or } \sigma(\theta) = -b\, db/\sin\theta\, d\theta \quad (1)$$

The angles of incidence and reflection are measured with respect to the normal at the point of scattering. From the geometry of the figure,

$$\theta = \pi - (i + r) = \pi - 2i \quad (2)$$

$$\sin\theta = \sin(\pi - 2i) = \sin 2i = 2\sin i\cos i \quad (3)$$

Since $r = i$ and $d\theta = -2di$
$b = R\sin i$ and $db = R\cos i\, di$

$$\text{We find } \sigma(\theta) = R^2/4 \quad (4)$$

The right hand side of (4) is independent of the scattering angle θ; that is the scattering is isotropic or equally in all directions.
The total cross-section

$$\sigma = \int \sigma(\theta)\,\mathrm{d}\,\Omega = \int_0^{\pi} (R^2/4).\ 2\pi\ \sin\theta\ \mathrm{d}\,\theta = \pi R^2$$

This is called geometric cross-section.

7.3.2 Rutherford Scattering

7.12 Rutherford law for scattered particles in the CMS is

$$\sigma(\theta^*) = \mathrm{d}\,\sigma(\theta^*)/\mathrm{d}\,\Omega^* = \frac{1}{16}\left(\frac{zZe^2}{T}\right)^2 \frac{1}{\sin^4\left(\frac{\theta^*}{2}\right)}$$

Now $\sin\,(\theta^*/2) = \sin((\pi - \varphi^*)/2) = \cos(\varphi^*/2) = \cos\varphi$
$\mathrm{d}\,\Omega^* = 2\,\pi\ \sin\,\theta^*\ \mathrm{d}\,\theta^* = 2\,\pi\ \sin(\pi - \varphi^*)\,\mathrm{d}\,\varphi^*$
$= 2\,\pi\ \sin\,2\,\varphi.2\,\mathrm{d}\,\varphi = 8\,\pi\ \sin\,\varphi\ \cos\,\varphi\ \mathrm{d}\varphi = 4\ \cos\,\varphi\ \mathrm{d}\Omega(\varphi)$
Therefore $\mathrm{d}\,\sigma(\varphi)/\mathrm{d}\Omega(\varphi) = (zZe^2/T)^2.1/\cos^3\varphi$

7.13 The Rutherford scattering cross-section for scattering between angle θ_1 and θ_2 is given by $\sigma(\theta_1,\ \theta_2) = \left(\pi\ {R_0}^2/4\right)[\cot^2(\theta_1/2) - \cot^2(\theta_2/2)]$
where $R_0 = zZe^2/4\pi\ \varepsilon_0\ T = 1.44zZ/T$
Therefore, $\frac{\sigma(60^\circ, 90^\circ)}{\sigma(90^\circ, 120^\circ)} = (\cot^2 30^\circ - \cot^2 45^\circ)/(\cot^2 45^\circ - \cot^2 60^\circ) = 3/1$

7.14 Since the atomic number and foil thickness and angles are the same, the probability for scattering will be inversely proportional to the square of energy and directly proportional to this square of the charge of the particle. Hence the probability for scattering of 5 MeV α-particles will be $\frac{2^2}{10^2} \times \frac{5^2}{1^2} \times 10^{-3} = 10^{-3}$

7.15 The minimum requirement is that the particle should be able to penetrate the nucleus of radius R. Thus
$T = 1.44zZ/R(\mathrm{fm}) = 1.44 \times 2 \times 60/10 \approx 17.3\,\mathrm{MeV}$

7.16 $\tan(\theta^*/2) = r(\mathrm{collision})/2b$
for $b = \frac{1}{2}\,r\,(\mathrm{collision}),\ \tan\,(\theta^*/2) = 1 = \tan 45^\circ$
Therefore, $\theta^* = 90^\circ$

7.17 Given $d = \frac{d_0}{2}(1 + \mathrm{cosec}\ (\theta/2))$
Here $d = r_{\min},\ d_0 = R_0$
Therefore, $R_0 = 1.44zZ/T = 1.44 \times 1 \times 82/11.8 = 10\,\mathrm{fm}$
$R_{\min} = (10/2)(1 + \mathrm{cosec}\ 26^\circ) = 16.4\ \mathrm{fm}$

7.18 Given $\theta = 2\tan^{-1}(a/2b)$

Therefore, $\tan\,(\theta/2) = R_0/2b$ (1)

$R_0 = 1.44zZ/T = 1.44 \times 1 \times 50/0.5 = 144\ \mathrm{fm}$

From (1)
$b = (R_0/2)\ \cot\,(\theta/2) = (144\ \cot\ 45^0)/2 = 72\ \mathrm{fm}$
Where we have used $\theta = 90^\circ$
$\sigma(90^\circ,\ 180^\circ) = \pi b^2 = \pi(72)^2 = 16,278\ \mathrm{fm}^2 = 1.628 \times 10^{-22}\ \mathrm{cm}^2$
Macroscopic cross-section
$\Sigma = \sigma\ N_0\,\rho/A = 1.628 \times 10^{-22} \times 6.02 \times 10^{23} \times 5/100 = 4.9\,\mathrm{cm}^{-1}$

Therefore, Mean free path $\lambda = 1/\Sigma = 0.2\,\text{cm}$
Fraction of scattered particles = Probability of scattering through more than 90^0 is $t/\lambda = 10^{-5}/0.2 = 5 \times 10^{-5}$.

7.19 $r_{\min} = \frac{R_0}{2}\left[1 + \left(+\frac{4b^2}{R_0^2}\right)^{\frac{1}{2}}\right]$
$R_0 = 1.44zZ/T = 1.44 \times 1 \times 50/(1 \times 10^{-3}) = 7.2 \times 10^4$ fm
$bp = L = \hbar$
$b = \hbar/p = \hbar c/cp = 197.3\,\text{MeV fm}/1 \times 10^{-3}\,\text{MeV} = 1.973 \times 10^5$ fm
$b/R_0 = 1.973 \times 10^5/(7.2 \times 10^4) = 2.74$
$r_{\min} = \frac{7.2}{2} \times 10^4\left[1 + (1 + 4 \times 2.74^2)^{\frac{1}{2}}\right]$
$= 2.365 \times 10^5$ fm

7.20 The mean free path $\lambda = 1/\Sigma = 1/N.\mathrm{d}\sigma$ (1)

where N is the number of atoms per cm^3

$N = N_0\rho/A = 6.02 \times 10^{23} \times 19.3/197 = 5.9 \times 10^{22}$ (2)

$\mathrm{d}\,\sigma = (1.44zZ/4T)^2\,\mathrm{d}\,\Omega/\sin^4(\theta/2)$ (3)

Put $z = 1$, $Z = 79$, $\theta = 60^\circ$, $\mathrm{d}\Omega = \text{area}/(\text{distance})^2 = 0.5/10^2 = 0.005$

Using these values in (3), $\mathrm{d}\,\sigma = 2.588\ \text{fm}^2$

$= 2.588 \times 10^{-26}\ \text{cm}^2$ (4)

Using (2) and (4) in (1) we find $\lambda = 655$ cm,
The probability of scattering at 60°,

$P = t/\lambda$ (5)

where t is the foil thickness
$t = p\lambda = (655 \times 1/5 \times 10^6)$ cm $= 1.31\ \mu$m.

7.21 The counting rate is dependent on the factor

$(zZ/T)^2\,(N_0/A)(\rho t/\sin^4(\theta/2))$

In the problem z, T, $(\rho\ t)$ and θ are unchanged. Hence, counting rate with platinum/counting rate with silver

$= (Z_{\text{pt}}{}^2/A_{\text{pt}})/(Z_{\text{Ag}}{}^2/A_{\text{Ag}}) = 1.52$

Substituting the known values: $Z_{\text{Ag}} = 47$, $A_{\text{Ag}} = 108.87$ and $A_{\text{pt}} = 195$, the above equation can be solved to yield $Z_{\text{pt}} = 77.55$ or 78.

7.22 Let the particle of mass m_1, charge z, velocity v and kinetic energy T collide elastically with an electron of mass $m_2 = m$. The electron velocity before and after the collision is $v_2{}^* = v_{\text{c}}$ in the CMS. The velocity $v_2{}^*$ is combined vectorially with the CMS velocity v_{c} to yield the LS velocity v_2 at angle φ with the incident direction

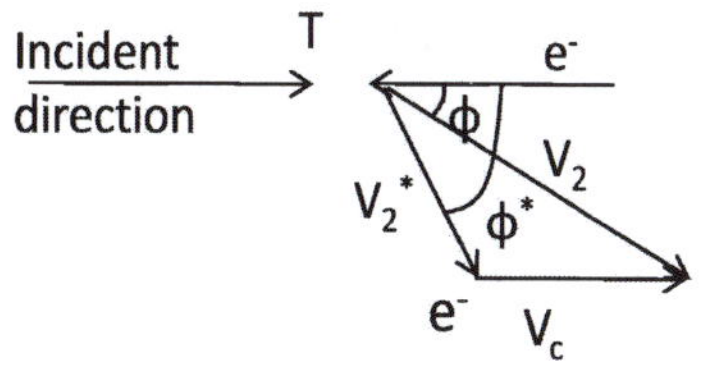

Fig. 7.14 Collision of a heavy charged particle with an electron

From the velocity triangle which is isosceles as in Fig. 7.14,

$v_2 = 2v_c \cos\varphi = 2m_1\, v \cos\varphi/(m_1 + m_2) \approx 2v \cos\varphi$

(Because $m_2 = m << m_1$)

The energy acquired by the electron

$$W = (m/2)\,(2v \cos\varphi)^2 = 2mv^2 \cos^2\varphi \quad (1)$$

If the recoil angle of the electron is φ^* in the CMS then $\varphi = \varphi^*/2$ and $\varphi^* = \pi - \theta^*$, where θ^* is the scattering angle of the incident particle in the CMS. And so

$$\cos^2\varphi = \sin^2(\theta^*/2)/2 \quad (2)$$

$$W = 2mv^2 \sin^2(\theta^*/2) \quad (3)$$

$$\mathrm{d}W = mv^2 \sin\theta^*\, \mathrm{d}\theta^* \quad (4)$$

But Rutherford's formula for scattering in the CMS is

$$\sigma(\theta^*) = \mathrm{d}\sigma/\mathrm{d}\Omega^* = z^2e^4/4\mu^2v^2 \sin^4\theta^*/2 \quad (5)$$

where we have put Z = −1 for electron. Since the electron mass is negligible compared to that of the incident particle, $\mu \approx m$. Further, the element of solid angle $\mathrm{d}\Omega^* = 2\pi \sin\theta^* \mathrm{d}\theta^*$.

Formula (5) becomes

$$\mathrm{d}\sigma = (2\pi \sin\theta^*\, \mathrm{d}\theta^*\, z^2\, e^4)/(4m^2v^4 \sin^4(\theta^*/2)) \quad (6)$$

Using (3) and (4) in (6)

$\mathrm{d}\sigma/\mathrm{d}W = 2\,z^2\, e^4/mv^2W^2$ (differential energy spectrum)

This gives us the cross-section for finding the delta rays(emitted electrons) of energy W per unit energy interval.

7.23 When the charged particle just grazes the nucleus

$$r_{\min} = R = ½R_0[1 + (1 + 4b^2/R_0{}^2)^{1/2}] \quad (1)$$

Solving for b, we obtain

$$b^2 = R^2 - RR_0 \quad (2)$$

Denoting the cross-section by $\sigma = \pi b^2$,

$\sigma/\sigma_g = \pi\, b^2/\pi\, R^2 = 1 - R_0/R$

7.24 The closest distance of approach r is given by

$r_{\min} = (R_0/2)(1 + \mathrm{cosec}\,(\theta/2)) = (R_0/2)(1 + 2) = 3R_0/2$

Now $R_0 = 1.44zZ/T = 1.44 \times 2 \times 13/4.5 = 8.32$
Hence $r_{\min} = 1.5 \times 8.32 = 12.48\,\text{fm}$

7.25 The minimum distance of approach $r_{\min}$ is obtained in the head-on collision when the initial α particle energy is entirely converted into the potential energy.

$T = zZe^2/4\pi\,\varepsilon_0 r_{\min}$

Putting $r_{\min} = R$, the nuclear radius, numerically
$\text{T(MeV)} = 1.44zZ/R(\text{fm}) = 1.44 \times 2 \times 47/7 = 19.34\,\text{MeV}$

7.26 For the same charges of interacting particles, target thickness and beam intensity and fixed scattering angle, the Rutherford scattering depends inversely as the square of particle energy. So long as the incident particle is outside the target nucleus, the Rutherford scattering is expected to be valid. But when the incident particle touches the nucleus, pure Rutherford scattering would be invalidated. We can assume that the observed counting rate N_8 is the expected counting rate for the lowest energy (8 MeV). For higher energy T the counting rate is expected to be $N_{\text{E}} = N_8(8/T)^2$. In the table below are displayed the calculated counting rates as well as the observed ones.

T(MeV)	8	12	18	22	26	27	30	34
N(Cal)	91,000	40,400	17,970	12,030	8,600	8,000	6,471	5,038
N(Obs)	91,000	40,300	18,000	12,000	8,400	100	12	1.1

Comparison between the calculated and observed counting rates indicates that Rutherford scattering begins to break down at 26 MeV. Since scattering angle is $\theta = 180°$, we are concerned with head-on collisions. Hence
$R_0 = R = 1.44\,zZ/T = 1.44 \times 2 \times 79/26 = 8.75\,\text{fm}$.
Hence the radius of gold nucleus is 8.75 fm.

7.27 When distance of closest approach $b = R_0$
$\tan(\theta/2) = R_0/2b = 0.5 = \tan 26.56°$
Therefore, $\theta/2 = 26.56°$ or $\theta = 53.1°$

7.28 We work out in the CMS because ^{7}Li being a light nucleus will recoil in the encounter. Equating the potential energy at the closest distance of approach R_0 to the initial kinetic energy,
$1.44zZ/R_0 = \mu\, v^2/2 = m_1 m_2 v^2/(m_1 + m_2) = m_2 T/(m_1 + m_2)$
where μ is the reduced mass. Solving for R_0
$R_0 = (1.44zZ/T)(1 + m_1/m_2) = (1.44\text{x}2\text{x}3/0.5)(1 + 4/7) = 27.15\,\text{fm}$

7.29 Given $nt = 1.0 \times 10^{19}\,\text{nuclei/cm}^2$
The fraction $\Delta N/N = 1 - \left(\pi\, R_0{}^2 \cot^2(\theta/2).nt\right)/4 = 1 - (\pi/4)(1.44zZ/T)^2 \cot^2(\theta/2).nt = 1 - \pi(1.44 \times 2 \times 79/0.5)^2 \cot^2(30/2) \times (1.0 \times 10^{19} \times 10^{-26})/4 = 0.82$
The factor 10^{-26} has been introduced to convert fm into cm^2.

7.30 The fraction of particles scattered in brass at angles exceeding θ is given by $\Delta N/N = (\pi/4)(1.44^2/T^2)\left(0.7Z_1{}^2/A_1 + 0.3Z_2{}^2/A_2\right)\rho t N_0 \cot^2\theta/2 \times 10^{-26}$ where $Z_1 = 29$ for copper and $Z_2 = 30$ for Zinc, $A_1 = 63.55$ and $A_2 = 65.38$ are the atomic masses, respectively, and $N_0 = 6.02 \times 10^{23}$ is the avagadro's number, $T = 1.5\,\text{MeV}$, $\rho t = 2 \times 10^{-3}\,\text{g cm}^{-2}$ and $\theta = 45^0$. The factor 10^{-26} is introduced to convert fm^2 into cm^2. Using these values in the above equation $\Delta N/N = 6.78 \times 10^{-4}$

7.31 (a) $\Delta\sigma = \sigma(90°,\ 180°) = (\pi/4)(1.44zZ/T)^2$
Using $\Delta\sigma = 0.6\,\text{kb} = 60\,\text{fm}^2,\ z = 2,\ Z = 79$ and solving for T, we get $T = 3.36\,\text{MeV}$

(b) $\sigma(\theta) = (1/16)(1.44zZ/T)^2 1/\sin^4(\theta/2)$
Put $\theta = 90°$ and $T = 3.36\,\text{MeV}$ to find $\sigma(90) = 1{,}146\,\text{fm}^2/\text{sr} = 11.46\,\text{b/sr}$.

7.32 First we work out in the CMS and then transform $\sigma(\theta^*)$ in the CMS to $\sigma(\theta)$ in the LS.
Rutherford's formula in CMS is

$$\sigma(\theta^*) = (1/4)(zZe^2/\mu v^2)^2 .1/\sin^4(\theta/2) \tag{1}$$

where the reduced mass

$$\mu = mM/(m+M) = m/(1+\gamma) \tag{2}$$

where m and M are the masses of the incident and target masses, respectively, and $\gamma = m/M$.
Further

$$\sin^4(\theta^*/2) = (1/4)(\sin^4\theta^*)/(1+\cos\theta^*)^2 \tag{3}$$

and

$$\tan\,\theta = \sin\,\theta^*/(\gamma + \cos\,\theta^*) \tag{4}$$

Squaring (4) and expressing it as a quadratic equation and solving it

$$\cos\,\theta = \gamma\ \sin^2\,\theta \pm \cos\,\theta(1-\gamma^2\sin^2\theta)^{1/2} \tag{5}$$

$$\text{Now } \sigma(\theta) = (1+\gamma^2+2\gamma\ \cos\,\theta^*)^{3/2}/|1+\gamma\ \cos\,\theta^*| \tag{6}$$

Combining (1), (2), (3), (5) and (6) and after some algebraic manipulations we get

$$\sigma(\theta) = \left(\frac{zZe^2}{2T}\right)^2 \frac{1}{\sin^4\theta} \frac{[\cos\theta \pm (1-\gamma^2\sin^2\theta)^{1/2}]^2}{(1-\gamma^2\sin^2\theta)^{1/2}}$$

If $\gamma < 1$, the positive sign only should be used before the square root. If $\gamma > 1$, the expression should be calculated for positive and negative signs and the results added to obtain $\sigma(\theta)$. For $\gamma = 1,\ \sigma(\theta) = \left(\frac{zZe^2}{T}\right)^2 \frac{\cos\theta}{\sin^4\theta}$

7.3.3 Ionization, Range and Straggling

7.33 The rate of loss of energy due to radiation is inversely proportional to the square of particle mass and directly proportional to the square of charge of the incident particle as well as the atomic number of the target nucleus and directly proportional to the kinetic energy.

$-(\mathrm{d}E/\mathrm{d}x)_{\mathrm{rad}} \propto z^2 Z^2 T/m^2$

As the medium is identical and both e and d are singly charged with the same kinetic energy the ratio of the radiation loss for deuteron and electron will be$(m_\mathrm{e}/m_\mathrm{d})^2 \approx (1/3,670)^2 = 7.4 \times 10^{-8}$ or $\approx 10^{-7}$

7.34 Ionization loss of muons in the rock $= 2\,\mathrm{MeV\,g^{-1}\,cm^2}$
$= 2\,\mathrm{MeV\,g^{-1}\,cm^2} \times \rho = 2\,\mathrm{MeV\,g^{-1}\,cm^2} \times 3.0\,\mathrm{g\,cm^{-3}}$
$= 6\,\mathrm{MeV/cm}$.

The depth of the rock which will reduce 60 GeV to zero $= 60 \times 10^3/6\,\mathrm{cm} = 10^4\,\mathrm{cm} = 100\,\mathrm{m}$.

7.35 $E_\mathrm{d} = m_\mathrm{d}\, v_\mathrm{d}{}^2/2$
$E_\mathrm{p} = m_\mathrm{p}\, v_\mathrm{p}{}^2/2$
$E_\mathrm{d}/E_\mathrm{p} = E/(E/2) = 2 = m_\mathrm{d}\, v_\mathrm{d}{}^2/m_\mathrm{p}\, v_\mathrm{p}{}^2 = 2v_\mathrm{d}{}^2/v_\mathrm{p}{}^2$
Therefore, $v_\mathrm{d} = v_\mathrm{p}$

Deuteron and proton having the same initial speed will have their ranges in the ratio of their masses. Therefore the deuteron has twice the range of proton.

7.36 $R = (M/z^2)\,\mathrm{f}(E/M)$
The E/M ratio for the three given particles is identical because $E_\mathrm{p}/M_\mathrm{p} = 10/1,\ E_\mathrm{d}/M_\mathrm{d} = 20/2$ and $E_\alpha/M_\alpha = 40/4$. Hence

$R_\mathrm{d} = (R_\mathrm{p} M_\mathrm{d} z_\mathrm{p}{}^2)/(z_\mathrm{d}^2 M_\mathrm{p}) = (0.316 \times 2 \times 1^2)/(1 \times 1^2) = 0.632\,\mathrm{mm}$

$R_\alpha = (R_\mathrm{p} M_\alpha z_\mathrm{p}{}^2)/(z_\alpha{}^2 M_\mathrm{p}) = (0.316 \times 4 \times 1^2)/(1 \times 2^2) = 0.316\,\mathrm{mm}$

7.37 Apply the Bragg–Kleeman formula

$R_\mathrm{Al}/R_\mathrm{air} = (\rho_\mathrm{air}/\rho_\mathrm{Al})\sqrt{A_\mathrm{Al}}/\sqrt{A_\mathrm{Air}}$

Substitute the values: $\rho_\mathrm{air} = 1.226 \times 10^{-3}\,\mathrm{g\,cm^{-3}},\ \rho_\mathrm{Al} = 2.7\,\mathrm{g\,cm^{-3}}$

$A_\mathrm{Al} = 27$ and $A_\mathrm{air} = 14.5$, we find $R_\mathrm{Al}/R_\mathrm{Air} = 1/1,614$

7.38 The straggling of charged particles is given by the ratio σ_R/R, where R is the mean range of a beam of particles in a given medium and σ_R is the standard deviation of the ranges.
Now $R = f(v_0,\ I)/z^2$ where v_0 is the initial velocity of beam of particles, I is the ionization of the atoms of the absorber. Further, $\sigma_\mathrm{R} = \sqrt{M}\ F(v_0,\ I)/z^2$. So

$\sigma_\mathrm{R}/R = \varphi(v_0, I)/\sqrt{M}$

As the straggling is inversely proportional to the square root of particle mass, the straggling for ^{3}He will be greater than that for ^{4}He of equal range.

7.39 $dE/dR = k/(\beta c)^2$

$dR = [(\beta c)^2/k]\,d(Mv^2/2) = k'M\,f(v)\,dv$

where k' is another constant
Integrating from zero to v_0

$$R = \int dR = k'M \int f(v)\,dv = k'M\,F(v_0)$$

If two single charge particles of masses M_1 and M_2, be selected so that their initial velocities are identical then their residual ranges
$R_1/R_2 = M_1/M_2$
$M_2 = R_2 M_1/R_1 = 165 \times 1,837\,m_e/1,100 = 275.5\,m_e$
It is a pion. Its accepted mass is $273\,m_e$

7.40 Balancing the centripetal force with the magnetic force at the point of extraction

$$mv^2/r = zevB \quad (1)$$

which gives us
$v = zeBr/m$
The ratio of velocities for α-particle and deuteron is
$v_d/v_\alpha = (z_d.m_\alpha)/(m_d.z_\alpha) = (1 \times 2)/(2 \times 1) = 1$
since $m_\alpha \approx 2m_d$
As the initial velocities are identical the ratio
$R_d/R_\alpha = (m_d {z_\alpha}^2)/(m_\alpha {z_d}^2) = 2^2/2 = 2$
Therefore, $R_d = 2R_\alpha$

7.41 Use Geiger's rule

$$R = K\,E^{3/2} \quad (1)$$

$8.6 = K\,(8.8)^{3/2}$

$K = 0.33$

At a distance of 4 cm from the source, residual range is 4.6 cm. At this point the energy can be found out by applying (1) again
$4.6 = 0.33 \times E_1^{3/2}$
Or $E_1 = 5.79\,\text{MeV}$
Differentiating (1)

$dR/dE = (3/2)K\,E^{1/2}$

$dE_1/dR = 2/3K\,{E_1}^{1/2} = 2/(3 \times 0.33 \times \sqrt{5.79})$

$= 0.84\,\text{MeV/cm}$

7.42 The proton velocity $v_p = (2E_p/m_p)^{1/2} = \sqrt{2 \times 4/m_p}$
The deuteron velocity $v_d = (2E_d/m_d)^{1/2} = \sqrt{2 \times 8/m_d} = (2 \times 4/m_p)^{1/2}$ (because $m_d \approx 2m_p$)

Thus, the proton and deuteron have the same velocity, and both of them are singly charged. Hence their stopping power is identical.

7.43 (a) $dE/dR \propto z^2/v^2$ or $\propto Mz^2/E$

$$(dE_\alpha/dR)/(dE_p/dR) = M_\alpha z_\alpha{}^2 E_p/M_p z_p{}^2 E_\alpha = (M_\alpha/M_p)(z_\alpha{}^2/z_p{}^2)$$
$$(E_p/E_\alpha) = 4 \times 4 \times 30/480 = 1$$

(b) The change in ionization over a given distance will be different for different particles in a medium. Calibration curves can be drawn for particles of different masses. The proton curve can be assumed to be the standard curve. This method is particularly useful for those particles which are not arrested within the emulsion stack or bubble chamber.

7.44 If S is the relative stopping power then
$R(\text{Al}) = R(\text{air})/S = 2/1,700\,\text{cm}$

The thickness of aluminum is obtained by multiplying the range in air by the density

Therefore, $R(\text{Al}) = (2.7 \times 2/1,700)\,\text{g cm}^{-2} = 3.18 \times 10^{-3}\,\text{g cm}^{-2}$

7.45 Apply the Bragg–Kleeman rule

$$R_a = R_s \rho_s \sqrt{A_a}/\rho_a \sqrt{A_s} \tag{1}$$

Now $R_s(\text{cm}) = R_s(\text{g cm}^{-2})/\rho_s = 2.5 \times 10^{-3}/\rho_s$
Hence (1) becomes
$R_a(\text{cm}) = 2.5 \times 10^{-3}\sqrt{14.5}/1.226 \times 10^{-3} \times \sqrt{56} = 1.04$
Apply Geiger's rule
$E = (R/0.32)^{2/3} = (1.04/0.32)^{2/3} = 2.19\,\text{MeV}$
α's of energy greater than 2.2 MeV will be registered.

7.46 $R(\text{Al}) = R(\text{air})/S = R(\text{air})/1{,}700$
We find $R(\text{air})$ by Geigers's rule
$R(\text{air}) = 0.32\,E^{3/2} = 0.32 \times 5^{3/2} = 11.18\,\text{cm}$
Therefore $R(\text{Al}) = 11.18/1,700 = 0.00658\,\text{cm} = 66\,\mu\text{m}$

7.47 Geiger's rule is
$R = 0.32(\text{E})^{3/2}$
where R is in cm and E in MeV
$3.8 = 0.32(5)^{3/2}$
$R = 0.32(10)^{3/2}$
Therefore, $R = 3.8 \times (10/5)^{3/2} = 10.75\,\text{cm}$

7.48 (a) $v_0 = (2T/m)^{1/2} = c(2T/mc^2)^{1/2} = c(2 \times 5/3728)^{1/2} = 0.0518\,c$
$R = 0.98 \times 10^{-27} \times (3 \times 10^{10} \times 0.0518)^3 = 3.678\,\text{cm}$

(b) Over the whole path total number of ion pairs
$=$ Total energy lost/Ionization energy of each pair $= 5.0 \times 10^6/34$
$= 1.47 \times 10^5$

For $R_1 = 1.839\,\text{cm}$, we can find the velocity at the middle of the path by the given formula
$R_1 = 1.839 = 0.98 \times 10^{-27}\, v_1{}^3$
or $v_1 = 1.233 \times 10^9$. The corresponding energy at the mid-path is
$E_1 = mv_1{}^2/2 = mc^2\, v_1{}^2/2c^2 = (1/2) \times 3728 \times (1.233 \times 10^9/3 \times 10^{10})^2 = 3.148\,\text{MeV}$
Energy lost in the first half of the path
$E = 5.000 - 3.148 = 1.852\,\text{MeV}$.

Number of ion pairs produced over the first half of the path $= 1.852 \times 10^6/34 = 5.45 \times 10^4$

7.49 $v_p^2 = 2E/m_p,\, v_d^2 = 2E/m_d = 2E/2m_p = v_p^2/2$
$= dE/dx \;\propto\; z^2/v^2$

$$\therefore\ \frac{(dE/dx)_p}{(dE/dx)_d} = \frac{v_d^2}{v_p^2} = 2(\because z = 1 \text{ for both } p \text{ and } d)$$

7.50 $(-\mathrm{d}E/\mathrm{d}x)_{\text{proton}} = (-\mathrm{d}E/\mathrm{d}x)_{\text{deuteron}}$

As their charges are identical, their velocity must be the same. Therefore, the ratio of their kinetic energies must be equal to the ratio of their masses.
$E_\mathrm{p}/E_\mathrm{d} = M_\mathrm{p}/M_\mathrm{d} = 1/2$

7.51 After crossing a radiation length of lead electrons emerge with an average energy of 2.7/e $=$ 2.7/2.71 $=$ 1.0 GeV. Thus, the average energy loss $= 1.7\,\text{GeV}$.

7.52 The root mean square multiple scattering angle is approximately given by
$(\overline{\theta^2})^{\frac{1}{2}} = \frac{20}{\beta p(MeV/c)}\sqrt{\frac{L}{L_\mathrm{R}}}$

For a traversal of a distance L. For $P = 400\,\text{MeV/c}$ and $L = L_\mathrm{R}/10$, this angle is 15.8 mr. An electron emitting a photon will be emitted at an angle of $m_\mathrm{e}c^2/E_\gamma$ with the direction of flight. This angle is 0.511/400 or 1.28 mr.

Thus, the angular distribution of Bremsstrahlung photons is determined mainly by the multiple scattering of electrons.

7.3.4 Compton Scattering

7.53 (a) Let $E = h\nu$ be the energy of the scattered photon and $h\nu/c$ be its momentum (Fig. 7.15).
Energy conservation gives

$$h\,\nu_0 = h\,\nu + T \tag{1}$$

where T is the electron's kinetic energy.
Balancing momentum along and perpendicular to the direction of incidence

Fig. 7.15

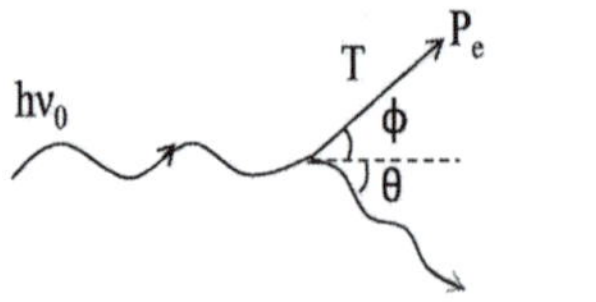

$$h\ \nu_0/c = h\ \nu\ \cos\ \theta/c + P_e\ cos\ \varphi \tag{2}$$

$$0 = -h\ \nu\ \sin\ \theta/c + P_e\ \sin\ \varphi \tag{3}$$

where P_e is the electron momentum
Re-arranging (2) and (3) and squaring

$$P_e{}^2\ \cos^2\ \varphi = \left(\frac{h\nu_0}{c} - \frac{h\nu\cos\theta}{c}\right)^2 \tag{4}$$

$$P_e{}^2\ \sin^2\ \varphi = \left(\frac{h}{c}\nu\sin\theta\right)^2 \tag{5}$$

Add (4) and (5) and using the relativistic equation

$$c^2 P_e{}^2 = T^2 + 2Tmc^2 = h^2(\nu_0{}^2 + \nu^2 - 2\nu_0\nu\ \cos\ \theta) \tag{6}$$

Eliminating T between (1) and (2) and simplifying we get

$$E = E_0/[1 + \alpha(1 - \cos\ \theta)] \tag{7}$$

(b) $T = E_0 - E = E_0 - E_0/[1 + \alpha(1 - \cos\theta)] = [\alpha E_0(1 - \cos\theta)]/[1 + \alpha(1 - \cos\theta)]$ (8)

(c) From (2) and (3), we get
$\text{Cot}\ \varphi = (\nu_0 - \nu\cos\theta)/\nu\sin\theta$
With the aid of (7), and re-arranging we find $\tan(\theta/2) = (1 + \alpha)\tan\varphi$ (9)

7.54 Fractional shift in frequency is

$$\Delta\nu/\nu_0 = \frac{1}{1 + \frac{Mc^2}{2h\nu_0\sin^2\left(\frac{\theta}{2}\right)}}$$

$h\nu_0 = E_\gamma = 1,241/\lambda_{\text{nm}} = 1,241/0.1 = 12,410\,\text{eV} = 0.01241\,\text{MeV}$
Put $\theta = 180^\circ$ and $MC^2 = 938.3\,\text{MeV}$ to obtain

$$(\Delta\nu/\nu_0)_{\max} = \frac{1}{1 + \frac{938.3}{2\times 0.01241}} = 2.645 \times 10^{-5}$$

7.55 The energy of scattered photon will be

$$h\ \nu = h\ \nu_0/[1 + \alpha(1 - \cos\theta)]$$

For $h\nu_0 = 30\,\text{keV},\ \alpha = h\nu_0/mc^2 = 30/511 = 0.0587$ and $\theta = 30°$
We find $h\nu = 29.766\,\text{keV}$
Therefore the kinetic energy of the electron

$$T = 30.0 - 29.766 = 0.234\,\text{keV}$$

The velocity $v = (2T/m)^{1/2} = c(2T/mc^2)^{1/2} = c(2\times 0.234/511)^{1/2} = 0.03\,c$

7.56 As the K-shell ionization energy for tantalum is under 80 keV, the ejected electrons due to photoelectric effect are expected to have kinetic energy little less than 1.5 MeV. Hence photoelectric effect is ruled out as the source of the observed electrons.

For the $e^+ e^-$ pair production the threshold energy is 1.02 MeV. The combined kinetic energy of the pair would be (1.5 − 1.02) or 0.48 MeV, a value which falls short of the observed energy of 0.7 MeV for the electron. Thus the observed electrons can not be due to this process.

In the Compton scattering the electrons can be imparted kinetic energy ranging from zero to $T_{max} = h\nu_0/(1 + 1/2\alpha)$, depending on the angle of emission of the electron. Substituting the values, $h\nu_0 = 1.5\,\text{MeV}$ and $\alpha = h\nu_0/\text{mc}^2 = 1.5/0.51 = 2.94$, we find $T_{max} = 1.28\,\text{MeV}$. Thus the Compton scattering is the origin of the observed electrons.

7.57 For Compton scattering, if λ_0 and λ are the wavelength of the incident photon and scattered photon,

$$\Delta\lambda = \lambda - \lambda_0 = \frac{h}{mc}(1 - \cos\theta) = 0.02425(1 - \cos\ 60°)\ \text{Å}$$
$$= 0.01212\ \text{Å}$$

Therefore, $\lambda_0 = \lambda - 0.01212 = 0.312 - 0.012 = 0.3\text{Å}$

7.58 (a) The energy of the scattered photon through an angle θ is

$$E_1 = E_0/[(1 + \alpha(1 - \cos\theta)] \qquad (1)$$

where $\alpha = E_0/mc^2$
For $\theta = 180^0,\ E_1 = E_0/(1 + 2\alpha)$

Loss of energy $\Delta E = E_0 - E_1 = 2\,\alpha\,E_0/(1 + 2\,\alpha)$ (2)

(b) The energy of photon after first scattering through 90° by the application of (1) is $E_1' = E_0/(1 + \alpha)$
The energy of this photon after the second scattering will be

$$E_2' = \frac{E_1'}{(1+\alpha)(1+\alpha')} = \frac{E_0}{1 + 2\,\alpha}\left(\because\ \alpha' = \frac{E_1'}{mc^2} = \frac{E_0}{1+\alpha}mc^2 = \frac{\alpha}{1+\alpha}\right)$$

The total energy loss $\Delta E' = E_0 - E_0/(1 + 2\alpha) = 2\alpha E_0/(1 + 2\alpha)$ (3)

(c) The energy of photon after the first scattering through 60° will be

$$E_1'' = 2E_0/(2 + \alpha)$$

The photon energy after second scattering through 60° will be

$$E_2'' = \frac{2E_1''}{2+\alpha''} = \frac{4E_0}{(2+\alpha)(4)(1+\alpha)(2+\alpha)} = \frac{E_0}{(1+\alpha)_0}$$

Photon energy after the third scattering through 60° will be

$$E_3'' = \frac{2E_2''}{2+\alpha''} = \frac{2\ E_0}{(1+\alpha)(2+3\alpha)/(1+\alpha)} = \frac{2E_0}{2+3\alpha}$$

∴ Total energy loss

$\Delta\ E'' = E_0 - 2E_0/(2+3\ \alpha) = (3\alpha E_0)/(2+3\ \alpha)$ (4)

Comparison of (2), (3) and (4) shows that the energy loss is equal for Case (a) and (b), and each is greater than in case (c)

7.59 $\mu_c = \sigma_c \rho N_0/A$

$\mu_c\, x = \sigma_c\, (N_0/A)\, (\rho\, x) = 8.1 \times 10^{-24} \times 6 \times 10^{23} \times 3.7/27 = 6.66$

$\mu_{ph} = 4 \times 10^{-24} \times 6 \times 10^{23} \times 3.7/27 = 3.288$

$\mu x = (\mu_c + \mu_{ph})\chi = 6.66 + 3.288 = 9.948$

The intensity will be reduced by

$I/I_0 = e^{-\mu x} = e^{-9.948} = 4.78 \times 10^{-5}$

Ratio of intensities absorbed due to the Compton effect and due to the photo-effect $= (1 - \exp(-\mu_c\, x))/(1 - \exp(-\mu_{ph}\, x)) = (1 - e^{-6.66})/(1 - e^{-3.288}) =$ 1.037

7.60 The change in wavelength in Compton scattering is given by

$\Delta\lambda = \frac{h}{mc}(1 - \cos\theta)$

where θ is the scattering angle. $\Delta\lambda$ will be maximum for $\theta = 180°$ in which case

$\Delta\lambda(\max) = 2h/mc = 4\hbar c/mc^2$

$= 4\pi \times 197.3$ MeV-fermi/0.511 MeV = 0.0485 Å

7.61 The energy E of the Compton scattered photon by the incident photon of energy E_0 is

$E = E_0/[(1 + \alpha\ (1 - \cos\theta)]$

where $\alpha = E_0/mc^2$ and θ is the scattering angle.

$E/E_0 = 1/2 = 1/[(1 + \alpha(1 - \cos\ 45°)]$

whence $\alpha = 3.415$

Or $E = (3.415((0.511) = 1.745$ MeV

7.62 Energy of the scattered gamma rays

$E = E_0/[1 + \alpha(1 - \cos\theta)]$ (1)

Where $\alpha = E_0/mc^2$

In $e^+ - e^-$ annihilation, each gamma ray has energy $E_0 = 0.511$ MeV

$\alpha = 0.511/0.511 = 1$

Therefore (1) becomes

$E = 0.511/(2 - \cos\theta)$

E_{max} is obtained by putting $\theta = 0$ and E_{min} by putting $\theta = 180°$. Thus

$E_{max} = 0.511$ MeV

$E_{min} = 0.17$ MeV

7.3.5 Photoelectric Effect

7.63 According to Einsteins's equation, kinetic energy of the photoelectron

$$T = h\,\nu - h\,\nu_0 \tag{1}$$

where ν is the frequency of the incident photon and ν_0 is the threshold frequency.

$\lambda = 2{,}536 \times 10^{-10}\,\text{m} = 253.6\,\text{nm}$

Corresponding energy $E = h\,\nu = 1{,}241/253.6\,\text{eV} = 4.894\,\text{eV}$

$\lambda_0 = 3{,}250 \times 10^{-10}\,\text{m} = 325\,\text{nm}$

$E = h\,\nu_0 = 1{,}241/325 = 3.818$

$T = 4.894 - 3.818 = 1.076\,\text{eV}$

$T = ½\,mv^2 = mc^2\,v^2/2c^2 = 0.511 \times 10^6 \times v^2/2\,c^2 = 1.076$

whence $v = 2.05 \times 10^{-3}\,c = 6.15 \times 10^5\,\text{m s}^{-1}$

7.64 The momentum $p = 300\,\text{Br MeV}/c$

$= 300 \times 3.083 \times 10^{-3}$

$= 0.925\,\text{MeV}/c$

$E^2 = (T + m)^2 = p^2 + m^2$

$T = (p^2 + m^2)^{1/2} - m$

Put $p = 0.925$ and $m = 0.511$

$T = 0.546\,\text{MeV}$

(a) The kinetic energy of the photoelectrons is 0.546 MeV

(b) The energy of the gamma ray photons is $0.546 + 0.116 = 0.662\,\text{MeV}$

7.65 $T = h\,\nu - h\,\nu_0$

$h\,\nu = 1{,}241/253.7 = 4.89\,\text{eV}$

$h\,\nu_0 = 1{,}241/325 = 3.81\,\text{eV}$

$T = eV = 4.89 - 3.81 = 1.08$

The required potential is 1.08 V

7.66 Suppose the photoelectric effect does take place with a free electron due to the absorption of a photon of energy T. The photoelectron must be ejected with energy in the incident direction. Energy and momentum conservation give

$$T = h\,\nu \tag{1}$$

$$P = h\,\nu\,/c \tag{2}$$

Equation (1) can be written as the relativistic relation connecting momentum and kinetic energy

$$T^2 = c^2\,p^2 = T^2 + 2Tmc^2 \tag{3}$$

Using (1) and (2) in (3), we get

$2h\,\nu\,.\,mc^2 = 0$

Neither h nor mc^2 is zero. We thus end up with an absurd situation. This only means that both energy and momentum can not be conserved for photoelectric effect with a free electron.

7.67 Number of gamma rays absorbed in the thickness x cm of lead
$N = N_0(1 - e^{-\mu x})$
where N_0 is the initial number and μ is the absorption coefficient expressed in cm^{-1}.
Now $\mu = \sigma N_0 \rho / A$
where N_0 is the Avagadro's number, ρ is the density of lead and A is its atomic weight.
$\mu = 20 \times 10^{-24} \times 6.02 \times 10^{23} \times 11.3/207 = 0.657\,\text{cm}^{-1}$
$n/n_0 = 90/100 = 1 - e^{-0.657x}$
$x = 3.5\,\text{cm}$

7.68 The K-shell absorption wavelength in Ag
$\lambda_k = 0.0424\,\text{nm}$
The corresponding energy
$E_K(\text{Ag}) = 1{,}241/0.0485 = 25{,}567\,\text{eV}$
We use the formula
$E_K(Z) = 13.6(Z - \sigma)^2$
For silver $25.567 \times 10^3 = 13.6(47 - \sigma)^2$
whence $\sigma = 3.64$
For the impurity X of atomic number $Z = 50$
$E_k(Z) = 13.6(50 - 3.64)^2$
$= 29{,}229\,\text{eV}$
$= 29.23\,\text{keV}$
Now the wavelength of 0.0424 nm corresponds to $E = 29.24\,\text{keV}$ which is in agreement with the calculated value. Thus the impurity is $_{50}\text{Sn}$

7.69 $eV = hc/\lambda - W$
A plot of V against $1/\lambda$ must be a straight line. The slope of the line gives hc/e, hence h can be determined. The intercept multiplied by hc give W, the work function. The threshold frequency is given by $\nu_0 = W/h$ (Fig. 7.16).

Fig. 7.16

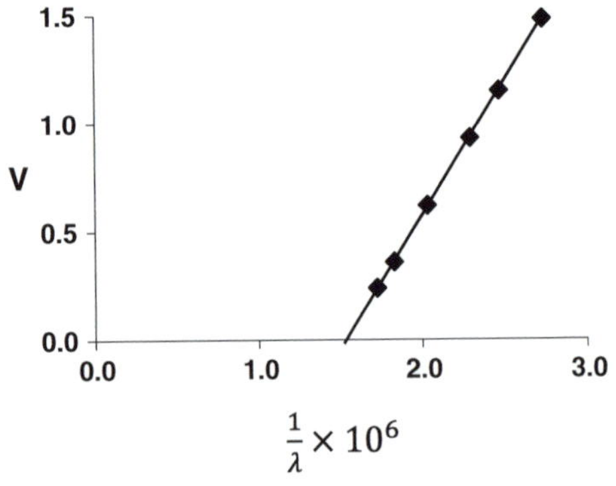

V	$\lambda \times 10^{-10}$ m	$(1/\lambda) \times 10^6$
1.48	3,660	2,732
1.15	4,050	2,469
0.93	4,360	2,293
0.62	4,920	2,032
0.36	5,460	1,831
0.24	5,790	1,727

Slope $= 1.24/10^6 = hc/e$
$h = 1.24 \times 10^{-6} \times e/c = (1.24 \times 10^{-6} \times 1.6 \times 10^{-19}/(3 \times 10^8) = 6.6 \times 10^{-34}$ J-s
Intercept $= 1.5 \times 10^{-6}\,\text{m}^{-1}$
$W = hc \times \text{intercept} = 6.6 \times 10^{-34} \times 3 \times 10^8 \times 1.5 \times 10^6$
$= 3 \times 10^{-19}\,\text{J} = 3 \times 10^{-19}/1.6 \times 10^{-19}\,\text{eV} = 1.9\,\text{eV}$
$h\nu_0 = W$
$\nu_0 = W/h = 3 \times 10^{-19}/6.6 \times 10^{-34} = 4.545 \times 10^{14}$
$\lambda_0 = 6.6 \times 10^{-7}\,\text{m}$

7.70 The solid angle subtended at the source
Ω = area of the face of the crystal/square of the distamce from the source
$= \pi R^2/d^2$

Assuming that the gamma rays are emitted isotropically, the fraction of gamma rays entering the crystal
$F = \Omega/4\pi = R^2/4d^2 = 2^2/4 \times 100^2 = 10^{-4}$

The number of photons entering the crystal per second from the source of strength S is
$N = SF = 100 \times 10^{-6} \times 3.7 \times 10^{10} \times 10^{-4}$
$= 370/\text{s}$

Number of photons absorbed in the crystal of 1 cm thickness due to photoelectric effect will be
$N_{ph} = N(1 - \exp(-\mu_{ph}\,x) = 370(1 - e^{-0.03 \times 1}) = 11$
and the number absorbed due to Compton scattering will be
$N_c = N(1 - \exp(-\mu_c\,x) = 370(1 - e^{-0.24 \times 1}) = 79$

Assuming that each photon that interacts in the crystal produces a photo electron, N_{ph} and N_c also denote the number of photoelectrons in the respective processes.
Number of photons that do not interact in 1sec in the crystal is $370 - (11 + 79) = 280$

7.71 In the first excited state the ionization energy is $13.6/2^2 = 3.4\,\text{eV}$
$T = h\nu - W$
$h\nu = T + W = 10.7 + 3.4 = 14.1\,\text{eV}$
In the ground state, $W = 13.6\,\text{eV}$
$T = 14.1 - 13.6 = 0.5\,\text{eV}$

7.72 $T = h\nu - W$ (Einstein's equation)
$T_1 = hc/\lambda_1 - W$

$T_2 = hc/\lambda_2 - W$
$T_2 - T_1 = hc(1/\lambda_2 - 1/\lambda_1)$
$(4.0 - 1.8) \times 1.6 \times 10^{-19}\,\text{J} = hc\left(\frac{1}{700\times 10^{-10}} - \frac{1}{800\times 10^{-10}}\right)$
Solving for h, we get $h = 6.57 \times 10^{-34}$ Js
The accepted value is $h = 6.625 \times 10^{-34}$ J-s

7.73 $h\nu_0 = W = 4.7\,\text{eV}$
$\lambda_0 = 1{,}241/4.7 = 264\,\text{nm}$.

7.74 $T_A = 4.25 - W_A$ (1)

$T_B = 4.7 - W_B$ (2)

$$\frac{T_A}{T_B} = \frac{P_A^2}{P_B^2} = \frac{\lambda_B^2}{\lambda_A^2} = 4 \quad (3)$$

$T_B = T_A - 1.5$ (4)

Solving the above equations,
$T_A = 2\,\text{eV}$ and $T_B = 0.5\,\text{eV}$
$W_A = 2.25\,\text{eV}$ and $W_B = 4.2\,\text{eV}$

7.3.6 Pair Production

7.75 The minimum photon energy required for the e^-e^+ pair production is $2mc$, where m is the mass of electron. Therefore
$h\,\nu = 2mc^2 = 2 \times 0.511 = 1.022\,\text{MeV}$
The corresponding wavelength
$\lambda = (1,241/1.022 \times 10^6\,\text{eV})\,\text{nm} = 1.214 \times 10^{-12}\,\text{m}$.

7.76 In the annihilation process energy released is equal to the sum of rest mass energy of positron and electron which is $2mc^2$. Because of momentum and energy conservation, the two photons must carry equal energy. Therefore each gamma ray carries energy
$E_\gamma = mc^2 = 0.511\,\text{MeV}$
The wavelength of each photon
$\lambda = 1,241/(0.511 \times 10^6)\text{nm} = 2.428 \times 10^{-3}\text{nm} = 2.428 \times 10^{-12}\,\text{m}$

7.77 Let us suppose that the $e^+ e^-$ pair is produced by an isolated photon of energy $E = h\,\nu$ and momentum $h\,\nu/c$. Let the electron and positron be emitted with momentum p_- and p_+, their total energy being E_- and E_+. Energy conservation gives

$h\,\nu_0 = E_+ + E_-$ (1)

The momentum conservation implies that the three momenta vectors, $\mathbf{p}_0$, $\mathbf{p}_+$ and $\mathbf{p}_-$ must form the sides of a closed triangle, as in Fig. 7.17. Now in any triangle, any side is equal or smaller than the sum of the other sides. Thus

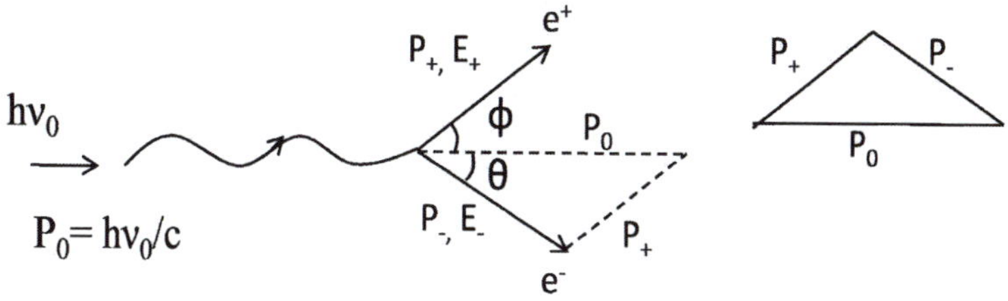

Fig. 7.17

$$h\,\nu/c \le p_+ + p_- \qquad (2)$$

Or $(h\,\nu)^2 \le (cp_+ + cp_-)^2$

By virtue of (1)

$(E_+ + E_-)^2 \le c^2 p_+{}^2 + c^2 p_-{}^2 + 2c^2 p_+ p_-$

Or $E_+{}^2 + E_-{}^2 + 2E_+E_- \le E_+{}^2 - m^2 c^4 + E_-{}^2 - m^2c^4 + 2[(E_+{}^2 - m^2c^4)$
$(E_-{}^2 - m^2 c^4)]^{1/2}$

$E_+E_- \le [(E_+{}^2 - m^2 c^4)(E_-{}^2 - m^2 c^4)]^{1/2} - mc^2$

Now the left hand side of the inequality is greater than the value of the radical on the right hand side. It must be still greater than the right hand side. We thus get into an absurdity, which has resulted from the assumption that both momentum and energy are simultaneously conserved in this process. Thus an electron–positron pair cannot be produced by an isolated photon.

7.3.7 Cerenkov Radiation

7.78 $P_\mu = 150\,\text{MeV/c}$
$E_\mu = (p^2{}_\mu + m^2{}_\mu)^{1/2} = (150^2 + 106^2)^{1/2} = 183.67\,\text{MeV}$
$\beta_\mu = P_\mu/E_\mu = 150/183.67 = 0.817$
$n_\mu = 1/\beta_\mu = 1.224$
$P_\pi = 150\,\text{MeV/c}$
$E_\pi = (P^2{}_\pi + m^2{}_\pi)^{1/2} = (150^2 + 140^2)^{1/2} = 205.18$
$\beta_\pi = P_\pi/E_\pi = 150/205.18 = 0.731$
$n_\pi = 1/\beta_\pi = 1.368$

Therefore the range of the index of refraction of the material over which the muons above give Cerenkov light is 1.224 – 1.368.

7.79 $\cos\theta = 1/\beta\,\mu$
$\beta = 1/\mu\cos\theta = 1/(1.8 \times \cos 11°) = 0.566$
$\gamma = (1 - \beta^2)^{-1/2} = 1.213$
Kinetic energy of proton $T = (\gamma - 1)\, m_p\, c^2$
$= (1.213 - 1) \times 938.3$
$= 200\,\text{MeV}$

7.80 For electron $z = -1$. The given integral is easily evaluated assuming that this refractive index μ is independent of frequency. Integrating between the limits ν_1 and ν_2.

$-dW/dl = (4\pi^2 e^2/c^2)(1 - 1/\beta^2\mu^2)[(\nu_2{}^2 - \nu_1{}^2)/2]$ (1)

Calling the average photon frequency as

$\nu = 1/2\,(\nu_1 + \nu_2)$ (2)

the average number of photons emitted per second is given by

$N = 1/h\,\bar{\upsilon} = (4\pi^2 e^2/hc)(1 - 1/\beta^2\mu^2)(\nu_2 - \nu_1)$

$= (2\pi/137)(1 - 1/\beta^2\mu^2)(1/\lambda_2 - 1/\lambda_1)$ (3)

where $\lambda_1 = c/\nu_1$ and $\lambda_2 = c/\nu_2$ are the vacuum wavelengths and μ is the average index of refraction over the wavelength interval $\lambda_2 = 4{,}000$ Å to $\lambda_1 = 8{,}000$ Å. Substituting $\beta\,\mu = 0.9 \times 1.33 = 1.197$,
$\lambda_1 = 8 \times 10^{-5}$ cm and $\lambda_2 = 4 \times 10^{-5}$ cm, in (3) we find the number of photons emitted per cm, $N = 173$.

7.3.8 Nuclear Resonance

7.81 The condition for resonance fluorescence is
$\Delta E_\gamma / E_\gamma = v/c$
Put $\Delta\, E_\gamma = \Gamma$
$\Gamma = vE_\gamma/c = (1\,\text{cm/s}/3 \times 10^{10}\,\text{cm/s}) \times (129\,\text{keV}) = 4.3 \times 10^{-6}\,\text{eV}$
The mean lifetime
$\tau = \hbar/\Delta\, E_\gamma = \hbar/\Gamma = 1.05 \times 10^{-34}/1.6 \times 10^{-19} \times 4.3 \times 10^{-6} = 1.5 \times 10^{-10}\,\text{s}$

7.82 Energy conservation gives

$h\,\nu = \Delta W - E_R$ (1)

Momentum conservation gives

$P_R = h\,\nu\,/c$ (2)

Therefore $E_R = P_R{}^2/2M = (h\,\nu)^2/2Mc^2$ (3)

Eliminating E_R between (1) and (3), we get a quadratic equation in $h\nu$.
$(h\,\nu)^2/2Mc^2 + h\,\nu - \Delta W = 0$
Which has the solution
$h\,\nu = -Mc^2 + Mc^2(1 + 2\Delta W/Mc^2)^{1/2}$
Expanding the radical binomially and retaining up to second power of ΔW, and simplifying
$h\,\nu = \Delta W(1 - \Delta W/Mc^2)$
$\nu = (\Delta W/h)(1 - \Delta W/Mc^2)$

7.83 The root mean square velocity of ^{137}Cs atoms
$\surd <v^2>= v = (3kT/m)^{1/2}$
Substituting $k = 1.38 \times 10^{-23}\,\text{J K}^{-1} = 0.8625 \times 10^{-10}\,\text{MeV K}^{-1}$
$T = 288\,\text{K}$, $mc^2 = 137 \times 931.5 = 1.276 \times 10^5$, we find $v/c = 7.64 \times 10^{-7}$

Optical Doppler effect is given by
$h\,\nu = h\,\nu_0(1 + \beta\cos\theta^*)$

The maximum and minimum energy of photons will be $h\,\nu_0(1 + \beta)$ and $h\,\nu_0(1 - \beta)$ or 661 keV $\pm$ 0.5 eV

7.84 The gravitational red shift is due to the change in the energy of a photon as it moves from one region of space to another differing gravitational potential. The photon carries an inertial as well as the gravitational mass given by $h\,\nu/c^2$. In its passage from a point where the gravitational potential is φ_1 to another point where the potential is φ_2 there will be expenditure of work given by $h\,\nu/c^2$ times the potential difference $(\varphi_2 - \varphi_1)$. This would result in an equivalent decrease in the energy content of the photon and hence its frequency.

$\Delta E = E(\varphi_2 - \varphi_1)/c^2$

A level difference of H near the earth's surface would result in the fractional shift of frequency

$\Delta\nu/\nu = gH/c$

Now, $\Delta\,\nu/\nu = v/c = gH/c^2$

or $v = gH/c = 9.8 \times 22.6/3 \times 10^8 = 7.38 \times 10^{-7}$ m/s $= 7.38 \times 10^{-4}$ mm/s

Thus resonance fluorescence would occur for downward velocity of the absorber of magnitude 7.38 mm/s.

7.3.9 Radioactivity (General)

7.85

T	0	4	8	12	16	20	24	28	32	36	40
dN/dt	18.59	13.27	10.68	9.34	8.55	8.03	7.63	7.30	6.99	6.71	6.44
$\ln(dN/dt)$	2.923	2.585	2.368	2.234	2.146	2.083	2.032	1.988	1.944	1.904	1.863

The log-linear plot of dN/dt versus time (t) is not a straight line because the source contains two types of radioactive material of different half-lives described by the sum of two exponentials. If the two half-lives are widely different then it is possible to estimate the half-lives by the following procedure.

Toward the end of the curve (Fig. 7.18), say for time 20–28 h, most of the atoms of the shorter-lived substance with half-life T_1 would have decayed and the curve straightens up, corresponding to the single decay of longer half-life of T_2. If this straight line is extrapolated back up to the y-axis, then the half-life T_2 can be estimated in the usual way as from the slope of the curve for a single source on the log-linear plot. In this example,

$$T_2 = \frac{0.693\,\Delta t}{\ln\left(\frac{dN}{dt}\right)_0 - \ln\left(\frac{dN}{dt}\right)t} = \frac{(0.693)(40)}{(2.083 - 1.863)} = 63 \text{ min}$$

For the shorter-lived substance, the contribution dN_2/dt of the source 2 can be subtracted from the observed values in the initial portion of the curve over suitable time interval and the procedure repeated. In this way we find $T_1 = 10$ min.

Fig. 7.18

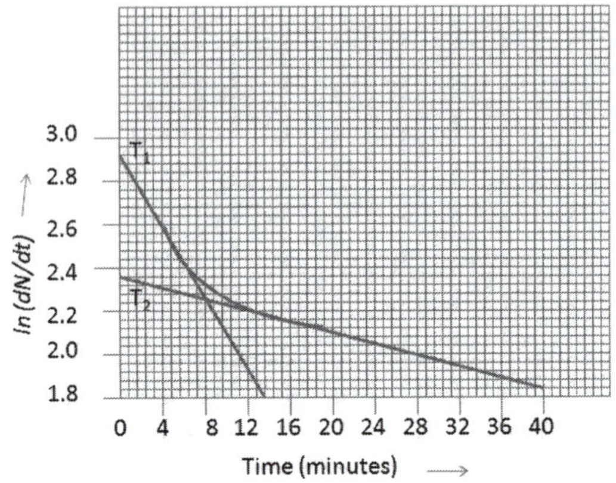

7.86 The flow of heat in a material placed between the walls of a coaxial cylinder is given by

$$\frac{dQ}{dt} = \frac{2\pi L}{\ln\left(\frac{r_2}{r_1}\right)}(T_1 - T_2) \tag{1}$$

Number of decays of radon atoms per second
$dN/dt = 100 \times 10^{-3} \times 3.7 \times 10^{10} = 3.7 \times 10^{9}$ disintegration/second
Energy deposited by α's $= 3.7 \times 10^{9} \times 5.5$ MeV/s
$= 2.035 \times 10^{10}$ MeV/s $= 3.256 \times 10^{-7}$ J $= 0.779 \times 10^{-3}$ Cal/s
Using the values, $k = 0.025$ Cal cm^{-2} s^{-1} C^{-1}, $L = 5$ cm, $r_1 = 2$ mm and $r_2 = 6$ mm in (1), and solving for $(T_1 - T_2)$ we find $(T_1 - T_2) = 1.09\,°$C

7.87 In the series decay A→B→C, if $\lambda_A < \lambda_B$ the transient equilibrium occurs when
$N_B/N_A = \lambda_A/(\lambda_B - \lambda_A)$
Here A = Uranium and B = Radium
$\lambda_A = 1/\tau_A = 0.693/4.5 \times 10^{9}$ year^{-1}, $\lambda_B = 1/\tau_B = 0.693/1{,}620$ year^{-1}
$N(\text{Rad})/N(\text{U}) \approx 1{,}620/4.5 \times 10^{9} = 1/2.78 \times 10^{6}$

7.88 Use the law of radioactivity

$$N_A = N_A{}^0 \exp(-\lambda_A t) \tag{1}$$

$$N_B = N_B{}^0 \exp(-\lambda_B t) \tag{2}$$

Dividing (2) by (1)
$N_B/N_A = \exp(\lambda_A - \lambda_B)t$ (Because $N_A{}^0 = N_B{}^0$)
Take $\log_e$ on both sides

$$t = \frac{1}{(\lambda_A - \lambda_B)\ln\left(\frac{N_B}{N_A}\right)}$$

Given $N_B/N_A = 2.2/0.53 = 4.15$
$\lambda_A = 0.693/T_{1/2}\,(A) = 0.693/12 = 0.05775\ \text{year}^{-1}$
$\lambda_B = 0.693/T_{1/2}\,(B) = 0.693/18 = 0.0385\ \text{year}^{-1}$
We find age of alloy $t = 73.93$ years.

7.89 Let W grams of ^{210}Po be required.

$$|dN/dt| = N\lambda \tag{1}$$

Required activity $|dN/dt| = 10\times10^{-3}\times3.7\times10^{10} = 3.7\times10^{8}$ disintegration/second

$$N = 6.02 \times 10^{23}\ W/210 = 2.867 \times 10^{21}\ W \tag{2}$$

$$\lambda = 0.693/T_{1/2} = 0.693/138 \times 86{,}400 = 5.812 \times 10^{-8} \tag{3}$$

Use (2) and (3) in (1) and solve for W to obtain $W = 2.22 \times 10^{-6} = 2.22\ \mu g$

7.90 Decay constant, $\lambda = 0.693/100 \times 86{,}400 = 8 \times 10^{-8}\ s^{-1}$
Let M be the gram molar weight of the substance.
Then number of atoms $N = N_0\ M = 6.02 \times 10^{23}$ M
$|dN/dt| = N\ \lambda = N_0\ M\lambda = 6.02 \times 10^{23} \times 8 \times 10^{-8}\ M = 4.816 \times 10^{16}\ M/s$
Power $P = 4.816 \times 10^{16}\ M \times 5 \times 10^{-14} = 2408$ M Watts
But 10% of this power is available, that is 240.8 M Watts. Equating this to the required power
$240.8\ M = 5$
Or $M = 0.02$ g-molecules

7.91 The present day activity

$$|dn/dt| = n_0\lambda\ e^{-\lambda t} \tag{1}$$

$n_0 = 6.02 \times 10^{23} \times 1.35 \times 10^{-10}/12 \times 100 = 6.77 \times 10^{10}$
$\lambda = 3.92 \times 10^{-10}\ s^{-1}$
$dn/dt = 12.9/60 = 0.215\ s^{-1}$
Using these values in (1) and solving for t, we get $t = 1.228 \times 10^{10}$ s or 390 years
$\lambda_E \quad \lambda_F$

7.92 RaE $\rightarrow$ RaF $\rightarrow$ RaG
The rate of decay of RaE is given by

$$dN_E/dt = -\lambda_E N_E \tag{1}$$

where N_E is the number of atoms of RaE and λ_E its decay constant.
The net change of RaF is given by

$$dN_F/dt = \lambda_E N_E - \lambda_F N_F \tag{2}$$

The first term on the right side represents the rate of increase of RaF (Note the positive sign) and the second term, the rate of decrease (Note the negative sign). Here N_E and N_F are the number of atoms of E and F respectively at time t.

Now at time t

$$N_E = N_E{}^0 \exp(-\lambda_E\, t) \tag{3}$$

Solution of (2) is

$$N_F = A \exp(-\lambda_E\, t) + B \exp(-\lambda_F\, t) \tag{4}$$

where A and B are constants which can be determined by the use of the initial conditions. At $t = 0$, the initial number of F is zero, that is $N_F{}^0 = 0$. Using this condition in (4) gives $B = -A$ and (4) becomes

$$N_F = A[\exp(-\lambda_E\, t) - \exp(-\lambda_F\, t)] \tag{5}$$

Further, $\mathrm{d}N_F/\mathrm{d}t = -\lambda_E\, A \exp(-\lambda_E\, t) + \lambda_F A \exp(-\lambda_F\, t)$
At $t = 0$

$$\mathrm{d}N_F{}^0/\mathrm{d}t = \lambda_E\, N_E{}^0 = -\lambda_E\, A + \lambda_F A$$

$$\text{or } A = \lambda_E\, N_A{}^0/(\lambda_B - \lambda_A) \tag{6}$$

Using (6) in (5)

$$N_F = \frac{\lambda_E N_E^0}{\lambda_F - \lambda_E}[\exp(-\lambda_E t) - \exp(-\lambda_F t)] \tag{7}$$

The time at which the greatest number of RaF atoms is obtained by differentiating N_F with respect to t in (7) and setting $\mathrm{d}N_F/\mathrm{d}t = 0$

$$\text{We find } t_{\max} = \frac{1}{\lambda_F - \lambda_E} \ln\left(\frac{\lambda_F}{\lambda_E}\right) \tag{8}$$

$\lambda_E = 0.693/5 = 0.1386\ \mathrm{day}^{-1}$, $\lambda_F = 0.693/138 = 0.0052\ \mathrm{day}^{-1}$

Number of bismuth atoms $= 6.02 \times 10^{23} \times 5 \times 10^{-10}/210 = 1.43 \times 10^{12}$

Using these values in (6) and (5) we find
$T = 24.6$ days
$N_F(\max) = 1.37 \times 10^{10}$

7.93 In the series decay, A → B → C, if $\lambda_A << \lambda_B$, that is $\tau_A >> \tau_B$, secular equilibrium is reached and $N_B/N_A = \lambda_A/\lambda_B = T_{1/2}(B)/T_{1/2}(A)$
Here A=Radium and B=Radon
$T_{1/2}(\text{Radium}) = N(\text{radium}).T_{1/2}(\text{Radon})/N(\text{radon})$
$= \left(\frac{1}{6.4 \times 10^{-6}}\right)\left(\frac{3.825}{365}\right) = 1{,}637$ years

7.94 Activity $|\mathrm{d}N_1/\mathrm{d}t| = \lambda N_1$
$|\mathrm{d}N_2/\mathrm{d}t| = \lambda\, N_2 = \lambda\, N_1 \exp(-\lambda t)$
Fractional decrease of activity
$= [\lambda N_1 - \lambda N_1 \exp(-\lambda t)]/\lambda N_1 = 1 - \exp(-\lambda t) = 4/100$
Or $\exp(-\lambda t) = 24/25$
$\lambda = \frac{1}{t} \ln\left(\frac{25}{24}\right)$
Put $t = 1$ h, $\lambda = 0.0408$
$\tau = 1/\lambda = 24.5$ h

7.95 Let the proportion of ^{235}U and ^{238}U at $t = 0$ be $x : (1 - x)$. Present day radioactive atoms for the two components after time t will be

$$N_{235} = xN_0 \exp(-\lambda_{235}t) \quad (1)$$

$$N_{238} = (1 - x)N_0 \exp(-\lambda_{238}t) \quad (2)$$

Dividing (1) by (2)

$$N_{235}/N_{238} = 1/140 = \frac{xe^{-(\lambda_{235}-\lambda_{238})t}}{1-x} \quad (3)$$

$\lambda_{235} = 0.693/T_{1/2}$, $T_{1/2} = 8.8 \times 10^8$ years
$\lambda_{238} = 0.693/T_{1/2}$, $T_{1/2} = 4.5 \times 10^9$ years
$t = 3 \times 10^9$ years
Substituting these values in (3) and solving for x we get $x = 1/22$

Therefore, the ratio of ^{235}U and ^{238}U atoms 3×10^9 years ago was 1/22:21/22 or 1:21

7.96 $|dN/dt| = N\lambda$
$N = 1.0 \times 10^{-6} \times 6.02 \times 10^{23}/242 = 2.487 \times 10^{15}$
$|dN/dt| = N\lambda$
$80 + 3/3{,}600 = 2.487 \times 10^{15} \times 0.693/T_{1/2}$
Solving for $T_{1/2}$, we find $T_{1/2} = 2.15 \times 10^{13}$ s or 6.8×10^5 years
For α-decay: $80 = N\lambda_\alpha = 2.487 \times 10^5\ \lambda_\alpha$
$\lambda_\alpha = 1.01 \times 10^{-6}$ year^{-1}
For fission $3/3{,}600 = 2.487 \times 10^{15}\lambda_f$
$\lambda_f = 1.05 \times 10^{-11}$ year^{-1}

7.97 $$N_{sr} = N_{sr}{}^0 \exp(-\lambda_{sr}t) \quad (1)$$

$$N_y = \frac{\lambda_{sr}N_{sr}^0}{\lambda_y - \lambda_{sr}}[\exp(\lambda_{sr}t) - \exp(-\lambda_y t)] \quad (2)$$

Dividing the two equations

$$\frac{N_y}{N_{sr}} = \frac{\lambda_{sr}}{\lambda_y - \lambda_{sr}}[1 - \exp(\lambda_{sr} - \lambda_y)t] \quad (3)$$

$\lambda_{sr} = 0.693/(28 \times 365 \times 24) = 2.825 \times 10^{-6}\,\mathrm{h}^{-1}$

$\lambda_y = 0.693/64 = 0.0108\,\mathrm{h}^{-1}$

(a) For $t = 1$ h and using the values for the decay constants $N_{sr}/N_y = 3.56 \times 10^5$
(b) For $t = 10$ years, $N_{sr}/N_y = 3{,}823$

7.98 If N is the number of atoms of each component at $t = 0$, then at time t the number of atoms of the two components will be

$$N_{235} = N_0 \exp(-\lambda_{235}\, t) \quad (1)$$

$$N_{238} = N_0 \exp(-\lambda_{238}\, t) \quad (2)$$

Dividing the two equations

$$N_{235}/N_{238} = (0.7/100) = \exp -(\lambda_{235} - \lambda_{238})t \quad (3)$$

Using $\lambda_{238} = 1/\tau_{238} = 1/6.52 \times 10^9,\ \lambda_{235} = 1/\tau_{235} = 1/1.02 \times 10^9$
And solving (3) we get the age of the earth as 5.26×10^9 years

7.99 Number of Th atoms in 2 μg is
$N = N_0\, W/A = 6.02 \times 10^{23} \times 2 \times 10^{-6}/224 = 5.375 \times 10^{15}$
$\lambda = 0.693/(3.64 \times 86,400) = 2.2 \times 10^{-6}$
Activity $A = |\mathrm{d}N/\mathrm{d}\,\lambda| = N\,\lambda = (5.375 \times 10^{15})(2.2 \times 10^{-6}) = 1.183 \times 10^{10}/\mathrm{s}$
$= (1.188 \times 10^{10}/3.7 \times 10^{10})\,\mathrm{ci} = 0.319\,\mathrm{ci}$

7.100 1 rad = 100 ergs g^{-1}
Energy received $= (40)(100)(20 \times 10^3)/10^7$ J
$= 8$ J
$= 8/4.18 = 1.91$ Cal

7.101 Let V cm^3 be the volume of blood, Initial activity is $16{,}000/V$ per minute per cm^3. $0.8 = \frac{16{,}000}{V} \exp(-0.693 \times 30/15)$
$= \frac{4,123}{V}$
$\therefore V = 5000\,\mathrm{cm}^3$

7.3.10 Alpha-Decay

7.102 $\lambda_1 = 10^{21}\ \exp(-2\pi\ zZ_1 e^2/\hbar\, v_1)$
$\lambda_2 = 10^{21}\ \exp(-2\,\pi\, zZ_2\, e^2/\hbar v_2)$
Given $\lambda_1 = \lambda_2$
$Z_1/\sqrt{E_1} = Z_2/\sqrt{E_2}$
$E_2 = E_1\ {Z_2}^2/{Z_1}^2 = 5.3(80/82)^2 = 5.04\,\mathrm{MeV}$
Thus energy from the second nuclei is 5.04 MeV.

7.103 The energy required to force an α-particle(classically) into a nucleus of charge Ze is equal to the potential energy at the barrier height and is given by $E = zZe^2/4\,\pi\,\varepsilon_0\,R$.

This is barely possible when α particle and the Uranium nucleus will just touch each other and the kinetic energy of the bombarding particle is entirely converted into potential energy.

$R = R_1 + R_2 = r(A_1^{1/3} + {A_2}^{1/3})$

$= 1.2(4^{1/3} + 238^{1/3}) = 9.34$ fm

$E = 1.44zZ/R = 1.44 \times 2 \times 92/9.34 = 28.37\,\mathrm{MeV}$

7.104 Geiger–Nuttal law is
$\log \lambda = k\ \log\ x + c$
where k and c are constants, λ is the decay constant and x is the range

$\log(0.693/T) = k \log x + c$

$k\ \log\ x + \log\ T = \log\ 0.693 - c = c_1$ (1)

where c_1 is another constant.

$$k \log 3.36 + \log(1622 \times 365) = c_1 \quad (2)$$

$$k \log 3.85 + \log 138 = c_1 \quad (3)$$

Solving (2) and (3), $k = 61.46$ and $c_1 = 36.5$. Using the values of k and c_1 in (1) $61.46 \log 6.97 + \log T = 36.5$
Solving for T, we find $T = 4.79 \times 10^{-16}$ days $= 4.14 \times 10^{-11}$ s

7.105 For α-decay we use the equation

$$\lambda = 1/\tau = 10^{21} \exp(-2\pi zZ/137\beta) \quad (1)$$

$$T_{1/2}(1)/T_{1/2}(2) = \tau_1/\tau_2 = \exp(2\pi zZ_1/137\beta_1)/\exp(2\pi zZ_2/137\beta_2) \quad (2)$$

For ${}_{88}Ra^{226}$ decay put $Z_1 = 86$ for the daughter nucleus, $z = 2$ for α-particle and $\beta_1 = (2E/Mc^2)^{1/2} = (2 \times 4.9/3728)^{1/2} = 0.05127$
For ${}_{90}Th^{226}$ decay, put $Z_2 = 88$ for daughter nucleus and $z = 2$ for α-particle and $\beta_2 = (2 \times 6.5/3728)^{1/2} = 0.05905$
Using the values in (2), we find
$\tau_1/\tau_2 = 5.19 \times 10^7$

7.3.11 Beta-Decay

7.106 The selection rules for allowed transitions in β-decay are:

$$\left.\begin{array}{ll} \Delta I = 0 & \\ I_i = 0 \rightarrow I_f = 0 & \text{allowed} \\ \Delta\pi = 0 & \end{array}\right\} \quad \text{Fermi Rule}$$

$$\left.\begin{array}{ll} \Delta I = 0, \pm 1 & \\ I_i = 0 \rightarrow I_i = 0 & \text{forbidden} \\ \Delta\pi = 0 & \end{array}\right\} \quad \text{G.T. Rule}$$

where I is the nuclear spin and π is the parity. In view of the above selection rules the first transition is Fermi transition, the second one Gamow–Teller transition. The third one occurs between two mirror nuclei ^{17}F and ^{17}O in which the proton number and neutron number are interchanged. The configuration of nucleus is very much similar in such nuclei, consequently the wave functions are nearly identical. This leads to a large value for the overlap integral. The log ft value for such transitions is small being in the range of 3–3.7. These are characterized by
$\Delta I = 0, \pm 1$ and $\Delta\pi = 0$. The given transition $^{17}F \rightarrow ^{17}O$ is an example of superallowed transition.

7.107 If the number of ^{23}Mg nuclides is N_0 at $t = 0$, then at time t the number decayed will be

$$N = N_0[1 - \exp(-\lambda t)] \quad (1)$$

$$\text{At } t = t_1, \; N_1 = N_0[1 - \exp(-\lambda t)] \quad (2)$$

$$\text{At } t = t_2, \; N_2 = N_0[1 - \exp(-\lambda t)] \quad (3)$$

Dividing (3) by (2)

$N_2/N_1 = [1 - \exp(-\lambda t_2)]/[1 - \exp(-\lambda\, t_1)]$

$N_2/N_1 = 2.66 = [1 - \exp(-6\,\lambda)]/[1 - \exp(-2\lambda)]$

$$= (1 - x^3)/(1 - x) = 1 + x + x^2 \quad (4)$$

where $x = \exp(-2\lambda)$
Solving the quadratic equation for x, we find $x = 0.885 = \exp(-2\lambda)$
whence the mean life time $\tau = 1/\lambda = 15.93$ s

7.108 According to Fermi's theory of β-decay, for $E_0 >> m_e c^2$, the decay constant

$$\lambda = \frac{G^2|M_{\mathrm{if}}|^2 E_0^5}{60\pi^3(\hbar c)^6\hbar} \quad (1)$$

So that with the value $G/\hbar^3 c^3 = 1.166 \times 10^{-5}$ GeV^{-2}, (1) becomes $\lambda = \frac{1.11 E_0^5 |M_{\mathrm{if}}|^2}{10^4} S^{-1} = \frac{1.11 \times (3.5)^5 \times 6}{10^4}$
$= 0.3466\,\mathrm{s}^{-1}$
Therefore, $T_{1/2} = 0.693/0.3466 = 2.0$ s. The experimental value is 0.8 s.

7.109 The mean energy of electrons

$$< E >= \int_0^{E\max} \mathrm{E\,n(E)dE} / \int_0^{E\max} n(E)\,\mathrm{d}E$$

Given $n(E)\mathrm{d}E = \mathrm{c}\sqrt{E}(E_{\max} - E)^2\,\mathrm{d}E$
where c = constant

$$< E >= \frac{c\int_0^{E\max} E\sqrt{E}(E_{\max} - E)^2 \mathrm{d}E}{c\int_0^{E\max} \sqrt{E}(E_{\max} - E)^2\,\mathrm{d}E} = \frac{E_{\max}}{3}$$

If all the electrons emitted are absorbed then the kinetic energy of the electrons is converted into heat.
Heat evolved/sec = (mean energy)(no. of electrons emitted/second) $= (0.156) \times (3.7 \times 10^7)/3$ MeV/s $= 1.92 \times 10^6$ MeV/s

7.110 Let the threshold energy be E_ν. In that case the particles in the CMS will be at rest. Now, in the neutron decay

$$n \to p + e^- + \bar{\nu} + 0.79\,\mathrm{MeV}$$

$$M_n - (M_p + M_e) = 0.79$$

as $\bar{\nu}$ has zero rest mass

$$M_n + M_e = M_p + 2m_e + 0.79 = M_p + 1.81$$

(the masses of electron and positron are identical, each equal to 0.511 MeV)
We use the invariance of
$(\Sigma\, E)^2 - |\Sigma \mathbf{P}|^2 = (\Sigma\, E^*)^2$ – zero
where (*) refers to the CMS and total momentum of particles in CMS is zero
$(E_\nu + M_p)^2 = (M_n + M_e)^2 = (M_p + 1.81)^2$
Or $E_\nu = 1.81$ MeV.

7.111 i. Momentum $= 300\,\mathrm{Br} = (300)(0.1)(0.061) = 1.83\,\mathrm{MeV/c}$
Use the relativistic equation to find total energy, $E = (p^2 + m^2)^{1/2} = (1.83^2 + 0.511^2)^{1/2} = 1.90\,\mathrm{MeV}$
Kinetic energy of electron
$T = E - m = 1.90 - 0.51 = 1.39\,\mathrm{MeV} = 1.39 \times 10^6\,\mathrm{eV}$
Neglecting the energy of the recoiling nucleus the neutrino energy $E_\nu = 2.4 - 1.39 = 1.01\,\mathrm{MeV}$

ii. Maximum kinetic energy will be carried by the nucleus when the nucleus recoils opposite to β-particle and the neutrino is at rest. Momentum conservation gives

$$P_\mathrm{N}{}^2 = 2M_\mathrm{N}T_\mathrm{N} = P_\mathrm{e}{}^2 = T_\mathrm{e}{}^2 + 2T_\mathrm{e}\,m \quad (1)$$

Energy conservation gives

$$T_\mathrm{N} + T_\mathrm{e} = 2.4 \quad (2)$$

Eliminating T_e, we find $T_\mathrm{N} = 5\,\mathrm{keV}$.

7.112 $n(E) = C\sqrt{E}(E_{\max} - E)^2$
where $C =$ constant.
Differentiate $n(E)$ with respect to E and set $\mathrm{d}n/\mathrm{d}E = 0$. We find that the maximum occurs at $E = E_{\max}/5$.

Chapter 8
Nuclear Physics – II

8.1 Basic Concepts and Formulae

Nuclear models

Fermi gas model

Total kinetic energy of all protons

$$E_{\rm Z} = \frac{3}{5} Z\, E_{\rm Z}(\max);\; E_{\rm Z}(\max) = \frac{K^2}{2M}\left(\frac{Z}{A}\right)^{\frac{2}{3}} \tag{8.1}$$

Total energy of all neutrons.

$$E_{\rm N} = \frac{3}{5} N\, E_N(\max);\; E_{\rm N}(\max) = \frac{K^2}{2M}\left(\frac{N}{A}\right)^{\frac{2}{3}} \tag{8.2}$$

$$\text{where } K = \frac{\hbar}{r_0}\left(\frac{9\pi}{4}\right)^{\frac{1}{3}} \tag{8.3}$$

Shell model Magic Numbers: 2, 8, 20, 28, 50, 82, 126. Table 8.1 shows the shell number (Λ) occupation number (N_s) and the Number of neutrons and protons in various shells (ΣN_s) under the spin – orbit coupling scheme.

Table 8.1

Λ	N_s	ΣN_s
5	$N'_\Lambda = 44$	126
4	$N'_\Lambda = 32$	82
3	$N'_\Lambda = 22$	50
3	$2(\Lambda + 1) = 8$	28
2	$N_\Lambda = 12$	20
1	$N_\Lambda = 6$	8
0	$N_\Lambda = 2$	2

The shell model energy levels are:

$$\left[1s_{\frac{1}{2}}\right]\left[1p_{\frac{3}{2}},\ 1p_{\frac{1}{2}}\right]\left[1d_{\frac{5}{2}},\ 2s_{\frac{1}{2}},1d_{\frac{3}{2}}\right]\left[1f_{\frac{7}{2}}\right]\left[2p_{\frac{3}{2}},\ 1f_{\frac{5}{2}},\ 2p_{\frac{1}{2}},\ 1g_{\frac{9}{2}}\right]$$
$$\left[1g_{\frac{7}{2}},\ 2d_{\frac{5}{2}},\ 2d_{\frac{3}{2}},\ 3s_{\frac{1}{2}},\ 1h_{\frac{11}{2}}\right]\ldots \tag{8.4}$$

Liquid drop model (8.5)

$$M(\text{atom}) = Z\,M_H + (A-Z)M_n - \Delta \tag{8.6}$$

$$\Delta = \text{mass defect}$$

$$P = \frac{M-A}{A} = \text{packing fraction.} \tag{8.7}$$

$$1\,\text{amu} = \frac{1}{12}\text{ of atomic mass of }{}^{12}\text{C atom} \tag{8.8}$$

$$\text{f} = \frac{B.E}{A} \tag{8.9}$$

$$1\,\text{amu} = 931.5\,\text{Mev} \tag{8.10}$$

$$1\,\text{amu} = 1.66\times10^{-27}\,\text{kg} \tag{8.11}$$

The f-A curve is shown in Fig. 8.1. A more detailed diagram is shown in Problem 8.40

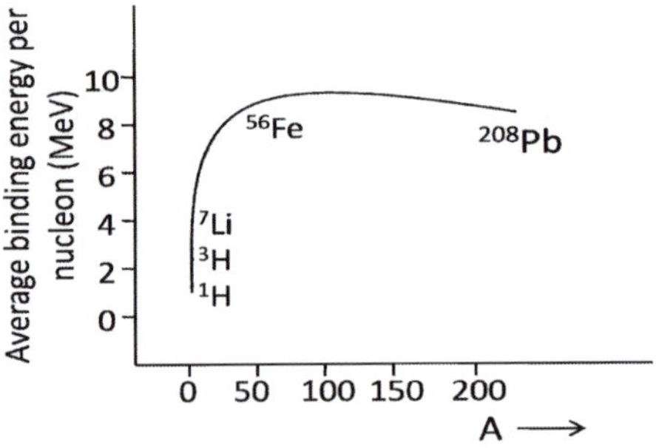

Fig. 8.1 BE/A Versus A

Stability against decay

$$\beta^- - \text{decay}: M(Z,\ A) \le M(Z+1,\ A) \tag{8.12}$$
$$\beta^+ - \text{decay}: M(Z+1,\ A) \le \text{M(Z, A)} + 2M_e \tag{8.13}$$
$$e^-\text{ capture}: M(Z+1, A) \le M(Z, A) \tag{8.14}$$
$$\alpha - \text{decay}: M(Z, A) \le M(Z-2,\ A-4) + M_{\text{He}^4} \tag{8.15}$$

Assuming that γ-ray precedes the decay, the energy released

$$Q_{\beta^-} = [M(Z, A) - M(Z+1, A)]\,c^2 = T_{\max} + T_\gamma \tag{8.16}$$
$$Q_{\beta^+} = [M(Z+1, A) - M(Z, A)]\,c^2 = 2M_e c^2 + T_{\max} + T_\gamma \tag{8.17}$$
$$Q_{\text{EC}} = [M(Z+1, A) - M(Z, A)]\,c^2 = T_\upsilon + T_\gamma \tag{8.18}$$

Charge symmetry of nuclear forces: p − p = n − n force
Charge Independence of nuclear forces: p − p = p − n = n − n force
Isospin: A fictitious quantum number (T) which is used in the formalism of charge independence The charge

$$\frac{Q}{e} = T_3 + \frac{B}{2} \tag{8.19}$$

where $\frac{Q}{e}$ is the charge of the particle in terms of electron charge, T_3 is the third component of T, and B is the baryon number.

Thus, n and p of similar mass form an isospin doublet of nucleon. For proton, $T_3 = +\frac{1}{2}$ and for neutron $T_3 = -\frac{1}{2}$. For strong interaction of other particles refer to Chap. 10. Although there is no connection between isospin and ordinary spin, the algebra is the same. For a system of particles, the notation for isospin will be I.

Nuclear spin (J)

$$\text{Odd A nuclei}: \; J = \frac{1}{2}, \frac{3}{2}, \frac{5}{2}, \ldots \tag{8.20}$$

$$\text{Even A nuclei}: \; J = 0, 1, 2, \ldots \tag{8.21}$$

Nuclear parity: By convention n and p are assigned even(+) intrinsic parity. In addition parity comes from the orbital angular momentum (l) and is given by $(-1)^l$.

Thus, for deuteron which is mainly in the s-state, this part of parity is +1. Parity is multiplicative quantum number, so that for deuteron, $P = +1$.

Hyperfine structure of spectral lines

Fine structure of spectral lines is explained by the electron spin, while the hyperfine structure is accounted for by the nuclear spin.

Nuclear magnetic moment (μ)

$$\mu = gJ\left(\frac{e\hbar}{2m_p c}\right) \tag{8.22}$$

where J is the nuclear spin and g is the nuclear g factor.

In Rabi's experiment the resonance technique is used. In a constant magnetic field B, the magnetic moment precesses with Larmor's frequency ν given by

$$\upsilon = \frac{\mu B}{Jh} \tag{8.23}$$

If an alternating magnetic field of frequency f is superimposed there will be a dip in the resonance curve when

$$\upsilon = f. \tag{8.24}$$

Electric quadrupole moment

For nuclei with $J = 0$ or $\frac{1}{2}$, electric dipole moment is zero. For spherical nuclei quadrupole moment is also zero. Only in non-spherical nuclei the quadrupole moment (Q) exists.

The concept of quadrupole comes from the classical electrostatic potential theory (Fig. 8.2).

$$\Phi(r, \theta) = \frac{1}{r}\Sigma_{n=0}^{\infty}\frac{a_n}{r^n}P_n(\cos\theta) \tag{8.25}$$

The quantity $2a_2/e$ is known as the quadrupole.

$$\frac{Q}{e} = \int \psi^* \,(3z^2 - r^2)\psi \,\mathrm{d}\tau \tag{8.26}$$

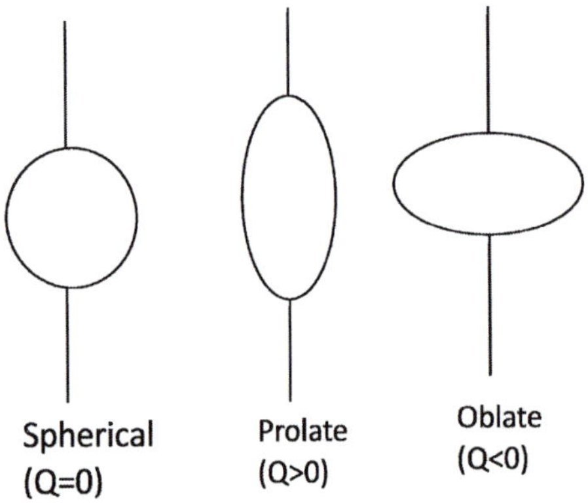

Fig. 8.2 Electric quadrupole moment for nuclei of various shapes

Nuclear reactions

$x(a, b)y \qquad\qquad a + x \rightarrow b + y$

$$Q = (m_x + m_{a-}m_y - m_a)c^2 = E_b + E_y - E_a \tag{8.27}$$

$$Q = E_b(1 + m_b/m_y) - E_a(1 - m_a/m_y) - \frac{2}{m_y}(m_a m_b E_a E_b)^{\frac{1}{2}}\cos\theta \tag{8.28}$$

Exoergic reactions: Q = +ve
Endoergic reactions: Q = −ve;

$$E_a\,(\text{threshold}) = |-Q|(1 + m_a/m_x) \tag{8.29}$$

Elastic Scattering: Q = zero

Table 8.2 Comparison of compound nucleus reactions and direct reactions

Feature	Compound nuclear reactions	Direct reactions
Times involved	10^{-14}–10^{-16} s	~10^{-20}–10^{-21} s
Dominance of reaction	Low energy	High energy
Nature of reaction	Surface phenomenon	Nuclear interior
Cross - section	~Barn	~Millibarn
Angular distribution	Isotropic in the CMS	Peaked in the forward hemisphere
Location of peaks	Energy dependent	Orbital angular momentum dependent

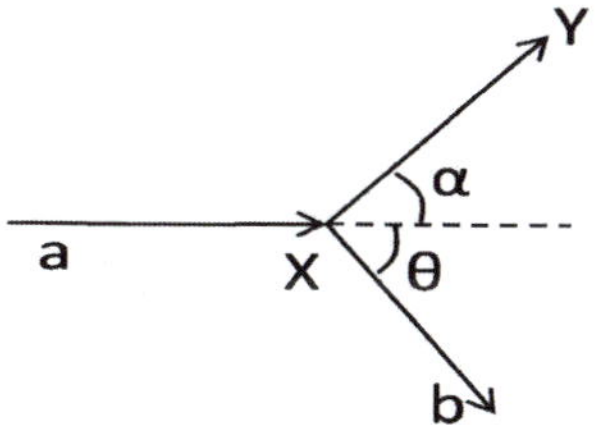

Fig. 8.3 Kinematics of a collision

Inverse reactions and the reciprocity theorem

$$\frac{\sigma(b \to a)}{\sigma(a \to b)} = \frac{(2I_x + 1)(2I_a + 1)}{(2I_y + 1)(2I_b + 1)} \cdot \frac{p_a^2}{p_b^2} \tag{8.30}$$

At the same CM energy.

$$\sigma_S^l = \frac{\pi}{k^2}(2l + 1)|1 - \eta_l|^2 \tag{8.31}$$

$$\sigma_r^l = \frac{\pi}{k^2}(2l + 1)(1 - |\eta_l|^2) \tag{8.32}$$

The Breit – Wigner formulae – Reactions via compound nucleus formation.

$$\sigma_S = \frac{\pi \lambda\!\!\!^{-2}\, \Gamma_s^2 . g}{(E - E_R)^2 + (\Gamma/2)^2} \tag{8.33}$$

$$\sigma_r = \frac{\pi \lambda\!\!\!^{-2} \quad \Gamma_R \Gamma_S . g}{(E - E_R)^2 + (\Gamma/2)^2} \tag{8.34}$$

$$\sigma_t = \sigma_s + \sigma_r = \frac{\pi \lambda\!\!\!^{-2}\, \Gamma_s \Gamma . g}{(E - E_R)^2 + (\Gamma/2)^2} \tag{8.35}$$

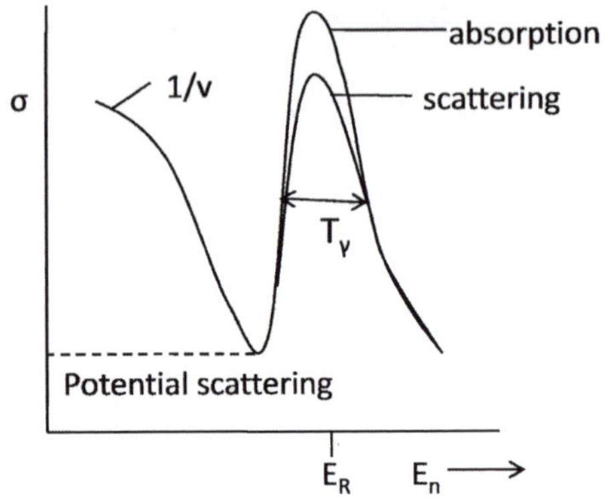

Fig. 8.4 Resonance and potential sacttering and absorption cross-sections as a function of neutron energy

$$\text{where } g = \frac{(2I_c + 1)}{(2I_a + 1)(2I_x + 1)} \tag{8.36}$$

$$\Gamma_a \tau_a = \hbar;\ \Gamma_b \tau_b = \hbar \tag{8.37}$$

$$\sigma_{\text{el}} = \frac{\pi}{k^2} \left| A_{\text{res}} + A_{\text{pot}} \right|^2 \tag{8.38}$$

$$\text{with } A_{\text{res}} = \frac{i\Gamma_n}{(E_n - E_R) + \frac{1}{2} i\Gamma} \tag{8.39}$$

$$\text{and} \quad A_{\text{pot}} = \exp(2ikR) - 1 \tag{8.40}$$

Optical model

$$V = -U - iW \tag{8.41}$$

$$U = \hbar v k_1 \tag{8.42}$$

$$W = \frac{\hbar}{2} v k \tag{8.43}$$

$$\text{where } v = \sqrt{2E/m} \tag{8.44}$$

is the velocity of the incident nucleon.

$$k = (2mE)^{1/2}/\hbar; \tag{8.45}$$

$$k_1 = [2m(E + U)]^{1/2}/\hbar \tag{8.46}$$

$$K = \frac{1}{\lambda} = \frac{A\,\sigma}{\text{vol}} = \frac{3\sigma}{4\pi r_0^3} \tag{8.47}$$

Direct reactions (Reactions without compound nucleus formation)

1. Inelastic scattering
2. Charge exchange reactions
3. Nucleon transfer reactions

4. Breakup reactions
5. Knock-out reactions

Pre-equilibrium reactions

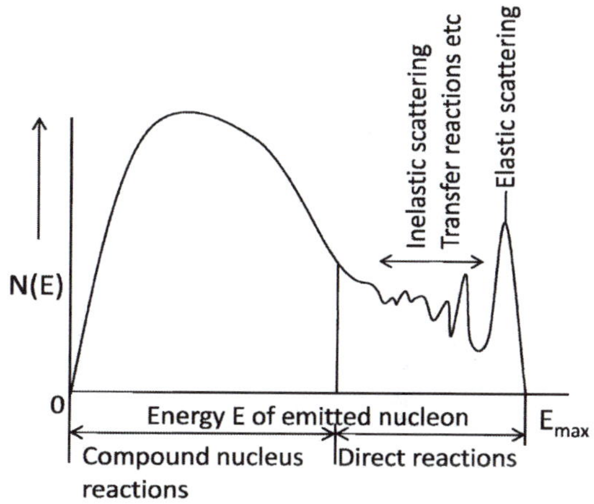

Fig. 8.5 Energy spectrum of particles emitted in various types of reactions

The pre-equilibrium reactions take place before the formation of the compound nucleus.

Heavy Ion reactions

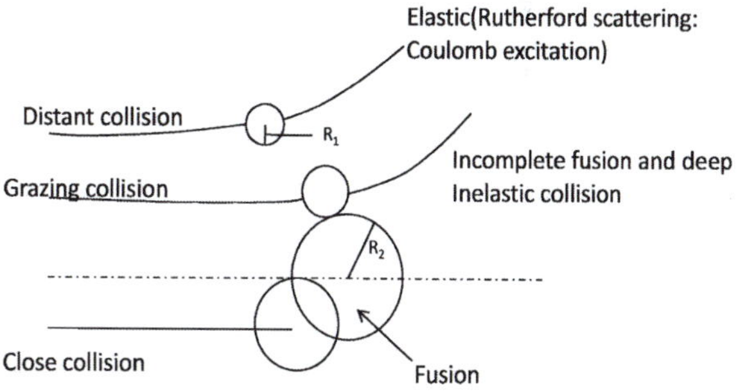

Fig. 8.6 Collision between heavy ions

Types of interactions

(a) The coulomb region with $r_{\min} > R_N$, where R_N is the distance for which the nuclear interactions are ineffective.
(b) The deep inelastic and the incomplete fusion region with $r_{\min} = R_1 + R_2$.
(c) The fusion region with $0 \leq r_{\min} \leq R_1 + R_2$.

If E_{cm} is the center of mass energy of the two interacting ions then the minimum distance of approach is given by

$$r_{min} = \frac{b}{\sqrt{1 - \frac{V(r_{min})}{E_{cm}}}} \tag{8.48}$$

8.2 Problems

8.2.1 Atomic Masses and Radii

8.1 Singly-charged lithium ions, liberated from a heated anode are accelerated by a difference of 400 V between anode and cathode. They then pass through a hole in the cathode into a uniform magnetic field perpendicular to their direction of motion. The magnetic flux density is 8×10^{-2} Wb/m^2 and the radii of the paths of the ions are 8.83 and 9.54 cm, respectively. Calculate the mass numbers of the lithium isotopes.

[Osmania University]

8.2 A narrow beam of singly charged ^{10}B and ^{11}B ions of energy 5 keV pass through a slit of width 1 mm into a uniform magnetic field of 1,500 gauss and after a deviation of 180° the ions are recorded on a photographic plate

(a) What is the spatial separation of the images.

(b) What is the mass resolution of the system?

[University of Manchester 1963]

8.3 Singly charged chlorine ions are accelerated through a fixed potential difference and then caused to travel in circular paths by means of a uniform field of magnetic induction of 1,000 gauss. What increase in induction is necessary to cause the mass 37 ion to follow the path previously taken by the mass 35 ion?

[University of London 1960]

8.4 $^{27}_{14}$Si and $^{27}_{13}$Al are mirror nuclei. The former is a positron emitter with $E_{max} =$ 3.48 MeV. Determine r_0.

8.5 Use the uncertainty relation to estimate the kinetic energy of the nucleon, the nuclear radius is about 5×10^{-13} cm and the mass of a proton is about 2×10^{-24} g.

[University of Bristol 1969]

8.6 ^{14}O is a positron emitter decaying to an excited state of ^{14}N. The ^{14}N γ - rays have an energy of 2.313 MeV and the maximum energy of the positron is 1.835 MeV. The mass of ^{14}N is 14.003074 amu and that of electron is 0.000548 amu. Find the mass of ^{14}O in amu.

8.2.2 Electric Potential and Energy

8.7 Derive an expression for the electrostatic energy of a spherical nucleus of radius R assuming that the charge $q = Ze$ is uniformly distributed homogeneously in the nuclear volume.

8.8 A charge q is uniformly distributed in a sphere of radius R. Obtain an expression for the potential $V(r)$ at a point distant r from the centre ($r < R$).

8.2.3 Nuclear Spin and Magnetic Moment

8.9 Suppose a proton is assumed to be a classical particle rotating with angular velocity of 2.6×10^{23} rad/s about its axis. If it posseses a rotational energy of 537.5 MeV, then show that it has angular momentum equal to h.

8.10 (a) A $D_{5/2}$ term in the optical spectrum of ${}^{39}_{19}$K has a hyperfine structure with four components. Find the spin of the nucleus.
(b) In (a) what interval ratios in the hyperfine quadruplet are expected?

8.11 Given that the proton has a magnetic moment of 2.79 magnetons and a spin quantum number of one half, what magnetic field strength would be required to produce proton resonance at a frequency of 60 MHz in a nuclear magnetic resonance spectrometer?

8.12 In a nuclear magnetic experiment for the nucleus ${}_{25}\text{Mn}^{55}$ of dipole moment $3.46\,\mu_N$, the magnetic field employed is 0.8 T. Find the resonance frequency. You may assume $J = 7/2$, $\mu_N = 3.15 \times 10^{-14}$ MeV T^{-1}

8.2.4 Electric Quadrupole Moment

8.13 Show that for a homogeneous ellipsoid of semi axes a, b the quadrupole moment is given by $Q = \frac{2}{5}\text{Ze}(a^2 - b^2)$

8.14 Estimate the ratios of the major to minor axes of ${}^{181}_{73}$Ta and ${}^{123}_{51}$Sb. The quadrupole moments are $+6 \times 10^{-24}$ cm^2 for Ta and -1.2×10^{-24} cm^2 for Sb (Take R $= 1.5\,\text{A}^{1/3}$ fm)

[Saha Institute 1964]

8.15 Show that the electric quadrupole moment of a nucleus vanishes for
(a) Spherically symmetric charge distribution
(b) Nuclear spin $I = 0$ or $I = 1/2$

8.16 Show that for a rotational ellipsoid of small eccentricity and uniform charge density the quadrupole moment is given by $Q = \frac{4}{5}ZR\Delta R$. Assuming that the quadrupole moment of ${}^{181}_{71}$Ta is 4.2 barns, estimate its size.

8.2.5 Nuclear Stability

8.17 In general only the heavier nuclei tend to show alpha decay. For large A it is found that $B/A = 9.402 - 7.7 \times 10^{-3}\,A$.

Given that the binding energy of alpha particles is 28.3 MeV, show that alpha decay is energetically possible for $A > 151$.

[University of Wales, Aberystwyth, 2003]

8.18 For the nucleus ^{16}O the neutron and proton separation energies are 15.7 and 12.2 MeV, respectively. Estimate the radius of this nucleus assuming that the particles are removed from its surface and that the difference in separation energies is due to the Coulomb potential energy of the proton.

[University of Wales, Aberystwyth 2004]

8.19 By considering the general conditions of nucleus stability show that the nucleus $^{229}_{90}$Th will decay and decide whether the decay will take place by α or β emission.

The atomic mass excesses of the relevant nuclei are:

Element	$^{4}_{2}$He	$^{225}_{88}$Ra	$^{229}_{89}$Ac	$^{229}_{90}$Th	$^{229}_{91}$Pa
Mass excess amu $\times 10^{-6}$	2,603	23,528	32,800	31,652	32,022

8.20 The masses of $^{64}_{28}$Ni, $^{64}_{29}$Cu and $^{64}_{30}$Zn are as tabulated below. It follows that $^{64}_{29}$Cu is radioactive. Detail the various possible decay modes to the ground state of the daughter nucleus and give the maximum energy of each component, ignoring the recoil energy. Compute the recoil energy in one of the cases and verify that it was justifiable to ignore it.

Nuclide	Atomic mass
$^{64}_{28}$Ni	63.927959
$^{64}_{29}$Cu	63.929759
$^{64}_{30}$Zn	63.929145

[University of Bristol 1969]

8.21 $^{28}_{13}$Al decays to $^{28}_{14}$Si via β^- emission with $T_{max} = 2.865$ MeV. $^{28}_{14}$Si is in the excited state which in turn decays to the ground state via γ-emission. Find the γ-ray energy. Take the masses ^{28}Al $= 27.981908$ amu, ^{28}Si $=$ 27.976929 amu.

8.22 $^{22}_{11}$Na decays to $^{22}_{10}$Ne via β^+ with $T_{max} = 0.542$ MeV followed by γ-decay with energy 1.277 MeV. If the mass of ^{22}Ne is 21.991385 amu, determine the mass of ^{22}Na in amu.

8.23 $^{7}_{4}$Be undergoes electron capture and decays to $^{7}_{3}$Li. Investigate if it can decay by the competitive decay mode of β^+ emission. Take masses $^{7}_{4}$Be $= 7.016929$ amu, $^{7}_{3}$Li $= 7.016004$ amu.

8.2.6 Fermi Gas Model

8.24 In the Fermi gas model the internal energy is given by $U = \frac{3}{5}AE_F$, where A is the mass number and E_F is the Fermi energy. For a nucleus of volume V with $N = Z = A/2$. $A = KVE_F{}^{3/2}$ where K is a constant. Using the thermodynamic relation, $p = -\left(\frac{\partial U}{\partial V}\right)_s$, show that the pressure is given by $p = \frac{2}{5}\rho_n E_F$, where ρ_n is the nucleon density.

8.25 Assuming that in a nucleus $N = Z = A/2$, calculate the Fermi momentum, Fermi energy E_F, and the well depth.

8.2.7 Shell Model

8.26 A certain odd-parity shell model state can accommodate up to a maximum of 12 nucleons. What are its j and l values?

8.27 The shell model energy levels are in the following way
$[1s_{1/2}][1p_{3/2}, 1p_{1/2}][1d_{5/2}, 2s_{1/2}, 1d_{3/2}][1f_{7/2}][2p_{3/2}, 1f_{5/2}, 2p_{1/2}, 1g_{9/2}]$
$[1g_{7/2}, 2d_{5/2}, 2d_{3/2}, 3s_{1/2}, 1h_{11/2}]\ldots$
Assuming that the shells are filled in the order written, what spins and parities should be expected for the ground state of the following nuclei?
${}^{7}_{3}\mathrm{Li}$, ${}^{16}_{8}\mathrm{O}$, ${}^{17}_{8}\mathrm{O}$, ${}^{39}_{19}\mathrm{K}$, ${}^{45}_{21}\mathrm{Sc}$.

8.28 Find the gap between the $1p_{1/2}$ and $1d_{5/2}$ neutron shells for nuclei with mass number $A \approx 16$ from the total binding energy of the $^{15}\mathrm{O}$ (111.9556 MeV), $^{16}\mathrm{O}$ (127.6193 MeV) and $^{17}\mathrm{O}$ (131.7627 MeV) atoms.

8.29 Compute the expected shell-model quadruple moment of $^{209}\mathrm{Bi}(9/2^-)$

8.30 From the shell model predictions find the ground state spin and parity of the following nuclides:
${}^{3}_{2}\mathrm{He}$; ${}^{20}_{10}\mathrm{Ne}$; ${}^{27}_{13}\mathrm{Al}$; ${}^{41}_{21}\mathrm{Sc}$;

8.31 Making use of the shell model, write down the ground state configuration of protons and neutrons for ${}^{12}_{6}\mathrm{C}$ and the next three isotopes of increasing A. Give the spin and parity assignment for the ground state of ${}^{13}_{6}\mathrm{C}$ and compare this with the equivalent assignments for the ground state of ${}^{15}_{6}\mathrm{C}$.

8.32 How does the shell model predict $\frac{7^-}{2}$ for the ground state spin parity of ${}^{41}_{20}\mathrm{Ca}$. What does the model predict about the spin and parities of the ground states of ${}^{30}_{14}\mathrm{Si}$ and ${}^{14}_{7}N$?

[University of Cambridge, Tripos 2004]

8.2.8 Liquid Drop Model

8.33 Deduce that with $a_c = 0.72\,\text{MeV}$ and $a_s = 23\,\text{MeV}$ the ratio z_{min}/A is approximately 0.5 for light nuclei and 0.4 for heavy nuclei.

[Royal Holloway, University of London 1998]

8.34 Determine the most stable isobar with mass number $A = 64$.

8.35 The masses (amu) of the mirror nuclei $^{27}_{13}\text{Al}$ and $^{27}_{14}\text{Si}$ are 26.981539 and 26.986704 respectively. Determine the Coulomb's coefficient in the semi emperical mass formula.

8.36 If the binding energies of the mirror nuclei $^{41}_{21}\text{Sc}$ and $^{41}_{20}\text{Ca}$ are 343.143 V and 350.420 MeV respectively, estimate the radii of the two nuclei by using the semi empirical mass formula $[e^2/4\pi\varepsilon_0 = 1.44\,\text{MeV fm}]$

8.37 The empirical mass formula (neglecting a term representing the odd – even effect) is $M(A, Z) = Z(m_p + m_e) + (A - Z)m_n - \alpha A + \beta A^{2/3} + \gamma(A - 2Z)^2/A + \varepsilon Z^2 A^{-1/3}$ where $\alpha,\ \beta,\ \gamma$ and ε are constants. By finding the minimum in $M(A, Z)$ for constant A obtain the expression $Z_{min} = 0.5A(1 + 0.25A^{2/3}\varepsilon/\gamma)^{-1}$ for the value of Z which corresponds to the most stable nucleus for a set of isobars of mass number A.

[Royal Holloway, University of London 1998]

8.38 The binding energy of a nucleus with atomic number Z and mass number A can be expressed by Weisacker's semi – empirical formula

$$B = a_v A - a_s A^{2/3} - \frac{a_c Z^2}{A^{1/3}} - a_a \frac{(N - Z)^2}{A} - \frac{a_p}{A^{1/2}}$$

where $a_v = 15.56\,\text{MeV},\ a_s = 17.23\,\text{MeV}$,

$a_c = 0.697\,\text{MeV},\ a_a = 23.285\,\text{MeV},\ a_p = -12, 0, 12\,\text{MeV}$

for even-even, odd-even (or even-odd) or odd-odd nucleus respectively.

Estimate the energy needed to remove one neutron from nucleus $^{40}_{20}\text{Ca}$.

8.39 (a) Consider the alpha particle decay $^{230}_{90}\text{Th} \rightarrow ^{226}_{88}\text{Ra} + \alpha$ and use the following expression to calculate the values of the binding energy B for the two heavy nuclei involved in this process.

$$B = a_v A - a_s A^{2/3} - a_c \frac{Z(Z-1)}{A^{1/3}} - a_a \frac{(N - Z)^2}{A} - a_p A^{-3/4}$$

where values for the constants $a_v,\ a_s,\ a_c,\ a_a$ and a_p are respectively 15.5, 16.8, 0.72, 23.0 and −34.5 MeV. Given that the total binding energy of the alpha particle is 28.3 MeV, find the energy Q released in the decay.

(b) This energy appears as the kinetic energy of the products of the decay. If the original thorium nucleus was at rest, use conservation of momentum and conservation of energy to find the kinetic energy of the daughter nucleus ^{226}Ra.

8.40 The following plot (Fig. 8.7) shows the average binding energy per nucleon for stable nuclei as a function of mass number.
Explain how the mass of a nucleus can be calculated from this plot and estimate the mass of ${}^{235}_{92}$U.

Briefly describe the main features of the plot in the context of nuclear models such as the liquid Drop Model, the Fermi Gas Model and the Nuclear Shell Model.

In terms of the Liquid Drop Model, explain why nuclear fission and fusion are possible and estimate the energy released when a nucleus of ${}^{235}_{92}$U undergoes fission into the fragments ${}^{87}_{35}$Br and ${}^{145}_{57}$La with the release of three prompt neutrons.

[University of Cambridge, Tripos 2004]

Fig. 8.7 B/A versus A plot

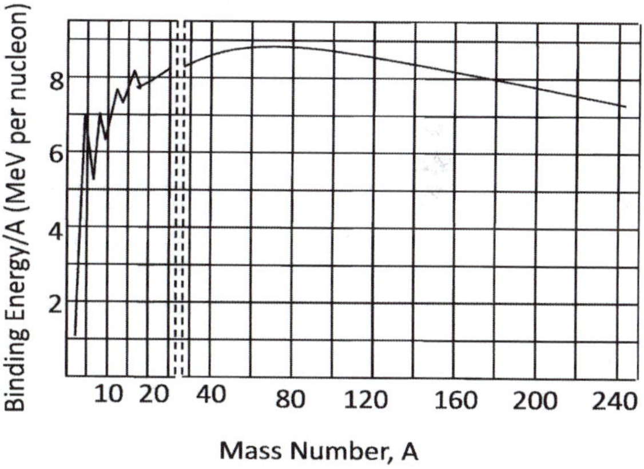

8.41 Investigate using liquid drop model, the β stability of the isobars ${}^{127}_{53}$I and ${}^{127}_{54}$Xe given that

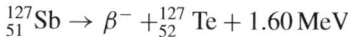

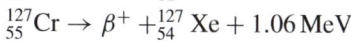

${}^{127}_{51}\text{Sb} \rightarrow \beta^- + {}^{127}_{52}\,\text{Te} + 1.60\,\text{MeV}$

${}^{127}_{55}\text{Cr} \rightarrow \beta^+ + {}^{127}_{54}\,\text{Xe} + 1.06\,\text{MeV}$

[University of London 1969]

8.42 The empirical mass formula is

${}^{A}_{Z}M = 0.99198\,\text{A} - 0.000841\,\text{Z} + 0.01968\,\text{A}^{2/3} + 0.0007668\,\text{Z}^2\,\text{A}^{-1/3}$
$+0.09966(\text{Z} - \text{A}/2)^2\text{A}^{-1} - \delta$

in atomic mass units, where $\delta = \pm 0.01204\,\text{A}^{-1/2}$ or 0.

Determine whether or not the nuclide ${}^{27}_{12}$Mg is stable to β decay.

[University of Newcastle 1966]

8.2.9 Optical Model

8.43 Show that the imaginary part of the complex potential $V = -(U + iW)$ in the optical model has the effect of removing particle flux from the elastic channel.

8.44 For neutrons with kinetic energy 100 MeV incident on nuclei with mass number $A = 120$, the real and imaginary parts of the complex potential are approximately -25 and -10 MeV, respectively. On the basis of these data, estimate

(i) the deBroglie wavelength of the neutron inside the nucleus
(ii) the probability that the neutron is absorbed in passing through the nucleus

8.2.10 Nuclear Reactions (General)

8.45 ^{13}N is a positron emitter with an end point energy of 1.2 MeV. Determine the threshold of the reaction $p + ^{13}\mathrm{C} \rightarrow ^{13}\mathrm{N} + n$, if the neutron – hydrogen atom mass difference is 0.78 MeV.

[Osmania University 1964]

8.46 The reaction, $\mathrm{p} + ^7_3\mathrm{Li} \rightarrow ^7_4\mathrm{Be} + \mathrm{n}$, is known to be endothermic by 1.62 MeV. Find the total energy released when ^{7}Be decays by K capture and calculate the energy carried off by the neutrino and recoil nucleus, respectively.

($M_\mathrm{p}c^2 = 938.23$ MeV, $M_\mathrm{n}c^2 = 939.52$ MeV, $M_\mathrm{e}c^2 = 0.51$ MeV)

[University of Bristol 1960]

8.47 If a target nucleus has mass number 24 and a level at 1.37 MeV excitation, what is the minimum proton energy required to observe scattering from this level.

[Osmania University 1966]

8.48 The nuclear reaction which results from the incidence of sufficiently energetic α-particles on nitrogen nuclei is $^4_2\mathrm{He} + ^{14}_7\mathrm{N} \rightarrow \mathrm{X} + ^1_1\mathrm{H}$. What is the decay product X? What is the minimum α-particle kinetic energy (in the laboratory frame) required to initiate the above reaction?
(Atomic masses in amu: ^{1}H = 1.0081; ^{4}He = 4.0039; ^{14}N = 14.0075; X = 17.0045)

[University of Manchester]

8.49 The Q values for the reactions ^{2}H$(d, n)\,^3$He and ^{2}H$(d, p)\,^3$H are 3.27 MeV and 4.03 MeV, respectively. Show that the difference between the binding energy of the ^{3}H nucleus and that of the ^{3}He nucleus is 0.76 MeV and verify that this is approximately the magnitude of Coulomb energy due to the two protons of the ^{3}He nucleus. (Distance between the protons in the nucleus $3^{1/3} \times 1.3$ fm).

[University of London 1968]

8.50 Thermal neutrons are captured by $^{10}_5$B to form $^{11}_5$B which decays by α-particle emission to Li. Write down the reaction equation and calculate

(a) The Q-Value of the decay in MeV
(b) The Kinetic energy of the α-particles in MeV.

(Atomic masses: $^{10}_5$B = 10.01611 amu; ^1_0n = 1.008987 amu; $^7_3$Li = 7.01822 amu; $^4_2$He = 4.003879 amu; 1 amu = 931 MeV)

[University of Bristol 1967]

8.51 Consider the reaction $^{27}\text{Al}(p, n)^{27}\text{Si}$. The positrons from the decay of ^{27}Si are observed and their spectrum is found to have an end – point energy of 3.5 MeV. Derive the Q value of the (p, n) reaction, and find the threshold proton energy for the reaction given that the neutron – proton mass difference is 0.8 MeV.

[University of Manchester 1963]

8.52 The Q value of the $^{3}\text{H}(p, n)^{3}\text{He}$ reaction is -0.764 MeV. Calculate (a) the threshold energy for appearance of neutrons in the forward direction (b) the threshold for the appearance of neutron in the 90° direction.

[University of Liverpool 1960]

8.53 When ^{30}Si is bombarded with a deuteron, ^{31}Si is formed in its ground state with the emission of a proton. Determine the energy released in this reaction from the following information:-

$$^{31}\text{Si} \rightarrow ^{31} p + \beta^{-} + 1.51\,\text{MeV}$$

$$^{30}\text{Si} + d \rightarrow ^{31} p + n + 5.10\,\text{MeV}$$

$$n \rightarrow p + \beta^{-} + \overline{\nu} + 0.78\,\text{MeV}$$

[University of London 1960]

8.54 The nucleus ^{12}C has an excited state at 4.43 MeV. You wish to investigate whether this state can be produced in inelastic scattering of protons through 90° by a carbon target. If you have access to a beam of protons of kinetic energy 15 MeV, what is the kinetic energy of the scattered protons for which you must look?

[University of Bristol 1966]

8.55 A thin hydrogenous target is bombarded with 5 MeV neutrons, and a detector is arranged to collect those protons emitted in the same direction as the neutron beam. The neutron beam is replaced by a beam of γ-rays; calculate the photon energy needed to produce protons of the same energy as with the neutron beam.

[Osmania University]

8.56 Protons of energy 5 MeV scattering from $^{10}_{5}\text{B}$ at an angle of 45° show a peak in the energy spectrum of the scattered protons at an energy of 3.0 MeV
(a) To what excitation energy of $^{10}_{5}\text{B}$ does this correspond?
(b) What is the expected energy of the scattered protons if the scattering is elastic?

8.57 Calculate the energy of protons detected at 90° when 2.1 MeV deutrons are incident on ^{27}Al to produce ^{28}Al with an energy difference $Q = 5.5$ MeV.

[Royal Holloway University of London 1998]

8.58 The reaction $^{2}\text{H} +^{1}\text{H} \rightarrow^{3}\text{He} + \gamma + 5.3\,\text{MeV}$ occurs with the deuterons and proton at rest. Estimate the energy of the helium nucleus.

If the reaction occurs in a region of the sun where the temperature is about 1.7×10^{7} K, estimate how close the deuteron and proton must approach for fusion to occur.

[Osmania University]

8.59 The Q- value of the reaction $^{16}O(d,n)^{17}F$ is -1.631 MeV, while that of the reaction $^{16}O(d,\ p)^{17}O$ is $+1.918$ MeV. Which is the unstable member of the pair $^{17}O-^{17}F$, and what is the maximum energy of β-particles it emits? (n$-^{1}$H mass difference is 0.782 MeV)

[University of Manchester 1958]

8.60 An aluminum target is bombarded by α-particles of energy 7.68 MeV, and the resultant proton groups at 90° were found to possess energies 8.63, 6.41, 5.15 and 3.98 meV. Draw an energy level diagram of the residual nucleus, using the above information.

[Osmania University 1963]

8.61 If the Q-value for the $^{3}H(p,n)^{3}He$ reaction is -0.7637 MeV and tritium (^{3}H) emits negative β-particles of end point energy 18.5 KeV, calculate the difference in mass between the neutron and the hydrogen atom.

[Andhra University]

8.62 The reaction $^{3}H(d,n)^{4}He$ has Q value of 17.6 MeV. What is the range of neutron energies that may be obtained from this reaction for an incident deuteron beam of 300 KeV?

[Osmania University 1970]

8.63 A target of ^{181}Ta is bombarded with 5 MeV protons to form ^{182}W in an excited state. Calculate the energy of the excited state (ignore the coulomb barrier and assume the target nuclei at rest). If ^{182}W in the same excited state were produced by bombarding a hydrogen target with energetic Ta nuclei, what energy would be needed? The atomic masses in amu are: $^{1}H_1 = 1.007825$; $^{181}_{73}Ta =$ 180.948007; $^{182}_{74}W = 181.948301$

[University of Durham 1963]

8.64 In the reaction $^{48}Ca + ^{16}O \rightarrow ^{49}Sc + ^{15}N$, the Q-value is -7.83 MeV. What is the minimum kinetic energy of bombarding ^{16}O ions to initiate the reaction? At this energy, estimate the orbital angular momentum in units of $\hbar$ of the ions for grazing collision. Take $R = 1.1\,A^{1/3}$ fm.

8.2.11 Cross-sections

8.65 Calculate the thickness of Indium foil which will absorb 1% of neutrons incident at the resonance energy for Indium (1.44 eV) where $\sigma = 28,000$ barns. At. Wt of Indium = 114.7 amu, density of Indium = 7.3 g/cm^3.

[Andhra University]

8.66 ^{60}Co is produced from natural cobalt in a reactor with a thermal neutron flux density of $5 \times 10^{12}\,n\ cm^{-2}\ s^{-1}$. Determine the maximum specific activity. Given $\sigma_{act} = 20$ b.

8.67 Natural Cobalt is irradiated in a reactor with a thermal neutron flux density of $3 \times 10^{12}\, n\ \text{cm}^{-2}\ \text{s}^{-1}$. How long an irradiation will be required to reach 20% of the maximum activity? Given $T_{1/2} = 5.3$ years

8.68 In a scattering experiment an aluminum foil of thickness 10 μm is placed in a beam of intensity 8×10^{12} particles per second. The differential scattering cross-section is known to be of the form

$$\frac{d\sigma}{d\Omega} = A + B\cos^2\theta$$

where A, B are constants, θ is scattering angle and Ω is the solid angle. With a detector of area 0.01 m^2 placed at a distance of 6 m from the foil, it is found that the mean counting rate is $50\,\text{s}^{-1}$ when θ is 30° and $40\,\text{s}^{-1}$ when θ is 60°. Find the values of A and B. The mass number of aluminum is 27 and its density is 2.7 g/cm^2.

8.69 A thin target of ^{48}Ca with 1.3×10^{19} nuclei per cm^2 is bombarded with a 10 nA beam of α particles. A detector, subtending a solid angle of 2×10^{-3} steradians, records 15 protons per second. If the angular distribution is measured to be isotropic, determine the total cross section for the $^{48}\text{Ca}(\alpha,\ p)$ reaction.

[University of Cambridge, Tripos 2004]

8.2.12 Nuclear Reactions via Compound Nucleus

8.70 Cadmium has a resonance for neutrons of energy 0.178 eV and the peak value of the total cross-section is about 7,000 b. Estimate the contribution of scattering to this resonance.

[Osmania University 1964]

8.71 A nucleus has a neutron resonance at 65 eV and no other resonances nearby. For this resonance, $\Gamma_n = 4.2$ eV, $\Gamma_\gamma = 1.3$ eV and $\Gamma_\alpha = 2.7$ eV, and all other partial widths are negligible. Find the cross-section for (n, γ) and (n, α) reactions at 70 eV.

[Osmania University]

8.72 Neutrons incident on a heavy nucleus with spin $J_\text{N} = 0$ show a resonance at an incident energy $E_\text{R} = 250$ eV in the total cross-section with a peak magnitude of 1,300 barns, the observed width of the peak being $\Gamma = 20$ eV. Find the elastic partial width of the resonance.

[University of Bristol 1970]

8.2.13 Direct Reactions

8.73 The reaction $d + {}^{14}\text{N} \rightarrow \alpha + {}^{12}\text{C}$ has been used to test the principle of detailed balance which relates the cross-section σ_{ab} for a reaction $a + x \rightarrow b + y$, to the cross-section for the inverse reaction and has the form
$(2S_a + 1)\,(2S_x + 1)\,{P_a}^2\sigma_{ab} = (2S_b + 1)\,(2S_y + 1)\,{P_b}^2\sigma_{ba}$

The reaction above will take place for low energy incident deuterons in a s-wave state leaving the ^{12}C nucleus in the ground state.

Given that the deuteron has spin 1 and positive parity while the alpha particle has zero spin and positive parity, estimate the spin of ^{14}N in the ground state.

Can the alpha particle come off with orbital angular momentum $l = 1$?

If the incident kinetic energy of the deuteron is 20 MeV in the laboratory frame, calculate the laboratory kinetic energy at which α's should scatter from ^{12}C to test the principle of detailed balance. What is the expected ratio between the cross-sections for the direct and inverse reactions? Atomic masses of ^{14}N, ^{2}H and ^{4}He are 14.003074, 2.014102, 4.002603 amu respectively. 1 amu = 931.44 MeV.

[University of Bristol 1969]

8.74 A beam of 460 MeV deuterons impinges on a target of bismuth. Given the binding energy of the deuteron is 2.2 MeV, compute the mean energy, spread in energy and the angle of the cone in which the neutrons are emitted.

[Osmania University 1975]

8.2.14 Fission and Nuclear Reactors

8.75 1.0 g of ^{23}Na of density 0.97 is placed in a reactor at a region where the thermal flux is $10^{11}/\text{cm}^2/\text{s}$. Set up the equation for the production of ^{24}Na and determine the saturation activity that can be produced. The half-life of ^{24}Na is 15 h, and the activation cross-section of ^{23}Na is 536 millibarns.

[Osmania University 1964]

8.76 Suppose 100 mg of gold ($^{197}_{79}Au$) foil are exposed to a thermal neutron flux of 10^{12} neutrons/cm^2/s in a reactor. Calculate the activity and the number of atoms of ^{198}Au in the sample at equilibrium [Thermal neutron activation cross-section for ^{197}Au is 98 barns and half-life for ^{197}Au is 2.7 h]

[Osmania University]

8.77 Estimate the energy released in fission of $^{238}_{92}U$ nucleus, given $a_c = 0.59$ MeV and $a_s = 14.0$ MeV.

[Osmania University 1962]

8.78 A small container of Ra–Be is embedded in the middle of a sphere of paraffin wax of a few cm radius so as to form a source of (predominantly) thermal neutrons. This source is placed at the centre of a very large block of graphite. Derive an expression for the density of thermal neutrons at a large distance r from the source in terms of the source strength Q, the diffusion coefficient D and diffusion length L.

A small BF_3 counter is placed in the graphite at a distance of 3 m from the above source contains 10^{20} atoms of ^{10}B. The cross-section of ^{10}B for the thermal neutron capture follows a $1/v$ law and has a magnitude of 3,000 barns for a neutron velocity $v = 2,200\,\text{m/s}$. If the counting rate is 250/min, calculate the value of Q. Given $L = 50\,\text{cm}$; $D = 5 \times 10^5\,\text{cm}^2/\text{s}$.

[University of Bristol 1959]

8.79 Calculate the thermal utilization factor for a heterogeneous lattice made up of cylindrical uranium rods of diameter 3 cm and pitch 18 cm in graphite
Take the flux ratio $\phi_\text{m}/\phi_\text{U}$ as 1.6
Densities : Uranium $= 18.7 \times 10^3\,\text{kg m}^{-3}$, Graphite $= 1.62 \times 10^3\,\text{kg m}^{-3}$
Absorption cross-sections $\sigma_\text{av} = 7.68\,\text{b}$; $\sigma_\text{am} = 4.5 \times 10^{-3}\,\text{b}$.

[University of Durham 1961]

8.80 Calculate approximately, using one-group theory results, the critical size of a bare spherical reactor, given $k_\infty = 1.54$ and Migration area $M^2 = 250\,\text{cm}^2$.

8.81 Assuming the energy released per fission of ^{235}U is 200 MeV, calculate the amount of ^{235}U consumed per day in Canada India reactor "Cirus" operating at 40 MW of power.

8.82 Assuming that the energy released per fission of $^{235}_{92}$U is 200 MeV, calculate the number of fission processes that should occur per second in a nuclear reactor to operate at a power level of 20,000 kW. What is the corresponding rate of consumption of $^{235}_{92}$U.

[University of London 1959]

8.83 (a) Assume that in each fission of ^{235}U, 200 MeV is released. Assuming that 5% of the energy is wasted in neutrinos, calculate the amount of ^{235}U burned which would be necessary to supply at 30% efficiency, the whole annual electricity consumption in Britain 50×10^9 kWh
(b) A thermal reactor contains 100 tons of natural uranium (density 19) and operates at a power of 100 MW (heat). Assuming that the thermal cross-section of ^{235}U is 550 barns and that the uranium contains 0.7 % of ^{235}U. Calculate the neutron flux near the centre of the reactor by neglecting neutron losses from the outside, and assuming flux constant through out the lattice.

[University of Liverpool 1959]

8.84 If the elastic scattering of neutrons by hydrogen nuclei is isotropic in the centre of mass system, show that

$$\overline{\ln(E_1/E_2)} = 1$$

where E_1 and E_2 are respectively the kinetic energies of a neutron before and after the collision.

[University of London 1969]

8.85 (a) Estimate the average number of collisions required to reduce fast fission neutrons of initial energy 2 MeV to thermal energy (0.025 eV) in graphite moderator.
(b) Calculate the corresponding slowing-down time given that $\Sigma_s = 0.385\,\text{cm}^{-1}$

8.86 Show that a homogeneous, natural uranium-graphite moderated assembly can not become critical. Use the following data:
400 moles of graphite per mole of uranium

Natural uranium	Graphite
$\sigma_a(\mathrm{U}) = 7.68\,\mathrm{b}$	$\sigma_a(\mathrm{M}) = 0.0032\,\mathrm{b}$
$\sigma_s(\mathrm{U}) = 8.3\,\mathrm{b}$	$\sigma_s(\mathrm{M}) = 4.8\,\mathrm{b}$
$\varepsilon = 1.0;\ \eta = 1.34$	$\xi = 0.158$

[Osmania University 1964]

8.87 A point source of thermal neutrons is placed at the centre of a large sphere of beryllium.

Deduce the spatial distribution of neutron density in the sphere. Estimate what its radius must be if less than 1% of the neutrons are to escape through the surface. Find also the neutron density near the surface in this case in terms of the source strength.

At. Wt of beryllium = 9
Density of beryllium = 1.85 g/cc
Avagadro number = 6×10^{23} atoms/g atom
Thermal neutron scattering cross-section on beryllium = 5.6 barns
Thermal neutron capture cross-section on beryllium = 10 mb (at velocity $v = 2{,}200\,\mathrm{m/s}$)

[University of Bristol 1961]

8.88 Calculate the steady state neutron flux distribution about a plane source emitting Q neutrons/s/cm^2 in an infinite homogeneous diffusion medium. Assume that neutrons are not produced in any region of interest.

8.89 Calculate the thermal diffusion time for graphite. Use the data:

$\sigma_a(\mathrm{C}) = 0.003$ b, $\rho_c = 1.62\,\mathrm{g\,cm^{-3}}$. Average thermal neutron speed $= 2{,}200\,\mathrm{m/s}$.

8.90 Estimate the generation time for neutrons in a critical reactor employing ^{235}U and graphite. Use the following data:

$\underset{a}{\Sigma} = 0.0006\,\mathrm{cm^{-1}};\ B^2 = 0.0003;\ L^2 = 870\,\mathrm{cm^2}\ ;\ \langle v \rangle = 2200\,\mathrm{ms}.$

8.91 The spatial distribution of thermal neutrons from a plane neutron source kept at a face of a semi-infinite medium of graphite was determined and found to fit $e^{-0.03x}$ law where x is the distance along the normal to the plane of the source. If the only impurity in the graphite is boron, calculate the number of atoms of boron per cm^3 in the graphite if the mean free path for scattering and absorption in graphite are 2.7 and 2,700 cm, respectively. The absorption cross-section of boron is 755 barns.

[Osmania University 1964]

8.92 Assuming that the elastic scattering of low energy neutrons is isotropic, show that the mean energy of the neutron after each collision will be

$E_f = (A^2 + 1)E_i/(A + 1)^2$

where A is the mass number of the target nucleus. Determine the number of collisions needed to thermalize fission neutrons (2 MeV) in graphite ($A = 12$).

8.2.15 Fusion

8.93 Determine the range of neutrino energies in the solar fusion reaction, $p + p \rightarrow d + e^+ + \nu$. Assume the initial protons have negligible kinetic energy and that the binding energy of the deuteron is 2.22 MeV, $m_p = 938.3\,\text{MeV}/\text{c}^2$ and $m_d = 1875.7\,\text{MeV}/\text{c}^2$ and $m_e = 0.51\,\text{MeV}/\text{c}^2$.

8.94 If the kinetic energy of the deuterons in the fusion reaction $D + D \rightarrow {}^3{}_2\text{He} + n + 3.2\,\text{MeV}$ can be neglected, what is the kinetic energy of the neutron?

8.95 (a) It is estimated that the deutrons have to come within 100 fm of each other for fusion to proceed. Calculate the energy that the deuterons must possess to overcome the electrostatic repulsion.
(b) If the energy is supplied by the thermal energy of the deutrons, what is the temperature of the deuteron?
[$e^2/4\pi\varepsilon_0 = 1.44\,\text{MeV fm}$, Boltzmann constant $k = 1.38 \times 10^{-23}\,\text{J K}^{-1}$]
(c) In (b) the actual required temperature is lower than the estimated value. Explain the mechanism by which the fusion reaction may proceed.

8.96 In a fusion reactor, the D-T reaction with Q value of 17.62 MeV is employed. Assuming that the deuteron density is $7 \times 10^{18}\,\text{m}^{-3}$ and the experimental value $< \sigma_{DT} \cdot v >= 10^{-22}\,\text{m}^3\,\text{s}^{-1}$ and that equal number of deuterons and tritons exist in the plasma at energy 10 keV, calculate the confinement time if the Lawson criterion is just satisfied.

8.3 Solutions

8.3.1 Atomic Masses and Radii

8.1 An ion of charge q will pick up kinetic energy, $T = qV$ in dropping through a P.D of V volts. In a magnetic induction B perpendicular to its path, the ion of momentum p will describe a circular path of radius r given by

$p = qBr = \sqrt{2MT} = \sqrt{2MqV}$

$$M = \frac{qB^2r^2}{2V} \qquad (1)$$

For the first ion, $q = 1.6 \times 10^{-19}\,\text{C}$, $B = 0.08$, $r = 0.0883\,\text{m}$ and $\text{V} = 400\,\text{V}$. Substituting these values in (1) we get

$M_1 = 9.98 \times 10^{-27}\,\text{kg}$

The mass of this ion is then $\dfrac{9.98 \times 10^{-27}\,\text{kg}}{1.66 \times 10^{-27}\,\text{kg/amu}} = 6.012\,\text{amu}$

Therefore the mass number is 6.

For the second ion, the only change is the radius of the orbit which is 0.0954 m. The mass of the second ion is

$$m_2 = m_1 \times \left(\frac{r_2}{r}\right)^2 = 6.012 \times \left(\frac{0.0954}{0.0883}\right)^2 = 7.0176\,\text{amu}$$

Therefore the mass number is 7.

8.2 The radius of curvature of an ion in the magnetic induction B, perpendicular to the orbit, with kinetic energy qV will be

$$r = \left(\frac{2MV}{qB^2}\right)^{1/2} \tag{1}$$

After a deviation of 180° the two ions will be separated by d, the difference between the diameters of the circular path.

(a) $$d = 2(r_{11} - r_{10}) = \left(\frac{8V}{qB^2}\right)^{1/2}\left(\sqrt{M_1} - \sqrt{M_2}\right) \tag{2}$$

Substitute $V = 5,000\,\text{V}$, $q = 1.6 \times 10^{-19}\,\text{C}$, $B = 0.15\,\text{T}$, $M_1 = 11 \times 1.66 \times 10^{-27}$ Kg and $M_2 = 10 \times 1.66 \times 10^{-27}$ Kg to find the separation $d = 0.021$ m or 2.1 cm.

(b) From (1), we get

$$\frac{\Delta M}{M} = 2\frac{\Delta r}{r}$$

The mass resolution is

$$\delta = \frac{M}{\Delta M} = \frac{r}{2\Delta r} = \frac{r}{d} = \frac{22\,\text{cm}}{2.1\,\text{cm}} = 10.5$$

where r is the mean radius of the ions, r_{11} and r_{10}, determined from (1) as 22.52 cm and 21.47 cm.

8.3 $p = \sqrt{2Tm} = qBr$

Here T, q and r are fixed.

$$\therefore \frac{B_2}{B_1} = \left(\frac{m_2}{m_1}\right) = \frac{37}{35}$$

Increase in induction

$$\Delta B = B_2 - B_1 = B_1\left(\frac{37}{35} - 1\right) = \frac{2B_1}{35} = \frac{2 \times 0.1}{35} = 0.57 \times 10^{-3}\,\text{T or } 5.7\,\text{G}$$

8.4 $^{27}\text{Si} \rightarrow {}^{27}\text{Al} + \beta^+ + \nu + T_{\max}$

$$M_{\text{Si}} - M_{\text{Al}} = 2m_e + T_{\max} = 2 \times 0.511 + 3.48 = 4.5\,\text{MeV}$$

The transition is between two mirror nuclei of charge $Z + 1$ and Z. The difference in Coulomb energy is

$$\Delta E_c = \frac{3}{5} \cdot \frac{1}{4\pi\varepsilon_0} \cdot \frac{e^2}{R}[Z(Z+1) - Z(Z-1)]$$
$$= \frac{6Ze^2}{5R} \cdot \frac{e^2}{4\pi\varepsilon_0} = 1.2 \times 1.44 \frac{Z}{R} \text{ MeV-fm}$$

Equating ΔE_c to $M_{Si} - M_A = 3.48 + 2 \times 0.51 = 4.5$ MeV, and $Z = 13$, we find R from which $r_o = R/A^{1/3}$ can be determined, where $A = 27$. Thus $r_o = 1.66$ fm.

8.5 $\Delta P_x . \Delta x = \hbar$ (uncertainty principle)

or $c\Delta P_x = \dfrac{\hbar c}{\Delta x} = \dfrac{197.3 \text{ MeV} - \text{fm}}{5 \text{ fm}} = 39.6 \text{ MeV}$

The kinetic energy $T = \dfrac{c^2 p^2}{2Mc^2} = \dfrac{(39.6)^2}{2 \times 940} = 0.83 \text{ MeV}$

8.6 The mass-energy equation for positron decay is

$$M(^{14}O) - M(^{14}N) = 2m_e + \frac{E_\beta(\max) + E_\gamma}{931.5}$$
$$= 2 \times 0.000548 + \frac{1.835 + 2.313}{931.5}$$
$$= 0.005549$$

or $\quad M(^{14}O) = 14.003074 + 0.005549 = 14.008623$ amu

8.3.2 Electric Potential and Energy

8.7 The charge density $\rho = 3ze/4\pi R^3$. Consider a spherical shell of radii r and $r + dr$.
The volume of the shell is $4\pi\ r^2\ dr$. The charge in the shell $q' = (4\pi r^2 dr)\rho$. The electrostatic energy due to the charge q' and the charge (q'') of the sphere of radius r, which is $\frac{4}{3}\pi r^3 \rho$, is calculated by imagining q'' to be deposited at the centre. The charge outside the sphere of radius r does not contribute to this energy (Fig. 8.8). Thus

$$dU = \frac{q'q''}{4\pi\varepsilon_0 r} = \frac{(4\pi r^2 dr \rho)(4\pi r^3 \rho/3)}{4\pi\varepsilon_0 r} = \frac{4\pi}{\varepsilon_0}\rho^2 r^4 dr$$

The total electrostatic energy is obtained by integrating the above expression in the limits 0 to R,

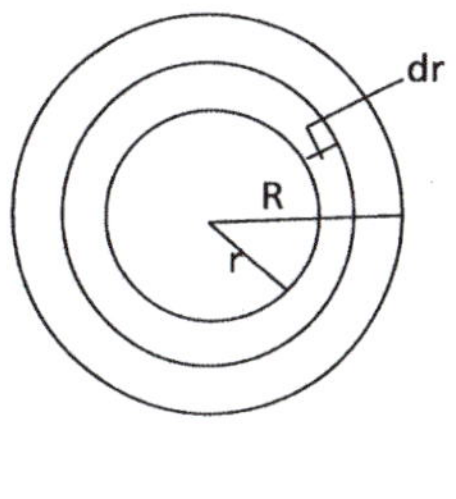

Fig. 8.8 Spherical shell of radius R

$$U = \int \mathrm{d}U = \frac{4\pi\rho^2}{3\varepsilon_0}\int_0^R r^4 \mathrm{d}r = \frac{4\pi\rho^2 R^5}{1.5\varepsilon_0} = \frac{3z^2e^2}{20\pi\varepsilon_0 R}$$

where we have substituted the value of ρ.

8.8 Choose a point at distance r ($<R$) from the centre of the nucleus. Let q' be the charge within the sphere of radius r.

Then $q' = q\left(\frac{r}{R}\right)^3$

The electric field will be $E = \frac{q'}{4\pi\varepsilon_0 r^2} = \frac{qr}{4\pi\varepsilon_0 R^3}$

The potential $V(r) = -\int E dr = -\int \frac{qr}{4\pi\varepsilon_0 R^3} + C$

$$= -\frac{qr^2}{8\pi\varepsilon_0 R^3} + C \qquad (1)$$

where C is a constant.

At $r = R$, the point is just on the surface and the potential will be given by Coulomb's law.

$$V(R) = \frac{q}{4\pi\varepsilon_0 R} \qquad (2)$$

Using (2) in (1), the value of C is determined as $C = \frac{3}{2}\frac{q}{4\pi\varepsilon_0 R}$ and (1) becomes

$$V(r) = \frac{q}{8\pi\varepsilon_0 R}\left(3 - \frac{r^2}{R^2}\right)$$

8.3.3 Nuclear Spin and Magnetic Moment

8.9 The rotational kinetic energy is given by

$$E_R = \frac{1}{2}I\omega^2 \qquad (1)$$

where E_R is the rotational energy, I the rotational inertia and ω is the angular velocity

The angular momentum is given by

$$J = I\omega \qquad (2)$$

Combining (1) and (2)

$$E_R = \frac{1}{2}J\omega$$

Therefore $J = \frac{2E_R}{\omega} = \frac{2 \times 537.5 \times 1.6 \times 10^{-13}}{2.6 \times 10^{23}} = 6.61 \times 10^{-34}\,\text{Js} = h$

8.10 (a) If J be the electronic angular momentum and I the nuclear spin, the multiplicity is given by $(2J + 1)$ or $(2I + 1)$, whichever is smaller.

Now, $2J + 1 = 2 \times \frac{5}{2} + 1 = 6$. But only four terms are found. Therefore, the multiplicity is given by $2I + 1 = 4$, whence $I = 3/2$.

(b) The magnetic field produced by the electron interacts with the nuclear magnetic moment resulting in the energy shift in hyperfine structure.

$\Delta E \approx 2\boldsymbol{I}.\boldsymbol{J} = F(F+1) - I(I+1) - J(J+1)$

where, $\boldsymbol{F} = \boldsymbol{I} + \boldsymbol{J}$ takes on integral values from 4 to 1.

$F = 4, \Delta E = 20 - 25/2$

$F = 3, \Delta E = 12 - 25/2$

$F = 2, \Delta E = 6 - 25/2$

$F = 1, \Delta E = 2 - 25/2$

The intervals are 8, 6 and 4 in the ratio 4:3:2

8.11 The resonance frequency ν is given by

$$\nu = \frac{\mu\beta}{Ih}$$

where μ is the magnetic moment, B the magnetic induction, I the nuclear spin in units of $\hbar$. A nuclear magneton $\mu_N = 5.05 \times 10^{-27}\,\text{JT}^{-1}$.

$$B = \frac{\nu I h}{\mu} = \frac{60 \times 10^6 \times (1/2) \times 6.625 \times 10^{-34}}{2.79 \times 5.05 \times 10^{-27}} = 1.4\,\text{T}$$

8.12 $f = \frac{\mu B}{Jh} = \frac{3.46 \times 3.15 \times 10^{-14} \times 1.6 \times 10^{-13} \times 0.8}{\left(\frac{7}{2}\right) \times 6.63 \times 10^{-34}}$

$$= 6.012 \times 10^6\,\text{Hz}$$

$$= 6.012\,\text{MHz}$$

8.3.4 Electric Quadrupole Moment

8.13 Quadrupole Moment

Assume that the charge is uniformly distributed over an ellipsoid of revolution, with the axis of symmetry along the z'-axis, the semi-major axis of length a, and the other two semi-axes of length b. The charge density ρ is

$$\rho = \frac{Ze}{\text{volume}} = \frac{Ze}{\frac{4\pi}{3}ab^2}$$

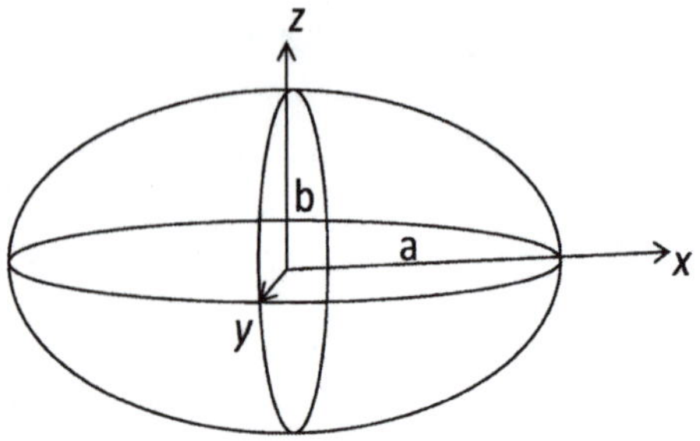

Fig. 8.9 Ellipsoid of revolution

The equation of the ellipsoid is

$$\left(\frac{x'^2}{b^2}\right)+\left(\frac{y'^2}{b^2}\right)+\left(\frac{z'^2}{a^2}\right)=1$$

In cylindrical coordinates (z', s', φ') the equation is

$$\left(\frac{s'^2}{b^2}\right)+\left(\frac{z'^2}{a^2}\right)=1, \text{ where } s'^2=x'^2+y'^2$$

$$Q=\int \rho(3z'^2-r'^2)\mathrm{d}\tau'$$

$$\rho=\frac{Ze}{\left(\frac{4\pi}{3}\right)ab^2}$$

$$\mathrm{Q}=\frac{3Ze}{4ab^2}\int_0^a \mathrm{d}z' \int_0^{b\sqrt{1-(z'^2/a^2)}}(2z'^2-s'^2)s'\mathrm{d}s' \int_0^{2\pi}\mathrm{d}\varphi'$$

$$=\frac{2}{5}Ze(a^2-b^2)$$

8.14 The quadrupole $Q=\frac{4}{5}\eta ZR^2$

$$\eta=\frac{a-b}{R}$$

Tantalum:

$$R=1.5\times(181)^{1/3}=8.48\text{fm}$$

$$\eta=\frac{5Q}{4ZR^2}=\frac{5\times 6\times 100\,\text{fm}^2}{4\times 73\times(8.48)^2\,\text{fm}^2}=0.143$$

The ratio of major to minor axes is

$$\frac{a}{b}\cong 1+\eta=1.143$$

Antimony: $R=1.5\times(123)^{1/3}=7.458\,\text{fm}$

$$\eta=\frac{5\,(-1.2\times 100\,\text{fm}^2)}{4\,51\times(7.458)^2\,\text{fm}^2}=-0.053$$

$$\frac{a}{b}=1-0.053=0.947$$

8.15 (a) The electrical quadrupole moment is the expectation value of the operator

$$Q_{ij} = \sum_{k=1}^{z} e_k (3x_i x_j - \delta_{ij} r^2)_k \tag{1}$$

where δ_{ij} is the kronecker delta.
The Q_{zz} – component,

$$Q_{zz} = \sum_{k=1}^{z} e_k (3z_k^2 - r_k^2) \tag{2}$$

is zero for a spherically charge distribution. This feature also becomes obvious from the formula

$Q = \frac{2}{5} ze(a^2 - b^2)$ for a homogenous ellipsoid of semi axes a, b (Problem 8.54).

For a sphere $a = b$ and therefore $Q = 0$.

(b) The quadrupole moment which is a tensor has the property that it is symmetric, that is $Q_{ij} = Q_{ji}$ and that its trace (the sum of the diagonal elements), $Q_{xx} + Q_{yy} + Q_{zz} = 0$. Using these two proiperties, Q_{ij} can be expressed in terms of the spin vector **I** which specifies the quantized state of the nucleus.

$$Q_{ij} = C\left(I_i I_j + I_j I_i - \frac{2}{3} \boldsymbol{I}^2 \delta_{ij}\right) \tag{3}$$

where C is a constant. Substituting $\boldsymbol{I}^2 = I(I+1)$ in (3)

$$Q = \frac{2}{3} CI(2I - 1) \tag{4}$$

which is zero for $I = 0$ or $I = 1/2$

8.16 From the results of Problem 8.15

$Q = \frac{2}{5}\left(a^2 - b^2\right) Z$

We can write

$$Q = \frac{4}{5} Z \left(\frac{a+b}{2}\right)(a-b)$$

Calling $R = \left(\frac{a+b}{2}\right)$ and $\Delta R = a - b$, we get

$$Q = \frac{4}{5} ZR\Delta R$$
$$Q = 4.2\,\text{barn} = 4.2 \times 10^{-24}\,\text{cm}^2 = 420\,\text{fm}^2$$
$$= \frac{2}{5} \times 71(a^2 - b^2)$$

Therefore $(a^2 - b^2) = 14.79\,\text{fm}^2$ (1)

Now, $A = \frac{4}{3}\pi ab^2\rho$ (2)

The nuclear charge density, $\rho = 0.17\,\text{fm}^{-3}$ (3)

Using the value of ρ and $A = 181$ in (2), we get

$ab^2 = 254.3\,\text{fm}^3$ (4)

Solving (2) and (4) we find $a = 7.1\,\text{fm},\ b = 6.0\,\text{fm}$

8.3.5 Nuclear Stability

8.17 Given $B/A = 9.402 - 7.7 \times 10^{-3}A$
For the parent nucleus

$$\text{B(A, Z)} = 9.402\text{A} - 7.7 \times 10^{-3}\,\text{A}^2$$

For the product nucleus

$$B(A-4, Z-2) = 9.402(A-4) - 7.7 \times 10^{-3}(A-4)^2$$
$$B(\propto) = 28.3$$

Condition that alpha decay is just energetically possible is

$$B(A, Z) = B(A-4, Z-2) + B(\propto)$$

Or $9.402A - 7.7 \times 10^{-3}A^2 = 9.402(A-4) - 7.7 \times 10^{-3}(A-4)^2 + 28.3$

Simplifying and solving for A, we find that $A = 153$. Thus, alpha decay is energetically possible for $A > 153$.

8.18 $S_n - S_p = 15.7 - 12.2 = 3.5 = \frac{3}{5} \times \frac{e^2}{4\pi\varepsilon_0 R}[Z^2 - (Z-1)^2]$
Substitute $e^2/4\pi\varepsilon_o = 1.44\,\text{MeV fm}$ and $Z = 8$ to find $R = 3.7\,\text{fm}$.

8.19 An atom of mass M_1 will decay into the product of mass M_2 and α particle of mass m_α if $M_1 > M_2 + m_\alpha$. Now the mass excess $\Delta = M - A$.

For ^{229}Th, $M_1 = 229 + 0.031652 = 229.031652$ amu. For α decay, the product atom would be ^{225}Ra, and $M_2 = 225 + 0.023528 = 225.023528$.

For α - particle, $m_\alpha = 4 + 0.002603 = 4.002603$

$M_2 + m_\alpha = 229.026131$

Since, $M_1 > M_2 + m_\alpha$, ^{229}Th will decay via α emission.

For β^- decay, $M_2 = 229 + 0.032022 = 229.032022$

As $M_1 < M_2$, decay via β^- emission is not possible. By the same argument the decay of ^{229}Th to ^{229}Ac is not possible via β^+ emission as $M_1 < M_2 + 2m_\text{e}$.

8.20 Three types of decays are possible

$$^{64}_{29}\text{Cu} \rightarrow ^{64}_{30}\text{Zn} + \beta^- + \bar{\nu}_e \quad (1)$$

$$^{64}_{29}\text{Cu} \rightarrow ^{64}_{28}\text{Ni} + \beta^+ + \nu_e \quad (2)$$

$$^{64}_{29}\text{Cu} + e^- \rightarrow ^{64}_{28}\text{Ni} + \nu_e \quad (3)$$

The energy released for the three processes are as follows:

$$Q_1 = m_{cu} - m_{zn} = (63.929759 - 63.929145) \times 931.5 = 0.572\,\text{MeV}$$
$$Q_2 = m_{cu} - m_{\text{Ni}} - 2m_e = (63.929759 - 63.927959) \times 931.5 - 2 \times 0.511$$
$$= 0.655\,\text{MeV}$$
$$Q_3 = m_{cu} - m_{\text{Ni}} = (63.929759 - 63.927959) \times 931.5 = 1.677\,\text{MeV}$$

We calculate the recoil energy in the process (3) and show that it is negligible.

$$T_{\text{Ni}} + T_\nu = Q = 1.677 \text{ (energy conservation)} \quad (4)$$
$$P_{\text{Ni}} = P_\nu$$
$$\text{or } P^2_{\text{Ni}} = 2M_{\text{Ni}}T_{\text{Ni}} = P^2_\nu = T^2_\nu \quad (5)$$

Solve (4) and (5) to obtain $T_{\text{Ni}} = 6\,\text{eV}$

8.21 $Q = (M_{\text{Al}} - M_{\text{Si}}) \times 931.5$
$= E_{\text{max}} + E_\gamma$
$= (27.981908 - 27.976929) \times 931.5$
$= 4.638\,\text{MeV}$
$= E_{\text{max}} + E_\gamma$
Therefore, $E_\gamma = Q - E_{\text{max}} = 4.638 - 2.865$
$= 1.773\,\text{MeV}$

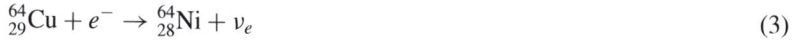

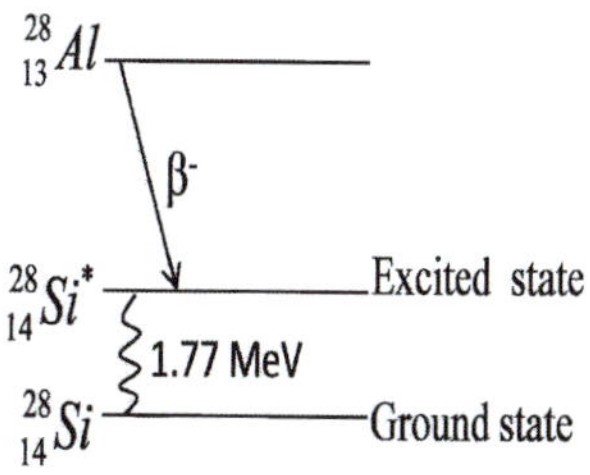

Fig. 8.10 β^- decay of $^{28}_{13}\text{Al}$ followed by γ decay

8.22 The mass-energy equation for the positron decay gives

$$M_{\mathrm{Na}} = M_{\mathrm{Ne}} + 2m_{\mathrm{e}} + \frac{T_{\max} + T_\gamma}{931.5}$$
$$= 21.991385 + 2 \times 0.511 + \frac{0.542 + 1.277}{931.5}$$
$$= 23.015338\,\mathrm{amu}$$

8.23 For electron capture $^7\mathrm{Be} + e^- \rightarrow ^7\mathrm{Li} + \nu_e$,

$$Q = (m_{\mathrm{Be}} - m_{\mathrm{Li}}) \times 931.5 = (7.016929 - 7.016004) \times 931.5 = 0.8616\,\mathrm{MeV}$$

which is positive

For positron emission the Q-value must be atleast 1.02 meV which is not available. Therefore, positron emission is not possible.

8.3.6 Fermi Gas Model

8.24 $$U = \frac{3}{5}AE_{\mathrm{F}} \tag{1}$$

$$p = -\frac{\partial U}{\partial V} = \frac{3}{5}A\frac{\partial E_{\mathrm{F}}}{\partial V} \tag{2}$$

From Fermi gas model

$$A = KVE_{\mathrm{F}}^{3/2} \tag{3}$$

Differentiating (3) with respect to V

$$\frac{3}{2}V\sqrt{E_{\mathrm{F}}}\frac{\partial E_{\mathrm{F}}}{\partial V} + E_{\mathrm{F}}^{3/2} = 0$$

whence $$\frac{\partial E_{\mathrm{F}}}{\partial V} = -\frac{2}{3}\frac{E_{\mathrm{F}}}{V} \tag{4}$$

Using (4) in (2)

$$p = \frac{2}{5}\frac{A}{V}E_{\mathrm{F}} = \frac{2}{5}\rho_{\mathrm{N}}E_{\mathrm{F}}$$

where $\rho_{\mathrm{N}} = A/V$, is the nucleon density.

8.25 The Fermi momentum for $N = Z = A/2$, is

$p_{\mathrm{F}}(n) = p_{\mathrm{F}}(p) = (\hbar/r_0)(9\pi/8)^{1/3}$

$cp_{\mathrm{F}} = (\hbar c/r_0)(9\pi/8)^{1/3} = (197.3\,\mathrm{MeV.fm}/1.3\,\mathrm{fm})(9\pi/8)^{1/3} = 231\,\mathrm{MeV}$

$p_{\mathrm{F}} = 231\,\mathrm{MeV/c}$

$E_{\mathrm{F}} = p_{\mathrm{F}}{}^2/2\mathrm{M} = (231)^2/(2 \times 940) = 28\,\mathrm{MeV}$

If B is the binding energy of a nucleon

$V = E_{\mathrm{F}} + B = 28 + 8 = 36\,\mathrm{MeV}$

8.3.7 Shell Model

8.26 A state with quantum number j can accommodate a maximum number of $N_j = 2(2j + 1)$ nucleons. Now $j = l \pm \frac{1}{2}$ and for $N_j = 12$, $j = 5/2$ and $l = 3$ or 2. However, because the parity is odd and $P = (-1)^l$, it follows that $l = 3$.

8.27 In the shell model the nuclear spin is predicted as due to excess or deficit of a particle (proton or neutron) when the shell is filled. Its parity is determined by the l value of the angular momentum, and is given by $(-1)^l$. For s- state, $l = 0$, p – state, $l = 1$, d-state $l = 2$, f-state $l = 3$ etc.

For the ground state of the nuclei:

$^7_3$Li: Spin is due to the third proton in $P_{3/2}$ state.

Therefore $J^\pi = (3/2)^-$ $\quad (\because\ l = 1)$

$^{16}_8$O: This is a doubly magic nucleus, and $J^\pi = 0^+$

$^{17}_8$O: Spin is due to the 9th neutron in $d_{5/2}$ state.

Therefore $J^\pi = (5/2)^+$ $\quad (\because\ l = 2)$

$^{39}_{19}$K: Spin is due to the proton hole in the $d_{3/2}$ state.

Therefore $J^\pi = (3/2)^+$ $\quad (\because\ l = 2)$

$^{45}_{21}$Sc: spin is due to the 21st proton in the $f_{7/2}$ state.

Therefore $J^\pi = (7/2)^-$ $\quad (\because\ l = 3)$

8.28 The ^{15}O nucleus in the $1p_{1/2}$ shell is an ^{16}O nucleus deficit in one neutron, its energy being B(15)–B(16), while ^{17}O in the $1f_{5/2}$ shell is an ^{16}O nucleus with a surplus neutron, its energy being B(16)–B(17). Thus the gap between the shells is

$$
\begin{aligned}
E(1\,f_{5/2}) - E(1\,p_{1/2}) &= B(16) - B(17) - [B(15) - B(16)] \\
&= 2\,B(16) - B(17) - B(15) \\
&= 2 \times 127.6193 - 131.7627 - 111.9556 \\
&= 11.52\,\text{MeV}
\end{aligned}
$$

8.29 $Q = -\left(\dfrac{2j-1}{2j+2}\right) < r^2 >$

For ^{209}Bi, $j = 9/2, < r^2 >= \frac{3}{5}R^2 = \frac{3}{5}r_0^2 A^{2/3} = \frac{3}{5}(1.2)^2(209)^{2/3}$ $= 30.42\,\text{fm}^2 = 0.3\,b$

$Q = -0.22\,b$

8.30 The spin and parity are determined as in Problem 8.27. The ground state spin and parity for the following nuclides are:

$^3_2$He : $J^\pi = (1/2)^+$ The state due to neutron hole is $1s_{1/2}$
$^{20}_{10}$Ne : $J^\pi = (0)^+$ The protons and neutrons complete the sub-shell.
$^{27}_{13}$Al : $J^\pi = (5/2)^+$ The state of an extra neutron is $1d_{5/2}$
$^{41}_{21}$Sc : $J^\pi = (7/2)^-$ The state of an extra proton is $1f_{7/2}$

8.31 For the nuclei $^{12}_6$C, $^{13}_6$C, $^{14}_6$C and $^{15}_6$C the ground state configuration of protons is $(1s_{1/2}, 1p_{3/2})$ For neutrons it is
$(1s_{1/2}, 1p_{3/2}), (1s_{1/2}1p_{3/2}, 1p_{1/2}), (1s_{1/2}, 1p_{3/2}, 1p_{1/2})$ and
$(1s_{1/2}, 1p_{3/2}, 1p_{1/2}, 1D_{5/2})$ respectively

The spin and parity assignment for $^{13}_6$C is $(1/2)^-$ and that for $^{15}_6$C, it is $(1/2)^+$.

8.32 Twenty neutrons and twenty protons fill up the third shell. The extra neutron goes into the $1f_{7/2}$ state. The spin and parity are determined by this extra neutron. Therefore the spin is 7/2. The parity is determined by the l - value which is 3 for the f-state. Hence the parity is $(-1)^l = (-1)^3 = -1$.

The model predicts $J^\pi = (0)^+$ for $^{30}_{14}$Si and $J^\pi = (1)^+$ for $^{14}_7$N nuclides.

8.3.8 Liquid Drop Model

8.33 Using the mass formula one can deduce the atomic number of the most stable isobar. It is given by

$$Z_{\min} = \frac{A}{2 + (a_c/2a_s)A^{2/3}}$$

$$\frac{Z_{\min}}{A} = \frac{1}{2 + 0.0156A^{2/3}}$$

For light nuclei, say $A = 10$, the second term in the denominators is small, and $z_{\min}/A = 0.48$. For heavy nuclei, say $A = 200$, $Z_{\min}/A = 0.4$

8.34 The most stable isobar is given by

$$Z_0 = \frac{A}{2 + 0.015\,\mathrm{A}^{2/3}} = \frac{64}{2 + 0.015 \times 64^{2/3}} = 28.57 \quad \text{or } 29$$

8.35 $\Delta m = (26.986704 - 26.981539) \times 931.5 = 4.76\,\text{MeV}$.

The difference in binding energy is due to mass difference of neutron and proton. (1.29 MeV) plus Δm, that is 6.05. The difference in the masses of mirror nuclei is assumed to be due to difference in proton number. Then equating the Coulomb energy difference to the mass difference

$$\Delta B = a_c \frac{[Z(Z+1) - Z(Z-1)]}{A^{1/3}} = \frac{2a_c Z}{A^{1/3}}$$

$$a_c = \frac{A^{1/3}\Delta B}{2Z} = \frac{3 \times 6.05}{2 \times 13} = 0.7$$

8.36 The difference in the binding energies of the two mirror nuclei is assumed to be the difference in the electrostatic energy, the mass number being the same.

$$B(\text{Ca}) - B(\text{Se}) = \frac{3}{5}\frac{[Z(Z+1) - Z(Z-1)]e^2}{4\pi\varepsilon_0 R}$$

$e^2/4\pi\varepsilon_0 = 1.44\,\text{MeV fm}$, and $Z = 20$

The left hand side is $350.420 - 343.143 = 7.227\,\text{MeV}$, R is calculated as 4.75 fm.

8.37 The empirical mass formula is

$$M(A, Z) = Z(m_p + m_e) + (A - Z)m_n - \alpha A + \beta A^{2/3} + \gamma(A - 2Z)^2/A + \varepsilon Z^2 A^{-1/3}$$

where α, β, γ and ε are constants. Holding A as constant, differentiate $M(A, Z)$ with respect to Z and set $\frac{\partial M}{\partial Z} = 0$.

$$\frac{\partial M}{\partial Z} = m_p + m_e - m_n - 4\gamma(A - 2Z)/A + 2\varepsilon Z A^{-1/3} = 0$$

The terms $m_p + m_e - m_n \cong m_H$, the mass of hydrogen atom which is neglected. Rearranging the remaining terms, we obtain

$$Z_{\text{min}} = \frac{A}{2 + (\varepsilon/2\gamma)A^{2/3}}$$

which is identical with the given expression.

8.38 $$B\left({}^{40}_{20}\text{Ca}\right) = 15.56 \times 40 - 17.23 \times 40^{2/3} - \frac{0.697 \times 20^2}{40^{1/3}} - 0 + \frac{12}{40^{1/2}} \quad (1)$$

$$B\left({}^{39}_{20}\text{Ca}\right) = 15.56 \times 39 - 17.23 \times 39^{2/3} - \frac{0.697 \times 20^2}{39^{1/3}} - \frac{23.285}{39} - 0 \quad (2)$$

Subtracting (2) from (1) gives us the binding energy of neutron $B(n) = 15.38\,\text{MeV}$.

This is the energy needed to separate one neutron from the nucleus.

8.39 (a) $B(\text{Th}) = 15.5 \times 230 - 16.8 \times 230^{2/3} - 0.72 \times \frac{90 \times 89}{230^{1/3}} - 23.0 \times \frac{50^2}{230} + \frac{34.5}{230^{3/4}}$

$= 1743.70\,\text{MeV}$

$B(\text{Ra}) = 15.5 \times 226 - 16.8 \times 226^{2/3} - 0.72 \times \frac{88 \times 87}{226^{1/3}} - 23.0 \times \frac{48^2}{226} + \frac{34.5}{226^{3/4}}$

$= 1740.88\,\text{MeV}$

$B(\alpha) = 28.3$

$Q = B(\text{Ra}) + B(\alpha) - B(\text{Th})$

$= 25.48\,\text{MeV}$

(b) $E(\text{Ra}) = \frac{Q \times 4}{4 + 226} = 0.44\,\text{MeV}$

8.40 For $A = 235$ the diagram (Fig. 8.7) indicates the binding energy per nucleon, $\frac{B}{A} = 7.6\,\text{MeV}$. Therefore, the total binding energy of ^{235}U nucleus $B = 7.6\,A = 7.6 \times 235 = 1{,}786\,\text{MeV}$.

Hence the rest mass energy of the nucleus

$$\begin{aligned} M(A, Z)c^2 &= Zm_pc^2 + (A - Z)m_nc^2 - B \\ &= 92 \times 938.3 + 143 \times 939.5 - 1,786 \\ &= 218,929\,\text{MeV} \end{aligned}$$

Therefore the mass is $\dfrac{218{,}929}{931.5} = 235.028$ amu

The plot is based on the semi-empirical mass formula obtained in the Liquid Drop Model. In this formula the lowering of binding energy at low mass numbers due to surface tension effects as well as at high Z (and hence large A) due to coulomb energy, are predicted by this model. The jumps in the curves at low mass numbers ($A = 2$–20) are attributed to the shell effects explained by the shell Model. The asymmetry term occurring in the mass formula is explained by the Fermi Gas Model. From the plot the binding energy per nucleon for ^{87}Br is found to be 8.7 MeV and that for ^{145}La it is 8.2 MeV. The energy released in the fission is

$$\begin{aligned} Q &= [B(\text{Br}) + B(\text{La})] - B(\text{U}) \\ &= 8.7 \times 87 + 8.2 \times 145 - 1786 \\ &= 160\,\text{MeV} \end{aligned}$$

8.41 The liquid drop model gives the value of Z for the most stable isobar of mass number A by

$$Z_0 = \frac{A}{2 + 0.015A^{2/3}}$$

For $A = 127$, $Z_0 = 53.38$, the nearest being $Z = 53$. Hence $^{127}_{53}$I is stable. But $^{127}_{54}$Xe is unstable against β^+ decay or e^- capture.

8.42 $^{27}_{12}\text{Mg} - ^{27}_{13}\text{Al} = -0.000841\ (12–13) + 0.0007668 \times 27^{-1/3}(12^2–13^2)$

$$+0.09966\left[\left(12 - \frac{27}{2}\right)^2 - \left(13 - \frac{27}{2}\right)^2\right]$$

$$= +0.225331\ \text{amu}$$

As the right hand side is positive $^{27}_{12}$Mg is heavier than $^{27}_{13}$Al, and therefore it is unstable against β^- decay.

8.3.9 Optical Model

8.43 Introduce the complex potential $V = -(U + iW)$ in the Schrodinger's equation

$$\nabla^2\psi + \frac{2m}{\hbar^2}(E + U + iW)\psi = 0 \tag{1}$$

Multiply (1) by ψ^*

$$\psi^* \nabla^2 \psi + \frac{2m}{\hbar^2}(E + U + iW)\psi^* \psi = 0 \tag{2}$$

Form the complex conjugate equation of (1) and multiply by ψ

$$\psi \nabla^2 \psi^* + \frac{2m}{\hbar^2}(E + U - iW)\psi^* \psi = 0 \tag{3}$$

Subtract (2) from (3)

$$\psi^* \nabla^2 \psi - \psi \nabla^2 \psi^* = \frac{-4imW}{\hbar^2}\psi\psi^* \tag{4}$$

Now the quantum mechanical expression for the current density is

$$\mathbf{j} = \frac{\hbar}{\mathbf{2im}}(\psi^* \nabla^2 \psi - \psi \nabla^2 \psi^*) \tag{5}$$

so that (4) becomes

$$\operatorname{div} \mathbf{j} = -\frac{2}{\hbar} W \psi^* \psi \tag{6}$$

Since $\psi^*\psi$ is the probability density and $W = \frac{1}{2}\hbar v K$ where K is the absorption coefficient, Eq. (6) is equivalent to the classical continuity equation

$$\frac{\partial \rho}{\partial t} + \operatorname{div} \boldsymbol{j} = -\frac{v}{\lambda}\rho \tag{7}$$

where v is the particle velocity and the mean free path $\lambda = 1/K$. When steady state has reached the first term on the LHS of (7) vanishes. Provided $W > 0$, the imaginary part of the complex potential has the effect of absorbing flux from the incident channel.

8.44 (i) $cp = \sqrt{2m(E - U)}$

$$\lambda = \frac{h}{p} = \frac{2\pi\hbar c}{\sqrt{2mc^2(E - U)}} = \frac{2\pi \times 197.3}{\sqrt{2 \times 939.6 \times (100 + 25)}} = 2.56\,\text{fm}$$

(ii) $W = \frac{1}{2}\hbar v K = \frac{1}{2}\hbar c \beta K$

$$\beta = \sqrt{\frac{2E}{mc^2}} = \sqrt{\frac{2 \times 100}{939.6}} = 0.46$$

$$K = \frac{2W}{\hbar c \beta} = \frac{2 \times 10}{197.3 \times 0.46} = 0.22\,\text{fm}^{-1}$$

$$2R = 2r_o A^{1/3} = 2 \times 1.3(120)^{1/3} = 12.82$$

$$2KR = 2.82$$

Probability that the neutron will be absorbed in passing diametrically through the nucleus $= (1 - e^{-2KR}) = (1 - e^{-2.82}) = 0.94$

8.3.10 Nuclear Reactions (General)

8.45 The given decay is

$$^{13}N \rightarrow {}^{13}C + \beta^{+} + \nu + 1.2\,\text{MeV} \quad (1)$$

$$M_{\text{N}} - M_{\text{C}} - 2m_{\text{e}} = 1.2\,\text{MeV} \quad (2)$$

where masses are atomic and $c = 1$
For the reaction

$$p + {}^{13}C \rightarrow {}^{13}N + n$$

$$Q = M_p + M_C - M_N - M_n$$

where the masses are nuclear.
Add and subtract $7m_{\text{e}}$ to get Q in atomic masses

$$Q = M_H + M_C - M_N - M_n = -[(M_n - M_H) + (M_N - M_C)]$$

$$Q = -[0.78 + 1.2 + 2m_{\text{e}}] = -3\,\text{MeV}$$

where we have used (2) and the mass difference $M_n - M_H = 0.78\,\text{MeV}$ and $m_e = 0.51\,\text{MeV}$

$$E_{\text{threshold}} = |Q|\left(1 + \frac{m_p}{m_c}\right) = 3 \times \left(1 + \frac{1}{13}\right) = 3.23\,\text{MeV}$$

8.46 Given reaction is

$$p + {}^{7}\text{Li} \rightarrow {}^{7}\text{Be} + n - 1.62\,\text{MeV} \quad (1)$$

$$M_{\text{Li}} + M_P - M_{\text{Be}} - M_n = -1.62\,\text{MeV} \quad (2)$$

where the masses are nuclear.
It follows that

$$M_{\text{Be}} - M_{\text{Li}} = 1.62 + 938.23 - 939.52 = 0.33\,\text{MeV}$$

Add and subtract $4m_{\text{e}}$ to get the Q mass difference of Be and Li

$$M_{\text{Be}} - M_{\text{Li}} - m_{\text{e}} = 0.33$$

or $M_{\text{Be}} - M_{\text{Li}} = 0.33 + m_{\text{e}} = 0.84\,\text{MeV}$

where $m_{\text{e}} = 0.51\,\text{MeV}$. Thus the total energy released in the electron capture $e^{-} + {}^{7}\text{Be} \rightarrow {}^{7}\text{Li} + \nu$ is 0.84 MeV

This energy is shared between the neutrino and the recoil nucleus. Energy and momentum conservation give

$$E_{\text{N}} + E_{\upsilon} = 0.84\,\text{MeV} \quad (1)$$

$$P_{\upsilon}^2 = E_{\upsilon}^2 = P_N^2 = 2\,M_{\text{N}} E_{\text{N}} \quad (2)$$

Using $M_{\text{N}} \cong 7\,\text{amu} = 6520.5\,\text{MeV}$, (1) and (2) can be solved to obtain $E_{\text{N}} = 7.5\,\text{keV}$ and $E_{\upsilon} \cong 0.84\,\text{MeV}$.

8.47 The Q-value is $-1.37\,\text{MeV}$. Minimum energy required is

$$E_a = |Q|\left(1 + \frac{m_a}{m_x}\right) = 1.37\left(1 + \frac{1}{24}\right) = 1.427\,\text{MeV}$$

8.48 Given reaction is

$${}_{2}^{4}\text{He} + {}_{7}^{14}\text{N} \rightarrow {}_{Z}^{A}\text{X} + {}_{1}^{1}\text{H}$$

As the atomic number (Z) and mass number (A) are conserved

$$2 + 7 = Z + 1, \text{ or } Z = 8$$
$$4 + 14 = A + 1 \text{ or } A = 17$$

Therefore the product X is ${}_{8}^{17}\text{O}$.

$$Q = [(4.0039 + 14.0075)–(17.0045 + 1.0081)] \times 931.5$$
$$Q = -1.118\,\text{MeV}$$

Thus it is an endoergic reaction for which the minimum α - particle kinetic energy required to initiate the above reaction is

$$E_{\text{threshold}} = |Q|\left(1 + \frac{M_{\text{He}}}{M_{\text{N}}}\right) = 1.118 \times \left(1 + \frac{4}{14}\right) = 1.437\,\text{MeV}$$

8.49 Given reactions are

$${}^{2}\text{H} + {}^{2}\text{H} \rightarrow \text{n} + {}^{3}\text{He} + 3.27\,\text{MeV}$$
$${}^{2}\text{H} + {}^{2}\text{H} \rightarrow \text{p} + {}^{3}\text{H} + 4.03\,\text{MeV}$$

It follows that

$$M_{\text{n}} + M_{\text{He}^3} + 3.27 = M_{\text{p}} + M_{\text{H}^3} + 4.03$$
$$M_{\text{H}^3} - M_{\text{He}^3} = \left(M_{\text{n}} - M_{\text{p}}\right) + 3.27 - 4.03$$
$$= 1.29 + 3.27 - 4.03 = 0.53\,\text{MeV}$$

Binding energies are given by

$$B(H^3) = m_{\text{p}} + 2m_{\text{n}} - M(H^3)$$
$$B(\text{He}^3) = 2m_{\text{p}} + m_n - M(\text{He}^3)$$
$$\therefore B(H^3) - B(\text{He}^3) = m_{\text{n}} - m_{\text{p}} - \left[M(H^3) - M(\text{He}^3)\right]$$
$$= 1.29 - 0.53 = 0.76\,\text{MeV}$$

The Coulomb energy of two protons

$$E_{\text{c}} = \frac{1.44 \times 1 \times 1}{3^{1/3} \times 1.3} = 0.769\,\text{MeV}$$

8.50 Given reaction can be written down as

$$\text{n} + {}_{5}^{10}\text{B} \rightarrow {}_{5}^{11}\text{B} \rightarrow \alpha + {}_{3}^{7}\text{Li} + Q$$

(a) $Q = (M_{\text{n}} + M_{\text{B}^{10}} - M_{\alpha} - M_{\text{Li}}) \times 931\,\text{MeV}$
$= (1.008987 + 10.01611 - 4.003879–7.01822) \times 931.5\,\text{MeV}$
$= 2.79\,\text{MeV}$

(b) The energy released is partitioned as follows

$$E_\alpha = \frac{QM_{\rm Li}}{M_\alpha + M_{\rm Li}} = \frac{2.79 \times 7.018}{4.004 + 7.018} = 1.78\,\text{MeV}$$

$$E_{\rm Li} = Q - E_\alpha = 2.79 - 1.78 = 1.01\,\text{MeV}$$

8.51 $p +^{27}\text{Al} \to n +^{27}\text{Si} + Q$ (1)

$^{27}\text{Si} \to^{27}\text{Al} + \beta^+ + \nu$ (2)

$$M_{\rm si} - M_{\rm Al} = E_{\max} + 2m_{\rm e}$$

where masses are atomic. In terms of nuclear masses

$$M_{\rm Si} - M_{\rm Al} = E_{\max} + 2m_{\rm e} = 3.5 + 0.51 = 4.01\,\text{MeV}$$

In (1), $Q = m_{\rm p} + m_{\rm Al} - m_{\rm n} - m_{\rm si}$
where masses are nuclear

$$Q = -\left(m_{\rm n} - m_{\rm p}\right) - (m_{\rm Si} - m_{\rm Al}) = -0.8 - 4.01 = -4.81\,\text{MeV}$$

$$E_{\rm Threshold} = |Q|\left(1 + \frac{m_{\rm p}}{m_{\rm Al}}\right) = 4.81 \times \left(1 + \frac{1}{27}\right) = 5.0\,\text{MeV}$$

8.52 (a) The threshold energy for appearance of neutron in the forward direction is

$$E_{\rm p}(\text{threshold}) = |Q|\left(1 + \frac{m_{\rm p}}{m_{\rm H^3}}\right)$$

$$= 0.764 \times \left(1 + \frac{1}{3}\right) = 1.019\,\text{MeV}$$

(b) The threshold for the appearance of neutrons in the 90° direction is

$$E_{\rm p}(\text{threshold}) = \frac{|Q|m_{\rm He}}{m_{\rm He} - m_{\rm p}}$$

$$= 0.764 \times \left(\frac{3}{3-1}\right) = 1.146\,\text{MeV}$$

8.53 $\text{d} +^{30}\text{Si} \to^{31}\text{Si} + \text{p}$ (1)

$Q = M_{\rm d} + M_{\rm Si^{30}} - M_{\rm Si^{31}} - M_{\rm p}$ (2)

Given

$M_{\rm Si^{30}} + M_{\rm d} = M_{\rm p^{31}} + M_{\rm n} + 5.1$ (3)

$M_{\rm Si^{31}} = M_{\rm p^{31}} + M_{\rm e} + 1.51$ (4)

Subtract (4) from (3)

$M_{\rm Si^{30}} + M_{\rm d} - M_{\rm Si^{31}} = M_{\rm n} - M_{\rm e} + 3.59$ (5)

Further

$M_{\rm n} = M_{\rm p} + M_{\rm e} + 0 + 0.78$ (6)

Add (5) and (6)

$$M_{\rm Si^{30}} + M_{\rm d} - M_{\rm Si^{31}} - M_{\rm p} = Q = 3.59 + 0.78 = 4.37\,\text{MeV}$$

8.54 Inelastic scattering is like an endoergic reaction except that the identity of particles is unchanged.

$$Q = E_b\left(1+\frac{m_b}{m_y}\right) - E_a\left(1-\frac{m_a}{m_y}\right) \quad \because \theta = 90^\circ$$

Substitute $Q = -4.4$, $m_a = 1$, $m_b = 1$, $m_y = 12$, $E_a = 15$ MeV to obtain $E_b = 8.63$ MeV

8.55 In the head-on collision the proton will receive full energy of the incident neutron, that is 5 MeV and neutron will stop. If the neutron was replaced by the gamma ray then the proton will be emitted in the forward direction as before but the gamma ray will be scattered backward at 180°. The kinetic energy imparted to the proton can be found out by the use of the formula employed for Compton scattering except now α would mean

$$\alpha = \frac{h\nu_o}{M_p c^2} = \frac{E_o}{Mc^2}.$$

$$T = \frac{\alpha E_o(1-\cos\theta)}{1+\alpha(1-\cos\theta)}$$

$$T_{max} = \frac{2\alpha E_o}{1+2\alpha} = \frac{2E_o}{2+Mc^2/E_o} = 5\,\text{MeV}$$

where we have put $\theta = 180°$. Putting $M = 938$ MeV, the above equation is easily solved to yield $E_o = h\nu = 51$ MeV.

8.56 (a) $Q = E_b\left(1+\frac{m_b}{m_y}\right) - E_a\left(1-\frac{m_a}{m_y}\right) - \frac{2}{m_y}\sqrt{m_a m_b E_a E_b}\cos\theta$

Here $m_a = m_b = 1$, $m_y = 10$, $E_a = 5$, $E_b = 3$ and $\theta = 45^0$. Substituting these values, we find $Q = -1.75$ MeV. Thus, the excitation energy of ^{10}B is 1.75 MeV.

(b) For elastic scattering $Q = 0$. Substituting the necessary values in the above equation which is quadratic in $\sqrt{E_b}$, we find $\sqrt{E_b} = 2.171$ so that $E_b = 4.715$ MeV.

Thus the expected energy of elastically scattered protons will be 4.715 MeV.

8.57 For the reaction $X(a, b)Y$,

$$Q = E_b\left(1+\frac{m_b}{m_Y}\right) - E_a\left(1-\frac{m_a}{m_Y}\right) - \frac{2}{m_Y}\sqrt{m_a m_b E_a E_b}Cos\theta$$

Here $b = p, a = d, Y = {}^{28}Al, \theta = 90^o$

$$5.5 = E_p\left(1+\frac{1}{28}\right) - 2.1\left(1-\frac{2}{28}\right)$$

$$E_p = 7.19\,\text{MeV}$$

8.58 $E_{(He)} + E_\gamma = 5.3\,\text{MeV}$ (energy conservation) (1)

$P_{He} = \sqrt{2M_{(He)}\,E_{(He)}} = P_\gamma = E_\gamma/c$ (momentum conservation) (2)

Or $E_\gamma = \sqrt{2M_{(He)}c^2 E_{(He)}}$ (3)

Use $M_{(He)}c^2 = 3,728\,\text{MeV}$ and solve (1) and (3) to find
$E_{(He)} = 3.775 \times 10^{-3}\,\text{MeV}$
$= 3.78\,\text{keV}$
Heat energy $= \frac{3}{2}\text{kT} = \frac{3}{2} \times 1.38 \times 10^{-23} \times 1.7 \times 10^7$
$= 3.5 \times 10^{-16}\,\text{J} = 2.19 \times 10^{-3}\,\text{MeV}$
Equating this to the electrostatic energy
$(1.44 \times 1 \times 1)/r = 2.19 \times 10^{-3}$
we get $r = 640\,\text{fm}$

8.59 $\text{d} + {}^{16}\text{O} \rightarrow \text{n} + {}^{17}\text{F} - 1.631$
$\text{d} + {}^{16}\text{O} \rightarrow \text{p} + {}^{17}\text{O} + 1.918$
It follows that

$$\text{m}_n + \text{m}_\text{F} - 1.631 = \text{m}_\text{p} + \text{m}_\text{O} + 1.918$$

$$\begin{aligned} \text{or } \text{m}_\text{F} - \text{m}_\text{O} &= 3.549 - (\text{m}_n - \text{m}_p) \\ &= 3.549 - (\text{m}_n - \text{m}_H + \text{m}_e) \\ &= 3.549 - (0.782 + 0.511) \\ &= 2.256\,\text{MeV} \end{aligned}$$

Therefore ^{17}F is heavier than ^{17}O. Actually ^{17}F decays to ^{17}O by β^+ emission.

$${}^{17}\text{F} \rightarrow {}^{17}\text{O} + \beta^+ + \nu$$

$$Q = E_\beta(\text{max}) + 2m_e = E_{\text{max}} + 1.022$$

$$E_{\text{max}} = \text{Q} - 1.022 = 2.256 - 1.022 = 1.234\,\text{MeV}$$

8.60 In the reaction $X(a, b)Y$ at 90°,

$$\begin{aligned} Q &= E_b\left(1 + \tfrac{m_b}{m_y}\right) - E_a\left(1 - \tfrac{m_a}{m_y}\right) \\ &= E_p\left(1 + \tfrac{1}{30}\right) - 7.68\left(1 - \tfrac{4}{30}\right) \end{aligned}$$

For $E_p = 8.63\,\text{MeV}$, $Q_0 = 2.262$

$E_p = 6.41\,\text{MeV}$, $Q_1 = -0.032$

$E_p = 5.15\,\text{MeV}$, $Q_2 = -1.334$

$E_p = 3.98\,\text{MeV}$, $Q_3 = -2.543$

The energy levels are determined by

$E_0 = Q_0 - Q_0 = 0$ (ground state)

$E_1 = Q_0 - Q_1 = 2.294$

$E_2 = Q_0 - Q_2 = 3.596$

$E_3 = Q_0 - Q_3 = 4.805$

The energy levels of ^{30}Si are shown in Fig. 8.11.

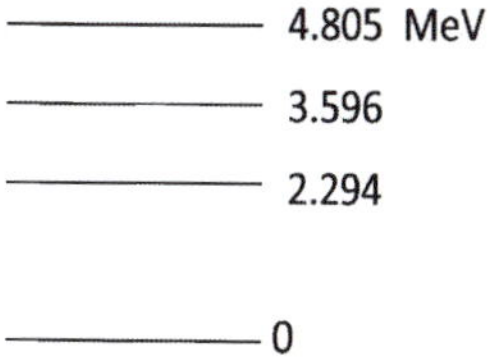

Fig. 8.11 Energy levels

8.61 For the nuclear reaction

$$^3\mathrm{H} + \mathrm{p} \rightarrow ^3\mathrm{He} + \mathrm{n} - 0.7637\,\mathrm{MeV} \quad (1)$$

$$^3\mathrm{H} -^3\mathrm{He} = \mathrm{n} - \mathrm{p} - 0.7637 \quad (2)$$

∴ Mass difference

$$m_\mathrm{n} - m_\mathrm{H} = m_{\mathrm{H}^3} - m_{\mathrm{He}^3} - m_\mathrm{e} + 0.7637 \quad (3)$$

where the masses are atomic.

Add and subtract m_e in the right hand side so that

$$m_\mathrm{n} - m_\mathrm{H} = m_{\mathrm{H}^3} - m_{\mathrm{He}^3} - m_\mathrm{e} + 0.7637\,\mathrm{MeV} \quad (4)$$

The masses are now atomic.

Now consider the decay

$$^3\mathrm{H} \rightarrow ^3\mathrm{He} + \beta^- + \nu + 18.5\,\mathrm{keV} \quad (5)$$

On the atomic scale

$$m_{\mathrm{H}^3} - m_{\mathrm{He}^3} = 18.5\,\mathrm{keV} = 0.0185\,\mathrm{MeV} \quad (6)$$

use (6) in (4) to find

$$m_\mathrm{n} - m_\mathrm{H} = (0.0185 + 0.7637)\,\mathrm{MeV}$$
$$= 0.7822\,\mathrm{MeV}.$$

8.62 For the reaction $X(a, b)Y$,

$$Q = E_b\left(1 + \frac{m_b}{m_Y}\right) - E_a\left(1 - \frac{m_a}{m_Y}\right) - \frac{2}{m_Y}\sqrt{m_a m_b E_a E_b}\cos\theta$$

For the reaction $^3\mathrm{H}(d, n)^4\mathrm{He}$, we identify $a = d,\ x =^3\mathrm{H},\ b = n,\ Y =^4\mathrm{He}$. Substitute $E_a = 0.3\,\mathrm{MeV}$. Maximum neutron energy is obtained by putting $\theta = 0^o$ and minimum energy for $\theta = 180°$ in the above equation.

$E_\mathrm{n}(\mathrm{max}) = 15.41\,\mathrm{MeV}$ and $E_\mathrm{n}(\mathrm{min}) = 13.08\,\mathrm{MeV}$.

Thus, the range of neutron energy will be 13.08 – 15.41 MeV, the energy at other angle of emission will be in between.

8.63 Energy available in the CMS

$$-W^* = \frac{E_\mathrm{p} m_\mathrm{Ta}}{m_\mathrm{Ta} + m_\mathrm{p}} \quad (1)$$

where E_p is the Lab proton kinetic energy. Using the values of the masses of E_a and m_p and E_p, we find the excited level $W^* = 4.972\,\mathrm{MeV}$. The energy of the excited state will be $m_\mathrm{w} + W^* = 169,490\,\mathrm{MeV}$.

When the target and the projectile are interchanged, the same excitation energy W^* produced with T_a is given by

$$-W^* = E_{Ta}\frac{m_p}{m_p + m_{Ta}} \quad (2)$$

Comparing (1) and (2)

$$E_{Ta} + E_p\frac{m_{Ta}}{m_p} = 897.7\,\text{MeV}$$

8.64 $E_{th} = |Q|\left(1 + \dfrac{m_o}{m_{ca}}\right) = 7.83\,(1 + \frac{16}{48}) = 10.44\,\text{MeV}$

The velocity, $\beta = \dfrac{v}{c} = \sqrt{\dfrac{2T}{mc^2}} = \sqrt{\dfrac{2 \times 10.44}{16 \times 931}} = 3.74 \times 10^{-2}$

The impact parameter, $b = R_1 + R_2 = 1.1\left(16^{1/3} + 48^{1/3}\right) = 6.153\,\text{fm}$

$$J = M_o\, vb = n\hbar$$

$$n = \frac{M_o\, vb}{\hbar} = \frac{M_o c^2 \beta b}{\hbar c} = \frac{16 \times 931 \times 3.74 \times 10^{-2} \times 6.153}{197.3} = 17.37 \text{ or } 17$$

8.3.11 Cross-sections

8.65 If 1% of neutrons are absorbed then 99% are transmitted. The transmitted number I are related to the incident number by

$$I = I_0 \exp(-\mu x) \quad (1)$$

where μ is the absorption coefficient and x is the thickness of the foil.

$$\mu = \Sigma = \sigma N = \sigma N_0 \rho / A$$

where σ is the microscopic cross-section, N_0 is the Avagardro's number, ρ the density and A the atomic weight.

$$\mu = \frac{28000 \times 10^{-24} \times 6.02 \times 10^{23} \times 7.3}{114.7}\,\text{cm}^{-1}$$

$$= 1{,}073\,\text{cm}^{-1}$$

$$I/I_0 = 99/100 = \exp(-1{,}073x)$$

$$x = 9.37 \times 10^{-6}\text{cm or } 9.37\,\mu\text{m}$$

8.66 Specific activity, that is activity per gram

$$|dQ/dt| = Q_{max}\lambda = \phi\,\Sigma_{act} = \phi\sigma_a N_0/A$$

$$= 5 \times 10^{12} \times 20 \times 10^{-24} \times 6.02 \times 10^{23}/60$$

$= 10^{12}$ disintegrations per second per gram.

8.67 $|dQ/dt| = Q\lambda = \phi\Sigma_{act}(1 - e^{-\lambda t})$

Given $|dQ/dt|/|dQ_s/dt| = 20/100 = 1 - e^{-0.693t/5.3}$

or $\exp(0.693t/5.3) = 1.25$

Take $\log_e$ to find $t = 1.7$ years.

8.68 $\frac{d\sigma}{d\Omega} = \frac{I}{I_0 N d\Omega} = A + B\cos^2\theta$

$I_0 = 8 \times 10^{12}/\text{m}^2 - \text{s}$

N = number of target atoms intercepting the beam

$$= \frac{N_0 \rho t}{A} = \frac{6.02 \times 10^{23} \times 2.7 \times 10^{-3}}{27} = 6.02 \times 10^{19}$$

$$d\Omega = \frac{0.01}{6^2} = 2.78 \times 10^{-4}$$

$$A + B\cos^2 30^\circ = \frac{50}{8 \times 10^8 \times 6.02 \times 10^{19} \times 2.78 \times 10^{-4}} = 3.73 \times 10^{-24}$$

$$A + B\cos^2 45^\circ = \frac{40}{8 \times 10^8 \times 6.02 \times 10^{19} \times 2.78 \times 10^{-4}} = 2.99 \times 10^{-24}$$

Solving the above equations we find

$A = 1.57\,\text{b/Sr}$

$B = 2.88\,\text{b/Sr}$

8.69 $\sigma(\text{total}) = \frac{\text{Total number of particles scattered/sec}}{\text{(beam intensity) (number of target particles within the beam)}}$

As the scattering is assumed to be isotropic total number of particles scattered $=$ (Observed number) $(4\pi/d\Omega) = 15 \times 4\pi/2 \times 10^{-3} = 9.42 \times 10^4/\text{s}$

Beam intensity, that is number of beam particles passing through unit area per second

$$= \frac{\text{beam current}}{\text{charge on each proton}} = \frac{10 \times 10^{-9} A}{1.6 \times 10^{-19} C} = 6.25 \times 10^{10}/\text{cm}^2\,\text{s}$$

Therefore $\sigma(\text{total}) = \frac{9.42 \times 10^4}{6.25 \times 10^{10} \times 1.3 \times 10^{19}} = 1.159 \times 10^{-25}\,\text{cm}^2$
$= 116\,\text{mb}$

8.3.12 Nuclear Reactions via Compound Nucleus

8.70 Breit–Wigner formulae are

$$\sigma_t = \pi \frac{\lambda\!\!\!^{-2} \Gamma_s \Gamma . g}{(E - E_R)^2 + \frac{\Gamma^2}{4}} \quad (1)$$

$$\sigma_s = \pi \frac{\lambda\!\!\!^{-2} \Gamma_s^2 . g}{(E - E_R)^2 + \frac{\Gamma^2}{4}} \quad (2)$$

Dividing (2) by (1)

$$\frac{\sigma_s}{\sigma_t} = \frac{\Gamma_s}{\Gamma} \quad (3)$$

Ignoring the statistical factor g, at resonance

$$\sigma_t = 7{,}000 \times 10^{-24}\,\text{cm}^2 = \frac{\lambda^2}{\pi}.\frac{\Gamma_s}{\Gamma} \tag{4}$$

But $\lambda = \dfrac{0.286}{\sqrt{E}} = \dfrac{0.286}{\sqrt{0.178}} = 0.678\text{Å} = 0.678 \times 10^{-8}\,\text{cm}$
From (3) and (4) we get

$$\Gamma_s/\Gamma = 47.815 \times 10^{-5} \tag{5}$$

Inserting (5) and the value of σ_t in (3), we find

$$\sigma_s = 3.35\text{b}$$

8.71 $\Gamma = \Gamma_n + \Gamma_\gamma + \Gamma_\alpha = 4.2 + 1.3 + 2.7 = 8.2\,\text{eV}$

$$\sigma(\text{n}, \gamma) = \frac{\lambda^2}{4\pi}.\frac{\Gamma_\gamma \Gamma_\text{n}}{(E - E_R)^2 + \frac{\Gamma^2}{4}}$$

$$\lambda = \frac{0.286}{\sqrt{70}} = 3.418 \times 10^{-10}\,\text{cm}$$

$\text{E}_R = 60\,\text{eV},\ \text{E} = 70\,\text{eV},\ \Gamma_\gamma = 1.3\,\text{eV}$ and $\Gamma_n = 4.2\,\text{eV}$
Ignoring the g - factor, we find $\sigma(n, \gamma) = 1215\,\text{b}$.

$$\sigma(n, \alpha) = \sigma(n, \gamma).\frac{\Gamma_\alpha}{\Gamma_\gamma} = 1215 \times \frac{2.7}{1.3} = 2523\,\text{b}$$

8.72 Breit–Wigner's formula is

$$\sigma_\text{total} = \frac{\pi\lambda^2\Gamma_s\Gamma g}{(E - E_\text{R})^2 + \frac{\Gamma^2}{4}} \tag{1}$$

For spin zero target nucleus, the statistical factor $g = 1$. At resonance energy (1) reduces to

$$\Gamma_s = \frac{\pi\Gamma\sigma_\text{total}}{\lambda^2} \tag{2}$$

$$\lambda = \frac{0.286}{\sqrt{E}}\text{Å} = \frac{0.286}{\sqrt{250}} = 0.018\text{Å} = 1.8 \times 10^{-10}\,\text{cm}$$

Substituting, $\sigma_\text{total} = 1300 \times 10^{-24}\,\text{cm}^2,\ \Gamma = 20\,\text{eV}$ and $\lambda = 1.8 \times 10^{-10}\,\text{cm}$ in (2), we find $\Gamma_s = 2.5\,\text{eV}$.

8.3.13 Direct Reactions

8.73 $Q = [(m_\text{d} + m_\text{N}) - (m_\alpha + m_\text{C})] \times 931.44\,\text{MeV}$
$= [(2.014102 + 14.003074) - (14.002603 + 12.0)] \times 931.44$
$= 13.57\,\text{MeV}$

For the forward reaction, energy available in the CMS is

$$E^* = Q + T_d^* + T_N^* = Q + \frac{T_d m_N}{m_N + m_d}$$
$$= 13.57 + \frac{20 \times 14}{14 + 2} = 31.07$$

The energy of 31.07 MeV is shared between α and ^{12}C. Using the energy and momentum conservation, we find $P_\alpha^* = 416.8\,\text{MeV}/c$.

The inverse reaction will be endoergic, and so an energy of $31.07 + 13.57 = 44.64$ MeV must be provided. This corresponds to the Lab kinetic energy for α given by

$$T_\alpha = 44.64 \times \frac{12 + 4}{12} = 59.52\,\text{MeV}$$

In the CMS the energy of 31.07 MeV is shared between ^2H and ^{14}N. The momentum of deuteron would be 441.9 MeV/c.

$$\frac{\sigma_{dN}}{\sigma_{\alpha C}} = \frac{(2S_\alpha + 1)(2S_C + 1)}{(2S_d + 1)(2S_N + 1)} \frac{P_\alpha^{*2}}{P_d^{*2}}$$
$$= \frac{1 \times 1}{3 \times (2S_N + 1)} \times \left(\frac{441.9}{319.4}\right)^2 = \frac{1.91}{2S_N + 1}$$

since $S_\alpha = S_c = 0$ and $S_d = 1$. The spin of ^{14}N can be determined from the experimental value of the cross-sections. For $S_N = 1$, the ratio of cross-sections is expected to be 0.64.

The intrinsic parities of all the four particles is positive. If the $\alpha's$ are captured in the s-state for which the parity will be positive as it is given by $(-1)^l$, the $\alpha's$ can not be produced in the $l = 1$ state for which the parity would be negative, resulting in the violation of parity.

8.74 According to Butler's theory, the neutron energy

$$E_n = \frac{1}{2} E_d = 0.5 \times 460 = 230\,\text{MeV}$$

The spread in energy

$$\Delta E_n = 1.5\sqrt{B_d E_d} = 1.5\sqrt{2.2 \times 460} = 47.7\,\text{MeV}$$

And the angular spread is

$$\Delta\theta = 1.6\sqrt{\frac{B_d}{E_d}} = 1.6\sqrt{\frac{2.2}{460}} = 0.11\,\text{radians}$$

8.3.14 Fission and Nuclear Reactors

8.75 If Q is the number of atoms of ^{23}Na per gram at any time t, the net rate of production of ^{24}Na is

$$\frac{dQ}{dt} = \phi\Sigma_a - \lambda Q$$

where the first term on the right hand side denotes the absorption rate of neutrons in ^{23}Na, and if it is assumed that each neutron thus absorbed produces a ^{24}Na atom, then this also represents the production rate of ^{24}Na. The second term represents the decay rate, so that $\mathrm{d}Q/\mathrm{d}t$ denotes the rate of change of atoms of ^{24}Na.

The saturation activity is obtained by setting $\mathrm{d}Q/\mathrm{d}t = 0$. Then

$$\lambda Q_s = \phi\Sigma_a = \phi\sigma_a \frac{N_o\rho}{A}$$

$$= \frac{10^{11} \times 536 \times 10^{-27} \times 6.02 \times 10^{23} \times 0.97}{23} = 1.36 \times 10^9\, s^{-1}$$

8.76 At equilibrium number of ^{198}Au atoms is

$$Q_s = \frac{\phi\sigma_a N_o W T_{1/2}}{0.0693A}$$

$$= \frac{10^{12} \times 98 \times 10^{-24} \times 6.02 \times 10^{23} \times 0.1 \times 2.7 \times 3,600}{0.693 \times 197}$$

$$= 4.2 \times 10^{14}$$

$$\text{Activity} = Q_s\lambda = \frac{Q_s \times 0.693}{T_{1/2}} = \frac{4.2 \times 10^{14} \times 0.693}{2.7 \times 3,600} = 3 \times 10^{12}\, s^{-1}$$

8.77 Consider a binary fission, that is a heavy nucleus of mass number A and atomic number Z breaking into two equal fragments each characterized by $\frac{A}{2}$ and $\frac{Z}{2}$. In this problem the only terms in the mass formula which are relevant are the Coulomb term and the surface tension term.

$$M(A, Z) = \frac{a_c Z^2}{A^{1/3}} + a_s A^{2/3}$$

Energy released is equal to the difference in energy of the parent nucleus and that of the two fragments

$$Q = M(Z, A) - 2M\left(\frac{Z}{2}, \frac{A}{2}\right)$$

$$= \left(a_s A^{2/3} + a_c \frac{Z^2}{A^{1/3}}\right) - 2\left[a_s\left(\frac{A}{2}\right)^{2/3} + a_c \frac{(Z/2)^2}{(A/2)^{1/3}}\right]$$

$$= a_s A^{2/3}(1 - 2^{1/3}) + a_c \frac{Z^2}{A^{1/3}}\left(1 - \frac{1}{2^{2/3}}\right)$$

Inserting $A = 238,\ Z = 92,\ a_c = 0.59$ and $a_s = 14$ we find $Q \cong 160\,\mathrm{MeV}$.

8.78 The diffusion equation for the steady state in the absence of sources at the point of interest is

$$\nabla^2\phi - K^2\phi = 0 \quad (1)$$

where $K^2 = 3\Sigma_a/\lambda_{tr}$

Writing the Laplacian for spherical geometry (1) becomes

$$\frac{d^2\phi}{dr^2} + \frac{2}{r}\frac{d\phi}{dr} - K^2\phi = 0 \tag{2}$$

Equation (2) is easily solved, the solution being

$$\phi = \frac{C_1 e^{Kr}}{r} + \frac{C_2 e^{-Kr}}{r} \tag{3}$$

As K is positive, the first term on the right hand side tends to ∞ as $r \to \infty$. Therefore, $C_1 = 0$ if the flux is required to be finite everywhere including at ∞.

$$\phi = \frac{C_2 e^{-Kr}}{r} \tag{4}$$

We can calculate the constant C_2 by considering the current J through a small sphere of radius r with its centre at the source.
The net current

$$J = -\frac{\lambda_{\text{tr}}}{3}\frac{\partial \phi}{\partial r} = \frac{\lambda_{\text{tr}}}{3r^2} C_2(Kr+1)e^{-Kr} \tag{5}$$

where we have used (4)

The net number of neutrons leaving the sphere per second is

$$4\pi r^2 J = \frac{4}{3}\pi\lambda_{\text{tr}} C_2(Kr+1)e^{-Kr} \tag{6}$$

But as $r \to 0$, the total number of neutrons leaving the sphere per second must be equal to the source strength Q. Thus from (6)

$$Q = \frac{4}{3}\pi\lambda_{\text{tr}} C_2$$

$$\text{or} \quad C_2 = \frac{3Q}{4\pi\lambda_{\text{tr}}} \tag{7}$$

The complete solution is

$$\phi = \frac{3Qe^{-Kr}}{4\pi r \lambda_{\text{tr}}} \tag{8}$$

Therefore the neutron density

$$n(r) = \frac{\phi}{v} = \frac{3Qe^{-Kr}}{4\pi\lambda_{\text{tr}} v r} = \frac{Qe^{-r/L}}{4\pi D r} \tag{9}$$

where $\frac{\lambda_{tr} v}{3} = D$, is the diffusion coefficient and $K = 1/L$, L being the diffusion length.

The counting rate,

$$R = N\sigma v n(r) \tag{10}$$

As the absorption obeys the $1/v$ law, the product $\sigma.v =$ constant. We then have

$$\frac{250}{60} = 10^{20} \times 3{,}000 \times 10^{-24} \times 2.2 \times 10^5 n(r)$$

Or $n(r) = 6.313 \times 10^{-5}/\mathrm{cm}^3$
Substituting the values: $D = 5 \times 10^5,\ r = 300,\ L = 50$ and $R = 250/60$ in (9), we find

$$Q = 4.8 \times 10^7/\mathrm{s}.$$

8.79 The absorption rate in the fuel is

$$\Sigma_{\mathrm{au}}\phi_{\mathrm{u}}V_{\mathrm{u}}$$

where V is the volume, and the absorption rate in the moderator is

$$\Sigma_{\mathrm{am}}\phi_{\mathrm{m}}V_{\mathrm{m}}$$

The fraction of thermal neutrons absorbed by the Uranium fuel as compared to the total number of thermal neutron absorptions in the assembly is known as the thermal utilization factor f and is given by

$$f = \frac{\Sigma_{\mathrm{au}}\phi_{\mathrm{u}}V_{\mathrm{u}}}{\Sigma_{\mathrm{au}}\phi_{\mathrm{u}}V_{\mathrm{u}} + \Sigma_{\mathrm{am}}\phi_{\mathrm{m}}V_{\mathrm{m}}} = \frac{1}{1 + \dfrac{\Sigma_{\mathrm{am}}\phi_{\mathrm{m}}V_{\mathrm{m}}}{\Sigma_{\mathrm{au}}\phi_{\mathrm{u}}V_{\mathrm{u}}}}$$

Fig. 8.12 shows a unit cell of a heterogeneous assembly in which the uranium rods of radius r, are placed at regular intervals (pitch). The equivalent cell radius r_1 is also indicated.

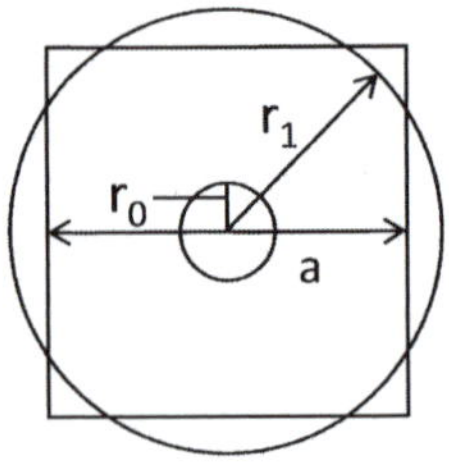

Fig. 8.12 Unit cell of a heterogeneous assembly

Area of the unit cell $A = \pi r_1^2$
Now,

$$\frac{\Sigma_{am}}{\Sigma_{au}} = \frac{N_m}{N_u} \times \frac{\sigma_{am}}{\sigma_{au}} = \frac{\rho_m/A_m}{\rho_u/A_u} \times \frac{\sigma_{am}}{\sigma_{au}}$$

$$= \frac{1.62/12}{18.7/238} \times \frac{4.5 \times 10^{-3}}{7.68} = 1.01 \times 10^{-3}$$

$$\frac{V_m}{V_u} = \frac{18^2 - \pi(1.5)^2}{\pi(1.5)^2} = 44.8$$

$$\frac{\phi_m}{\phi_u} = 1.6$$

Hence, $f = \dfrac{1}{1 + (1.01 \times 10^{-3}) \times 44.8 \times 1.6} = 0.933$

8.80 The equation for a critical reactor is

$$\nabla^2\phi + B^2\phi = 0 \tag{1}$$

Where ϕ is the neutron flux and B^2 is the buckling.
For spherical geometry, Eq. (1) becomes

$$\frac{d^2\phi}{dr^2} + \frac{2}{r}\frac{d\phi}{dr} + B^2\phi = 0 \tag{2}$$

which has the solution

$$\phi = \frac{A}{r}\sin(\pi r/R) \tag{3}$$

where A is the constant of integration and R is the radius of the bare reactor

$$\frac{d\phi}{dr} = -\frac{A}{r^2}\sin\left(\frac{\pi r}{R}\right) + \frac{\pi A}{Rr}\cos\left(\pi\frac{r}{R}\right) \tag{4}$$

$$\frac{d^2\phi}{dr^2} = \frac{2A}{r^3}\sin\left(\pi\frac{r}{R}\right) - \frac{\pi A}{Rr^2}\cos\left(\frac{\pi r}{R}\right) - \frac{A\pi^2}{R^2 r}\sin\left(\pi\frac{r}{R}\right) \tag{5}$$

Therefore (2) becomes

$$-\frac{A\pi^2}{dr^2}\sin\left(\frac{\pi r}{R}\right) + \frac{B^2 A}{r}\sin\left(\pi\frac{r}{R}\right) = 0$$

Therefore, $B^2 = \frac{\pi^2}{R^2}$
Or the critical radius,

$$R = \frac{\pi}{B} \tag{6}$$

$$B^2 = \frac{k_\infty - 1}{M^2} = \frac{1.54 - 1}{250} = 2.16 \times 10^{-3}\,\text{cm}^{-2}$$

$$B = 0.04647$$

$$R = \frac{\pi}{0.04647} = 67\,\text{cm}$$

The actual radius will be shorter by $d = 0.71\,\lambda_{tr}$ where d = extrapolated distance.

8.81 Let n fissions take place per second.
Energy released per second = $200\,n$ MeV = $(200\,n)(1.6 \times 10^{-13})$ J
$200 \times 1.6 \times 10^{-13} n = 40 \times 10^6$
Or $\quad n = 1.25 \times 10^{18}/\text{s}$

$$\text{Number of atoms in 1.0 g of } ^{235}\text{U} = \frac{6.02 \times 10^{23}}{235} = 2.562 \times 10^{21}$$

$$\text{Mass of uranium consumed per second} = \frac{1.25 \times 10^{18}}{2.562 \times 10^{21}} = 4.88 \times 10^{-4}\,\text{g}$$

Mass consumed in 1 day = $(4.88 \times 10^{-4})(86, 400) = 42.16$ g

8.82 Let n be the number of fissions occurring per second in the nuclear reactor.
Energy released = $200\,n$ MeV s^{-1}

$$= (200\,n)(1.6 \times 10^{-13})\text{J-s}^{-1}$$

$$= \text{power} = 2 \times 10^7\ \text{W}$$

Therefore, $n = 6.25 \times 10^{17}\ \text{s}^{-1}$
In 1 g there are $N_0/A = 6.02 \times 10^{23}/235 = 2.56 \times 10^{21}$ atoms of ^{235}U. Therefore consumption rate of ^{235}U will be

$$\frac{6.25 \times 10^{17}}{2.56 \times 10^{21}} = 2.44 \times 10^{-4}\ \text{g s}^{-1}$$

8.83 (a) Let n fissions occur per second. Then energy available will be 200 n MeV or $200\,n \times 1.6 \times 10^{-13}$J. Allowing for 5% wastage and 30% efficiency, net power used is $P = 200\,n \times 1.6 \times 10^{-13} \times 0.95 \times 0.3 = 9.12 \times 10^{-12} n$ W.

Required energy in 1 s that is power $= \dfrac{50 \times 10^9 \times 10^3 \times 3, 600}{3.15 \times 10^7} =$ 5.714×10^9

Equating the two powers

$$9.12 \times 10^{-12} n = 5.714 \times 10^9$$

$$n = 6.26 \times 10^{20}/\text{s}$$

$$\text{Number of atoms in 1 g of } ^{235}\text{U} = \frac{6.02 \times 10^{23}}{235} = 2.56 \times 10^{21}$$

$$\text{Mass consumed per second} = \frac{6.26 \times 10^{20}}{2.56 \times 10^{21}} = 0.244\,\text{g}$$

Mass consumed in 1 year = $0.244 \times 3.15 \times 10^7\ g = 7.7 \times 10^6\ g = 7.7$ tons

(b) Volume, $V = \dfrac{M}{\rho} = \dfrac{10^8}{19000} = 526.3\,\text{m}^3$

Power density (Power per unit volume)

$$P/V = \frac{100 \times 10^6}{526.3} = 1.9 \times 10^5\text{W/m}^3$$

Number of ^{235}U atoms/cm^3,

$$N = \frac{N_0 \rho}{A} \times \frac{0.7}{100} = \frac{6 \times 10^{23} \times 19}{238} \times \frac{0.7}{100} = 3.353 \times 10^{20}$$

$$\Sigma_a = \sigma_a N = 550 \times 10^{-24} \times 3.353 \times 10^{20} = 0.1844\,\text{cm}^{-1}$$

If ϕ is the neutron flux and 200 MeV is released per fission, then energy released in 1 cm^3/s will be $200\phi\Sigma_a$ MeV or $200 \times 0.1844 \times 1.6 \times 10^{-13}\phi = 5.9 \times 10^{-12}\phi = 7.648$ Therefore $\phi = 1.3 \times 10^{12}/\text{cm}^2 - \text{s}$

8.84 Let the Lab kinetic energy of neutron be E_1 and E_2 before and after the scattering. The neutron and proton mass is approximately identical.

The neutron velocity in the CMS, is $v_1^* = v_1/2$ as the masses of projectile and target are nearly identical. As the scattering is elastic, the velocity of the neutron $v_2^* = v_1^*$ in magnitude. Let the neutron be scattered at angle θ^* in the CMS. The velocity v_2^* is combined with v_c to yield v_2 in the Lab. From Fig. 8.13,

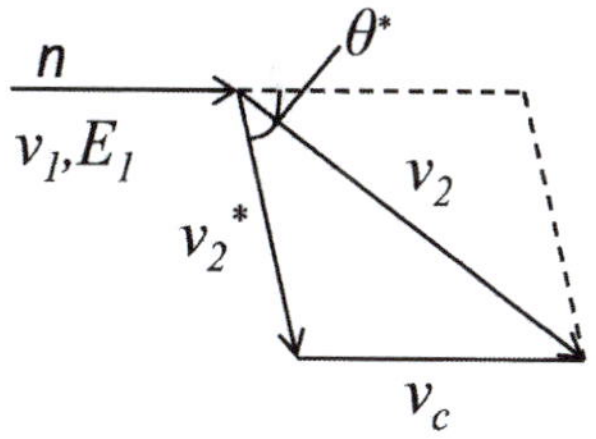

Fig. 8.13 Kinematics of scattering

$$v_2^2 = v_2^{*2} + v_c^2 + 2v_2^* v_c \cos\theta^* \tag{1}$$

$$v_2^2 = v_1^2(1 + \cos\theta^*)/2$$

$$(\because v_c = v_2^* = v_1/2)$$

or

$$E_2 = \frac{E_1}{2}(1 + \cos\theta^*) \tag{2}$$

Let the neutrons scattered between the angles θ^* and $\theta^* + \mathrm{d}\theta^*$ in the CMS appear with energy between E_2 and $E_2 + \mathrm{d}E$ in the LS.
Differentiating (2), holding E_1 as constant.

$$\mathrm{d}E_2 = \frac{E_1}{2}\mathrm{d}\cos\theta^* \tag{3}$$

The mean value

$$\left\langle \ln\frac{E_1}{E_2} \right\rangle = \frac{\int \ln\left(\frac{E_1}{E_2}\right) 2\pi \sin\theta^*\,\mathrm{d}\theta^*}{\int 2\pi \sin\theta^*\,\mathrm{d}\theta^*} \tag{4}$$

Writing $\sin\theta^* d\theta^* = -d(\cos\theta^*)$, and using (3), (4) becomes

$$\left\langle \ln\frac{E_1}{E_2}\right\rangle = \int_0^{E_1} \ln\left(\frac{E_1}{E_2}\right)\frac{dE_2}{E_1} = -\int \ln\left(\frac{E_2}{E_1}\right)\frac{dE_2}{E_1}$$

$$= -\frac{E_2}{E_1}\ln\frac{E_2}{E_1} + \frac{E_2}{E_1}\Big|_0^{E_1}$$

At the upper limit the value is 1. At the lower limit, the second term gives 0. The first term also contributes zero because $x\ln x$ in the limit $x \to 0$ gives zero. Thus,

$$\left\langle \ln\frac{E_1}{E_2}\right\rangle = 1.$$

8.85 (a) The number of collisions required is

$$n = \frac{1}{\xi}\ln\frac{E_0}{E_n}$$

The average logarithmic energy decrement

$$\xi = 1 + \frac{(A-1)^2}{2A}\ln\frac{A-1}{A+1}$$

For graphite ($A = 12$), $\xi = 0.158$

$$\therefore n = \frac{1}{0.158}\ln\left(\frac{2\times 10^6}{0.025}\right) = 115$$

(b) Slowing down time

$$t = \frac{\sqrt{2m}}{\xi\Sigma_s}[E_f^{-1/2} - E_i^{-1/2}]$$

$$\cong \frac{\sqrt{2mc^2/E_f}}{c\xi\Sigma_s} \qquad (\because E_i >> E_f)$$

Inserting the values, $mc^2 = 940\times 10^6$ eV, $E_f = 0.025$ eV, $\xi = 0.158$ and $\Sigma_s = 0.385\,\text{cm}^{-1}$ for graphite, we find the slowing down time $t = 1.5\times 10^{-4}\,s$.

8.86 If N_u is the number of Uranium atom per cm^3 and N_0 is the number of ^{238}U atoms per cm^3, then $N_0 = \frac{139}{140}N_u$. Further, $N_m/N_u = 400$.
Therefore, $N_m/N_0 = 400\times 140/139 = 402.9$
Thermal utilization factor (f);

$$f = \frac{\Sigma_a(U)}{\Sigma_a(U) + \Sigma_a(m)} = \frac{N_u\sigma_a(U)}{N_u\sigma_a(U) + N_m\sigma_a(m)}$$

$$= \frac{1}{1 + \frac{N_m\sigma_a(m)}{N_u\sigma_a(U)}} = \frac{1}{1 + \frac{402.9\times 0.0032}{7.68}} = 0.857$$

We can now calculate p, the resonance escape probability.

$$\Sigma_s/N_0 = \frac{\Sigma_s(U) + \Sigma_s(m)}{N_0} = \sigma_s(U) + \frac{N_m}{N_u}\sigma_a(M)$$
$$= 8.3 + 402.9 \times 4.8 = 1942 \text{ b}$$

We can use the empirical relation for the effective resonance integral (ERI)

$$\int_E^{E_0} (\sigma_a)_{eff} \frac{dE}{E} = 3.85\,(\Sigma_s/N_0)^{0.415} = 3.85(1942)^{0.145} = 89 \text{ b}$$

We have ignored the contribution of Uranium to the scattering as its inclusion hardly changes the result. Thus $\Sigma_0/N_0 \simeq \Sigma_s/N_0 = \frac{N_m}{N_0}\sigma_s(M) = 402.9 \times 4.8 = 1934$

$$p = \exp\left[-(\text{ERI}) \Big/ \frac{\Sigma_0 \xi}{N_0}\right] = \exp\left(-\frac{89}{1934 \times 0.158}\right) = 0.747$$

The reproduction factor

$k_\infty = \xi n f p = (1.0)(1.34)(0.857)(0.747) = 0.858$

Thus, $k_\infty < 1$, and so the reactor cannot go critical.

8.87 The spatial distribution was derived for Problem 8.78.

$$\phi(r) = \frac{3Q}{4\pi \lambda_{tr}} \frac{e^{-r/L}}{r} \quad (1)$$

If 1% of the neutrons are to escape then

$$\frac{\phi(r)}{Q} = \frac{1}{100} = \frac{3}{4\pi \lambda_{tr}} \frac{e^{-r/L}}{r} \quad (2)$$

$$L = \left(\frac{\lambda_{tr}\lambda_a}{3}\right)^{1/2}$$

$$\lambda_s = \frac{1}{\Sigma_s} = \frac{A}{\sigma_s N_0 \rho} = \frac{9}{5.6 \times 10^{-24} \times 6.02 \times 10^{23} \times 1.85} = 1.443 \text{ cm}$$

$$\lambda_s = \frac{1}{\Sigma_a} = \frac{A}{\sigma_a N_0 \rho} = \frac{9}{10 \times 10^{-27} \times 6.02 \times 10^{23} \times 1.85} = 808 \text{ cm}$$

$$\lambda_{tr} = \frac{\lambda_s}{1 - \frac{2}{3A}} = \frac{1.443}{1 - \frac{2}{3 \times 9}} = 1.564 \text{ cm}$$

$$L = \left(\frac{1.564 \times 808}{3}\right)^{1/2} = 20.52 \text{ cm}$$

Inserting the values of λ_{tr} and L in (2) and solving for r, we get $r = 9.6$ cm. Thus the radius ought to be greater than 9.6 cm.

At the surface the neutron density corresponding to $r = 9.6\,\text{cm}$ and mean neutron velocity $2.2 \times 10^5\ \text{cm s}^{-1}$

$$n(r) = \frac{\phi}{v} = \frac{3Q}{4\pi\lambda_{\text{tr}} v}\frac{e^{-9.6/L}}{9.6} = 2.93 \times 10^{-9} Q\,\text{cm}^{-3}$$

8.88 Consider the diffusion equation

$$\frac{\partial n}{\partial t} = S + \frac{\lambda_{\text{tr}}}{3}\nabla^2\phi - \phi\Sigma a \tag{1}$$

where n is the neutron density, S is the rate of production of neutrons/ cm^3/s, $\phi\Sigma a$ is the absorption rate/ cm^3/s and $\frac{\lambda_{\text{tr}}\nabla^2\phi}{3}$ represents the leakage of neutrons. Σa is the macroscopic cross-section, ϕ is the neutron flux and λ_{tr} is the transport mean free path.

Since it is a steady state, $\frac{\partial n}{\partial t} = 0$. Further $S = 0$ because neutrons are not produced in the region of interest. As we are interested only in the x-direction the Laplacian reduces to $\frac{\text{d}^2}{\text{d}x^2}$. Thus (1) becomes

$$\frac{\lambda_{\text{tr}}}{3}\frac{\text{d}^2\phi}{\text{d}x^2} - \phi\Sigma a = 0 \tag{2}$$

or

$$\frac{\text{d}^2\phi}{\text{d}x^2} - K^2\phi = 0 \tag{3}$$

where

$$K^2 = \frac{3\Sigma a}{\lambda_{\text{tr}}} = \frac{3}{\lambda_a\lambda_{\text{tr}}} \tag{4}$$

λ_a being the absorption mean free path.The solution of (3) is

$$\phi = C_1 e^{Kx} + C_2 e^{-Kx} \tag{5}$$

where C_1 and C_2 are constants of integration. The condition that the flux should be finite at any point including at infinity means that $C_1 = 0$. Therefore, (5) becomes

$$\phi = C_2 e^{-Kx} \tag{6}$$

We can now determine C_2. Consider a unit area located in the YZ plane at a distance x from the plane source as in Fig. 8.14. On an average half of the neutrons will be travelling along the positive x-direction. As $x \to 0$, the net current flowing in the positive x-direction would be equal to $\frac{1}{2}$Q; the diffusion of neutrons through unit area at $x = 0$ would have a cancelling effect because from symmetry equal number of neutrons would diffuse in the opposite direction.

Fig. 8.14

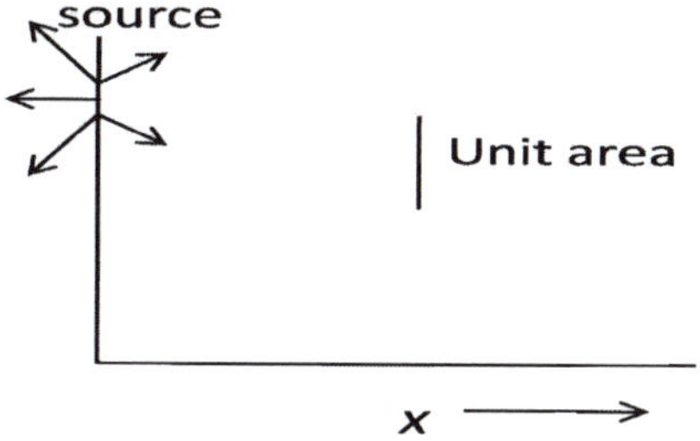

The net current,

$$J = -\frac{\lambda_{tr}}{3}\frac{\partial \phi}{\partial x} = A_2 K \frac{\lambda_{tr}}{3} e^{-kx}$$

As $x \to 0$, $J = \frac{Q}{2} = A_2 K \frac{\lambda_{tr}}{3}$

whence $A_2 = \frac{3Q}{2K\lambda_{tr}}$

The complete solution is

$$\phi = \frac{3Q}{2K\lambda_{tr}} e^{-Kx} \tag{7}$$

8.89 Thermal diffusion time is given by

$$t = \frac{\lambda_a}{v} = \frac{1}{v\Sigma_a}$$

Now $\Sigma_a = \sigma_a \frac{N_0 \rho}{A} = \frac{0.003 \times 10^{-24} \times 6.02 \times 10^{23} \times 1.62}{12}$

$= 2.44 \times 10^{-3}\,\text{cm}^{-1}$

$$t = \frac{1}{2.2 \times 10^3 \times 2.44 \times 10^{-3}} = 0.19\,\text{s}$$

8.90 The generation time for neutrons in a critical reactor is calculated from the formula

$$t = \frac{1}{\Sigma_a < v > (1 + L^2 B^2)}$$

$$= \frac{1}{0.0006 \times 2.2 \times 10^5 (1 + 870 \times 0.0003)}$$

$= 6 \times 10^{-3}\,\text{s}$

8.91 Let N be the number of boron atoms/cm^3. Ignore the scattering of neutrons in boron.

$$\Sigma_a(\text{graphite}) = \frac{1}{\lambda_a(C)} = \frac{1}{2,700} = 3.7 \times 10^{-4}\,\text{cm}^{-1}$$

$$\Sigma_a = \Sigma_a(\text{C}) + \Sigma_a(\text{B}) = \Sigma_a(\text{C}) + \sigma_a(\text{B})\text{N}$$

$$= 3.7 \times 10^{-4} + 755 \times 10^{-24}\,\text{N} \quad (1)$$

$$\lambda_{tr} = \frac{\lambda_s}{1 - \frac{2}{3A}} = \frac{2.7}{1 - \frac{2}{3\times 12}} = 2.859\,\text{cm} \quad (2)$$

The attenuation dependence of $e^{-0.03x}$ implies that the diffusion length

$$L = \frac{1}{0.03} = 33.33\,\text{cm} \quad (3)$$

$$\text{But } L^2 = \frac{\lambda_{tr}\lambda_a}{3} = \frac{\lambda_{tr}}{3\Sigma_a} \quad (4)$$

Combining (1), (2), (3), and (4) and solving for N, we find
$N = 5.83 \times 10^{17}$ boron atoms/cm^3

8.92 Refer to solution of Problem 8.84 with the change of notation.

$$v_1{}^2 = v_1{}^{*2} + v_c{}^2 + 2\,v_1{}^*\,v_c\,\cos\,\theta^*$$

$$= \frac{v_0^2}{(A+1)^2}(A^2 + 1 + 2A\cos\theta^*)$$

where we have used the relations

$v_1{}^* = A\,V_0/(A+1)$ and $v_c = v_0/(A+1)$

We can then write

$$\frac{E_1}{E_0} = \frac{A^2 + 2A\cos\theta^* + 1}{(A+1)^2}$$

$$< E_1/E_0 >= \frac{1}{(A+1)^2}\int_{-1}^{+1}(A^2 + 2A\cos\theta^* + 1)\frac{1}{2}\text{d}(\cos^*)$$

$$= \frac{(A^2+1)}{(A+1)^2}$$

Therefore, $< E_f >= E_i\frac{(A^2+1)}{(A+1)^2}$

Let the neutron energy be E_n after n collisions

$$\frac{E_n}{E_0} = \frac{E_1}{E_0}\cdot\frac{E_2}{E_1}\cdot\frac{E_3}{E_2}\cdots\frac{E_n}{E_{n-1}} = \left(\frac{E_1}{E_0}\right)^n$$

Therefore $\frac{0.025\,\text{eV}}{2\times 10^6\,\text{eV}} = \left[\frac{A^2+1}{(A+1)^2}\right]^n = \left(\frac{145}{169}\right)^n$

where we have put $A = 12$ for graphite. Taking logarithm on both sides and solving for n, we obtain $n = 119$. Compare this value with $n = 115$ obtained from the average logarithmic decrement (Problem 8.85).

8.3.15 Fusion

8.93 The minimum energy of neutrino is zero when d and e^+ are emitted in opposite direction and the neutrino carries zero energy. The maximum energy of neutrino corresponds to a situation in which e^+ is at rest and d moves in a direction opposite to neutrino.

$Q = 2 \times 938.3 - 1875.7 - 0.51 = 0.39\,\text{MeV}$

$T_d + T_\nu = Q = 0.39$ (energy conservation) (1)

${P_{ed}}^2 = {P_\nu}^2$ (momentum conservation) (2)

$2m_d T_d = {T_\nu}^2$ (3)

Equations (1) and (3) can be solved to give $T_\nu(\max) = 0.38996\,\text{MeV}$. Thus neutrino energy will range from zero to 0.39 MeV.

8.94 $E_n = \frac{Qm_{He}}{m_{He}+m_n} = \frac{3.2\times 3}{3+1} = 2.4\,\text{MeV}$

8.95 (a) $E = \frac{1.44 z_1 z_2}{r} = \frac{1.44\times 1\times 1}{80} = 0.018\text{MeV} = 18\,\text{keV}$

(b) Equating the kinetic energy to heat energy

$E = \frac{3}{2}kT$

$18 \times 1.6 \times 10^{-16}\text{J} = \frac{3}{2} \times 1.38 \times 10^{-23}\text{T}$

$\text{T} = 1.39 \times 10^8\,\text{K}$

(c) The temperature can be lowered because with smaller energy Coulomb barrier penetration becomes possible.

8.96 The Lawson criterion is just satisfied if

$$L = \frac{\text{energy output}}{\text{energy input}} = \frac{n_d^2 < \sigma_{dt} v > t_c Q}{6 n_d kT} = \frac{n_d < \sigma_{dt} v > t_c Q}{6kT} = 1$$

Substitute $n_d = 7 \times 10^{18}\text{m}^{-3}$, $kT = 10\,\text{keV} = 10^4\,\text{eV}$

$< \sigma_{dT}.v \geq 10^{-22}\text{m}^3\text{s}^{-1}$, $Q = 17.62 \times 10^6\,\text{eV}$ to find $t_c = 4.86\,\text{s}$

Chapter 9
Particle Physics – I

9.1 Basic Concepts and Formulae

Interactions and decay

Probability for scattering

$$P = t/\lambda \tag{9.1}$$

where t is the thickness and λ is the mean free path.

Attenuation of beam intensity due to interaction

$$I = I_o e^{-\mu x} \tag{9.2}$$

The absorption coefficient μ is in metre^{-1} if x is in metre.

$$\begin{aligned}
&\mu \text{ is in cm}^2/\text{g if } x \text{ is in g/cm}^2.\\
&x \text{ in g/cm}^2 = (x \text{ in cm})(\text{density in g/cm}^3)\\
&\mu = \Sigma(\text{macroscopic cross - section})\\
&\Sigma = \sigma N_0 \rho / A
\end{aligned} \tag{9.3}$$

where σ = microscopic cross-section, N_o = Avogadro's number, ρ = density and A = Atomic or molecular weight.

Attenuation of beam intensity due to decay

$$I = I_o e^{-t/\tau} = I_o e^{-s/v\gamma\tau_0} \tag{9.4}$$

where $v = \beta c$ is the particle velocity, γ is the Lorentz factor, s is the distance travelled and τ_o is the proper life time, that is the life time in the rest frame of the particle.

Energy in the center of mass system (CMS)

In the collision of particle of mass m_1, of total energy E_1 in the lab system with m_2 initially at rest, total energy E^* available in the CMS is given by

$$E^* = \left(m_1^2 + m_2^2 + 2m_2E_1\right)^{1/2} \tag{9.5}$$

Division of energy in the decay $A \to B + C$, at rest.

Total energy carried by B,

$$E_B = \frac{m_A^2 + m_B^2 - m_C^2}{2m_A} \tag{9.6}$$

Circular accelerators

In static magnetic field, a charged particle is not accelerated but is only bent into a circular path if the field is perpendicular to the plane of the path. Otherwise the particle goes into a helical path.

The radius of curvature r is related to the momentum p by

$$P = 0.3\,Br \tag{9.7}$$

where p is in Gev/c, r in meters, and the field B in Tesla. (1 Tesla = 10^4 gauss).

$$\text{If } n = \gamma - 1 \text{ then } R = \frac{mc}{qB}(n^2 + 2n)^{1/2} \tag{9.8}$$

Betatron

Energy gained

$$\Delta T = \mathrm{e}\frac{\Delta\Phi}{\Delta t} \tag{9.9}$$

where $\Delta\varphi/\Delta t$ is the rate of change of flux.

Cyclotron

$\omega_0 = \frac{qB}{m}$ (Resonance condition)
where $\omega_0 = 2\pi f_0$
Energy at the extraction point

$$T = \frac{(qBR)^2}{2m} \tag{9.10}$$

Synchrocyclotron

$$\omega_0 = \frac{qB}{m_0} \tag{9.11}$$

$$\omega = \frac{qB}{m} \tag{9.12}$$

$$\frac{\omega}{\omega_0} = \frac{m_0}{m} = \frac{1}{\gamma} \tag{9.13}$$

Synchrotron

The synchrotron radiation loss per turn

$$\Delta E = \frac{4\pi}{3R} \frac{e^2}{4\pi\epsilon_0} \left(\frac{E}{mc^2}\right)^4 \tag{9.14}$$

Linear Accelerator

Total length

$$L = \frac{1}{2f} \left(\frac{2\,\mathrm{eV}}{m}\right)^{1/2} \Sigma\sqrt{n} \tag{9.15}$$

where n is the number of drift tubes.

Colliders

Luminosity

$$L = \frac{nN_1N_2f}{A} \tag{9.16}$$

The number of interactions per second

$$N = L\sigma \tag{9.17}$$

where σ is the interaction cross-section of a given type

N_1 or N_2 = the number of particles/bunch in each beam
A = area of cross-section of intersecting beams
n = number of bunches/beam
f = frequency of revolution

Detectors

Proportional counter

Gas multiplication factor

$$M = \frac{CV}{Ne} \tag{9.18}$$

where N is number of ion pairs released.

G.M counter

$$E = \frac{V}{r\ \ln(b/a)} \tag{9.19}$$

where E is the electric field, V is the applied voltage, r is the distance from the anode, b and a are the diameters of the cathode and anode.

Dead time of a counter(τ)

$$n = \frac{n_0}{1 - n_0 \tau} \tag{9.20}$$

where n = true counts, n_0 = observed counts.

Double source method

$$\tau = \frac{N_1 + N_2 - N_{12} + B}{2(N_1 - B)(N_2 - B)} \tag{9.21}$$

where N_1 and N_2 are the counting rates from individual sources, N_{12} is the counting rate from the combined sources and B is the background counting rate.

Cerenkov counter

$$\cos\theta = \frac{1}{\beta n} \tag{9.22}$$

where θ is the opening angle, n = index of refraction

$$\text{Threshold velocity } \beta = \frac{1}{n} \tag{9.23}$$

9.2 Problems

9.2.1 System of Units

9.1 Show that 1 kg = 5.6×10^{26} GeV/c^2

9.2 In the natural system of units ($\hbar = c = 1$) show that
(a) 1 m = 5.068×10^{15} GeV^{-1}
(b) 1 GeV^{-2} = 0.389 mb
(c) 1 s = 1.5×10^{24} GeV^{-1}

9.3 In the natural system of units show that
(a) The Compton wavelength for an electron is $\lambda_c = 1/m_e$
(b) The Bohr radius of a hydrogen atom is $1/\alpha\ m_e$
(c) The velocity of an electron in the ground state of hydrogen atom is αc

Also, calculate the numerical values for the above expressions where $\alpha = 1/137$ is the fine structure constant. The electron mass is $m_e = 0.511\,\text{MeV}$.

9.4 One of the bound states of positronium has a lifetime given in natural units by $\tau = 2/m\alpha^5$ where m is the mass of the electron and α is the fine structure constant. Using dimensional arguments introduce the factors $\hbar$ and c and determine τ in seconds.

9.5 The V-A theory gives the formula for the width (Γ_μ) of the muon decay in natural units.

$\Gamma_\mu = \hbar/\tau = G_F{}^2\, m_\mu{}^5/192\pi^3$

Convert the above formula in practical units and calculate the mean life time of muon

$[(G_F/(\hbar c)^3 = 1.116 \times 10^{-5}\,\text{GeV}^{-2},\; m_\mu\, c^2 = 105.659\,\text{MeV}]$

9.2.2 Production

9.6 An ultra high energy electron ($\beta \approx 1$) emits a photon. (a) Derive an expression to express the emission angle θ in the lab system in terms of θ^*, the angle of emission in the rest frame of the electron. Also, (b) Show that half of the photons are emitted within a cone of half angle $\theta \approx 1/\gamma$.

9.7 Show that in a fixed target experiments, the energy available in the CMS goes as square root of the particle energy (relativistic) in the Lab system.

9.8 A positron with laboratory energy 50 GeV interacts with the atomic electrons in a lead target to produce $\mu^+\mu^-$ pairs. If the cross-section for this process is given by $\sigma = 4\pi\alpha^2\, \hbar^2 c^2/3(E_{CM})^2$, calculate the positron's interaction length. The density of lead is $\rho = 1.14 \times 10^4\,\text{kg m}^{-3}$

9.9 It is desired to investigate the interaction of e^+ and e^- in flight, yielding a nucleon-antinucleon pair according to the equation of $e^+ + e^- \rightarrow p + p^-$. (a) To what energy must the positrons be accelerated for the reaction to be energetically possible in collisions with stationary electrons. (b) How do the energy requirements change if the electrons are moving, for example in the form of a high energy beam? (c) What is the minimum energy requirement?

$(m_e\, c^2 = 0.51\,\text{MeV},\; M_p\, c^2 = 938\,\text{MeV})$

[University of Bristol 1967]

9.2.3 Interaction

9.10 A proton with kinetic energy 200 MeV is incident on a liquid hydrogen target. Calculate the centre-of-mass energy of its collision with a nucleus of hydrogen. What kinds of particles could be produced in this collision?

[University of Wales, Aberystwyth 2003]

9.11 Estimate the thickness of iron through which a beam of neutrinos with energy 200 GeV must travel if 1 in 10^9 of them is to interact. Assume that at high energies the neutrino-nucleon total cross-section is 10^{-42} E_ν m^2, where E_ν is the neutrino energy is in GeV. The density of iron is 7,900 kg m^{-3}. [Mass of nucleon = 1.67×10^{-27} kg.]

9.12 (a) The cross-section for scattering of muons with air at atmospheric pressure is 0.1 barns, and the natural lifetime of muons 2.2×10^{-6} s. Explain what is meant by the terms elastic scattering and lifetime. Which of these factors limits the distance over which a beam of muons can travel in air in a laboratory, if the muon velocity is 10^6 ms^{-1}?
(Assume number density of air at STP = 2.69×10^{25} m^{-3})
(b) Given your answer to (a), why is it possible to detect showers of muons at ground level caused by the impact of primary cosmic ray particles with air at around 12 km altitude? Given that the mean energy of muons detected at ground level is $\approx$ 4 GeV, calculate the distance (in air) over which the number of such muons in a beam would reduce by a factor of e. What kind of interactions contribute to the scattering cross-section for these particles? [1 barn = 10^{-28} m^2; mass of muon = 105 MeV/c^2; number density of air at STP = 2.69×10^{25} m^{-3}]

9.13 Obtain an approximate value for the interaction length (in cm) of a fast proton in lead from the following data: $r_0 = 1.3 \times 10^{-13}$ cm, Atomic weight of lead = 207, Density of lead = 11.3 g cm^{-3}, Avogadro's number = 6×10^{23} molecules mole^{-1}

[University of Durham 1962]

9.14 A liquid hydrogen target of volume 125 cm^3 and density 0.071 g cm^{-3} is bombarded with a mono-energetic beam of negative pions with a flux 2×10^7 m^{-2} s^{-1} and the reaction $\pi^- + p \rightarrow \pi^0 + n$ observed by detecting the photons from the decay of the π^0. Calculate the number of photons emitted from the target per second if the cross-section is 40 mb.

9.15 A beam of π^+ mesons contaminated with μ^+ mesons is passed through an iron absorber. Given that the interaction cross-section of π^+ mesons with iron is 600 mb/nucleus, calculate the thickness of iron necessary to attenuate the π^+ beam by a factor of 500. Explain why muons will be reduced to a much less extent. (The density of iron is 7,900 kg m^{-2}, the atomic mass being 55.85 amu)

9.16 A neutrino of high energy ($E_0 >> m$) undergoes an elastic scattering with stationary electron of mass m. The electron recoils at an angle ϕ with energy T. Show that for small angle, $\phi = (2m/T)^{1/2}$

9.17 Why the following reactions can not proceed as strong interactions

(a) $\pi^- + p \rightarrow K^- + \Sigma^+$
(b) $d + d \rightarrow {}^4\text{He} + \pi^0$?

9.18 From the data given calculate the mass of the hyper-fragment ${}_{\Lambda}\text{He}^5$
Mass of Λ hyperon is 1115.58 MeV/c^2, Mass of He^4 is 3727.32 MeV/c^2
Binding energy B_{Λ} for He^5 is 3.08 MeV

[University of Dublin 1968]

9.19 Show that for a high energy electron scattering at an angle θ, the value of Q^2 is given approximately by

$$Q^2 = 2E_{\text{i}}E_{\text{f}}(1 - \cos\theta)$$

where E_{i} and E_{f} are the initial and final values of the electron energy and Q^2 is the four-momentum transfer squared. State when this approximation is justified.

9.2.4 Decay

9.20 Consider the decay process $K^+ \to \pi^+\pi^0$ with the K^+ at rest. Find
(a) the total energy of the π^0 meson
(b) its relativistic kinetic energy
The rest mass energy is 494 MeV for K^+, 140 MeV for π^+, 135 MeV for π^0.

9.21 A collimated beam of π-mesons of 100 MeV energy passes through a liquid hydrogen bubble chamber. The intensity of the beam is found to decrease with distance s along its path as exp(-as) with $a = 9.1 \times 10^{-2}\,\text{m}^{-1}$. Hence calculate the life time of the π-meson (Rest energy of the π-meson 139 MeV)

[University of Durham 1960]

9.22 One of the decay modes of K^+ mesons is $K^+ \to \pi^+ + \pi^+ + \pi^-$. What is the maximum kinetic energy that any of the pions can have, if the K^+ decays at rest? Given $m_{\text{K}} = 966.7\,m_{\text{e}}$ and $m_{\pi} = 273.2\,\text{m}_{\text{e}}$

9.23 Show that the energy of the neutrino in the pion rest frame, E'_{ν}, can be written in natural units as

$$E'_{\nu} = \left(m_{\pi}^2 - m_{\mu}^2\right)/2m_{\pi}$$

where m_{π} and m_{μ} are the masses of the charged pion and the muon, respectively. (You may assume here that the neutrino mass is negligible)

[University of Cambridge, Tripos 2004]

9.24 Find the maximum energy of neutron in the decay of Σ^+ hyperon at rest via the scheme, $\Sigma^+ \to n + \mu^+ + \nu_{\mu}$. The masses of Σ^+, n, μ^+ and ν_{μ} are 1,189 MeV, 939 MeV, 106 MeV and zero, respectively.

9.25 A charged kaon, with an energy of 500 MeV decays into a muon and a neutrino. Sketch the decay configuration which leads to the neutrino having the maximum possible momentum, and calculate the magnitude of this value (mass of charged kaon is 494 MeV/c^2, mass of muon is 106 MeV/c^2)

9.26 Heavy mesons, $M = 950\,m_e$ produced in nuclear interactions initiated by a more energetic beam of π-meson, have an energy of 50 MeV. Their tracks up to the point of decay have a mean length of 1.7 m. Calculate their mean life time.

[University of Durham 1960]

9.27 A pion beam from an accelerator target has momentum 10 GeV/c. What fraction of the particles will not have decayed into muons in a pathlength of 100 m?
Out of the pions decays one muon of a 8 GeV/c and a neutrino are produced at the beginning of the flight path. Assuming that the decay particles follow the same path, calculate the difference in arrival times at the end of the path. (Pion mass = 139.6 MeV, muon mass = 105.7 MeV. Mean life time of pion $= 2.6 \times 10^{-8}$ s, $c = 3 \times 10^{8}$ m s^{-1})

[University of Durham 1972]

9.28 Pions in a beam of energy 5 GeV decay in flight. What are the maximum and minimum energies of the muons from these decays? ($m_\pi = 139.5\,\mathrm{MeV/c^2}$; $m_\mu = 105.7\,\mathrm{MeV/c^2}$)

9.29 Assuming that a drop in intensity by a factor less than 10 is tolerable, show that a 1 GeV/c K^+ meson beam can be transported over 10 m without a serious loss of intensity due to decay, while a Λ -hyperon beam of the same momentum after the same distance will not have useful intensity (Take masses of K^+ meson and Λ-hyperon to be 0.5 and 1 GeV/c^2, respectively and their lifetimes 10^{-8} and 2.5×10^{-10} s respectively.

[University of Durham 1970]

9.30 If a particle has rest mass m_0 and momentum p, show that the distance traveled in one lifetime is $d = pT_0/m_0$ where T_0 is the life time in the frame of reference in which the particle is at rest.

[University of Dublin 1968]

9.31 A beam of muon neutrinos is produced from the decay of charged pions of $E_\pi = 20$ GeV. Show that the relationship between the neutrino energy in the laboratory frame, E_ν, and its angle relative to the pion beam θ, for small θ, is

$$E_\nu = \frac{E_\pi}{(1+\gamma^2\theta^2)}\left\{1 - {m_\mu}^2/{m_\pi}^2\right\}$$

where E_π is the energy of the pion and $\gamma = E_\pi/m_\pi >> 1$

[University of Cambridge, Tripos 2004]

9.32 It is intended to use a charged mono-energetic hyperon beam to perform scattering experiments off liquid hydrogen. Assuming that the beam transport system must have a minimum length of 20 m, calculate the minimum momentum of a Σ beam such that 1% of the hyperons accepted by the transport system arrive at the hydrogen target ($\tau = 0.8 \times 10^{-10}$ s, $m_\Sigma = 1.19\,\mathrm{GeV/c^2}$). What measurements would you make to identify the elastic scattering in a $(\Sigma^- p)$ collision?

[University of Manchester]

9.33 The energy spectrum for the electrons emitted in muon decay is given by

$$\frac{d\omega}{dE_e} = \frac{2G_F^2(m_\mu c^2)^2 E_e^2}{(2\pi)^3(\hbar c)^6}\left(1 - \frac{4E_e}{3m_\mu c^2}\right)$$

where the electron mass is neglected. Calculate the most probable energy for the electron. Show on a diagram the orientation of momenta of the decay product particles and their helicitis when $E_e \approx m_\mu c^2/2$. Furthermore, show the helicity of the muon. Integrate the energy spectrum to find the total decay width of the muon. Hence compute the muon mean lifetime in seconds $[G_F/(\hbar c)^3 = 1.166 \times 10^{-5}\,\text{GeV}^{-2}]$

9.34 (a) Explain the following statements. The mean lifetime of the π^+ meson is 2.6×10^{-8} s while that of π^0 is 0.8×10^{-16} s.
(b) The π^+ and π^- mesons are of equal mass, but the Σ^+ and Σ^- baryon masses differ by 8 MeV/c^2
(c) The mean life of the Σ^0 baryon is many orders of magnitude smaller than those of the Λ and Ξ^0 baryons.

9.35 In a bubble chamber two tracks originate from a common point, one caused by a proton of 440 MeV/c and the other one by π^- meson of momentum 126 MeV/c. The angle between the tracks is 64°. Determine the mass of the unknown particle and identify it.

9.2.5 Ionization Chamber, GM Counter and Proportional Counters

9.36 The dead time of a counter system is to be determined by taking measurements on two radioactive sources individually and collectively. If the pulse counts over a time interval t are, respectively, N_1, N_2 and N_{12}, what is the value of the dead time?

9.37 An ionization chamber is connected to an electrometer of capacitance 0.5 μμ F and voltage sensitivity of 4 divisions per volt. A beam of α-particles causes a deflection of 0.8 divisions. Calculate the number of ion pairs required and the energy of the source of the α-particles (1 ion pair requires energy of 35 eV, $e = 1.6 \times 10^{-19}$ Coulomb)

[Osmania University]

9.38 An ionization chamber is used with an electrometer capable of measuring 7×10^{-11} A to assay a source of 0.49 MeV beta particles. Assuming saturation conditions and that all the particles are stopped within the chamber, calculate the rate at which the beta particles must enter the chamber to just produce a measurable response. Given the ionization potential for the gas atoms is 35 eV.

9.39 Estimate the gas multiplication required to count a 2 MeV proton which gives up all its energy to the chamber gas in a proportional counter. Assume that the amplifier input capacitance in parallel with the counter is 1.5×10^{-9} F and that its input sensitivity is 1 mV. Energy required to produce one ion pair is 35 eV.

9.40 Calculate the pulse height obtained from a proportional counter when a 14 keV electron gives up all its energy to the gas. The gas multiplication factor of the proportional counter is 600, capacitance of the circuit is 1.0×10^{-12} F and energy required to produce an ion pair is 35 eV

9.41 The plateau of a G.M. Counter working at 800 V has a slope of 2.0% count rate per 100 V. By how much can the working voltage be allowed to vary if the count rate is to be limited to 0.1%?

9.42 An organic-quenched G.M. tube has the following characteristics.
Working voltage 1,000 V, Diameter of anode 0.2 mm, Diameter of cathode 2.0 cm. Maximum life time 10^9 counts.

What is the maximum radial field and how long will it last if used for 30 h per week of 3,000 counts per minute?

[University of New Castle]

9.43 A G.M. tube with a cathode and anode of 2 cm and 0.10 mm radii respectively is filled with Argon gas to 10 cm Hg pressure. If the tube has 1.0 kV applied across it, estimate the distance from the anode, at which electron gains just enough energy in one mean free path to ionize Argon. Ionization potential of Argon is 15.7 eV and mean free path in Argon is 2×10^{-4} cm at 76 cm Hg pressure.

9.44 A S^{35} containing solution had a specific activity of 1m Curie per ml. A 25 ml sample of this solution mass was assayed.

(a) in a Geiger-Muller counter, when it registered 2,000 cpm with a background count of 750 in 5 min; and

(b) in a liquid scintillation counter, where it registered 9,300 cpm with a background count of 300 in 5 min.

Calculate the efficiency as applied to the measurement of radioactivity and discuss the factors responsible for the difference in efficiency of these two types of counters.

[University of Dublin]

9.45 The dead time of a G.M. counter is 100 μs. Find the true counting rate if the measured rate is 10,000 counts per min.

[Osmania University]

9.46 A pocket dosimeter has a capacitance of 5.0 PF and is fully charged by a potential of 100 V. What value of leakage resistance can be tolerated if the meter is not to lose more than 1% of full charge in 1 day?

9.47 A G.M tube with a cathode 4.0 cm in diameter and a wire diameter of 0.01 cm is filled with argon in which the mean-free-path is 8×10^{-4} cm. Given that the ionizing potential for Argon is 15.7 V, calculate the maximum voltage which must be applied to produce the avalanche.

9.48 An electrometer with a capacitance 0.4 PF and 5.0 PF chamber are both charged to a potential of 600 V. Calculate the potential of the combined system after 8×10^9 ions have been collected in the chamber.

9.2.6 Scintillation Counter

9.49 Pions and protons, both with momentum 4 GeV/c, travel between two scintillation counters distance L m apart. What is the minimum value of L necessary to differentiate between the particles if the time of flight can be measured with an accuracy of 100 ps?

9.50 In the historical discovery of antiproton, negatively charged particles of protonic mass had to be isolated from a heavy background of K^-, π^- and μ^-. The negatively charged particles which originated from the bombardment of a target with 5.3 GeV protons, were subject to momentum analysis in a magnetic field which permitted only those particles with momentum $p = 1.19\,\text{GeV/c}$ to pass through a telescope system comprising two scintillation counters in coincidence with a separation of d=12 metres between the detectors. Identify the particles whose time of flight in the telescope arrangement was determined as $t = 51 \pm 1$ ns. What would have been the time of flight of π mesons?

9.51 A scintillation spectrometer consists of an anthracene crystal and a 10-stage photomultiplier tube. The crystal yields about 15 photons for each 1 keV of energy dissipated. The photo-cathode of the photomultiplier tube generates one photo-electron for every 10 photons striking it, and each dynode produces 3 secondary electrons. Estimate the pulse height observed at the output of the spectrometer if a 1 MeV electron deposits its energy in the crystal. The capacitance of the output circuit is 1.2×10^{-10} F.

9.52 A sodium iodide crystal is used with a ten-stage photomultiplier to observe protons of energy 5 MeV. The phosphor gives one photon per 100 eV of energy loss. If the optical collection efficiency is 60% and the conversion efficiency of the photo-cathode is 5%, calculate the average size and the standard deviation of the output voltage pulses when the mean gain per stage of the multiplier is 3 and the collector capacity is 12 PF.

[University of Manchester]

9.53 A 400-channel pulse-height analyzer has a dead time $\tau = (17 + 0.5\,K)\,\mu\text{s}$ when it registers counts in channel K. How large may the pulse frequencies become if in channel 100 the dead time correction is not to exceed 10%? Repeat the calculations for the channel 400.

9.54 An anthracene crystal and a 12-stage photomultiplier tube are to be used as a scintillation spectrometer for β-rays. The phototube output circuit has a combined capacitance of 45 PF. If an 8-mV output pulse is desired whenever a 55-keV beta particle is incident on the crystal, calculate the electron multiplication required per stage. Assume perfect light collection and a photo-cathode efficiency of 5% (assume 550 photons per beta particle)

9.55 Figure 9.1 shows the gamma-ray spectrum of ^{22}Na in NaI scintillator. Indicate with explanation the origin of the parts labeled as A, B, C, D and E, Given that ^{22}Na is a positron emitter and emits a γ-ray of 1.275 MeV.

Fig. 9.1

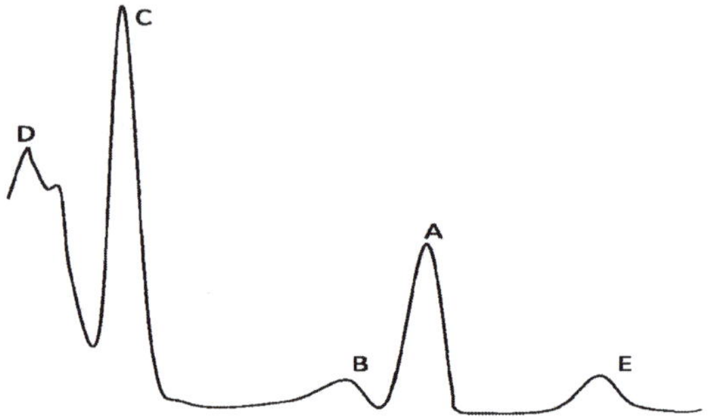

9.56 Assuming that the shape of the photpeak in the scintillation counter is described by the normal distribution, show that the half width at half-maximum is HWHM = 1.177 σ

9.57 The peak response to the 661 keV gamma rays from ^{137}Cs occurs in energy channel 298 and 316. Calculate the standard deviation of the energy, and the coefficient of variation of the energy determination, assuming the pulse analyzer to be linear.

9.2.7 Cerenkov Counter

9.58 Explain what is meant by Cerenkov radiation. How may a Cerenkov detector distinguish between a kaon and a pion with the same energy? A pion of energy 20 GeV passes through a chamber containing CO_2 at STP. Calculate the angle to the electron's path with which the Cerenkov radiation is emitted. (Use the result that the velocity of a relativistic particle $v = c(1 - \gamma^{-2})^{1/2}$. [Mass of pion = 140 MeV, $M_p = 938$ MeV; refractive index of CO_2 at STP = 1.0004]

9.59 An electron incident on a glass block of refractive index 1.5 emits Cerenkov radiation at an angle 45° to its direction of motion. At what speed is the electron travelling?

[University of Cambridge, Tripos 2004]

9.60 Consider Cerenkov radiation emitted at angle θ relative to the direction of a charged particle in a medium of refractive index n. Show that its rest mass energy mc^2 is related to its momentum by $mc^2 = pc(n^2 \cos^2 \theta - 1)^{1/2}$

9.61 In an experiment using a Cerenkov counter, one measures the kinetic energy of a given particle species as $E(\text{kin}) = 420\,\text{MeV}$ and observes that the Cerenkov angle in flint glass of refractive index ($\mu = 1.88$) is $\theta = \text{arc}\cos(0.55)$. What particles are being detected (calculate their mass in m_e units)

9.62 Calculate the number of Cerenkov photons produced by a particle travelling at $\beta = 0.95$ in water ($n = 1.33$) in the response range (3500–5500 A) per unit path length.

9.63 Estimate the minimum length of a gas Cerenkov counter that could be used in threshold mode to distinguish between charged pions and charged kaons with momentum 15 GeV/c. Assume that a minimum of 175 photons need to be radiated to ensure a high probability of detection. Assume also that the radiation covers the whole visible spectrum between 400 and 700 nm and neglect the variation with wavelength of the refractive index of the gas ($n = 1.0004$)

9.64 What type of material would you choose for a threshold Cerenkov counter which is to be sensitive to 900 MeV/c pions but not to 900 MeV/c protons.

9.2.8 Solid State Detector

9.65 A depletion-layer detector has an electrical capacitance determined by the thickness of the insulating dielectric. Estimate the capacitance of a silicon detector with the following characteristics: area $1.5\,\text{cm}^2$, dielectric constant 10, depletion layer 40 μm. What potential will be developed across the capacitance by the absorption of a 5.0 MeV alpha particle which produces one ion pair for each 3.5 eV dissipated?

9.2.9 Emulsions

9.66 The range of protons in C_2 emulsion is given in the following table (Range in microns, energy in MeV). Draw a graph of the Range – Energy-Relation for Deuterons and ^{3}He particles.

R	0	50	100	150	200	250	300	350	400	450	500
E	0	2.32	3.59	4.61	5.48	6.27	7.01	7.69	8.32	8.91	9.47

9.2.10 Motion of Charged Particles in Magnetic Field

9.67 For a relativistic particle of charge e moving in a circular orbit of radius r in a magnetic field B perpendicular to the orbital plane, show that $p = 0.3\,\text{Br}$, where the momentum p is in GeV/c and B in Tesla and r in metres.

9.68 A 10 GeV proton is directed toward the centre of the earth when it is at distance of 10^3 earth radius away. Through what mean transverse magnetic field (assumed uniform) must it move if it is to just miss the earth?

9.69 A proton of kinetic energy 20 keV enters a region where there is uniform electric field of 500 V/cm, acting perpendicularly to the velocity of the proton. What is the magnitude of a superimposed magnetic field that will result in no deflection of the proton?

[University of Manchester]

9.70 A particle of known charge but unknown rest mass is accelerated from rest by an electric field E into a cloud chamber. After it has traveled a distance d, it leaves the electric field and enters a magnetic field B directed at right angles to E. From the radius of curvature of its path, determine the rest mass of the particle and the time it had taken to traverse the distance d.

[University of Durham]

9.71 A proton accelerated to 500 keV enters a uniform transverse magnetic field of 0.51 T. The field extends over a region of space $d = 10$ cm thickness. Find the angle α through which the proton deviates from the initial direction of its motion.

9.72 An electron is projected up, at an angle of 30° with respect to the x-axis, with a speed of 8×10^5 m/s. Where does the electron recross the x-axis in a constant electric field of 50 N/C, directed vertically upwards?

9.73 A proton moving with a velocity 3×10^5 ms^{-1} enters a magnetic field of 0.3 T, at an angle of 30° with the field. What will the radius of curvature of its path be (e/m for proton $\approx 10^8$ C kg^{-1})?

9.74 What energy must a proton have to circle the earth at the magnetic equator? Assume a 1 G magnetic field.

9.75 The average flux of primary cosmic rays over earth's surface is approximately 1 cm^{-2} s^{-1} and their average kinetic energy is 3 GeV. Calculate the power delivered to the earth from cosmic rays in giga watts. (Earth's radius = 6,400 km)

9.76 Kaons with momentum 25 GeV/c and deflected through a collimator slit at a distance of 10 m from a bending magnet 1.5 m long which produces a field of 1.2 T. What should be minimum width of the slit so that it accepts particles of momenta within 1% of the central value?

9.2.11 Betatron

9.77 The orbit of electrons in a betatron has a radius of 1m. If the magnetic field in which the electrons move is changing at the rate of 50 W m^{-2} s^{-1}, calculate the energy acquired by an electron in one rotation. Express your answer in electron volts.

[University of New Castle 1965]

9.78 In a betatron of diameter 72 in operating at 50 c/s the maximum magnetic field is $1.2\,\mathrm{Wb/m^2}$. Calculate
(a) the number of revolutions made each quarter cycle
(b) the maximum energy of the electrons
(c) the average energy gained per revolution

[University of New Castle 1965]

9.2.12 Cyclotron

9.79 A cyclotron is designed to accelerate protons to an energy of 6 MeV. Deduce the expression for the kinetic energy in terms of radius, magnetic field etc and calculate the maximum radius attained.

[University of Bristol 1967]

9.80 If the acceleration potential (peak voltage, 50,000) has a frequency of 1.2×10^7 c/s, find the field strength for cyclotron resonance when deuterons and alpha particles, respectively are accelerated. For how long are the particles accelerated if the radius of orbit at ejection is 30.0 cm? Deuteron rest mass = 2.01 amu. α-particle rest mass = 4.00 amu

[University of New Castle 1965]

9.81 The original (uniform magnetic field) type of fixed frequency cyclotron was limited in energy because of the relativistic increase of mass of the accelerated particle with energy. What percentage increment to the magnetic flux density at extreme radius would be necessary to preserve the resonance condition for protons of energy 20 MeV?

[University of Durham 1972]

9.82 A cyclotron is powered by a 50,000 V, 5 Mc/s radio frequency source. If its diameter is 1.524 m. what magnetic field satisfies the resonance conditions for (a) protons (b) deuterons (c) alpha-particles? Also what energies will these particles attain?

[University of Durham 1961]

9.83 In a synchrocyclotron, the magnetic flux density decreases from 15,000 G at the centre of the magnet to 14,300 G at the limiting radius of 206 cm. Calculate the range of frequency modulation required for deuteron acceleration and the maximum kinetic energy of the deuteron. (The rest mass of the deuteron is 3.34×10^{-24} g)

[University of London 1959]

9.84 Calculate the magnetic field, B and the Dee radius of a cyclotron which will accelerate protons to a maximum energy of 5 MeV if a radio frequency of 8 MHz is available.

9.85 When a cyclotron shifts from deuterons to alpha particles it is necessary to drop the magnetic field slightly. If the atomic masses of ^{2}H and ^{4}He are 2.014102 amu and 4.002603 amu, what is the percentage decrease in field strength that is required?

[Osmania University]

9.86 A synchro-cyclotron has a pole diameter of 4 m and a magnetic field of $1.5\,\mathrm{W\,m^{-2}}$ (15,000 G). What is the maximum energy that can be transmitted to electrons which are struck by protons extracted from this accelerator?

[University of Bristol 1968]

9.87 If the frequency of the dee voltage at the beginning of an accelerating sequence is 20 Mc/s, what must be the final frequency if the protons in the pulse have an energy of 469 MeV?

[University of Durham 1963]

9.2.13 Synchrotron

9.88 At what radius do 30 GeV protons circulate in a synchrotron if the guide field is $1\,\mathrm{W\,m^{-2}}$

9.89 Calculate the orbit radius for a synchrotron designed to accelerate protons to 3 GeV assuming a guide field of 14 kG

[University of Durham 1962]

9.90 What percentage depth of modulation must be applied to the dee voltage of a synchrotron in order to accelerate protons to 313 MeV assuming that the magnetic field has a 5% radial decrease in magnitude.

[University of Durham 1962]

9.91 An electron synchrotron with a radius of 1 m accelerates electrons to 300 MeV. Calculate the energy lost by a single electron per revolution when it has reached maximum energy.

[Andhra University 1966]

9.92 Show that the radius R of the final orbit of a particle of charge q and rest mass m_0 moving perpendicular to a uniform field of magnetic induction B with a kinetic energy n times its own rest mass energy is given by
$R = m_0\, c\, (n^2 + 2n)^{1/2}/qB$

9.93 Protons of kinetic energy 50 MeV are injected into a synchrotron when the magnetic field is 147 G. They are accelerated by an alternating electric field as the magnetic field rises. Calculate the energy at the moment when the magnetic field reaches 12,000 G (rest energy of proton = 938 MeV)

[University of Bristol 1962]

9.94 A synchrotron (an accelerator with an annular magnetic field) accelerates protons (mass number $A = 1$) to a kinetic energy of 1,000 MeV. What kinetic energy could be reached by deuteron ($A = 2$) or ^{3}He ($A = 3,\ Z = 2$) when accelerated in this machine? Take the proton mass to be equivalent to 1,000 MeV.

[University of Durham 1970]

9.95 A synchrocyclotron accelerates protons to 500 MeV. $B = 18\,\text{kG},\ V = 10\,\text{kV}$ and $\varphi_s = 30^0$. To find

(a) the radius of the orbit at extraction
(b) Energy of ions for acceptance
(c) the initial electric frequency limits
(d) the range of frequency modulation

9.96 Explain how a synchrotron accelerates particles. What is the main energy loss mechanism in these devices? How much more power is needed to maintain a beam of 500 GeV electrons in a synchrotron of radius 1 km than to maintain a beam of protons of the same energy? Is this feasible?

[University of Aberyswyth 2003]

9.97 Protons are accelerated in a synchrotron in the orbit of 10 m. At one moment in the cycle of acceleration, protons are making one revolution per microsecond. Calculate the value at this moment of the kinetic energy of each proton in MeV.

[University of Bristol 1961]

9.98 Electrons are accelerated to an energy of 10 MeV in a linear accelerator, and then injected into a synchrotron of radius 15 m, from which they are accelerated with an energy of 5 GeV. The energy gain per revolution is 1keV

(a) Calculate the initial frequency of the RF source. Will it be necessary to change this frequency?
(b) How many turns will the electron make?
(c) Calculate the time between injection and extraction of the electrons
(d) What distance do the electrons travel within the synchrotron?

9.2.14 Linear Accelerator

9.99 Protons of 2 MeV energy enter a linear accelerator which has 97 drift tubes connected alternately to a 200 MHz oscillator. The final energy of the protons is 50 MeV (a) What are the lengths of the second cylinder and the last cylinder (b) How many additional tubes would be needed to produce 80 MeV protons in this accelerator?

9.100 The Stanford linear accelerator produces 50 pulses per second of about 5×10^{11} electrons with a final energy of 2 GeV. Calculate (a) the average beam current (b) the power output.

9.101 A section of linear accelerator has five drift tubes and is driven by a 50 Mc oscillator. Assuming that the protons are injected into the first drift tube at 100 kV and gain 100 kV in every gap crossing

(a) What is the output energy after the fifth drift tube?
(b) What is the total length of the whole section? (Ignore the gap length)

[AEC 1966 Trombay]

9.102 What is the length L of the longest drift tube in a linac which operating at a frequency of $f = 25\,\mathrm{MHz}$ is capable of accelerating ^{12}C ions to a maximum energy of $E = 80\,\mathrm{MeV}$?

9.2.15 Colliders

9.103 Two beams of particles consisting of n bunches with N_1 and N_2 particles in each circulate in a collider and make head-on-collisions. A, the cross-sectional area of the beam and f is the frequency with which the particles circulate, obtain an expression for the luminosity L.

9.104 In an electron-positron collider the particles circulate in short cylindrical bunches of radius 1.2 mm. The number of particles per bunch is 6×10^{11} and the bunches collide at a frequency of 2 MHz. The cross-section for $\mu^+ \mu^-$ creation at 8 GeV total energy is, $1.4 \times 10^{-33}\,\mathrm{cm}^2$; how many $\mu^+ \mu^-$ pairs are created per second?

9.105 (a) Show that in a head-on-collision of a beam of relativistic particles of energy E_1 with one of energy E_2, the square of the energy in the CMS is 4 E_1 E_2 and that for a crossing angle θ between the beam this is reduced by a factor $(1+\cos\theta)/2$, (neglect the masses of beam particles in comparison with the energy)

(b) Show that the available kinetic energy in the head-on-collision with two 25 GeV protons is equal to that in the collision of a 1,300 GeV proton with a fixed hydrogen target. (Courtesy D.H. Perkins, Cambridge University Press)

9.106 Head-on collisions are observed between protons each moving with velocity (relative to the fixed observer) corresponding to 10^{10} eV. If one of the protons were to be at rest relative to the observer, what would the energy of the other need to be so as to produce the same collision energy as before. The rest energy of the proton is 10^9 eV.

[University of Manchester 1958]

9.107 It is required to cause protons to collide with an energy measured in their centre of mass frame, of 4 M_0c^2 in excess of their rest energy 2 M_0c^2. This can be achieved by firing protons at one another with two accelerators each of which imparts a kinetic energy of 2 M_0c^2. Alternatively, protons can be fired from an accelerator at protons at rest. How much energy must this single machine be capable of imparting to a proton? What is the significance of this result for experiments in high energy nuclear physics?

[University of New Castle 1966]

9.108 The HERA accelerator in Hamburg provided head-on collisions between 30 GeV electrons and 820 GeV protons. Calculate the centre of mass energy that was produced in each collision.

[Manchester 2008]

9.109 At a collider, a 20 GeV electron beam collides with a 300 GeV proton beam at a crossing angle of 10°. Evaluate the total centre of mass energy and calculate what beam energy would be required in a fixed-target electron machine to achieve the same total centre-of-mass energy.

9.3 Solutions

9.3.1 System of Units

9.1 $E = Mc^2 = 1 \times (3 \times 10^8)^2 = 9 \times 10^{16}$ J
$= 9 \times 10^{16}$ J/$(1.6 \times 10^{-10}$ J/GeV$) = 5.63 \times 10^{26}$ GeV

9.2 (a) $\hbar c = 197.3$ MeV-fm
In natural units, $\hbar = c = 1$
Therefore $1 = 197.3$ MeV-fm $= 0.1973$ GeV-10^{-15} m
Therefore 10^{-15} m $= 1/0.1973$ GeV^{-1}
or 1 m $= 10^{15}/0.1973 = 5.068 \times 10^{15}$ GeV^{-1}
(b) From (a) we have
1 m$^2 = (5.068)^2 \times 10^{30}$ GeV$^{-2} = 25.6846 \times 10^{30}$ GeV^{-2}
Therefore 1 GeV$^{-2} = 10^{-30}/25.6846$ m$^2 = 0.389$ mb
(c) $\hbar = 1.055 \times 10^{-34}$ J-s
$1 = 1.055 \times 10^{-34}$ J-s$/1.6 \times 10^{-10}$ J/GeV
Therefore 1 s $= (1.6 \times 10^{-10}/1.055 \times 10^{-34})$ GeV$^{-1} = 1.5 \times 10^{24}$ GeV^{-1}

9.3 (a) In practical units $\lambda_c = \hbar/m_e c$
Put $\hbar = c = 1$
In natural units $\lambda_c = 1/m_e$
(b) In practical units Bohr's radius of hydrogen atom is
$a_0 = \varepsilon_0\, \hbar^2/\pi\, me^2 = \varepsilon_0\, \hbar\hbar c/\pi m\, ce^2$
In natural units $a_0 = 1/\alpha m_e$
(c) In the ground state velocity
$v = \hbar/ma_0 = \hbar\alpha$
where we have used the results of (b)
Put $\hbar = 1$ to find $v = \alpha$ in natural units.
Numerical values:
$\lambda_c = \hbar/m_e c = (\hbar c/m_e c^2)$ MeV.fm/MeV $= 197.3/0.511$ fm $= 385$ fm
$= 0.00385 \times 10^{-10}$ m
$= 0.00385$ Å
$a_0 = 4\pi\varepsilon_0\, \hbar^2/e^2\, m = (4\,\pi\,\varepsilon_0/e^2)\,(\hbar\, c)^2/mc^2$
$= [1/(1.44$ MeV-fm$)] \times (197.3$ MeV-fm$)^2/0.511$ MeV
$= 53{,}000$ fm $= 0.53 \times 10^{-10}$ m
$v = \alpha c = (3 \times 10^8/137)$ ms^{-1}
$= 2.19 \times 10^6$ ms^{-1}

9.4 In natural units $\tau = 2/m\alpha^5$. Introduce the factors $\hbar$ and c

Let $\tau = 2\hbar^x\, c^y/m\,\alpha^5$

Find the dimensions: $[\hbar] = [\text{Energy} \times \text{Time}]$, $[c] = [\text{velocity}]$

$[\tau] = [1/m][ML^2\,T^{-1}]^x[LT^{-1}]^y$

$[T] = [M]^{x-1}[L]^{2x+y}[T]^{-x-y}$

Equating the coefficients on both sides of the equation,

$-x - y = 1,\ 2x + y = 0,\ x - 1 = 0$

we get $x = 1,\ y = -2$

Therefore $\tau = 2\hbar c^{-2}/m\alpha^5 = 2\,\hbar/mc^2\alpha^5$

$\tau = 2 \times (0.659 \times 10^{-21}\,\text{MeV-s}) \times (137)^5/0.511\,\text{MeV}$

$= 1.245 \times 10^{-10}\,\text{s}$

9.5 $\Gamma_\mu = G_F{}^2\, m_\mu{}^5/192\pi^3$ (1)

Now, the dimensional formula for G_F is

$[\text{Energy}]^{-2}\,[\hbar\, c]^3$ and $[\Gamma_\mu] = [\text{Energy}]$

Introduce $\hbar$ and c in (1) and take dimensions on both sides.

$\Gamma_\mu = G_F{}^2\, m\mu^5\,\hbar^x\, c^y/192\pi^3$ (2)

$[M\,L^2\,T^{-2}] = [M\,L^2\,T^{-2}]^{-4}\,[M]^5\,[M\,L^2\,T^{-1}]^{x+6}\,[L\,T^{-1}]^{y+6}$

or $[M\,L^2\,T^{-2}] = [M]^{7+x}\,[L]^{10+2x+y}\,[T]^{-4-x-y}$

Equating powers of M, L and T, we find

$x = -6$ and $y = 4$

and (2) becomes

$\Gamma_\mu = G_F{}^2\, m_\mu{}^5\,\hbar^{-6}\, c^4/192\pi^3$

$\Gamma_\mu = \hbar/\tau = G_F{}^2\,(m_\mu c^2)^5/192\,(\hbar c)^6\pi^3$

$= (1.116 \times.10^{-5}\,\text{GeV}^{-2})^2\,(105.659 \times 10^{-3})^5/192\pi^3$

$\tau = 2.39 \times 10^{-6}\,\text{s}$

9.3.2 Production

9.6 (a) The relation for lab angle θ and the C.M.S. angle θ^* is given by

$\tan\theta = \sin\theta^*/\gamma_c(\cos\theta^* + \beta_c/\beta^*)$ (1)

where $\beta_c = v_c/c$ is the CMS velocity and γ_c is the corresponding Lorentz factor (see summary of Chap. 6). For photon $\beta^* = 1$ and $\beta_c = 1$ as the electron is ultra relativistic. Dropping off the subscript c, (1) becomes

$\tan\theta = \sin\theta^*/\gamma(\cos\theta^* + 1) = \tan(\theta^*/2)/\gamma$ (2)

(b) Assuming that the photons are emitted isotropically, half of the photons will be contained in the forward hemisphere in the CMS, that is within $\theta^* = 90^\circ$. Substituting $\theta^* = 90^\circ$ in (2)

$\tan\theta = 1/\gamma$

As the incident electron is ultrarelativistic the photons in the lab would come off at small angles so that $\tan\theta \approx \theta = 1/\gamma$. Thus half of the photons will be emitted within a cone of half angle $\theta \approx 1/\gamma$.

9.7 If a particle m_1 moving with total energy E_1 collides with the target of mass m_2, then the total energy in the CMS will be
$E^* = \left(m_1{}^2 + m_2{}^2 + 2\,m_2\,E_1\right)^{1/2}$
If $E_1 >> m_1$ or m_2, then $E^* \propto \sqrt{E_1}$

9.8 $E_{CM} = E^* \approx (2mE_{Lab})^{1/2} = (2 \times 0.511 \times 10^{-3} \times 50)^{1/2}$
$= 0.226\,\text{GeV} = 226\,\text{MeV}$
$\sigma = 4\pi\alpha^2\hbar^2 c^2/3(E_{CM})^2$
$= (4\pi/137^2) \times (197.3\,\text{MeV-fm})^2/(3 \times 226^2) = 1.7 \times 10^{-4}\,\text{fm}^2$
$= 1.7 \times 10^{-30}\,\text{cm}^2$
Macroscopic cross-section per atom
$\Sigma = \sigma N_0\rho/A = 1.7 \times 10^{-30} \times 6.02 \times 10^{23} \times 11.4/207 = 0.5636 \times 10^{-7}\,\text{cm}^{-1}$
$\Sigma_e = Z.\Sigma_{atom} = 82 \times 0.5636 \times 10^{-7} = 4.62 \times 10^{-6}\,\text{cm}^{-1}$
Interaction mean free path $\lambda = 1/\Sigma_e = 1/4.62 \times 10^{-6} = 2.16 \times 10^5\,\text{cm}$
$= 2.16\,\text{km}$

9.9 (a) Minimum energy required in the C.M.S. is
$E^* = M + M = 2 \times 0.938 = 1.876\,\text{GeV}$
If E is the positron energy required in the LS, then
$E = E^{*2}/2m_e = (1.876)^2/2 \times 0.511 \times 10^{-3} = 3{,}444\,\text{GeV}$
(b) Energy requirements are drastically reduced
(c) Each beam of energy $m_p - m_e \approx 938\,\text{MeV}$, need to be oppositely directed.

9.3.3 Interaction

9.10 In the collision of a particle of mass m_1 of total energy E_1 with m at rest, the centre-of-mass energy is

$$E^* = \left(m_1{}^2 + m_2{}^2 + 2\,m_2\,E_1\right)^{1/2} \qquad (1)$$

Here, $m_1 = 938\,\text{MeV},\ m_2 = 938\,\text{MeV},\ E_1 = 938 + 200 = 1{,}138\,\text{MeV}$
Using these values in (1), we find $E^* = 1973\,\text{MeV}$
Useful energy for particle production
$= E^* - (m_1 + m_2) = 1,973 - (938 + 938) = 97\,\text{MeV}$
Pion can not be produced as the threshold energy is 290 MeV. Muons and electron – positron pairs are not produced in strong interactions. The resonance $\Delta(1, 236\,\text{MeV})$ also can not be produced.

9.11 $\sigma = 10^{-42}E = 10^{-42} \times 200 = 2 \times 10^{-40}\,\text{m}^2$
Number of nucleons$/m^3$, $n = \rho/m_p = 7,900/1.67 \times 10^{-27} = 4.7 \times 10^{30}$
$\Sigma = n\sigma = 4.7 \times 10^{30} \times 2 \times 10^{-40} = 9.4 \times 10^{-10}\,\text{m}^{-1}$
The mean free path $\lambda = 1/\Sigma = 1/9.4 \times 10^{-10} = 1.06 \times 10^9\,\text{m}$.
If t is the thickness of iron then the probability for interaction
$P = t/\lambda = t/1.06 \times 10^9 = 1/10^9$
Required thickness of iron, $t \approx 1.0\,\text{m}$.

9.12 (a) In elastic scattering, total kinetic energy is conserved. Life-time means mean-life-time i.e., the time in which the population of unstable particles is reduced by a factor e. The important interactions that the muons undergo are elastic scattering and the reaction, $\mu^- + p \to n + \nu_\mu$.

Let the muons travel a distance d metres. Then their intensity will be reduced due to decay by a factor

$$I/I_0 = \exp(-d/v\gamma\tau_0) \quad (1)$$

$$\gamma = 1/(1 - v^2/c^2)^{1/2} = 1/\left[1 - (10^6/3 \times 10^8)^2\right]^{1/2}$$
$$= 1.0033$$

$$I/I_0 = \exp -(d/(1.0033 \times 10^6 \times 2.2 \times 10^{-6}))$$
$$= \exp -(0.453\,d) \quad (2)$$

The intensity reduction due to interaction will be

$$I/I_0 = \exp(-\mathrm{d}\Sigma) \quad (3)$$

$$\Sigma = \sigma\, n = 0.1 \times 10^{-28} \times 2.69 \times 10^{25} = 2.69 \times 10^{-4}$$

$$I/I_0 = \exp(-2.69 \times 10^{-4})\, d \quad (4)$$

Comparing (2) and (4), reduction due to decay will be by far greater than by interaction for any value of d.

(b) Repetition of calculation for $T = 4\,\mathrm{GeV}$, $\gamma = 38.7$, $\beta \approx 1$, and $d =$ 12,000 m, gives $I/I_0 = e^{-0.47} = 0.625$, so that a large number of muons survive at the ground level. The distance at which muons of energy 4 GeV will be reduced by e is given by $d = v\gamma\tau_0 = 3 \times 10^8 \times 38.7 \times 2.2 \times 10^{-6} = 2.55 \times 10^4\,\mathrm{m} = 25.5\,\mathrm{km}$.

9.13 The geometrical cross-section is
$\sigma_g = \pi R^2 = \pi (r_0 A^{1/3})^2 = \pi r_0{}^2 A^{2/3}$
$= \pi (1.3 \times 10^{-13})^2 (208)^{2/3} = 1.86 \times 10^{-24}\,\mathrm{cm}^2$
Number of lead atoms/cm^3, $n = N_0\, \rho/A = 6 \times 10^{23} \times 11.3/207 = 3.275 \times 10^{22}$

Macroscopic cross-section
$\Sigma = n\,\sigma = (3.275 \times 10^{22})\,(1.86 \times 10^{-24}) = 0.06\,\mathrm{cm}^{-1}$
The interaction length, $\lambda = 1/\Sigma = 1/0.06 = 16.7\,\mathrm{cm}$

9.14 Number of interactions per second in volume V
$I = \Sigma V I_0$
$= \sigma\, N_0\, \rho V\, I_0/A$
$= (40 \times 10^{-27}\,\mathrm{cm}^2)(6.02 \times 10^{23} \times 0.071/2) \times 125 \times 2 \times 10^3$
$= 213.7$ neutral pions/s are produced.

Each pion decays into two photons. Therefore number of photons produced per second = 427

9.15 $I/I_0 = \exp(-\Sigma x)$
$\Sigma = n\sigma = N\rho\sigma/A$

$= 6.02 \times 10^{23} \times 7.9 \times 600 \times 10^{-27}/55.95 = 0.051\ \mathrm{cm}^{-1}$
$I/I_0 = 1/500 = \exp(-0.051\,x)$
Take $\log_e$ on both sides of the equation and solve for x.
We find $x = 122\ \mathrm{cm} = 1.22\ \mathrm{m}$.

Thus a thickness of 1.22 m iron will reduce the pion beam by a factor of 500. This does not include the reduction due to the decay of pions. Muons have much less interaction cross-section so that much greater thickness is required.

9.16 From the momentum triangle (Fig. 9.2)
$P_\nu{}^2 = p_0{}^2 + p_e{}^2 - 2\,p_0\,p_e\,\cos\phi$ (momentum conservation)
$E_\nu = E_0 - T$ (energy conservation)
Since $m_\nu = 0,\ p_\nu = E_\nu$
Therefore

$$P_\nu{}^2 = E_\nu{}^2 = (E_0 - T)^2 = E_0{}^2 + T^2 + 2mT - 2E_0(T^2 + 2mT)^{1/2}\cos\phi$$

Simplifying,

$$E_0(T^2 + 2mT)^{1/2}\cos\phi = T(E_0 - m) \approx T\,E_0\ (\text{because } E_0 >> m)$$

Squaring and simplifying
$\tan\phi \approx \phi = (2m/T)^{1/2}$

Fig. 9.2

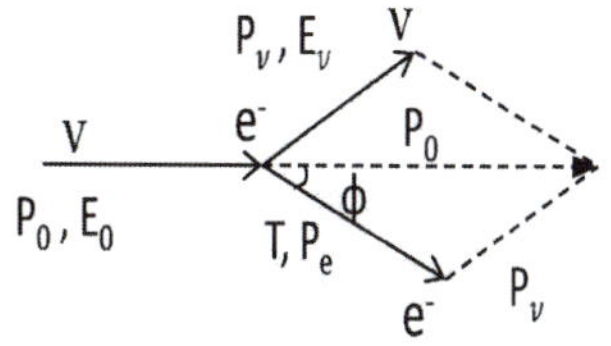

9.17 (a) In this reaction initially the total strangeness quantum number
$S_i = S_\pi + S_p = 0 + 0 = 0$, while for the final state
$S_f = S_K + S_\Sigma = -1 + (-1) = -2$
$\Delta S = -2$. Therefore, the rule $\Delta S = 0$ for strong interactions is violated.
(b) The total isospin for the initial state $I = 0 + 0 = 0$, while for the final state $I = 0 + 1 = 1$.
$\Delta I = 1$. Therefore, the rule $\Delta I = 0$ for strong interactions is violated.

9.18 $\mathrm{M}(^5\mathrm{He}_\Lambda) = \mathrm{M}(\Lambda) + \mathrm{M}(\mathrm{He}^4) - \mathrm{B}(\Lambda)$
$= 1{,}115.58 + 3{,}727.32 - 3.08 = 4{,}839.82\ \mathrm{MeV}/c^2$

9.19 Let P_i and P_f be the initial and final four-momentum transfer.
$\boldsymbol{P}_i = (\boldsymbol{p}_i,\ iE_i)$
$\boldsymbol{P}_f = (\boldsymbol{p}_f,\ iE_f)$
$\boldsymbol{Q} = \boldsymbol{P}_i - \boldsymbol{P}_f = (\boldsymbol{p}_i - \mathbf{p}_f) + i(E_i - E_f)$
$Q^2 = p_i{}^2 + p_f{}^2 - 2\boldsymbol{p}_i.\boldsymbol{p}_f - E_i{}^2 - E_f{}^2 + 2\,E_i\,E_f$
$= -m^2 - m^2 + 2\,E_i\,E_f - 2p_i\,p_f\cos\theta$
If $E_i >> m$ and $E_f >> m$, then $p_i \approx E_i$, and $p_f \approx E_f$.

$Q^2 = 2\,E_\mathrm{i}\,E_\mathrm{f}(1 - \cos\theta)$

Thus, the approximation is valid when the particle mass is much smaller than the energies.

9.3.4 Decay

9.20 (a) If a particle A at rest decays into B and C then the total energy of B will be
$E_\mathrm{B} = \left(m_\mathrm{A}{}^2 + m_\mathrm{B}{}^2 - m_\mathrm{C}{}^2\right)/2m_\mathrm{A}$
Inserting the values $m_\mathrm{A} = 494\,\mathrm{MeV}$, $m_\mathrm{B} = 135\,\mathrm{MeV}$ and $m_\mathrm{C} = 140\,\mathrm{MeV}$, we find $\mathrm{E}(\pi^0) = 245.6\,\mathrm{MeV}$.
(b) Kinetic energy of π^0 is $E_{\pi 0} - m_{\pi 0} = 245.6 - 135 = 110.6\,\mathrm{MeV}$

9.21 If I_0 is the original intensity and I the observed intensity, then
$I = I_0\,e^{-s/v\tau} = I_0\,e^{-as}$
$\gamma = 1 + T/m = 1 + 100/139 = 1.719$
$\beta = (\gamma^2 - 1)^{1/2}/\gamma = (1.719^2 - 1)^{1/2}/1.719 = 0.813$
$a = 1/v\tau \rightarrow \tau = 1/va = 1/\beta\,ca = 1/(0.813 \times 3 \times 10^8 \times 9.1 \times 10^{-2})$
$\tau = 4.5 \times 10^{-8}\,\mathrm{s}$
Proper life time $\tau_0 = \tau/\gamma = 4.5 \times 10^{-8}/1.719 = 2.62 \times 10^{-8}\,\mathrm{s}$

9.22 $m_\mathrm{K}\,\mathrm{c}^2 = 966.7 \times 0.511 = 494\,\mathrm{MeV}$
$m_\pi\,\mathrm{c}^2 = 273.2 \times 0.511 = 139.5\,\mathrm{MeV}$
Energy released in the decay
$Q = m_\mathrm{K} - 3m_\pi = 494.0 - 3 \times 139.5 = 75\,\mathrm{MeV}$

Maximum kinetic energy of one pion will occur when the other two pions go together in the opposite direction, that is energy is shared between one π and 2 π's
Non-relativistically
$T_\pi\ (\mathrm{max}) = Q \times 2m_\pi/(2m_\pi + m_\pi) = 75 \times 2/3 = 50\,\mathrm{MeV}$
Relativistically
$T_\pi\ (\mathrm{max}) = E_\pi - m_\pi = \left[\left(m_K{}^2 + m_\pi{}^2 - (2m_\pi)^2\right)/2m_\mathrm{K}\right] - m_\pi$
Inserting the values of m_K and m_π, we find $T_\pi\ (\mathrm{max}) = 48.4\,\mathrm{MeV}$. The difference in the two results is 3.3%.

9.23 In the decay $A \rightarrow B + C$,
$E_B = \left(m_A{}^2 + m_B{}^2 - m_C{}^2\right)/2m_A$
Here $B = \nu,\ A = \pi,\ C = \mu$. It follows that
$E'_\nu = \left(m_\pi^2 - m_\mu^2\right)/2m_\pi$ (because $m_\nu = 0$)

9.24 The maximum energy of neutron is obtained when the neutrino has zero energy, neutron and muon being emitted in the opposite direction.

The energy released in the decay is
$Q = m_\Sigma - (m_\mathrm{n} + m_\mu + m_\nu) = 1{,}189 - (939 + 106 + 0) = 144\,\mathrm{MeV}$
Non-relativistically,
$T_\mathrm{n} = Q\,m_\mu/(m_\mu + m_\mathrm{n}) = 144 \times 106/(106 + 939) = 14.6\,\mathrm{MeV}$
Relativistically,

$E_n = \left(m_\Sigma{}^2 + m_n{}^2 - m_\mu{}^2\right)/2m_\Sigma = (1{,}189^2 + 939^2 - 106^2)/2 \times 1{,}189 =$ 960.6 MeV

Therefore $T_n = E_n - m_n = 960.6 - 939 = 21.6$ MeV

9.25 Maximum neutrino momentum in the L-system is obtained when the neutrino is emitted in the forward direction, that is at $\theta^* = 0$ in the rest system of K-meson (Fig. 9.3).
The energy of the neutrino in the rest frame of K-meson is
$E_\nu{}^* = \left(m_K{}^2 - m_\mu{}^2\right)/2m_K = 235.6\,\text{MeV}$
$P_\nu{}^* = 235.6\,\text{MeV/c}$ (because $m_\nu = 0$)
$P_\nu = \gamma_K\, p_\nu{}^*\ (1 + \beta_K \beta_\nu{}^*\ \cos\theta^*)$
$\beta_\nu{}^* = 1;\ \theta^* = 0;\ p_\nu{}^* = 235.6\,\text{MeV/c}$
$\gamma_K = 1 + T_K/m_K = 1 + 500/494 = 2.012$
$\beta_K = 0.868$
$p_\nu = (2.012)(235.6)(1 + 0.868) = 885.5\,\text{MeV/c}$

Fig. 9.3 Decay configuration

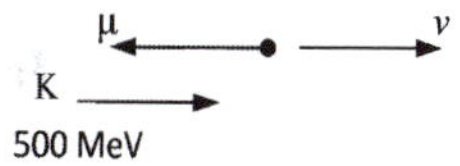

9.26 Rest mass energy of the heavy meson, $M = 965 \times 0.511 = 493$ MeV
$\gamma = 1 + T/M = 1 + 50/493 = 1.101$
$\beta = (\gamma^2 - 1)^{1/2}/\gamma = (1.101^2 - 1)^{1/2}/1.101 = 0.418$
Observed mean lifetime τ is related to proper lifetime by
$\tau = \tau_0 \gamma$
$\tau = d/\beta c = 1.7/0.418 \times 3 \times 10^8 = 1.355 \times 10^{-8}\,\text{s} = 1.101\,\tau_0$
Therefore $\tau_0 = 1.23 \times 10^{-8}$ s

9.27 Fraction of pions that will not decay
$I/I_0 = \exp(-t/\tau_0) = \exp(-d/\beta\, c\, \gamma \tau_0)$
$\gamma = E/m = (p^2 + m^2)^{1/2}/m = [(p^2/m^2) + 1]^{1/2} = [(10/0.1396)^2 + 1]^{1/2} =$ 71.64
$\beta \approx 1$
Therefore $I/I_0 = \exp(-100/1 \times 3 \times 10^8 \times 71.64 \times 2.6 \times 10^{-8}) = \exp(-0.179) =$ 0.836
$\gamma_\mu = ((p^2/m^2) + 1)^{1/2} = ((8^2/0.1057^2) + 1)^{1/2} = 75.69$
$\beta_\mu \approx 1$
$t_\mu = d\gamma/\beta c = 100 \times 75.69/1 \times 3 \times 10^8 = 25.23 \times 10^{-6}$ s
$t_\nu = d/c = 100/3 \times 10^8 = 0.333 \times 10^{-6}$ s
Therefore $\Delta t = t_\mu - t_\nu = (25.23 - 0.33) \times 10^{-6}$ s
$= 24.9\,\mu$s

9.28 In the rest system of pion the muon is emitted with kinetic energy
$T_\mu{}^* = 4$ MeV (see Problem 6.54)
Its energy in the LS will be maximum when $\theta^* = 0$ and minimum when $\theta^* = 180$ in the CMS. Applying the relativistic transformation equation

$E_\mu = E_\mu{}^* \gamma_\pi (1 + \beta_\mu{}^* \beta_\pi \cos\theta^*)$
$E_\mu{}^* = 105.7 + 4.0 = 109.7\,\text{MeV};\ \gamma_\mu{}^* = 1.0378;\ \beta_\mu{}^* = 0.2676$
$\gamma_\pi = 5/0.1395 = 35.84;\ \beta_\pi \approx 1$
$E_\mu\,(\text{max}) = 4.98\,\text{GeV};\ E_\mu\,(\text{min}) = 2.88\,\text{GeV}$

9.29 The decay equation is

$$I = I_0 \exp(-t/\gamma\tau) = I_0 \exp(-d/\beta\, c\, \gamma\tau) \quad (1)$$

For K^+ meson beam
$\beta_K = 1/\left[1 + m_K{}^2/p_K{}^2\right]^{1/2}$
Substituting $m_K = 0.5\,\text{GeV/c}^2$ and $p_K = 1\,\text{GeV/c}$
$\beta_K = 0.894;\ \gamma_K = 2.23;\ \tau_K = 10^{-8}\,\text{s};\ d = 10\,\text{m}$
Using these values in Eq. (1)
$I/I_0 = \exp(-1.672) = 0.188$
which is tolerable.

For Λ-hyperon beam
$\beta_\Lambda = 1/\left[1 + m_\Lambda{}^2/p_\Lambda{}^2\right]^{1/2} = 0.7$
$\gamma_\Lambda = 1.96,\ \tau_\Lambda = 2.5 \times 10^{-10}\,\text{s};\ d = 10\,\text{m}$

Using the above values in (1) we find
$I/I_0 = \exp(-97) \approx \text{zero}$
which is not at all useful.

9.30

$$d = vt \quad (1)$$

$$t = \gamma T_0 \quad (2)$$

$$p = m_0\, v\, \gamma \quad (3)$$

Combining (1), (2) and (3)
$d = pT_0/m_0$

9.31 Inverse Lorentz transformations gives

$$E_\nu{}^* = \gamma E_\nu(1 - \beta\cos\theta) = E_\nu[\gamma - (\gamma^2 - 1)^{1/2}\cos\theta]$$

$$= E_\nu[\gamma - \gamma(1 - 1/2\gamma^2)(1 - \theta^2/2)]$$

$$E_\nu{}^* \approx (E_\nu/2\gamma)(1 + \gamma^2\theta^2) \quad (1)$$

where we have neglected terms $\theta^2/2\gamma$ for $\gamma >> 1$ and considered small θ.

But $E_\nu{}^* = \left(m_\pi{}^2 - m_\mu{}^2\right)/2m_\pi$ and $E_\pi = \gamma\, m_\pi$ (2)

Using (2) in (1) we get the desired result.

9.32 $I/I_0 = 1/100 = \exp(-d/\beta c\,\gamma\tau)$
Put $d = 20\,\text{m}$, $c = 3 \times 10^8\,\text{m/s}$ and $\tau = 0.8 \times 10^{-10}\,\text{s}$ and take $\log_e$ on both sides to solve for $\beta\gamma$. We find $\beta\gamma = 181$. Therefore, the momentum of hyperon,
$cp = m\beta\gamma = 1.19 \times 181 = 215\,\text{GeV}$
Thus, the minimum momentum required is $P = 215\,\text{GeV/c}$

The elastic scattering of Σ^- hyperons with protons can be recorded in the hydrogen bubble chamber from the kinematical fits of events.

It may be noted that the beams of sigma hyperons are available only in recent times. This has been possible because of time dilation for high momenta particles. Even beams of omega minus have been used.

9.33 $$\frac{d\omega}{dE_e} = \frac{2G_F^2 \left(m_\mu c^2\right)^2 E_e^2}{(2\pi)^3(\hbar c)^6}\left(1 - \frac{4E_e}{3m_\mu c^2}\right)$$

This is the differential energy distribution of electrons from the decay of muon. The distribution has a maximum, which is easily found out by differentiating the right hand side of the above expression with respect to E_e and setting it equal to zero. We easily find the maximum of distribution to occur at $E_e = m_\mu c^2/2$.

Now, if we regard the electron mass to be negligible then the maximum value of the electron energy will occur when it is emitted in a direction opposite to the two neutrinos, its maximum energy being also $m_\mu c^2/2$. A rough plot of the electron energy spectrum is shown in Fig. 9.4.

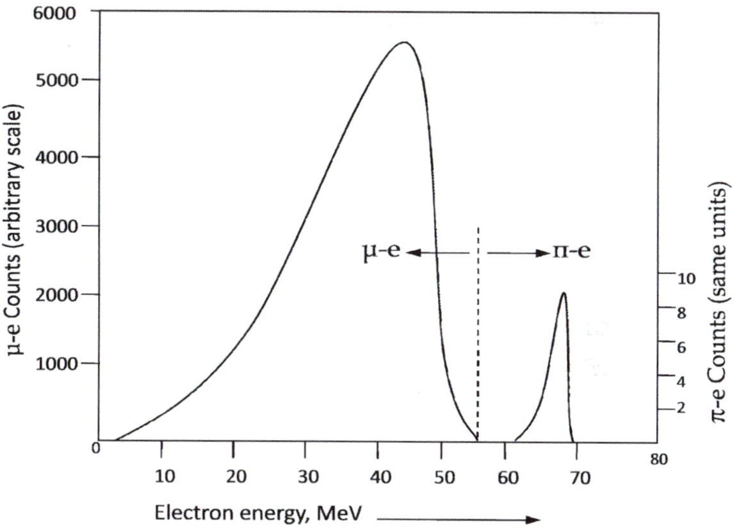

Fig. 9.4

The orientation of momenta and helicities of the particles in the decay are shown in Fig. 9.5.

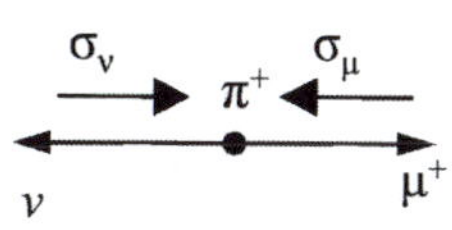

Fig. 9.5 Pion rest-frame sketch indicating sense of spin polarization in pion

The integration of the expression

$$\int_0^{\frac{m_\mu c^2}{2}} E_e^2\left(1 - 4E_e/3m_\mu\, c^2\right) dE_e$$

gives $(m_\mu\, c^2)^3/48$ so that the full expression for full width is

$\Gamma = G_F{}^2 (mc^2)^5/(\hbar c)^6\; 192\pi^3$

which is identical with the one stated in Problem 9.5. The mean life time is obtained from $\tau = \hbar/\Gamma$ as in the solution of 9.5.

9.34 (a) The charged pions $\pi^\pm$ decay by weak interactions $\pi \rightarrow \mu + \nu$, and so their mean lifetime is relatively longer, while the neutral pion π^0 decays via $\pi^0 \rightarrow 2\gamma$, electromagnetically and therefore their mean life time is shorter.

(b) The π^+ and π^- are particle and antiparticle pair and they are expected to have the same mass by the CPT theorem. On the other hand Σ^+ and Σ^- hyperons are not particle-antiparticle. Actually $\bar{\Sigma}^-$ is the antiparticle of Σ^+. Σ^+ and Σ^- are the members of the isospin triplet $(\Sigma^+,\ \Sigma^0,\ \Sigma^-)$ and because of difference in their charges can slightly differ in their mass similar to the masses of neutron and proton for the isospin doublet of nucleon.

(c) Λ-hyperon decays via $\Lambda \rightarrow p + \pi^-$ or $n + \pi^0$, the interaction is weak. Similarly the hyperon Ξ^0 decays via $\Xi^0 \rightarrow \Lambda + \pi^0$, which is also a weak decay. In both the cases the lifetimes are relatively long on the nuclear scale. On the other hand the decay of Σ^0-hyperon via $\Sigma^0 \rightarrow \Lambda + \gamma$ is electromagnetic. The explanation is the same as in (a)

9.35 Apply the formula

$$M^2 = m_p{}^2 + m_\pi{}^2 + 2(E_p E_\pi - P_p P_\pi\ \cos\theta) \qquad (1)$$

Use the values:

$m_p = 0.939\,\text{GeV},\ m_\pi = 0.139\,\text{GeV},\ P_p = 0.44\,\text{GeV}$
$P_\pi = 0.126\,\text{GeV},\ E_p = 1.036\,\text{GeV},\ E_\pi = 0.188\,\text{GeV},\ \theta = 64^0$ and find $M = 1.114\,\text{GeV}/\text{c}^2$. The particle is Λ hyperon

9.3.5 Ionization Chamber, GM Counter and Proportional Counters

9.36 The problem is based on the double source method for the determination of the dead time of a G.M. counter. Two radioactive sources of comparable strength are chosen. In all four counts are taken. First, the background rate B per second is found out when neither source is present. One of the sources is placed in a suitable position so that a high counting rate N_1 is registered. While this counter is in the same position, the second source is placed by its side to get the counting rate N_{12}. Finally, the first source is removed so that the second counter above gives a count of N_2. If $n_1,\ n_2$ and n_{12} are the true counting rates then we expect

$(n_1 + B) + (n_2 + B) = n_{12} + B$

or

$n_1 + n_2 = n_{12} - B$ (1)

Now, for each particle that is counted, on an average there will be a dead time τ during which particles are not counted. Then if N is the counting rate (number of counts/second) then the time lost will be $N\tau$, so that the true counting rate

$n = N/(1 - N\tau)$

Thus $n_1 = N_1/(1 - N\tau)$ etc.
We can then write (1) as

$$N_1/(1 - N_1\,\tau) + N_2/(1 - N_2\,\tau) = N_{12}/(1 - N_{12}\,\tau) - B/(1 - B\tau) \quad (2)$$

Now, in practice N_1 and N_2 will be of the order of 100 per second, N_{12} is of 200 per second, $B \approx 1$ per second and $\tau \approx 10^{-4}$ second, so that $N_1\,\tau << 1$, etc. We can then expand the denominators binomially and write to a good approximation

$N_1(1 + N_1\tau) + N_2(1 + N_2\tau) = N_{12}(1 + N_{12}\tau) - B(1 + B\tau)$

$N_1 + N_2 - N_{12} + B = \tau\left[N_{12}{}^2 - N_1{}^2 - N_2{}^2\right] \approx \tau\left[(N_1 + N_2)^2 - N_1{}^2 - N_2{}^2\right]$

Or $\tau = (N_1 + N_2 - N_{12} + B)/2N_1\,N_2$

9.37 The voltage sensitivity is 4 divisions per volt.
So 0.8 divisions correspond to 0.2 V.
The charge deposited, $Q = CV = 0.5 \times 10^{-12} \times 0.2 = 10^{-13}$ Coulomb
If n ion pairs are released then $ne = Q$
Therefore $n = Q/e = 10^{-13}/1.6 \times 10^{-19} = 6.25 \times 10^5$ ion pairs.
Energy of alpha particles = (number of ion pairs) × (Ionization energy)
$= (6.25 \times 10^5)(35) = 21.87 \times 10^6$ eV = 21.87 MeV.

9.38 Let n beta particles enter the ionization chamber per second and stopped. Number of ion pairs released when ionization energy is I is given by
$n = N/I = 0.49 \times 10^6/35 = 1.4 \times 10^4$
The ionization current, i = Nne
Therefore, $N = i/ne = 7 \times 10^{-11}/1.4 \times 10^4 \times 1.6 \times 10^{-19} = 3.125 \times 10^4$ beta particles/second

9.39 If M is the gas multiplication factor and N is the number of ion pairs released, V the voltage developed, C the capacitance and e the electronic charge then the charge Q deposited will be
$Q = CV = MNe$
Or $M = CV/Ne = 1.5 \times 10^{-9} \times 10^{-3}/(2 \times 10^6/35) \times 1.6 \times 10^{-19} = 164$

9.40 Number of primary ions produced by a 14 keV electron, $N = 14{,}000/35 = 400$
Charge collected $Q = MNe$
where M is the gas multiplication and $e = 1.6 \times 10^{-19}$ Coulomb the elementary charge. If C is the capacitance of the circuit the pulse height will be
$V = Q/C = MNe/C = 600 \times 400 \times 1.6 \times 10^{-19}/10 \times 10^{-12} = 3.84 \times 10^{-3}$ V = 3.84 mV.

9.41 Slope $= \dfrac{2\%}{100\,V} = \dfrac{0.1\%}{5\,V}$

The voltage should not vary more than ±5 V from the operating voltage of 800 V.

9.42 For the cylindrical geometry of the G.M. tube the electric field is given by
$E = V/r \ln(b/a)$
where V is the applied voltage, r is the distance of a point from the anode, b and a are the diameters of the cathode and the anode wire.

The field will be maximum close to the anode.
$r = 0.1\,\text{mm} = 10^{-4}\,\text{m}$
$E_{\text{max}} = 1,000/10^{-4}\ \ln(20/0.2) = 2.17 \times 10^6\,\text{V/m}$

The lifetime of G.M. tube in years is given by dividing total number of possible counts by counts per year.
$t = 10^9/(52 \times 30 \times 60 \times 3,000) = 3.56$ years.
In practice, the G.M. tube will not work properly long before the above estimate.

9.43 As the mean free path is inversely proportional to the pressure, the mean free path at 10 cm pressure will be $2 \times 10^{-4} \times (76/10)$ or 1.52×10^{-3} cm.

At a distance r = mean free path = 1.52×10^{-3} cm. from the anode, the electric field should be such that the electron acquires sufficient energy to ionize Argon for which the ionization energy is 15.7 eV. The required value of E is
$E = 15.7/1.52 \times 10^{-3} = 1.03 \times 10^4\,\text{V/cm}$
$r = V/E\ \ln(b/a) = 1000/1.03 \times 10^4\ \ln(20/0.1) = 0.0183\,\text{cm} = 0.183\,\text{mm}$

9.44 The source gives $3.7 \times 10^7 \times 25 = 9.25 \times 10^8$ disintegrations/second
G.M. Counting rate of beta rays plus gamma rays
= (2,000/60) – (750/300) = 30.83/s
Therefore, efficiency = $30.83/(9.25 \times 10^8) = 3.33 \times 10^{-8}$
Scintillation counter counting rate of beta particles = (9,300/60) – (300/300) = 154/s
Therefore, efficiency = $154/(9.25 \times 10^8) = 16.65 \times 10^{-8}$

For both the counters the efficiency is low because of small solid angle of acceptance. Both the counters would register beta particles as well as gamma rays. But the efficiency for counting gamma rays in G.M. counter will be quite low, being of the order of 1%. This is because gamma rays cause ionization only indirectly by hitting the walls of the GM counter and ejecting electrons. The efficiency is relatively higher in the scintillation counter. For this reason, in the given situation the efficiency for scintillation counter is approximately five times greater.

9.45 The true counting rate n is related to the observed counting rate n_0 by
$n = n_0/(1 - n_0\,\tau)$

where τ is the dead time. $\tau = 100\,\mu\text{s} = 10^{-4}$ s.
$n_0 = 10^4/\text{min} = 166.7/\text{s}$
$n = (10^4/\text{min})/(1 - 166.7 \times 10^{-4}) = 10{,}169/\text{min}$

9.46 $V = V_0\, e^{-t/RC}$
$e^{t/RC} = V_0/V = 100/99$
$R = t/C \ln(100/99) = 86{,}400/(5 \times 10^{-12})(0.01) = 1.73 \times 10^{18}$ ohm.

9.47 $V = Er\ \ln(d_1/d_2)$ (1)
Put $r = d_2/2 = 0.01/2 = 0.005$ cm
$E = 15.7/\lambda = 15.7/8 \times 10^{-4}$ V/cm $= 19{,}625$ V/cm
$d_1/d_2 = 4/0.01 = 400$
$V = (19{,}625)(0.005) \ln\ 400 = 588$ V.

9.48 $Q = CV$
Initial chamber charge:
$Q_{\text{ch}} = 5 \times 10^{-12} \times 600 = 3.0 \times 10^{-9}$ Coulomb
Initial electrometer charge:
$Q_{\text{el}} = 0.4 \times 10^{-12} \times 600 = 0.24 \times 10^{-9}$ Coulomb
Chamber charge lost:
$8 \times 10^9 \times 1.6 \times 10^{-19} = 1.28 \times 10^{-9}$ Coulomb
Final chamber charge:
$Q_{\text{ch}}{}'(3.0 - 1.28) \times 10^{-9} = 1.72 \times 10^{-9}$
Final electrometer charge:
$Q'_{\text{el}} = 0.24 \times 10^{-19}$ Coulomb
Final system charge
$\text{Q} = (1.72 + 0.24) \times 10^{-9} = 1.96 \times 10^{-9}$ Coulomb
Total capacitance $C = (5.0 + 0.4) \times 10^{-12}\,\text{F} = (5.4 \times 10^{-12})$
Final potential$= 1.96 \times 10^{-9}/5.4 \times 10^{-12} = 363\,V$.

9.3.6 Scintillation Counter

9.49 The velocity of a particle
$\beta = p/E = p/(p^2 + m^2)^{1/2}$
The time taken to cross a distance L is
$t = L/\beta c = (L/c)[(p^2 + m^2)/p^2]^{1/2} = (L/c)[(1 + (m^2/p^2)]^{1/2}$
The difference in time taken for proton and pion is
$\Delta t = t_{\text{p}} - t_\pi = (L/c)\left\{\left[1 + \left(m_{\text{p}}{}^2/p^2\right)\right]^{1/2} - \left[1 + \left(m_\pi{}^2/p^2\right)\right]^{1/2}\right\}$
Substituting, $\Delta t = 100 \times 10^{-12}$ s, $c = 3 \times 10^8\text{ms}^{-1}$
$m_{\text{p}} = 0.94\,\text{GeV/c}^2$, $m_\pi = 0.14\,\text{GeV/c}^2$ and $p = 4\,\text{GeV/c}$, and solving for L, we find the minimum value of $L = 1.14$ m

9.50 $t = \dfrac{d}{c}\left(1 + \dfrac{m^2}{p^2}\right)^{1/2}$
Put $t = 51 \times 10^{-9}$ s, $c = 3 \times 10^8$ m/s, $d = 12$ m and $p = 1.19$ GeV/c and solve for m to find $m_{p-} = 0.94\,\text{GeV/c}^2$ for the anti proton mass.

$E_\pi = \left(P^2 + m_\pi{}^2\right)^{1/2} = (1.19^2 + 0.14^2)^{1/2} = 1.198\,\text{GeV}$
$\beta_\pi = P/E_\pi = 1.19/1.198 = 0.9933$
$t_\pi = d/\beta_\pi\, c = 12/(0.9933 \times 3 \times 10^8) = 40.26 \times 10^{-9}\,\text{s}$
$= 40.26\,\text{ns}$

9.51 Number of photons emitted due to absorption of 1 MeV electron is $15 \times 10^6/10^3 = 15{,}000$.
Number of photo – electrons emitted $= 15{,}000/10 = 1{,}500$

The electron multiplication factor $M = 3^{10}$ because the photomultiplier tube has 10 dynodes and each dynode produces 3 secondary electrons. The charge collected at the output is
$q = 1{,}500 \times 3^{10} \times 1.6 \times 10^{-19} = 1.417 \times 10^{-11}$ Coulomb

The pulse height will be
$V = q/C = 1.417 \times 10^{-11}/1.2 \times 10^{-10} = 0.118\,\text{V}$

9.52 The number of electrons liberated by the phosphor when 5 MeV proton is stopped
$n = 5 \times 10^6/100 = 5 \times 10^4$

Allowing for light collection efficiency (η) and conversion efficiency (ε) of the photocathode, number of electrons released from the cathode
$N = n\eta\varepsilon = 5 \times 10^4 \times (60/100)(5/100) = 1{,}500$

After going through 10 stages the number of electrons reaching the anode becomes with a gain (G) of 3/stage
$N\ G = 1,500 \times 3^{10} = 8.85 \times 10^7$

The charge collected at the anode
$q = NGe = 8.85 \times 10^7 \times 1.6 \times 10^{-19} = 14.16 \times 10^{-12}$ Coulomb

The voltage developed
$V = q/C = 14.16 \times 10^{-12}/12 \times 10^{-12} = 1.18\,\text{V}$.

9.53 $\tau = (17 + 0.5\,\text{k})\mu\text{s}$
For channel 100, $\tau = 17 + 0.5 \times 100 = 67\,\mu\text{s}$
True counting rate $N = N_0/(1 - 67 \times 10^{-6}\text{f})$
$N_0/N = 90/100 = 1 - 67 \times 10^{-6}\text{f}$
Or $f = 1{,}490\,\text{s}^{-1}$
Similarly for channel 400, $f = 460\,\text{s}^{-1}$

9.54 Let the gain/stage be G, so that the net gain due to electron multiplication will be G^{12} (because there are 12 stages).
Number of photons producing electrons from the cathode for each beta particle absorbed, with 5% photo-cathode efficiency $= 550 \times 5/100 = 27.5$
Number of electrons reaching the anode $= 27.5 \times \text{G}^{12}$
Charge collected, $q = 27.5 \times G^{12} \times 1.6 \times 10^{-19}$
The voltage developed, $V = q/C = 27 \times G^{12} \times 1.6 \times 10^{-19}/45 \times 10^{-12} = 8 \times 10^{-3}$
Solving for G, we get the gain/stage, $G = 2.567$

9.55 Label A is due to photopeak of 1.275 MeV γ-rays. Some of the primary photons will be absorbed within the crystal after undergoing Compton scattering. Such events merely enhance the photopeak. In other cases, the Compton-scattered photon will escape from the crystal, the light output will now be proportional to the energy of the recoil electron which will be absorbed in a large crystal. There will be a energy continuum of the recoil electrons with energy ranging from zero to maximum. The label B represents the Compton shoulder. The strong peak labeled C marks the photopeak at 0.511 MeV due to electron-positron annihilation leading to absorption of one of the photons. The annihilation may take place from the positron emitted by the source or by the pair production caused by the primary photon. Now the total kinetic energy available by the electron-positron pair will be $(h\nu - 1.02)$ MeV. When all of the energy is dissipated, the positron will be annihilated. If both the annihilation photons escape from the crystal, a peak (called Escape peak) will occur at an energy $(h\nu - 1.02)$ MeV $= 1.275 - 1.02 = 0.255$ MeV, represented by label D. If one of the photons escapes, another peak will occur at an energy $(h\nu - 0.511)$ MeV $= 1.275 - 0.511 = 0.764$ MeV (not shown in the Fig. 9.6). If both the photons are absorbed, full output will be realized and this will be added up to the photo peak.

Fig. 9.6

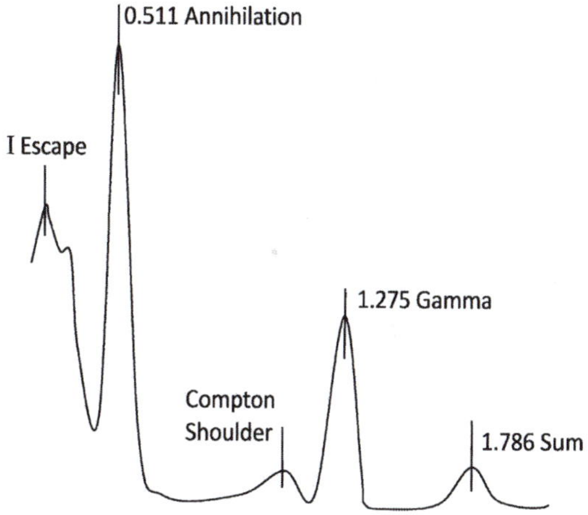

If the annihilation peak at 0.511 MeV and the photo peak due to primary photon occur simultaneously within the resolving time of the instrument then their energies are added and the events are recorded as a single event, known as sum peak at energy $1.275 + 0.511 = 1.786$ MeV (label E), usually with small amplitude.

9.56 The normal distribution is

$$p(x) = \left(\sigma\sqrt{2\pi}\right)^{-1} \exp[-(\bar{x} - x)/2\sigma^2] \tag{1}$$

At $\bar{x}$,

$$p(\bar{x}) = \left(\sigma\sqrt{2\pi}\right)^{-1} \quad (2)$$

Let the HWHM points be located at

$$\bar{x} - x = k\sigma \quad (3)$$

$$\text{Then } p(k\sigma) = \left(\sigma\sqrt{2\pi}\right)^{-1} \exp(-k^2/2) \quad (4)$$

By definition,

$$\frac{p(k\sigma)}{p(\bar{x})} = \frac{1}{2} = \exp\left(-\frac{k^2}{2}\right) \quad (5)$$

where we have used (2) and (4). Taking $\log_e$ on both sides of (5)
$k^2/2 = \ln 2 = 0.69315$
$\therefore k = 1.1774$
Therefore HWHM $= 1.177\sigma$

9.57 Channel 298 corresponds to 661 keV energy (Fig. 9.7)
FWHM (full width at half maximum)$= 316 - 281 = 35$ channels
HWHM (half width at half maximum)$= 35/2 = 17.5$ channels
HWHM corresponds to (661 keV)(17.5)/298 $= 38.8$ keV
Now HWHM $= 1.177\,\sigma$
Therefore $\sigma =$ HWHM/1.177
$= 38.8/1.177 = 33$ keV
The coefficient of the energy determination $= (33)(100)/661 = 5\%$

Fig. 9.7 Photo peak 661 keV γ-rays

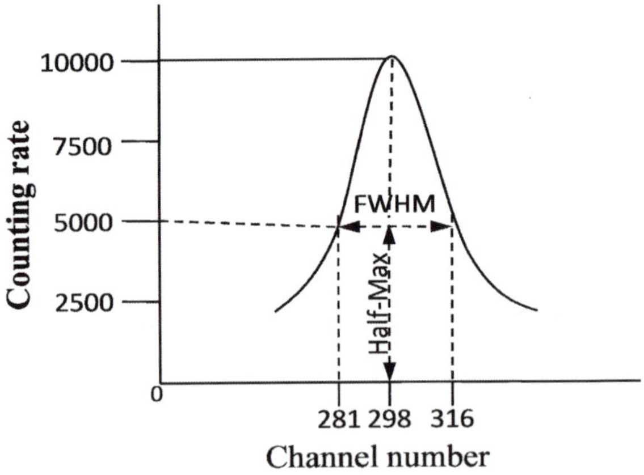

9.3.7 Cerenkov Counter

9.58 When a charged particle moves through a medium with velocity greater than the light velocity in the medium then light is emitted known as Cerenkov radiation, after the discoverer. The radiation is emitted around the surface of a cone with its axis along the particle's path and with a half-angle given by

$$\cos\,\theta = 1/\beta\, n \tag{1}$$

where n is the refractive index and βc is the particle velocity. The threshold corresponds to $\theta = 0^0$ and is given by

$$\beta = 1/n \tag{2}$$

Kaon and pion of the same energy would have different velocities, the pion velocity being higher than the kaon velocity. If the medium be chosen such that $\beta_{\mathrm{k}} < 1/n$ but $\beta_\pi > 1/n$, then the pion will be counted by the Cerenkov counter but not the kaon: In practice there can be several modifications of this principle as by providing coincidence/anticoincidence with a scintillation counter, and by restricting the emitted light to a prescribed angular interval.
$\gamma_\pi = 1 + (T/m_\pi) = 1 + (20/0.14) = 143.857$
$\beta_\pi = \left[\left(1 - (1/\gamma_\pi{}^2)\right)\right]^{1/2} = 0.9999758$
$\cos\,\theta = 1/\beta_\pi\, n = 0.999624$
$\theta = 1.57^\circ$

9.59 $\cos\,\theta = 1/\beta\, n$
$\beta = 1/n\,\cos\,\theta = 1/(1.5 \times \cos\,45^\circ) = 0.943$
$v = \beta c = 0.943 \times 3 \times 10^8$
$= 2.828 \times 10^8\,\mathrm{ms}^{-1}$

9.60 $\beta = 1/n\,\cos\,\theta$ (1)
Also $\beta = cp/E$ (2)

where E is the total energy.
Combining (1) and (2)

$$E = cpn\,\cos\,\theta \tag{3}$$

The relativistic equation is

$$E^2 = c^2\,p^2 + m^2\,c^4 \tag{4}$$

Eliminating E between (3) and (4), we get $mc^2 = pc\,(n^2\cos^2\theta - 1)^{1/2}$

9.61 $\theta = 0.55$ radians $= 0.55 \times 57.3^\circ = 31.515^\circ$
$\beta = 1/n\,\cos\,\theta = 1/1.88 \times 0.55 = 0.967$
$\gamma = 1/(1-\beta^2)^{1/2} = 3.925$
$\gamma = 1 + (T/m) \rightarrow m = T/(\gamma - 1)$
$m = 420/(3.925 - 1) = 107\,\mathrm{MeV/c^2}$
$= (107/0.511)\,m_{\mathrm{e}} = 209\,m_{\mathrm{e}}$
It is a muon

9.62 The number of photons $N(\lambda)\,\mathrm{d}\lambda$ radiated per unit path in a wavelength interval $\mathrm{d}\lambda$ can be shown to be
$N(\lambda)\,\mathrm{d}\lambda = 2\pi\alpha\,(1 - 1/\beta^2\,n^2)\,(1/\lambda_1 - 1/\lambda_2)$
where $\alpha = 1/137$, is the fine structure constant.
Inserting $\beta = 0.95$, $n = 1.33$, $\lambda_1 = 3500\times10^{-8}$ cm and $\lambda_2 = 5500\times10^{-8}$ cm in the above expression, we find $N(\lambda)\,\mathrm{d}\lambda = 178$ per cm.

9.63 Number of photons emitted per unit length

$$N = 2\pi\alpha(1 - 1/\beta^2 n^2)(1/\lambda_1 - 1/\lambda_2) \qquad (1)$$

Pions: $\beta_\pi = p/(p^2 + m^2)^{1/2} = 15/(15^2 + 0.14^2)^{1/2} = 0.999956$
Kaons: $\beta_k = 15/(15^2 + 0.494^2)^{1/2} = 0.999458$
If the signal is to be given by pions but not kaons, the condition on the refractive index is
$\beta_\pi > 1/n > \beta_k$
$0.999956 > 1/\mathrm{n} > 0.999458$
For a value of $n = 1.0004$, $1/n = 0.9996$, the above condition is satisfied. Inserting
$\alpha = 1/137$, $\beta_\pi = 0.999956$, $n = 1.0004$,
$\lambda_1 = 4 \times 10^{-7}$ m, and $\lambda_2 = 7 \times 10^{-7}$ m in (1), we find $N = 35$ photons/m. Therefore to obtain 175 photons a length of 5.0 m is required.

9.64 Pions: $E = (p^2 + m^2)^{1/2} = (900^2 + 140^2)^{1/2} = 911$ MeV
$\beta_\pi = p/E_\pi = 900/911 = 0.9879$
For threshold, $n_\pi = 1/\beta = 1/0.9879 = 1.012$
Protons: $E = (900^2 + 938^2)^{1/2} = 1{,}300$ MeV
$\beta_\mathrm{p} = 900/1300 = 0.6923$
For threshold $n_\mathrm{p} = 1.444$
The material chosen must have the refraction index $1.012 < n < 1.444$
For $n = 1.012$, Cerenkov light will come off at 0° with the path. If a higher index of refraction is chosen, light will come off at wider angle. The n must be less than 1.444, otherwise protons will be counted.

9.3.8 Solid State Detector

9.65 If A is the area, d the thickness of depletion layer and K the dielectric constant then the capacitance is
$C = \varepsilon AK/d = 8.8 \times 10^{-12} \times 1.5 \times 10^{-4} \times 10/40 \times 10^{-6} = 3.3 \times 10^{-10}$ F
The charge liberated
$q = 5 \times 10^6 \times 1.6 \times 10^{-19}/3.5 = 2.286 \times 10^{-13}$ Coulomb
Potential developed
$V = q/C = 2.286 \times 10^{-13}/3.3 \times 10^{-10} = 0.69 \times 10^{-3}$ V
$= 0.69$ mV

9.3.9 Emulsions

9.66 The Range-Energy-Relation must be such that it ensures that the ionization $-\mathrm{d}E/\mathrm{d}R$ is a function of $z^2\,\mathrm{f}(R/M)$. One such relation is $E = K\,z^{2\mathrm{n}}\,M^{1-\mathrm{n}}\,R^{\mathrm{n}}$ where E is in MeV and R in microns, K and n are empirical constants which depend on the composition of emulsions, z is the charge of the ion and M its mass in terms of proton mass. The data on R and E for protons have been used to determine n and R. We find $n = 0.6$ and $K = 0.2276$. Using these values of n and K, the energy for various values of R have been determined for deuterons and ^{3}He as tabulated below and the corresponding graphs are drawn, Fig. 9.8.

Fig. 9.8

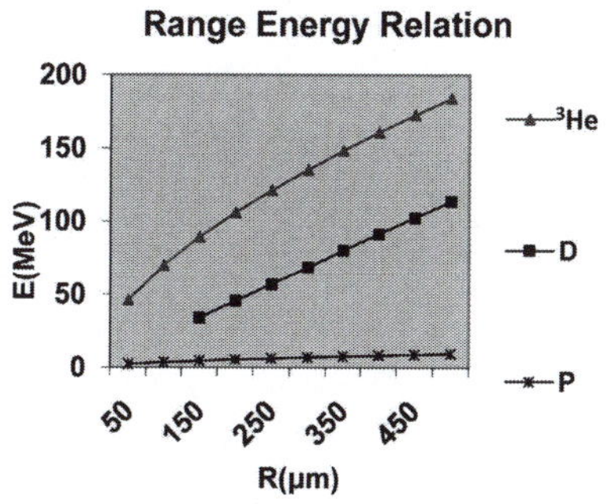

R(μm)	0	50	100	150	200	250	300	350	400	450	500
D E(MeV)	0	3.06	6.31	34.2	45.6	57.0	68.6	80.1	91.4	102.6	113.7
^{3}He	0	46.2	70.0	89.3	106.2	121.4	135.4	148.5	160.9	172.7	184.0

9.3.10 Motion of Charged Particles in Magnetic Field

9.67 For a circular orbit of a charged particle of charge q and momentum p moving in a magnetic field of B Tesla perpendicular to the orbit, the radius r in metres is related by the formula

$P = qBr \rightarrow cp = qBrc$ J

$cp = (1.6 \times 10^{-19})(3 \times 10^8)\, Br$ J

$= 4.8 \times 10^{-11}\, Br$ J

$= (4.8 \times 10^{-11}/1.6 \times 10^{-10})\, Br$ GeV

$P = 0.3\, Br$ GeV/c

9.68 Let the proton be at a distance d from S, the earth's centre. Under the influence of magnetic field it will describe an arc of a circle of radius r.

From Fig. 9.9, it is clear that

$(R + r)^2 = d^2 + r^2$

Or $2\,R\,r = d^2 - R^2 \approx d^2$ (Because $R << d$)

Therefore $r = d^2/2R = (1{,}000\,R)^2/2R$

$= 5 \times 10^5\, R$

The momentum, $p = 0.3\, Br$

$P = (E^2 - m^2)^{1/2} = (10.94^2 - 0.94^2)^{1/2}$

$= 10.9$ GeV/c

$B = p/0.3r = 10.9/0.3 \times 5 \times 10^5 \times 6.4 \times 10^6$

$= 1.13 \times 10^{-11}\,\mathrm{T}$
$= 1.13 \times 10^{-7}\,\mathrm{G}$

Fig. 9.9

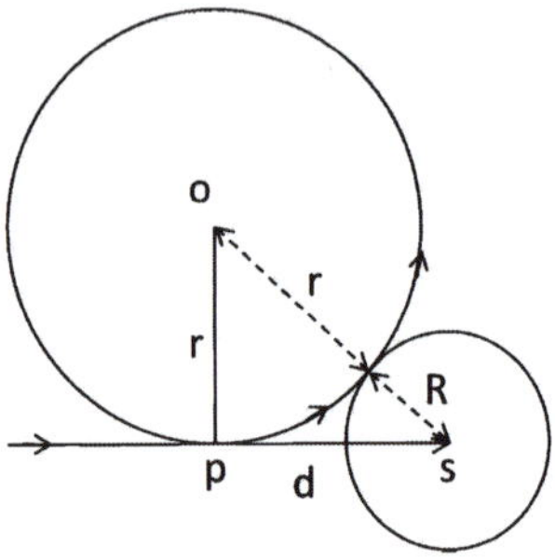

9.69 Electric force $= qE$
Magnetic force $= qvB$
If these two forces cross, that is
$qE = qvB$
then the condition for null deflection is
$v = E/B$
Now $v = (2T/m)^{1/2} = c(2T/mc^2)^{1/2} = c(2 \times 20/940 \times 1{,}000)^{1/2} = 0.00652\,c$
$E = 500\,V/\mathrm{cm} = 5 \times 10^4\,\mathrm{V/m}$
Therefore $B = E/v = 5 \times 10^4/0.00652 \times 3 \times 10^8 = 0.0256\,\mathrm{T} = 256\,\mathrm{G}$

9.70 $p^2 = 2m\ T = 2m\ qV = 2mqEd$ (1)
Also $p = qBr$ (2)

Combining (1) and (2)

$m = qB^2 r^2/2Ed$ (3)

Further $d = at^2/2 = qE\ t^2/2m$ (4)

Using (3) in (4) and solving for t,
$t = Br/E$

9.71 Drop a perpendicular DA at D and extend the path DP to meet the extension of the initial path HC in E. From the geometry of the figure (Fig. 9.10) angle CAD $= \theta$. Drop a perpendicular DG on CA. In the triangle AGD, $\sin\theta = \mathrm{GD/AD} = d/R$, where R is the radius of curvature.

$$R = \frac{\sqrt{2mK}}{qB}$$
$$= \frac{\sqrt{2 \times 1.67 \times 10^{-27} \times 5 \times 10^5 \times 1.6 \times 10^{-19}}}{1.6 \times 10^{-19} \times 0.51}$$
$\approx 0.2\,\mathrm{m}.$
$\sin\theta = d/R = 0.1/0.2 = 0.5$
$\theta = 30^\circ$

Fig. 9.10

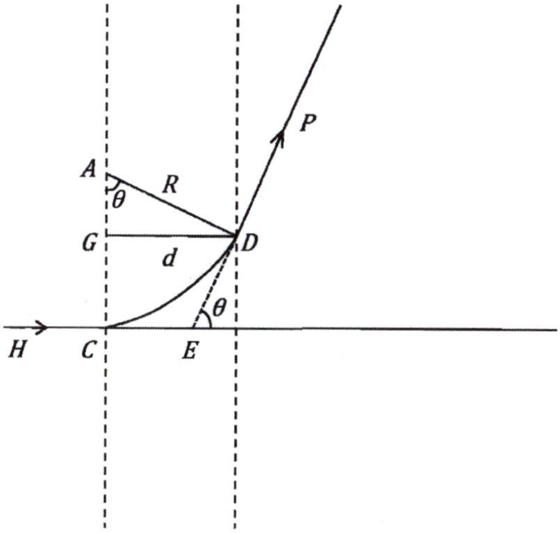

9.72 $t = \dfrac{2u \sin \alpha}{\alpha} = \dfrac{2u \sin \alpha}{Ee}.m$

Where the acceleration $\alpha = Ee/m$

$$t = \frac{2 \times 8 \times 10^5 \sin 30^\circ \times 9.1 \times 10^{-31}}{50 \times 1.6 \times 10^{-19}}$$

$= 91 \times 10^{-9}\,\text{s} = 91\,\text{ns}$

9.73 The component of the velocity, $\perp$ to the field is $v \sin \theta$. Equating the centripetal force to the magnetic force

$mv_\perp^2 = qv_\perp BR$

$R = \text{m}v_\perp/qB = v \sin \theta/(q/m)B$

$= \dfrac{3 \times 10^5 \sin 30^0}{10^8 \times 0.3}$

$= 0.5 \times 10^{-2}\,\text{m}$

$= 0.5\,\text{cm}$

9.74 $p = 0.3BR = (0.3)(10^{-4})(6.4 \times 10^6) = 192\,\text{GeV/c}$

$T \approx 192\,\text{GeV}$

9.75 Cosmic ray flux $= 1\,\text{cm}^{-2}\,\text{s}^{-1} = 10^4\,\text{m}^{-2}\,\text{s}^{-1}$

Earth's surface area, $A = 4\,\pi R^2 = 4\pi(64 \times 10^5)^2$

$= 5.144 \times 10^{14}\,\text{m}^2$

Cosmic rays incident on earth's surface

$= 5.144 \times 10^{14} \times 10^4$

$= 5.144 \times 10^{18}\,\text{m}^{-2}\,\text{s}^{-1}$

Cosmic rays energy delivered to earth

$= 5.144 \times 10^{18} \times 3\,\text{GeV}\,\text{s}^{-1}$

$= 1.5432 \times 10^{19}\,\text{GeV}\,\text{s}^{-1}$

$= 1.5432 \times 10^{19} \times 1.6 \times 10^{-10}\,\text{J}\,\text{s}^{-1}$

$= 2.47 \times 10^9\,\text{W}$

$= 2.47\,\text{GW}.$

9.76 $p = 0.3BR$ (1)

where p is in GeV/c, B in Tesla and R in metres.

Now $\theta \approx L/R = 0.3\ BL/p$ (2)

where L is the length of the magnet, θ is the angle of deflection, R is the radius of curvature of the circular arc of the path in the magnetic field, and l is the length of the straight path from the slits (Fig. 9.11).

$\Delta\theta = s/l = 0.3\ BL\Delta\ p/p^2$

$\Delta p/p = 1/100$

$S = 0.3 \times 1.2 \times 1.5 \times 10 \times (1/100)(1/25)$

$= 2.16 \times 10^{-3}\ \text{m} = 2.16\ \text{mm}$

Fig. 9.11

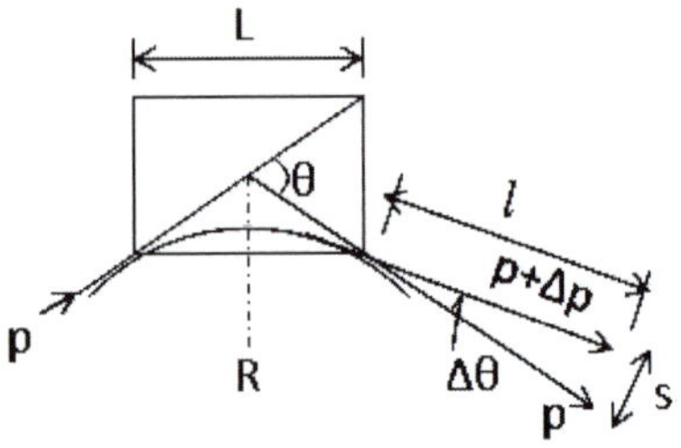

9.3.11 Betatron

9.77 $\Delta T = e\Delta\varphi/\Delta t = 1.6 \times 10^{-19} \times 50\ \text{J}$

$= 50\ \text{eV}$

9.78 (a) If N is the number of revolutions then

$(2\pi RN)(4f) = c$

The factor 4 arises due to the fact that the duty cycle is over a quarter of a period.

$N = c/8\pi\ Rf = 3 \times 10^8/8\pi \times 0.9 \times 50 = 2.65 \times 10^5$

(b) Radius, $R = 0.9\ \text{m}$

$T_{max} = BRec = 1.2 \times 0.9 \times 1.6 \times 10^{-19} \times 3 \times 10^8$

$= 5.184 \times 10^{-11}\ \text{J}$

$= 5.184 \times 10^{-11}/1.6 \times 10^{-13} = 324\ \text{MeV}$

(c) The average energy gained per revolution

$\Delta T = T_{max}/N = 324\ \text{MeV}/2.65 \times 10^5$

$= 122.2 \times 10^{-5}\ \text{MeV}$

$= 1.222\ \text{keV}$

9.3.12 Cyclotron

9.79 The resonance condition is

$\omega = qB/m$

Further $p = qBR$
$T = p^2/2m = q^2B^2R^2/2m$
Maximum radius
$R(\max) = (2mT_{\max})^{1/2}/qB$
$= (2 \times 1.67 \times 10^{-27} \times 6 \times 1.6 \times 10^{-13})^{1/2}/(1.6 \times 10^{-19}\ B)$
$= 0.354/B$ m

9.80 The resonance condition is
$\omega = 2\pi\ f = qB/m$
$B = 2\pi\ fm/q$
For deuterons, $B_d = 2\,\pi \times 1.2 \times 10^7 \times 2.01 \times 1.66 \times 10^{-27}/1.6 \times 10^{-19}$
$= 1.5715\,\text{T}$
For alpha particles, $B_\alpha = 2\,\pi \times 1.2 \times 10^7 \times 4.00 \times 1.66 \times 10^{-27}/2 \times 1.6 \times 10^{-19}$
$= 1.5637\,\text{T}$
Deuterons: At ejection kinetic energy
$E_f = (qBR)^2/2m = (1.6 \times 10^{-19} \times 1.5715 \times 0.3)^2/(2 \times 2.01 \times 1.66 \times 10^{-27})\text{J}$
$= 0.085 \times 10^{-11}\,\text{J}$
$= 5.3\,\text{MeV}$
If N is the number of orbits, t total time, and T_0 the time period then the mean energy increment per orbit is E/N and the average time for each orbit (supposed to be constant)
$T_0 = 2\pi/\omega_0 = 1/f_0$
$N = t/T_0 = t\,f_0$
Therefore $E_f/N = E_f/tf_0 = 2\,\text{eV}$ (the factor 2 is introduced because there are two gaps)
$t = E_f/(2\text{eV})f_0 = 5.3 \times 10^6/(2 \times 5 \times 10^4) \times 1.2 \times 10^7 = 4.42 \times 10^{-6}\,\text{s}$
$= 4.42\,\mu\text{s}$
α-particles: At ejection $E_f = (\text{qBR})^2/2\text{m}$
$= (2 \times 1.6 \times 10^{-19} \times 1.5637 \times 0.3)^2/(2 \times 4.0 \times 1.66 \times 10^{-27})$
$= 0.1697 \times 10\,\text{J}$
$= 10.6\,\text{MeV}$
Total time $t = E_f/(2 \times 2\,\text{eV})f_0$
$= 10.6 \times 10^6/(4 \times 5 \times 10^4 \times 1.2 \times 10^7)$
$= 4.42 \times 10^{-6}\,\text{s} = 4.42\,\mu\text{s}$

9.81 Cyclotron resonance condition is
$\omega = qB/m$
Because of relativistic increase of mass the resonance condition would be
$\omega' = qB'/m\gamma$
If $\omega' = \omega$, then $B' = B\gamma$
Fractional increase of magnetic flux density required is
$(B' - B)/B = \Delta B/B = \gamma - 1$
For protons of 20 MeV. $\gamma = (T/m) + 1 = (20/940) + 1 = 1.0213$
Therefore percentage increase of B is $(\gamma - 1) \times 100$
$= (1.0213 - 1) \times 100 = 2.13$

9.82 (a) Resonance condition for protons is
$B = 2\pi fm/q = (2\pi \times 5 \times 10^6)(1.6726 \times 10^{-27})/(1.6 \times 10^{-19})$
$= 0.3284\,\text{T}$
$T = (1/2)(Bqr)^2/m = (0.3284)^2(1.6 \times 10^{-19} \times 0.762)^2/(2 \times 1.6726 \times 10^{-27})$
$= 4.79 \times 10^{-13}\,\text{J} = 3\,\text{MeV}$

(b) For deuteron, the charge is the same as that of proton but mass is approximately double, the required magnetic field will be that for (a). So $B = 0.655\,\text{T}$.
The kinetic energy $\propto B^2/m$, so that it will be $(2^2/2) \times 3$ or 6 MeV

(c) For alpha particle the mass is approximately four times and charge is double compared to proton, so that the required magnetic field is twice that for proton, that is $B = 0.655\,\text{T}$, and kinetic energy will be $(2 \times 2)^2/4$, that is, four times the proton energy or 12 MeV.

9.83 The resonance frequency at the beginning
$f_0 = B_0\, q/2\pi\, m = 1.5 \times 1.6 \times 10^{-19}/(2\pi \times 3.34 \times 10^{-27}) = 11.44 \times 10^6\, c/s$
$= 11.44\,\text{Mc/s}$
The resonance frequency at the limiting radius is
$f = qB/2\pi\, m = 1.43 \times 1.6 \times 10^{-19}/2\,\pi \times 3.34 \times 10^{-27}$
$= 10.91 \times 10^6\,\text{c/s} = 10.91\,\text{Mc/s}$
Range of frequency modulation is 11.44–10.91 Mc/s.
$T_{\text{max}} = q^2B^2r^2/2\,m = (1.6 \times 10^{-19})^2(1.43)^2(2.06)^2/2 \times 3.34 \times 10^{-27}$
$= 3.3256 \times 10^{-11}\,\text{J}$
$= 3.3256 \times 10^{-11}/1.6 \times 10^{-13}\,\text{MeV} = 207.8\,\text{MeV}$

9.84 $B = \omega\, m/q = (2\pi \times 8 \times 10^6)(1.66 \times 10^{-27})/(1.6 \times 10^{-19})$
$= 0.52\,\text{T}$
$r = (2T/m\omega^2)^{1/2} = c(2T/mc^2\omega^2)^{1/2}$
$= 3 \times 10^8(2 \times 5/938 \times 4\pi^2 \times 8^2 \times 10^{12})^{1/2}$
$= 0.616\,\text{m}$

9.85 The cyclotron resonance condition is $\omega = qB/m$
For deuterons, $\omega_{\text{d}} = 1 \times B_{\text{d}}/m_{\text{d}}$
For alpha particles, $\omega_\alpha = 2 \times B_\alpha/m_\alpha$
If the resonance frequency is to remain unaltered $\omega_{\text{d}} = \omega_\alpha$
$B_{\text{d}}/B_\alpha = 2 \times (2.014102/4.002603) = 1.006396$
Fractional decrease of magnetic field
$(B_\alpha - B_{\text{d}})/B_{\text{d}} = -0.006355$
Percentage decrease $= 0.6355\%$

9.86 The energy of protons extracted from the accelerator is calculated from the equations
$E^2 = p^2 + m^2$
$p = 0.3\;BR = 0.3 \times 1.5 \times 2 = 0.9\,\text{GeV/c}$
$= 900\,\text{MeV/c}.$

$E = (p^2 + m^2)^{1/2} = (900^2 + 938^2)^{1/2}$
$= 1,300\,\text{MeV}$
$\gamma = 1,300/938 = 1.386$
$\beta = 0.692$
Maximum energy transferred to electron is
$E(\max) = 2m\beta^2\gamma^2 = 2 \times 0.511 \times 0.692^2 \times 1.386^2$
$= 0.94\,\text{MeV}$

9.87 Initially $\omega_0 = qB/m$
$\omega = qB/m\gamma = \omega_0/\gamma$
$\gamma = 1 + T/m = 1 + 469/938 = 1.5$
$f = f_0/\gamma = 20/1.5 = 13.33\,\text{Mc/s}$

9.3.13 Synchrotron

9.88 $P = 0.3\,BR$
$P = (T^2 + 2\,Tm)^{1/2} = (30^2 + 2 \times 30 \times 0.938)^{1/2}$
$= 30.924\,\text{GeV/c}$
$R = p/0.3\,\text{B} = 30.924/0.3 \times 1 = 103.08\,\text{m}$

9.89 $p = 0.3\,BR$
$P = (T^2 + 2Tm)^{1/2} = [3^2 + (2 \times 3 \times 0.938]^{1/2} = 3.825\,\text{GeV/c}$
$R = p/0.3\,\text{B} = 3.825/(0.3 \times 1.4) = 9.1\,\text{m}$

9.90 Initially $\omega_0 = B_0\, e/m$
Finally $\omega = 0.95\,B_0\,e/(m + T)$
$\omega/\omega_0 = 0.95m/(m + T)$

$$\frac{\omega_0 - \omega}{\omega_0} = \frac{T + 0.05\,m}{T + m} = \frac{313 + 0.05 \times 938}{313 + 938} = 0.288$$

Therefore Depth of modulation is 28.8 %

9.91 Radiation loss per revolution
$\Delta E = (4\pi\, e^2/3R)(E/mc^2)^4.1/4\,\pi\varepsilon_0$
$= (1.44 \times 4\pi/3R)(E/mc^2)^4\,\text{MeV-fm}$
Substituting $E = 300\,\text{MeV},\ mc^2 = 0.511\,\text{MeV}$
$R = 1.0 \times 10^{15}\,\text{fm}$
$\Delta E = 716 \times 10^{-6}\,\text{MeV} = 716\,\text{eV}$

9.92 $p = qBR$
$R = p/qB = m_0\gamma\beta\, c/qB = m_0\, c\,(\gamma^2 - 1)^{1/2}/qB$
But $\gamma = (T/m_0c^2) + 1 = \text{n} + 1$
$\therefore \gamma^2 - 1 = n^2 + 2n$
$\therefore R = (m_0\, c/qB)(n^2 + 2n)^{1/2}$

9.93 Using the result of Problem 9.92

$$R = \frac{m_0 c}{qB}(n^2 + 2n)^{\frac{1}{2}}$$

$$= \frac{m_0 c}{0.0147q}\left[\left(\frac{50}{940}\right)^2 + \left(\frac{2 \times 50}{940}\right)\right]^{1/2} = 22.48\frac{m_0 c}{q} \quad (1)$$

As the radius of synchrotron does not change, we can use the same relation at higher energy

$$R = \left(\frac{m_0 c}{1.2q}\right)(N^2 + 2N)^{1/2} \quad (2)$$

Combining (1) and (2) and solving for n, we find $N = 26$

$\therefore T = Nm_0\, c^2 = 26 \times 0.938 = 24.39\,\text{GeV}$

9.94 Using the results of Problem 9.92

$$q/m_0 = (c/BR)(n^2 + 2n)^{1/2} \quad (1)$$

Proton: $q/m_0 = 1/1 = 1$

$n = T/m_\text{p} = 1{,}000/1{,}000 = 1$

Using the above values in (1)

$$c/BR = 1/\sqrt{3} \quad (2)$$

Deuteron: $q/m_0 = 1/2$

Therefore $\frac{1}{2} = (n^2 + 2n)^{1/2}/\sqrt{3}$

Solving for n, we find $n = 0.3229$

Kinetic energy of deuteron $= \text{nm}_\text{d}$

$= 0.3229 \times 2{,}000\,\text{MeV}$

$= 646\,\text{MeV}$

$^3\text{He}: \ q/m_0 = 2/3$

Therefore $2/3 = (1/\sqrt{3})(n^2 + 2n)^{1/2}$

Solving for n, we find $n = 0.527$

Therefore Kinetic energy of $^3\text{He} = nm_{\text{He}^3}$

$= 0.527 \times 3{,}000\,\text{MeV}$

$= 1{,}583\,\text{MeV}$.

9.95 (a) $p = 0.3$ BR (GeV/c if B is in Tesla and R in metres)

$$P = (T^2 + 2T\ mc^2)^{1/2} = (0.5^2 + 2 \times 0.5 \times 0.938)^{1/2} = 1.09\,\text{GeV/c}$$

$$B = 18\,\text{kG} = 1.8\,\text{T}$$

$$R = p/0.3\,B = 1.09/(0.3 \times 1.8) = 2.02\,\text{m}.$$

(b) Energy of ions for acceptance

$$T_\text{s} = \pm(2\text{eV}.mc^2/\pi)^{1/2}[(\varphi_\text{s} - \pi/2)\sin\varphi_\text{s} + \cos\varphi_\text{s}]^{1/2} \quad (1)$$

Substitute in (1), $10\,\text{keV} = 0.01\,\text{MeV}$

$mc^2 = 938\,\text{MeV};\ \varphi_\text{s} = 30^0 = 0.5236$ radians

$T_\text{s} = \pm 2.5\,\text{MeV}$

(c) The initial frequency

$$f = Bqc^2/2\pi\ (mc^2 + T_\text{s}) \quad (2)$$

$B = 1.8\,\mathrm{T}$; $q = 1.6 \times 10^{-19}$ C; $c = 3 \times 10^8$ m/s; $mc^2 = 938 \times 1.6 \times 10^{-13}$ J

$T_s = \pm 2.5\,\mathrm{MeV} = \pm 2.5 \times 1.6 \times 10^{-13}$ J

Substituting the above values in (2) we find

$f_1 = 27.43$; $f_2 = 27.51$ Mc

(d) The required electrical frequency for 500 MeV protons is

$$f = Bqc^2/2\pi(mc^2 + T) \tag{3}$$

Put $(mc^2+T) = (938+500) \times 1.6 \times 10^{-13}$ J, B $= 1.8$ T, $c = 3 \times 10^8$ m/s and $q = 1.6 \times 10^{-19}$ C, in (3) to obtain $f = 17.94$ Mc.

Thus the range of frequency modulation is 27.51 – 17.94 Mc

9.96 The particles are constrained to move in a vacuum pipe bent into a torus that threads a series of electromagnets, providing a field normal to the orbit. The particles are accelerated once or more per revolution by radio frequency cavities. Both the magnetic field and the R.F. frequency must increase and the synchronized with the particle velocity as it increases.

The major energy loss is caused by the emission of synchrotron radiation.

The synchrotron radiation loss per turn is

$\Delta E = (4\pi/3R)(e^2/4\pi\varepsilon_0)(E/mc^2)^4$

For electron

$$\Delta E = \left(\frac{4\pi}{3}\frac{(1.44\,\mathrm{MeV.fm})}{(10^3 \times 10^{15}\,\mathrm{fm})}\right)\left(\frac{500 \times 10^3}{0.511}\right)^4$$

$= 5{,}526 \times 10^3$ MeV

$= 5{,}526$ GeV

an energy loss which is an order of magnitude greater than the electron energy to which the electrons are to be accelerated, which is impossible. On the other hand for protons the energy loss per turn will be smaller by a factor $(1,836)^4$ or 1.1×10^{13}

Thus for protons $\Delta E = 5,526/1.1 \times 10^{13} = 5 \times 10^{-10}$ GeV $= 0.5$ eV which is quite small.

As the synchrotron radiation losses for electrons in circular machines are much beyond tolerable limits, linear accelerators are employed which are capable of accelerating electrons up to 30–40 GeV.

9.97 Orbital frequency $f = 1/10^{-6} = 10^6$ c/s

$P = 0.3\,Br$

$f = qB/2\pi\,m$

$\therefore\ p = 0.3 \times 2\pi\,m\,f\,r/q = 0.3 \times 2\pi \times 1.67 \times 10^{-27} \times 10^6 \times 10/(1.6 \times 10^{-19})$

$= 0.1966$ GeV/c

$T^2 + 2mT = p^2$

Using $m = 0.938$ GeV/c^2 and solving for T, we find $T = 0.02$ GeV or 20 MeV

9.98 (a) At injection $\gamma = 1 + \frac{10}{0.511} = 20.57$
$\beta = 0.9988$
At extraction $\gamma = 1 + \frac{5{,}000}{0.511} = 9,786$
$\beta \approx 1$
Initial frequency $f_1 = \dfrac{\beta c}{2\pi r} = \dfrac{(0.9988) \times 3 \times 10^8}{2\pi \times 15} c = 3.1809\,\text{Mc}$
Final frequency $f_2 = \dfrac{1 \times 3 \times 10^8}{2\pi \times 15} = 3.1847\,\text{Mc}$
As the initial and final frequencies are nearly the same there is hardly any need to change the R.F. frequency.
(b) Total energy gain= $E_f - E_i = 5{,}000 - 10 = 4{,}990$ MeV
Energy gain per turn = 1 keV
$\therefore$ Number of turns, $n = \frac{4{,}990}{10^{-3}} = 4.99 \times 10^6$
(c) The period of revolution
$T_0 = 1/f = 1/3.18 \times 10^6 = 3.14 \times 10^{-7}$ s
Time between injection and extraction is
$T = nT_0 = 4.99 \times 10^6 \times 3.14 \times 10^{-7} = 1.567$ s
(d) The total distance traveled by the electron is
$d = 2\pi rn = 2\pi \times 15 \times 4.99 \times 10^6$
$= 4.7 \times 10^8\ \text{m} = 4.7 \times 10^5\ \text{km}.$

9.3.14 Linear Accelerator

9.99 (a) As there are 97 drift tubes, there will be 96 gaps. The energy gain per gap is $\Delta E = (50–2)/96 = 0.5\,\text{MeV/gap}$
After crossing the first gap, the protons are still non-relativistic and their velocity will be in the second drift tube will be

$v_2 = (2T/m)^{1/2} = \text{c}\,[2 \times (2 + 0.5)/938]^{1/2} = 0.073\,c$

The length of the second tube

$l_2 = v_2/2f = (0.073 \times 3 \times 10^{10}/2 \times 2 \times 10^8)\,\text{cm}$
$= 5.48\,\text{cm}$

In the last tube v is calculated relativistically.

$\gamma = 1 + T/m = 1 + 50/938 = 1.053$

$\beta = (\gamma^2 - 1)^{1/2}/\gamma = 0.313$

$l_f = \beta c/2f = 0.313 \times 3 \times 10^{10}/2 \times 2 \times 10^8 = 23.48\,\text{cm}$

(b) To produce 80 MeV protons, number of additional tubes required is $(80 - 50)/0.5 = 60$

9.100 (a) Beam current $i = q/t = 50 \times 5 \times 10^{11} \times 1.6 \times 10^{-19}$
$= 4 \times 10^{-6}$ amp
$= 4\,\mu\text{A}$

(b) Power output $W = iV$
The current is obtained from (a) and the voltage V corresponding to final voltage of 2 GeV would be 2×10^9.
Therefore $W = 4 \times 10^{-6} \times 2 \times 10^9$
$= 8{,}000$ W

9.101 There are five drift tubes in the section but only four gaps. Initial energy is 100 keV. Therefore, the final energy will be

(a) $E = 100\,\text{keV} + n\,\text{eV}$
$= 100 + 4 \times 100 = 500\,\text{keV}$
(b) $L = (1/2f)(2\text{eV}/m)^{1/2} \sum_1^4 \sqrt{n}$
$= \frac{1}{2\times 50\times 10^6}\left(\frac{2\times 1.6\times 10^{-19}\times 10^5}{1.67\times 10^{-27}}\right)^{1/2}(1+\sqrt{2}+\sqrt{3}+\sqrt{4})$
$= 0.269\,\text{m} = 26.9\,\text{cm}$

9.102 The longest drift tube will be in the end of the linear accelerator where maximum ion energy has been achieved.
$L = v/2f$
We can do non-relativistic calculations for v as the ion energy is not large
$v = (2T/M)^{1/2} = c\,(2T/Mc^2)^{1/2} = c\,(2 \times 80/12 \times 938)^{1/2} = 0.119\,c$
$L = 0.119 \times 3 \times 10^8/2 \times 25 \times 10^6 = 0.714\,\text{m}.$

9.3.15 Colliders

9.103 In the colliding beam experiments, for two colliding beams the reaction rate is written in terms of the "luminosity" L. The number of interactions per second is
$N = L\sigma$
where σ is the interaction cross-section in question.
If N_1 or N_2 = number of particles/bunch in each beam
A = area of cross-section of intersecting beams
N = number of bunches/beam
and f = frequency of revolution
then luminosity is simply given by
$L = n\,N_1\,N_2\,f/A$

9.104 Number of beam electrons/cm^2/s, $N_1 = nf/\pi r^2$
$= 6 \times 10^{11} \times 2 \times 10^6/\pi(0.12)^2 = 2.65 \times 10^{19}$
Number of beam positrons/s, $N_2 = 6 \times 10^{11}$
Expected production rate of $\mu^+\,\mu^-$ pairs/second
$= N_1\,N_2\sigma(e^+e^- \to \mu^+\mu^-)$
$= (2.65 \times 10^{19})(6 \times 10^{11})(1.4 \times 10^{-33})$
$= 0.022$

9.105 (a) $E^{*2} = (E_1 + E_2)^2 - (p_1 - p_2)^2$
$= E_1{}^2 + 2\,E_1\,E_2 + E_2{}^2 - p_1{}^2 + 2\,p_1\,p_2 - p_2{}^2$
$= \left(E_1{}^2 - p_1{}^2\right) + \left(E_2{}^2 - p_2{}^2\right) + 2(E_1\,E_2 + p_1\,p_2)$
$= m_1{}^2 + m_2{}^2 + 2(E_1\,E_2 + p_1\,p_2)$
If $E_1 >> m_1$ and $E_2 >> m_2$, $m_1{}^2$ and $m_2{}^2$ can be neglected and $p_1 \approx E_1$, $p_2 \approx E_2$
Therefore $E^{*2} \approx 4E_1\,E_2$
If the beams cross at an angle θ then
$E^{*2} = (E_1 + E_2)^2 - \left(p_1{}^2 + p_2{}^2 - 2\,p_1\,p_2\,\cos\theta\right)$
$\approx 2E_1\,E_2\,(1 + \cos\theta) = 4\,E_1\,E_2\,(1 + \cos\theta)/2$
Thus the available energy in the CMS is reduced by a factor of $(1 + \cos\theta)/2$ compared to head-on-collision.
(b) In the CMS $E^* = 25 + 25 = 50\,\text{GeV}$
With the fixed proton target
$E^* = (m^2 + m^2 + 2\,T_1\,m)^{1/2} = 50\,\text{GeV}$
Substituting $m = 0.94\,\text{GeV}$, we find $T_1 = 1329\,\text{GeV}$

9.106 When the protons travel toward each other with equal energy, and therefore with the same speed, their net momentum is zero. In that case the Lab system is reduced to the C-M system, and the observer is watching the events sitting in the C.M. system. The total energy is then,
$E^* = 10^{10}\,\text{eV} + 10^{10}\,\text{eV} + 10^9\,\text{eV} + 10^9\,\text{eV} = 22 \times 10^9$ or 22 GeV

Let E_1 be the energy of a proton in the lab system moving toward the target proton originally at rest, then if the total energy available in the CMS has to be the same as $E^* = 22\,\text{GeV}$,

$$\left(m_1{}^2 + m_2{}^2 + 2\,E_1\,M_1\right)^{1/2} = E^* = 22\,\text{GeV}$$

$$(1^2 + 1^2 + 2\,E_1 \times 1)^{1/2} = 22$$

Or $E_1 = 241$ GeV

Therefore required kinetic energy $= 241 - 1 = 240\,\text{GeV}$

9.107 Total energy available in the CMS is
$E^* = \left(m_1{}^2 + m_2{}^2 + 2\,E_1\,m_2\right)^{1/2}$
where E_1 is the total energy of projectile of mass m_1 and m_2 is the mass of the target.

$$E^* = 4M_0 + 2\,M_0 = 6M_0$$

$$m_1 = m_2 = M_0$$

$$6M_0 = \left(M_0{}^2 + M_0{}^2 + 2\,E_1\,M_0\right)^{1/2}$$

whence $E_1 = 17\,M_0$
Or $T_1 = E_1 - M_0 = 16\,M_0$

The significance of this result is that in the colliders a lot more energy is available than in fixed target experiments.

9.108 The CMS energy is calculated from the invariance of $E^2 - p^2$

$$E^{*2} = E^2 - p^2 = \left(E_p + E_e\right)^2 - \left(p_p - p_e\right)^2$$
$$= E_p^2 - p_p^2 + E_e^2 - p_e^2 + 2\left(E_p E_e + p_p p_e\right)$$
$$\approx m_p{}^2 + m_e{}^2 + 4E_p E_e$$

Since $m_p << E_p$ and $m_e << E_e$

$$E^* \approx \sqrt{4E_p E_e} = \sqrt{4 \times 820 \times 30} = 314\,\text{GeV}$$

Note that the HERA accelerator is different from other colliders in that the energy of the colliding particles (protons and electrons) is quite asymmetrical. It has been possible to achieve high momentum transfer in the CMS (20,000 GeV^2), necessary for the studies of proton structure.

9.109 Total CMS energy, $E^* \approx [2\ E_1\ E_2\ (1 + \cos\theta)/2]^{1/2}$
Substituting $E_1 = 20\,\text{GeV}$, $E_2 = 300\,\text{GeV}$, and $\theta = 10^0$
we find $E^* = 109\,\text{GeV}$

If the same energy ($E^* = 109\,\text{GeV}$) is to be achieved in a fixed target experiment, then the electron energy in the lab-system would be
$(2\ E\ M + M^2 + m^2)^{1/2} = E^*$
Neglecting M^2 and m^2
$E = (E^*)^2/2M = (109)^2/(2 \times 0.94) \approx 6,300\,\text{GeV}$

Chapter 10
Particle Physics – II

10.1 Basic Concepts and Formulae

Classification of particles

Table 10.1 gives the mass, mean lifetimes (τ) and common decay modes of elementary particles excluding resonances. Their classification into hadrons, photon and leptons is also indicated. Further subdivision of hadrons into mesons (pions and kaons) and baryons (nucleons and hyperons) is also shown. Electron (e^-), muon (μ^-), Tauon (τ^-), and the three neutrinos ν_e, ν_μ, and ν_τ constitute the class of leptons. A hadron stands for a strongly interacting particle distinguished from lepton which has only weak or electromagnetic interactions. Photon is the massless carrier of the electromagnetic field.

In the fourth family, graviton a massless particle of spin 2, the quantum of gravitation is not yet discovered.

Mesons and photon are Bosons (a particle of integral spin, $0\hbar, 1\hbar, 2\hbar, \ldots$). Bosons obey Bose-Einstein statistics, the wave function describing two identical bosons is symmetric under particle exchange. The baryons and leptons are Fermions (a particle with half integral spin, $\frac{1}{2}\hbar, \frac{3}{2}\hbar, \ldots$). Fermions obey Fermi-Dirac statistics, for which the wave function of two identical particle is anti symmetric (changes sign under particle exchange).

Antiparticle: Every particle has in association an antiparticle, with exactly the same mass and lifetime but opposite values of electric charge, magnetic moment, baryon number, lepton number, and flavor. Thus positron (e^+) is the anti particle of electron (e^-), antiproton (p^-) that of proton (p), $\bar{\nu_e}$ that of ν_e etc. Photon is the antiparticle of itself, so also π^0.

Fundamental interactions

1. Strong (nuclear) interaction
2. Electromagnetic interaction
3. Weak (nuclear) interaction
4. Gravitational interaction.

Table 10.1

			Particle	Mass (MeV/c^2)	$\tau(s)$	Common decay mode
Hadrons	Mesons	Pions	π^-, π^+	139	2.5×10^{-8}	$\mu\nu$
			π^0	135	1.8×10^{-16}	$\gamma\gamma$
		Kaons	K^-, K^+	494	1.2×10^{-8}	$\mu\nu$
			K^0	498		$\pi^\pm\pi^0$
			Mixture of K_1, K_2			
			K_1		0.89×10^{-10}	$\pi^+\pi^-$
						$\pi^0\pi^0$
			K_2		5.18×10^{-8}	$\pi^0\pi^0\pi^0$
						$\pi^+\pi^-\pi^0$
						$\pi\mu\nu$
						$\pi\mu e\bar{\nu}$
			η	550	10^{-18}	$\gamma + \gamma$
						$\pi^+ + \pi^- + \pi^0$
	Baryons	Nucleons	p	938.2	$> 10^{38}$	stable
			n	939.5	10^3	$pe^-\nu$
		Hyperons	Λ	1,115	2.6×10^{-10}	$p\pi^-$, $n\pi^0$
			Σ^+	1,189	0.8×10^{-10}	$p\pi^0$, $n\pi^+$
			Σ^0	1,192	10^{-20}	$\Lambda\gamma$
			Σ^-	1,197	1.6×10^{-10}	$n\pi^-$
			Ξ^0	1,314	3×10^{-10}	$\Lambda\pi^0$
			Ξ^-	1,321	1.8×10^{-10}	$\Lambda\pi^-$
			Ω^-	1,675	1.3×10^{-10}	$\Xi\pi$
						ΛK^-
		Photon	γ	0	∞	Stable
		Leptons	τ^-	1,784	3.4×10^{-13}	Electrons and neutrinos
			μ^-	105	2×10^{-6}	$e\nu\bar{\nu}$
			e^-	0.51	∞	Stable
			ν_e	0	∞	Stable
			ν_μ	<0.5		Stable
			ν_τ	<164	∞	Stable
		Graviton	?	0	–	Stable

Here we will be concerned with only the first three types. Table 10.2 summarizes the characteristics of the interactions.

Coupling constant: Particles interact through strong electromagnetic or weak charges. The square of the charge is known as the coupling constant. It enters the interaction matrix which determines the cross-sections and decay rates. Strictly speaking, the coupling constants are not constant but vary very gradually with the particle energy. They are called running constants.

QED (quantum electrodynamics) is the quantum field theory of the electro magnetic interaction whose predictions have been verified to a precision of one part in a billion.

QCD (quantum chromo dynamics) is the field theory of the strong color interaction between quarks.

Table 10.2 Characteristics of the fundamental interactions

	Strong	Electromagnetic	Weak	Gravitational
Carrier of field spin-parity(J^P) of quantum	Gluon 1^-	Photon 1^-	$W^{\pm}, Z^0 1^-, 1^+$	Gravitation ? 2^+
Coupling constant	$\alpha_s \leq 1$	$\alpha = \frac{e^2}{4\pi\hbar c} = \frac{1}{137}$	$\frac{G\left(Mc^2\right)^2}{(\hbar c)^3}$	$\frac{G_N M^2}{4\pi\hbar c} = 5 \times 10^{-40}$
Mass	0	0	80, 90 GeV	0
Relative strength	1	10^{-2}	$\leq 10^{-5}$	10^{-38}
Time scale	10^{-23}s	$10^{-18} - 10^{-20}$s	$>10^{-13}$s	
Range	$\leq 10^{-15}$m	∞	10^{-18}m	∞
Source	Colour charge	Electric charge	Weak charge	Mass

Standard model: As of today, the physics embodied in electrodynamics, chromo dynamics and electro-weak interaction is termed as the standard model of elementary particles.

Resonances or resonant states are the analogs of excited states of atoms. They are the excited states of familiar hadrons. Some of them are so short lived ($\sim 10^{-23} - 10^{-24}$s) that their direct detection is not possible. They ultimately decay into more familiar particles, like nucleons, mesons, leptons and photons. Because of their short lives their energy (mass) spread is enormous, due to uncertainty principle.

Baryon number (B) is the generalization of mass number. For nucleons and hyperons $B = +1$, for anti baryons, $B = -1$, for pions, kaons and other particles $B = 0.B$ is an additive quantum and is conserved in all the three types of interactions.

Isospin (T or I) is a quantum number applicable to hadrons and is conserved in strong interactions. It results from the near equality of u and d-quarks. This is reflected in the near equality of masses of charged multiplets such as (n, p), (π^+, π^0, π^-), (K^+, K^0) etc, as well as for the atomic nuclei once the coulomb interaction is removed. It is thus named because its mathematical description is entirely analogous to ordinary spin or angular momentum in quantum mechanics. The tables for Clebsch – Gordon coefficients for the addition of angular momenta (displayed in the summary of Chap. 3) can be directly used for the isospins. T is the additive quantum number. The charge multiplicity is given by $2T + 1$. The antiparticle has the same T as the particle but opposite T_3. T_3 is analogous to L_z for angular momentum. Total isospin (I) is conserved in strong interactions but breaks down in em and weak interactions. The third component (I_3) of a system of hadrons is conserved in strong and em interactions but is violated in weak interactions $\left(\Delta I_3 = \pm^1/_2\right)$.

The generalized pauli principle

$$(-1)^{l+s+I} = -1 \tag{10.1}$$

where l is the orbital angular momentum, s the spin and I the isospin.

Strangeness and strange particles: Heavy unstable particles such as kaons and hyperons which are produced copiously but decay slowly are named as strange particles. A new quantum number S, strangeness is introduced to distinguish them from other particles. Gellmann's formula

$$\frac{Q}{e} = T_3 + \frac{S + B}{2} \tag{10.2}$$

where B is the baryon number. K^+ and K^0 are assigned $S = +1$, while K^- and $\overline{K^0}$ have $S = -1$. The Σ hyperons and Λ hyperons have $S = -1$, Ξ^- and Ξ^0 have $S = -2$, Ω^- has $S = -3$. The ordinary particles, n, p, π^+, π^0, π^- have $S = 0$. The anti particles have opposite strangeness. Strangeness S is an additive quantum number. Table 10.3 summarises the strangeness S for various hadron multiplets.

Table 10.3 T and S assignments

T/S	−3	−2	−1	0	1	2	3
0	Ω^-		Λ		$\overline{\Lambda}$		$\overline{\Omega}^+$
				p			
		Ξ^0	K^-	n	K^+	$\overline{\Xi}^0$	
½		Ξ^-	K^0	$\overline{p}$	K^0	$\overline{\Xi}^+$	
				$\overline{n}$			
			Σ^+	π^+	$\overline{\Sigma}^-$		
1			Σ^0	π^0	$\overline{\Sigma}^0$		
			Σ^-	π^-	$\overline{\Sigma}^+$		

Strangeness S is conserved in strong and electromagnetic interactions, that is $\Delta S = 0$, but breaks down in weak interactions, such as decays, the rule being $\Delta S = \pm 1$.

Leptons: The electron, the muon and the tauon and their respective neutrinos as well as their antiparticles constitute the family of leptons. Leptons are assigned lepton number, $L = +1$ and antileptons, $L = -1$. The numbers L_e, L_μ and L_τ are separately conserved in all the three types of interactions. The lepton numbers are shown in Table 10.4.

Table 10.4 Lepton numbers

Q/e	$L_e = 1$	$L_\mu = 1$	$L_\tau = 1$
0 −1	$\begin{pmatrix} \nu_e \\ e^- \end{pmatrix}$	$\begin{pmatrix} \nu_\mu \\ \mu^- \end{pmatrix}$	$\begin{pmatrix} \nu_\tau \\ \tau^- \end{pmatrix}$
Q/e	$L_e = -1$	$L_\mu = -1$	$L_\tau = -1$
0 +1	$\begin{pmatrix} \overline{\nu_e} \\ e^+ \end{pmatrix}$	$\begin{pmatrix} \overline{\nu_\mu} \\ \mu^+ \end{pmatrix}$	$\begin{pmatrix} \overline{\nu_\tau} \\ \tau^+ \end{pmatrix}$

Helicity or handedness: The helicity H is defined as the ratio J_z/J where J_z is the component of spin along the momentum vector of the particle and J is the total spin. Massless particles have spin components $J_z = +J$ only. Thus $H = +1$ or -1. Neutrinos have $H = -1$ (left-handed) and anti-neutrinos have $H = +1$ (right-handed). Massive particles are not in pure helicity eigen states and contain both LH and RH components.

Parity (p or π): The concept of parity was mentioned in Chap. 3. The absolute intrinsic parity cannot be determined. Parity of a particle can be stated only relative to another particle. By convention baryons are assigned positive parity. All the antifermions have parity opposite to the fermions. On the other hand, bosons have the same parity for particle and antiparticle. Pions and Kaons have odd parity.

Parity is a multiplicative number, so that the parity of a composite system is equal to the parities of the parts. Thus, for a system comprising of particles A and B,

$$P(\text{AB}) = p(\text{A})\,.\,p(\text{B})\,.\,p(\text{orbital motion}) \tag{10.3}$$

where $p(\text{orbital motion}) = (-1)^l$ is the parity associated with the relative motion of the particles, l being the orbital angular momentum quantum number $(0, 1, 2, \ldots)$.

Overall parity is conserved in strong and em interactions but is violated in weak interactions.

Charge – conjugation: (C-parity) is the process of replacing a particle by an antiparticle or a system of particles by the anti particle (s).

In general, a system whose charge is not zero cannot be an eigen function of C. However if $Q = B = S = 0$, the effect of C is to produce eigen value ± 1.

C is conserved in strong and em interactions but not in weak interactions. For π^0, $C = +1$. For photon $C = -1$ and for n-photons

$$C = (-1)^n \tag{10.4}$$

G-parity: The operation G consists of rotation of 180^0 about the y-axis or z-axis in isospin space followed by charge conjugation.

G-parity for the pion is -1 and for baryon it is zero. It is a multiplicative quantum number. For a system of n pions

$$G = (-1)^n \tag{10.5}$$

G - parity is conserved in strong interaction and is a good quantum number for non-strange mesons. For a $N - \overline{N}$ system,

$$G = (-1)^{l+S+I} \tag{10.6}$$

Time reversal means changing the sign of time. Strong interactions are invariant under time reversal as evidenced by the absence of electric dipole moment of neutron and verification of the predicted ratio of forward and backward reactions at the same energy in the CMS.

If $A + B \to C + D$,

$$\text{then } \frac{\sigma_{AB \to CD}}{\sigma_{CD \to AB}} = \frac{(2s_A + 1)(2s_B + 1)P_C^{*2}}{(2s_C + 1)(2s_D + 1)P_A^{*2}} \tag{10.7}$$

where s is the spin of the particles, and p^* is the momentum, the particle beams being unpolarised.

The TCP theorem: A lagrangian which is invariant under proper Lorentz transformation is invariant with respect to the combined operation CPT, taken in any order. The predictions of the TCP theorem which have been verified are

(i) the existence of an anti particle for every particle.
(ii) The equality of masses, lifetimes, and magnetic moments of particles and anti particles.

Table 10.5 Conservation laws for the three types of interactions

	Quantity	Strong	Electromagnetic	Weak
1.	Q (Charge)	Yes	Yes	Yes
2.	B (Baryon no.)	Yes	Yes	Yes
3.	J (Angular momentum)	Yes	Yes	Yes
4.	Mass + Energy	Yes	Yes	Yes
5.	Linear momentum	Yes	Yes	Yes
6.	I (Isospin)	Yes	No	No
7.	I_3 (Third component of I)	Yes	Yes	No $\Delta I_3 = \pm 1/2$
8.	S (Strangeness)	Yes	Yes	No $\Delta S = \pm 1$
9.	p (parity)	Yes	Yes	No
10.	C (Charge conjugation)	Yes	Yes	No
11.	G (G-parity)	Yes	No	No
12.	L (Lepton number)	Yes	Yes	Yes

Quarks are the structureless fermions from which all the strongly interacting particles (hadrons) are built. The quarks occur with fractional baryon number, $B = \frac{1}{3}$ and charges $+\frac{2}{3}e$ or $-\frac{1}{3}e$. Baryons are built with three quarks (u, d, s) and mesons with a quark-antiquark pair. Their characteristics are shown in Table 10.6.

Table 10.6 Characteristics of quarks

Quark	Symbol	Mass(GeV/c^2)	Q/e	I	S	C	B^*	T
Down	d	~ 0.3	$-\frac{1}{3}$	½	0	0	0	0
Up	u	~ 0.3	$+\frac{2}{3}$	½	0	0	0	0
Strange	s	0.5	$-\frac{1}{3}$	0	−1	0	0	0
Charmed	c	1.6	$+\frac{2}{3}$	0	0	+1	0	0
Bottom	b	4.5	$-\frac{1}{3}$	0	0	0	−1	0
Top	t	175	$+\frac{2}{3}$	0	0	0	0	+1

For all the quarks spin-parity $J^{\mathrm{P}} = ½^+$. The quark structure of some of the hadrons is as follows.

$$p = uud; n = \text{udd}; \pi^+ = \text{u}\bar{d}; \pi^- = \bar{u}d; \Sigma^+ = \text{uus};\ \bar{k}^\circ = \bar{d}\text{s};$$

$$\Xi^- = \text{dss}; \Omega^- = sss; \text{K}^+ = \text{u}\bar{\text{s}}; \text{D}^+ = \text{c}\bar{d}; \pi^0 = \frac{u\bar{u} - \overline{dd}}{\sqrt{2}}; \Delta^{++} = uuu$$

Gellman's equation is generalized as:

$$Q/e = I_3 + 1/2(B + S + C + B^* + T) \qquad (10.8)$$

where B denotes Baryon number, C the charm, B^* the beauty or bottom and T the top.

Quarks are not observed as free particles as they are confined in hadrons.

In order to save pauli's principle, a new quantum number called "colour" is assigned to quarks. This has nothing to do with ordinary colour. The quarks appear in three colours, red, blue, and green. The antiquarks have anticolour. The observed hadrons are colourless. Color plays a role in strong interactions similar to charge in electromagnetic interaction.

The strong color field between quarks is mediated by massless gluons analogous to electro-magnetic field mediated by photons. While a photon does not carry electric charge, gluon itself carries color charge. There are eight types of gluons.

Charmonium ($c\bar{c}$) is the state formed from the charmed anti-charmed quark pair.
D mesons (D^0, $D^{\pm}$) contain a charmed quark or antiquark. They are pseudoscalar like pions ($J^P = 0^-$) and decay weakly predominantly into non-charmed strange mesons.
Flavor is a generic name to describe different types of quark and lepton.
Generation: The six flavors of quarks and of leptons are grouped into three generations or families. The quarks (*d, u*), (*s, c*) and (*b, t*) are of first, second and third generations; the corresponding leptons being (e^-, ν_e), (μ^-, ν_μ) and (τ^-, ν_τ)

Cabibbo – Kobayashi – Maskowa (CKM) matrix

$$V_{ij} = \begin{pmatrix} V_{ud} & V_{us} & V_{ub} \\ V_{cd} & V_{cs} & V_{cb} \\ V_{td} & V_{ts} & V_{tb} \end{pmatrix}$$

The probability for a transition from a quark q to a quark q' is proportional to $|V_{qq'}|^2$, the square of the magnitude of the matrix element. The diagonal elements of this matrix, V_{ud}, V_{cs}, V_{tb} which correspond to transitions within a family are short of unity by only a few percent. Hence, transitions $u \to d$, $c \to s$, and $t \to b$ are Cabibbo favoured.

The elements V_{us}, V_{cd}, V_{cb}, and V_{ts} are small but not zero. Hence transitions, $s \to u$, $c \to d$, $b \to u$ and $t \to s$ are Cabibbo suppressed.

The elements V_{ub} and V_{td} are nearly zero. Hence transitions $b \to u$ and $t \to d$ are Cabibbo forbidden.
The boson propagator: The rate of a particular reaction mediated by boson exchange is proportional to the square of the amplitude $f(q^2)$ multiplied by a phase factor and determines the cross-section or the decay of an unstable particle. Here q^2 is the square of the four-momentum transfer.

$$f(q^2) \propto \frac{1}{q^2 + m^2} \qquad (10.9)$$

where m is the mass of the exchanged boson. The quantity $(q^2 + m^2)^{-1}$ is known as the propagator term. For low momentum transfers the propagator is insensitive to q, but in high energy collisions for large momentum transfer, $f(q^2)$ decreases with increasing q.
Spurion is a hypothetical particle which is introduced into the initial state to convert the weak decay into a strong interaction.

Weak interaction- characteristics

Weak decays have long lifetimes ($>10^{-13}$s) and small interaction cross-sections, typically $\sim10^{-39}\,cm^2$. Charged leptons experience both weak and em interaction while neutrinos only weak interaction. Strangeness and parity are not conserved.

Depending on the extent to which leptons are involved, the weak decays are divided into three classes.

(1) *Leptonic decays* in which the decay products are leptons only, as in the decay $\mu^- \to e^- + \overline{\nu_e} + \nu_\mu$
(2) *Semi leptonic decays* which involve both hadrons and leptons. Examples are

(a) $n \to p + e^- + \overline{\nu_e}$ for $\Delta S = 0$
(b) $K^+ \to \pi^0 + e^+ + \nu_e$ for $|\Delta S| = 1$

(3) *Non-leptonic decays* which do not involve leptons. Parity or strangeness are not conserved, the selection rules being $\Delta S = \pm 1$ and $\Delta I = \pm\frac{1}{2}$ as in the decay, $\Lambda \to p + \pi^-$

In weak decays the flavors of the quark changes in contrast with strong or electromagnetic decays where the flavour is conserved. For example the decay of neutron, $n \to p + e^- + \overline{\nu_e}$ is represented by $udd \to udu + e^- + \overline{\nu_e}$ in which a d-quark is converted into a u-quark, $d \to u + e^- + \overline{\nu_e}$.
Charge current weak interaction is mediated by massive bosons $W^\pm$. The $W^\pm$ exchange results in the change of the lepton charge as in the anti neutrino absorption

$$\overline{\nu_e} + p \to n + e^+.$$

Neutral current weak interaction is mediated by the massive boson Z^0. The exchange of Z^0 does not cause the change of lepton as in $\nu_\mu + e^- \to \nu_\mu + e^-$.

Electro – Weak interaction

The electro magnetic interaction and weak interaction are two aspects of a single interaction called electro-weak interaction. The corresponding charges are related by θ_w the Weinberg angle

$$e = g\sin\theta_w \tag{10.10}$$

$$\sin^2\theta_w = 0.2319 \tag{10.11}$$

$$M_w/M_z = 0.88 \tag{10.12}$$

Feynman diagrams are short-hand for writing down individual terms in the calculation of transition matrix elements in various processes pictorially. As an example, consider the diagram in Fig. 10.1. The solid lines are the fermion lines. The convention used here is that time runs from left to right. The top solid lines represent electron and bottom lines μ^-. The diagram represents the scattering of a muon with electron, $\mu^- + e^- \to \mu^- + e^-$ by electromagnetic interaction. The dots corresponding to vertices are points where interactions occur. The wriggly line is the photon line. At the vertex, the electron emits a photon which is absorbed by the muon or viceversa. This corresponds to the lowest order of perturbation theory and is known as the first order Feynman diagram or leading Feynman diagram.

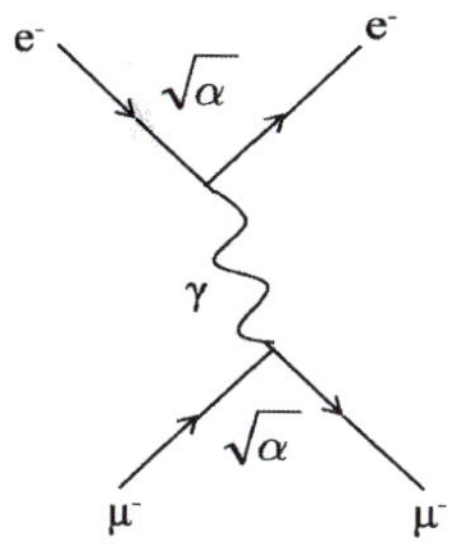

Fig. 10.1 Muon-electron scattering

The arrows on the solid lines towards the vertices indicate the direction of Fermions in time, the antifermions are indicated by reversed arrows, moving backward in time. The photon being an antiparticle of itself does not need any arrow on the photon line. The lines which begin and end within the diagrams are the internal lines correspond to virtual particles that is those which are not observed. The lines which enter or leave the diagram are the external lines which represent the observed or real particles. The external lines show the physical process of an event, while the internal lines indicate its mechanism. The direction of fermions is such that charge is conserved on each vertex. Also, the four-momentum is conserved at each vertex. The virtual particles which are exchanged (photon in Fig. 10.1) are not present in the initial or final state. They exist briefly during the interaction. In this short time τ energy can be violated compatible with the uncertainty principle, $\tau \,.\, \Delta E \approx \hbar$. Consequently, a virtual particle is not required to satisfy the relativistic relation, $E^2 = p^2 + m^2$. A virtual particle can be endowed with any mass which is different from that of a free particle. In Fig. 10.1, the exchanged photon couples to the charge of one electron at the top vertex and the second one at the bottom vertex. For each vertex the transition amplitude carries a factor which is proportional to e that is $\sqrt{\alpha}$ (square-root of fine structure constant). The transition matrix will be proportional to $\sqrt{\alpha}\sqrt{\alpha}$ or α. The exchanged particle also introduces a propagation term in the matrix element, the general form being $(q^2 + m^2)^{-1}$, where m is the mass of the exchanged particle and q is the four-momentum transfer.

In general, the same end result is obtained from a number of Feynman diagrams. The transition matrix element includes the superposition of amplitudes of all such diagrams.

For the weak interaction the exchanged particle is a massive boson, $W^{\pm}$ or Z^0, indicated by a wavy or broken line. For the strong interaction, the exchanged particle is a gluon indicated by a spring.

10.2 Problems

10.2.1 Conservation Laws

10.1 Are the following particle interactions allowed by the conservation rules? If so, state which force is involved and draw a Feynman diagram for the process.

[University of Aberystwyth 2003]

(i) $\mu^- \to e^- + \nu_\mu + \overline{\nu_e}$
(ii) $\Lambda \to \pi^+ + \pi^-$
(iii) $\nu_e + n \to p + e^-$
(iv) $\pi^0 \to \tau^+ + \tau^-$
(v) $e^+ + e^- \to \mu^+ + \mu^-$

10.2 Indicate, with an explanation, whether the following interactions proceed through the strong, electromagnetic or weak interactions, or whether they do not occur.

(i) $\pi^- \to \mu^- + \overline{\nu_\mu}$
(ii) $\tau^- \to \mu^- + \nu_\tau$
(iii) $\Sigma^0 \to \Lambda + \gamma$
(iv) $p \to n + e^+ + \nu_e$
(v) $\pi^- + p \to \pi^0 + \Sigma^0$
(vi) $\pi^- + p \to K^0 + \Sigma^0$
(vii) $e^+ + e^- \to \mu^+ + \mu^-$

10.3 Consider the decay of K^0 meson of momentum P_0 into π^+ and π^- of momenta p_+ and p_- in the opposite direction such that $p_+ = 2p_-$. Determine p_0.

[Mass of K^0 is 498 MeV/c^2; mass of $\pi^{\pm}$ is 140 MeV/c^2]

10.4 Indicate, with an explanation, whether the following interactions proceed through the strong, electromagnetic or weak interactions, or whether they do not occur.

(i) $\Xi^- \to \Sigma^- + \pi^0$
(ii) $\tau^- \to e^- + \overline{\nu_e} + \nu_\tau$
(iii) $\tau^+ \to \mu^+ + \gamma$

(iv) $\mu^+ + \mu^- \rightarrow \tau^+ + \tau^-$
(v) $p \rightarrow e^+ + \pi^0$
(vi) $\pi^0 \rightarrow \gamma + \gamma$
(vii) $\pi^- + p \rightarrow K^+ + \Sigma^-$
(viii) $\pi^- + p \rightarrow K^- + \Sigma^+$

10.5 The ρ^0 meson is known to have an intrinsic spin of $\hbar$, and pions zero spin. Show that the requirements of symmetry on the total wave function of the final state permit the decay

$\rho^0 \rightarrow \pi^+\pi^-$ but not $\rho^0 \rightarrow \pi^0\pi^0$

10.6 The baryon Ω^- has a mass 1,672 MeV/c^2 and strangeness $s = -3$. Which of the following decay modes are possible?

(a) $\Omega^- \rightarrow \Xi^- + \pi^0 (m_{\Xi^-} = 1{,}321\,\text{MeV}/c^2,\ S = -2,\ m_{\pi^0} = 135\,\text{MeV}/c^2)$
(b) $\Omega^- \rightarrow \Sigma^0 + \pi^- (m_{\Sigma^0} = 1{,}192\,\text{MeV}/c^2,\ S = -1,\ m_{\pi^-} = 139\,\text{MeV}/c^2)$
(c) $\Omega^- \rightarrow \Lambda + K^- \begin{pmatrix} m_\Lambda = 1{,}115\,\text{MeV}/c^2,\ S = -1 \\ m_{K^-} = 494\,\text{MeV}/c^2,\ S = -1 \end{pmatrix}$
(d) $\Omega^- \rightarrow n + K^- + \overline{K^0} (m_{\overline{K^0}} = 498\,\text{MeV}/c^2,\ S = -1)$

10.7 The following transitions have Q-values and mean lifetimes as indicated

	Transition	Q-value (MeV)	Lifetimes(s)
(a)	$\Delta^{++} \rightarrow p + \pi^+$	120	10^{-23}
(b)	$\pi^0 \rightarrow 2\gamma$	135	10^{-17}
(c)	$\mu^+ \rightarrow e^+ + \nu_e + \overline{\nu_\mu}$	105	2.2×10^{-6}
(d)	$\mu^+ + {}^{12}\text{C} \rightarrow {}^{12}\text{B} + \nu_\mu$	93	2×10^{-6}

State which interactions are responsible in each case and estimate the relative coupling strengths.

10.8 Indicate how the following quantities will transform under the P (space inversion) and T(time reversal) operation:

(a) Position coordinate $\mathbf{r}$
(b) Momentum vector $\mathbf{P}$
(c) Spin or angular momentum vector $\boldsymbol{\sigma} = \mathbf{r} \times \mathbf{P}$
(d) Electric field $\mathbf{E} = -\nabla V$
(e) Magnetic field $\mathbf{B} = \mathbf{i} \times \mathbf{r}$
(f) Electric dipole moment $\boldsymbol{\sigma}.\mathbf{E}$
(g) Magnetic dipole moment $\boldsymbol{\sigma}.\mathbf{B}$

10.9 The deuteron is a bound state of neutron and proton and has spin 1 and positive parity. Prove that it can exist only in the $^3S_1$ and $^3D_1$ states.

10.10 Show that (a) $\rho \to \eta + \pi$ is forbidden as a decay through strong interaction.
(b) $\omega \to \eta + \pi$ is forbidden as an electro-magnetic or strong decay.

10.11 (a) The ρ^0 and K^0 mesons both decay predominantly to $\pi^+ + \pi^-$. Explain why the mean lifetime of the ρ^0 is 10^{-23} s, while that of K^0 is 0.89×10^{-10} s.
(b) The Δ^0 and the Λ both decay to proton and π^- meson. Explain why the Δ^0 meson lifetime is $\sim 10^{-23}$ s while that of Λ is 2.6×10^{-10} s.

10.12 State and give reasons for the following decay modes of the ρ-meson ($J^P = 1^-, I = 1$) which are allowed by the strong or electromagnetic interaction

(a) $\rho^0 \to \pi^+\pi^-$
(b) $\rho^0 \to \pi^0\pi^0$
(c) $\rho^0 \to \eta^0\pi^0$
(d) $\rho^0 \to \pi^0\gamma$

10.13 Conventionally nucleon is given positive parity. What does one say about deuteron's parity and the intrinsic parities of u and d-quarks?

10.14 Which of the following hyperon decays are allowed in lowest order weak interactions?

(a) $\Xi^0 \to \Sigma^- + e^+ + \nu_e$
(b) $\Xi^0 \to p + \pi^- + \pi^0$
(c) $\Omega^- \to \Xi^0 + e^- + \overline{\nu_e}$
(d) $\Omega^- \to \Xi^- + \pi^+ + \pi^-$

10.15 Explain how the parity of K^- meson has been determined.

10.16 (a) Briefly define the following terms giving two examples of each:

(i) Hadron
(ii) Lepton
(iii) Baryon
(iv) Meson

(b) A π^0 meson at rest decays into two photons of equal energy. What is the wavelength (in m) of the photons? [The mass of the π^0 is 135 MeV/c^2]
[University of London]

10.2.2 Strong Interactions

10.17 (a) The Δ^{++} Resonance has a full width of $\Gamma = 120$ MeV. How far on average would such a particle of energy 200 GeV travel before decaying?
(b) Given that the width for W-boson decay is less than 6.5 MeV, estimate the limit for the corresponding lifetime.

10.18 Analyze the pion – proton scattering data in terms of isospin amplitudes $a_{1/2}$ and $a_{3/2}$ for the reactions:

$$\pi^+ + p \to \pi^+ + p \quad (1)$$

$$\pi^- + p = \pi^- + p \quad (2)$$

$$\pi^- + p = \pi^0 + n \quad (3)$$

Show that if $a_{1/2} \ll a_{3/2}$ then $\sigma_1 : \sigma_2 : \sigma_3 = 9 : 1 : 2$ and if $a_{1/2} \gg a_{3/2}$, then $\sigma_1 : \sigma_2 : \sigma_3 = 0 : 2 : 1$

10.19 Use the results of π-N scattering at the same energy,

$$\sigma^+(\pi^+ p \to \pi^+ p) = |a_{3/2}|^2$$

$$\sigma^-(\pi^- p \to \pi^- p) = \frac{1}{9}|a_{3/2} + 2a_{1/2}|^2$$

$$\sigma^0(\pi^- p \to \pi^0 n) = \frac{2}{9}|a_{3/2} - a_{1/2}|^2$$

to deduce the inequality $\sqrt{\sigma^+} + \sqrt{\sigma^-} - \sqrt{2\sigma^0} \geq 0$

10.20 Calculate the branching ratio for the decay of the resonance Δ^+ (1232) which has two decay modes

$$\Delta^+ \to p\pi^0$$
$$\to n\pi^+$$

10.21 A resonance X^+(1520) decays by the strong interaction to the final states $n\pi^+$ and $p\pi^0$ with branching ratios of approximately 36 and 18% respectively. What is its isospin?

10.22 Given that the ρ-meson has a width of 158 MeV/C^2 in its mass, how would you classify the interaction for its decay?

10.23 In which isospin states can (a) $\pi^+\pi^-\pi^0$ (b) $\pi^0\pi^0\pi^0$ exist?

10.24 The particles X and Y can be produced by strong interaction

$$K^- + p \to K^+ + X$$

$$K^- + p \to \pi^0 + Y$$

Identify the particles X (1,321 MeV) and Y(1,192 MeV) and deduce their quark content. If their decay schemes are $X \to \Lambda + \pi^-$ and $Y \to \Lambda + \gamma$, give a rough estimate of their lifetime.

10.25 The scattering of pions by proton shows evidence of a resonance at a centre of mass system momentum of 230 MeV/c. At this momentum, the cross-section for scattering of positive pions reaches a peak cross-section of 190 mb while that of negative pions is only 70 mb. What can you deduce about the properties of the resonance (a) from the ratio of the two cross-sections (b) from the magnitude of the larger?

[University of Bristol]

10.26 Consider the formation of the resonance Δ (1,236) due to the incidence of π^+ and π^- on p. Assuming that at the resonance energy the $I = 1/2$ contribution to the $\pi^- + p$ interaction is negligible show that at the resonance peak

$$\frac{\sigma(\pi^+ + p \to \Delta)}{\sigma(\pi^- + p \to \Delta)} = 3$$

10.27 Consider the reactions at the same energy

$$\pi^+ + p \to \Sigma^+ + K^+$$
$$\pi^- + p \to \Sigma^- + K^+$$
$$\pi^- + p \to \Sigma^0 + K^0$$

Assuming that the isospin amplitude $a_{1/2} \ll a_{3/2}$, show that the cross sections for the reactions will be in the ratio 9:1:2

10.28 Calculate the ratio of the cross sections for the reaction $\pi^- p \to \pi^- p$ and $\pi^- p \to \pi^0 n$ on the assumption that the two I spin amplitudes are equal in magnitude but differ in phase by 30°.

10.29 Negative pions almost at rest are absorbed by deuterium atoms and undergo the following reaction

$$\pi^- + d \to n + n$$

which is established by the direct observation of the neutrons which have a unique energy for this process. Assuming that the parity of neutron and deuteron is positive, show how the existence of the above reaction affords the determination of parity of negative pion.

10.30 The cross-section for $K^- + p$ shows a resonance at $P_K \approx 400\,\text{MeV/c}$. This resonance appears in the reactions

$$K^- + p \to \Sigma + \pi$$
$$\to \Lambda + \pi + \pi$$

But not in the reaction

$$K^- + p \to \Lambda + \pi^0$$

What conclusion can you draw on the isospin value of the resonance?

10.31 K^- mesons are incident with equal frequency on protons and neutrons and the following reactions are observed:

$$K^- + p \to \Sigma^+ + \pi^-$$
$$\to \Sigma^0 + \pi^0$$
$$\to \Sigma^- + \pi^+$$
$$K^- + n \to \Sigma^- + \pi^0$$
$$\to \Sigma^0 + \pi^-$$

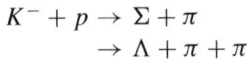

Show that the number of charged Σ's will be equal to twice the number of neutral Σ's.

10.32 In the rest system of the B^+-meson, the products of the strong interaction decay, $B^+ \rightarrow \omega^0 + \pi^+$ are found to be formed in an s-state. Deduce the spin, parity and isospin of the B-meson. What difference would it make to your conclusion if the decay took place by the week interaction (spin and parity are respectively 0^- for the charge triplet pion and 1^- for the ω^0).

[University of Durham]

10.33 The Δ (1,232) is a resonance with $I = 3/2$. What is the predicted branching ratio for $(\Delta^0 \rightarrow p\pi^-)/(\Delta^0 \rightarrow n\pi^0)$? What would be the ratio for a resonance with $I = 1/2$?

10.34 Show the position of pseudo scalar mesons $\pi^+, \pi^-, \pi^0, K^0, \overline{K^0}, K^+, K^-$ and η on the S – I_3 diagram.

10.35 Show the position of vector mesons $\rho^-, \rho^0, \rho^+, \varphi, \omega, k^{*0}, k^{*+}, k^{*-}$ and $\overline{k^{*0}}$ on S . I_3 diagram.

10.36 Show the position of Baryons $p, n, \Xi^-, \Xi^0, \Lambda, \Sigma^+, \Sigma^-$, and Σ^0 particles with spin-parity $\left(\frac{1}{2}\right)^+$ on the Y – I_3 diagram.

10.37 Describe the $(3/2)^+$ baryon decuplet on Y – I_3 diagram.

10.38 (a) Explain why at the same energy the total cross-sections

$$\sigma(\pi^- + p) \cong \sigma(\pi^+ + n), \text{ while } \sigma(K^- + p) \neq \sigma(K^+ + n)$$

(b) How can the neutral K-mesons, K^0 and $\overline{K^0}$ be distinguished?

10.39 A hyper nucleus is formed when a neutron is replaced by a Λ-hyperon. In the reactions of K^- in a helium bubble chamber, the mirror hyper nuclei ${}^4_\Lambda\text{He}$ and ${}^4_\Lambda\text{H}$ are produced

$$K^- + {}^4\text{He} \rightarrow {}^4_\Lambda\text{He} + \pi^-$$
$$\rightarrow {}^4_\Lambda\text{H} + \pi^0$$

Determine the ratio of the cross sections.

10.40 In the reaction $K^- + {}^4\text{He} \rightarrow {}^4_\Lambda\text{H} + \pi^0$, the isotropy of the decay products has established $J({}^4_\Lambda\text{H}) = 0$. Show that this implies a negative parity for the K^-- meson, regardless of the angular momentum state from which the K^--meson is captured.

10.41 Show that the reaction $\pi^- + d \rightarrow n + n + \pi^0$ cannot occur for pions at rest.

10.42 At 600 MeV the cross sections for the reactions $p + p \rightarrow d + \pi^+$ and $p + n \rightarrow d + \pi^0$ are $\sigma^+ = 3.15$ mb and $\sigma^0 = 1.5$ mb. Show that the ratio of the cross sections is in agreement with the iso-spin predictions.

[Osmania University]

10.43 Explain which of the following combination of particles can or cannot exist in $I = 1$ state

(a) $\pi^0\pi^0$
(b) $\pi^+\pi^+$
(c) $\pi^+\pi^-$
(d) $\Lambda\pi^0$
(e) $\Sigma^0\pi^0$
(f) $\pi^-\pi^-$?

10.2.3 Quarks

10.44 Describe the phenomena when a quark is struck by a high energy electron with a high enough momentum transfer.

10.45 The B^- meson is the lightest particle consisting of a b quark and $\overline{u}$ antiquark. Which type of interaction causes its decay. Describe with explanation its decay chain.

10.46 The 3γ decay of positronium (the bound state of e^+e^-) has a width that in QED is predicted to be $\Gamma(3\gamma) = 2(\pi^2 - 9)\alpha^6 m_e c^2/9\pi$, where α is the fine structure constant.

(a) If the hadronic decay of the $c\overline{c}$ bound state J/ψ (3,100) proceeds via an analogous mechanism, but involving three gluons, use the experimental hadronic width (fragmentation into hadrons occurring with probability unity) $\Gamma(3g) = 80\,\text{keV}$ to estimate the effective strong interaction coupling constant α_s.

(b) Determine α_s from the radiative width $\Gamma(gg\gamma) = 0.16\,\text{keV}$ of the $b\overline{b}$ bound state $\gamma(9{,}460)$

10.47 Calculate the ratio R of the cross section for $e^+e^- \rightarrow Q\overline{Q} \rightarrow$ hadrons to that for the reaction $e^+e^- \rightarrow \mu^+\mu^-$ as a function of increasing CMS energy up to 400 GeV. Assume the quark masses in GeV/c^2 up or down 0.31, strange 0.5, charm 1.6, bottom 4.6 and top 175.

10.48 Show that the quark model predicts the following cross-section relation

$$\sigma(\Sigma^- n) = \sigma(pp) + \sigma(K^- p) - \sigma(\pi^- p)$$

10.49 Using the additive quark model the total interaction cross-section is assumed to result from the sum of the cross-sections of various pairs. Assuming that $\sigma(qq) = \sigma(q\overline{q})$ prove the relation $\sigma(\Lambda p) = \sigma(pp) + \sigma(K^- n) - \sigma(\pi^+ p)$

10.50 The coulomb self- energy of a hadron with charge $+e$ or $-e$ is about 1 MeV. The quark content and rest energies (in MeV)of some hadrons are

$n(udd)940$, $p(uud)938$, $\Sigma^-(dds)1197$, $\Sigma^0(uds)1192$, $\Sigma^+(uus)1189$,

$K^0(d\overline{s})498$, $K^+(u\overline{s})494$

The u and d quarks make different contribution to the rest energy. Estimate this difference.

10.51 (a) What are the quark constituents of the states Δ^-, Δ^0, Δ^+, Δ^{++}?
(b) Assuming the quarks are in states of zero angular momentum, what fundamental difficulty appears to be associated with the Δ states, which have $I = 3/2$ and how is it resolved?
(c) How do you explain the occurrence of excited states of the nucleons with the higher values of J. What parities would the higher states have?

10.52 Use the quark model to determine the quark composition of (a) Σ^+, Σ^-, n and p (b) K^+, K^-, π^+, π^- mesons.

10.53 (a) What are the three particles described by taking the three identical quarks?
(b) What are the quantum numbers of the b quark?

10.54 Draw the quark flow diagrams for the decays (a) $\varphi \rightarrow K^+K^-$ (b) $\omega \rightarrow \pi^+\pi^-\pi^0$ (c) Show that the decay $\varphi \rightarrow \pi^+\pi^-\pi^0$ is suppressed.

10.55 At a beam energy of 60 GeV, $\sigma(\pi^+ p) \cong \sigma(\pi^- p) = 25$ mb while $\sigma(pp) \cong \sigma(pn) = 38$ mb. Show that the ratio of cross sections $\sigma(\pi N)/\sigma(NN)$ can be explained by simple quark model.

10.56 The production of a leptonic pair in a pion–nucleon collision is explained by the Drell–Yan mechanism which consists of the annihilation of the anti-quark from the pion with a quark from the nucleon, producing a virtual photon that transforms to a muon pair. Show that the cross section in π-^{12}C collisions away from heavy meson resonances is predicted as

$\sigma(\pi^- C)/\sigma(\pi^+ C) = 4 : 1$

[Courtesy D.H. Perkins, Introduction to High Energy Physics, University of Cambridge Press]

10.57 Below the production threshold of charm particles, the cross-section for the reaction $e^+e^- \rightarrow \mu^+\mu^-$ is 20 nb. Estimate the cross-section for hadron production

10.58 In the quark model, a meson is described as a bound quark-antiquark state. It is usual to represent the potential energy between q and $\overline{q}$ by $V(r) = -\frac{A}{r} + Br$ where A and B are positive constants. At $r >$ few fermis, A is negligible. Use the method of variation, with the trial function $\psi(r) = e^{-r/a}$ to show that the ground state energy is given by $E_0 = 2.96\left(\frac{B^2\hbar^2}{m_q}\right)^{1/3}$, where m_q is the quark mass.

10.2.4 Electromagnetic Interactions

10.59 The reaction $e^+e^- \rightarrow \mu^+\mu^-$ is studied using colliding beams each of energy 10 GeV and at these energies the reaction is predominantly electromagnetic. Draw the lowest order Feynman diagram. The differential cross-section is given by

$$\frac{d\sigma}{d\Omega} = \frac{\alpha^2\hbar^2c^2}{4E_{CM}^2}(1+\cos^2\theta)$$

where E_{CM} is the total centre of mass energy and θ is the scattering angle with respect to the beam direction. Calculate the total cross-section at this energy.

10.60 (a) The Σ^0 hyperon decays to $\Lambda + \gamma$ with a mean lifetime of 7.4×10^{-20} s. Estimate its width.

(b) Explain why the absence of the decay $K^+ \rightarrow \pi^+ + \gamma$ can be considered an argument in favor of spin zero for K^+ meson.

10.61 Positronium is the bound state of positron and electron. It is found either in the singlet s-state (para-positronium) or in a triplet s-state (ortho-positronium). Show how the C-invariance restricts the number of photons into which the positronium can annihilate for these two types of systems

10.62 Which of the following processes are allowed in electro-magnetic interactions, and which are allowed in weak interactions via the exchange of a single $W^\pm$ or Z^0?

(a) $K^+ \rightarrow \pi^0 + e^+ + \nu_e$
(b) $\Sigma^0 \rightarrow \Lambda + \nu_e + \overline{\nu_e}$

10.2.5 Weak Interactions

10.63 Estimate the rate of decay for $D^+(1{,}869) \rightarrow e^+ +$ anything, $D^0(1{,}864) \rightarrow e^+ +$anything. Given the branching fractions $B = 19\%$, and 8% respectively, $\tau_{D^+} = 10.6 \times 10^{-13}$ s, $\tau_{D^0} = 4.2 \times 10^{-13}$ s

10.64 It is observed that the cross section for neutrino-electron scattering falls by 20% as the momentum transfer increases from very small values to 30 GeV/c. Deduce the mass of the exchanged boson.

10.65 Estimate the number of $W^+ \rightarrow e^+\nu_e$ events produced in 10^9 pp^- interactions [The cross-sections $\sigma(pp^- \rightarrow W^+) = 1.8$ nb and $\sigma(pp^- \rightarrow \text{anything}) = 70$ mb]

(University of Cambridge, Tripos 2004)

10.66 Use Cabibbo theory to explain the difference in the decays $D^+ \rightarrow \overline{K^0}\mu^+\nu_\mu$ and $D^+ \rightarrow \pi^0\mu^+\nu_\mu$. Given that the D^+ consists of a c quark and $\overline{d}$ antiquark.

10.67 Show that the ratio of decay rates

$$R \equiv \frac{\Gamma(\Sigma^- \rightarrow n + e^- + \overline{\nu_e})}{\Gamma(\Sigma^- \rightarrow \Lambda + e^- + \overline{\nu_e})} \cong 17$$

10.68 Use lepton universality and lepton-quark symmetry to estimate the branching fraction for the decay $\tau^- \to e^- + \overline{\nu_e} + \nu_\tau$. Ignore final states that are cabibbo-suppressed relative to the lepton modes.

10.69 The following decays are all ascribed to the weak interaction, resulting in three final state particles. For each process, the available energy Q in the decay is given as well as the mean lifetime.

	Q(MeV)	τ(s)
(a) $\mu^+ \to e^+ + \nu_e + \overline{\nu_\mu}$	106	2.19×10^{-6}
(b) $n \to p + e^- + \overline{\nu_e}$	0.78	900
(c) $\tau^+ \to e^+ + \nu_e + \overline{\nu_\tau}$	1776	3.4×10^{-13}
(d) $\pi^+ \to \pi^0 + e^+ + \nu_e$	4.1	2.56
(e) $^{14}O \to ^{14}N^* + e^+ + \nu_e$	1.81	198

Apply Sergent's law of radioactivity to show that the Q-values and mean lifetimes are compatible with the same weak coupling.

10.70 Which of the following reactions are allowed and which forbidden as under weak interactions?

(a) $\nu_\mu + p \to \mu^+ + n$
(b) $\nu_e + p \to e^- + \pi^+ + p$
(c) $K^+ \to \pi^0 + \mu^+ + \nu_\mu$
(d) $\Lambda \to \pi^+ + e^- + \overline{\nu_e}$

10.71 Introduce a fictitious particle called Spurion to calculate the ratio of decay rates $\Xi^- \to \Lambda + \pi^- / \Xi^0 \to \Lambda + \pi^0$

10.72 Consider the following decays of Σ hyperons

$$\Sigma^+ \to n + \pi^+$$

$$\Sigma^- \to n + \pi^-$$

$$\Sigma^+ \to p + \pi^0$$

Using the $\Delta I = ½$ rule, show that the triangle relation for the amplitudes is $a_+ + \sqrt{2}a_0 = a_-$

10.73 Λ hyperon can decay predominantly through the non-leptonic modes $\Lambda \to p + \pi^-$ and $\Lambda \to n + \pi^0$. Introduce a fictitious particle called spurion of isospin ½ to convert the decay into a reaction and determine the branching ratios for these modes.

10.74 Assuming that the entire energy resulting from the p–p chain reaction escapes from the sun's surface, calculate the flux of neutrinos received on earth. Take the earth-sun distance as 1.5×10^8 km. Assume that the total energy output of the sun is $L_\odot = 3.83 \times 10^{26}$ Js^{-1}, and that each α-particle produced implies the generation of 26.72 MeV

10.75 The observation of neutrinos emitted by the supernova SN 1987A 170,000 years ago provided a rough estimate of neutrino's mass. Assume that the neutrino energy was spread between 5 and 15 MeV over a period of 4 s. Estimate the upper limit for the mass of neutrino.

10.76 A particle X decays at rest weakly as follows

$X \rightarrow \pi^0 + \mu^+ + \nu_\mu$

Determine the following properties of X
(a) Charge
(b) baryon number
(c) Lepton number
(d) Isospin
(e) Strangeness
(f) spin
(g) boson or fermion
(h) lower limit on its mass in MeV/c^2
(i) Identity of X

10.77 The α-decay of an excited 2^- state in ^{16}O to the ground 0^+ state of ^{12}C is found to have a width $T_\alpha \cong 1.0 \times 10^{-10}$ eV. Explain why this decay indicates a parity-violating potential.

10.78 Given the mean life time of μ^+ meson is 2.197 μs and its branching fraction for $\mu^+ \rightarrow e^+ + \nu_e + \overline{\nu_\mu}$ is 100%, estimate the mean lifetime of τ^+ if the branching fraction B for the decay $\tau^+ \rightarrow e^+ + \nu_e + \overline{\nu_\tau}$ is 17.7%. The masses of muon and τ-lepton are 105.658 and 1,784 MeV.

10.79 State with reasoning which of the following particles may undergo two-pion decay?
(a) $\omega^0(J^{PI} = 1^{-0})$
(b) $\eta^0(J^{PI} = 0^{-0})$
(c) $f^0(J^{PI} = 2^{+0})$

10.80 Why is the decay $\eta \rightarrow 4\pi$ not observed?

10.81 Van Royen-Weisskopf proposed a formula for the partial width of the leptonic decays of the vector mesons. For the vector mesons $\rho^0(765)$, $\omega^0(785)$ and $\Phi^0(1{,}020)$ which have similar masses, the partial width $\Gamma \propto Q^2$ where $Q^2 = \left|\sum a_i Q_i\right|^2$ is the squared sum of the charges of the quarks in the meson. Show that

$\Gamma(\rho^0) : \Gamma(\omega^0) : \Gamma(\Phi^0) = 9 : 1 : 2$

[Courtesy D.H. Perkins, Introduction to High Energy Physics, University of Cambridge Press]

10.82 Classify the following semi-leptonic decays of the $D^+(1{,}869) = c\bar{d}$ meson as Cabibbo-allowed, Cabibbo-suppressed or forbidden in lowest order weak interactions.

(a) $D^+ \rightarrow K^+ + \pi^- + e^+ + \nu_e$
(b) $D^+ \rightarrow \pi^+ + \pi^- + e^+ + \nu_e$

10.83 Classify the following semileptonic decays of the $D^+(1{,}869) = c\bar{d}$ meson as Cabibbo-allowed, Cabibbo-suppressed or forbidden in lowest order weak interactions.

(a) $D^+ \to K^- + \pi^+ + e^+ + \upsilon_e$
(b) $D^+ \to \pi^+ + \pi^+ + e^- + \overline{\nu_e}$

10.84 Which of the following decays are allowed in lowest order weak interactions?

(a) $K^- \to \pi^+ + \pi^- + e^- + \overline{\nu_e}$
(b) $\Xi^0 \to \Sigma^- + e^+ + \nu_e$
(c) $\Omega^- \to \Xi^- + \pi^+ + \pi^-$

10.85 Which of the following decays are allowed and which are forbidden ?

(a) $K^0 \to \pi^- e^+ \nu_e$
(b) $K^0 \to \pi^+ e^- \overline{\nu_e}$
(c) $\overline{K^0} \to \pi^+ e^- \overline{\nu_e}$
(d) $\overline{K^0} \to \pi^- e^+ \nu_e$

10.86 A muon neutrino is generated at time $t = 0$ at a particle accelerator. Show that at a later time t the probability that it is still a muon neutrino is, in natural units and in the neutrino rest frame

$$P_\mu(t) = 1 - \sin^2 2\theta \sin^2 \left[\frac{(E_2 - E_1)t}{2} \right]$$

[Courtesy D.H. Perkins, Introduction to High Energy Physics, University of Cambridge Press]

10.87 (a) In Problem 10.86, write down an expression for the probability $P_e(t)$ that at the same time t, the neutrino has oscillated into an electron neutrino.
(b) Derive the expression for the time at which the probabilities $P_\mu(t)$ and $P_e(t)$ are first equal. Assuming that $m_{\nu_e} = 2\,\text{eV}$ and $m_{\nu_\mu} = 3\,\text{eV}$ and $\theta = 34^0$; find time t when beam energy is 1 GeV

10.88 Show how the following data prove the universality of the weak coupling constant. $\tau_\mu = 2.197 \times 10^{-6}\,\text{s}$, $\tau_\tau = 2.91 \times 10^{-13}\,\text{s}$, the branching fraction of the tauon, $\text{B}(\tau^+ \to e^+ \nu_e \overline{\nu_\tau}) = 0.178$, $m_\mu = 105.658\,\text{MeV/c}^2$, $m_\tau = 1{,}777\,\text{MeV}$. Note that $\Gamma(\mu \to e \nu_e \overline{\nu_\mu}) = \dfrac{G^2 m^5 \mu}{192\pi^3}$ in natural units and the Fermi constant $G \propto g^2$, where g is the weak charge also known as the coupling amplitude.

10.89 Consider the semi leptonic weak decays (a) $\Sigma^- \to n + e^- + \overline{\nu_e}$ (b) $\Sigma^+ \to n + e^+ + \nu_e$ Explain why the reaction (a) is observed but (b) is not.

10.90 The D^+ meson ($c\bar{d}$) decays via the weak interaction to $\overline{K^0}\, \mu^+ \nu_\mu$ Alternatively the D^+ can decay to $\pi^0 \mu^+ \nu_\mu$. What are the predictions of Cabibbo's theory for the relative rates of the two decays?

10.91 The neutral kaons K^0 and $\overline{K^0}$ are the charge conjugate of each other and are distinguished by their strangeness. However, they decay similarly and mixing can occur by a virtual process like $K^0 \Leftrightarrow \pi^+ + \pi^- \Leftrightarrow \overline{K^0}$. Starting with a pure beam of K^0's at $t = 0$ obtain the intensity of K^0's and $\overline{K^0}$'s at time t, in terms of the mean lifetimes $\tau_L(0.9 \times 10^{-10}\ s)$ and $\tau_s(0.5 \times 10^{-7}\ s)$ for the components K_L and K_s, long lived and short lived respectively.

10.92 Suppose one starts with a pure beam of K^0's which traverses in vacuum for a time of the order of 100 K_s mean lives so that all the K_s- component has decayed and one is left with K_L only. If now the K_L traverses a carbon screen some of the K_S states are regenerated. Explain this phenomenon of regeneration.

10.2.6 Electro-Weak Interactions

10.93 The observation of the process $\overline{\nu_\mu} e^- \rightarrow \overline{\nu_\mu} e^-$, signifies the presence of a neutral current interaction. Similarly, why does the process $\overline{\nu_e} e^- \rightarrow \overline{\nu_e} e^-$, not indicate the presence of such an interaction?

10.94 From the data on the partial and full decay width of Z^0 boson show that the number of neutrino generations is 3 only.

$$\Gamma_z(\text{total}) = 2.534\,\text{GeV},\ \Gamma(z^0 \rightarrow \text{hadrons}) = 1.797\,\text{GeV},$$

$$\Gamma(z^0 \rightarrow l^+ l^-) = 0.084\,\text{GeV}.\text{Theoretical value for } \Gamma(z^0 \rightarrow \nu_l \overline{\nu_l}) = 0.166\,\text{GeV}.$$

10.95 (a) What are the experimental signatures and with what detectors would one measure (a) $W \rightarrow e\nu$ and $W \rightarrow \mu\nu$ (b) $Z^0 \rightarrow e^+e^-$ and $Z^0 \rightarrow \mu^+\mu^-$

(b) The weak force is due to W, Z exchange, mass $\cong$ 100 GeV. Give the range in meters.

[University of London 2000]

10.96 Using the results of electro-weak theory of Salam and Weinberg, calculate the masses of W and Z bosons. Take the fine-structure constant $\alpha = 1/128$ and Fermi's constant $G_F/\hbar^3 c^3 = 1.166 \times 10^{-5}\,\text{GeV}^{-2}$ and Weinberg angle $\theta_W = 28.17°$

10.2.7 Feynman Diagrams

10.97 Explain which force is responsible for the following particle interactions and draw a Feynman diagram for each:

[University of Wales, Aberystwyth 2004]

(i) $\tau^+ \rightarrow \mu^+ + \nu_\mu + \overline{\nu_\tau}$

(ii) $K^- + p \rightarrow \Omega^- + K^+ + K^0$ $(\text{K}^- = \text{s}\overline{\text{u}};\ \text{K}^+ = \text{u}\overline{\text{s}};\ \text{K}^0 = \text{d}\overline{\text{s}};\ \Omega^- = \text{sss})$

(iii) $\overline{D^0} \rightarrow K^+ + \pi^-$ $(\overline{D^0} = \text{u}\overline{\text{c}};\ \text{K}^+ = \text{u}\overline{\text{s}};\ \pi^- = \text{d}\overline{\text{u}})$

10.98 Sketch the Feynman diagrams and replace the symbols l with the correct leptons or anti-leptons in the following:

[University of Cambridge, Tripos 2004]

(i) $l + n \rightarrow e^- + p$
(ii) $\tau^- \rightarrow \mu^- + \ell + \bar{l}$
(iii) $B^0 \rightarrow D^- + \mu^+ + l$
[The quark content of B^0 is $\bar{b}$d and D^- is $\bar{c}$d]

10.99 Draw the lowest order Feynman diagram for the following decays
(a) $\Delta^0 \rightarrow p\pi^-$
(b) $\Omega^- \rightarrow \Lambda K^-$

10.100 (a) Draw the lowest order Feynman diagram for $e^+e^- \rightarrow \nu_\mu \overline{\nu_\mu}$
(b) Draw the lowest order Feynman diagram for $D^0 \rightarrow K^-\pi^+$ and estimate the ratio of the transition rates.
[Quark contents (masses in MeV/c^2): $\Delta^+ = uud(1,232)$, $\Omega^- = sss(1,672)$, $\Xi^0 = uss(1,315)$, $p = uud(938)$, $n = udd(940)$, $\pi^0 = \frac{1}{\sqrt{2}}\left[u\bar{u} + d\bar{d}\right](135)$, $\pi^+ = u\bar{d}(140)$, $D^0 = c\bar{u}(1,865)$, $K^- = \bar{u}s(494)$]

[Adapted from University of Cambridge, Tripos 2004]

10.101 Draw the lowest order Feynman diagram for the following processes:
(a) $e^- - e^-$ elastic scattering (Moller scattering)
(b) $e^+e^- \rightarrow e^+e^-$ (Bhabha scattering)

10.102 Draw the lowest order Feynman diagram at the quark level for the following decays
(a) $\Lambda \rightarrow p + e^- + \overline{\nu_e}$
(b) $D^- \rightarrow K^0 + \pi^-$

10.103 Draw the Feynman diagrams at the quark level for the reactions:
(a) $\pi^- + p \rightarrow K^0 + \Lambda$
(b) $e^+ + e^- \rightarrow \overline{B^0} + B^0$, where B is a meson containing a b-quark.

10.104 Draw the lowest order Feynman diagram for the decay $K^- \rightarrow \mu^- + \bar{\nu}_\mu + \gamma$ and hence deduce the form of the overall effective coupling.

10.105 Explain with the aid of Feynman diagrams, why the decay $D^0 \rightarrow K^- + \pi^+$ can occur as a charged-current weak interaction at lowest order, but the decay $D^+ \rightarrow K^0 + \pi^+$ cannot.

10.106 Why is the mean lifetime of the charged pion much longer than that of the neutral pion? Draw Feynman diagram to illustrate your answer.

10.107 Draw Feynman diagrams for
(a) Bremsstrahlung
(b) pair production.

10.108 Draw Feynman diagrams for
(a) photo electric effect
(b) Compton scattering

10.109 Draw Feynman diagrams for the processes
(a) $e^+e^- \to q\overline{q}$
(b) $\nu_\mu + N \to \nu_\mu + X$.

10.110 Draw Feynman diagrams for the decays
(a) $\Xi^- \to \Lambda + \pi^-$
(b) $K^+ \to \pi^+\pi^+\pi^-$

10.111 (a) Draw the Feynman diagram for the semi leptonic decay of $D^+ \to K^0 + l + \overline{l}$.
(b) Draw Feynman diagram for the tauon decay $\tau^- \to \pi^- + \nu_\tau$

10.112 Draw Feynman diagrams for (a) Two-photon annihilation (b) Three-photon annihilation of positronium.

10.113 Draw the Feynman diagram for the decay $\Lambda \to p + \pi^-$

10.3 Solutions

10.3.1 Conservation Laws

10.1 (i) Weak decay

Fig. 10.2(a)

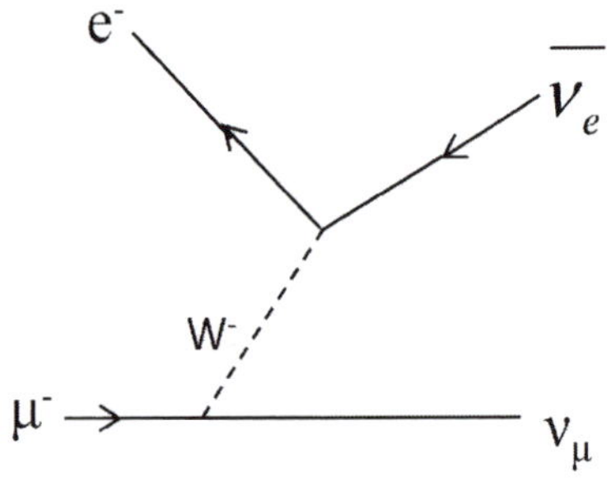

$\mu^- \to \bar{e}\, \nu_\mu \overline{\nu}_e$

(ii) Forbidden because Baryon number is violated
(iii) Charge current weak interaction

Fig. 10.2(b)

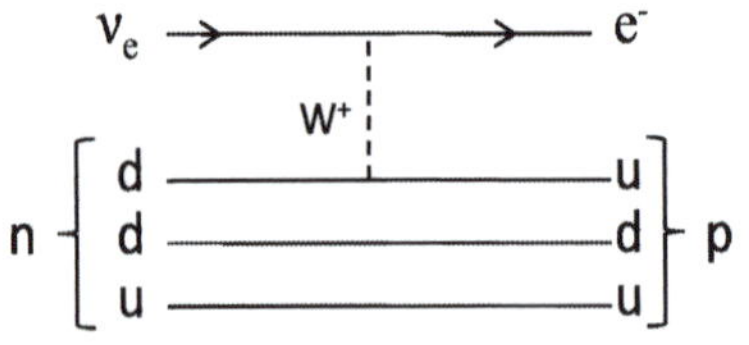

$\nu_e + n \to pe^-$

(iv) Forbidden because energy is not conserved

(v)

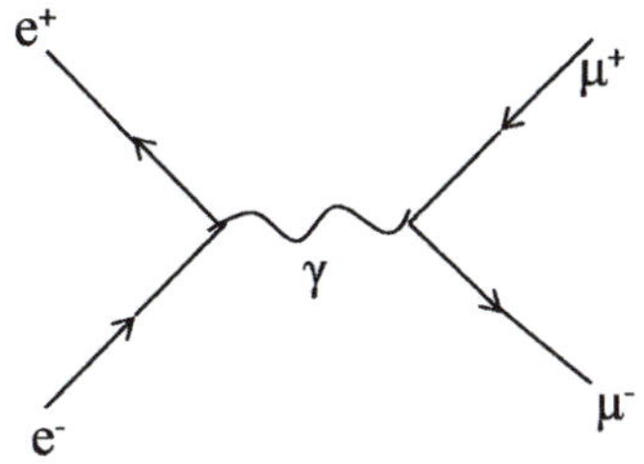

Fig. 10.2(c) A EM interaction

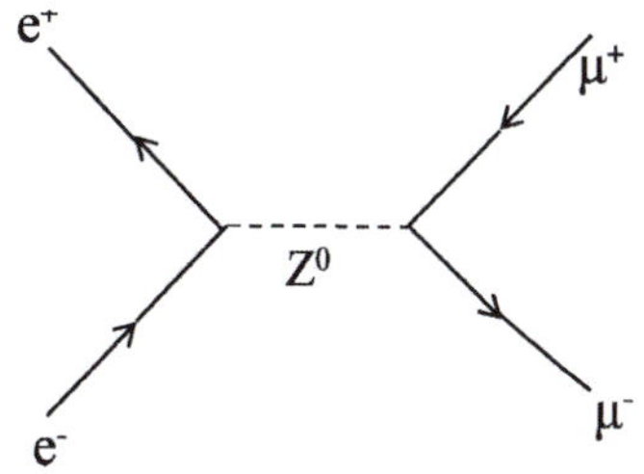

Fig. 10.2(d) Weak interaction

10.2 (i) Weak interaction because neutrino is involved
(ii) Does not occur because lepton number is violated
(iii) Electromagnetic interaction as gamma ray is involved and $\Delta S = 0$
(iv) Allowed as a weak decay if proton is bound but forbidden when proton is free because proton is lighter than the sum of masses of the product particles.
(v) Does not occur as a strong or electromagnetic interaction because $\Delta S \neq 0$
(vi) Strong interaction because $\Delta S = 0$ and other quantum numbers are conserved.
(vii) Weak interaction, because a lepton- antilepton pair is involved.

10.3 If P_0 is the momentum of Kaon, p_1 and p_2 the momenta of the pions, then momentum conservation requires

$$p_0 = p_1 - p_2 = 2p_2 - p_2 = p_2$$

Energy conservation requires

$$E_2 + E_1 = E_0$$

or

$$\sqrt{p_2^2 + m_\pi^2} + \sqrt{p_1^2 + m_\pi^2} = \sqrt{p_0^2 + m_K^2} \tag{1}$$

But $p_2 = p_0$ and $p_1 = 2p_2 = 2p_0$ (2)

Using (2) in (1) and solving the resultant equation

$$p_0 = \frac{m_{\text{K}}}{2}\left(\frac{m_{\text{K}}^2 - 4m_\pi^2}{2m_{\text{K}}^2 + m_\pi^2}\right)^{1/2} \tag{3}$$

Substituting $m_K = 498\,\mathrm{MeV/c^2}$ and $m_\pi = 140\,\mathrm{MeV/c^2}$, we get
$p_0 = 142.8\,\mathrm{MeV/c^2}$

10.4 (i) Does not occur because energy is not conserved
(ii) Weak interaction because neutrinos are involved
(iii) Does not occur because lepton number is not conserved
(iv) Weak interaction because leptons are involved
(v) Does not occur because of non-conservation of baryon number and lepton number
(vi) Electromagnetic interaction because γ-rays are involved, charge and c-parity are conserved.
(vii) Occurs as strong interaction because strangeness is conserved
(viii) Does not occur because strangeness is not conserved.

10.5 ρ^0 has $J^P = 1^-$. The conservation of angular momentum requires that the two π^0's are in $l = 1$ state of orbital angular momentum. This state is anti-symmetric and is therefore forbidden for identical bosons which require the state to be symmetric with respect to the exchange of two bosons.

10.6 (a) This decay mode is allowed and is observed. $\Delta S = 1$ as required for the weak decays of strange particles.
(b) The decay is forbidden as $\Delta S = 2$
(c) The decay is allowed as $\Delta S = 1$. Further, the rest mass energy of Ω^- is greater than the sum of the energies of decay products. Also Q/e and B are conserved.
(d) The decay is forbidden although $\Delta S = 1$ and Q/e and B are conserved. But E is violated.

10.7 (a) Strong interaction
(b) Electromagnetic interaction
(c) Weak interaction
(d) Weak interaction
Relative strength:$1 : 10^{-2} : 10^{-7} : 10^{-7}$

10.8 (a) $\mathbf{r} \to -\mathbf{r}$ under P- operation as $x \to -x,\ y \to -y$ and $z \to -z$ but $\mathbf{r} \to \mathbf{r}$ under T- operation.
(b) **P** reverses its sign under both P and T operation, $P \to -P$. Both **r** and **p** are known as polar vectors.
(c) $\boldsymbol{\sigma}$ or **L** are axial vectors. $\boldsymbol{\sigma} = \mathbf{r} \times \mathbf{p}$. Since both **r** and **p** change their sign under P-operation, **L** does not. However under T-operation **r** does not change sign but **p** does and so $\boldsymbol{\sigma}$ changes its sign.
(d) $E = -\partial V/\partial r$ for the above argument changes sign under P-operation as **r** changes its sign and does not under T-operation as **r** does not.
(e) The magnetic field like angular momentum is an axial vector $\mathbf{B} = \mathbf{i} \times \mathbf{r}$. Under p-operation $\mathbf{B} \to \mathbf{B}$ because $\mathbf{i} \to -\mathbf{i}$ and $\mathbf{r} \to -\mathbf{r}$ but under T-operation because $\mathbf{r} \to \mathbf{r}$ and $\mathbf{i} \to -\mathbf{i}$ so that $\mathbf{B} \to -\mathbf{B}$

(f) $\boldsymbol{\sigma}.\mathbf{E} \rightarrow -\boldsymbol{\sigma}.\mathbf{E}$ under both P and T-operations.
(g) Similarly $\sigma.B \rightarrow -\sigma.B$ under both p-operation and under T-operation.

10.9 The deuteron which is the nucleus of deuterium (heavy hydrogen) consists of one proton and one neutron. Since the parity of neutron and proton are +1, that of deuteron is also +1. Spin of deuteron is 1, and $l = 0$ mostly, with 4% admixture of $l = 2$ so that the parity determined by $(-1)^l = +1$ for $l = 0$ or 2. The deuteron is in a state of total angular momentum $J = 1$. Thus $J^P = 1^+$. Using the spectroscopic notation $^{2s+1}L_J$, deuteron's state is described by $^3S_1$ and $^3D_1$.

10.10 (a) For ρ, η and π, the I^G values are 1^+, 0^+ and 1^- respectively. Therefore G-parity is violated. Therefore, the decay of ρ is forbidden by strong interaction.
(b) For ω, η and π, the parities are −1. The overall parity of η and π will be $(-1)(-1) = +1$, as they are emitted in the s-state of relative angular momentum. Therefore, both strong and em interactions will be forbidden. The strong interaction will be forbidden also because of non conservation of isospin.

10.11 (a) In the decay $\rho^0 \rightarrow \pi^+\pi^-$, all the quantum numbers (Q/e, B, I, G, π) are conserved as required by a strong interaction. Therefore its lifetime ($\sim 10^{-23}$ s) is characteristic of a strong interaction. On the other hand the decay $K^0 \rightarrow \pi^+ \pi^-$, violates strangeness. It is therefore a weak decay, characterized by relatively long lifetime of the order of 10^{-10} s.
(b) $\Delta^0 \rightarrow p + \pi^-$
$\rightarrow n + \pi^0$
The Δ^0 has $T = 3/2$ and $T_3 = -1/2$. In both the decays T_3 is conserved (for the first one $p + \pi^-$ system has $T_3 = +1/2 - 1 = -1/2$, and for the second one $n + \pi^0$ system has $T_3 = -1/2 + 0 = -1/2$). Therefore, the decay proceeds through strong interaction with a characteristic life time of $\sim 10^{-23}$ s.
In the case of Λ,

$$\Lambda \rightarrow p + \pi^-$$
$$\rightarrow n + \pi^0$$

Λ has $T_3 = 0$, and therefore T_3 is violated. The decay being weak has the characteristic life time of 10^{-10} s.

10.12 (a) For ρ^0(770 MeV), $J^P = 1^-$ and $I^G = 1^+$ while for π (139 MeV), $J^P = 0^-$ and $I^G = 1^-$. In the decay $\rho^0 \rightarrow \pi^+ \pi^-$, all the quantum numbers (Q/e, B, I, G, π) are conserved. Note that the decay involves a large Q-value (491 MeV) so that the pions will come off with relative angular momentum of $l = 1$, contributing (−1) to overall parity. Thus, the overall parity is conserved (each pion has intrinsic parity −1).
(b) Decay is forbidden because of Bose symmetry.

(c) Decay via strong or electromagnetic interaction is forbidden because of violation of cp invariance. For ρ^0, $cp = -1$. But $c_\eta = c_{\pi^0} = +1$ because both decay to two gamma rays.
(d) The decay is allowed via electromagnetic interaction.

10.13 Parity of deuteron $\pi_{\rm d} = \pi_{\rm p}\,\pi_{\rm n}.\,(-1)^l$. As $\pi_{\rm p} = \pi_{\rm n} = +1$ for s or d state $l = 0$ or 2.

$\pi_{\rm d} = +1$

The intrinsic parity of quarks is assumed to be positive because the intrinsic parity of a nucleon (+) comes from the parities of three quarks and $l = 0$.

10.14 Decay (a) is forbidden by the $\Delta S = \Delta Q$ rule for the semi – leptonic decays and (b) is forbidden by the $\Delta S = 0, \pm 1$ rule for the hadronic weak decays. (c) and (d) are allowed and both have been experimentally observed.

10.15 Since strange particles are always produced in pairs, as in the reaction $\pi^- + p \to \Lambda + K^0$, the intrinsic parity of a strange particle can only be determined relative to that of another. Thus, for example, one can determine the kaon parity relative to that of Λ, which by convention is assigned a positive parity. Consider the reaction

$$
\begin{aligned}
K^- + He^4 &\to {}_\Lambda He^4 + \pi^- \\
&\to {}_\Lambda H^4 + \pi^0
\end{aligned}
$$

These reactions are known to occur in a helium bubble chamber. Now, the Λ is bound in an s-state relative to the nuclear core He^3 or H^3 which have positive parity. Furthermore, all the participants in the reaction are spinless. $L_{\rm initial} = L_{\rm final}$. The orbital angular momentum does not contribute to the parity because of s-state. The only relevant parities in the above reaction are $P_{\rm K} = P_\Lambda.\, P_{\pi^-} = -P_\Lambda$, as $P_{\pi^-} = -1$

The validity of the argument obviously hinges on the hyper-nuclei having zero spin. If the spin were 1, for example, angular momentum conservation would require $l = 1$ in the final state, thus reversing the conclusion. The spin of ${}_\Lambda H^4$ has also been experimentally determined to be indeed zero. It is concluded that the relative parity of K^- is negative.

10.16(a) (i) *Hadron* is an elementary particle which participates in strong interactions, examples being neutron and pion.
(ii) *Lepton* participates in weak interaction and if charged in em interaction as well, examples being electron and muon. Lepton number is universally conserved.
(iii) *Baryon* comprises nucleons and hyperons which participate in strong interaction, examples being proton and Σ-hyperon. Baryons are fermions and baryon number is universally conserved.
(iv) *Mesons* are the carrier of strong forces, examples being pion and kaon.

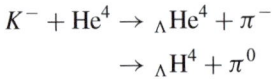

(b) Each photon will carry 67.5 MeV

$$\lambda = \frac{1241}{67.5 \times 10^6}\text{nm} = 18.385 \times 10^{-6}\text{nm} = 1.84 \times 10^{-14}\text{m}$$

10.3.2 Strong Interactions

10.17 (a) $\tau = \frac{\hbar}{\Gamma} = \frac{\hbar c}{\Gamma c} = \frac{197.3 \times 10^{-15}(\text{MeV} - \text{m})}{(120\text{MeV})(3 \times 10^8\ \text{m/s})} = 5.5 \times 10^{-24}\ s$

$$\gamma = 1 + \frac{200}{1.236} \cong 163$$

$$\beta \cong 1$$

$$d = \beta c \gamma \tau = 1 \times 3 \times 10^8 \times 163 \times 5.5 \times 10^{-24}\ \text{m}$$

$$\cong 2.7 \times 10^{-13}\ \text{m}$$

$$\cong 0.0003\ \text{nm}$$

a distance which is much less than the resolution obtainable by the available techniques. The best resolution obtained in photographic emulsions is only 1 μm.

(b) Uncertainty Principle:

$$\Gamma.\tau \geq \hbar$$

$$\tau \geq \frac{\hbar}{\Gamma} = \frac{\hbar c}{\Gamma c} = \frac{197.3\ \text{MeV} - \text{fm}}{6.5(\text{MeV}) \times 3 \times 10^8(\text{m/s})} = \frac{197.3 \times 10^{-15}}{19.5 \times 10^8}$$

$$\tau \geq 10^{-22}\ s$$

10.18 $\pi^+ + p \rightarrow \pi^+ + p$ (1)

$\pi^- + p \rightarrow \pi^- + p$ (2)

$\pi^- + p \rightarrow \pi^0 + n$ (3)

The cross-section is proportional to the square of the matrix element M_{if} connecting the initial and final states

$$M_{\text{if}} = \langle \psi_f\, |H|\, \psi_i \rangle$$

where H is the isospin operator, and

$$\sigma \propto |M_{if}|^2$$

As pion has $T = 1$ and proton $T = 1/2$, the reactions can proceed either through $I = 1/2$ or $I = 3/2$ channels. Designating the corresponding operators by H_1 and H_3 and the matrix elements for the reactions by M_1 and M_3, we can write

$$M_1 = \left\langle \psi_f \left(\frac{1}{2}\right) \middle| H_1 \middle| \psi_i \left(\frac{1}{2}\right) \right\rangle$$

$$M_3 = \left\langle \psi_f \left(\frac{3}{2}\right) \middle| H_3 \middle| \psi_i \left(\frac{3}{2}\right) \right\rangle$$

As I-spin is conserved there is no operator connecting different isospin states. Reaction (1) involves a pure state of $I = 3/2,\ I_3 = +3/2$.

Therefore, $\sigma_1 = C\,|M_3|^2$
where C = constant
For reaction (2) we have the mixture of $I = 1/2$ and $I = 3/2$ states.
Refering to the table for $3/2 \times 1/2$ C.G. coefficients we can write

$$|\psi_i\rangle = |\psi_f\rangle = \sqrt{\frac{1}{3}}\left|\phi\left(\frac{3}{2}, -\frac{1}{2}\right)\right\rangle - \sqrt{\frac{2}{3}}\left|\phi\left(\frac{1}{2}, -\frac{1}{2}\right)\right\rangle$$

Therefore, $\sigma_2 = C\left\langle\psi_f\,|H_1 + H_3|\,\psi_i\right\rangle^2 = C\left|\frac{1}{3}M_3 + \frac{2}{3}M_1\right|^2$

For the reaction (3) we have

$$|\psi_i\rangle = \sqrt{\frac{1}{3}}\left|\phi\left(\frac{3}{2}, -\frac{1}{2}\right)\right\rangle - \sqrt{\frac{2}{3}}\left|\phi\left(\frac{1}{2}, -\frac{1}{2}\right)\right\rangle$$

$$|\psi_f\rangle = \sqrt{\frac{2}{3}}\left|\phi\left(\frac{3}{2}, -\frac{1}{2}\right)\right\rangle + \sqrt{\frac{1}{3}}\left|\phi\left(\frac{1}{2}, -\frac{1}{2}\right)\right\rangle$$

Therefore, $\sigma_3 = C\left|\sqrt{\frac{2}{9}}M_3 - \sqrt{\frac{2}{9}}M_1\right|^2$
The ratio of cross-sections are

$$\sigma_1 : \sigma_2 : \sigma_3 = |M_3|^2 : \frac{1}{9}\,|M_3 + 2M_1|^2 : \frac{2}{9}\,|M_3 - M_1|^2$$

If $a_{1/2} << a_{3/2},\ M_1 << M_3,\ \sigma_1 : \sigma_2 : \sigma_3 = 9 : 1 : 2$
And if $a_{1/2} >> a_{3/2},\ M_1 >> M_3,\ \sigma_1 : \sigma_2 : \sigma_3 = 0 : 2 : 1$

10.19 From the known $\pi - N$ scattering cross-sections in terms of the amplitudes $a_{3/2}$ and $a_{1/2}$ obtained in Problem 10.18 we can construct a diagram for the amplitudes in the complex plane, Fig. 10.3

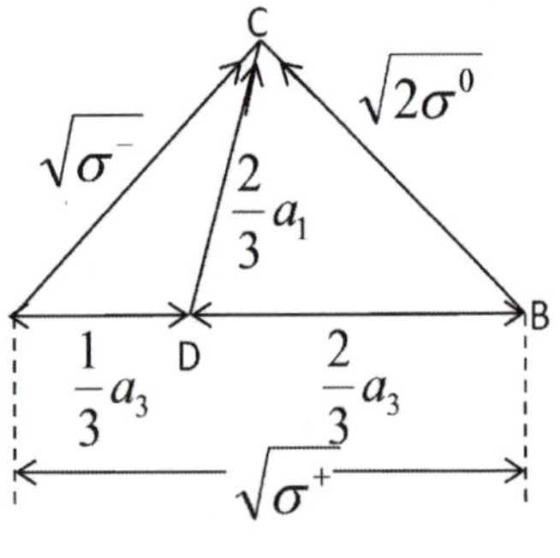

Fig. 10.3 Diagram of amplitudes

$$\sqrt{\sigma^+} = |a_3|$$

$$\sqrt{\sigma^-} = \frac{1}{3}\,|a_3 + 2a_1|$$

$$\sqrt{2\sigma^0} = \frac{2}{3}\,|a_3 - a_1|$$

where for brevity we have written a_1 for $a_{1/2}$ and a_3 for $a_{3/2}$
From the triangle the required inequality follows from the fact that the sum of two sides is equal to or greater than the third side.

10.20 The nucleon has $T = 1/2$ and pion $T = 1$. The Δ^+ has $T = 3/2$ and $T_3 = +1/2$. Using the C.G.C. for $1 \times 1/2$ (Table of Chap. 3), we have

$$\left|\frac{3}{2}, \frac{1}{2}\right\rangle = \sqrt{\frac{2}{3}}\,|1, 0\rangle \left|\frac{1}{2}, \frac{1}{2}\right\rangle + \sqrt{\frac{1}{3}}\,|1, 1\rangle \left|\frac{1}{2}, -\frac{1}{2}\right\rangle$$

$$\left|N\pi; \frac{3}{2}, \frac{1}{2}\right\rangle = \sqrt{\frac{2}{3}}\left|P\pi^0\right\rangle + \sqrt{\frac{1}{3}}\left|n\pi^+\right\rangle$$

$$\therefore \frac{\Gamma(\Delta^+ \to \pi^0 + p)}{\Gamma(\Delta^+ \to \pi^+ + n)} = \left(\sqrt{2/3}\right)^2 \Big/ \left(\sqrt{1/3}\right)^2 = 2$$

Note that for the charge states Δ^{++}, the strong decay is through only one channel ($\Delta^{++} \to p + \pi^+$), so also for Δ^-, viz $\Delta^- \to n + \pi^-$.

10.21 $X^+ \to n + \pi^+$

$\to p + \pi^0$

X^+ can have either $T = 3/2$ or 1/2. For $T = 3/2$ the predicted ratio $\Gamma(X^+ \to n\,\pi^+)/\Gamma(X^+ \to p\pi^0) = 1/2$ (as in Problem 10.20) which is in disagreement with the experimental ratio of 36/18 or 2.

If we assume the value $T = 1/2$ for X^+ the state $\left|\frac{1}{2}, \frac{1}{2}\right\rangle$ must be orthogonal to the state $\left|\frac{3}{2}, \frac{1}{2}\right\rangle$. Therefore

$$X^+\left|N\pi, \frac{1}{2}, \frac{1}{2}\right\rangle = -\sqrt{\frac{1}{3}}\left|p\pi^0\right\rangle + \sqrt{\frac{2}{3}}\left|n\pi^+\right\rangle$$

upto an overall phase factor.
The branching ratio
$\therefore\ \Gamma(X^+ \to n\pi^+)/\Gamma(X^+ \to p\pi^0) = 2/1$, which is in agreement with experimental ratio of 36/18 or 2/1.

10.22 $\tau = \frac{\hbar}{\Gamma} = \frac{\hbar c}{\Gamma c} = \frac{197.3\ \text{MeV-fm}}{158\ (\text{MeV}) \times 3 \times 10^8\ (\text{m/s})} = \frac{197.3 \times 10^{-15}\ (\text{MeV-m})}{474\ (\text{MeV}) \times 10^8\ (\text{m/s})} = 4 \times 10^{-24}$ s a lifetime which is characteristic of a strong interaction. Therefore, the ρ-meson decays via strong interaction.

10.23 We first write down the isospin for a pair of pions and then combine the resultant with the third pion. (a) Each pion has $T = 1$, so that the π^+, π^- combination has $I = 2, 1, 0$. When π^0 is combined, possible values are $I = 3, 2, 1, 0$ (b) Two π^0's give $I = 2$. When the third π^0 is added total $I = 3$ or 1.

10.24 From the conservation laws for strong interactions the quantum numbers for X are $B = +1$, $Q/e = -1$, $S = -2$. It is Xi hyperon (Ξ^-) and decays weakly with lifetime $\tau \sim 10^{-13}$ s. Its quark structure is dss; the quantum numbers for Y are $B = +1$, $Q/e = 0$, $S = -1$. It is a sigma hyperon (Σ^0) which decays electromagnetically with lifetime of the order of 10^{-20} s. Its quark structure is uds.

10.25 (a) From Problem 10.18, we have the result

$$|a_{3/2}|^2 \propto 190\,\text{mb}$$

$$\left\{\frac{1}{\sqrt{3}}|a_{3/2} + 2a_{1/2}|\right\}^2 \propto 70\,\text{mb}$$

Dividing one by the other, and solving we find $a_{1/2} = 0.0256\,a_{3/2}$. Thus the amplitude $a_{1/2}$ is negligible. The resonance is therefore characterized by $I = 3/2$.

(b) The fact that $\pi^+ p$ scattering cross-section has a fairly high value at $p = 230\,\text{MeV/c}$ implies that $a_{\frac{3}{2}}$ amplitude is dominant.

10.26 From the analysis of π-N scattering (Problem 10. 18) the ratio

$$\frac{\sigma(\pi^+ + p \to \Delta)}{\sigma(\pi^- + p \to \Delta)} = \frac{\left|a_{\frac{3}{2}}\right|^2}{\left\{\frac{1}{\sqrt{3}}\left|a_{\frac{3}{2}} + 2a_{\frac{1}{2}}\right|\right\}^2}$$

If we put $a_{1/2} = 0$, we get the ratio as 3.

10.27 The analysis is identical with that for πp reactions (problem 10.18). The Σ's are isospin triplet ($T = 1$) with the third components $T_3 = +1, 0, -1$ for Σ^+, Σ^0, Σ^- similar to π^+, π^0, π^-. Further, the K^+ and K^0 form a doublet ($T = 1/2$) with $T_3 = +1/2$ and $-1/2$, analogous to p and n. Therefore the result on the ratio of cross-section will be identical with that for pion – nucleon reactions as in Problem 10.18

$$\begin{aligned}&\sigma(\pi^+ + p \to \Sigma^+ + K^+) : \sigma(\pi^- + p \to \Sigma^- + K^+) :\\ &\quad \sigma(\pi^- + p \to \Sigma^0 + K^0)\\ &= |a_{3/2}|^2 : \tfrac{1}{9}|a_{3/2} + 2a_{1/2}|^2 : \tfrac{2}{9}|a_{3/2} - a_{1/2}|^2\end{aligned}$$

If $a_{1/2} << a_{3/2}$, then the ratio becomes 9:1:2.

10.28 Pions have $T = 1$ and nucleons $T = ½$, so that the resultant isospin both in the initial state and the final state can be $I = ½$ or 3/2. Looking up the table for Clebsch – Gordon Coefficients of 1 × 1/2 (Table 3.3) we can write

$$|\pi^- p\rangle = \sqrt{\frac{1}{3}}\left|\frac{3}{2}, -\frac{1}{2}\right\rangle - \sqrt{\frac{2}{3}}\left|\frac{1}{2}, -\frac{1}{2}\right\rangle$$

$$|\pi^0 n\rangle = \sqrt{\frac{2}{3}}\left|\frac{3}{2}, -\frac{1}{2}\right\rangle + \sqrt{\frac{1}{3}}\left|\frac{1}{2}, -\frac{1}{2}\right\rangle$$

Therefore, $\langle \pi^- p\,|H|\,\pi^- p\rangle = \frac{1}{3}a_3 + \frac{2}{3}a_1$

$$\langle \pi^0 n|\,H\,|\pi^- p\rangle = \frac{\sqrt{2}}{3}a_3 - \frac{\sqrt{2}}{3}a_1$$

The ratio of cross sections is

$$\frac{\sigma^-}{\sigma^0} = \frac{|a_3 + 2a_1|^2}{\left|\sqrt{2}\,(a_3 - a_1)\right|^2}$$

Put $a_3 = a_1 e^{i\Phi}$ with $\Phi = \pm 30^\circ$

$$\frac{\sigma_1}{\sigma_2} = \frac{1}{2}\frac{\left|e^{i\phi} + 2\right|^2}{\left|e^{i\phi} - 1\right|^2} = \frac{1}{2}\frac{\left(e^{i\phi} + 2\right)\left(e^{-i\phi} + 2\right)}{\left(e^{i\phi} - 1\right)\left(e^{-i\phi} - 1\right)} = \frac{1}{2}\frac{(5 + 4\cos\phi)}{(2 - 2\cos\phi)} = 15.8$$

10.29 Deuteron has spin 1 and is in the s-state ($l = 0$) mostly, with an admixture of $l = 2$. The total angular momentum $J = 1$ so that $J^p = 1^+$. Thus deuteron's state is described by $^3S_1$ and $^3D_1$. Next we consider the parity arising from the angular momentum of the π^-–d system. The time for a π^- to reach the $K-$ orbit in the deuterium mesic atom is estimated as $\sim 10^{-10}$ s. Furthermore, direct capture of π^- from 2p level is negligible compared to the transition from 2p to 1s level. It is therefore concluded that all π^-'s will be captured from the s-state of the deuterium atom before they decay. Therefore, the parity arising from the angular momentum is $(-1)^l = (-1)^0 = +1$. If p_{π^-} is the parity of π^- then the parity of the initial state will be

$$\pi(\text{initial}) = p_d . p_{\pi^-} .1 = p_{\pi^-}$$

In the final state the neutrons obey Fermi – Dirac statistics and must be in the anti symmetric state. The two neutrons can have total spin $S =$ zero (singlet, antisymmetric) or $S = 1$ (triplet, symmetric). Now the total wave function which is the product of space and spin parts must be antisymmetric.

$$\psi = \varphi(\text{space})\chi(\text{spin})$$

Now the symmetry of the spin function is $(-1)^{s+1}$ and that for the spatial part it is $(-1)^L$. Hence the overall symmetry of the wavefunction ψ under the interchange of both space and spin will be $(-1)^{L+S+1}$.

The initial state of $\pi^- - d$ system is characterized by $L = 0$ and $S = 1$ ($\because$ the deuteron spin $= 1$ and the pion spin is zero) and $L = 0$. Hence the total angular momentum of the initial state is 1. By conservation of angular momentum, final state must also have $J = 1$. The requirement $J = 1$ implies $L = 0,\ S = 1$ or $L = 1,\ S = 1$ or $L = 1,\ S = 0$ or 1, or $L = 2,\ S = 1$. The possible final states are enumerated $^3S_1$(symmetrical); $^3P_1$(antisymmetrical); $^3D_1$(symmetrical); $^1P_1$(symmetrical). Of these combinations $L = S = 1$ alone gives $L + S$ even.

Thus the only correct state is $^3P_1$ with parity $(-1)^l = -1$, which also requires negative parity for the initial state as parity is conserved in strong interactions. It follows that $p_{\pi^-} = -1$.

It may be pointed out that the assumed parity of $+1$ for neutron and deuteron is immaterial because the parities of baryon particles get cancelled in any reaction due to conservation of baryon number.

10.30 In the initial state $T(K) = ½$ and $T(p) = ½$, so that $I = 0$ or 1. In the first two reactions $I = 2$, 1 or 0; so that these reactions can proceed through $I = 1$ or 0. In the third reaction $T(\Lambda) = 0$ and $T(\pi^0) = 1$ so that $I = 1$ only. Since the resonance does not go through, the conclusion is that the resonance has $T = 0$.

10.31
$$K^- + p \to \Sigma^+ + \pi^- \quad (1)$$
$$K^- + p \to \Sigma^0 + \pi^0 \quad (2)$$
$$K^- + p \to \Sigma^- + \pi^+ \quad (3)$$
$$K^- + n \to \Sigma^- + \pi^0 \quad (4)$$
$$K^- + n \to \Sigma^0 + \pi^- \quad (5)$$

The initial $K^- + p$ system of two particles of $T = ½$ has $I_3 = 0$ and consists equally of $I = 0$ and $I = 1$ state. The final $\Sigma + \pi$ state will be

$$|\psi\rangle = \frac{1}{\sqrt{2}}(a_0\,|\phi\,(0,0)\rangle + a_1\,|\phi\,(1,0)\rangle)$$

For $I_1 = I_2 = 1$, referring to the C.G. Coefficients (Table 3.3) we can write

$$|\phi\,(0,0)\rangle = \frac{1}{\sqrt{3}}\left(\left|\Sigma^+\pi^-\right\rangle - \left|\Sigma^0\pi^0\right\rangle + \left|\Sigma^-\pi^+\right\rangle\right)$$

$$|\phi\,(1,0)\rangle = \frac{1}{\sqrt{2}}\left(\left|\Sigma^-\pi^+\right\rangle - \left|\Sigma^+\pi^-\right\rangle\right)$$

$$|\psi\rangle = \frac{1}{\sqrt{2}}\left[\left(\frac{a_0}{\sqrt{3}} - \frac{a_1}{\sqrt{2}}\right)\left|\Sigma^+\pi^-\right\rangle - \frac{a_0}{\sqrt{3}}\left|\Sigma^0\pi^0\right\rangle + \left(\frac{a_0}{\sqrt{3}} + \frac{a_1}{\sqrt{2}}\right)\left|\Sigma^-\pi^+\right\rangle\right]$$

$$\therefore (\Sigma^+\pi^-) : (\Sigma^0\pi^0) : (\Sigma^-\pi^+) = \frac{1}{2}\left(\frac{a_0}{\sqrt{3}} - \frac{a_1}{\sqrt{2}}\right)^2 : \frac{a_0^2}{6} : \frac{1}{2}\left(\frac{a_0}{\sqrt{3}} + \frac{a_1}{\sqrt{2}}\right)^2$$

The reactions of K^- with n go through $I = 1$ only. Since $I_3 = -1$ and the final $\Sigma\pi$ state is $a_1\,|\psi\,(1,-1)\rangle = \frac{a_1}{\sqrt{2}}\left[\left|\Sigma^-\pi^0\right\rangle - \left|\Sigma^0\pi^-\right\rangle\right]$

$$(\Sigma^-\pi^0) : (\Sigma^0\pi^-) = \frac{a_1^2}{2} : \frac{a_1^2}{2}$$

If K^- is incident with equal frequency on p and n, then

$$\Sigma^- + \Sigma^+ = \frac{1}{2}\left(\frac{a_0}{\sqrt{3}} - \frac{a_1}{\sqrt{2}}\right)^2 + \frac{1}{2}\left(\frac{a_0}{\sqrt{3}} + \frac{a_1}{\sqrt{2}}\right)^2 + \frac{a_1^2}{2} = \frac{a_0^2}{3} + a_1^2$$

$$\Sigma^0 = \frac{a_0^2}{6} + \frac{a_1^2}{2}$$

It follows that $\Sigma^- + \Sigma^+ = 2\Sigma^0$

10.32 $B^+ \to \omega^0 + \pi^+$

J^P	1^-	0^-
I	0	1

Assuming that the decay proceeds via strong interaction, the parity
$P_B = P_\omega P_\pi = (-1)(-1)(-1)^0 = +1$
$J_\omega = 1 + 0 = 1$ (because $l = 0$)
$I = 0 + 1 = 1$
For the B^+ meson, $J^P = 1^+$ and $I = 1$

In case of weak decay, the spin would still be 1 but it would not be meaningful to talk about I or P.

10.33 The nucleon has $T = 1/2$ and pion $T = 1$. The Δ^0 has $T = 3/2$ and $T_3 = -1/2$. Using Clebsch – Gordon coefficients (C.G.C) for $1 \times 1/2$ (Table 3.3 of Chap. 3), we have

$$\underset{\Delta^0}{\left|\frac{3}{2}, -\frac{1}{2}\right\rangle} = \sqrt{\frac{2}{3}}\,\underset{\pi^0}{|1, 0\rangle}\,\underset{n}{\left|\frac{1}{2}, -\frac{1}{2}\right\rangle} + \sqrt{\frac{1}{3}}\,\underset{\pi^-}{|1, -1\rangle}\,\underset{p}{\left|\frac{1}{2}, \frac{1}{2}\right\rangle}$$

The ratio of the amplitudes for the decays $\Delta^0 \to \pi^- p$ and $\Delta^0 \to \pi^0 n$ is given by the ratio of the corresponding C.G.C. and the ratio of the cross sections by the squares of the C.G.C. Thus, the branching ratio would be

$$\left(\sqrt{\frac{2}{3}}\right)^2 \Big/ \left(\sqrt{\frac{1}{3}}\right)^2 = 2$$

10.34

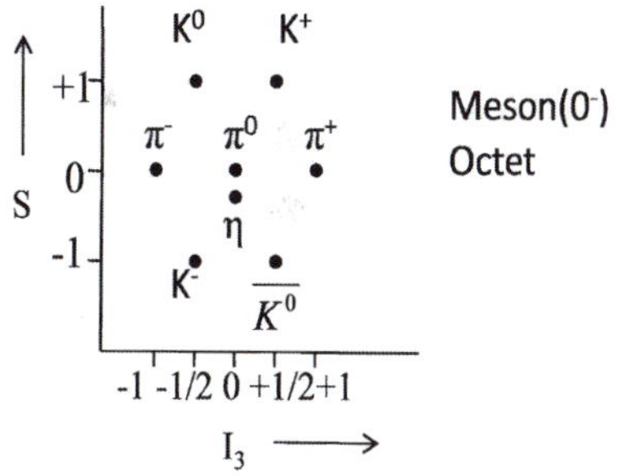

Fig. 10.4 S – I_3 plot for pseudoscalar meson octet

10.35

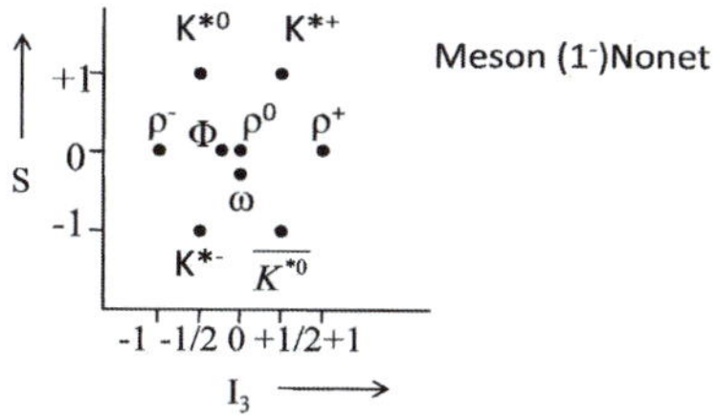

Fig. 10.5 S – I_3 plot for vector meson nonet

10.36

Fig. 10.6 Y – I_3 plot for baryon octet

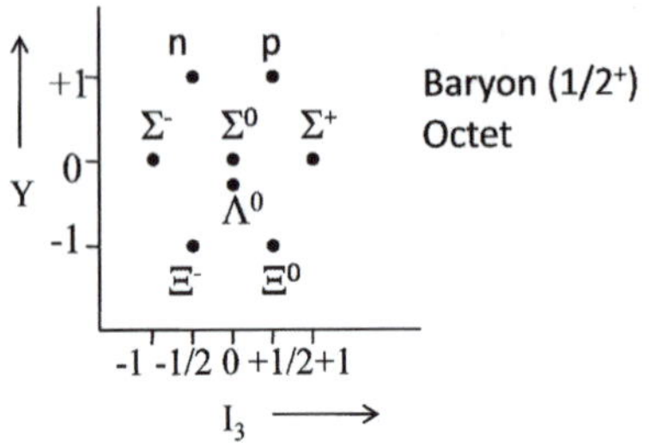

10.37

Fig. 10.7 Y – I_3 plot for baryon decuplet

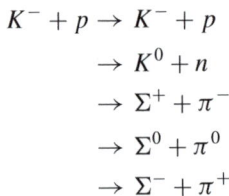

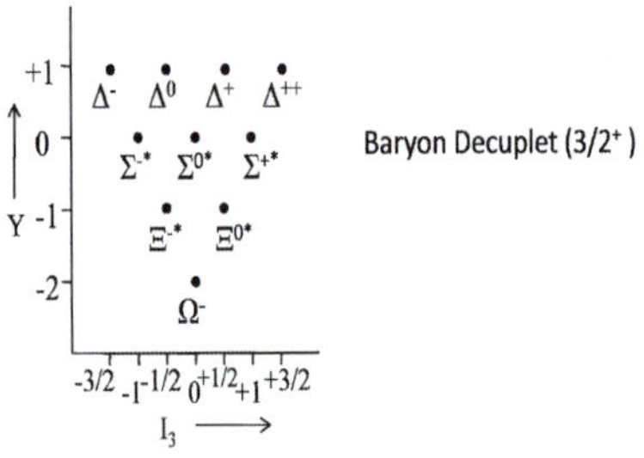

10.38 (a) The reaction channels for the pion reactions $\pi^- + p \rightarrow \pi^- + p$ and $\pi^- + p \rightarrow \pi^0 + n$ are charge symmetric to the reactions, $\pi^+ + n \rightarrow \pi^+ + n$ and $\pi^+ + n \rightarrow \pi^0 + p$. We can therefore expect the reaction cross sections with pions to be nearly identical. On the other hand, the reactions with kaons are not charge symmetric because K^+ and K^- do not constitute an iso spin doublet but are particle–antiparticle. Typical reactions with K^+ are $K^+ + n \rightarrow K^+ + n$ and $K^+ + n \rightarrow K^0 + p$
However for K^- many more channels are open.

$$
\begin{aligned}
K^- + p &\rightarrow K^- + p \\
&\rightarrow K^0 + n \\
&\rightarrow \Sigma^+ + \pi^- \\
&\rightarrow \Sigma^0 + \pi^0 \\
&\rightarrow \Sigma^- + \pi^+
\end{aligned}
$$

So, there is no reason for $\sigma(K^- + p)$ to be equal to $\sigma(K^+ + n)$.

(b) Because of the CPT theorem, K^0 and $\overline{K^0}$ which are particle and anti-particle, cannot be distinguished by their decay modes, but can be identified through their distinct strong interaction process. Thus

$$K^0 + p \to K^+ + n \quad \text{allowed}$$

$$\left.\begin{array}{l} K^0 + p \to \Sigma^+ + \pi^0 \\ K^0 + n \to K^- + p \end{array}\right\} \text{forbidden because of strangeness non-conservation}$$

$$\left.\begin{array}{l} \overline{K^0} + p \to \Sigma^+ + \pi^0 \\ \overline{K^0} + n \to K^- + p \end{array}\right\} \text{allowed}$$

$$\overline{K^0} + p \to K^+ + n \quad \text{forbidden}$$

10.39 ^{4}He is singlet and so $T = 0$, while K^- has $T = 1/2$. Therefore, the initial state is a pure $I = 1/2$ with $I_3 = -1/2$. In the final state $^4_\Lambda$He and $^4_\Lambda$H form a doublet, with $T = 1/2$. As pion has $T = 1$, the final state will be a mixture of $I = 3/2$ and 1/2. Looking up the Clebsch – Gordon Coefficients $1 \times 1/2$ (given in Table 3.3 of Chap. 3), we find the

$$\text{final state} \sim \frac{1}{\sqrt{3}} \left|\pi^0 \mathrm{H}_\Lambda\right\rangle - \sqrt{\frac{2}{3}} \left|\pi^- \mathrm{He}_\Lambda\right\rangle$$

$$\frac{\sigma({}_\Lambda{}^4\mathrm{He})}{\sigma({}_\Lambda{}^4\mathrm{H})} = \frac{\left(\sqrt{2/3}\right)^2}{\left(1/\sqrt{3}\right)^2} = \frac{2}{1}$$

10.40 As both K^- and π^0 have zero spin, l is conserved. Thus the orbital angular momentum l must be the same for the initial and final states. Since this is a strong interaction, conservation of parity requires that

$$(-1)^l P_k = (-1)^l P_\pi = (-1)^l(-1)$$

Therefore, $P_k = -1$

10.41 The deuteron spin $J_\mathrm{d} = 1$ and the capture of pion is assumed to occur from the s-state. Particles in the final state must have $J = 1$. As the Q- value is small (0.5 MeV), the final $nn\pi^0$ must be an s-state. It follows that the two neutrons must be in a triplet spin state (symmetric) which is forbidden by Pauli's principle.

10.42 Referring to the two nucleons by 1 and 2, + for proton and – for neutron, and using the notation $|I, I_3\rangle$ for the isotopic state and the Clebsch–Gordon Coefficients for $1/2 \times 1/2$ we can write the p–p state:

$$|p\rangle\,|p\rangle \equiv |1+\rangle\,|2+\rangle = |1, 1\rangle$$

The p–n state

$$|p\rangle\,|n\rangle \equiv |1+\rangle\,|2-\rangle = \frac{1}{\sqrt{2}}\left[|1, 0\rangle + |0, 0\rangle\right]$$

While the p–p system is a pure $I = 1$ state the p-n system is a linear superposition with equal statistical weight of $I = 1$ and $I = 0$ states. Therefore if isospin is conserved only the $I = 1$ state can contribute since the final state is a pure $I = 1$ state. The expected ratio of cross-sections at a given energy is

$$\frac{\sigma^+}{\sigma^0} = \frac{1}{\left(1/\sqrt{2}\right)^2} = 2$$

The observed ratio $3.15/1.5 = 2.1$ is consistent with the expected ratio of 2.0.

10.43 Pions obey Bose statistics and the system of pions must be symmetric upon interchange of space and isospin coordinates of any two pions. For each pion $T = 1$ and for a system of two pions, total isospin, $I = 2, 1, 0$. The corresponding states are

$|2, \pm 2\rangle, |2, \pm 1\rangle, |2, 0\rangle$ (Symmetric)

$|1, \pm 1\rangle, |1, 0\rangle$ (Anti symmetric)

$|0, 0\rangle$ (Symmetric)

Now $\psi(\text{total}) = \psi(\text{space})\psi(\text{isospin})$

Thus for the symmetric states like $|2, \pm 1\rangle$ or $|0, 0\rangle$, ψ (space) must be symmetric because of Bose symmetry. But for two pions the interchange of the spatial coordinates introduces a factor $(-1)^l$ so that only even l states are allowed. Thus, (π^0, π^0), (π^+, π^+), (π^-, π^-) in (a), (b) and (f) will be found in the states $L = 0, 2, \ldots$ However, the states $|1, \pm 1\rangle, |1, 0\rangle$ are antisymmetric and this requires, $L =$ odd, that is $L = 1, 3, \ldots$ (c) being antisymmetric can have odd values for L, that is $L = 1, 3, \ldots$ (d) and (e) can exist in $I = 1$ regardless of the L values because of two particles in these two cases are fermion and boson, so that the previous considerations are inconsequential.

10.3.3 Quarks

10.44 High energy collisions can cause the quarks within hadrons, or newly created quark–antiquark pairs, to fly apart from each other with very high energies. As the distance from the origin increases the colour force also increases, and it is energetically more favourable for the quarks to be fragmented, that is transformed into a jet of hadrons, mostly pions. Gluons may also radiate a separate jet. However, the quarks can not be dislodged as free particles.

10.45

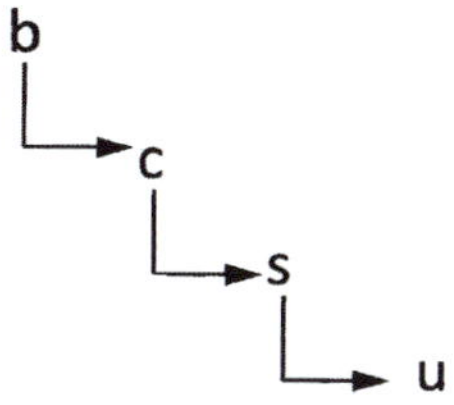

Fig. 10.8 Chain decay of B-meson

The decays are weak at each stage as the quantum numbers b, c and S change by one unit and is evidenced by narrow decay widths characteristic of weak interactions. The route for the chain decays is in accordance with the Cabibbo scheme as explained in the Sect. 10.1.

10.46 The QED formula is

$$\Gamma(3\gamma) = \frac{(2\pi^2 - 9)\alpha^6 m_e c^2}{9\pi} \tag{1}$$

(a) By analogy the QCD formula for charmonium ($c\bar{c}$ bound state) would be

$$\Gamma(3g) = \frac{(2\pi^2 - 9)\alpha_s^6 m_c c^2}{9\pi} \tag{2}$$

where α_s is the coupling constant for strong interaction and $m_c c^2 \approx 1.55\,\text{GeV}$ is the constituent mass of the c-quark. Substituting $\Gamma(3g) = 8 \times 10^{-5}\,\text{GeV}$ in (2), we find $\alpha_s = 0.307$

(b) For the radiative decay of the $b\bar{b}$ bound state the analogous formula would be

$$\Gamma(gg\gamma) = \frac{2(\pi^2 - 9)\alpha_s^4 \alpha^2 m_b c^2}{9\pi}$$

Substituting $\Gamma(gg\gamma) = 1.6 \times 10^{-7}\,\text{GeV}$, $\alpha = 1/137$, $m_b c^2 = 4.72\,\text{GeV}$, we find $\alpha_s = 0.319$

10.47 The ratio of the cross sections

$$R = \frac{3\sigma(e^+e^- \to Q\overline{Q})}{\sigma(e^+e^- \to \mu^+\mu^-)} = \frac{3\sum_i (q_i/e)^2}{1} \tag{1}$$

where q_i/e is the charge of the quark in terms of the charge of the electron. The charges and masses for the quarks are tabulated below

	q/e	M (GeV/c^2)
u-quark	+2/3	0.31
d-quark	−1/3	0.31
s-quark	−1/3	0.5
c-quark	+2/3	1.6
b-quark	−1/3	4.6
t-quark	+2/3	175

The sum of the squares of the quark charges refer to those quark pairs which can possibly contribute to the reactions at a given CMS energy ($\sqrt{s}$). The factor 3 in (1) arises due to 3 colors. At the low s-values, below the $c\bar{c}$ threshold, only u, d, s quarks are involved, the expected ratio being

$$R(\sqrt{s} < 3\,\text{GeV}) = 3 \times [(2/3)^2 + (1/3)^2 + (1/3)^2] = 2$$

$$R(3.2 < \sqrt{s} < 9.2\,\text{GeV}) = 3 \times [(2/3)^2 + (1/3)^2 + (1/3)^2 + (2/3)^2] = 10/3$$

$$R(9.2 < \sqrt{s} < 350\,\text{GeV}) = 3 \times [(2/3)^2 + (1/3)^2 + (1/3)^2 + (2/3)^2 + (1/3)^2] = 11/3$$

$$R(\sqrt{s} > 350\,\text{GeV}) = 3 \times [(2/3)^2 + (1/3)^2 + (1/3)^2 + (2/3)^2 + (1/3)^2 + (2/3)^2] = 5$$

10.48 In the quark model $\Sigma^- = dds,\ P = uud,\ n = udd,\ K^- = s\bar{u},\ \pi^- = d\bar{u}$

$$\sigma(\Sigma^- n) = 9\sigma(qq)$$
$$\sigma(pp) = 9\sigma(qq)$$
$$\sigma(K^- p) = 3\sigma(qq) + 3\sigma(q\bar{q})$$
$$\sigma(\pi^- p) = 3\sigma(qq) + 3\sigma(q\bar{q})$$

It follows that

$$\sigma(\Sigma^- n) = \sigma(pp) + \sigma(K^- p) - \sigma(\pi^- p)$$

10.49 In the quark model, $\Lambda = uds,\ p = uud,\ n = udd,\ K^- = s\bar{u}, \pi^+ = u\bar{d}$

$$\sigma(\Lambda p) = \sigma(uds)(uud)$$
$$= \sigma(uu + uu + ud + du + du + dd + su + su + sd) = 9\sigma(qq)$$

Similarly $\sigma(pp) = 9\sigma(qq)$

$\sigma(K^- n) = 3\sigma(qq) + 3\sigma(q\bar{q})$

$\sigma(\pi^+ p) = 3\sigma(qq) + 3\sigma(q\bar{q})$

Therefore $\sigma(\Lambda p) = \sigma(pp) + \sigma(K^- n) - \sigma(\pi^+ p)$

where we have assumed that $\sigma(qq) = \sigma(q\bar{q})$

10.50 Substract 1 MeV from the rest energies of the charged particles and take the difference in the masses.

$$n(940) - p(938 - 1) = 3\,\text{MeV}$$
$$= udd - uud = d - u$$

$$\Sigma^-(1{,}197 - 1) - \Sigma^0(1{,}192) = 4\,\text{MeV}$$
$$= dds - uds = d - u$$

$$\Sigma^0(1192) - \Sigma^+(1189 - 1) = 4\,\text{MeV}$$
$$= uds - uus = d - u$$

$$K^0(498) - K^+(494 - 1) = 5\,\text{MeV}$$
$$= d\bar{s} - u\bar{s} = d - u$$

Thus the mean difference in the masses of d-quark and u-quark is 4 MeV.

10.51 (a) $\Delta^- : ddd,\ \Delta^0 : ddu,\ \Delta^+ : uud,\ \Delta^{++} : uuu$

(b) The fundamental difficulty is that Pauli's principle is violated. For example consider the spin of Δ^-. All the three d-quarks have to be aligned with the same $j_z = -1/2$. The same difficulty arises for Δ^{++}. The difficulty is removed by endowing a new intrinsic quantum number (colour) to the quarks. Thus, the three d-quarks in Δ^- or the three u-quarks in Δ^{++} differ in colour, green, blue, and red.

(c) The higher values of J are accounted for by endowing higher orbital angular momentum values to the quarks. The parity is determined by the value of $(-1)^l$, where l is the orbital angular momentum quantum number.

10.52 (a) $\Sigma^+ : uus,\ \Sigma^- : dds,\ n : udd,\ p : uud$

(b) $K^+ : u\bar{s},\ K^- : ds,\ \pi^+ : u\bar{d},\ \pi^- : \bar{u}d$

10.53 (a) $\Delta^{++}, \Delta^-, \Omega^-$

(b) The quantum numbers of the b-quark are spin = ½ $\hbar$, charge $Q/e = -1/3$, mass $\sim$4.5 GeV/c^2, and Beauty $= -1$.

10.54 Figure 10.9a, b are the quark flow diagrams for the decays $\Phi \to K^+K^-$ and $\omega \to \pi^+\pi^0\pi^-$, respectively. Figure 10.9 c shows that the decay $\Phi \to \pi^+\pi^0\pi^-$ is suppressed because of unconnected quark lines.

10.55 $\sigma(\pi N) = \sigma(q\bar{q})(qqq) = 6\,\sigma(qq)$
$\sigma(NN) = \sigma(qqq)(qqq) = 9\,\sigma(qq)$
where we have assumed $\sigma(qq) = \sigma(q\bar{q})$
$\therefore \sigma(\pi N)/\sigma(NN) = 6/9 = 2/3$
which is in agreement with the ratio 25/38.

10.56 The cross section for this electromagnetic process is proportional to the square of the quark charge. In the annihilation of $\pi^-(=\bar{u}d)$ with ^{12}C nucleus $(= 18u + 18d)$, $18\bar{u}$ are involved, the cross section being proportional to $18Q_U^2$ or 18(4/9) or 8. In the annihilation of $\pi^+(= u\bar{d})$ with ^{12}C nucleus, $18\bar{d}$ are involved, the cross section being proportional to $18Q_d^2$ or 18(1/9) or 2.
Therefore the cross section ratio $\sigma(\pi^-\text{C})/\sigma(\pi^+\text{C}) = 8/2$ or 4:1

10.57 According to the quark model, the ratio

$$R = \frac{\sigma\left(e^+e^- \to \text{hadrons}\right)}{\sigma\left(e^+e^- \to \mu^+\mu^-\right)} = \frac{3\sum Q_i^2}{1}$$

where Q_i is the charge of the quark and the factor 3 arises due to the three colours in which the quarks can appear. Now the charges of the three quarks, u, d, and s are $+2/3$, $-1/3$, and $-1/3$
Therefore $\Sigma Q_i^2 = \left(\frac{2}{3}\right)^2 + \left(-\frac{1}{3}\right)^2 + \left(-\frac{1}{3}\right)^2 = \frac{2}{3}$
and $R = 3 \times 2/3 = 2$
Hence $\sigma\left(e^+e^- \to \text{hadrons}\right) = 3 \times \frac{2}{3} \times \sigma\left(e^+e^- \to \mu^+\mu^-\right) = 2 \times 20\,\text{nb} = 40\,\text{nb}$

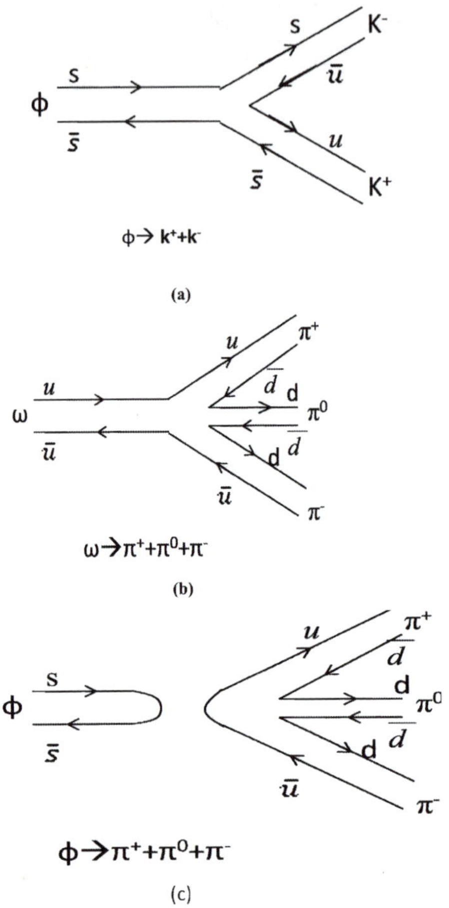

Fig. 10.9

Actually the value of R will be slightly greater than 2 because of an increased value of α_s, the running coupling constant.

10.58 $$E = \langle \psi | H | \psi \rangle / \langle \psi \mid \psi \rangle = \frac{\int_0^\infty e^{-r/a} \left\{ -\frac{\hbar^2}{2\mu} \left[\frac{\mathrm{d}^2}{\mathrm{d}r^2} + \frac{2}{r}\frac{\mathrm{d}}{\mathrm{d}r} \right] + \mathrm{B}r \right\} e^{-r/a} . 4\pi r^2 \mathrm{d}r}{\int_0^\infty e^{-2r/a} . 4\pi r^2 \mathrm{d}r}$$

where $\mu = m_q/2$ is the reduced mass

$$E = \frac{\hbar^2}{2\mu}\frac{1}{a^2} + \frac{3}{2}\mathrm{Ba} \quad (1)$$

The ground state energy is obtained by minimizing E, $\frac{\partial E}{\partial a} = 0$;

$$-\frac{\hbar^2}{\mu a^3} + \frac{3}{2}B = 0 \to a = \left(\frac{2\hbar^2}{3\mu B}\right)^{1/3} \quad (2)$$

Substituting (2) in (1) and putting $\mu = m_q/2$

$$E = E_0 = 2.45\left(\frac{B^2\hbar^2}{m_q}\right)^{1/3}$$

10.3.4 Electromagnetic Interactions

10.59

$$\begin{aligned}\sigma &= \int \frac{d\sigma}{d\Omega}.d\Omega = \int \frac{d\sigma}{d\Omega}.2\pi \sin\theta \, d\theta \\ &= \frac{2\pi\alpha^2\hbar^2c^2}{4E_{cM}^2}\int_{-1}^{1}(1+\cos^2\theta)d\cos\theta \\ &= \frac{2\pi}{4} \times \frac{1}{137^2} \times \frac{(197.3\,\text{MeV}-\text{fm})^2}{(2\times 10^4\,\text{MeV})^2}\left[\cos\theta + \frac{\cos^3\theta}{3}\right]_{-1}^{1} \\ &= 2.17 \times 10^{-8}\text{fm}^2 \\ &= 0.217\text{nB}\end{aligned}$$

The Feynman diagram is given in Fig. 10.10

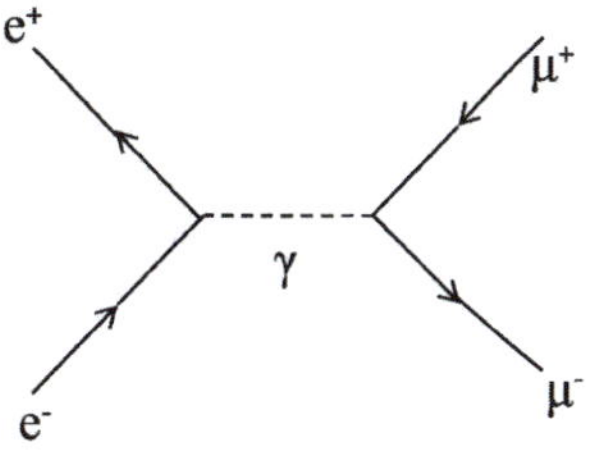

Fig. 10.10 $e^+e^- \to \gamma \to \mu^+\mu^-$

10.60 (a) $\Gamma = \frac{\hbar}{\tau} = \frac{\hbar c}{\tau c} = \frac{197.3\,\text{MeV}-\text{fm}}{(7.4\times 10^{-20}\text{s})(3\times 10^8 \times 10^{15}\text{fm/s})}$
$= 8.89 \times 10^{-3}$ MeV $= 8.89$ keV

Note that because of the inverse dependence of decay width on lifetime, weak decays with relatively long lives ($> 10^{-3}$ s) have widths of the order of a small fraction of eV, while em decays ($\tau \sim 10^{-19}$–10^{-20} s) have widths of the oder a kV, and strong decays ($\tau \sim 10^{-23} - 10^{-24}$ s) have widths of several MeV.

(b) Photon has spin 1 and pion zero and K^+ meson is known to have zero spin. Pion will be emitted with $l = 0$ as its energy is small. Therefore the occurrence of the radiative decay would constitute the violation of angular momentum conservation.

10.61 Neither e^+ nor e^- is an eigen state of C. However, the system $e^+ e^-$ in a definite (l, s) state is an eigen state of C. According to the generalized Pauli Principle under the total exchange of particles consisting of changing Q, $\mathbf{r}$ and s labels, the total function Ψ must change its sign.
Space exchange gives a factor $(-1)^l$ as this involves parity operation.
Spin exchange gives a factor $(-1)^{S+1}$.
Charge exchange gives a factor C.
The condition becomes

$$(-1)^l(-1)^{S+1}C = -1 \text{ or } C = (-1)^{l+S}$$

where S is the total spin.

We shall now show how C-invariance restricts the number of photons into which the positrinium annihilates. Let n be the number of photons in the final state. Conservation of C-parity gives

$$(-1)^{l+S} = (-1)^n$$

Two cases arise

(i) Singlet s state ${}^1S_0$; $l = S = 0$ (para – positronium)

$$e^+e^- \to 2\gamma \text{ allowed with lifetime } 1.25 \times 10^{-7}\text{s.}$$
$$\to \quad 3\gamma \text{ forbidden}$$

(ii) Triplet s state ${}^3\mathrm{S}_1$; $l = 0,\ S = 1$ (ortho – positronium)

$$e^+e^- \to 3\gamma \text{ allowed with lifetime } 1.5 \times 10^{-7}\text{s.}$$
$$\to \quad 2\gamma \text{ forbidden}$$

Note that annihilation into a single photon is not possible as it would violate the conservation of linear momentum.

10.62 (a) It is forbidden as electromagnetic interaction because $\Delta S \neq 0$ and also forbidden as weak interaction because there is no strangeness changing current

(b) It is allowed as an electromagnetic process because $\Delta S = 0$ (both Σ^0 and Λ have $S = -1$. S does not apply to e^+, and e^-)

10.3.5 Weak Interactions

10.63 The decay rate W is given by the inverse of lifetime multiplied by the branching fraction B, that is $W = B/\tau$

$$W(D^+) = \frac{0.19}{10.6 \times 10^{-13}} = 1.8 \times 10^{11}\text{s}^{-1}$$

$$W(D^0) = \frac{0.08}{4.2 \times 10^{-13}} = 1.9 \times 10^{11}\text{s}^{-1}$$

10.64 $\sigma \propto \dfrac{1}{\left[q^2 + m^2\right]^2}$

At low energies $q \to 0$

$$\frac{\sigma}{\sigma_0} = \frac{80}{100} = \left[\frac{m^2}{30^2 + m^2}\right]^2$$

$$m = 87\,\mathrm{GeV/c^2}$$

10.65 Number of $W^+ \to e^+ \nu_e = \dfrac{\sigma(pp^- \to W^+)}{\sigma(pp^- \to \text{anything})} \times 10^9$

$$= \frac{1.8 \times 10^{-9}}{70 \times 10^{-3}} \times 10^9 = 26$$

10.66 The decay $D^+(= c\bar{d}) \to \overline{K^0}(= s\bar{d}) + \mu^+ + \nu_\mu$ involves the transformation of a c-quark to an s-quark which is Cabibbo favored. In the case of the decay

$$D^+(= c\bar{d}) \to \pi^0(= \frac{u\bar{u} - d\bar{d}}{\sqrt{2}}) + \mu^+ + \nu_\mu$$

the transformation of a c-quark to a u-quark is Cabibbo suppressed. This explains why the former process is more likely than the latter.

10.67 The difference in the decay rates is due to two reasons (i) the Q - values in the decays are different. For $\Sigma^- \to n + e^- + \overline{\nu_e},\ Q_1 = 257\,\mathrm{MeV}$, while for $\Sigma^- \to \Lambda + e^- + \overline{\nu_e},\ Q_2 = 81\,\mathrm{MeV}$, so that by Sargent's law, the decay rate (ω) will be proportional to Q^5. (ii) The first decay involves $|\Delta S| = 1$, so that ω will be proportional to $\sin^2\theta_c$, where θ_c is the Cabibbo angle. The second one involves $|\Delta S| = 1$, so that ω is proportional to $\cos^2\theta_c$. In all

$$R = \frac{\omega\left(\Sigma^- \to n + e^- + \overline{\nu_e}\right)}{\omega\left(\Sigma^- \to \Lambda + e^- + \overline{\nu_e}\right)} = \frac{\sin^2\theta_c}{\cos^2\theta_c}\left(\frac{Q_1}{Q_2}\right)^5 = 0.0533\left(\frac{257}{81}\right)^5 = 17.14$$

where we have used $\theta_c = 13°$. The experimental value for R is 17.8

10.68 The leptonic decays of τ^- are

$$\tau^- \to e^- + \overline{\nu_e} + \nu_\tau \quad (1)$$

$$\tau^- \to \mu^- + \overline{\nu_\mu} + \nu_\tau \quad (2)$$

which from lepton – quark symmetry have equal probability. In addition, the possible hadronic decays are of the form $\tau^- \to \nu_\tau + X$

where X can be $d\bar{u}$ or $s\bar{c}$ so that X may have negative charge. However the latter possibility is ruled out because $m_\tau < m_s + m_c$. We are then left with only one possibility for X so that the hadronic decay will be

$$\tau^- \to \nu_\tau + d\bar{u} \quad (3)$$

Reaction (3) has relative weightage of 3 because of three colours. Thus the branching fraction of (1) is predicted to be 1/5 or 0.2, a value which is in agreement with the experimental value of 0.18.

10.69 According to Sargent's law, $\Gamma \propto Q^5$

	Decay	Q (MeV)	τ (s)	$Q^5\tau$ (MeV^5 – s)
(a)	$\mu^+ \to e^+ + \nu_e + \overline{\nu_\mu}$	105.15	2.197×10^{-6}	2.82×10^4
(b)	$n \to p + e^- + \overline{\nu_e}$	0.782	900	2.63×10^2
(c)	$\tau^+ \to e^+ + \nu_e + \overline{\nu_\tau}$	1784	3.4×10^{-13}	6.14×10^3
(d)	$\pi^+ \to \pi^0 + e^+ + \nu_e$	4.08	2.56	2.89×10^3
(e)	$^{14}O \to {}^{14}N^* + e^+ + \nu_e$	1.81	198	3.85×10^3

The numbers in the last column vary only over one order of magnitude which is small compared to the lifetimes which span over 15 orders of magnitude, thereby conforming to the Sargents' law ($\tau \propto 1/T^5{}_{\max}$) and therefore favoring the same coupling constant for the weak decays.

10.70 (a) Forbidden because lepton number is not conserved, left side has $L_\mu = -1$ while right side has $L_\mu = +1$

(b) Forbidden because charge is not conserved.

(c) Allowed because $B,\ L_\mu,\ Q$ etc are conserved. Also the selection rule, $\Delta Q = \Delta S = \pm 1$ is obeyed.

(d) Allowed for reason stated in (c).

10.71 Introduce the spurion, (Sp) a hypothetical particle of spin ½ and isospin 1/2 and convert the weak decay into a strong interaction

$$\mathrm{Sp} + \Xi^- \to \Lambda + \pi^-$$

	Sp	Ξ^-	Λ	π^-
I	½	½	0	1
I_3	– 1/2	– 1/2	0	– 1

The reaction must occur in pure $I = 1$ state. Looking up the Clebsch – Gordon coefficients for ½× ½, in Table 3.3,

$$|\Lambda, \pi^-\rangle = a_1\,|1, -1\rangle$$

For the second decay, we associate again a spurion and consider the reaction

$$\mathrm{Sp} + \Xi^0 \to \Lambda + \pi^0$$

	Sp	Ξ^0	Λ	π^0
I	½	½	0	1
I_3	– 1/2	+ 1/2	0	0

The initial state is a mixture of $I = 0$ and $I = 1$ states. The final state can exist only in $I = 1$.

$$|\Lambda, \pi^0> = \frac{1}{\sqrt{2}}[a_1|1,0> + a_0|0,0>]$$

$$\frac{\lambda(\Xi^- \to \Lambda + \pi^-)}{\lambda(\Xi^0 \to \Lambda + \pi^0)} = \frac{1}{(1/\sqrt{2})^2} = 2$$

10.72 Let us introduce a fictitious particle called spurion of $I = 1/2,\ I_3 = -1/2$, and neutrally charged and add it on the left hand side so that the weak decay is converted into a strong interaction in which I is conserved. The reactions with Σ^+ can proceed in $I = 1/2$ and 3/2 channels while that with Σ^- through pure $I = 3/2$ channel. We can then write down the amplitudes for the initial state by referring to the C.G. coefficients for $1 \times 1/2$:

$$\Sigma^+ + s \to \sqrt{\frac{1}{3}}a_3 + \sqrt{\frac{2}{3}}a_1$$

$$\Sigma^- + s \to a_3$$

where a_1 and a_3 are the $I = 1/2$ and $I = 3/2$ contributions. Similarly, for the final state the amplitudes are

$$n\pi^+ \to \sqrt{\frac{1}{3}}a_3 + \sqrt{\frac{2}{3}}a_1$$

$$n\pi^- \to a_3$$

$$p\pi^0 \to \sqrt{\frac{2}{3}}a_3 - \sqrt{\frac{1}{3}}a_1$$

Therefore,
$$a_+ = \langle \Sigma^+ \mid n\pi^+ \rangle = \tfrac{1}{3}a_3^2 + \tfrac{2}{3}a_1^2$$
$$a_- = \langle \Sigma^- \mid n\pi^- \rangle = a_3^2$$
$$a_0 = \langle \Sigma^+ \mid p\pi^0 \rangle = \tfrac{\sqrt{2}}{3}a_3^2 - \tfrac{\sqrt{2}}{3}a_1^2$$

giving $a_+ + \sqrt{2}\, a_0 = a_-$

10.73 Converting the decay into a reaction

$$S + \Lambda \to p + \pi^-$$
$$\to n + \pi^0$$

where s is the fictitious particle of isospin $T = ½$ and $T_3 = -½$. Because Λ has $T = 0$ and $T_3 = 0$, the initial state will be a pure $I = ½$ state with $I_3 = -1/2$. In the final state the nucleon has $T = ½$ and pion has $T = 1$, so that $I = 3/2$ or ½. I and I_3 conservation require that the final state must be characterized by $I = ½$ and $I_3 = -½$. Looking up the table for C.G. Coefficients for $1 \times 1/2$ (given in Table 3.3) we can write down

$$|N, \pi, 1/2, -1/2\rangle = -\sqrt{\frac{2}{3}}\,|p\pi^-\rangle + \sqrt{\frac{1}{3}}\,|n\pi^0\rangle$$

Therefore, $\dfrac{\omega\left(\Lambda \to p\pi^-\right)}{\omega\left(\Lambda \to n\pi^0\right)} = \left(\sqrt{\dfrac{2}{3}}\right)^2 \Big/ \left(\sqrt{\dfrac{1}{3}}\right)^2 = 2$

10.74 $$L_\odot = 3.83 \times 10^{26}\,\mathrm{Js^{-1}} = \frac{3.83 \times 10^{26}}{1.6 \times 10^{-13}}\,\mathrm{MeV\ s^{-1}}$$

$$= 2.39 \times 10^{39}\,\mathrm{MeV\ s^{-1}}$$

Number of α's produced, $N_\alpha = \dfrac{2.39 \times 10^{39}}{26.72} = 8.94 \times 10^{37}\,\mathrm{s^{-1}}$

As the number of neutrinos produced is double the number of α's,

$N_\nu = 2N_\alpha = 1.788 \times 10^{38}\,\mathrm{s^{-1}}$

Number of neutrinos received per square metre on earth's surface, that is flux

$$\phi = \frac{N_\upsilon}{4\pi r^2} = \frac{1.788 \times 10^{38}}{(4\pi)\left(1.5 \times 10^{11}\right)^2} = 6.33 \times 10^{14}\,\mathrm{m^{-2}\,s^{-1}}$$

10.75 Using the formula $E = mc^2\gamma$, the speed of antineutrinos of mass m

$$\upsilon = c\left[1 - \frac{m^2c^4}{E^2}\right]^{1/2}$$

The time taken to travel to earth

$$t = \frac{d}{\upsilon} = \frac{d}{c}\left[1 - \frac{m^2c^4}{E^2}\right]^{-1/2} \cong \frac{d}{c}\left[1 + \frac{1}{2}\frac{m^2c^4}{E^2}\right]$$

The difference in travel time for two neutrinos of energies E_1 and E_2 where $E_2 > E_1 >> m$,

$$\Delta t = \frac{d}{2c}\left(mc^2\right)^2\left[\frac{1}{E_1^2} - \frac{1}{E_2^2}\right]$$

Substituting $\Delta t = 4\,\mathrm{s}$, $E_1 = 5\,\mathrm{MeV}$ and $E_2 = 15\,\mathrm{MeV}$, $\frac{d}{c} = 17 \times 10^4 \times 3.15 \times 10^7$ seconds, and solving for mc^2 we find $mc^2 = 6.48 \times 10^{-6}\,\mathrm{MeV} = 6.5\,\mathrm{eV}$

10.76 (a) $Q/e = +1$ $\because \Delta Q = 0$
(b) $B = 0$ $\because \Delta B = 0$
(c) $l_\mu = 0$ $\because \Delta l_\mu = 0$
(d) $T = 1/2$ $\because \Delta T_3 = \pm 1/2$
(e) $s = \pm 1$ $\because \Delta S = \pm 1$
(f) spin = 0 or 1 $\because$ spin of π^0, μ and ν are 0, 1/2 and 1/2 respectively
(g) Boson $\because$ on right hand side there are two fermions (μ, ν) and one boson (π)

(h) Mass $m_x \geq (m_\pi + m_\mu + m_\upsilon)/c^2 = (135 + 106 + 0)/c^2 = 241\,\text{MeV/c}^2$
(i) X is identified as K^+-meson.

10.77 The initial state $^{16}O^*$ has odd parity, while the parity of ^{12}C is even and so also that of α. If α is emitted with $l = 0$ the parity of the final state is even. Clearly the decay goes through a weak interaction because parity is violated and the observed width is consistent with a weak decay. On the other hand the width of the electro-magnetic decay $^{16}O^* \rightarrow ^{16}O + \gamma$, is 3×10^{-3} eV.

10.78 $\tau_\tau = B \times \tau_\mu \left(\frac{m_\mu}{m_\tau}\right)^5 = \frac{17.7}{100} \times 2.197 \times 10^{-6} \times \left(\frac{105.658}{1784}\right)^5 = 0.28 \times 10^{-12} s$ a value which is in excellent agreement with the experimental value of 0.3×10^{-12} s

10.79 Pions have $T = 1$. A system of two pions can exist in the isospin state $I = 0$, 1, and 2 state. Therefore, isospin is conserved in all the three cases.

Now pions obey Bose statistics and so Ψ_{total} is symmetrical. As $J = 0$ for pions, the spatial part of pions system must be symmetrical. The intrinsic parity of each pion being negative, does not contribute to the parity of the state. The parity of the state is mainly determined by the L-value; $p = (-1)^l$. Therefore, allowed states correspond to $L = 0, 2, 4, \ldots$ It follows that particles with $J^P = 0^+, 2^+ \ldots$ can decay to two pions, so that only possible decay is $f^0 \rightarrow \pi^+ + \pi^-$; the other two particles ω^0 and η^0 actually decay into three pions.

10.80 The mass of η – meson is $549\,\text{Mev/c}^2$ while the mass of four pions is $558\,\text{MeV/c}^2$ if all the four pions are charged and $540\,\text{MeV/c}^2$ if uncharged. In the first state the decay cannot occur because of energy conservation and in the second case the decay will be heavily suppressed because of a small phase space on account of small Q-value. Apart from this the only possible pionic decay is the 3-body mode, the strong decay into two pions or four pions being forbidden by parity conservation.

10.81 The leptonic decays are assumed to proceed via exchange of a single virtual photon, Fig. 10.11

Fig. 10.11 Leptonic decay of vector meson

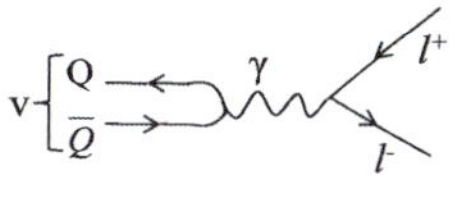

Apart from numerical factors the partial width $\Gamma(V \rightarrow l^+ l^-)$ is proportional to the squared sum of the charges of the quarks in the meson, that is $\Gamma \propto \sum |a_i\, Q_i|^2$, where a_i the amplitudes from all the quarks in the meson are superposed. The approach is similar to the Rutherford scattering where the cross-sections are assumed to be proportional to the sum of squares of the charges of quarks. The quark wave functions and the quantity $\left|\sum a_i\, Q_i\right|^2$ are tabulated below

Meson	Quark wave function	$\lvert\Sigma a_i\, Q_i\rvert^2$
ρ^0	$(u\bar{u} - d\bar{d})/\sqrt{2}$	$\left[\frac{1}{\sqrt{2}}\left(\frac{2}{3} - \left(-\frac{1}{3}\right)\right)\right]^2 = \frac{1}{2}$
ω^0	$(u\bar{u} + d\bar{d})/\sqrt{2}$	$\left[\frac{1}{\sqrt{2}}\left(\frac{2}{3} - \frac{1}{3}\right)\right]^2 = \frac{1}{18}$
Φ^0	$s\bar{s}$	$\left(\frac{1}{3}\right)^2 = \frac{1}{9}$

The expected leptonic widths are in the ratio

$$\Gamma(\rho^0) : \Gamma(\omega^0) : \Gamma(\varphi^0) = 9 : 1 : 2$$

in agreement with the observed ratios.

10.82 For the decay of charmed particles, the selection rules are (i) $\Delta C = \Delta S = \Delta Q = \pm 1$, for Cabibbo allowed and (ii) $\Delta C = \Delta Q = \pm 1,\ \Delta S = 0$, for Cabibbo suppressed decays
(ΔQ applies to hadrons only). Using these rules (a) is Cabibbo allowed and (b) is Cabibbo suppressed

10.83 For the charmed particles, the selection rule for the Cabibbo allowed decay is $\Delta C = \Delta S = \Delta Q = \pm 1$ and for the Cabibbo suppressed decays the selection rule is $\Delta C = \Delta Q = \pm 1,\ \Delta S = 0$. Using these rules we infer that the decays are (a) Cabibbo allowed (b) forbidden

10.84 For semileptonic decays the rule is $\Delta S = \Delta Q = \pm 1$,. In (a) both Q and S increase by one unit. Hence it is allowed and experimentally observed. (b) is also a semileptonic decay in which Q decreases by one unit, but S increases (from -2 to -1) by one unit. Therefore, the decay is forbidden. (c) is a non-leptonic weak decay in which $\Delta S = +1$ and the energy is conserved. It is allowed and experimentally observed.

10.85 All the decays are semi-leptonic. For Cabibbo allowed decay,
$\Delta Q = \Delta S = \pm 1$.
In (a) $\Delta Q = -1$ and $\Delta S = -1$ ($\because$ S changes from $+1$ to 0) Therefore it is allowed.
In (b) $\Delta Q = +1$ while $\Delta S = -1$. Therefore, forbidden.
In (c) $\Delta Q = +1$ and $\Delta S = +1(\because S = -1 \rightarrow S = 0)$, therefore allowed.
In (d) $\Delta Q = -1$ and $\Delta S = +1$, therefore forbidden.

10.86 If lepton number is not absolutely conserved and neutrinos have finite masses, then mixing may occur between different types of neutrinos ($\nu_e,\ \nu_\mu,\ \nu_\tau$) In what follows consider only two types of neutrinos ν_e and ν_μ. The neutrino states ν_μ and ν_e which couple to the muon and electron, respectively could be linear combinations

$$\nu_\mu = \nu_1 \cos\theta + \nu_2 \sin\theta \tag{1}$$

and $\nu_e = -\nu_1 \sin\theta + \nu_2 \cos\theta$ (2)

which form a set of ortho normal states. ν_μ and ν_e are the sort of states which are produced in charged pion decay ($\pi \to \mu + \nu_\mu$) and neutron decay ($n \to p + e^- + \overline{\nu_e}$); ν_1 and ν_2 are the mass eigen states, corresponding to the neutrino masses m_1 and m_2. In the matrix form

$$\begin{pmatrix} \nu_\mu \\ \nu_e \end{pmatrix} = \begin{pmatrix} \cos\theta & \sin\theta \\ -\sin\theta & \cos\theta \end{pmatrix} \begin{pmatrix} \nu_1 \\ \nu_2 \end{pmatrix} \tag{3}$$

The difference in the masses leads to difference in the characteristic frequencies with which the neutrinos are propagated. Here θ is known as the mixing angle which is analogous to the Cabibbo angle θ_c. Using natural units ($\hbar = c = 1$) at time $t \neq 0$,

$$\nu_1(t) = \nu_1(0)e_1^{-iE_1\,t}$$

$$\nu_2(t) = \nu_2(0)e_2^{-iE_2\,t} \tag{4}$$

where the exponentials are the usual oscillating time factors associated with any quantum mechanical stationary state. Since the momentum is conserved the states $\nu_1(t)$ and $\nu_2(t)$ must have the same momentum p. If the mass $m_i << E_i (i = 1, 2)$

$$E_i = (p^2 + m_i^2)^{1/2} \cong p + \frac{m_i^2}{2p} \tag{5}$$

Suppose at $t = 0$, we start with muon type of neutrinos so that, $\nu_\mu(0) = 1$ and $\nu_e(0) = 0$ then by (3),

$$\nu_2(0) = \nu_\mu(0) \sin\theta \tag{6}$$

$$\nu_1(0) = \nu_\mu(0) \cos\theta$$

and $\nu_\mu(t) = \cos\theta\,\nu_1(t) + \sin\theta\,\nu_2(t)$ (7)

Thus at time $t \neq 0$, the muon–neutrino beam is no longer pure but develops an electron neutrino component. Similarly, an electron–neutrino beam would develop a muon–neutrino beam component.

Using (4) and (6) in (7)

$$\frac{\nu_\mu(t)}{\nu_\mu(0)} = \cos^2\theta . e^{-iE_1 t} + \sin^2\theta . e^{-iE_2 t}$$

and the intensity

$$\frac{I_\mu(t)}{I_\mu(0)} = \left|\frac{\nu_\mu(t)}{\nu_\mu(0)}\right|^2 = \cos^4\theta + \sin^4\theta + \sin^2\theta\cos^2\theta \left[e^{i(E_2-E_1)t} + e^{-i(E_2-E_1)t}\right]$$

$$P(\nu_\mu \to \nu_e) = 1 - \sin^2 2\theta \sin^2\left[\frac{(E_2 - E_1)}{2} t\right] \tag{8}$$

where we have used $(\cos^4\theta + \sin^4\theta) = (\cos^2\theta + \sin^2\theta)^2 - 2\cos^2\theta\sin^2\theta$ in simplifying the above equation.

10.87 (a) $P_e(t) = 1 - P_\mu(t) = \sin^2 2\theta \sin^2\left[\frac{(E_2-E_1)}{2}t\right]$

(b) $P_\mu(t) = P_e(t)$

$$1 - \sin^2 2\theta \sin^2\left[\frac{(E_2 - E_1)}{2}t\right] = \sin^2 2\theta \sin^2\left[\frac{(E_2 - E_1)}{2}t\right]$$

$$2\sin^2 2\theta \sin^2\left[\frac{(E_2 - E_1)}{2}t\right] = 1$$

Restoring to practical units the above equation becomes

$$2\sin^2 2\theta \sin^2\left[\frac{\Delta m^2 c^4}{2E\hbar}t\right] = 1$$

where $\Delta m^2 = {m_2}^2 - {m_1}^2$ and $\theta = 34^0$

If m_1 and m_2 are in $\mathrm{eV/c^2}$, and E in MeV and L the distance from the source, then the last equation becomes

$$2\sin^2 2\theta \sin^2\left(\frac{1.27\Delta m^2.L}{E}\right) = 1$$

Inserting $\theta = 34^0$, $\Delta m^2 = 5^2 - 3^2 = 16$ and $E = 1{,}000\,\mathrm{MeV}$, we find $L = 426\,\mathrm{m}$, giving $t = L/C = 1.42 \times 10^{-6}\mathrm{s}$.

10.88 $$\tau(\mu^+ \to e^+ \nu_e \overline{\nu_\tau}) = \frac{G^2}{(\hbar c)^6}\frac{m_\mu^5}{192\pi^3} \qquad (1)$$

$G^2 \sim g^2 / M_\mathrm{w}^2$, where $\mathrm{M_w}$ is the mass of W-boson.

From the τ lepton lifetime and formula (1) for the dependence of parent particle mass, we can test the universality of the couplings g_μ and g_τ to the W – boson

$$\left(\frac{g_\tau}{g_\mu}\right)^4 = B\left(\tau^+ \to e^+\nu_e\overline{\nu_\tau}\right)\left(\frac{m_\mu}{m_\tau}\right)^5\left(\frac{\tau_\mu}{\tau_\tau}\right)$$

Inserting $B = 0.178$, $m_\mu = 105.658\,\mathrm{MeV/c^2}$, $m_\tau = 1777.0\,\mathrm{MeV/c^2}$, $\tau_\mu = 2.197 \times 10^{-6}\mathrm{s}$ and $\tau_\tau = 2.91 \times 10^{-13}$ s, we find

$$\frac{g_\tau}{g_\mu} = 0.987$$

Comment: From the branching fractions for $\tau^+ \to e^+\nu_e\overline{\nu_\tau}$ and $\tau^+ \to \mu^+\nu_\mu\overline{\nu_\tau}$ the ratio $g_\mu/g_e = 1.001$. A similar result is obtained from the branching ratio of $\pi \to e\nu$ and $\pi \to \mu e$, proving thereby different flavours of leptons have identical couplings to the $\mathrm{W}^\pm$ bosons.

The principle of universality is equally valid for the Z^0 coupling. Thus, the branching fractions are predicted as

$$Z^0 \to e^+e^- : \mu^+\mu^- : \tau^+\tau^- = 1:1:1$$

in agreement with the experimental ratios. Formula (1) affords the most accurate determination of G, the Fermi constant because the mass and lifetime of muon are precisely known by experiment.

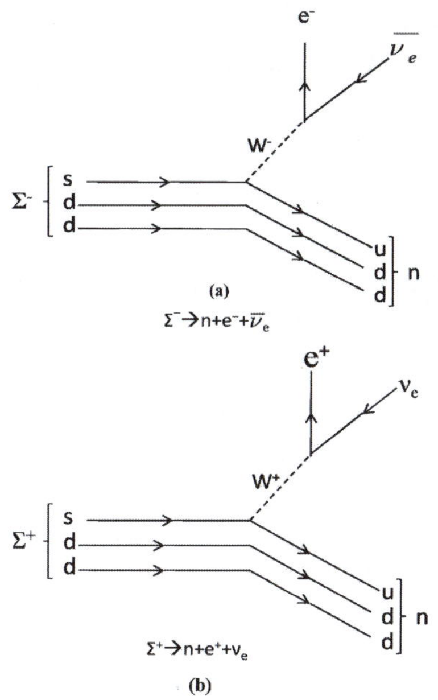

Fig. 10.12

10.89 The reaction (a) can go via the mechanism of the diagram shown in Fig. 10.12a. However, no diagram with single W exchange can be drawn for the reaction (b) which at the quark level implies the transformation

$$uus \rightarrow udd + e^+ + \nu_e$$

as in Fig. 10.12b and would require two separate quark transitions which involve the emission and absorption of two W bosons – a mechanism which is of higher order and therefore negligibly small.

The above conclusion can also be reached by invoking for the selection rule for semi leptonic decays. Reaction (a) obeys the rule $\Delta S = \Delta Q = +1$ and is therefore allowed, while in reaction (b) we have $\Delta S = +1$, but $\Delta Q = -1$, and therefore forbidden.

10.90 The difference in the two decay rates is due to two factors (i) The decay $D^+ \rightarrow \overline{K^0}\mu^+\nu_\mu$ involves $|\Delta s| = 1$ and hence proportional to $\sin^2\theta_c$, where $\theta_c = 12.9^0$ is the Cabibbo angle, while $D^+ \rightarrow \pi^0\mu^+\nu_\mu$ involves $|\Delta s| = 0$ and is proportional to $\sin^2\theta_c$ (ii) The Q-values are different for these decays. For the first one $Q_1 = 1870 - (498 + 106) = 1{,}266$ MeV and for the second one $Q_2 = 1{,}870 - (135 + 106) = 1629$MeV. Thus using Sargent's rule

$$R \approx \frac{\sin^2\theta_c}{\cos^2\theta_c}\left(\frac{Q_1}{Q_2}\right)^5 = \tan^2 12.9^0 \left(\frac{1266}{1629}\right)^5 \cong 0.015$$

10.91 Let a K^0 beam be formed through a strong interaction like $\pi^- + p \to K^0 + \Lambda$. Neither $\left|K^0\right\rangle$ nor $\left|\overline{K^0}\right\rangle$ is an eigen state of $|cp\rangle$. However, linear combinations can be formed.

$$|K_s\rangle = \frac{1}{\sqrt{2}}\left(\left|K^0\right\rangle + \left|\overline{K^0}\right\rangle\right)$$

$$cp\,|K_s\rangle = \frac{1}{\sqrt{2}}\left(\left|\overline{K^0}\right\rangle + \left|K^0\right\rangle\right) = |K_s\rangle$$

$$cp = +1$$

$$|K_L\rangle = \frac{1}{\sqrt{2}}\left(\left|K^0\right\rangle - \left|\overline{K^0}\right\rangle\right)$$

$$cp\,|K_L\rangle = \frac{1}{\sqrt{2}}\left(\left|\overline{K^0}\right\rangle - \left|K^0\right\rangle\right) = -\,|K_L\rangle$$

$$cp = -1$$

while K^0 and $\overline{K^0}$ are distinguished by their mode of production, K_s and K_L are distinguished by the mode of decay. Typical decays are $K_s \to \pi^0\pi^0,\ \pi^+\pi^-,\ K_L \to \pi^+\pi^-\pi^0,\ \pi\mu\nu$.
At $t = 0$, the wave function of the system will have the form

$$\psi(0) = \left|K^0\right\rangle = \frac{1}{\sqrt{2}}\left(|K_s\rangle + |K_L\rangle\right)$$

As time develops K_s and K_L amplitudes decay with their characteristic lifetimes. The intensity of K_s or K_L components can be obtained by squaring the appropriate coefficient in $\Psi(t)$. The amplitudes therefore contain a factor $e^{-iEt/\hbar}$ which describes the time dependence of an energy eigen function in quantum mechanics.

In the rest frame of the K^0 we can write the factor $e^{-iEt/\hbar}$ as $e^{-imc^2t/\hbar}$, where m is the mass. The complete wavefunction for the system can therefore be written as

$$\psi(t) = \frac{1}{\sqrt{2}}\left[|K_s\rangle\, e^{-t\left(\frac{1}{2\tau_s} + \frac{im_S c^2}{\hbar}\right)} + |K_L\rangle\, e^{-t\left(\frac{1}{2\tau_L} + \frac{im_L c^2}{\hbar}\right)}\right]$$

$$= \frac{1}{\sqrt{2}} e^{-im_s c^2 t/\hbar}\left[|K_s\rangle\, e^{-\frac{t}{2\tau_s}} + |K_L\rangle\, e^{i\Delta mc^2 t/\hbar}\right]$$

where $\Delta m = m_s - m_L$ and we have neglected the factor $e^{-t/2\tau_L}$ which varies slowly ($\tau_L \cong 70\tau_s$). Reexpressing $|K_1\rangle$ and $|K_2\rangle$ in terms of $\left|K^0\right\rangle$ and $\left|\overline{K^0}\right\rangle$

$$|K_s\rangle = \frac{1}{\sqrt{2}}\left(|K^0\rangle + \left|\overline{K^0}\right\rangle\right)$$

$$|K_L\rangle = \frac{1}{\sqrt{2}}\left(|K^0\rangle - \left|\overline{K^0}\right\rangle\right)$$

$$\psi(t) = \frac{1}{2}e^{-im_s c^2 t/\hbar}\left[e^{-t/2\tau_s}\left(|K^0\rangle + \left|\overline{K^0}\right\rangle\right) + e^{i\Delta mc^2 t/\hbar}\left(|K^0\rangle - \left|\overline{K^0}\right\rangle\right)\right]$$

$$= \frac{1}{2}e^{-im_s c^2 t/\hbar}\left[|K^0\rangle\left(e^{-t/2\tau_s} + e^{i\Delta mc^2 t/\hbar}\right) + \left|\overline{K^0}\right\rangle\left(e^{-t/2\tau_s} - e^{i\Delta mc^2 t/\hbar}\right)\right]$$

The intensity of the component is obtained by taking the absolute square of the coefficient of $\left|\overline{K^0}\right\rangle$

$$I\left(|K^0\rangle\right) = \frac{1}{4}\left[e^{-t/\tau_s} + 1 + 2e^{-t/2\tau_s}\cos\left(\Delta mc^2 t/\hbar\right)\right]$$

Similarly,

$$I\left|\left(\overline{K^0}\right\rangle\right) = \frac{1}{4}\left[e^{-t/\tau_s} + 1 - 2e^{-t/2\tau_s}\cos\left(\Delta mc^2 t/\hbar\right)\right]$$

10.92 Refering to Problem 10.91, the K_L state can be written as

$$|K_L\rangle = \frac{1}{\sqrt{2}}\left(|K^0\rangle - \left|\overline{K^0}\right\rangle\right) \quad (1)$$

When K_L enters the absorber, strong interactions would occur with $K^0(S = +1)$ and $\left|\overline{K^0}\right\rangle(S = -1)$ components of the beam of the original K^0 beam intensity, 50% has disappeared by K_S-decay. The remaining K_L component consists of 50% $\overline{K^0}$. Upon traversing the material the existence of $\overline{K^0}$ with $S = -1$ is revealed by the production of hyperons in a typical reaction, $\overline{K^0} + p \rightarrow \Lambda + \pi^+$

While K^0 components can undergo elastic and charge-exchange scattering only, the $\overline{K^0}$ component can in addition participate in absorption processes resulting in the hyperon production. The emergent beam from the slab will then have the K^0 amplitude $f\,|K^0\rangle$ and $\overline{K^0}$ amplitude $\overline{f}\left|\overline{K^0}\right\rangle$ with $\overline{f} < f < 1$. The composition of the emergent beam from the slab is given by modifying (1).

$$\frac{1}{\sqrt{2}}\left(f\,|K^0\rangle - \overline{f}\left|\overline{K^0}\right\rangle\right) = \frac{(f+\overline{f})}{2\sqrt{2}}\left(|K^0\rangle - \left|\overline{K^0}\right\rangle\right) + \frac{(f-\overline{f})}{2\sqrt{2}}\left(|K^0\rangle + \left|\overline{K^0}\right\rangle\right)$$

$$= \frac{1}{2}(f+\overline{f})\,|K_L\rangle + \frac{1}{2}(f-\overline{f})\,|K_S\rangle$$

Since $f \neq \overline{f}$, we conclude that some of the K_S-state has regenerated. The regeneration of the short lived K_1-component in a long-lived K_L beam was experimentally confirmed from observation of two-pion decay mode (1956).

10.3.6 Electro-weak Interactions

10.93

Fig. 10.13(a)

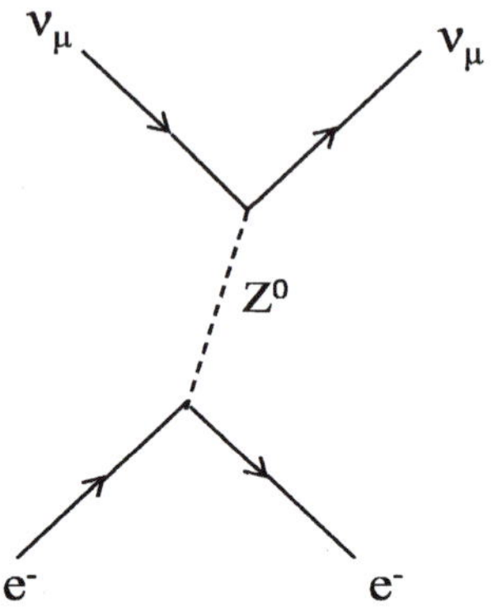

(a) $\nu_\mu + e^- \to z^0 \to \nu_\mu + e^-$

A charged current weak interaction is mediated by the exchange of $W^\pm$ boson, as for the decay $\mu^- \to e^- + \overline{\nu_e} + \nu_\mu$. Neutral current interaction is mediated by the exchange of Z^0, as in the scattering, $\nu_\mu + p \to \nu_\mu + p$. For $\nu_\mu + e^- \to \nu_\mu + e^-$, the Feynman diagram for weak neutral current shown in Fig. 10.13a is unambiguous.

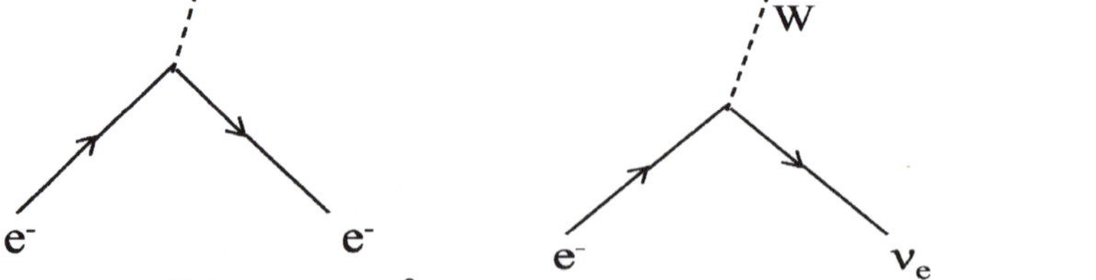

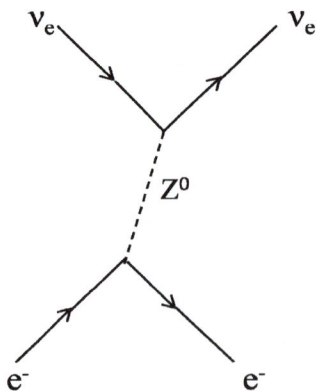

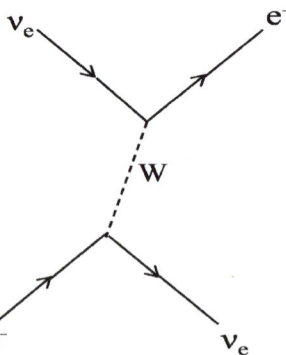

(b) $\nu_e + e^- \to z^0 \to \nu_e + e^-$ (c) $\nu_e + e^- \to \overline{w} \to \nu_e + e^-$

Fig. 10.13(b), (c)

However, for $\nu_e + e^- \rightarrow \nu_e + e^-$, there are two diagrams shown in Fig. 10.13b and Fig. 10.13c.

Thus the reaction can be described by neutral or charged current and therefore does not constitute an unequivocal evidence for neutral current.

10.94 If N_ν is the number of neutrino types in the sequence $\nu_e,\ \nu_\mu,\ \nu_\tau$ and assuming that there are only three charged leptons $e,\ \mu$ and τ, the balance equation for the decay rate can be written as

$$\Gamma_z(\text{total}) = \Gamma(Z^0 \rightarrow \text{hadrons}) + 3\,\Gamma(Z^0 \rightarrow l^+l^-) + N_\nu \Gamma(Z^0 \rightarrow \nu_l \overline{\nu_l})$$

The factor 3 is for the three charged leptons. Substituting the given data
$2.534 = 1.797 + 3 \times 0.084 + 0.166 \mathrm{N}_\nu$
and solving, we get $N_\nu = 2.92$ or 3.

10.95 (a) $W^\pm$ and Z^0 bosons are produced in the annihilation of $p^- - p$ at high energies (Fig. 10.14). The elementary production and decay processes are

$$u + \overline{d} \rightarrow W^+ \rightarrow e^+ + \nu_e,\ \mu^+ + \nu_\mu$$

$$\overline{u} + d \rightarrow W^- \rightarrow e^- + \overline{\nu_e},\ \mu^- + \overline{\nu_\mu}$$

$$\left.\begin{matrix} u + \overline{u} \\ d + \overline{d} \end{matrix}\right\} \rightarrow Z^0 \rightarrow e^+e^-,\ \mu^+\mu^-$$

Detector: It consists of calorimeter detectors, the central tracking chamber to detect individual secondary particles, surrounded by an electromagnetic calorimeter to detect electron – photon showers, a much larger calorimeter to detect and measure hadron jets and an outside muon detector.

Signature for $W \rightarrow e\nu$ event:

(i)An isolated single electron track with high transverse momentum (P_T) in the central track detector.

(ii)The electron track points to a shower in the electromagnetic calorimeter with appreciable energy deposition in the nearby hadronic calorimeter in the neighborhood

(iii)There should be missing P_T overall when summation is made over all the secondaries. The missing P_T is attributed to the unseen neutrino from the W-decay.

Signature of $Z^0 \rightarrow e^+e^-$ events:

Two isolated tracks with large P_T values and invariant mass $M_{e^+e^-} >$ 50 GeV, pointing to localized track clusters in the electromagnetic calorimeter.

For muonic decays, $W^\pm \rightarrow \mu\nu,\ Z^0 \rightarrow \mu\mu$ are observed by imposing high P_T requirements on the P_T values for muons which are able to penetrate hadron calorimeter and observed in the external muon chambers.

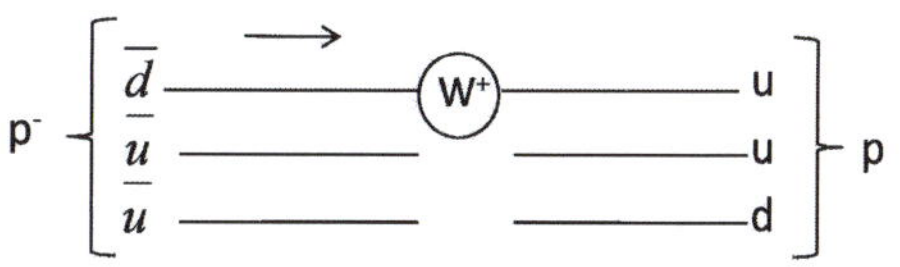

Fig. 10.14 Annihilation process

(b) $R \sim \dfrac{\hbar c}{mc^2} = \dfrac{197\,\text{MeV} - \text{fm}}{100 \times 10^3\,\text{MeV}} = 2 \times 10^{-3}\,\text{fm}$

10.96 $\text{M}_\text{W}{}^2 = \dfrac{4\pi\alpha}{8\sin^2\theta_W} \cdot \sqrt{2}\dfrac{(\hbar c)^3}{G_F}$

$= \dfrac{(4\pi)}{8\sin^2 28.17^0} \cdot \dfrac{\sqrt{2}}{128} \times \dfrac{1}{1.166 \times 10^{-5}} = 6,670$

$M_\text{W}c^2 = 81.67\,\text{GeV}$

the experimental value being 81 GeV

$M_\text{W}/M_\text{Z} = \cos\theta_\text{W} \to M_\text{Z} = M_\text{W}/\cos\theta_\text{W}$

$M_\text{Z} = 81.67/\cos 28.17^0$

$= 92.64\,\text{GeV}$

the experimental value being 94 GeV.

10.3.7 Feynman Diagrams

10.97 (i) Weak interaction

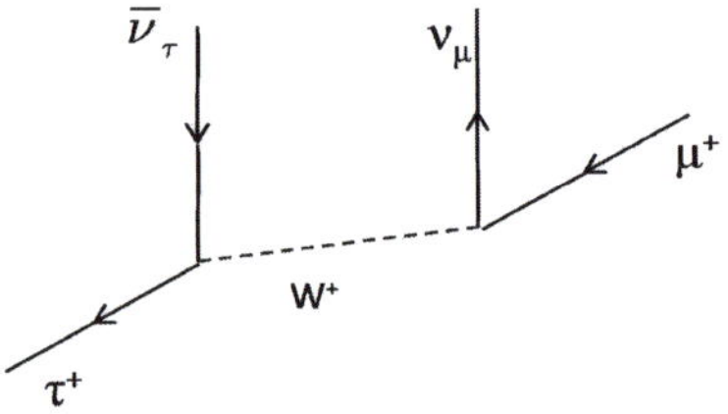

Fig. 10.15(i) τ^+ decay
$\tau^+ \to \mu^+ + \nu_\mu + \overline{\nu}_\tau$

(ii) Strong interaction

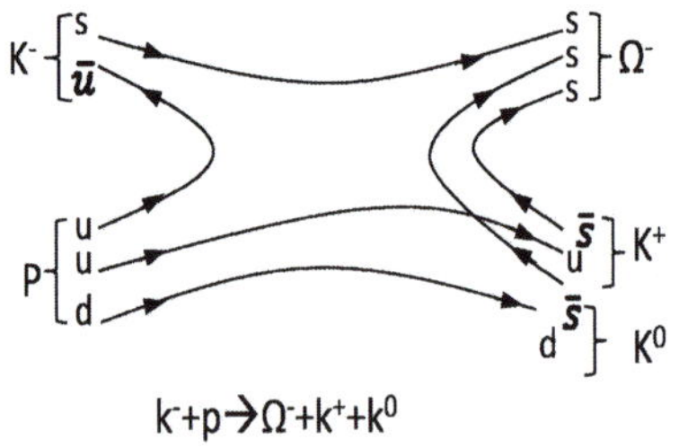

Fig. 10.15(ii) Ω^- production in K^- interaction
$K^- + p \to \Omega^- + K^+ + K^0$

(iii)

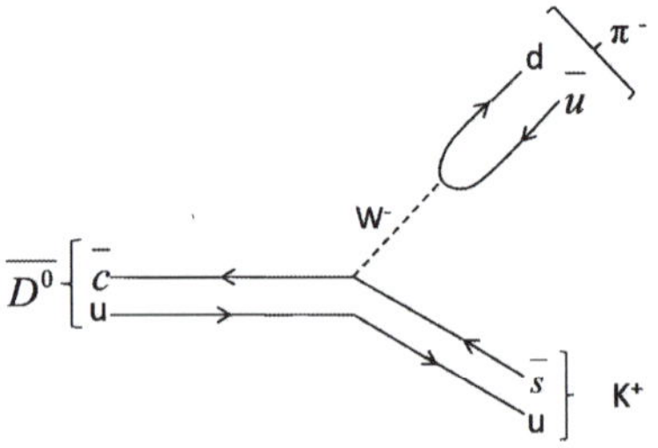

Fig. 10.15(iii) Decay of $\overline{D^0}$
$\overline{D^0} \to K^+ + \pi^-$

10.98 (i) $l = \nu_e$

Diagram is the same as in Problem 10.2(iii)

(ii) $\tau^- \rightarrow \mu^- + l + l$

Fig. 10.16(a) Decay of τ^- $\tau^- \rightarrow \nu_\tau + \mu^- + \overline{\nu}_\mu$

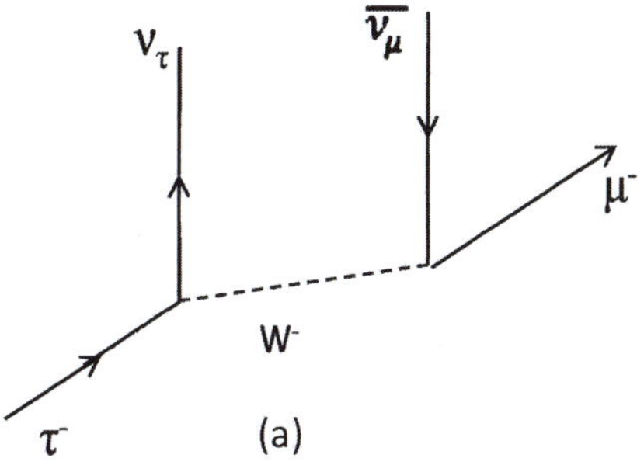

(iii) $l = \nu_\mu$

Fig. 10.16(b) Decay of B^0 meson

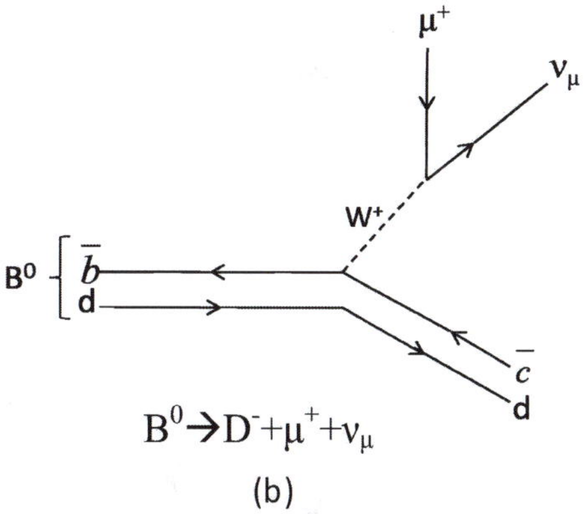

10.99 (a) $\Delta^0 \rightarrow p\pi^-$

The decay $\Delta^0 \rightarrow p + \pi^-$ proceeds via strong interaction in which a gluon is involved. $\Delta^0 = udd; p = udu; \pi^- = d\overline{u}$

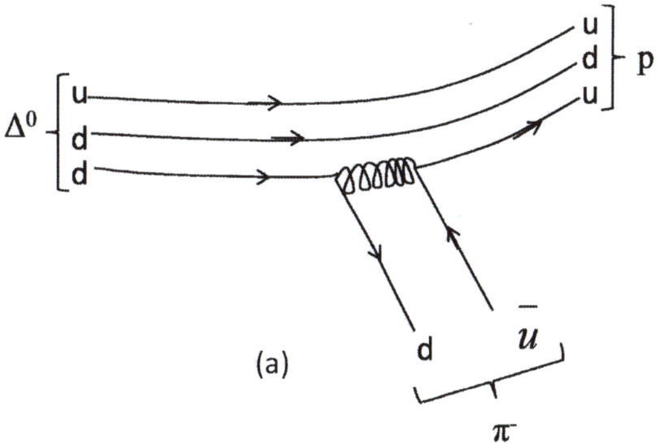

Fig. 10.17 (a)

(b) $\Omega^- \to \Lambda K^-$

Fig. 10.17 (b)

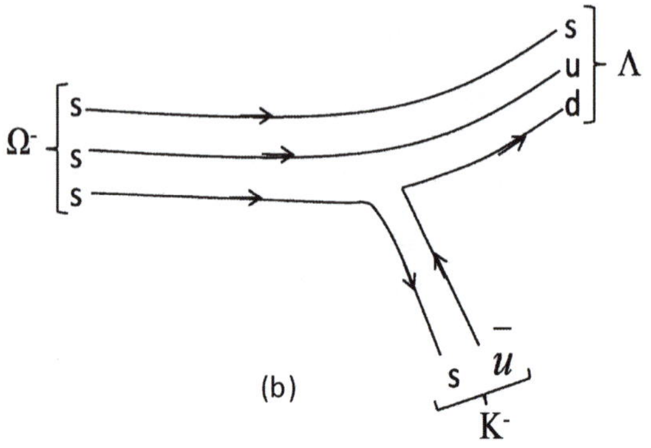

10.100 (a) $e^+e^- \to \nu_\mu \overline{\nu_\mu}$

Fig. 10.18 (a)

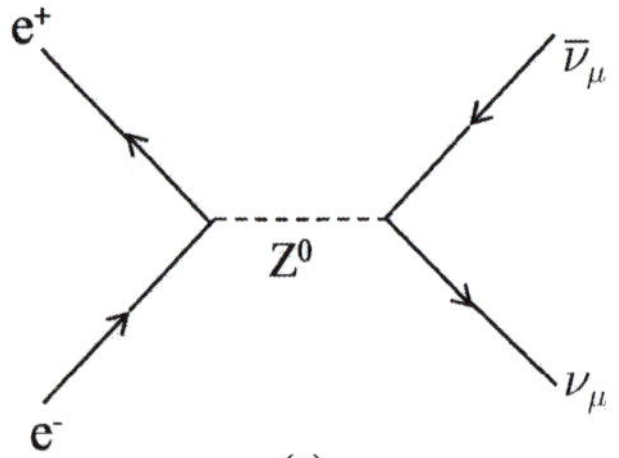

(b) $D^0 \to K^-\pi^+$

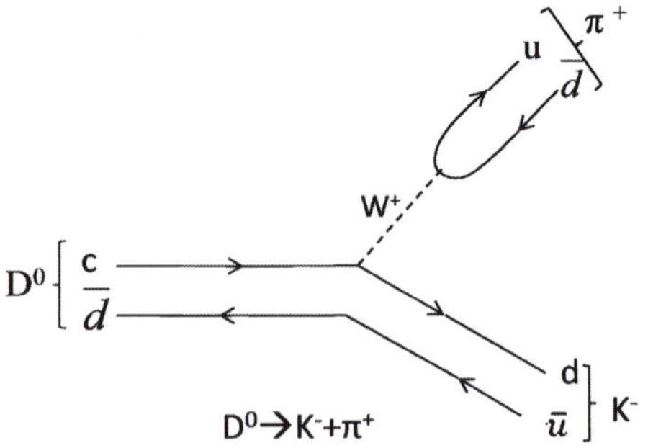

Fig. 10.18 (b)

(b)

10.101 (a)

Fig. 10.19(a) Moller scattering

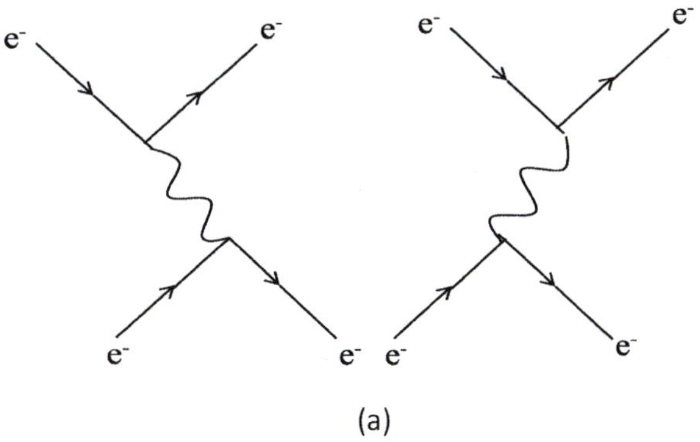

(b)

Fig. 10.19(b) Bhabha scattering

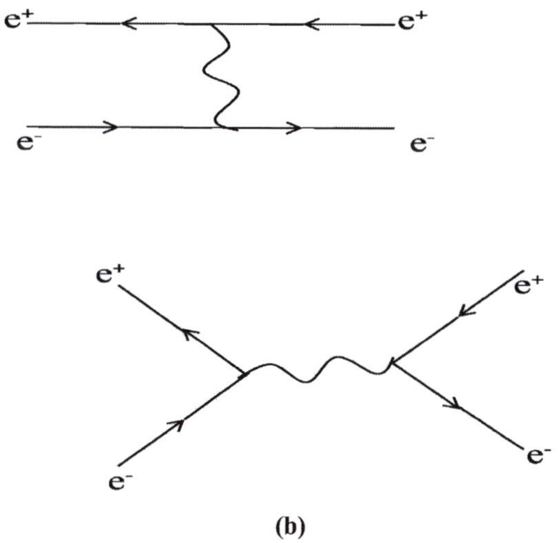

10.102 (a) $\Lambda \to p + e^- + \overline{\nu}_e$

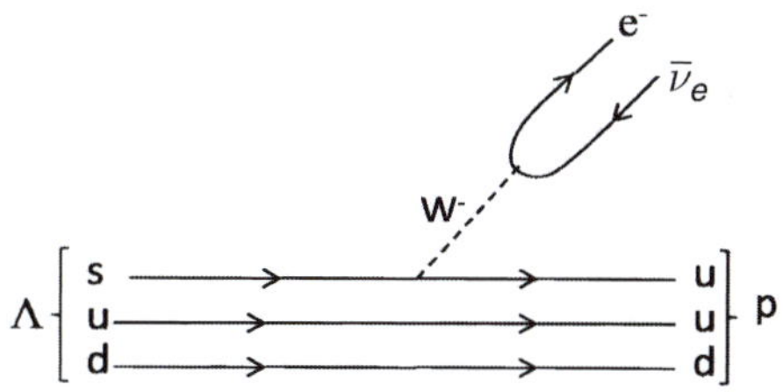

Fig. 10.20(a)

(a)

(b)

Fig. 10.20(b)

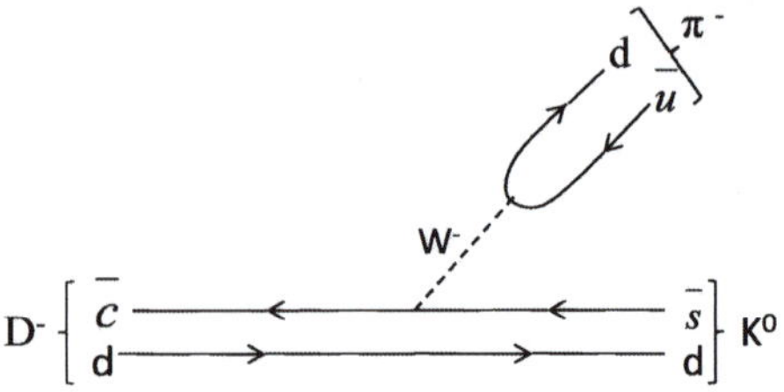

$D^- \rightarrow K^0\pi^-$

(b)

10.103 (a)

Fig. 10.21(a)

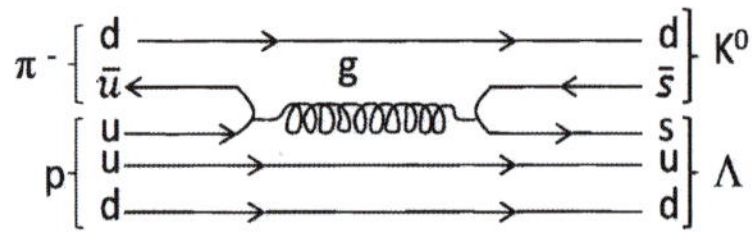

$\pi^- + p \rightarrow \Lambda + k^0$

(a)

(b)

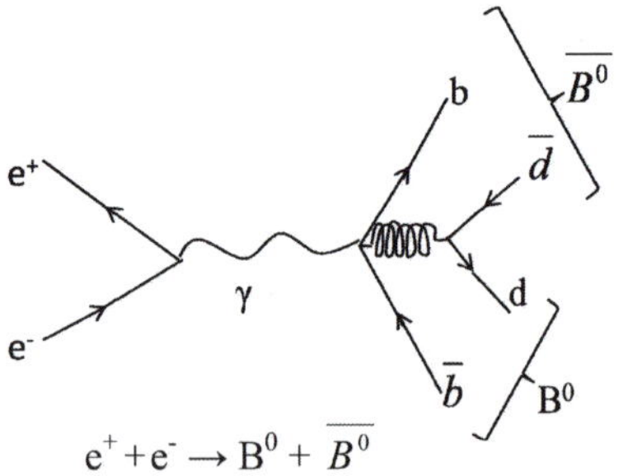

$e^+ + e^- \rightarrow B^0 + \overline{B^0}$

Fig. 10.21(b)

(b)

10.104 The Feynman diagram is

Fig. 10.22

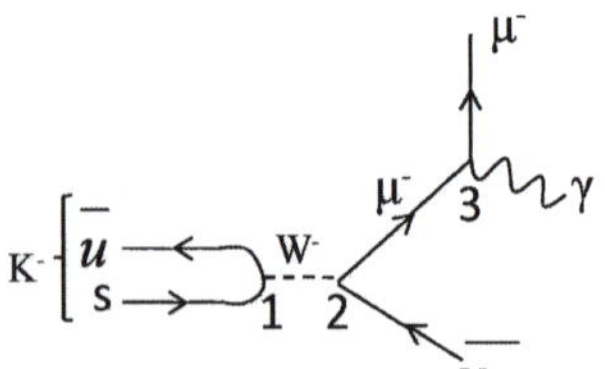

$K^- \rightarrow \mu^- \overline{\nu_\mu} \gamma$

The two vertices 1 and 2 where the W boson couples pertain to weak interactions and have strengths $\sqrt{\alpha_{\mathrm{w}}}$. Vertex 3 is electromagnetic and has strength $\sqrt{\alpha_{\mathrm{em}}}$. The overall strength of the diagram is $\sqrt{\alpha_{\mathrm{w}}}.\sqrt{\alpha_{\mathrm{w}}}.\sqrt{\alpha_{\mathrm{em}}}$ or $\alpha_{\mathrm{w}}\sqrt{\alpha_{\mathrm{em}}}$

10.105 The decay of $D^0 \rightarrow K^- + \pi^+$ is accomplished by the exchange of W^+ boson as illustrated by the Feynman diagram Fig. 10.23.

Fig. 10.23

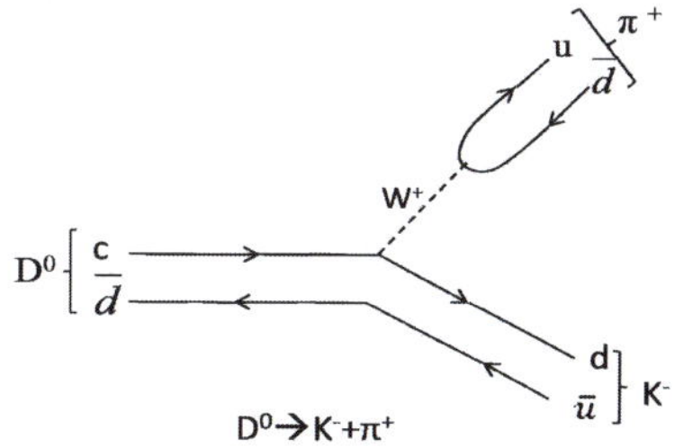

The quark composition ($D^0 = c\overline{u},\ K^- = s\overline{u}, \pi^+ = u\overline{d}$) is also indicated.

The favoured route for the decay is via $c \rightarrow s$. Hence the decay occurs via lowest order charge current weak interaction. However, for $D^+ \rightarrow K^0 + \pi^+$ the c-quark is required to decay to d-quark via W emission and the subsequent decay of W^+ to π^+. This would mean that the $\overline{d}$ quark in the D^+ decay to an $\overline{s}$ quark in the K^0 which is not possible as they both have the same charge. Furthermore the transition $c \rightarrow d$ is not favoured in the Cabibbo scheme.

10.106 Consider the Feynman diagrams for the decays $\pi^+ \rightarrow \mu^+ + \nu_\mu$ and $\pi^0 \rightarrow \gamma + \gamma$

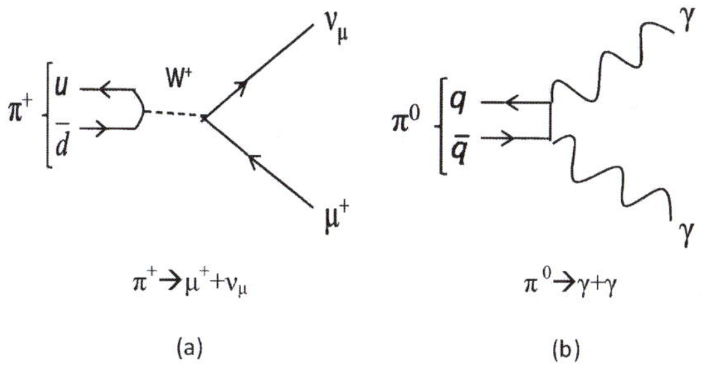

Fig. 10.24

For charged pion there are two vertices of strength $\sqrt{\alpha_{\mathrm{w}}}$, and a propagator

$$\frac{1}{Q^2 + M_{\mathrm{w}}^2 c^2} \cong \frac{1}{M_{\mathrm{w}}^2 c^2}$$

In the limit $Q << M_{\mathrm{w}} c$ momentum transfer squared Q^2 carried by W boson is quite small. Therefore, the decay rate will be proportional to

$$\left(\frac{\sqrt{\alpha_{\mathrm{w}}}\sqrt{\alpha_{\mathrm{w}}}}{M_{\mathrm{w}}^2}\right)^2 = \frac{\alpha_{\mathrm{w}}^2}{M_{\mathrm{w}}^4}$$

For the neutral pion there are two vertices of strength $\sqrt{\alpha_{\text{em}}}$ but the propagator term is absent because photon has zero rest mass. Thus the decay rate will be proportional to α_{em}^2. But $\alpha_{\text{w}} \cong \alpha_{\text{em}}$ according to the electro-weak unified theory. Therefore, the ratio of the decay rates $\frac{\omega(\pi^{\pm}\to\mu\nu)}{\omega(\pi^0\to 2\gamma)} \approx \frac{1}{M_{\text{W}}^4}$ which is quite small and so charged pion lifetime ($\sim 10^{-8}$ s) is much larger than that of neutral pion ($\sim 10^{-17}$ s).

10.107 (a)

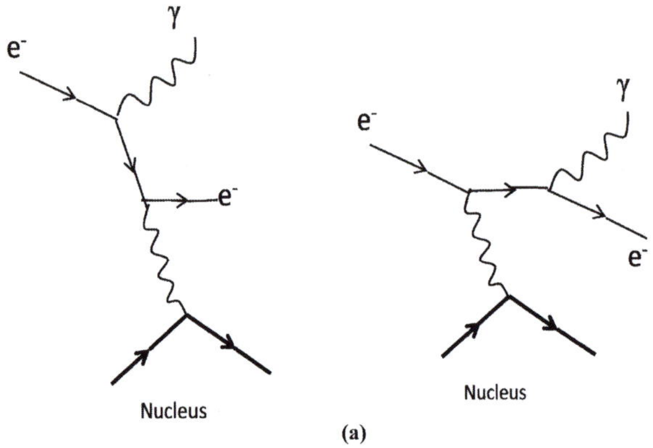

(a)

Fig. 10.25(a) Dominant Feynman diagrams for the Bremsstrahlung process $e^- + (Z, A) \to e^- + \gamma + (Z, A)$

(b) The pair production process $\gamma + (Z, A) \to e^- + e^+ + (Z, A)$

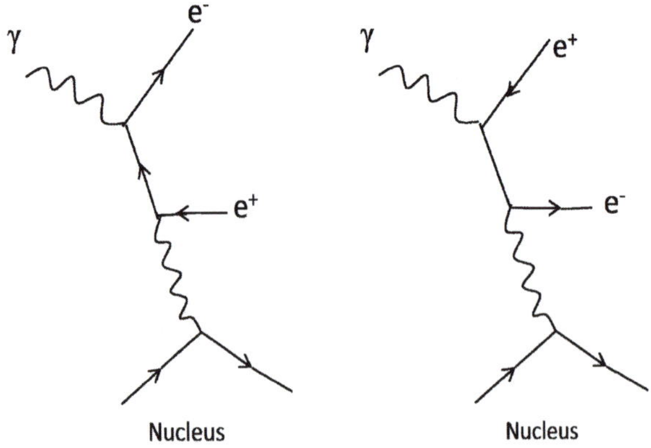

(b)

Fig. 10.25(b)

10.108 (a)

Fig. 10.26(a) Photo electric effect

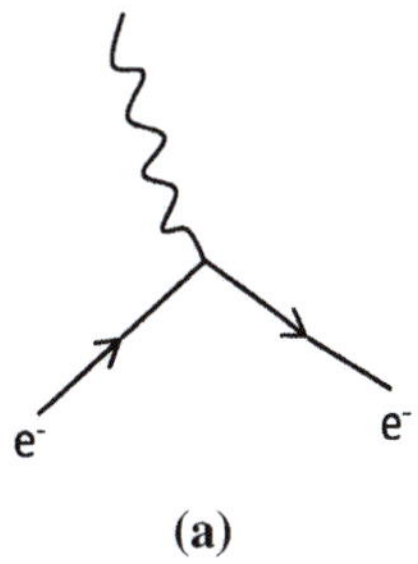

(b)

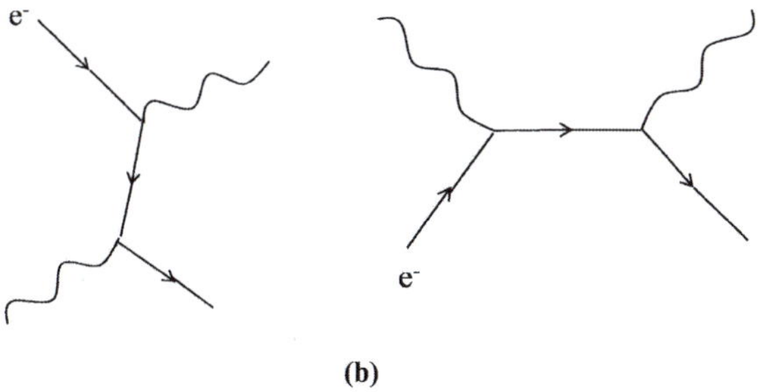

Fig. 10.26(b) Lowest order Feynman diagrams for the Compton scattering

10.109

Fig. 10.27(a)

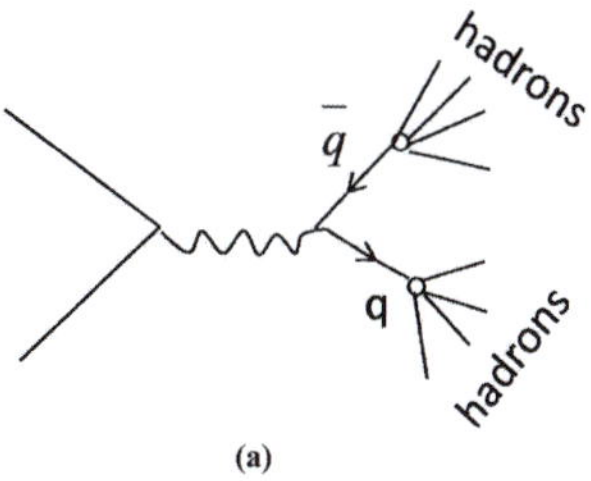

Fig. 10.27(b)

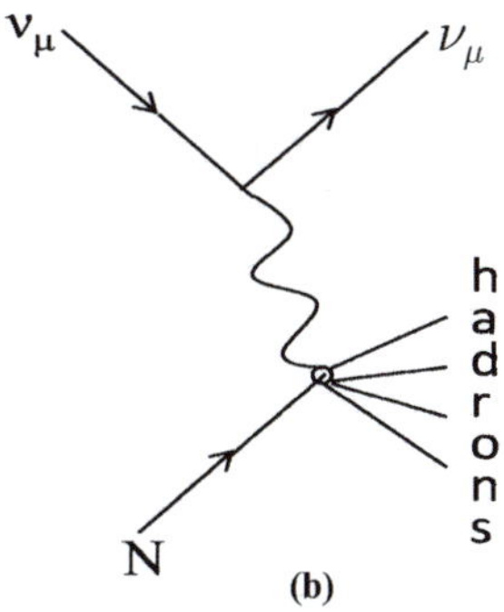

Neutral current reaction

10.110

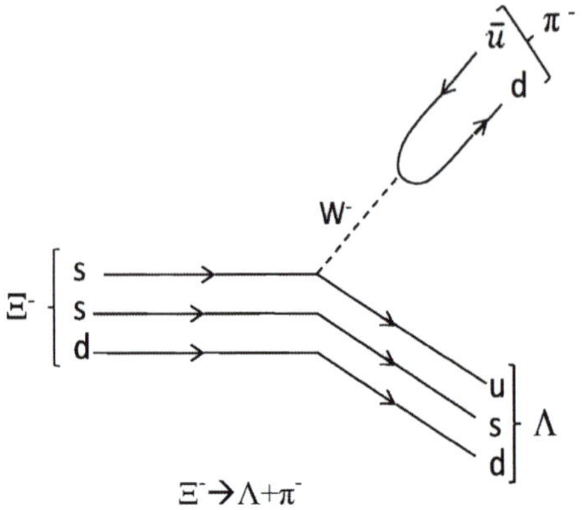

Fig. 10.28(a)

(a)

Fig. 10.28(b)

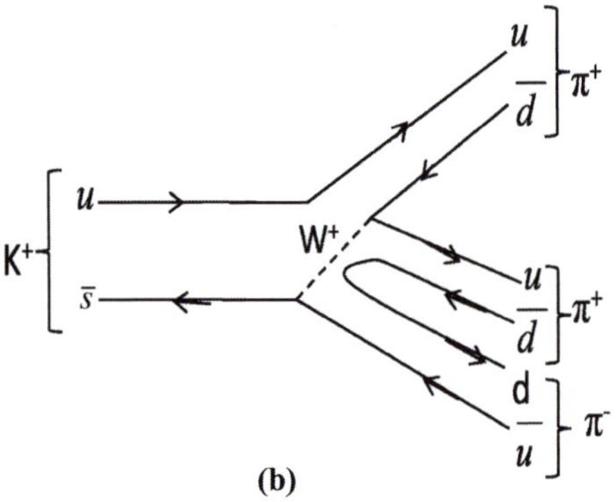

(b)

The decay $K^+ \to \pi^+\pi^+\pi^-$. In all hadrons, gluon interactions keep the gluon bound. In this example a $d\bar{d}$ pair is created.

10.111 (a) There are five basic mechanisms for the D^+ decay

Fig. 10.29(a)
$D^+ \to K^0 + l + \bar{l}$

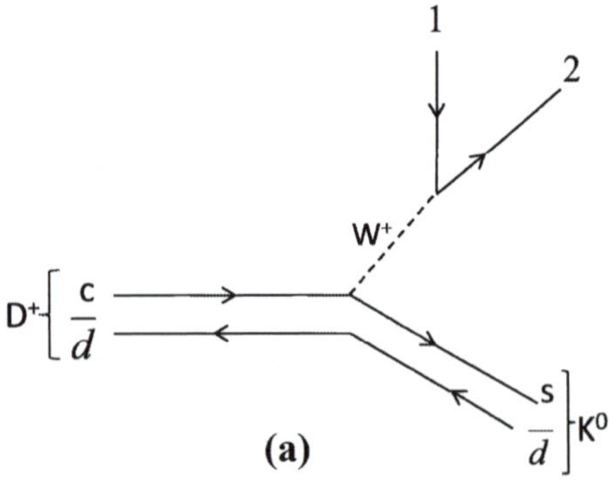

(a)

where the combinations (1,2) are $(e^+,\ \nu_e)$ or (μ^+, ν_μ) or $(\bar{d}, u)$, the last one three times for three colors. The combinations (τ^+, ν_τ) and $(\bar{s}, c)$ are ruled out as energy is violated.

(b) Hadronic decay of τ^-

$\tau^- \to \pi^- \nu_\tau$

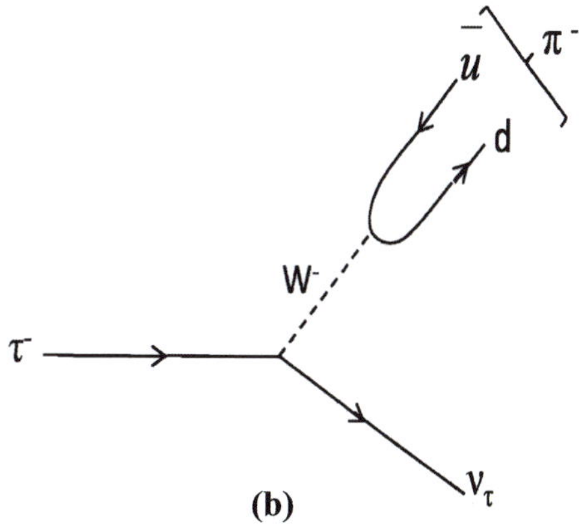

Fig. 10.29(b)

10.112 (a)

Fig. 10.30(a)

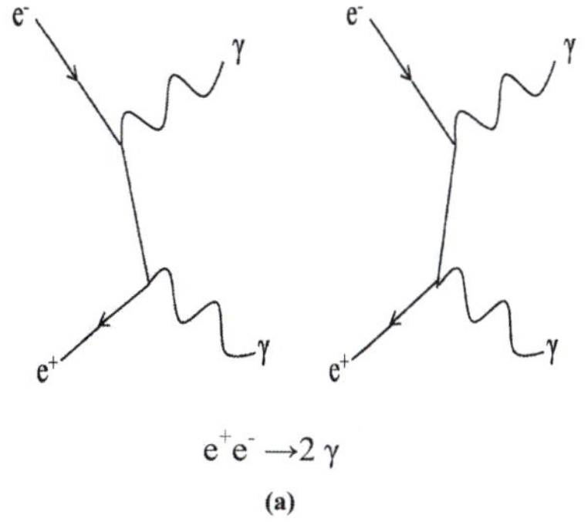

Two-photon annihilation; two ordered diagrams

(b)

Fig. 10.30(b)

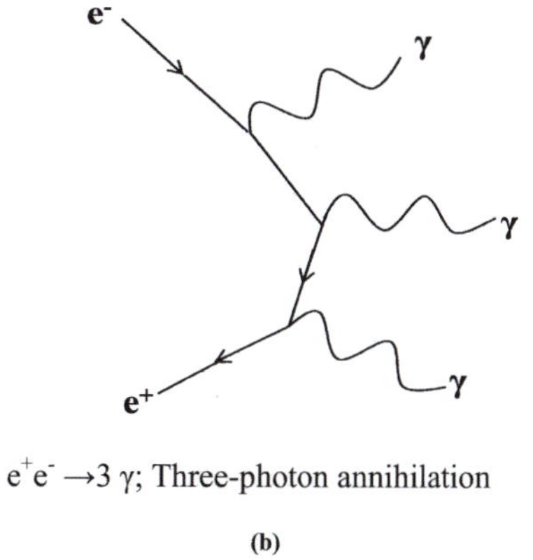

10.113

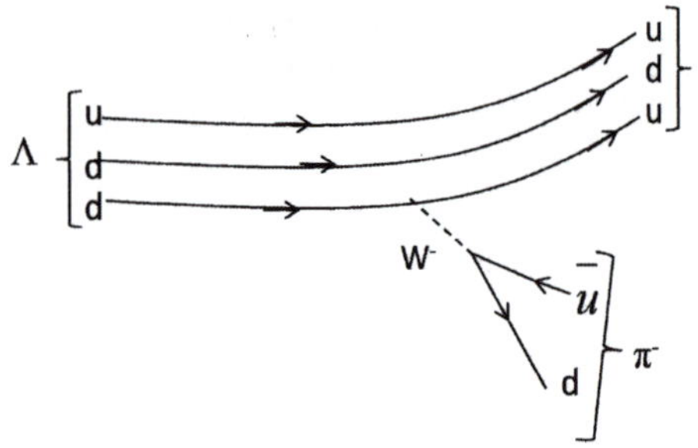

Fig. 10.31

$$\Lambda \rightarrow p\pi^-$$
$$\Lambda = uds;\ p = udu;\ \pi^- = \bar{u}d$$

Appendix: Problem Index

Chapter 1 Mathematical Physics

1.2.1 Vector Calculus

1.2.2 Fourier Series and Fourier Transforms

Chapter 2 Quantum Mechanics-1

Chapter 3 Quantum Mechanics 2

3.2.5 Hydrogen Atom

3.2.6 Angular Momentum

3.2.7 Approximate Methods

4.2.2 Thermodynamic Relations

4.2.3 Statistical Distributions

Chapter 6 Special theory of relativity

6.2.2 Length, Time, Velocity

6.2.3 Mass, Momentum, Energy

7.2.4 Compton Scattering

7.2.5 Photoelectric Effect

7.2.6 Pair Production

7.2.7 Cerenkov Radiation

7.2.8 Nuclear Resonance

7.2.9 Radioactivity (General)

Chapter 8 Nuclear Physics II

8.2.15 Fusion

Chapter 9 Particle Physics I

9.2.1 System of Units

9.2.2 Production

9.2.3 Interaction

9.2.4 Decay

10.2.3 Quarks

10.2.4 Electromagnetic Interaction

10.2.5 Weak Interaction

Index